Recent Advances in Indoor Localization Systems and Technologies

Recent Advances in Indoor Localization Systems and Technologies

Editors

Gyula Simon
László Sujbert

MDPI • Basel • Beijing • Wuhan • Barcelona • Belgrade • Manchester • Tokyo • Cluj • Tianjin

Editors
Gyula Simon
Óbuda University
Hungary

László Sujbert
Budapest University of Technology and Economics
Hungary

Editorial Office
MDPI
St. Alban-Anlage 66
4052 Basel, Switzerland

This is a reprint of articles from the Special Issue published online in the open access journal *Applied Sciences* (ISSN 2076-3417) (available at: https://www.mdpi.com/journal/applsci/special_issues/indoor_localization_systems_and_technologies).

For citation purposes, cite each article independently as indicated on the article page online and as indicated below:

LastName, A.A.; LastName, B.B.; LastName, C.C. Article Title. *Journal Name* **Year**, *Volume Number*, Page Range.

ISBN 978-3-0365-1483-3 (Hbk)
ISBN 978-3-0365-1484-0 (PDF)

Contents

About the Editors

Gyula Simon received his M.Sc. and Ph.D. degrees in electrical engineering from the Budapest University of Technology, Hungary, in 1991 and 1998, respectively, and his Dr. habil. degree from the University of Pannonia in 2017. Currently, he is a professor at Óbuda University, Székesfehérvár, Hungary. His research interests include adaptive signal processing, localization, and sensor networks.

László Sujbert received his M.Sc., Ph.D., and Dr. habil. degrees in electrical engineering from the Budapest University of Technology and Economics, Hungary, in 1992, 1998, and 2017, respectively. He has been with the Department of Measurement and Information Systems, the Budapest University of Technology and Economics, since 1992, where he is the head of the Digital Signal Processing Laboratory. He is active in the fields of measurement, signal processing and embedded systems.

Editorial

Special Issue on "Recent Advances in Indoor Localization Systems and Technologies"

Gyula Simon [1,*] and László Sujbert [2]

1 Alba Regia Technical Faculty, Óbuda University, 8000 Székesfehérvár, Hungary
2 Department of Measurement and Information Systems, Budapest University of Technology and Economics, 1111 Budapest, Hungary; sujbert@mit.bme.hu
* Correspondence: simon.gyula@amk.uni-obuda.hu

Abstract: Despite the enormous technical progress seen in the past few years, the maturity of indoor localization technologies has not yet reached the level of GNSS solutions. The 23 selected papers in this special issue present recent advances and new developments in indoor localization systems and technologies, proposing novel or improved methods with increased performance, providing insight into various aspects of quality control, and also introducing some unorthodox positioning methods.

Keywords: indoor localization technologies; hybrid positioning; model based techniques; quality control

Citation: Simon, G.; Sujbert, L. Special Issue on "Recent Advances in Indoor Localization Systems and Technologies". *Appl. Sci.* **2021**, *11*, 4191. https://doi.org/10.3390/app11094191

Received: 26 April 2021
Accepted: 1 May 2021
Published: 5 May 2021

While outdoor localization solutions mainly depend on services offered by well-established Global Navigation Satellite Systems (e.g., Global Positioning System–GPS), indoor localization technologies have not yet reached this level of maturity. There has been enormous technical progress in the past decade, resulting in various competing techniques and technologies for sensing, positioning, and tracking. The two main directions of indoor localization methods contain reference-based and reference-free methods. For references, various types of anchor nodes (e.g., WiFi hotspots, light sources, or pseudo-satellites–pseudolites) [1–3], or various maps (e.g., WiFi or magnetic fingerprint maps of a building) [4,5] can be used. Reference-free approaches do not use such external aids but rather estimate the position of the tracked entity from its measured inertial properties (e.g., Pedestrian Dead Reckoning–PDR) [6].

A wide range of technologies is used to create the reference system, including audio signals [7], magnetic signals [8], visible light [9], and radio communication techniques (e.g., WiFi, Bluetooth, Bluetooth Low Energy–BLE, Radio-Frequency Identification–RFID, and Ultra-wideband–UWB), [1,10–12]. For sensing, various metrics have been proposed, including the Received Signal Strength (RSS), the Time of Flight (ToF) or Time Difference of Arrival (TDoA) of signals, the Angle of Arrival (AoA) or Angle Difference of Arrival (ADoA) of signals, or the Phase of Arrival (PoA) of signals [13]. From the measured signals' properties, the position is estimated using various analytical or iterative methods.

To increase accuracy and performance of indoor localization systems, hybrid solutions have been proposed (e.g., PDR combined by WiFi [14]). Machine learning methods (Neural Networks and Deep Learning) and various model-based techniques are widely used (e.g., various Kalman Filters, Particle Filters) [15,16].

Despite advances in technology, the main challenges in indoor localization are still present: Our current sensing systems are sensitive to the state of the environment (e.g., non-Line of Sight conditions, fading, multipath effects, reflections, noise) and also changes in the environment (e.g., the presence or absence of people, change of furniture), the detection and adaptation to such circumstances are essential in order to maintain a high quality of localization services. Energy efficiency and low cost are further enabling factors of successful localization technologies [14].

Papers in this special issue focus on recent advances and new developments in indoor localization systems and technologies. A collection of 23 papers has been selected, which address various problems of indoor positioning, propose novel or improved methods with increased performance, and also provide insights into various aspects of quality control. The papers of this special issue can be classified into the following four research areas:

Hybrid Positioning: The papers in this section provide enhanced positioning performance by combining some of the available sensing modalities. In [17], WiFi RSS is used to provide both fingerprinting and trilateration, and PDR is used to enhance the position estimation accuracy. In [18], a genetic-particle filter (GPF) algorithm is proposed to integrate PDR and geomagnetic positioning. In [19], common infrastructure elements and services, already present in smart environments (e.g., WiFi, access-control services, and BLE beaconing), are used to enhance the quality of indoor positioning. The solution proposed in [20] combines BLE and proximity sensors to increase coverage and accuracy in an Ambient and Assisted Living environment. The hybrid system of [21] combines PDR with occasional correctional steps using Visual Place Recognition, to eliminate the cumulative error of PDR and enhance its accuracy when vertical movement between different floors is possible. In [22], a closed-form localization algorithm is introduced for scenarios where multiple TDOA measurements and a single TOA measurement are also available. The review paper [23] introduces and compares the performance of various integrated remote sensing technologies, which combine, e.g., GPS, UWB, RFID, sensor networks, and digital imaging.

Techniques to enhance the performance of positioning systems: In addition to the combination of various sensing modalities, other techniques have been proposed to improve the performance of indoor localization systems. Papers [24–26] propose enhancements to systems using RSS-based WiFi fingerprinting. In [24], a dimensionality reduction technique is proposed, where non-informative anchor nodes are eliminated to reduce computation time and also increase system performance. In [25], methods are proposed to enhance the performance of the feature extraction and the fingerprinting processes, and extensions are introduced to handle the unknown-map situation. In [26], Deep Learning methods are used to enhance the performance of WiFi fingerprinting: A stacked denoising auto-encoder is applied to extract robust features from noisy measurements and other model-based techniques (Kalman filters and Hidden Markov Models) are used to improve the estimate. In [27] an Extended Kalman Filter-based solution, utilizing Channel State Information, is proposed to enhance the accuracy of RSS-based ranging in challenging indoor environments. In [28], a unified state-estimation algorithm is proposed applying finite memory structure filtering and smoothing. The proposed unified approach has attractive properties, e.g., it provides robust, unbiased, and dead-beat state estimates (e.g., position, speed, and acceleration) in noise-free cases. To improve the accuracy of PDR systems, in [29], a novel adaptive Kalman filter-based heading estimation method is proposed.

Quality control: The actual state of the environment may strongly influence the accuracy of measurements used for positioning, thus affect the quality of the positioning process as well. High precision localization techniques must monitor the state of both the environment and the measurement process, detect potential anomalies and adapt the estimation process accordingly. Papers in this section propose various solutions to problems emerging in laser- and radio-based positioning systems. Paper [30] investigates positioning systems using laser sensors, being sensitive to refracted and reflected beams when transparent objects are present. The authors propose a novel algorithm to detect the presence of transparent objects, by analyzing the pattern of the reflected noise. In [31], a novel method is proposed to identify the indoor/outdoor environment using GPS signals and machine learning classification techniques. Paper [32] proposes a method to ensure the accuracy and reliability of positioning results in indoor pseudolite systems: An adaptive fault-detection method is applied to find and exclude potentially faulty pseudolite measurements, influenced by multipath effects, clock drift, or noise. In papers [33–35], problems of line-of-sight (LOS) and non-line-of-sight (NLOS) situations are studied in radio-based systems, where accurate ranging is possible with LOS. These papers propose various NLOS identification methods, which help to eliminate the influence of NLOS errors on positioning accuracy. In [33], a NLOS tracer method is proposed based on the improved Modified Probabilistic Data Association and Interacting Multiple Model algorithms. In [34], a Gaussian model is proposed to identify NLOS signals, while [35] applies machine learning methods to identify not only LOS and NLOS, but also multipath situations.

Unorthodox Positioning and Sensing methods: Some of the papers proposed new or unorthodox approaches both for sensing and positioning. In [36], a novel inexpensive sensing method is proposed using a Circular Photodiode Array (for sensor) along with modulated light sources (for anchor nodes), to provide an ADoA-based positioning system. Paper [37] proposes a single-camera trilateration method, which is able to estimate the 3D pose of a camera from a single image, using landmarks at known positions. The systems of [38,39] offer help for the visually impaired, using portable cameras. In [38], a marker-based localization and navigation system is proposed, using convolutional neural networks to identify markers on the camera stream. In [39], a system is proposed, which helps to identify objects on the camera image, using a novel multi-label convolutional support vector machine.

Funding: This research received no external funding.

Institutional Review Board Statement: Not applicable.

Informed Consent Statement: Not applicable.

Data Availability Statement: Not applicable.

Acknowledgments: Not applicable.

Conflicts of Interest: The authors declare no conflict of interest.

References

1. Geok, T.K.; Aung, K.Z.; Aung, M.S.; Soe, M.T.; Abdaziz, A.; Liew, C.P.; Hossain, F.; Tso, C.P.; Yong, W.H. Review of Indoor Positioning: Radio Wave Technology. *Appl. Sci.* **2020**, *11*, 279. [CrossRef]
2. Maheepala, M.; Kouzani, A.Z.; Joordens, M.A. Light-Based Indoor Positioning Systems: A Review. *IEEE Sens. J.* **2020**, *20*, 3971–3995. [CrossRef]
3. Li, X.; Zhang, P.; Huang, G.; Zhang, Q.; Guo, J.; Zhao, Y.; Zhao, Q. Performance analysis of indoor pseudolite positioning based on the unscented Kalman filter. *GPS Solutions* **2019**, *23*, 79. [CrossRef]
4. Zhu, X.; Qu, W.; Qiu, T.; Zhao, L.; Atiquzzaman, M.; Wu, D.O. Indoor Intelligent Fingerprint-Based Localization: Principles, Approaches and Challenges. *IEEE Commun. Surv. Tutorials* **2020**, *22*, 2634–2657. [CrossRef]
5. Ashraf, I.; Kang, M.; Hur, S.; Park, Y. MINLOC:Magnetic Field Patterns-Based Indoor Localization Using Convolutional Neural Networks. *IEEE Access* **2020**, *8*, 66213–66227. [CrossRef]
6. Kuang, J.; Niu, X.; Chen, X. Robust Pedestrian Dead Reckoning Based on MEMS-IMU for Smartphones. *Sensors* **2018**, *18*, 1391. [CrossRef]
7. Leonardo, R.; Barandas, M.; Gamboa, H. A Framework for Infrastructure-Free Indoor Localization Based on Pervasive Sound Analysis. *IEEE Sens. J.* **2018**, *18*, 4136–4144. [CrossRef]
8. Subbu, K.P.; Gozick, B.; Dantu, R. LocateMe. *ACM Trans. Intell. Syst. Technol.* **2013**, *4*, 1–27. [CrossRef]
9. Guan, W.; Chen, S.; Wen, S.-S.; Tan, Z.; Song, H.; Hou, W. High-Accuracy Robot Indoor Localization Scheme Based on Robot Operating System Using Visible Light Positioning. *IEEE Photon J.* **2020**, *12*, 1–16. [CrossRef]
10. Giuliano, R.; Cardarilli, G.C.; Cesarini, C.; Di Nunzio, L.; Fallucchi, F.; Fazzolari, R.; Mazzenga, F.; Re, M.; Vizzarri, A. Indoor Localization System Based on Bluetooth Low Energy for Museum Applications. *Electronics* **2020**, *9*, 1055. [CrossRef]
11. Hatem, E.; Abou-Chakra, S.; Colin, E.; Laheurte, J.-M.; El-Hassan, B. Performance, Accuracy and Generalization Capability of RFID Tags' Constellation for Indoor Localization. *Sensors* **2020**, *20*, 4100. [CrossRef]
12. Gezici, S.; Tian, Z.; Giannakis, G.B.; Kobayashi, H.; Molisch, A.F.; Poor, H.V.; Sahinoglu, Z. Localization via ultra-wideband radios: A look at positioning aspects for future sensor networks. *IEEE Signal Process. Mag.* **2005**, *22*, 70–84. [CrossRef]
13. Zafari, F.; Gkelias, A.; Leung, K.K. A Survey of Indoor Localization Systems and Technologies. *IEEE Commun. Surv. Tutorials* **2019**, *21*, 2568–2599. [CrossRef]
14. Liu, X.; Zhou, B.; Huang, P.; Xue, W.; Li, Q.; Zhu, J.; Qiu, L. Kalman Filter-Based Data Fusion of Wi-Fi RTT and PDR for Indoor Localization. *IEEE Sens. J.* **2021**, *21*, 8479–8490. [CrossRef]
15. Roy, P.; Chowdhury, C. A Survey of Machine Learning Techniques for Indoor Localization and Navigation Systems. *J. Intell. Robot. Syst.* **2021**, *101*, 63. [CrossRef]

16. Gu, F.; Hu, X.; Ramezani, M.; Acharya, D.; Khoshelham, K.; Valaee, S.; Shang, J. Indoor Localization Improved by Spatial Context—A Survey. *ACM Comput. Surv.* **2019**, *52*, 1–35. [CrossRef]
17. Poulose, A.; Kim, J.; Han, D.S. A Sensor Fusion Framework for Indoor Localization Using Smartphone Sensors and Wi-Fi RSSI Measurements. *Appl. Sci.* **2019**, *9*, 4379. [CrossRef]
18. Sun, M.; Wang, Y.; Xu, S.; Cao, H.; Si, M. Indoor Positioning Integrating PDR/Geomagnetic Positioning Based on the Genetic-Particle Filter. *Appl. Sci.* **2020**, *10*, 668. [CrossRef]
19. Fernández, P.J.; Santa, J.; Skarmeta, A.F. Hybrid Positioning for Smart Spaces: Proposal and Evaluation. *Appl. Sci.* **2020**, *10*, 4083. [CrossRef]
20. Kolakowski, M. Improving Accuracy and Reliability of Bluetooth Low-Energy-Based Localization Systems Using Proximity Sensors. *Appl. Sci.* **2019**, *9*, 4081. [CrossRef]
21. Qian, J.; Cheng, Y.; Ying, R.; Liu, P. A Novel Indoor Localization Method Based on Image Retrieval and Dead Reckoning. *Appl. Sci.* **2020**, *10*, 3803. [CrossRef]
22. Deng, Z.; Wang, H.; Zheng, X.; Fu, X.; Yin, L.; Tang, S.; Yang, F. A Closed-Form Localization Algorithm and GDOP Analysis for Multiple TDOAs and Single TOA Based Hybrid Positioning. *Appl. Sci.* **2019**, *9*, 4935. [CrossRef]
23. Moselhi, O.; Bardareh, H.; Zhu, Z. Automated Data Acquisition in Construction with Remote Sensing Technologies. *Appl. Sci.* **2020**, *10*, 2846. [CrossRef]
24. Abed, A.; Abdel-Qader, I. RSS-Fingerprint Dimensionality Reduction for Multiple Service Set Identifier-Based Indoor Positioning Systems. *Appl. Sci.* **2019**, *9*, 3137. [CrossRef]
25. Yoo, J.; Park, J. Indoor Localization Based on Wi-Fi Received Signal Strength Indicators: Feature Extraction, Mobile Fingerprinting, and Trajectory Learning. *Appl. Sci.* **2019**, *9*, 3930. [CrossRef]
26. Wang, Y.; Gao, J.; Li, Z.; Zhao, L. Robust and Accurate Wi-Fi Fingerprint Location Recognition Method Based on Deep Neural Network. *Appl. Sci.* **2020**, *10*, 321. [CrossRef]
27. Wang, J.; Park, J.G. A Novel Indoor Ranging Algorithm Based on a Received Signal Strength Indicator and Channel State Information Using an Extended Kalman Filter. *Appl. Sci.* **2020**, *10*, 3687. [CrossRef]
28. Kim, P.S. Finite Memory Structure Filtering and Smoothing for Target Tracking in Wireless Network Environments. *Appl. Sci.* **2019**, *9*, 2872. [CrossRef]
29. Chai, D.; Chen, G.; Wang, S. A Novel Method of Adaptive Kalman Filter for Heading Estimation Based on an Autoregressive Model. *Appl. Sci.* **2019**, *9*, 3727. [CrossRef]
30. Jung, J.-W.; Park, J.-S.; Kang, T.-W.; Kang, J.-G.; Kang, H.-W. Mobile Robot Path Planning Using a Laser Range Finder for Environments with Transparent Obstacles. *Appl. Sci.* **2020**, *10*, 2799. [CrossRef]
31. Bui, V.; Le, N.T.; Vu, T.L.; Nguyen, V.H.; Jang, Y.M. GPS-Based Indoor/Outdoor Detection Scheme Using Machine Learning Techniques. *Appl. Sci.* **2020**, *10*, 500. [CrossRef]
32. Zhao, Y.; Guo, J.; Zou, J.; Zhang, P.; Zhang, D.; Li, X.; Huang, G.; Yang, F. A Holistic Approach to Guarantee the Reliability of Positioning Based on Carrier Phase for Indoor Pseudolite. *Appl. Sci.* **2020**, *10*, 1199. [CrossRef]
33. Cheng, L.; Xue, M.; Liu, Z.; Wang, Y. A Robust Tracking Algorithm Based on a Probability Data Association for a Wireless Sensor Network. *Appl. Sci.* **2019**, *10*, 6. [CrossRef]
34. Si, M.; Wang, Y.; Xu, S.; Sun, M.; Cao, H. A Wi-Fi FTM-Based Indoor Positioning Method with LOS/NLOS Identification. *Appl. Sci.* **2020**, *10*, 956. [CrossRef]
35. Sang, C.L.; Steinhagen, B.; Homburg, J.D.; Adams, M.; Hesse, M.; Rückert, U. Identification of NLOS and Multi-Path Conditions in UWB Localization Using Machine Learning Methods. *Appl. Sci.* **2020**, *10*, 3980. [CrossRef]
36. Zachár, G.; Vakulya, G.; Simon, G. Bearing Estimation for Indoor Localization Systems Using Planar Circular Photodiode Arrays. *Appl. Sci.* **2020**, *10*, 3683. [CrossRef]
37. Zhou, Y.; Liu, W.; Lu, X.; Zhong, X. Single-Camera Trilateration. *Appl. Sci.* **2019**, *9*, 5374. [CrossRef]
38. Elgendy, M.; Guzsvinecz, T.; Sik-Lanyi, C. Identification of Markers in Challenging Conditions for People with Visual Impairment Using Convolutional Neural Network. *Appl. Sci.* **2019**, *9*, 5110. [CrossRef]
39. Bazi, Y.; Alhichri, H.; Alajlan, N.; Melgani, F. Scene Description for Visually Impaired People with Multi-Label Convolutional SVM Networks. *Appl. Sci.* **2019**, *9*, 5062. [CrossRef]

Article

A Sensor Fusion Framework for Indoor Localization Using Smartphone Sensors and Wi-Fi RSSI Measurements

Alwin Poulose, Jihun Kim and Dong Seog Han *

School of Electronics Engineering, Kyungpook National University, 80 Daehak-ro, Buk-gu, Daegu 41566, Korea; alwinpoulosepalatty@knu.ac.kr (A.P.); soji@knu.ac.kr (J.K.)

* Correspondence: dshan@knu.ac.kr; Tel.: +82-53-950-6609

Received: 7 September 2019; Accepted: 14 October 2019; Published: 16 October 2019

Abstract: Sensor fusion frameworks for indoor localization are developed with the specific goal of reducing positioning errors. Although many conventional localization frameworks without fusion have been improved to reduce positioning error, sensor fusion frameworks generally provide a further improvement in positioning accuracy. In this paper, we propose a sensor fusion framework for indoor localization using the smartphone inertial measurement unit (IMU) sensor data and Wi-Fi received signal strength indication (RSSI) measurements. The proposed sensor fusion framework uses location fingerprinting and trilateration for Wi-Fi positioning. Additionally, a pedestrian dead reckoning (PDR) algorithm is used for position estimation in indoor scenarios. The proposed framework achieves a maximum of 1.17 m localization error for the rectangular motion of a pedestrian and a maximum of 0.44 m localization error for linear motion.

Keywords: indoor localization; smartphone sensors; pedestrian dead reckoning (PDR); Wi-Fi indoor positioning; sensor fusion frameworks; Kalman filter; location fingerprinting; trilateration; received signal strength indication (RSSI)

---

1. Introduction

Accurate positioning for indoor or outdoor scenarios requires that a positioning system's displacement error be minimized. For locating user position in outdoor environments, localization systems such as global positioning system (GPS) [1] and base transceiver station (BTS) [2] based approaches exist. Both GPS and BTS face significant challenges when used for indoor localization. These include signal interference and spatial coverage limitations. The BTS cell phone technology does not achieve accurate results for indoor localization due to constrained signal coverage of the target area and dense urban environments characterized by high-rise buildings. To overcome these challenges, it is necessary to develop alternative localization systems for indoor localization. Existing indoor localization systems are based on ultra-wideband (UWB) technology [3], pedestrian dead reckoning (PDR) [4], radio frequency identification (RFID) [5], Bluetooth [6], visible light communication (VLC) [7], Zigbee [8] and Wi-Fi systems [9]. Each of these techniques achieves high position accuracy for indoor localization. However, combined systems give better performance compared to these individual systems. PDR systems use data from smartphone accelerometers, magnetometers, and gyroscopes for localization.

It is possible to use received signal strength indication (RSSI) signals for indoor localization with Wi-Fi access points (APs) in indoor environments. Many studies [10–12] have used smartphones equipped with a Wi-Fi receiving module for the RSSI data collection. The fusion of smartphone sensor data and Wi-Fi generated data for indoor localization has become prominent in indoor localization studies [13–15]. This paper proposes a sensor data fusion framework for both PDR and Wi-Fi systems. The proposed sensor fusion framework combines the location fingerprinting and trilateration algorithms for Wi-Fi indoor positioning and it fuses with PDR position results. The experiment results demonstrate that the proposed fusion framework reduced the localization errors when compared with conventional localization approaches.

The conventional approaches used for sensor fusion are PDR+Wi-Fi trilateration and PDR+Wi-Fi fingerprint algorithms. The most popular and easiest conventional approach used for indoor localization is the PDR+Wi-Fi trilateration algorithm. However, this algorithm does not provide very good accuracy (average error of 1.5–2 m) [16]. The multipath effects and non-line of sight (NLOS) conditions in the experiment areas reduce the trilateration algorithm performance. In trilateration algorithm, we use a free space propagation model for estimating the user distance from the Wi-Fi AP. The estimated user distance depends on the RSSI measurements. However, the measured RSSI signals tend to fluctuate according to the indoor environment's physical characteristics and contained objects. Wi-Fi fingerprint algorithms can solve the multipath fading effect by creating a fingerprint map, but changes in the indoor environment affect the fingerprint maps and require updates to the fingerprint database. To overcome the challenges faced by conventional approaches, we present a sensor fusion framework for indoor localization by utilizing the PDR and Wi-Fi positioning results.

A recent trend in indoor localization is to use the hybrid localization system for locating the user position with minimum errors. The hybrid localization systems give better performance than individual localization systems. In hybrid localization systems, we combine multiple localization technologies with the help of sensor fusion frameworks. The most common hybrid localization system is the PDR with Wi-Fi localization system. In this paper, we propose a PDR with Wi-Fi localization system with high position accuracy for indoor localization. The proposed system uses a sensor fusion framework for combining the PDR results with Wi-Fi localization system results. For PDR approach, we used our previous model proposed in [17]. Our previous PDR model reduced the smartphone sensor errors and provides accurate localization results. In the case of Wi-Fi localization, we proposed a fusion algorithm and it uses the results from Wi-Fi trilateration and Wi-Fi fingerprint algorithms to enhance the position accuracy. Finally, we proposed a sensor fusion framework to combine the Wi-Fi fusion algorithm and PDR for indoor localization using the Kalman filter. Through extensive experiments, we demonstrated that a higher indoor positioning accuracy based on the framework is achievable.

The rest of the paper is organized as follows: The related works on the sensor fusion for PDR and Wi-Fi localization systems are discussed in Section 2. The proposed sensor fusion framework model is presented in Section 3. In the Section 4, the experiment results and analysis are discussed. Finally, the conclusions and future work are presented in Section 5.

2. Related Works

Sensor data fusion frameworks have been studied in the past for PDR and Wi-Fi localization systems. In this section, we discuss the related work on sensor fusion frameworks. The proposed sensor fusion framework model depends on three research areas, which are PDR, Wi-Fi localization systems and sensor data fusion frameworks for combined PDR and Wi-Fi localization systems. The performance of a PDR system depends on a smartphone's sensor data error. When pedestrian walking distance increases, a PDR system will drift, and this error reduces the location accuracy. PDR algorithms depend on step detection [18], heading [19] and position estimation [20]. The localization accuracy of a Wi-Fi system depends on RSSI fluctuation. In Wi-Fi localization systems, the distance between a user and APs is estimated from free space path loss model. The Wi-Fi signal fluctuation affects distance estimation and degrades system performance. To overcome the problems related to PDR and Wi-Fi localization systems, a more accurate sensor data fusion framework for those systems is proposed in this paper.

The first sensor data fusion framework for Wi-Fi and PDR was introduced by Evennou and Marx [21]. In [21], Evennou and Marx proposed a sensor fusion framework which uses a Kalman filter and a particle filter. The benefits of the Evennou and Marx architecture were evaluated and compared with pure Wi-Fi localization systems and inertial navigation system (INS) positioning systems. A sensor fusion framework using a particle filter is explained in [22–24]. The real-world experiment results from [22–24] indicate remarkable performance improvements for indoor localization. An effective implementation of the extended Kalman filter (EKF) for sensor fusion is explained in [25–27]. The experiment results from [25–27] show that the author's proposed sensor fusion frameworks achieve high localization accuracy when compared to individual localization systems. A complementary extended Kalman filter for the sensor fusion framework is explained by Leppäkoski et al. [28]. The results from [28] show that both the map information and wireless local area network (WLAN) signals can be used to improve the PDR system accuracy. The idea of using an unscented Kalman filter (UKF) algorithm for a sensor fusion framework is introduced by Chen et al. [29]. In [29], an integrated technique for merging Wi-Fi localization system, PDR and smartphone sensor data using a UKF algorithm for 3D indoor localization is proposed. In [30,31], a detailed analysis of sensor fusion frameworks of Wi-Fi fingerprint with PDR systems are discussed and the results indicate that a hybrid localization system's performance is better than that of individual localization systems. An adaptive and robust filter for indoor localization using Wi-Fi and PDR is proposed by Li et al. [32]. Results from [32] show that the sensor fusion framework reduced the gross errors in the Wi-Fi localization system and gives accurate position results for indoor localization.

So far, we have discussed different types of sensor fusion frameworks used for Wi-Fi and PDR localization systems fusion. These sensor fusion frameworks reduced the localization errors; however, these require further position accuracy improvement for indoor localization. The proposed sensor fusion framework of this paper uses the position results from Wi-Fi localization and PDR systems with the help of a linear Kalman filter (LKF). For Wi-Fi localization systems, the proposed model uses a fusion algorithm to combine the trilateration and fingerprint algorithms. The fusion algorithm compensates for the line of sight (LOS) problem by taking advantage of fingerprinting to enhance the Wi-Fi system's performance. The trilateration algorithm achieves better results when applied in different environments and the system is free from a calibration phase . The Wi-Fi fingerprint technology is useful for non-line of sight (NLOS) conditions when the APs and the user face interference from other objects. Combining these two technologies offers better performance than the individual systems. The PDR algorithm in the proposed model uses two sensor fusion techniques to reduce the smartphone sensor errors. The accelerometer and gyroscope sensors in the smartphone are used for pitch and roll estimation. The accelerometer and gyroscope sensor fusion reduces the accumulated errors and drift errors from the sensors and gives better results for pitch and roll estimation. In the case of heading estimation in the PDR algorithm, the proposed model uses another sensor fusion which combines the magnetometer and gyroscope headings together for better heading estimation. The step length from pitch-based step detector and heading results from heading estimator are for position estimation and the position results from PDR is free from smartphone sensor errors and gives better results than conventional localization systems.

3. Proposed Sensor Fusion Framework Model Using PDR and W-Fi Localization Systems

The proposed sensor fusion framework model uses a Wi-Fi fusion algorithm results with PDR to enhance the position accuracy for indoor localization. The Wi-Fi fusion model in the proposed algorithm utilizes the effective features of two classical localization algorithms such as trilateration and fingerprint for user position estimation. The Wi-Fi fusion results combined with PDR results and the results from the proposed model are free from smartphone sensor errors and Wi-Fi RSSI signal strength problems. Figure 1 shows the proposed sensor fusion framework model.

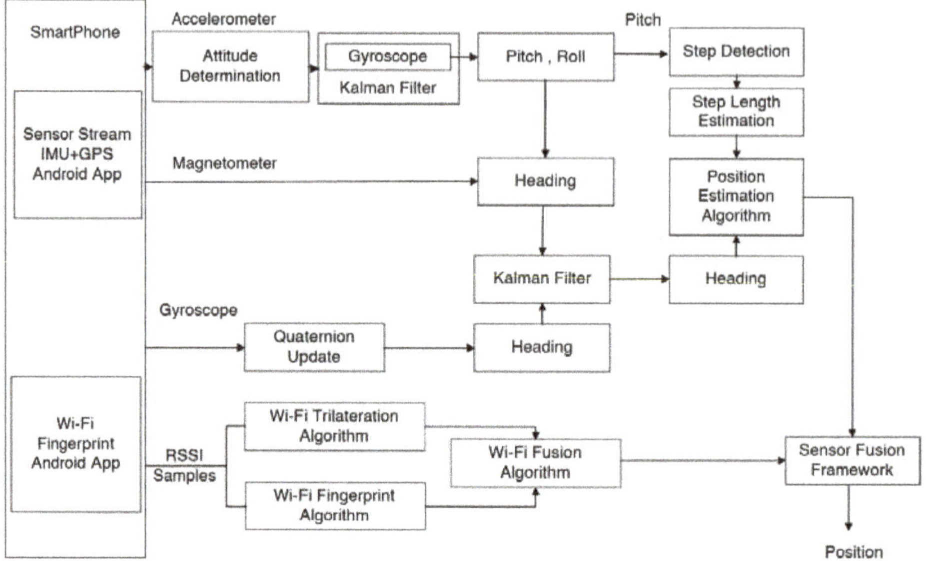

Figure 1. Proposed sensor fusion framework for PDR and Wi-Fi.

The position results from the PDR algorithm in the proposed model utilizes different smartphone sensors such as the accelerometer, gyroscope and magnetometer for position estimation. The data from accelerometer and gyroscope is used for pitch and roll estimation and the step detector uses the pitch values for step detection. The pitch and roll values together are used for heading estimation from the magnetometer. The data from gyroscope is also estimating the user heading and an LKF filter combines the gyroscope and magnetometer heading together for better performance. The step length is estimated from step detector and the position estimation algorithm uses the step length values and heading together for identifying the current user position. The results from PDR algorithm used in the sensor fusion framework for further position improvements. In the case of Wi-Fi fusion algorithm in the proposed model, it uses the results from trilateration and fingerprint algorithms together for user position estimation. The results of the Wi-Fi fusion algorithm are used in the sensor fusion framework for combining with PDR results. The linear Kalman filter explained in [33] is used for sensor fusion framework implementation.

3.1. PDR Positioning

The PDR positioning in the proposed model uses our previous work [17] for user position estimation. In [17], we proposed a sensor fusion technique for pitch and roll estimation, a pitch-based step detector algorithm for step detection, step length estimation from step detector, a sensor fusion technique for heading estimation and a position estimator. The pitch values from the proposed sensor fusion technique are used for user step detection. The pitch-based step detector identifies the user steps from pitch amplitude. A step is detected when the pitch amplitude crosses the threshold level. The step length estimator uses the results from step detector and the step length estimator follows a model presented in [34]. The sensor fusion technique for heading estimation uses the heading results from the magnetometer and gyroscope sensors. The individual heading results of the magnetometer and gyroscope are not free from smartphone sensor errors and these heading results are not sufficient for position estimation. To compensate the magnetometer and gyroscope sensor errors, the PDR model combines the gyroscope heading with the magnetometer heading and the combined results are more accurate than individual sensor heading results. The LKF used in [33] is used for heading sensor fusion. The position estimation algorithm proposed in [35] is used for current user position estimation.

The position estimation algorithm takes the user step length and heading values and estimates the current user position. For more details of PDR implementation, refer to our previous work in [17].

3.2. Wi-Fi Positioning

The classical localization approaches for Wi-Fi positioning are Wi-Fi trilateration and Wi-Fi fingerprinting. Of these approaches, Wi-Fi fingerprinting is the most popular and it does not require the coordinates of Wi-Fi APs in the experiment area. However, the fingerprint approach has two drawbacks when used in practical applications. The fingerprint approach needs a *priori* knowledge of the local area and thus requires a lot of time for location survey and manpower to generate the fingerprints. The second drawback is the position accuracy. As opposed to other localization approaches, the position accuracy of the fingerprint approach depends on the amount of fingerprint data that is generated for a specific coverage area. If the amount of data used in the fingerprint maps is not sufficient, it is difficult to estimate the user position with high accuracy. The computation time for estimating user position in fingerprinting algorithm is very high as compared to other Wi-Fi localization approaches. In the case of Wi-Fi trilateration algorithm, we estimated the distance from the Wi-Fi AP to the user using a free-space path loss model [36]. The localization algorithm uses the distance values from path loss model to estimate the current user position. The performance of trilateration algorithm depends on the channel conditions between APs and the user. For position estimation using trilateration algorithm, the user should be within a limited Wi-Fi signal coverage area with LOS conditions. In order to improve the localization accuracy of Wi-Fi systems and to compensate the position errors of classical Wi-Fi localization approaches, we propose a Wi-Fi fusion algorithm for indoor localization. The proposed Wi-Fi fusion algorithm improves the indoor position accuracy by utilizing the advantages of fingerprint and trilateration algorithms.

3.2.1. Wi-Fi Trilateration Algorithm

The model presented in our previous work [16] is used for trilateration algorithm implementation. For accurate position estimation using trilateration algorithm, the experiment area should contain at least three APs. In our experiment we use four APs and these APs are placed at four corners of the experiment area. A model for Wi-Fi localization using trilateration is shown in Figure 2.

A free space path loss, F is used to estimate the user position from APs and is expressed as [37]:

$$\text{F [dBm]} = 20\log(f) - 22.55 + 20\log_{10}(d) \tag{1}$$

where d is the user distance from AP in meters, f is the signal frequency expressed in megahertz. The trilateration algorithm follows the models explained in [38,39]. The distance d_i between smartphone (x_p, y_p) and the APs $A_i(x_i, y_i)$ is expressed as:

$$d_i^2 = (x_i - x_p)^2 + (y_i - y_p)^2 \tag{2}$$

The expanded form of the equation is;

$$d_i^2 = x_i^2 + x_p^2 - 2x_ix_p + y_i^2 + y_p^2 - 2y_iy_p \tag{3}$$

The above equation can be rewritten as

$$d_k^2 = x_k^2 + x_p^2 - 2x_kx_p + y_k^2 + y_p^2 - 2y_ky_p \tag{4}$$

Subtracting Equation (3) from Equation (2), then the equation can be expressed as;

$$d_i^2 - d_k^2 + x_k^2 + y_k^2 - x_i^2 - y_i^2 = 2(x_k - x_i)x_p + 2(y_k - y_i)y_p \tag{5}$$

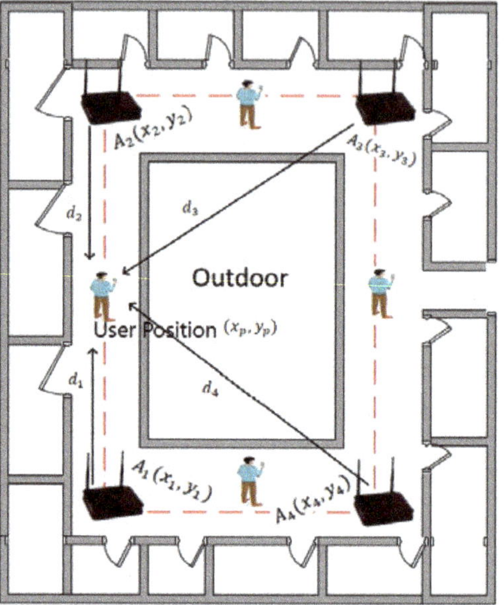

Figure 2. Wi-Fi localization system using trilateration.

With $i = 1$ and by varying index $k = 2, 3, 4$ we obtain Equation (6).

$$\begin{bmatrix} 2(x_2 - x_1) & 2(y_2 - y_1) \\ 2(x3 - x_1) & 2(y_3 - y_1) \\ 2(x4 - x_1) & 2(y_4 - y_1) \end{bmatrix} \begin{bmatrix} x_p \\ y_p \end{bmatrix} = \begin{bmatrix} d_1^2 - d_2^2 + x_2^2 + y_2^2 - x_1^2 - y_1^2 \\ d_1^2 - d_3^2 + x_3^2 + y_3^2 - x_1^2 - y_1^2 \\ d_1^2 - d_4^2 + x_4^2 + y_4^2 - x_1^2 - y_1^2 \end{bmatrix} \tag{6}$$

The coordinates (x_p, y_p) of the smartphone is obtained from the above system of equations and we express the equation in a linear form with three unknowns, $Ax = b$ where

$$A = \begin{bmatrix} 2(x_2 - x_1) & 2(y_2 - y_1) \\ 2(x_3 - x_1) & 2(y_3 - y_1) \\ 2(x_4 - x_1) & 2(y_4 - y_1) \end{bmatrix}, \tag{7}$$

$$x = \begin{bmatrix} x_p \\ y_p \end{bmatrix} \tag{8}$$

and

$$b = \begin{bmatrix} d_1^2 - d_2^2 + x_2^2 + y_2^2 - x_1^2 - y_1^2 \\ d_1^2 - d_3^2 + x_3^2 + y_3^2 - x_1^2 - y_1^2 \\ d_1^2 - d_4^2 + x_4^2 + y_4^2 - x_1^2 - y_1^2 \end{bmatrix} \tag{9}$$

The solution of the equations can be the $\left(x_p^*, y_p^*\right)$ that minimizes the δ defined by the following:

$$\delta = (Ax^* - b)^T (Ax^* - b) \tag{10}$$

$$x^* = \begin{bmatrix} x_p^* \\ y_p^* \end{bmatrix} \tag{11}$$

Applying Minimum Mean Square Error (MMSE) method in the above equation, we express user position, x^* in the following form.

$$x^* = \left(A^T A\right)^{-1} A^T b \quad (12)$$

3.2.2. Wi-Fi Fingerprint Algorithm

The basic idea of creating fingerprint maps for Wi-Fi positioning is explained in [40,41]. In this localization approach, we use a grid-based representation of the indoor environment. The fingerprint algorithm consists of two stages. The creation of fingerprint maps of the localization area from the sampled RSSI values is the first stage of the fingerprint algorithm. To accomplish this, we divide the location area into different zones and collect RSSI samples at all the grid points. In the second stage, we estimate the position of the receiving module from the fingerprint maps using the nearest neighbor method. Other methods used in position estimation are the support vector machine (SVM) and the hidden Markov model. Figure 3 shows the Wi-Fi fingerprint flow chart.

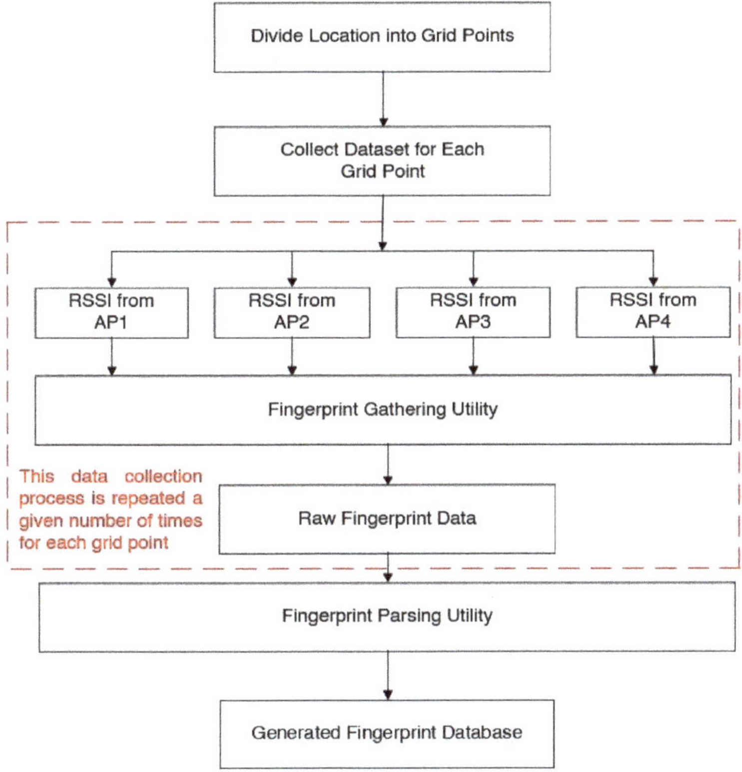

Figure 3. Wi-Fi fingerprint flow chart [41].

The first stage of Wi-Fi fingerprint positioning is explained in Figure 3. The localization area is divided into grid points and RSSI samples are collected from each grid point. The fingerprint gathering utility is used for data collection at each access point. The fingerprint parsing utility takes the RSSI samples as input and generates the data necessary to build the RSSI probability distribution for each reference point. The location estimation procedure for the Wi-Fi fingerprint algorithm is shown in Figure 4. The location estimation procedure uses the localization algorithm on the fingerprint maps of the four APs and the received RSSI samples on-the-fly. The nearest neighbor method is used to determine the user location based on the fingerprint data. The nearest neighbor approach calculates the Euclidean distance between the live RSSI sample and each reference point fingerprint. The minimum Euclidean distance is the approximated user location.

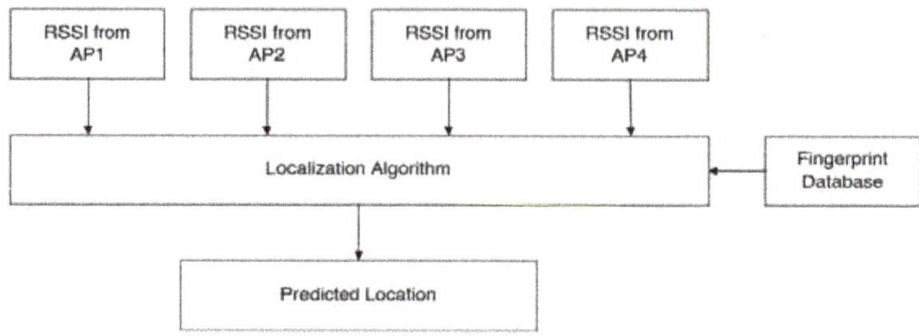

Figure 4. Location estimation procedure.

Let the RSSI samples from N Wi-Fi APs in the fingerprint maps be expressed as a vector $\rho = \{\rho_1, \rho_2, \rho_3, \ldots\ldots, \rho^N | \rho_i \in R^N\}$ to generate fingerprints database and $i = 1, 2, 3, \ldots, N$ be the number of grid points. The measured RSSI of a user from N Wi-Fi access points at a particular location during the experiment time is expressed as the vector $\delta = \left\{\delta_1, \delta_2, \delta_3, \ldots\ldots, \delta^N | \delta_k \in R^k\right\}$. The Euclidean distance D_i between fingerprint maps and the corresponding measured values can be expressed as in [42]:

$$D_i = |\rho_i - \delta| = \sqrt{\sum_{i=1}^{N} (\rho_i - \delta)^2} \tag{13}$$

From the resulting matrix D_i, the minimum value represents the approximated user location and is expressed as

$$[x, y] = \arg\min_i D_i \tag{14}$$

3.2.3. Proposed Wi-Fi Fusion Algorithm

The idea of combining Wi-Fi fingerprint algorithm with trilateration algorithm is explained in [43,44]. The position results from the Wi-Fi fusion algorithm show better position accuracy than individual Wi-Fi position estimation algorithm results. A state dynamic model and a measurement model are used for the Wi-Fi fusion process. The Wi-Fi fusion algorithm uses a simple linear model and it combines the position results for better user position estimation. The Wi-Fi fusion algorithm utilizes the features of LKF, and its implementation is shown in Figure 5.

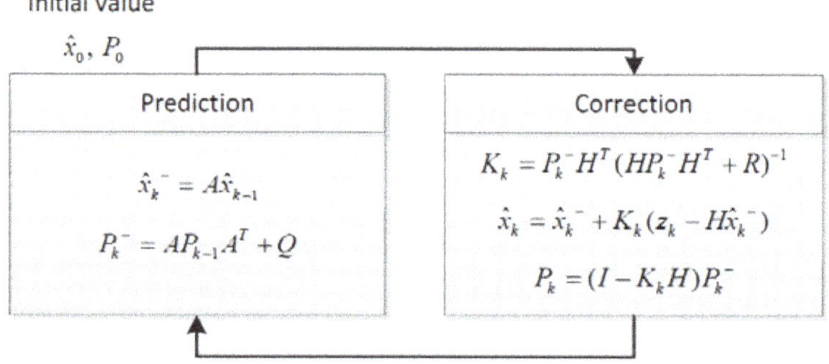

Figure 5. LKF implementation.

The unique characteristics of LKF give Wi-Fi fusion results which are free from localization errors and it improves the user position accuracy. The variables used for LKF implementation are shown in Table 1.

Table 1. Variables used in the LKF implementation.

$\hat{x}_k$	State variable
P_k	Error covariance matrix
A	State transition matrix
k	Variable used for the recursive execution of Kalman filter
$\hat{x}_{k-1}, P_{k-1}$	For internal computation
Q	Noise covariance of the process
K_k	Kalman gain
z_k	Measurement
H	Matrix for calculating the predicted value in the form of a measured value
R	Covariance of the measurement noise

3.3. Proposed Sensor Fusion Framework Algorithm using Wi-Fi Fusion and PDR Position Results

The proposed sensor fusion framework as well uses the advantages of the LKF for combining the Wi-Fi fusion and PDR position results. As compared to other sensor fusion frameworks such as EKF, complementary EKF, UKF and Markov model [45], the LKF is simple and easy to use for the fusion process. In our experiment scenarios, we tackled the problem formulation in the linear perspective and the LKF is the best option for combining the user position results. The proposed sensor fusion framework uses the same LKF implementation explained in Figure 5 for Wi-Fi and PDR systems fusion. The LKF is a recursive filter that estimates the state in a linear system. The estimation in LKF is repeated to minimize errors from inaccurate measurements and observations and optimum performance can be achieved when the noise is Gaussian noise. The LKF implementation consists of two phases, prediction and correction as illustrated in Figure 5.

The position results from Wi-Fi and PDR are fused by LKF filter and the results from LKF shows better performance than individual systems. In the LKF prediction phase, the state variable $\hat{x}_k$ and error covariance matrix P_k are estimated using state transition matrix (A). It also predicted the corrected state $\hat{x}_{k-1}$ and the error covariance P_{k-1} at previous time. The Gaussian noise (Q) is used for error covariance prediction. In the correction phase, the Kalman gain represents the reliability of the predicted value and the measured value. The larger the Kalman gain, the more reliable the measured value, and the smaller the Kalman gain, the more reliable the predicted value. The final output value P_{k-1} of the LKF is determined according to the Kalman gain.

4. Experiment and Result Analysis

The performance of the proposed sensor fusion framework is evaluated by two experiment scenarios based on the user motion. We used two user motions; rectangular and linear motion with predefined paths. For collecting data, a user of age 27 and height 172 cm walked on the reference path in the experiment areas with a smartphone in his hand. Figure 6 shows the experiment scenarios and the red line indicates the reference paths used for the experiment.

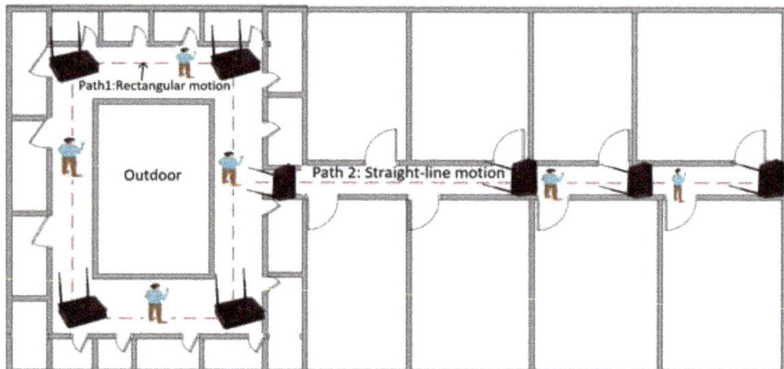

Figure 6. Experiment scenarios.

For collecting the smartphone sensor and RSSI data, we used a Samsung Galaxy Note 8 smartphone with a Snapdragon 835 processor and 6 GB RAM. The experiment area uses a LG U plus Wi-Fi APs for providing RSSI signals. In the first experiment, the user took a rectangular motion with a test area of 45 m × 37 m. The test area was divided into 4 × 20 grid points and the Wi-Fi RSSI for each of the points was recorded. To collect the Wi-Fi RSSI, we used a fingerprint Android application which can store the RSSI measurements in form of comma-separated values (CSV). This application provides an option for selecting the required APs for the experiment. Four APs were chosen for localization based on their signal strength in the experiment area. For the second experiment, a linear motion of the user was considered in a 75 m × 3 m corridor as shown in Figure 6. The corridor was divided into 2 × 60 grid points and the user walked in the reference paths. For PDR data collection, the sensor stream inertial measurement unit and global positioning system Android application were used. The Android application can access the accelerometer, gyroscope, and magnetometer sensors in the smartphone. The user can select any of the sensors and analyze the current values of the sensors. The application has an option for adjusting sensor frequency based on the experiment requirements.

The performance of the proposed sensor fusion framework was analyzed using the data from paths 1 and 2. Figures 7 and 8 show the results from the proposed sensor fusion framework and conventional sensor fusion frameworks for paths 1 and 2. From Figures 7 and 8, it can be seen that the proposed sensor fusion framework can minimize localization errors when compared to conventional sensor fusion approaches. In these experiments, the starting position was the origin and the experiments were carried out strictly along the predefined paths. The red line in Figure 7 shows the reference path for rectangular pedestrian motion and the dashed red line in Figure 8 shows the reference path for linear user motion. The accuracy of the proposed sensor fusion framework was evaluated by computing the average localization error and probability distribution of the localization errors. The average localization errors for the proposed sensor fusion framework and conventional sensor fusion frameworks for paths 1 and 2 are shown in Figures 9 and 10.

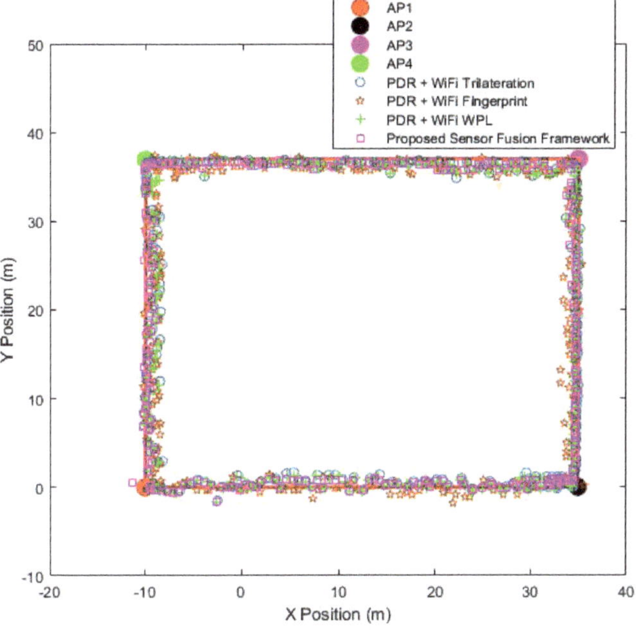

Figure 7. Trajectories of ground truth, proposed sensor fusion framework, conventional sensor fusion frameworks from rectangular motion.

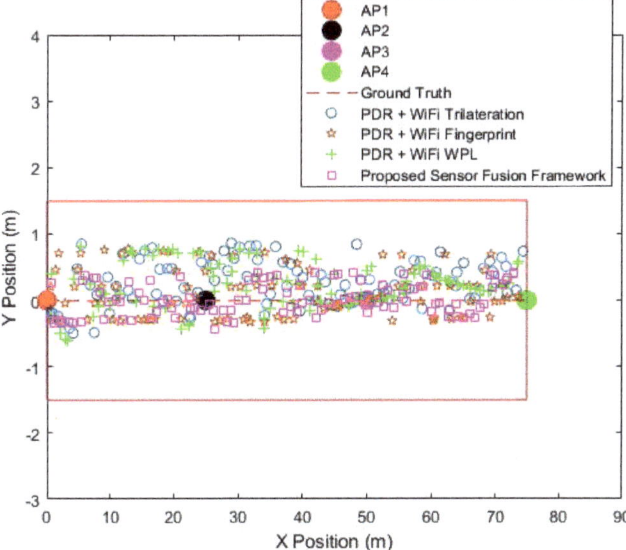

Figure 8. Trajectories of ground truth, proposed sensor fusion framework, conventional sensor fusion frameworks from linear motion.

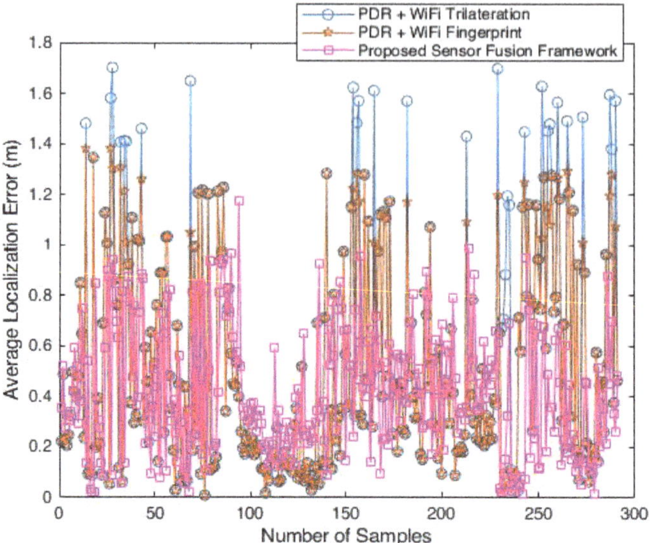

Figure 9. Average localization error from rectangular motion.

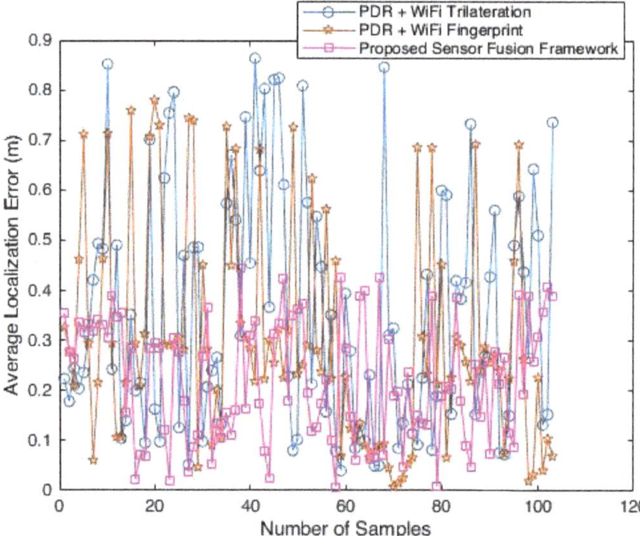

Figure 10. Average localization error from linear motion.

From Figures 9 and 10, the proposed sensor fusion framework has less position errors and gives better performance when compared to conventional sensor fusion approaches. The average localization error in the proposed sensor fusion framework is lower than in conventional sensor fusion frameworks. The average localization error from the PDR+Wi-Fi fingerprint approach shows a constant average localization error for most of the experiment time. This is due to the fingerprint map approximation between online data and offline data. The PDR+Wi-Fi trilateration approach shows the worst performance compared to other sensor fusion approaches. The localization accuracy of PDR+Wi-Fi trilateration approach depends on the multipath effects. The NLOS conditions and free space path loss model in the PDR+Wi-Fi trilateration approach yields larger position errors than other sensor fusion approaches. The probability distribution of localization errors for sensor fusion frameworks in both experiment scenarios is shown in Figures 11 and 12.

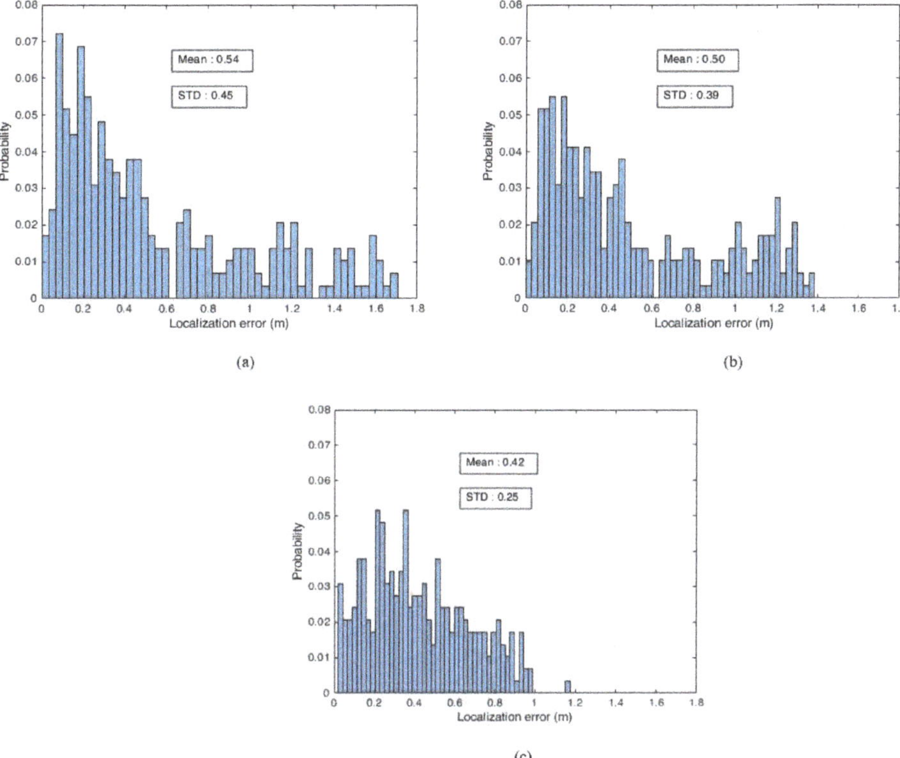

Figure 11. Probability distribution of the localization error for rectangular motion . (**a**) PDR+Wi-Fi trilateration. (**b**) PDR+Wi-Fi fingerprint. (**c**) Proposed Sensor Fusion Framework.

From probability distribution of the localization error plots, it is clear that the proposed sensor fusion framework is better than other methods and gives high position accuracy for indoor localization. When the experiment starts, the proposed sensor fusion framework and PDR+Wi-Fi fingerprint approach show the same localization errors. When the experiment time increases the proposed sensor fusion framework shows the highest position accuracy as compared to the other sensor fusion frameworks. At some point in the experiment, the PDR+Wi-Fi fingerprint and PDR+Wi-Fi trilateration approaches show the same localization errors. Tables 2 and 3 summarize the performance of the proposed sensor fusion framework versus conventional sensor fusion frameworks in terms of mean error, maximum error, minimum error, and standard deviation of error (STD).

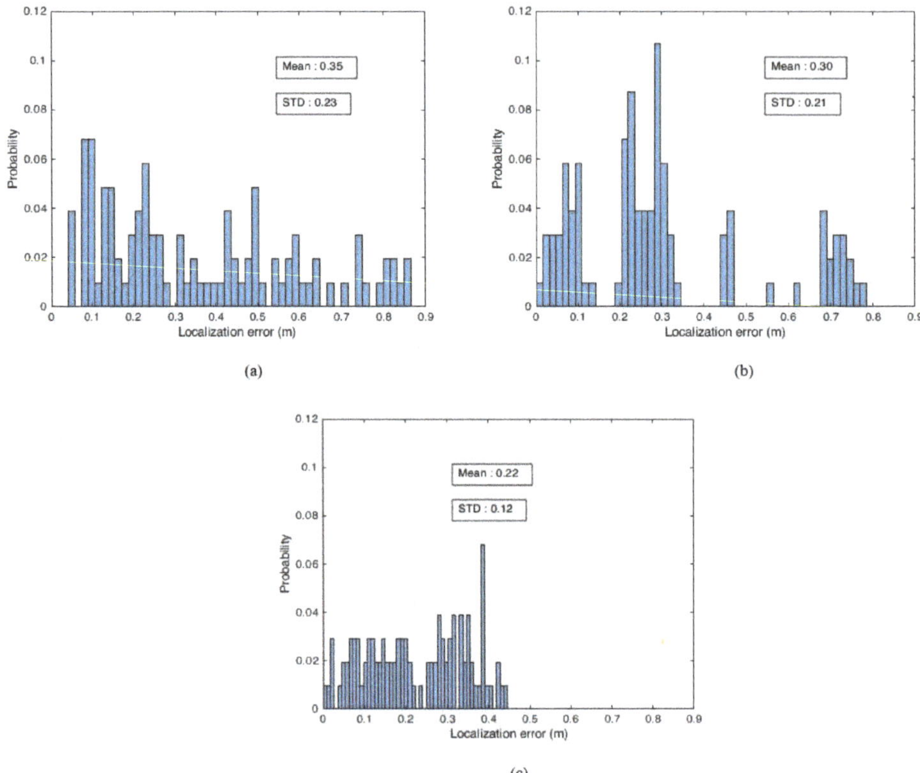

Figure 12. Probability distribution of the localization error for linear motion. (**a**) PDR+Wi-Fi trilateration. (**b**) PDR+Wi-Fi fingerprint. (**c**) Proposed Sensor Fusion Framework.

Table 2. Performance of different sensor fusion frameworks for path 1 rectangular motion.

Position Estimation Systems	Mean Error (m)	Max. Error (m)	Min. Error (m)	Standard Deviation of Error (m)
PDR+Wi-Fi Trilateration	0.54	1.7	0.01	0.45
PDR+Wi-Fi Fingerprint	0.50	1.38	0.5	0.39
Proposed sensor fusion framework	0.42	1.17	0.02	0.25

From Tables 2 and 3, the proposed sensor fusion framework has a lower mean error and maximum error than conventional sensor fusion approaches. The mean error results show that the PDR+Wi-Fi trilateration and PDR+Wi-Fi fingerprint approaches have almost similar error results compared to the proposed sensor fusion approach. However, the maximum error from PDR+Wi-Fi trilateration is very high compared to the other two sensor fusion approaches. The minimum error results indicate that the PDR+Wi-Fi trilateration approach has the least minimum error compared to the other two approaches. The standard deviation of error results validates the significance of the proposed sensor fusion framework with respect to conventional sensor fusion frameworks. From all the experiment results and analysis, we conclude that the proposed sensor fusion approach significantly outperforms that of conventional sensor fusion approaches and shows significant position accuracy for indoor localization.

Table 3. Performance of different sensor fusion frameworks for path 2 straight-line motion.

Position Estimation Systems	Mean Error (m)	Max. Error (m)	Min. Error (m)	Standard Deviation of Error (m)
PDR+Wi-Fi Trilateration	0.35	0.86	0.04	0.23
PDR+Wi-Fi Fingerprint	0.30	0.78	0.008	0.21
Proposed sensor fusion framework	0.22	0.44	0.007	0.12

5. Conclusions and Future Work

This paper proposed a sensor fusion framework for indoor localization using Wi-Fi RSSI signals and smartphone sensors. In the proposed sensor fusion framework model, we used a PDR algorithm which reduces the smartphone sensor errors and gives accurate position results for indoor localization. In the case of Wi-Fi localization systems, we proposed a Wi-Fi fusion algorithm for localization that achieves better performance than conventional Wi-Fi localization systems. To combine the position results from PDR and Wi-Fi systems, we proposed a sensor fusion framework model that shows better indoor localization accuracy than conventional Wi-Fi localization systems. The results from the rectangular motion experiment shows that the proposed sensor fusion framework has a maximum of 1.17 m error when compared to the predefined path. In the case of linear motion, the results show that the proposed sensor fusion framework has a maximum of 0.44 m errors when compared to predefined path. From the experiments and results, the proposed sensor fusion framework showed reasonable localization accuracy for indoor localization with fewer IMU sensor errors and solved the Wi-Fi RSSI signal fluctuation problems. However, in the future, we will mainly focus on multiple pedestrian localization. We will carry out experiments for multiple users in indoor scenarios.

Author Contributions: Writing—original draft, A.P. and J.K.; Writing—review & editing, D.S.H.

Funding: This work was supported by Institute of Information & Communications Technology Planning & Evaluation (IITP) grant funded by the Korea government (MSIT) (2016-0-00564, Development of Intelligent Interaction Technology Based on Context Awareness and Human Intention Understanding).

Conflicts of Interest: The authors declare no conflict of interest.

References

1. Xu, R.; Chen, W.; Xu, Y.; Ji, S. A new indoor positioning system architecture using GPS signals. *Sensors* **2015**, *15*, 10074–10087. [CrossRef]
2. Oussar, Y.; Ahriz, I.; Denby, B.; Dreyfus, G. Indoor localization based on cellular telephony RSSI fingerprints containing very large numbers of carriers. *EURASIP J. Wirel. Commun. Network.* **2011**, *2011*, 81. [CrossRef]
3. Alarifi, A.; Al-Salman, A.; Alsaleh, M.; Alnafessah, A.; Al-Hadhrami, S.; Al-Ammar, M.; Al-Khalifa, H. Ultra wideband indoor positioning technologies: Analysis and recent advances. *Sensors* **2016**, *16*, 707. [CrossRef]
4. Kuang, J.; Niu, X.; Zhang, P.; Chen, X. Indoor positioning based on pedestrian dead reckoning and magnetic field matching for smartphones. *Sensors* **2018**, *18*, 4142. [CrossRef]
5. Saab, S.S.; Nakad, Z.S. A standalone RFID indoor positioning system using passive tags. *IEEE Trans. Ind. Electron.* **2011**, *58*, 1961–1970. [CrossRef]
6. Zhuang, Y.; Yang, J.; Li, Y.; Qi, L.; El-Sheimy, N. Smartphone-based indoor localization with bluetooth low energy beacons. *Sensors* **2016**, *16*, 596. [CrossRef] [PubMed]

7. Guo, X.; Shao, S.; Ansari, N.; Khreishah, A. Indoor localization using visible light via fusion of multiple classifiers. *IEEE Photon. J.* **2017**, *9*, 1–16. [CrossRef]
8. Uradzinski, M.; Guo, H.; Liu, X.; Yu, M. Advanced indoor positioning using Zigbee wireless technology. *Wirel. Pers. Commun.* **2017**, *97*, 6509–6518. [CrossRef]
9. Husen, M.; Lee, S. Indoor location sensing with invariant Wi-Fi received signal strength fingerprinting. *Sensors* **2016**, *16*, 1898. [CrossRef]
10. Jiang, P.; Zhang, Y.; Fu, W.; Liu, H.; Su, X. Indoor mobile localization based on Wi-Fi fingerprint's important access point. *Int. J. Distrib. Sens. Netw.* **2015**, *11*, 429104. [CrossRef]
11. Liu, H.; Yang, J.; Sidhom, S.; Wang, Y.; Chen, Y.; Ye, F. Accurate WiFi based localization for smartphones using peer assistance. *IEEE Trans. Mob. Comput.* **2014**, *13*, 2199–2214. [CrossRef]
12. Kim, Y.; Chon, Y.; Cha, H. Smartphone-based collaborative and autonomous radio fingerprinting. *IEEE Trans. Syst. Man Cybern. Part C (Appl. Rev.)* **2012**, *42*, 112–122. [CrossRef]
13. Xiao, W.; Ni, W.; Toh, Y.K. Integrated Wi-Fi fingerprinting and inertial sensing for indoor positioning. In Proceedings of the 2011 International Conference on Indoor Positioning and Indoor Navigation (IPIN), Guimaraes, Portugal, 21–23 September 2011; pp. 1–6.
14. Tian, Z.; Fang, X.; Zhou, M.; Li, L. Smartphone-based indoor integrated WiFi/MEMS positioning algorithm in a multi-floor environment. *Micromachines* **2015**, *6*, 347–363. [CrossRef]
15. Tarrío, P.; Besada, J.A.; Casar, J.R. Fusion of RSS and inertial measurements for calibration-free indoor pedestrian tracking. In Proceedings of the 16th International Conference on Information Fusion, Istanbul, Turkey, 9–12 July 2013; pp. 1458–1464.
16. Poulose, A.; Eyobu, O.S.; Han, D.S. A Combined PDR and Wi-Fi Trilateration Algorithm for Indoor Localization. In Proceedings of the 2019 International Conference on Artificial Intelligence in Information and Communication (ICAIIC), Okinawa, Japan, 11–13 February 2019; pp. 72–77.
17. Poulose, A.; Eyobu, O.S.; Han, D.S. An Indoor Position-Estimation Algorithm Using Smartphone IMU Sensor Data. *IEEE Access* **2019**, *7*, 11165–11177. [CrossRef]
18. Ho, N.-H.; Truong, P.; Jeong, G.-M. Step-detection and adaptive step-length estimation for pedestrian dead-reckoning at various walking speeds using a smartphone. *Sensors* **2016**, *16*, 1423. [CrossRef] [PubMed]
19. Yuan, X.; Yu, S.; Zhang, S.; Wang, G.; Liu, S. Quaternion-based unscented Kalman filter for accurate indoor heading estimation using wearable multi-sensor system. *Sensors* **2015**, *15*, 10872–10890. [CrossRef] [PubMed]
20. Kang, W.; Han, Y. SmartPDR: Smartphone-based pedestrian dead reckoning for indoor localization. *IEEE Sens. J.* **2015**, *15*, 2906–2916. [CrossRef]
21. Evennou, F.; Marx, F. Advanced integration of WiFi and inertial navigation systems for indoor mobile positioning. *Eur. J. Appl. Signal Process.* **2006**, *2006*, 164. [CrossRef]
22. Wang, H.; Lenz, H.; Szabo, A.; Bamberger, J.; Hanebeck, U.D. WLAN-based pedestrian tracking using particle filters and low-cost MEMS sensors. In Proceedings of the 2007 4th Workshop on pOsitioning, Navigation and Communication (WPNC), Hannover, German, 22 March 2007; pp. 1–7.
23. Carrera, J.L.; Zhao, Z.; Braun, T.; Li, Z. A real-time indoor tracking system by fusing inertial sensor, radio signal and floor plan. In Proceedings of the 2016 International Conference on Indoor Positioning and Indoor Navigation (IPIN), Alcala de Henares, Spain, 4–7 October 2016; pp. 1–8.
24. Knauth, S. Smartphone PDR positioning in large environments employing WiFi, particle filter, and backward optimization. In Proceedings of the 2017 International Conference on Indoor Positioning and Indoor Navigation (IPIN), Sapporo, Japan, 18–21 September 2017; pp. 1–6.
25. Frank, K.; Krach, B.; Catterall, N.; Robertson, P. Development and evaluation of a combined WLAN & inertial indoor pedestrian positioning system. In Proceedings of the ION GNSS, Savannah, Georgia, 22–25 September 2009; pp. 1–9.
26. Zhuang, Y.; El-Sheimy, N. Tightly-coupled integration of WiFi and MEMS sensors on handheld devices for indoor pedestrian navigation. *IEEE Sens. J.* **2016**, *16*, 224–234. [CrossRef]
27. Deng, Z.-A.; Hu, Y.; Yu, J.; Na, Z. Extended Kalman filter for real time indoor localization by fusing WiFi and smartphone inertial sensors. *Micromachines* **2015**, *6*, 523–543. [CrossRef]
28. Leppäkoski, H.; Collin, J.; Takala, J. Pedestrian navigation based on inertial sensors, indoor map, and WLAN signals. *J. Signal Process. Syst.* **2013**, *71*, 287–296. [CrossRef]

29. Chen, G.; Meng, X.; Wang, Y.; Zhang, Y.; Tian, P.; Yang, H. Integrated WiFi/PDR/Smartphone using an unscented kalman filter algorithm for 3D indoor localization. *Sensors* **2015**, *15*, 24595–24614. [CrossRef] [PubMed]
30. Waqar, W.; Chen, Y.; Vardy, A. Incorporating user motion information for indoor smartphone positioning in sparse Wi-Fi environments. In Proceedings of the 17th ACM international conference on Modeling, Analysis and Simulation of Wireless and Mobile Systems (ACM), Montreal, QC, Canada, 21–26 September 2014; pp. 267–274.
31. Li, Y.; Zhuang, Y.; Lan, H.; Zhou, Q.; Niu, X.; El-Sheimy, N. A hybrid WiFi/magnetic matching/PDR approach for indoor navigation with smartphone sensors. *IEEE Commun. Lett.* **2016**, *20*, 169–172. [CrossRef]
32. Li, Z.; Liu, C.; Gao, J.; Li, X. An improved WiFi/PDR integrated system using an adaptive and robust filter for indoor localization. *ISPRS Int. J. Geo-Inf.* **2016**, *5*, 224. [CrossRef]
33. Kim, P. *Kalman Filter for Beginners: With MATLAB Examples*; CreateSpace: Scotts Valley, CA, USA, 2011.
34. Diaz, E.M.; Gonzalez, A.L.M. Step detector and step length estimator for an inertial pocket navigation system. In Proceedings of the 2014 International Conference on Indoor Positioning and Indoor Navigation (IPIN), Busan, Korea, 27–30 October 2014; pp. 105–110.
35. Zhang, M.; Wen, Y.; Chen, J.; Yang, X.; Gao, R.; Zhao, H. Pedestrian dead-reckoning indoor localization based on OS-ELM. *IEEE Access* **2018**, *6*, 6116–6129. [CrossRef]
36. Shchekotov, M. Indoor localization method based on wi-fi trilateration technique. In Proceedings of the 16th Conference of Fruct Association, Oulu, Finland, 27–31 October 2014; pp. 177–179.
37. Ilci, V.; Alkan, R.M.; Gülal, V.E.; Cizmeci, H. Trilateration technique for WiFi-based indoor localization. In Proceedings of the 11th International Conference on Wireless and Mobile Communications (ICWMC), St. Julians, Malta, 11–16 October 2015; p. 36.
38. Yim, J.; Jeong, S.; Gwon, K.; Joo, J. Improvement of Kalman filters for WLAN based indoor tracking. *Expert Syst. Appl.* **2010**, *37*, 426–433. [CrossRef]
39. Yim, J. Development of Web Services for WLAN-based Indoor Positioning and Floor Map Repositories. *Int. J. Control Autom.* **2014**, *7*, 63–74. [CrossRef]
40. Navarro, E.; Peuker, B.; Quan, M.; Clark, C.; Jipson, J. *Wi-Fi Localization Using RSSI Fingerprinting*; Citeseer: 2010. Available online: http://digitalcommons.calpoly.edu/cgi/viewcontent.cgi?article=1007&context=cpesp (accessed on 18 March 2019).
41. Zegeye, W.K.; Amsalu, S.B.; Astatke, Y.; Moazzami, F. WiFi RSS fingerprinting indoor localization for mobile devices. In Proceedings of the 2016 IEEE 7th Annual Ubiquitous Computing, Electronics & Mobile Communication Conference (UEMCON), New York, NY, USA, 20–22 October 2016; pp. 1–6.
42. Khan, M.; Kai, Y.D.; Gul, H.U. Indoor Wi-Fi positioning algorithm based on combination of location fingerprint and unscented Kalman filter. In Proceedings of the 2017 14th International Bhurban Conference on Applied Sciences and Technology (IBCAST), Islamabad, Pakistan, 10–14 January 2017; pp. 693–698.
43. Retscher, G. Fusion of location fingerprinting and trilateration based on the example of differential wi-fi positioning. *ISPRS Ann. Photogramm. Remote Sens. Spat. Inf. Sci.* **2017**, *4*, 377–384. [CrossRef]
44. Chan, S.; Sohn, G. Indoor localization using wi-fi based fingerprinting and trilateration techiques for lbs applications. *Int. Arch. Photogramm. Remote Sens. Spat. Inf. Sci.* **2017**, *38*, C26. [CrossRef]
45. Elliott, R.J.; Aggoun, L.; Moore, J.B. *Hidden Markov Models: Estimation And Control*; Springer Science & Business Media: New York, NY, USA, 28 July 2008; Volume 29.

Article

Indoor Positioning Integrating PDR/Geomagnetic Positioning Based on the Genetic-Particle Filter

Meng Sun [1,2], Yunjia Wang [1,2,*], Shenglei Xu [2], Hongji Cao [2] and Minghao Si [2]

[1] Key Laboratory of Land Environment and Disaster Monitoring, MNR, China University of Mining and Technology, Xuzhou 221116, China; msun@cumt.edu.cn

[2] School of Environment Science and Spatial Informatics, China University of Mining and Technology, Xuzhou 221116, China; cumtxsl@163.com (S.X.); hjcao@cumt.edu.cn (H.C.); 13218021689@163.com (M.S.)

* Correspondence: wyjc411@163.com; Tel.: +86-139-5220-7068

Received: 29 November 2019; Accepted: 16 January 2020; Published: 17 January 2020

Abstract: This paper proposes a fusion indoor positioning method that integrates the pedestrian dead-reckoning (PDR) and geomagnetic positioning by using the genetic-particle filter (GPF) algorithm. In the PDR module, the Mahony complementary filter (MCF) algorithm is adopted to estimate the heading angles. To improve geomagnetic positioning accuracy and geomagnetic fingerprint specificity, the geomagnetic multi-features positioning algorithm is devised and five geomagnetic features are extracted as the single-point fingerprint by transforming the magnetic field data into the geographic coordinate system (GCS). Then, an optimization mechanism is designed by using gene mutation and the method of reconstructing a particle set to ameliorate the particle degradation problem in the GPF algorithm, which is used for fusion positioning. Several experiments are conducted to evaluate the performance of the proposed methods. The experiment results show that the average positioning error of the proposed method is 1.72 m and the root mean square error (RMSE) is 1.89 m. The positioning precision and stability are improved compared with the PDR method, geomagnetic positioning, and the fusion-positioning method based on the classic particle filter (PF).

Keywords: indoor positioning; PDR; geomagnetic positioning; particle filter; genetic algorithm

---

1. Introduction

With the development of modern society, people's demands for location services is becoming increasingly urgent [1]. The Global Navigation Satellite System (GNSS) [2] can provide users with a high-precision positioning service in the outdoor environment. However, in the indoor environment, because the GNSS signal is blocked by humans, buildings and many other factors, it becomes so weak that the positioning accuracy will be greatly reduced. In order to solve indoor positioning problems, many kinds of method have been studied, such as Wi-Fi positioning [3,4], Bluetooth positioning [5,6], ultra-wideband (UWB) [7,8], radio frequency identification (RFID) [9], infrared [10], ZigBee [11], etc. Every positioning method is capable of providing an indoor location service for people, especially the UWB technology which can reach sub-meter accuracy in the indoor open areas [8]. But these approaches require additional signal transmitting devices, which means that people cannot obtain their position if there is no such equipment in the environment. Moreover, these devices are usually expensive which will make the cost very high if we apply these technologies to a large-scale environment.

Pedestrian dead-reckoning (PDR) [12,13] technology can use an accelerometer, gyroscope, and magnetometer to obtain continuous positions instead of installing the signal transmitting stations. But the positioning error will accumulate quickly because of the initial position error, heading angle estimation error and other measurement errors. Therefore, it is necessary to integrate PDR with other positioning methods to restrict error accumulation.

In 2000, Suksakulchai et al. [14] conducted an experimental analysis in the corridor environment and proposed that the indoor magnetic field can be used for indoor positioning. Then, experiments in [15] studied in detail the stability and distribution of an indoor magnetic field, pointing out that it is stable and reliable. All of these related research works provide a solid theoretical basis for the application of the magnetic field for indoor positioning. In recent years, researchers have found that utilizing indoor magnetic field can achieve high positioning precision at almost no investment that are safe and infrastructure-free compared with other technologies [16]. These characteristics make indoor geomagnetic positioning has great research value and also gradually become the research hotspot.

Today there are many kinds of geomagnetic positioning methods, which can be divided into two main types: sequence-based magnetic matching positioning (SBMP) and single-point-based magnetic matching positioning (SPMP). The dynamic time-wrapping algorithm (DTW) [17] is a representative positioning algorithm of SBMP. In [18,19], experimenters applied the DTW algorithm to magnetic matching by comparing the similarity of different trajectories and employed this magnetic positioning method on smartphones. The results show that the accuracy of 80% test positions is approximately 2 m. Generally speaking, the SBMP method has high positioning precision, but it relies on stable magnetic data sequences. However, the magnetic data sequence will generate large fluctuations due to the device heterogeneity. Even though some methods have been studied to solve this problem, they cannot be implemented in real-time positioning [20–22]. What is worse is that SBMP is restricted by the trajectory and walking speed [16,23]. In other words, if participants want to get their position, they should walk on the trajectory where they collect magnetic field data before, which means that this method will perform poorly in the open area.

By contrast with SBMP, the SPMP method uses the geomagnetic features data at the signal points rather than the magnetic sequence data and there is also no limitation to the speed and trajectory of pedestrian walking. Therefore, it is more flexible and easier to implement compared with SBMP method. This method usually does not have high positioning precision if there are few geomagnetic features. But the particle filter algorithm makes up for this shortcoming and has high positioning accuracy, which is also a typical single-point-based fingerprint positioning algorithm [24,25]. In related research, Huang et al. [26] used the Hausdorff distance and Pearson correlation coefficient [27] to control the initial position error and accelerate the convergence speed of the filter. In [28], an adaptive particle filter algorithm was proposed in geomagnetic positioning. The number of particles in the experiment can adjust according to the different features and the average positioning accuracy was 0.51 m. Shu et al. [29] combined the Wi-Fi signal with magnetic field data and designed a two-pass bidirectional particle-filtering process to improve precision, which achieved a mean location accuracy of 0.9 m in the office environment. Although geomagnetic positioning technology based on the particle filter can achieve high accuracy, it suffers from the particle degradation problem. Increasing particle numbers is one of the approaches to solve this problem, for example, the number of particles in [28] even reaches 3000. A large number of particles will result in a large computational load and limit its application on smart terminals. Therefore, how to better solve the particle degradation problem is the key to improving the performance of the particle filter.

In this paper, we choose the SPMP method for geomagnetic positioning because it is more flexible than the SBMP method. Five magnetic features data are extracted to ensure the reliability of geomagnetic fingerprint and the accuracy of the geomagnetic multi-features positioning algorithm. Moreover, the Mahony complementary filter (MCF) algorithm is applied to calculate the heading angles in PDR technology. In order to optimize the particle degradation problem, we use the gene mutation method in the genetic algorithm [30] and the method of reconstructing the particle set by combining the geomagnetic positioning results to update the particle set. Lastly, we integrate the PDR and the geomagnetic positioning by using the genetic-particle filter (GPF) algorithm and carry out several experiments to evaluate its performance. The contributions can be summarized as follows:

1. We transform the geomagnetic data into the geographic coordinate system (GCS) and extract five geomagnetic features data. These features data can improve the specificity of geomagnetic fingerprint and the accuracy of the devised geomagnetic multi-features positioning algorithm.
2. We design the optimization mechanism for the particle degradation problem, which utilizes the genetic mutation method to increase the diversity of the resampled particles and uses the particle set reconstruction method by combining the geomagnetic positioning results to improve the particles' reliability. The proposed mechanism better ameliorates the particle degradation problem;
3. We propose the fusion-positioning method that utilizes the PDR positions as the system state and uses the geomagnetic positions as the system observation based on the genetic-particle filter. The mean positioning error is 1.72 m and the root mean square error is 1.89 m.

The remainder of the paper is organized as follows: Section 2 gives a detailed description of each module, such as PDR technology; Section 3 evaluates the performance of the proposed methods; Section 4 summarizes the conclusions and discussion.

2. Materials and Methods

2.1. Pedestrian Dead-Reckoning (PDR) Module

PDR technology mainly includes three parts: heading angle estimation, step detection, and step length estimation. As illustrated in Figure 1, if the initial position is known, PDR technology can utilize the measured heading angles and step length to reckon positions by using Equation (1). However, due to the factors like the accuracy of heading estimation and the error of the initial position, system error will accumulate rapidly with time, which directly causes the navigation trajectory to deviate seriously from the actual path. Therefore, other positioning methods are usually needed to weaken the influence of error accumulation on the positioning results.

$$\begin{cases} N_{k+1} = N_k + l_k \bullet \cos \psi_k \\ E_{k+1} = E_k + l_k \bullet \sin \psi_k \end{cases} \tag{1}$$

where N_k and E_k represent the coordinate values of the north and east at time k, respectively; l_k and ψ_k represent the step length and heading angle at time k, whose calculation methods will be described in the following sections.

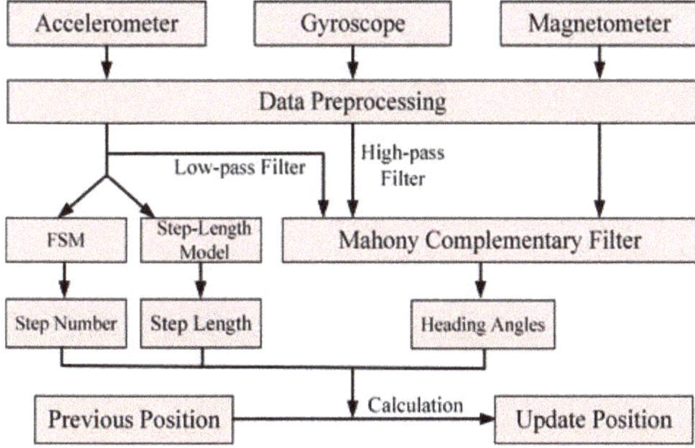

Figure 1. The framework of pedestrian dead-reckoning (PDR) technology in this research.

2.1.1. Heading Angle Estimation

Estimating heading angle is important for the PDR method to calculate position coordinates. In general, the heading angle can be measured by using the gyroscope. But there are many integration calculations that will make the measured heading angle inaccurate. In our research, the Mahony complementary filter [31] is implemented on the smartphone and used to compute the heading angle. It should be clear that the MCF algorithm mainly utilizes the gyroscope data to calculate the attitude of carrier, but the accelerometer and magnetometer will correct the gyroscope data so that the estimation error will not accumulate.

Generally, the whole producer of the MCF algorithm contains error compensation calculation and gyroscope data correction. The gyroscope data will be expressed as the quaternion vector, which represents current attitude prediction. Assuming that the quaternion vector is $\boldsymbol{Q} = [q_0\ q_1\ q_2\ q_3]^T$, it is defined as follows:

$$\begin{cases} q_0 = \cos(\beta/2) \\ q_1 = \omega'_x \sin(\beta/2) \\ q_2 = \omega'_y \sin(\beta/2) \\ q_3 = \omega'_z \sin(\beta/2) \end{cases} \tag{2}$$

where β is the rotation angle, ω'_x, ω'_y and ω'_z are the normalized instantaneous gyroscope data. Then, in the error compensation calculation stage, two error items will be calculated by using the accelerometer and magnetometer data. If the gyroscope error correction is $\boldsymbol{e} = \begin{bmatrix} e_x & e_y & e_z \end{bmatrix}^T$, it can be defined as:

$$\boldsymbol{e} = \boldsymbol{e}_a + \boldsymbol{e}_m \tag{3}$$

where $\boldsymbol{e}_a$ and $\boldsymbol{e}_m$ are the error correction items calculated by using the accelerometer and magnetometer data, respectively. They can be expressed as $\boldsymbol{e}_a = \begin{bmatrix} e_{ax} & e_{ay} & e_{az} \end{bmatrix}^T$ and $\boldsymbol{e}_m = \begin{bmatrix} e_{mx} & e_{my} & e_{mz} \end{bmatrix}^T$. Their calculation methods are as follows:

$$\begin{cases} \boldsymbol{e}_a = (\boldsymbol{C}_n^b \bullet \boldsymbol{g}_a) \times \boldsymbol{a} \\ \boldsymbol{e}_m = (\boldsymbol{C}_n^b \bullet \boldsymbol{b}_m) \times \boldsymbol{m} \end{cases} \tag{4}$$

where $\boldsymbol{g}_a$ is the gravity vector in the geographic coordinate system, $\boldsymbol{g}_a = \begin{bmatrix} 0 & 0 & g \end{bmatrix}^T$, b_m is the geomagnetic vector when x-axis of device points to the north, $\boldsymbol{b}_m = \begin{bmatrix} b_{mx} & 0 & b_{mz} \end{bmatrix}^T$; $\boldsymbol{a}$ and $\boldsymbol{m}$ are the normalized measured accelerometer and magnetometer data; $\boldsymbol{C}_n^b$ is the rotation matrix from the geographic coordinate system to the carrier coordinate system and it will be described in Section 2.2; "×" represents the vector cross product, respectively. After getting the error correction, the corrected gyroscope data can be calculated as:

$$\boldsymbol{\omega} = \boldsymbol{\omega}_g + K_P \boldsymbol{e} + K_I \int \boldsymbol{e} \tag{5}$$

where $\boldsymbol{\omega}_g = \begin{bmatrix} \omega_{gx} & \omega_{gy} & \omega_{gz} \end{bmatrix}^T$ and $\boldsymbol{\omega} = \begin{bmatrix} \omega_x & \omega_y & \omega_z \end{bmatrix}^T$ are the normalized gyroscope raw data and the corrected gyroscope data, respectively; K_P and K_I are the error control items, which are 2.0 and 0.001 based on our test, respectively. After getting the corrected gyroscope data, we should substitute $\boldsymbol{\omega}$ into the quaternion differential equation and solve this equation by using the first-order Runge–Kutta method [32] to obtain the quaternion update equation as Equation (6). The quaternion values can be calculated by this equation:

$$
\begin{cases}
q_0(t+T) = q_0(t) - \frac{T}{2}\left[\omega_x q_1(t) + \omega_y q_2(t) + \omega_z q_3(t)\right] \\
q_1(t+T) = q_1(t) + \frac{T}{2}\left[\omega_x q_0(t) + \omega_z q_2(t) - \omega_y q_3(t)\right] \\
q_2(t+T) = q_2(t) + \frac{T}{2}\left[\omega_y q_0(t) - \omega_z q_1(t) + \omega_x q_3(t)\right] \\
q_3(t+T) = q_3(t) + \frac{T}{2}\left[\omega_z q_0(t) + \omega_y q_1(t) - \omega_z q_2(t)\right]
\end{cases}
\tag{6}
$$

where $q_i(t)$ and $q_i(t+T)$ are the quaternion values at time t and $(t+T)$, $i = 1,2,3,\cdots$, T is the sampling time interval. Then, normalizing the updated quaternion values and substituting them into Equation (7), the heading angle can be obtained. In our research, the first-order low-pass filter [33] and the first-order high-pass filter [34] are applied to process the observation noise of the acceleration raw data and the gyroscope raw data, respectively. Because there are serval heading angles in one second, the median filter [35] is used to reduce the measurement error and obtain more accurate results.

$$
\begin{cases}
\psi = \mathrm{tg}^{-1}\frac{2(q_1q_2+q_0q_3)}{1-2(q_2^2+q_3^2)}, \psi \in (0, 2\pi) \\
\theta = -\sin^{-1}2(q_1q_3 - q_0q_2), \theta \in (-\frac{\pi}{2}, \frac{\pi}{2}) \\
\gamma = \mathrm{tg}^{-1}\frac{2(q_2q_3+q_0q_1)}{1-2(q_1^2+q_2^2)}, \gamma \in (-\pi, \pi)
\end{cases}
\tag{7}
$$

where ψ, θ, and γ are the heading angle, pitch, and roll, respectively, ψ is also called yaw.

2.1.2. Step Detection

There are many step detection methods in PDR technology, such as peak detection [36], zero velocity update [37], etc. As shown in Figure 2a, the acceleration changes regularly during the walking process and the acceleration cure has peaks and valleys. The peak detection method realizes the step detection by detecting the peaks and valleys. This method is easy and has a low calculation burden. However, the measurement error will affect it when recognizing steps. In our research, we employ the finite-state machine (FSM) algorithm to the step detection process because this algorithm is easy to be implemented and more resistant to error interference than the peak detection method.

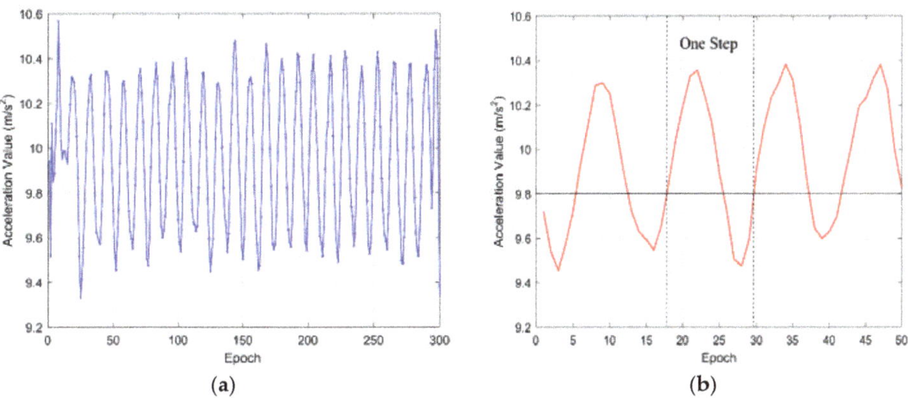

Figure 2. (**a**) Acceleration curve of the walking process; (**b**) acceleration curve of one step.

In fact, FSM is a special peak detection algorithm. It divides the walking process into five states, which can be recorded as $S_0\sim S_4$. S_0 represents the steady state and the acceleration at this time is approximately equal to 9.8 m/s^2; S_1 is the taking step state and the acceleration gradually increases; S_2 is the finding peaks state after entering the state S_1; S_3 is the peak threshold judgment state and filter out the false peaks; S_4 represents the falling state and the acceleration gradually decreases. As shown in Figure 3, the current walking state is transferred in the above five states and one step will be recognized if the five states all satisfy the threshold conditions. When a large measurement noise appears during the algorithm execution, it will return to the S_0 state and re-detect the walking state. Therefore, FSM can effectively eliminate the interference of false peaks and the measurement errors in the step detection process.

Table 1 gives all the variable names of FSM. It should be noted that P_{up} and P_{down} will be reset as zero at S_0 state or there exist large measurement errors. They can be calculated as follows:

$$\begin{cases} P_{up} = P_{up} + 1, a_i - a_{i-1} > acc\_diff \\ P_{down} = P_{down} + 1, a_i - a_{i-1} < acc\_diff \end{cases} \tag{8}$$

where a_i and a_{i-1} are the acceleration values at time i and $i-1$, respectively.

Table 1. Variable names of finite-state machine (FSM).

Notation	Description of Notation
acc_diff	Acceleration difference threshold
acc_thr	Entering the FSM algorithm threshold
up_thr	Maximum value judged as a rising state
$down\_thr$	Minimum value judged as a falling state
$peak\_thr$	Peak threshold
D_thr	Max interference value
P_{up}	Times of acceleration greater than acc_diff
P_{down}	Times of acceleration smaller than acc_diff
a_i	Acceleration value

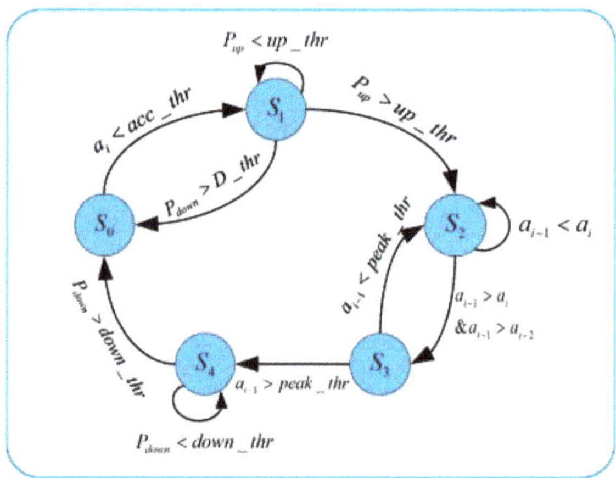

Figure 3. The execution process of the FSM algorithm.

2.1.3. Step Length Estimation

When people are walking, the step-length can be estimated by using the acceleration data. The existing step length models include linear model [38], constant models [39], nonlinear models [40], etc. Because of the difference between people's height and stride, the step length is also different. Therefore, the application of the constant model and linear model has certain limitations. We use the Weinberg model [41] that is a non-linear model to estimate step length. During the one-step process, if the maximum and minimum of the acceleration are $A_{\max}$ and $A_{\min}$, step length can be calculated as:

$$StepLength = K \bullet \sqrt[4]{A_{\max} - A_{\min}} \tag{9}$$

where K is a constant value, which can be obtained by training a large number of acceleration data.

2.2. Geomagnetic Positioning Module

2.2.1. Geomagnetic Multi-Features Data Extraction

It will be better to describe the distribution of the magnetic field at a single point by utilizing more geomagnetic features data. In addition to the three-axis components $\begin{pmatrix} B_x & B_y & B_z \end{pmatrix}$, the horizontal intensity B_h and the total intensity B represent the magnitude of the magnetic field in the horizontal plane and the 3-dimensional space. If these two components are utilized for geomagnetic matching positioning, they can provide additional distance information for the matching process. Therefore, we extracted these two features and their calculation methods are as follows:

$$\begin{cases} B_h = \sqrt{B_x^2 + B_y^2} \\ B = \sqrt{B_x^2 + B_y^2 + B_z^2} \end{cases} \tag{10}$$

However, the three-axis components of the magnetic field measured by smartphones are in the carrier coordinate system (CCS), whose x, y and z axes point to the right, front and top of the device as shown in Figure 4a. These three components vary with the attitudes of smartphones and it is meaningless to directly use the untransformed three-axis magnetic data. Therefore, it is necessary for us to transform the three-axis geomagnetic data to the geographic coordinate system (GCS) because they are stable in the GCS.

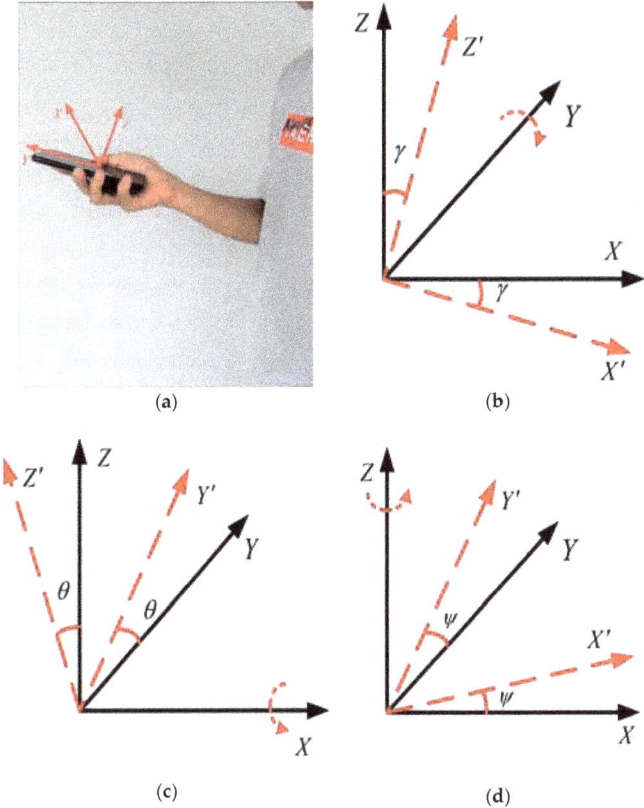

Figure 4. (**a**) The carrier coordinate system (CCS); (**b**) roll; (**c**) pitch; (**d**) yaw.

The GCS and the CCS can be rotated into each other through attitude angles, which contain yaw, pitch, and roll. Their calculation methods have been described particularly in Section 2.1.1. The rotation matrix C_n^b between the two systems can be defined as:

$$C_n^b = C_n^3 C_n^2 C_n^1 = \begin{bmatrix} \cos\gamma & 0 & -\sin\gamma \\ 0 & 1 & 0 \\ \sin\gamma & 0 & \cos\gamma \end{bmatrix} \begin{bmatrix} 1 & 0 & 0 \\ 0 & \cos\theta & \sin\theta \\ 0 & -\sin\theta & \cos\theta \end{bmatrix} \begin{bmatrix} \cos\psi & -\sin\psi & 0 \\ \sin\psi & \cos\psi & 0 \\ 0 & 0 & 1 \end{bmatrix} \tag{11}$$

where C_n^1, C_n^2 and C_n^3 are the corresponding matrices for yaw, pitch, and roll. Yaw, pitch, and roll respectively represent rotation around the z, x, and y axes as shown in Figure 4. The rotation matrix C_b^n from the CCS to the GCS can be calculated as:

$$C_b^n = (C_n^b)^{-1} = (C_n^b)^T \tag{12}$$

where $(C_n^b)^{-1}$ and $(C_n^b)^T$ are the inverse matrix and transposed matrix of C_n^b. Supposing that the three-axis components vector of magnetic data in the CCS is $\boldsymbol{B}_M = \begin{bmatrix} B_x & B_y & B_z \end{bmatrix}^T$ and $\boldsymbol{B}_S = \begin{bmatrix} B_x{}' & B_y{}' & B_z{}' \end{bmatrix}^T$ is the magnetic data vector in the GCS, the relationship between them is defined as:

$$\boldsymbol{B}_S = \left(C_n^b\right)^T \boldsymbol{B}_M = \begin{bmatrix} \cos\gamma\cos\psi + \sin\gamma\sin\psi\sin\theta & \sin\psi\cos\theta & \sin\gamma\cos\psi - \cos\gamma\sin\psi\sin\theta \\ -\cos\gamma\sin\psi + \sin\gamma\cos\psi\sin\theta & \cos\psi\cos\theta & -\sin\gamma\sin\psi - \cos\gamma\cos\psi\sin\theta \\ -\sin\gamma\cos\theta & \sin\theta & \cos\gamma\cos\theta \end{bmatrix} \begin{bmatrix} B_x \\ B_y \\ B_z \end{bmatrix} \tag{13}$$

We can obtain three-axis stable geomagnetic data by using Equation (13). As shown in Figure 5, the transformed data are stable and the x-axis components data are close to zero, which are consistent with theoretical results. However, we found that the x-axis components data of some experimental areas were not close to zero in our research. Therefore, we retain the x-axis component in this paper. Eventually, we can extract five geomagnetic features data as the fingerprint data by using the above method.

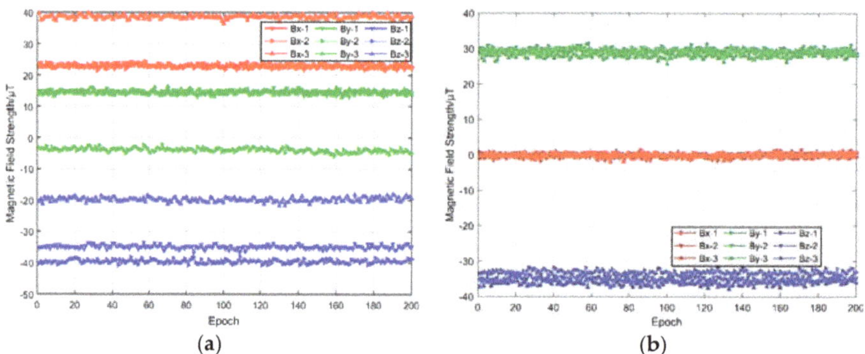

Figure 5. Comparison of the untransformed and transformed three-axis geomagnetic data. (**a**) Untransformed three-axis geomagnetic data; (**b**) Transformed three-axis geomagnetic data

2.2.2. Geomagnetic Multi-Features Positioning Algorithm

The geomagnetic multi-features positioning algorithm includes the offline database construction stage and online positioning stage. During the database construction stage, the experimenter uses mobile phones to collect the geomagnetic data on each selected point for 15 s, then extracts five geomagnetic features data. The fingerprint of a single point can be expressed as $\left(B_x, B_y, B_z, B_h, B, x, y\right)$

by combining the actual geographic coordinates. For the points without data collection, a linear interpolation [42] is used to obtain the geomagnetic data of these points.

In the online positioning stage, the real-time collected geomagnetic data will be firstly transformed into the geographic coordinate system, and five geomagnetic features data will be extracted. Then the k-nearest neighbor (KNN) method [43] is employed to obtain the positioning results by traversing the geomagnetic database. If one piece of the real-time collected data is $\left(B_x{}', B_y{}', B_z{}', B_h{}', B'\right)$, the distance between the measured and fingerprint data can be calculated as:

$$d = \sqrt{(B_x - B_x{}')^2 + \left(B_y - B_y{}'\right)^2 + (B_z - B_z{}')^2 + (B_h - B_h{}')^2 + (B - B')^2} \tag{14}$$

If there are n fingerprint data in the database, n distances can be obtained by traversing the geomagnetic database. We should find the k minimum distances, in the meantime, the coordinate points corresponding to these k minimum distances should be recorded. If the current position vector is $\boldsymbol{X}_m = \begin{bmatrix} x_{mx} & y_{my} \end{bmatrix}^T$, it can be calculated as:

$$\boldsymbol{X}_m = \begin{bmatrix} x_{mx} & y_{my} \end{bmatrix}^T = \begin{bmatrix} \frac{1}{k}\bullet\sum\limits_{i=1}^{k} x_i & \frac{1}{k}\bullet\sum\limits_{i=1}^{k} y_i \end{bmatrix}^T \tag{15}$$

where x_i and y_i are the coordinate values of the k nearest points, $i = 1,2, \cdots k$.

In one second, m pieces of geomagnetic data can be collected. All of the collected geomagnetic data will be involved in the positioning calculation. Thus, there will be m positioning results in one second. These results are essential for fusion positioning in Section 2.3. In order to better understand the geomagnetic multi-features positioning algorithm, Algorithm 1 gives the detailed process of it.

Algorithm 1 Geomagnetic multi-features positioning algorithm

Input: magnetic database ***MD***, Coordinate data ***CD***, measured magnetic data $\boldsymbol{B}_{measure}$, positive integer k
$\boldsymbol{B}_{measure} \leftarrow normalize(\boldsymbol{B}_{measure})$
for $i \leftarrow 1$ to length $(\boldsymbol{B}_{measure})$ **do**
  for $j \leftarrow 1$ to length(MD) **do**
  calculate the distance d_j
  add to $Dist$, $Dist \leftarrow d_j$
  end
  find k minimum values of the $Dist$, $mindist \leftarrow min(Dist)$
  index$\leftarrow find(Dist == mindist)$
  $current\_coordinate \leftarrow sum(CD(index))/k$
  add $current\_coordinate$ to pos_result, $pos\_result \leftarrow current\_coordinate$
end
Output: pos_result

2.3. Fusion Positioning Module

2.3.1. Classical Particle Filter

The particle filter (PF) is an algorithm based on the Monte Carlo method [44] and the Bayesian approach [45], which has good performance in solving non-linear problems. Generally, a dynamic system can be assumed that it obeys the Markov process of order one, it can be modeled as follows:

$$\begin{cases} \boldsymbol{X}_k = \boldsymbol{f}_k(\boldsymbol{X}_{k-1}) + \delta_{k-1} \\ \boldsymbol{Z}_k = \boldsymbol{h}_k(\boldsymbol{X}_k) + \sigma_k \end{cases} \tag{16}$$

where $\boldsymbol{X}(k)$ and $\boldsymbol{Z}(k)$ are the system state and observation at time k, δ_{k-1} and σ_k are the process noise and measurement noise at time $k-1$ and k, respectively. $\boldsymbol{f}_k$ reflects the relationship between the current and previous state. $\boldsymbol{h}_k$ expresses the relationship between states and observations. The posterior probability density function (pdf) $p(\mathbf{X}_k|\mathbf{Z}_{1:k})$ is usually used to calculate the optimal estimation and predict the appearance probability of $\boldsymbol{X}(k)$, which is called as the prediction in PF. $p(\mathbf{X}_k|\mathbf{Z}_{1:k})$ relies on the previous posterior pdf $p(\mathbf{X}_k|\mathbf{Z}_{1:k-1})$ based on the Bayesian theory, which can be calculated by using Equation (17) if there are observations available:

$$P(\mathbf{X}_k|\mathbf{Z}_{1:k-1}) = \int p(\mathbf{X}_k, \mathbf{X}_{k-1}|\mathbf{Z}_{1:k-1})d\mathbf{X}_{k-1} = \int p(\mathbf{X}_k|\mathbf{X}_{k-1})p(\mathbf{X}_{k-1}|\mathbf{Z}_{1:k-1})d\mathbf{X}_{k-1} \quad (17)$$

where $p(\mathbf{X}_k|\mathbf{X}_{k-1})$ is the transition probability distribution defined by $\boldsymbol{X}(k)$, $p(X_{k-1}|Z_{1:k-1})$ is the previous posterior, respectively. Assuming the previous posterior is known, the current posterior is defined as follows:

$$P(\mathbf{X}_k|\mathbf{Z}_{1:k}) = \frac{p(\mathbf{Z}_k|\mathbf{X}_k, \mathbf{Z}_{1:k-1})p(\mathbf{X}_k|\mathbf{Z}_{1:k-1})}{p(\mathbf{Z}_k|\mathbf{Z}_{1:k-1})} = \frac{p(\mathbf{Z}_k|\mathbf{X}_k)p(\mathbf{X}_k|\mathbf{Z}_{1:k-1})}{p(\mathbf{Z}_k|\mathbf{Z}_{1:k-1})} \quad (18)$$

where $p(\mathbf{Z}_k|\mathbf{Z}_{1:k-1})$ is the normalizing constant, it is calculated by using Equation (19):

$$p(\mathbf{Z}_k|\mathbf{Z}_{1:k-1}) = \int p(\mathbf{Z}_k|\mathbf{X}_k)p(\mathbf{X}_k|\mathbf{Z}_{1:k-1})d\mathbf{X}_k \quad (19)$$

We can use Equation (18) to update the current posterior probability and this is an important step in PF. But it is difficult for us to solve Equations (17) and (19) due to the complex integral operation. The Monte Carlo method provides an effective approach for solving this problem. This method uses a sequence of weighted samples to represent the posterior pdf and the mean value of these samples is substituted for the integration. If the sample set has N samples, $p(\mathbf{X}_k|\mathbf{Z}_{1:k})$ can be calculated by:

$$p(X_k|Z_{1:k}) \approx \frac{1}{N}\sum\nolimits_{i=1}^{N} \omega_k^i \delta(X_k - X_k^i) \quad (20)$$

where ω_k^i and $\mathbf{X}_k^i$ are the weight and the state of i-th sample, $\delta(\cdot)$ is the Dirac delta function. The update formula of ω_k^i is defined as follows:

$$\omega_k^i \propto \omega_{k-1}^i \frac{p(\mathbf{Z}_k|\mathbf{X}_k^i)p(\mathbf{X}_k^i|\mathbf{X}_{k-1}^i)}{q(\mathbf{X}_k^i|\mathbf{X}_{k-1}^i, \mathbf{Z}_k)} \quad (21)$$

where $q(\cdot)$ is the importance density. When $N \to \infty$, the result of Equation (20) approaches the true posterior density [45]. PF will resample a different set at each timestamp k based on the updated posterior probability, which can also be called weight. Resampling will copy large-weight samples and abandon the small-weight ones to obtain the optimal estimation. But this will decrease the diversity of samples if the system runs for a long time. This phenomenon is particle degradation which can affect the performance of PF and waste computing resources. Therefore, ensuring the diversity of samples and optimizing particle degradation are the key factors in determining the performance of PF. We concentrate on these problems and design the optimization mechanism for particle degradation in our research work.

2.3.2. Gene Mutation

The gene mutation method of the genetic algorithm (GA) [30] is essential to genetic algorithm optimization. GA follows the "survival of the fittest" principle and it uses the genetic operators of natural genetics to cross and mutate the population to create new generations. The information of each individual is expressed as the binary code, which is called the DNA. The whole process of gene mutation is as shown in Figure 6.

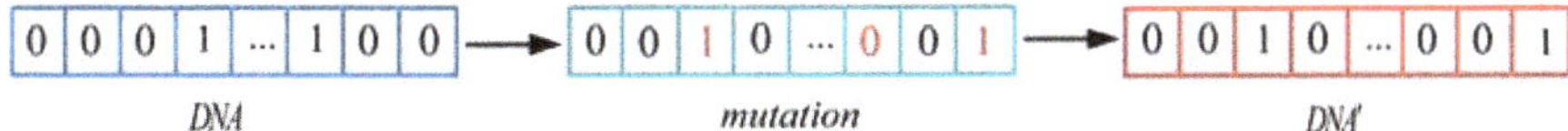

Figure 6. Gene mutation.

Assuming that the information of an individual is I and it can be described in the decimal system. In the mutation process, I will be transformed into the binary code vector $\boldsymbol{M}$, which can be described as:

$$\boldsymbol{M} = \begin{bmatrix} b_i & b_{i-1} & \cdots & b_0 \end{bmatrix}_2 \tag{22}$$

where b_i is the binary code, its value is 0 or 1. And then, the binary code will change randomly to imitate the DNA mutation and individuals with different characteristics can be obtained by decompiling the mutated binary codes using Equation (23):

$$I = \left(\sum_{j=0}^{i} b_j \bullet 2^j \right)_{10} \tag{23}$$

2.3.3. Optimization Mechanism for Classical Particle Filter

The optimization mechanism aims at improving the diversity and reliability of the resampled particles to ameliorate the particle degradation. Therefore, we utilize the gene mutation method to increase the diversity and use the reconstruction particle set method by combining the geomagnetic positioning results to improve the reliability of particles. We can assume that the characteristics of each resampled particle are its coordinate values. Firstly, we should utilize the gene mutation method to transform the coordinates into binary codes. Since the coordinate values are decimal, the problem becomes the decimal-to-binary conversion question [46]. As shown in Algorithm 2, it gives the detailed transformation process of x value. It should be clear that we should output $\boldsymbol{M}$ in the inverse order to obtain the binary codes.

Algorithm 2 decimal-to-binary of x value

```
Input: coordinate value x, empty set M, Boolean variable isRun
get the integer part of x, x_in ← floor(x); isRun ← true
while (isRun) do
    calculate the remainder r, r ←x%2
    calculate the quotient q, q ←(x − r)/2
    if q == 0 do
    calculate the remainder r, r ←x%2
    add r to M, M ← r; isRun ←false
    else do
    add r to M, M ← r
    end
    x_in← q
end
binarycode← reverse (M)
Output: binarycode
```

The binary codes of y value can be obtained by using Algorithm 2. As shown in Figure 7, if we change the binary codes to imitate the gene mutation, we can get the particle swarms with different characteristics. Therefore, the resampled particles become various and the diversity is improved.

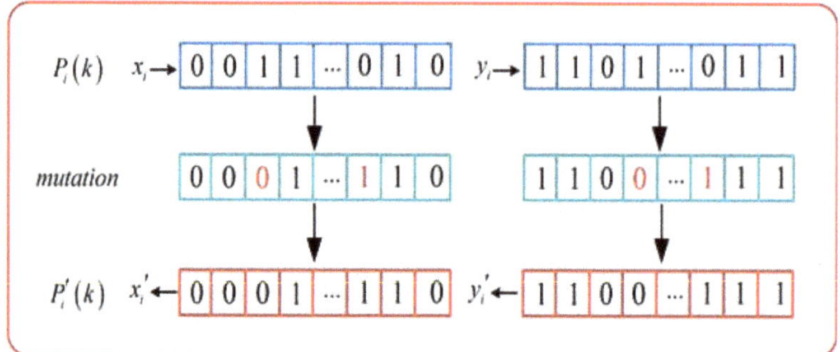

Figure 7. Particle mutation.

Apart from increasing the diversity, we should also assure the reliability of particles so that PF abandon fewer particles in the resampling step. In Section 2.2.2, we have pointed out that there are many geomagnetic positioning results rather than one result. In these results, we should select s results that are close to the PDR position. If the PDR positioning results are $\mathbf{X}(k) = \{x_k, y_k\}$ and the geomagnetic positioning results are $\mathbf{Z}(k) = \{\mathbf{Z}_1, \mathbf{Z}_2, \cdots, \mathbf{Z}_m\}, \mathbf{Z}_1 = (x_1, y_1), \cdots$, the Euclidean distance between $\mathbf{Z}(k)$ and $\mathbf{X}(k)$ can be computed by using Equation (24):

$$D_i = \sqrt{(x_i - x_k)^2 + (y_i - y_k)^2} \tag{24}$$

Then, based on the system state $\mathbf{X}(k)$, we can also use the KNN method in Section 2.2.2 to find s geomagnetic results that are close to the PDR position to replace the s particles. After this optimization, each generation particles are composed of the previous generation "excellent" particles after mutation and the geomagnetic positioning results of the current generation. These two methods assure the diversity and reliability of the particle set and avoid the "excellent" particles reduction of each generation. The particles of each generation seem to be in evolution rather than degradation in classical PF. Figure 8 shows the detailed process of the genetic-particle filter algorithm in this paper, and the size of different balls represents different characteristics of particles.

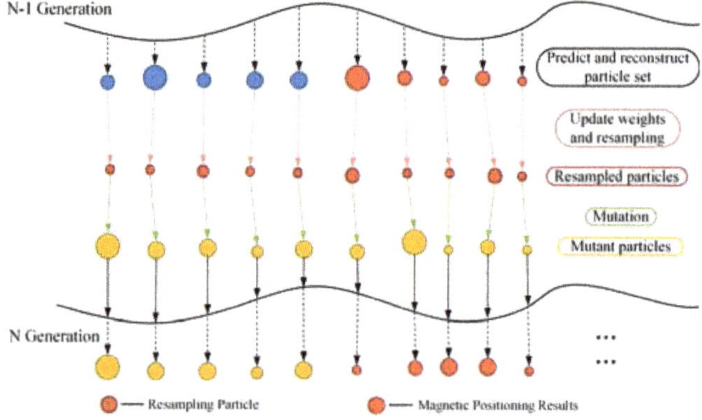

Figure 8. Process of genetic-particle filter.

2.3.4. Fusion-Positioning Algorithm based on Genetic-Particle Filter

In this section, we will describe the detailed process of the fusion positioning algorithm based on the G-Particle Filter. We use PDR as the system state $\boldsymbol{X}(k)$ and the geomagnetic positioning results

$Z(k) = \{Z_1, Z_2, \cdots, Z_m\}, Z_1 = (x_1, y_1), \cdots$, as the system observation; the PDR position will be corrected in real-time by the fusion positioning results. The general process is as follows:

Step 1, predict the system state $\boldsymbol{X}(k)$ at time k and reconstruct the particle set based on $\boldsymbol{X}(k)$. We assume the particle set is $\boldsymbol{P}(k) = \{\boldsymbol{P}_1, \boldsymbol{P}_2, \cdots, \boldsymbol{P}_n\}$ and the weight of each particle is $1/N$. Then, we should select the s observations with the smallest distance to replace the last s particles to reconstruct the particle set. The new particle set is $\boldsymbol{P}(k) = \{\boldsymbol{P}_1, \boldsymbol{P}_2, \cdots \boldsymbol{P}_{n-s}', \cdots, \boldsymbol{P}_n'\}$.

Step 2, update and normalize the weights of particles. After getting the reconstructed particle set, each particle's weight should be updated by using the Gaussian function:

$$\omega_i(k) = \frac{1}{\sigma_r \sqrt{2\pi}} exp\left[-\frac{\|\boldsymbol{P}_i - \boldsymbol{X}(k)\|^2}{2{\sigma_r}^2}\right] \tag{25}$$

where σ_r is the observation noise. The normalized weight of each particle can be calculated by:

$$\omega'_i(k) = \frac{\omega_i(k)}{\sum\limits_{i=1}^{n} \omega_i(k)} \tag{26}$$

where $\omega_i(k)$ and $\omega'_i(k)$ are the un-normalized and normalized weights of i-th particle, respectively.

Step 3, resample particles and calculate the fusion-positioning results. Based on each particle's normalized weight, the particles with small weights are abandoned and the particles with large weights are copied to replace them. After resampling, the average coordinates of the resampled particles are computed by using Equation (27) and setting as the fusion position $\boldsymbol{R}(k)$:

$$\boldsymbol{R}(k) = \frac{1}{n}\left(\sum_{i=1}^{n} x_i, \sum_{i=1}^{n} y_i\right) \tag{27}$$

Step 4, correct the PDR positions. After the three steps above, we should correct the PDR coordinates by using the fusion position $\boldsymbol{R}(k)$.

Step 5, particles mutation. In this step, we should convert the coordinates of the resampled particles into the binary code by using Algorithm 2. Gene mutation is used to change the coordinate values of the particles and the mutated particles will be passed to the next generation.

To summarize, when the system is running continuously, the above steps will be executed cyclically. At each moment, the system will geneticize the best quality particles to the next moment and the geomagnetic positioning results at the next moment will reconstruct the particle set. Particle mutation and the reconstruction of particle set methods ensure the reliability and the diversity of particles. Figure 9 shows the process of the fusion-positioning system.

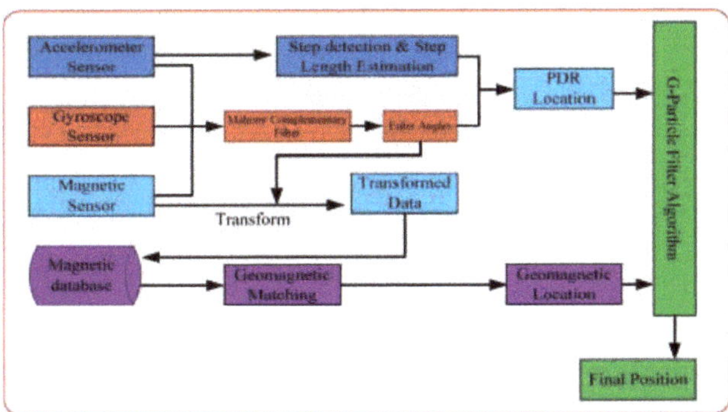

Figure 9. Fusion-positioning algorithm based on genetic-particle filter.

3. Experiments and Results

In this section, several experiments are designed to verify the methods mentioned in the above sections. As shown in Figure 10, they are the experimental area scene. The test path connects area A and B and the total length is approximately 130 m. There is a square area which is about 54 square meters between the two areas and the total experimental area is about 360 square meters. We use a Xiaomi 6 mobile phone that has installed our developed data acquisition software to collect data with a sampling rate of 50 HZ. In order to test the performance of the MCF algorithm, the actual azimuth angles from A_1 to A_2, A_2 to B_1, and B_1 to B_2 are measured beforehand, which are 88.1°, 357.17°, and 268.1°, respectively. The size of the particle set is 100 in the genetic-particle filter and all the data processing and simulation work are done on the computers.

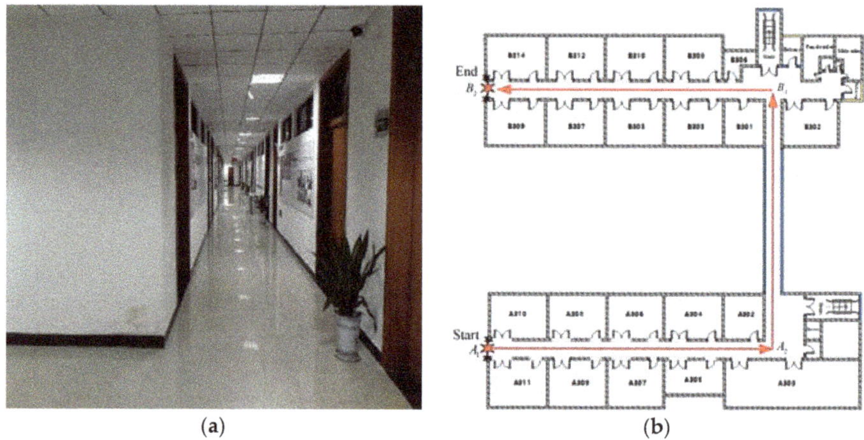

Figure 10. (**a**) Experimental area; (**b**) experimental path.

3.1. Geomagnetic Fingerprint Database Construction

It is essential for geomagnetic positioning to build a high precision database. In our construction work, we set the reference point every 1.2 m, and then collected the geomagnetic data for 15 s on these points. Five geomagnetic features are extracted by using the method in Section 2.2.1. As for the uncollected area, the linear interpolation method is used to obtain the geomagnetic data. Five geomagnetic maps are generated as illustrated in Figure 11. From these figures, we can find that geomagnetic characteristics of different regions are very obvious, especially Figure 11a which shows the x-axis component is not close to zero in some experimental areas and this is contrary to theory. We analyzed that ferrous materials lead to the anomalous magnetic field. But they can better describe the magnetic field distribution and improve the specificity of the geomagnetic fingerprint. Therefore, x-axis components are extracted and combine the other four features to construct the fingerprint database.

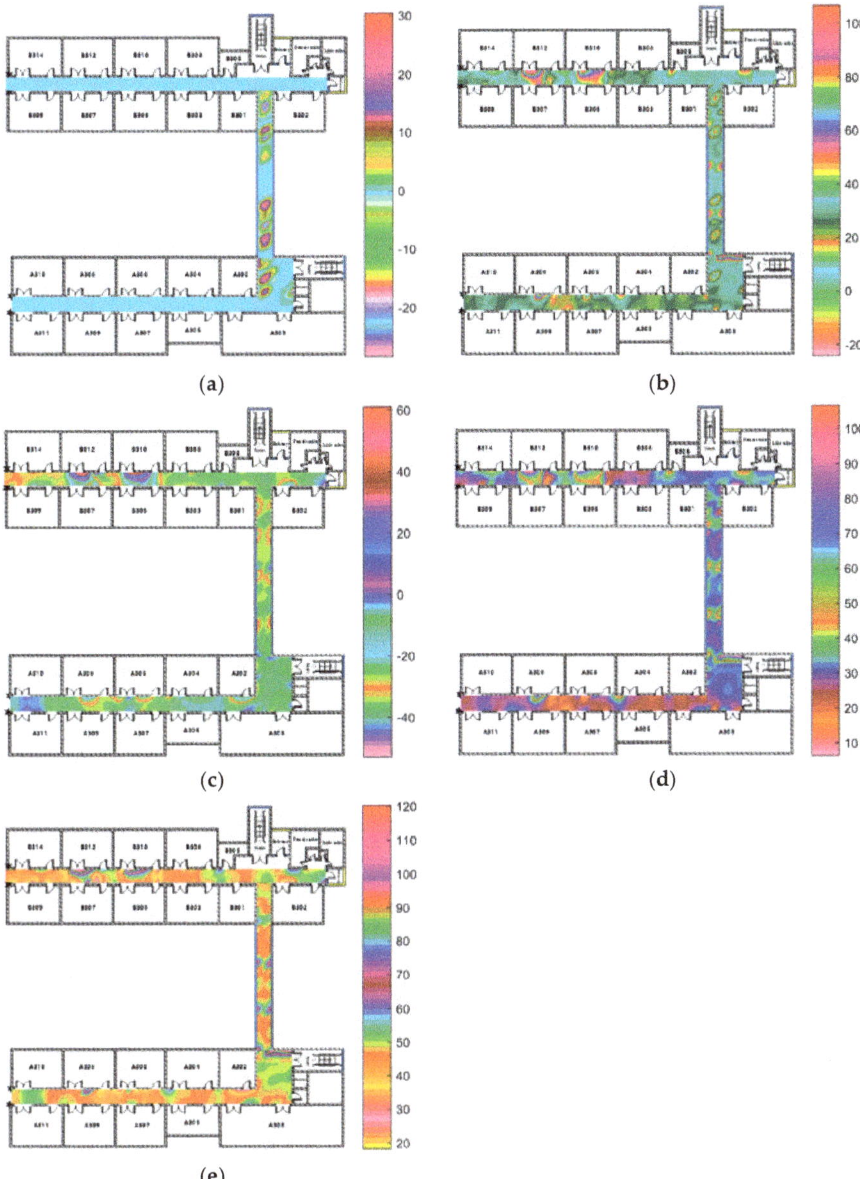

Figure 11. Geomagnetic characteristics interpolation map. (**a**) X-axis component interpolation map; (**b**) Y-axis component interpolation map; (**c**) Z-axis component interpolation map; (**d**) horizontal component interpolation map; (**e**) magnetic field strength interpolation map.

3.2. Geomagnetic Positioning Experiment

In this section, we apply one feature (total intensity of magnetic field), two features (horizontal and total components), three features (three-axis components), four features (three-axis components and horizontal intensity) and five features to geomagnetic positioning. Positioning errors are presented in Figure 12. We also counted the positioning error of 50% and 80% of the test points, and compute the mean positioning error and root mean square error (RMSE) in Table 2.

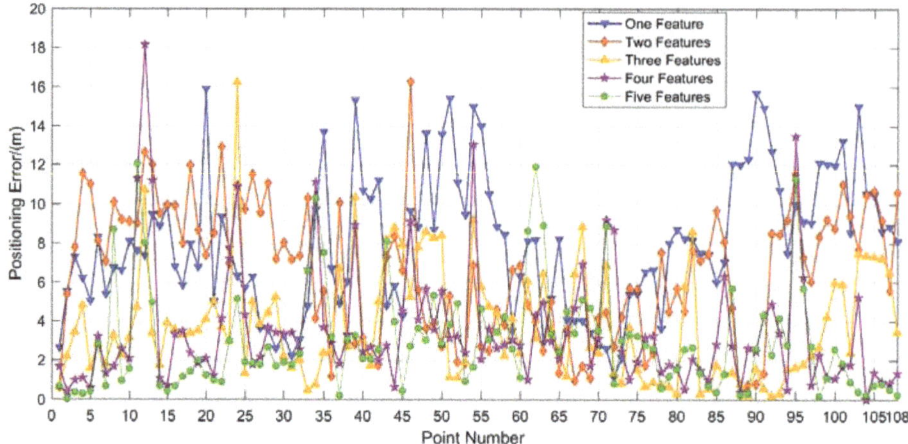

Figure 12. Geomagnetic positioning comparison by using different features data.

Table 2. Positioning errors comparison of different geomagnetic features (m).

Features	Min	Max	Mean	RMSE	50% Error	80% Error
One feature	1.16	15.91	8.01	8.74	7.98	10.69
Two features	0.27	16.29	6.62	7.41	7.06	9.54
Three features	0.19	16.27	3.80	4.77	3.15	6.08
Four features	0.02	18.16	3.38	4.66	2.68	4.65
Five features	0.04	12.07	2.98	4.01	2.56	4.51

As shown in Table 2 and Figure 12, one-feature or two-features data can differentiate positions with lower accuracy. When five features data are used, the positioning accuracy and the RMSE are 2.98 m and 4.01 m, which are better than those of using four or fewer features. The positioning accuracy gradually increases with the number of features of data increasing. Based on the experimental results in Section 3.1, five-features data have different distributions in different areas and can better describe the magnetic field at different points. Therefore, we can conclude that five geomagnetic features could improve the geomagnetic positioning accuracy and geomagnetic fingerprint specificity.

3.3. Heading Angle Experiment

In order to evaluate the performance of the Mahony complementary filter (MCF) algorithm, we utilized the mobile phone to collect the heading angles by using the MCF algorithm, electronic compass (EC) and the uncorrected gyroscope on the same path. The estimation errors of these three methods were compared in Table 3 and Figure 13. The mean estimation accuracy and the root mean square error (RMSE) of the MCF algorithm are 7.21 ° and 6.58°, respectively. Compared with the estimation errors of EC and the uncorrected gyroscope, we can conclude that the MCF algorithm can provide more accurate and stable heading angles.

Table 3. Heading angles estimation error (°).

Algorithm	Min	Max	Mean	RMSE
MCF	0.04	50.78	7.21	6.58
EC	0.05	59.03	11.46	9.81
Gyroscope	0.18	88.05	16.61	13.39

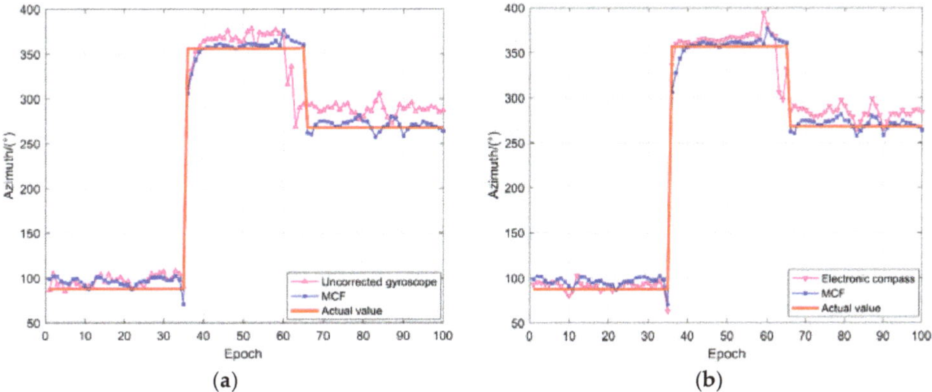

Figure 13. Heading angles comparison by using different methods (small angles added 360°). (**a**) Uncorrected gyroscope VS MCF; (**b**) Electronic compass VS MCF

3.4. Fusion-Positioning Experiment

In this section, we analyzed the mean positioning accuracy and root mean square error (RMSE) of the pedestrian dead-reckoning (PDR) method, geomagnetic positioning, fusion positioning based on the classic particle filter (PF) which has no gene mutation and particle set reconstruction, and the proposed fusion method. The computational cost (CC) obtained by calculating the average positioning time of different methods is also counted for a comparison.

As shown in Table 4, compared with the PDR method, geomagnetic positioning and fusion positioning by using PF, the mean positioning error of the proposed method is 1.72 m and the RMSE is 1.89 m, respectively. Moreover, 80% of the test points have the positioning error within 2.45 m. But under the same confidence level, the positioning accuracy of geomagnetic positioning, the PDR method and PF algorithm are only 4.51 m, 4.80 m and 3.22 m, respectively. In terms of the computational cost, we can find that our proposed algorithm can give positions within 0.3 s. This is slightly larger than PF, but it is still low because the size of the particle set is just 100. Table 4 shows the positioning accuracy and stability of our proposed algorithm are better improved and the computational cost is low. Moreover, the points positioning errors and the cumulative distribution of positioning errors are presented in Figure 14.

Table 4. Positioning error and computational cost (CC) table of different methods.

Method	Min/m	Max/m	Mean/m	RMSE/m	50% Error/m	80% Error/m	CC/s
PDR	0.12	6.29	3.14	3.43	2.54	4.80	0.004
Geomagnetic	0.04	12.07	2.98	4.01	2.56	4.51	0.252
PF	0.10	4.83	1.96	2.28	1.84	3.22	0.285
Proposed	0.13	3.42	1.72	1.89	1.75	2.45	0.296

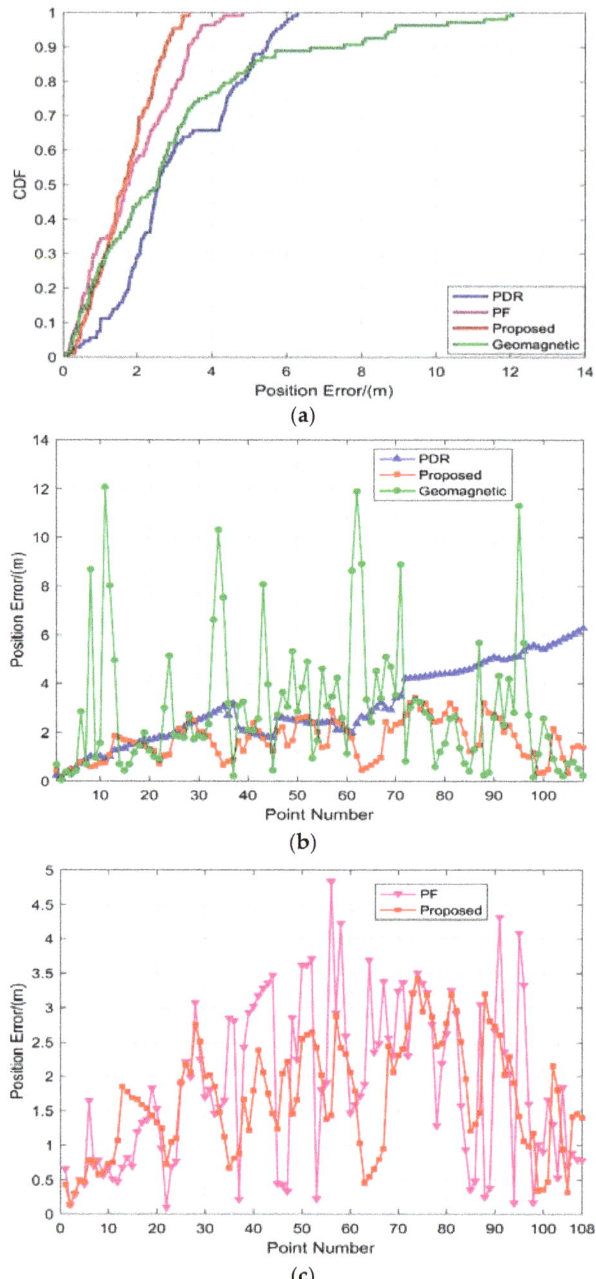

Figure 14. Positioning cumulative error curve and point error curves of different methods. (**a**) Positioning cumulative error curves; (**b**) points error curves of PDR, geomagnetic positioning and the proposed method; (**c**) points error curves of fusion positioning based on PF and the proposed method.

From Figure 14a we can find that the positioning error of all the test points using the proposed fusion-positioning method are within 4 m and more than 70% of test points have the positioning error better than 2.5 m. These are much better than the other three positioning methods. In terms of the points positioning errors, as shown in Figure 14b,c, the positioning error of the PDR method gradually accumulates with time and the error of final point accumulates to more than 6 m. Geomagnetic positioning and the fusion positioning method based on PF are both unstable and several large positioning errors appear in the error curves. Conversely, our proposed fusion-positioning method has neither a large positioning error nor the error accumulation phenomenon. The positioning accuracy and stability are improved.

Based on the above error analysis, we have concluded that our proposed algorithm has better positioning performance. Figure 15 shows the positioning trajectories of different methods, the PDR method can provide more reliable positioning results at first, but the trajectory begins to shift after the first turn. What is worse, the positioning trajectory directly passes through the room after the second turn, which seriously deviates from the actual path. The fusion-positioning method based on PF also shows the instability after the first turn, which suggests that particle degradation affects the positioning stability and accuracy as the time grows. Contrary to the instability and the navigation path drift, the navigation trajectory indicated by our proposed method has no large drift error and is closer to the actual experimental path. There is no doubt that the mutation and reconstruction of the particle set methods improve the performance of the particle filter and ensure the stability and high precision of the fusion-positioning algorithm.

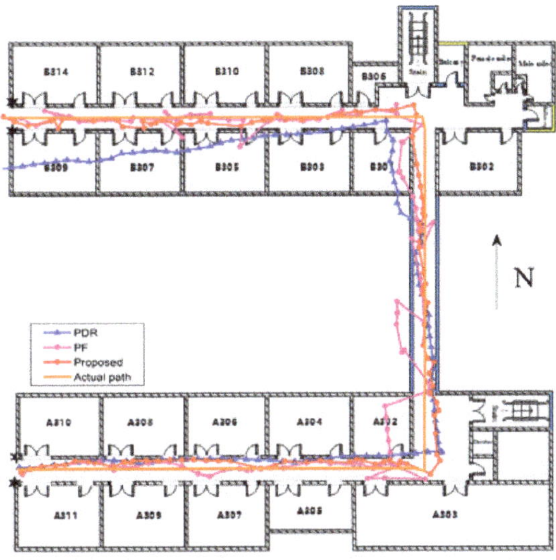

Figure 15. Positioning trajectories of different methods.

4. Discussion and Conclusions

In this paper, we propose a fusion positioning method based on the genetic-particle filter. In order to optimize the particle degradation problem, we design an optimization mechanism that utilizes the gene mutation method and the method of reconstructing particle set to ensure the reliability and diversity of particles. The experiment results show that the average position estimation error is 1.72 m and the RMSE is 1.89 m, the positioning accuracy and stability are improved. Contrary to the high computational burden caused by increasing the number of particles in papers [26,28,29], the size of the particle set in this paper is 100 and the computational load is lower compared with them. Furthermore, the single-point-based magnetic positioning (SPMP) method is more flexible and easier to be implemented and its computational load is also not very high. Therefore, it is possible for us to implement this fusion positioning algorithm on smartphones. Moreover, five geomagnetic features are extracted in our work and they can improve the fingerprint specificity of a single point and geomagnetic positioning accuracy. Besides, better heading angle estimation performance of the Mahony complementary filter (MCF) algorithm also provides another effective approach for us to calculate the heading angle by using smartphones. All of these results can support that our proposed fusion-positioning method has application value in indoor positioning and navigation.

However, there are also some limitations in our fusion-positioning method. The linear interpolation method can reduce data collection work. Whether there is another effective approach that can reduce more geomagnetic database construction work is still needed to explore in depth. We did not consider the influence of different walking speed on fusion positioning. The effect of different motion states on heading angle estimation is not studied. All of these limitations are our future research work.

Author Contributions: Conceptualization, M.S. (Meng Sun); Formal analysis, M.S. (Meng Sun) and S.X.; Funding acquisition, Y.W.; Methodology, H.C.; Project administration, Y.W.; Data curation, H.C., M.S. (Minghao Si); Writing-original draft preparation, M.S. (Meng Sun). All authors have read and agreed to the published version of the manuscript.

Funding: This research was funded by the National Key Research and Development Program of China [grant number 2016YFB0502102], in part by the State Key Laboratory of Satellite Navigation System and Equipment Technology [grant number CEPNT-2018KF-03].

Conflicts of Interest: The authors declare no conflict of interest. The founding sponsors had no role in the design of the study; in the writing of the manuscript; and in the decision to publish the results.

References

1. Davidson, P.; Piché, R. A survey of selected indoor positioning methods for smartphones. *IEEE Commun. Surv. Tutor.* **2017**, *19*, 1347–1370. [CrossRef]
2. Li, X.; Zhang, X.; Ren, X.; Fritsche, M.; Wickert, J.; Schuh, H. Precise positioning with current multi-constellation global navigation satellite systems: GPS, GLONASS, Galileo and BeiDou. *Sci. Rep.* **2015**, *5*, 8328. [CrossRef]
3. Lin, K.; Chen, M.; Deng, J.; Hassan, M.M.; Fortino, G. Enhanced fingerprinting and trajectory prediction for IoT localization in smart buildings. *IEEE Trans. Autom. Sci. Eng.* **2016**, *13*, 1294–1307. [CrossRef]
4. He, S.; Chan, S.-H.G. Wi-Fi fingerprint-based indoor positioning: Recent advances and comparisons. *IEEE Commun. Surv. Tutor.* **2016**, *18*, 466–490. [CrossRef]
5. Yoon, P.K.; Zihajehzadeh, S.; Kang, B.-S.; Park, E.J. Adaptive Kalman filter for indoor localization using Bluetooth Low Energy and inertial measurement unit. In Proceedings of the 2015 37th Annual International Conference of the IEEE Engineering in Medicine and Biology Society (EMBC), Milan, Italy, 25–29 August 2015; pp. 825–828.
6. De Blasio, G.; Quesada-Arencibia, A.; García, C.R.; Rodríguez-Rodríguez, J.C.; Moreno-Díaz, R. A protocol-channel-based indoor positioning performance study for Bluetooth low energy. *IEEE Access* **2018**, *6*, 33440–33450. [CrossRef]
7. Schroeer, G. A Real-Time UWB Multi-Channel Indoor Positioning System for Industrial Scenarios. In Proceedings of the 2018 International Conference on Indoor Positioning and Indoor Navigation (IPIN), Danvers, MA, USA, 24–29 September 2018; pp. 1–5.
8. Chen, P.; Ye, K.; Chen, X. A UWB/Improved PDR Integration Algorithm Applied to Dynamic Indoor Positioning for Pedestrians. *Sensors* **2017**, *17*, 2065. [CrossRef]
9. Bekkali, A.; Sanson, H.; Matsumoto, M. RFID indoor positioning based on probabilistic RFID map and Kalman filtering. In Proceedings of the Third IEEE International Conference on Wireless and Mobile Computing, Networking and Communications (WiMob 2007), New York, NY, USA, 8–10 October 2007; p. 21.
10. Yucel, H.; Edizkan, R.; Ozkir, T.; Yazici, A. Development of indoor positioning system with ultrasonic and infrared signals. In Proceedings of the 2012 International Symposium on Innovations in Intelligent Systems and Applications, Trabzon, Turkey, 2–4 July 2012; pp. 1–4.
11. Huang, C.-N.; Chan, C.-T. ZigBee-based indoor location system by k-nearest neighbor algorithm with weighted RSSI. *Procedia Comput. Sci.* **2011**, *5*, 58–65. [CrossRef]
12. Davidson, P.; Takala, J. Algorithm for pedestrian navigation combining IMU measurements and gait models. *Gyroscopy Navig.* **2013**, *4*, 79–84. [CrossRef]

13. Zheng, L.; Zhou, W.; Tang, W.; Zheng, X.; Peng, A.; Zheng, H. A 3D indoor positioning system based on low-cost MEMS sensors. *Simul. Model. Pract. Theory* **2016**, *65*, 45–56. [CrossRef]
14. Siiksakulchai, S.; Thongchai, S.; Wilkes, D.M.; Kawamura, K. Mobile robot localization using an electronic compass for corridor environment. In Proceedings of the 2000 IEEE International Conference on Systems, Man, and Cybernetics, Nashville, TN, USA, 8–11 October 2000; pp. 3354–3359.
15. Li, B.; Gallagher, T.; Dempster, A.G.; Rizos, C. How feasible is the use of magnetic field alone for indoor positioning? In Proceedings of the 2012 International Conference on Indoor Positioning and Indoor Navigation (IPIN), Nashville, TN, USA, 8–10 October 2000; pp. 1–9.
16. He, S.; Shin, K.G. Geomagnetism for smartphone-based indoor localization: Challenges, advances, and comparisons. *ACM Comput. Surv. (CSUR)* **2018**, *50*, 97. [CrossRef]
17. Senin, P. Dynamic time warping algorithm review. *Inf. Comput. Sci. Dep. Univ. Hawaii Manoa Honol. USA* **2008**, *855*, 40.
18. Subbu, K.P.; Gozick, B.; Dantu, R. LocateMe: Magnetic-fields-based indoor localization using smartphones. *ACM Trans. Intell. Syst. Technol. (TIST)* **2013**, *4*, 73. [CrossRef]
19. Subbu, K.P.; Gozick, B.; Dantu, R. Indoor localization through dynamic time warping. In Proceedings of the 2011 IEEE International Conference on Systems, Man, and Cybernetics, Anchorage, AK, USA, 9–12 October 2011; pp. 1639–1644.
20. Ashraf, I.; Hur, S.; Park, Y. mPILOT-Magnetic Field Strength Based Pedestrian Indoor Localization. *Sensors* **2018**, *18*, 2283. [CrossRef] [PubMed]
21. Ashraf, I.; Hur, S.; Shafiq, M.; Park, Y. Floor Identification Using Magnetic Field Data with Smartphone Sensors. *Sensors* **2019**, *19*, 2538. [CrossRef] [PubMed]
22. Hui, L.; Luo, H.; Fang, Z.; Li, X. *TACO: A Traceback Algorithm Based on Ant Colony Optimization for Geomagnetic Positioning*; Springer: Berlin/Heidelberg, Germany, 2014; pp. 208–224.
23. Shu, Y.; Shin, K.G.; He, T.; Chen, J. Last-mile navigation using smartphones. In Proceedings of the 21st Annual International Conference on Mobile Computing and Networking, Paris, France, 7–15 September 2015; pp. 512–524.
24. Le Grand, E.; Thrun, S. 3-axis magnetic field mapping and fusion for indoor localization. In Proceedings of the 2012 IEEE International Conference on Multisensor Fusion and Integration for Intelligent Systems (MFI), Hamburg, Germany, 13–15 September 2012; pp. 358–364.
25. Kim, B.; Kong, S.H. A Novel Indoor Positioning Technique Using Magnetic Fingerprint Difference. *IEEE Trans. Instrum. Meas.* **2016**, *65*, 2035–2045. [CrossRef]
26. Huang, H.; Li, W.; Luo, D.A.; Qiu, D.W.; Gao, Y. An Improved Particle Filter Algorithm for Geomagnetic Indoor Positioning. *J. Sens.* **2018**, *2018*, 5989678. [CrossRef]
27. Benesty, J.; Chen, J.; Huang, Y.; Cohen, I. Pearson correlation coefficient. In *Noise Reduction in Speech Processing*; Springer: Berlin, Germany, 2009; pp. 1–4.
28. Zheng, M.; Zhao, Y.; He, Y.; Xu, C.-Z. Sensitivity-based adaptive particle filter for geomagnetic indoor localization. In Proceedings of the 2017 IEEE/CIC International Conference on Communications in China (ICCC), Qingdao, China, 22–27 October 2017; pp. 1–6.
29. Shu, Y.; Bo, C.; Shen, G.; Zhao, C.; Li, L.; Zhao, F. Magicol: Indoor localization using pervasive magnetic field and opportunistic WiFi sensing. *IEEE J. Sel. Areas Commun.* **2015**, *33*, 1443–1457. [CrossRef]
30. Srinivas, M.; Patnaik, L.M. Genetic algorithms: A survey. *Computer* **2002**, *27*, 17–26. [CrossRef]
31. Mahony, R.; Hamel, T.; Pflimlin, J.-M. Nonlinear Complementary Filters on the Special Orthogonal Group. *IEEE Trans. Autom. Control* **2008**, *53*, 1203–1218. [CrossRef]
32. Butcher, J.C.; Butcher, J. *The Numerical Analysis of Ordinary Differential Equations: Runge-Kutta and General Linear Methods*; Wiley-Interscience: New York, NY, USA, 1987; Volume 512.
33. Ahn, D.; Park, J.-S.; Kim, C.-S.; Kim, J.; Qian, Y.; Itoh, T. A design of the low-pass filter using the novel microstrip defected ground structure. *IEEE Trans. Microw. Theory Tech.* **2001**, *49*, 86–93. [CrossRef]
34. Keselbrener, L.; Keselbrener, M.; Akselrod, S. Nonlinear high pass filter for R-wave detection in ECG signal. *Med. Eng. Phys.* **1997**, *19*, 481–484. [CrossRef]
35. Zhang, P.; Li, F. A new adaptive weighted mean filter for removing salt-and-pepper noise. *IEEE Signal Process. Lett.* **2014**, *21*, 1280–1283. [CrossRef]

36. Brajdic, A.; Harle, R. Walk detection and step counting on unconstrained smartphones. In Proceedings of the 2013 ACM International Joint Conference on Pervasive and Ubiquitous Computing, Zurich, Switzerland, 8–12 September 2013; pp. 225–234.
37. Zhuang, Y.; Lan, H.; Li, Y.; El-Sheimy, N. PDR/INS/WiFi integration based on handheld devices for indoor pedestrian navigation. *Micromachines* **2015**, *6*, 793–812. [CrossRef]
38. Ladetto, Q. On foot navigation: Continuous step calibration using both complementary recursive prediction and adaptive Kalman filtering. In Proceedings of the ION GPS, Salt Lake City, UT, USA, 19–22 September 2000; pp. 1735–1740.
39. Vildjiounaite, E.; Malm, E.-J.; Kaartinen, J.; Alahuhta, P. Location estimation indoors by means of small computing power devices, accelerometers, magnetic sensors, and map knowledge. In *Pervasive Computing*; Springer: Berlin, Germany, 2002; pp. 211–224.
40. Tian, Q.; Salcic, Z.; Kevin, I.; Wang, K.; Pan, Y. A multi-mode dead reckoning system for pedestrian tracking using smartphones. *IEEE Sens. J.* **2015**, *16*, 2079–2093. [CrossRef]
41. Weinberg, H. Using the ADXL202 in pedometer and personal navigation applications. *Analog Devices AN-602 Appl. Note* **2002**, *2*, 1–6.
42. Blu, T.; Thévenaz, P.; Unser, M. Linear Interpolation Revitalized. *IEEE Trans. Image Process. A Publ. IEEE Signal Process. Soc.* **2004**, *13*, 710–719. [CrossRef]
43. Deng, Z.; Zhu, X.; Cheng, D.; Zong, M.; Zhang, S. Efficient kNN classification algorithm for big data. *Neurocomputing* **2016**, *195*, 143–148. [CrossRef]
44. Ma, Y.; Dou, Z.; Jiang, Q.; Hou, Z. Basmag: An optimized HMM-based localization system using backward sequences matching algorithm exploiting geomagnetic information. *IEEE Sens. J.* **2016**, *16*, 7472–7482. [CrossRef]
45. Arulampalam, M.S.; Maskell, S.; Gordon, N.; Clapp, T. A tutorial on particle filters for online nonlinear/non-Gaussian Bayesian tracking. *IEEE Trans. Signal Process.* **2002**, *50*, 174–188. [CrossRef]
46. Gay, D.M. Correctly rounded binary-decimal and decimal-binary conversions. *Numer. Anal. Manuscr. 90–10*. 1990. Available online: https://ampl.com/REFS/rounding.pdf (accessed on 29 November 2019).

Article

Hybrid Positioning for Smart Spaces: Proposal and Evaluation

Pedro J. Fernández [1,*], José Santa [2] and Antonio F. Skarmeta [1]

1 Department of Information and Communications Engineering, University of Murcia, 30100 Murcia, Spain; skarmeta@um.es
2 Department of Electronics and Computer Technology, Technical University of Cartagena, 30202 Cartagena, Spain; jose.santa@upct.es
* Correspondence: pedroj@um.es; Tel.: +34-868-88-4644

Received: 30 April 2020 ; Accepted: 10 June 2020; Published: 13 June 2020

Abstract: Positioning capabilities have become essential in context-aware user services, which make easier daily activities and let the emergence of new business models in the trendy area of smart cities. Thanks to wireless connection capabilities of smart mobile devices and the proliferation of wireless attachment points in buildings, several positioning systems have appeared in the last years to provide indoor positioning and complement GPS for outdoors. Wi-Fi fingerprinting is one of the most remarkable approaches, although ongoing smart deployments in the area of smart cities can offer extra possibilities to exploit hybrid schemes, in which the final location takes into account different positioning sources. In this paper we propose a positioning system that leverages common infrastructure and services already present in smart spaces to enhance indoor positioning. Thus, GPS and WiFi are complemented with access control services (i.e., ID card) or Bluetooth Low Energy beaconing, to determine the user location within a smart space. Better position estimations can be calculated by hybridizing the positioning information coming from different technologies, and a handover mechanism between technologies or algorithms is used exploiting semantic information saved in fingerprints. The solution implemented is highly optimized by reducing tedious computation, by means of opportunistic selection of fingerprints and floor change detection, and a battery saving subsystem reduces power consumption by disabling non-needed technologies. The proposal has been showcased over a smart campus deployment to check its real operation and assess the positioning accuracy, experiencing the noticeable advantage of integrating technologies usually available in smart spaces and reaching an average real error of 4.62 m.

Keywords: hybrid positioning; fingerprinting; indoor positioning; smart buildings; mobile devices

---

1. Introduction

In the last decades, boosted by the appearance and proliferation of smartphones, many indoor positioning systems have been developed. The point in common in all of them has been the usage of existing infrastructures and technologies integrated in these devices for multiple purposes, facilitating the deployment and reducing costs. As a result, many positioning systems can coexist in the same terminal implementing hybrid approaches. In addition, when a base technology or algorithm must be chosen, a handover policy schema should be defined to get the best approximation of the real position. Technologies like WiFi, Bluetooth Low Energy (LE), and built-in movement sensors have been widely exploited to implement indoor positioning systems.

In recent years, new technologies related to smart spaces and Internet of Things (IoT) have appeared, presenting potential positioning capabilities. This strong relation between indoor positioning and IoT is stated in the work of Macagnano et al. [1], for instance. Access control systems and face recognition cameras are two representative examples that take profit of these new technologies.

However, existing positioning systems do not exploit these new infrastructures to better estimate the location. Hence, one of our main contributions in this paper is the presentation of a hybrid and modular positioning system that uses, apart from the most common technologies present in smartphones, new technologies available in smart spaces.

Proximity and triangulation based algorithms using radio-frequency requires sensing the medium to compute location. Moreover, two-phase positioning algorithms like fingerprinting [2] also requires the creation of a radio-map in a first phase. At this stage a database is created storing reference points linking a determined position with sensed signal strength indicators (RSSIs) from surrounding access points or tags. In a second phase, the fingerprinting algorithm compares the current received RSSI using a matching algorithm to determine the closest reference point. However, the usage of all available technologies in smartphones and smart spaces for positioning purposes is not straightforward, and a proper platform able to integrate measurements and account for the error committed by each technology is necessary. This issue is even more complex when considering power constrains and mobility patterns of smartphones.

Our solution is designed to be used as a library to be easily incorporated in different applications and services. Due to the two-phase nature of some algorithms like Wi-Fi fingerprinting or even generic proximity to base stations, a training application has to be developed to prepare the target indoor spaces to let the system work properly. Here it is necessary to register the fingerprints and the installed base stations, such as WiFi access points or Bluetooth tags. In our case, this application has been developed for Android platform, as reference app that integrates and tests our positioning library developed in Java. An important contribution to the solution presented in this paper is the way different positioning technologies integrate to get not only better position estimations, but also a reduction of required computation effort and, hence, energy consumption, prolonging the life of smartphone batteries. This is implemented through an intelligent handover mechanism between technologies or algorithms, which is fed by extra semantics included in the fingerprints. Apart from supporting the position calculation, registered fingerprints indicate the indoor-outdoor transition areas, floor changing locations, and extensible fields to further suggest positioning technologies are not needed.

The paper is organized in the following parts. Section 2 presents most prominent related papers in the literature. An small survey of indoor positioning technologies is detailed in Section 3. The proposed architecture of our approach is given in Section 4. Section 5 showcases our implementation in a real-world testbed within a smart building and analyses the obtained results. Finally, Section 6 presents the conclusions and future research lines.

2. Related Work

Considering several technologies for indoor positioning are quite common in the recent literature, given that indoor spaces are gaining momentum with the relevance of smart buildings. For example, in [3] it is presented a framework where the system accepts new technologies in a modular fashion, but this work lacks full implementation and no evaluation tests are carried out. In addition, due to the age of the proposal (2012), new communication technologies such as Bluetooth LE or NFC are considered. Similar issues are found in the more recent contribution in [4], where NFC for indoor support is included, with the bid drawback of needing an important deployment of NFC tags to approximate indoor positions. Regarding switching between technologies, a handover procedure is also addressed in [5], presenting a similar approach to the one followed in the current paper, which is based on specific border zones that prepare the decision algorithm to select the next and most promising technology. This allows a seamless transition, since technologies like GPS need more time to warm-up. Another handover approach is introduced in [6], but focusing more on the accuracy estimation. Although GPS provides dilution of precision (DOP) information, for the case of Wi-Fi only the amount of available APs is used as accuracy measurement. A simpler approach is presented in [7], in which the handover triggering is performed taking into account the number of visible GPS satellites.

In our case, regarding indoor fingerprinting, the accuracy is computed on the basis of how the nearest fingerprints are distributed in the space, once the pattern matching algorithm is performed. According to [8], this accuracy estimation method has a good trade-off between error and computational cost.

Dead reckoning (DR) algorithms are very extended in the literature, but here some works that deal with multiple information sources are highlighted. For example, in [9] the DR algorithm is improved by the addition of landmarks to eliminate the cumulative error. No communication technologies are taken into account in this work, limiting the scope of the solution. The work presented in [10] suffers from the same limitation, but at least it proposes using digital maps and Bluetooth LE to perform DR drift corrections. These pedestrian DR proposals do not have the need to train the system, as it is the case of fingerprinting; however, they are not properly prepared to switch between other positioning technologies. Nevertheless, in [11] a handover mechanism similar to ours is considered, but technologies and algorithms are too coupled with the solution and the proposed DR algorithm is based on a good initial location approximation.

The solution proposed in [12] uses a promising handover algorithm for switching between indoor and outdoor technologies. This is achieved thanks to the light and magnetic measurements made by the smartphone sensors, selecting GPS technology for outdoors and the pedestrian dead reckoning (PDR) algorithm for indoors. Our approach also offers this differentiation between outdoor and indoor operation, but it is achieved by the addition of extra semantics in fingerprints that easily allow the identification of indoor and outdoor zones.

The issue with PDR approaches is the complexity of calculations and the need of continuously compensating cumulative errors. Due to this, PDR is not initially taken into account in our solution, but thanks to its modular design, it can be easily added in the future. This is a distinctive factor of our positioning system. It is considered as an open framework where multiple technologies and algorithms can coexist and contribute to improve accuracy. Moreover, infrastructure-based positioning technologies available in smart spaces, such us card-based authentication, are considered in our approach. Some extra technical advances are also included to provide history-based selection of potential fingerprints, multi-floor support, high-accuracy positioning using proximity to Bluetooth LE tags and WiFi APs, weighted positioning, and support of different mobile terminals with different radio sensibilities, among others.

3. Review of Indoor Positioning Technologies

Indoor positioning systems have been implemented thanks to different technologies. First, indoor systems started to use infrared signals [13], in which users are provided with badges that emit diffuse infrared signals and pre-installed infrared sensors detect emitters and report its approximate position. However, the usage of infrared implies interference from florescent lights and sunlight. To avoid these drawbacks, systems based in ultrasound signals [14,15] started to replace them, using ultrasound tags in a similar fashion. Due to ultrasounds are unable to penetrate walls and suffer from reflection interference, radio-frequency signals were proposed as an improvement, sometimes in combination with ultrasound signals, in order to implement time-of-flight trilateration techniques. With these, user terminals send both kinds of signal simultaneously and usually the infrastructure is in charge of computing the difference in arrival times to each receiver to determine a quite accurate position.

The previous positioning systems require the user to acquire additional hardware such as budges or tags. Therefore, more recent positioning systems [16] proposed the usage of hardware already present. Nowadays, everybody has a smartphone in the pocket, which are extremely equipped devices with good communication possibilities. Despite global navigation satellite systems like GPS, GLONASS, GALILEO and BEIDOU are also available in smartphones and offer enough accuracy for outdoors applications, satellite signals are too week to reach indoor spaces. Other radio-frequency technologies addressing general-purpose data communications, like Bluetooth, WiFi or 4G, among others, can be used for navigation purposes using a wide range of positioning algorithms.

The most relevant algorithms exploiting wireless communications capabilities of mobile devices are summarized next:

- Proximity: This is the simplest way to calculate the position. It uses RSSI values sensed from well-known base stations, such as WiFi access points or Bluetooth emitters, which are compared to a reference signal level to estimate its proximity. Usually the $RSSI_{1m}$ value measured at one meter of distance is considered as the baseline. This algorithm has a very low computational cost, but offers low accuracy depending on the distance and obstruction to the access points or tags.
- Triangulation: Triangulation uses the geometric properties of triangles to determine the location. It has two derivations: lateration and angulation. Techniques based on the measurement of the propagation time of signals (e.g., Time of Arrival (TOA), Rount-Trip Time of Flight (RTOF), Time Difference of Arrival (TDOA) and even hybrid Time of Flight with Direction of Arribal (TOF-DOA) studied in [17]) and RSS-based methods are called lateration techniques [10,11]. The Angle of Arrival (AOA) estimation technique is considered a representative example of an angulation technique. Jun-Ho et al. [18] uses trilateration with BLE beacons distributed in hexagonal cells, for instance.
- Fingerprinting: This is one of the most used algorithms in the literature. It is a two-phase algorithm that consists of, first, generating a radio-map with a set of reference points (fingerprints) with their RSSI measurements to surrounding transmitter like WiFi access points or Bluetooth LE tags. The second stage, which involves the real-time operation of the system, imply sensing the RSSI in communications with surrounding transmitters and then comparing with the fingerprints to find the nearest. Euclidean distance is the most usual method to search among fingerprints [19]. The output of the algorithm, i.e., the most similar fingerprint to current RSSI values, determines the position of the device. The density of reference points should consider a trade-off between accuracy and computational cost, as more reference points imply a better approximation at the expense of extra computational cost. Although the infrastructure can be already present for the case of WiFi, the high cost of the radio-map generation and maintenance should be also taken into account.
- Dead Reckoning: Given that smartphones are usually provided with built-in gyroscopes and accelerometers, they can use orientation and acceleration data, respectively. A representative algorithm exploiting these capabilities is Dead Reckoning (DR). It consists of estimating the current position based on the last known position, movement direction and speed. The main disadvantage here is the accumulative error when computing new positions based on previous ones. This technique should be combined with another one that helps to eliminate the accumulated error and estimate and initial location. For instance, the work presented in Kang et al. [20] is a good representative of this positioning method. We expect to introduce similar PDR technology in our solution soon.

Emerging methods of measuring reference distances for indoor positioning are introduced in [21,22], where the usage of OFDM signals in combination with the properties of the Zadoff–Chu sequences achieves promising results. Another example is the work in [23], which exploits millimeter-wave (mmWave) combined with Machine Learning (ML) procedures to obtain excellent results.

With the deployment of new smart spaces crowded of sensors, cameras and services, new data sources for positioning are available. For example, in Anca Morar et al. [24] it is summarized some computer vision methods using cameras and QR codes. This is the line followed in the current paper: exploiting smart infrastructure to support indoor navigation. In particular, our implementation integrates an access control system based in ID cards to improve accuracy when card readers registered in the platform are used. In general, our work combines traditional radio-frequency positioning technologies with capabilities of smart environments and enrich fingerprinting knowledge with semantics about particular locations, to provide extra continuity, accuracy and support to reduce energy consumption of mobile devices using the system.

4. Hybrid Positioning Approach

4.1. Overall Architecture

A collaborative/hybrid positioning approach has been designed on the basis of usual positioning sources (like cellular, GPS and wireless proximity) and some extra indoor positioning technologies and algorithms coming from smart spaces. This way, the most accurate location source will be chosen or a combination of them, improving the accuracy and availability of the system across different places. A graphical representation of how the main functions of the positioning system work is shown in Figure 1. Data retrieved from radio-frequency communication technologies coming from the same mobile device are combined with data received by smart infrastructures to compute the final position.

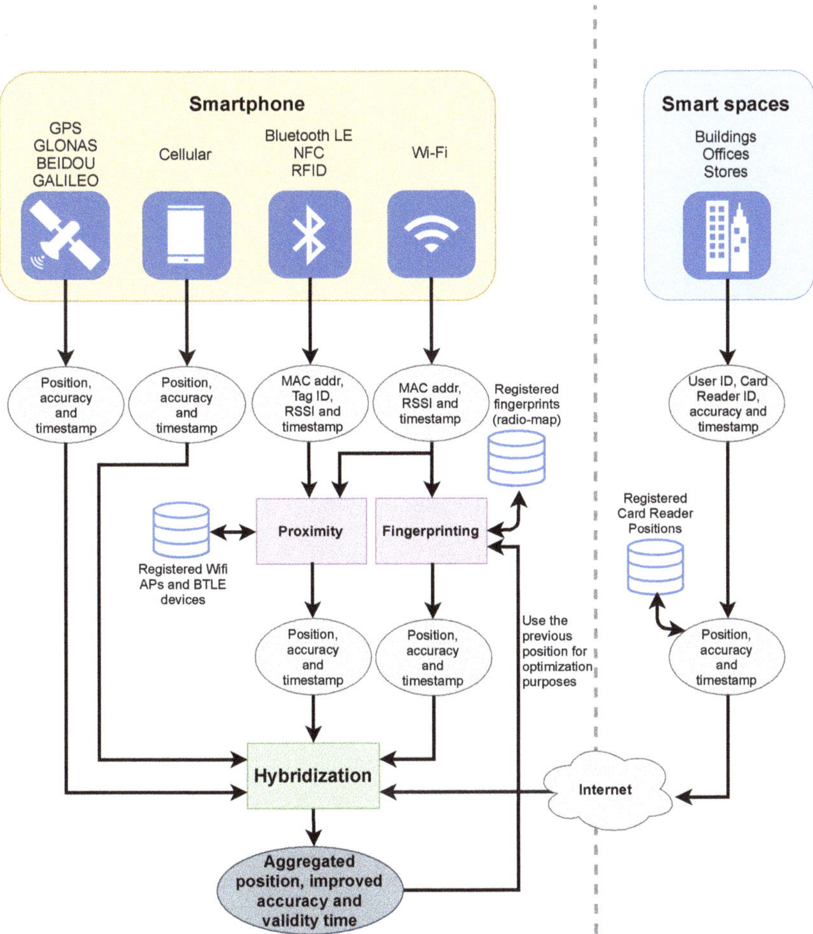

Figure 1. Positioning subsystem.

Positioning technologies such as GPS or cellular, which provide a final position, directly input the decision module with the hybridization algorithm. In this case an accuracy approximation is usually provided by the same technology, such as the position quality indicators of GPS using the geometry of satellite signals. Cell ID carried out using cellular networks, as they are integrated in Android devices at least, also provide an accuracy estimation in meters. However, regarding measurements coming from short and medium-range communication technologies, the signal strength (RSSI) and ID of base stations (Wi-Fi access points and BLE tags) sensed from the surroundings are the input for state-of-the-art fingerprinting and proximity algorithms improved as explained in the next parts. These provide an estimated location together with a position accuracy indicator, which is used to check the most appropriate technology or algorithm.

Our architecture also puts the focus on energy saving, due to devices like smartphones are power constrained and use short-life batteries. Thanks to the hybridization carried out between GPS/Celular, with good performances outdoors, and WiFi/BLE based algorithms, the positioning system disable or enable GPS/Cellular positioning methods when the user is detected inside a building or near the exit, respectively. This can be detected using special fingerprints/BLE beacons located near the entry points of the building. Another energy saving measure have been added in the fingerprinting matching calculation, which is a CPU-intensive task consuming much energy. It is focused on filtering as many fingerprints as possible using the last reported position to only consider closer fingerprints of the same floor. This introduces the extra issue of detecting floor changes, but this is efficiently solved using special fingerprints/BLE beacons located close to stairs or lifts.

Once the position and accuracy information coming from all technologies and algorithms and smart infrastructure devices is received, a hybridization process is carried out to compute a final position and provide a combined estimation of the accuracy.

4.2. Fingerprinting Algorithm

The fingerprinting algorithm performs a pattern matching search comparing previously collected fingerprints (in a prior training stage) with current measurements. For the moment, a state-of-the-art deterministic approach [2] has been used for the matching method, using the Euclidean distance, but instead of a single fingerprint output, a set of the most similar fingerprints are considered, ordered by distance. This search is bounded within a configurable threshold radio (default value is 20 m) around the previous location, in order to offer significant results and save computational cost and energy consumption. The searching process is also carried out over fingerprints of the same floor when the user is indoors. As has been said, with the aim of supporting the change between floors, handover areas are modelled with flagged fingerprints. When the user is detected on these fingerprints, a vertical cylinder-based search is performed using the previously configured radio. Once the list of candidate fingerprints is obtained, a weighted position is calculated on the basis of its Euclidean closeness to the sensed radio environment. Finally, an estimation of the position accuracy is calculated using the "Best Candidate Set" method proposed in [8]. Taking into account only a set of closest fingerprints, this accuracy estimation is calculated as the mean of the euclidean distances between the closest fingerprint and the rest.

4.3. Proximity Algorithm

The proximity algorithm detects the closeness to a radio tag or a short/medium-range base station. In a previous phase, all base stations were registered in the system database, taking into account its geographical position and the signal strength received at a one-meter of distance (this is not required for BLE tags due to this 1-meter reference is included in the beacon messages). Depending on the technology and the communication range configured, when the signal strength of the detected base-stations/beacons are obtained in a scan process, the distance is approximated using the 1-meter reference. This calculation is finally translated into distances in meters, which let us know which base-station is the closest one. At this point we can estimate that the mobile device is positioned at the installation point of the base station or tag, with an estimation error equal to this calculated distance. The stronger the detected base-station signal, the more accurately the mobile device is located.

4.4. Smart Space Integration

From smart deployments including ID-card readers for door opening systems, a momentary but very accurate position can be obtained by knowing the installation point. In our case, the accuracy estimation is reduced proportionally to the average walking speed. Since people walk calmly between 1 and 2 m per second, we consider that the accuracy estimation is reduced by 1 m per second in the next 30 s after opening a door. As the access control and positioning systems are independent, the interaction between them is not straightforward. In order to let the smartphone know about a door opening event, a position monitoring online service has been deployed to receive the current position of different users with a certain periodicity. This could be useful for centralized control purposes. As depicted in Figure 1, when a door opening event occurs, the access control system notifies the position monitoring system with the user ID, the card reader ID and the timestamp of the event. The associated location is retrieved from the registered card readers database. Then, the ID-card driven location source returns a new position and its associated accuracy in order to be taken into account in the hybridization module to be combined with the rest of positioning data coming from the other technologies and algorithms.

4.5. Final Accuracy and Position Estimation

Position estimations coming from the different data sources are provided with a timestamp, an identifier of the technology used and the associated estimated accuracy. This way it is possible to know the accuracy level and validity time of the estimated position. Depending on hybridization policy scheme, the position estimations are combined to get a final location.

Our hybrid approach denotes the usage of different ways of approximating the user position, but the most important part is how they are combined to get a better approximation minimizing the error. It is determined by a set of predefined handover policies. Initially the handover policy was very simple, selecting the most accurate technology at each moment and discarding the rest. However, in a second stage of the development we realized that technologies can support each other to improve accuracy. The strategy followed is illustrated next, with a representative example about how fingerprinting can provide a valuable help to the proximity algorithm to reduce the error.

Figure 2 shows a first case without combining technologies on the left. Here, the positioning system uses the Bluetooth beacon depicted in the figure to locate the person. The distance between the person and the beacon can be determined using the perceived signal strength. However, the positioning system is only able to figure out that the user is probably located at any point of the dashed blue circumference surrounding the beacon. As the mean position of all possibilities in a circumference is the middle point, the algorithm returns this location as result, with an accuracy/error equals to the calculated distance between the person and the beacon. In contrast, in the second case illustrated in Figure 2, due to the fact that Wi-Fi fingerprinting is integrated with Bluetooth proximity, in addition to the distance between the person and the beacon, the location of the nearest fingerprint can be used to choose a best estimate location within the circumference. Hence, as can be seen, the final error is greatly reduced. As we cannot know the exact value of this new error, the same accuracy/error blue circumference is also shown in Figure 2b, but centered in this case in the new approximated position. This assumption is valid if we have at least three surrounding fingerprints. This hybridization is just an example of the possibilities offered by the platform, which is open to new merging possibilities.

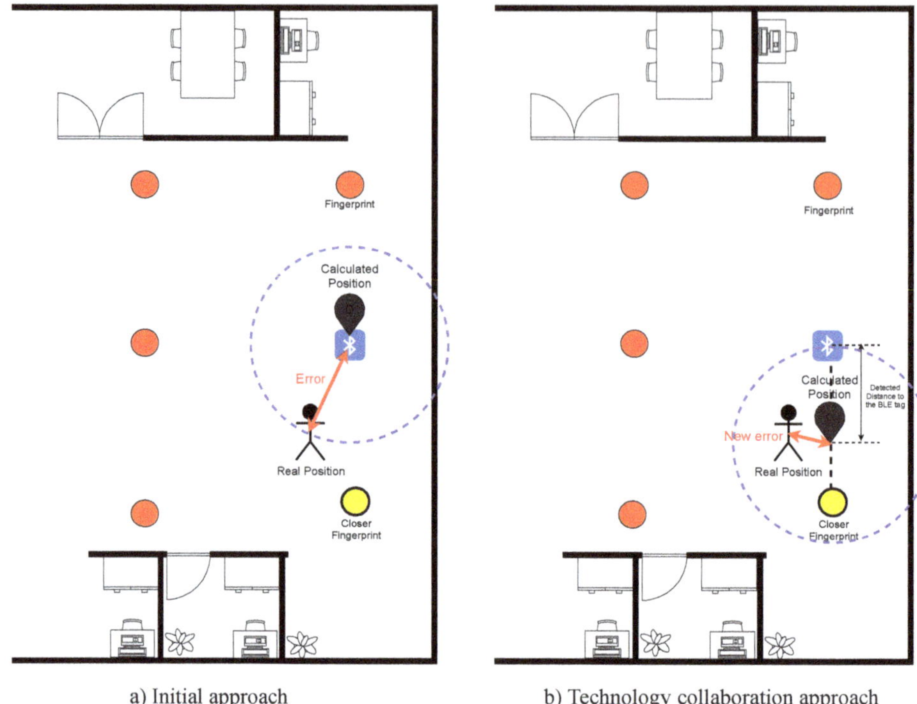

Figure 2. Technology collaboration strategy to improve positioning accuracy.

In addition, the handover mechanism has to support the effective change between technologies, in order to determine what is the main and supporting technologies at each moment. This approach can be used for other purposes like battery saving and preparing the technology beforehand. This is particularly interesting in our case when using GPS, since it can be disabled indoors and activated when the mobile device is detected at especially identified handover fingerprints close to the main door of the building.

4.6. Managing Hardware Diversity

Positioning systems using radio-map databases like fingerprinting have a notorious problem when different devices are used for both training and matching procedures, especially if RSSI-based matching algorithms are used. Under this situation, radio sensing varies with the hardware, so different RSSI levels are perceived by different devices. As not a single smartphone model is going to be used, it is necessary to scale measurements.

In our case, the application used to sense the radio environment is firstly configured with a simple test near well-known base stations or tags, in order to compute a correcting factor for the signal strength perceived. This scaling solution is proposed in [25], based on the assumption that the order of magnitude among RSSI measurements does not change between different smartphones. In other words, two different terminals at the same place will sense the signals of the same access points, not necessarily with the same intensity, but maintaining relative RSSI differences among measurements. Due to this, RSSI perceived from each access point or tag is only used to rank them. The fingerprinting matching algorithm uses the similarity of the rankings to estimate the distance of each fingerprint to the current position. An advantage of this approach is that the whole radio-map is valid for all devices, and only a minor calibration of the mobile device is necessary the first time the positioning system is used.

4.7. Modular Design

The presented framework is designed in a modular fashion, so the supported technologies and algorithms are considered as positioning sources. More technologies and algorithms can be easily incorporated to the system. For example, PDR algorithms offer good results and could complement the presented implementation based on proximity and fingerprinting.

As some positioning algorithms need the previous estimated position in order to apply optimization measures or calculate the next one, the rest of positioning sources allow to retrieve it properly from the system. This would be the case if PDR-based methods were developed. In our case, the fingerprinting algorithm uses the previous position to consider only a reduced subset of potentially matching fingerprints, reducing computational effort and saving energy, as it is shown in Figure 1.

5. Implementation and Evaluation of the Solution

A real testbed was deployed at the Computer Science Faculty (University of Murcia). As can be seen in Figure 3, real plans of the building floors are used to show the main architectural components and one of the validation tests. Prior to the tests, 26 WiFi APs were registered with their positions (latitude and longitude), as well as their measured RSSI at 1 meter of distance using the smartphone app. In addition, five BLE tags were installed at strategic points. For the sake of clarity, these tags were depicted with the technology symbol, while WiFi APs were included as points surrounded by waves. Only some of the doors of the first floor of the building were provided with contact-less card readers; that supposes approximately 34 of them. However we only used one in the trials in order not to compromise workmates privacy. It is shown in purple in the map on the right of the first floor, as can better be seen in the zoomed part of the map. The door opening system of the building was prepared to send positions to the positioning system when the user was identified by the ID card readers.

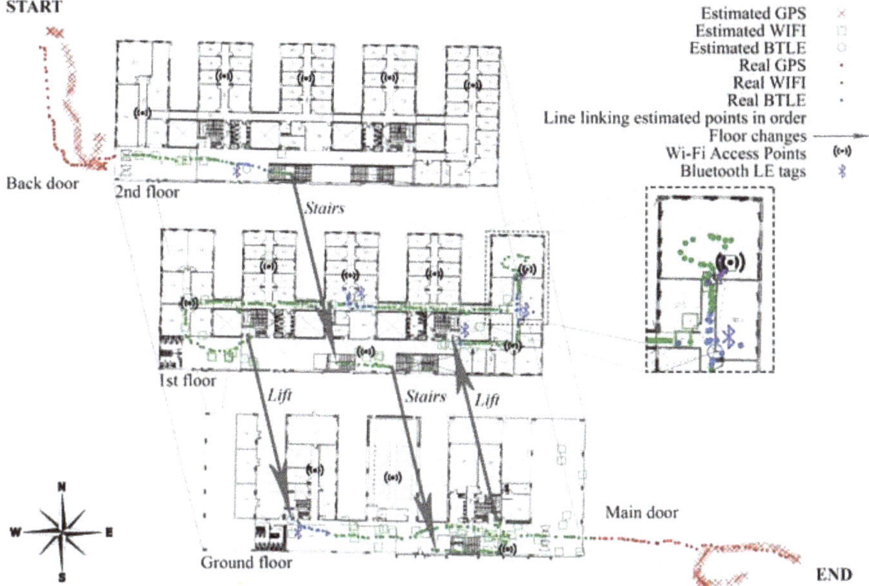

Figure 3. Deployment and operation test of the positioning solution.

The main part of the positioning system was implemented as a Java library used in an smartphone app, and a small server with a REST interface developed in PHP, which worked as a proxy between the door opening system of the building and our positioning system. The app could continuously collect radio measurements and generate position estimates. Since the scan of access points and BLE devices took approximately five seconds, a new position could not be outputted at a higher frequency, so the system provided the same estimated location several times if a higher frequency was requested. The system always responded immediately with the last estimated position available. Every estimation was accompanied with a time-stamp, so the user could estimate the age of every received approximation by the positioning system.

5.1. Validation

One of the trials performed using the positioning system is plotted on the building plans in Figure 3. A prior training for WiFi fingerprinting was performed using the smartphone app, creating a three-meter resolution grid (443 fingerprints in total) shown in the screen shots in Figure 4. Both real and estimated positions are shown. The real position is represented by a black X on the center of the screen, which was manually set constantly during the test by dragging the background map while the user walks. This way, the actual positionwas taken from the graphical interface at the same time the positioning system worked. The estimated position was marked with a message indicating the main technology used, being the center of a circumference that showed the estimated error. Other possible solutions with their associated error are also shown. For each estimated position, the actual location was also saved for later comparison and error calculation.

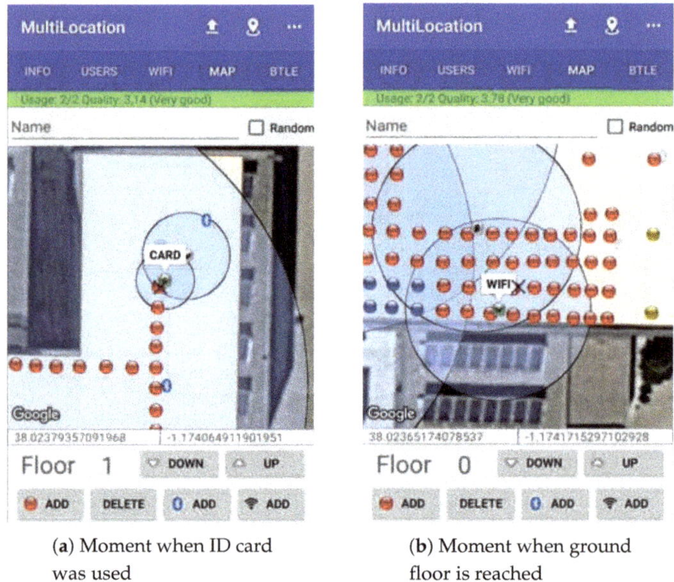

(**a**) Moment when ID card was used

(**b**) Moment when ground floor is reached

Figure 4. Two representative screen-shots of the app used to perform the tests.

The test in Figure 3 started outside the building, and then the user entered the second floor to go downstairs. In this first stretch GPS was initially used, presenting a great error due to multi-path and signal obstruction of the metal facade of the building. Although enough satellites were received, between four and six, measurements still suffered from multi-path interference from the metal facade. The same phenomenon happened at the end of the path, when the user went outside the building. All the accuracy estimations returned by the different technologies were all expressed in meters, so they could be easily compared. Since the accuracy perceived by using WiFi fingerprinting was better than GPS in this moment, a technology handover was performed. Then the position accuracy was clearly better. Just before the stairs, a Bluetooth LE tag was detected and used with the proximity algorithm for a while, since it presented an error estimation better than WiFi fingerprinting. At the first floor the user went downstairs again to the ground floor, the moment captured in Figure 4b. At this floor, the user took the lift and returned to the first floor to walk within our department. Just going out the lift, a Bluetooth LE tag was detected, improving the floor change detection.

At the middle of this stretch the user entered the central corridor and was located with another Bluetooth LE tag. Then he entered a laboratory, a moment also captured in Figure 4a. The door opening infrastructure informed the positioning system about the user location. Since this position had an optimum accuracy level, it was selected as the most accurate position, but only for a while. The error produced by this position source was incremented by 1 m/s. Later, after continuing walking through the first floor, the user entered another lift to come back to the ground floor and go straight forward to the main door exit. Close to the lift, another Bluetooth LE tag helped in the user location and floor change detection. Finally, when the user was near the main door, the positioning system selected flagged fingerprints represented as yellow balls in Figure 4b, that notified an exit proximity. This event was used by the system to turn on the GPS positioning system in the user smartphone. Usually, smartphones automatically turn off the GPS capabilities to save battery life if no GPS signals are received, so our app had to turn it on again.

5.2. Accuracy Analysis

The distribution of errors committed by the positioning system in the previous test is plotted in Figures 5 and 6. Figure 5 includes all the position estimations returned by each technology during the test. On the contrary, Figure 6 only includes the most accurate position approximations at each moment, i.e., the final estimation chosen. The histogram labeled with the hybrid solution in Figure 5 includes the final error produced by the hybridization process, while the rest of histograms show the error produced by each technology itself. Hence, all data included in Figure 6 belong to the hybrid part of Figure 5.

As can be seen in Figure 6, most of the calculated positions by the hybridization process had an error below 10 m, and a great part below 5 m. Most of the high-impact errors were due to GPS, since it was used near the building with a metal-made facade that reflected part of the satellite signals. However, there were some errors between 10 and 20 m that were attributed to WiFi fingerprinting, which presented deviations when the user moved near the floor edges. This was due to the bad distribution of fingerprints surrounding the user at these locations. Although accuracy was highly impacted by the wide usage of WiFi fingerprinting, overall results were improved by the small areas where Bluetooth LE proximity and ID card readers were used. Here the errors detected were usually below 5 m. The Android Network positioning source was not used, since the accuracy value obtained for this technology (based on cell ID and WiFi proximity) was worse than the accuracy estimations of the rest of technologies. This is the reason why in Figure 6 there are not values in the NET histogram.

A summary of the error results for each technology is included in Table 1. From these results we can estate that the mean error of the system was 4.62 ± 0.31 m for our testbed involving a mix of technologies. It is important to remark the challenging scenario considered, where the GPS signal is partially disrupted by the metal-made building facade, increasing the overall error. These results could be further improved with a lower resolution grid for WiFi fingerprinting, deploying more Bluetooth LE tags and entering more spaces with a card reader available. Many rooms in the building provide such capability and we have also disabled many other Bluetooth LE tags in order to present a realistic and cost-efficient setup.

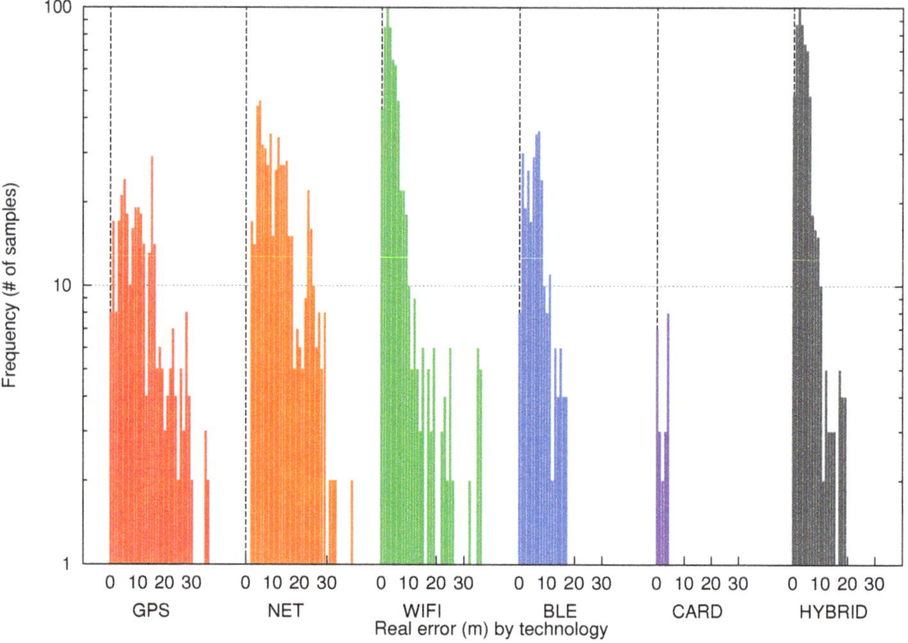

Figure 5. Histograms of the error committed for each technology (all the samples).

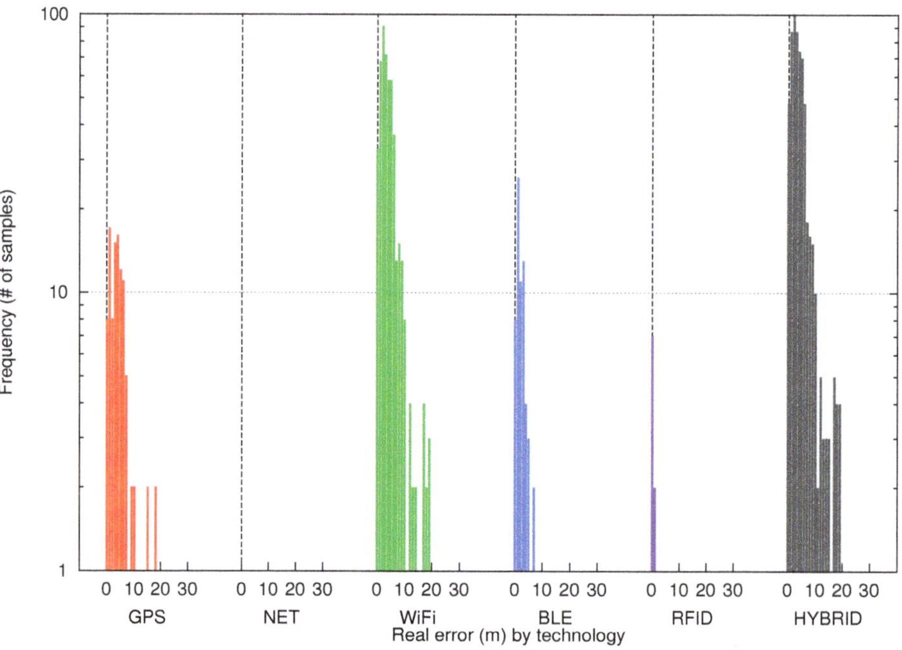

Figure 6. Histograms of the error committed for each technology (only the selected samples).

Table 1. Statistics obtained from the error values collected for each technology.

Stats	HYBRID	GPS	NET	WIFI	BTLE	CARD (NFC)
Mean	**4.62** m	12.71 m	15.04 m	8.45 m	6.53 m	2.61 m
Maximum	**19.23** m	38.40 m	40.85 m	37.56 m	17.69 m	4.78 m
Minimum	**0.12** m	0.53 m	2.53 m	0.12 m	0.51 m	0.65 m
Standard deviation	**4.08** m	9.39 m	13.19 m	13.78 m	3.91 m	1.63 m
Confidence interval 95%	**0.31** m	0.99 m	1.07 m	1.03 m	0.45 m	0.66 m
Samples	**664**	342	573	671	279	23

6. Conclusions

A holistic positioning framework, which is open to multiple indoor and outdoor technologies and algorithms, is presented in the paper. Smart deployments are especially considered and provide an essential support to improve performance. The system maintains accuracy by selecting the most appropriate solution every time based on the estimated error. The fingerprinting algorithm and model are essential in the system, by providing not only position estimates, but also including extended semantics with information that supports the handover between technologies and algorithms used, and the detection of especial situation such as floor changing or building entering/exiting. Moreover, the advantages of the architecture to hybridize position estimates is showed with a reference development of improving the base proximity algorithm with information coming from nearby fingerprints.

The actual implementation considers GPS, WiFi, Bluetooth LE and card ID technologies, together with fingerprinting and proximity positioning algorithms. All of them have been used in a reference testbed aimed at presenting a realistic deployment scenario. This involves not only indoor situations, but also outdoor parts surrounding the building and different handover scenarios in which it is necessary to change between different technologies and algorithms to provide the best position estimation. Challenging scenarios when entering and exiting the building, using the lift for floor changes and walking near walls are considered in the tests. The results obtained in the indoor/outdoor scenario present an average real error of 4.62 ± 0.31 m. This implies that our system provides good position estimations even under challenging settings.

It is planned to extend the positioning system with new technologies and algorithms, considering our background gained in the past [26,27]. The hybridization approach among technologies to compute a merged solution paves the way to find more accurate and cost-efficient solutions for mobile devices. Next steps also include the evolution of our parallel line about reducing computational cost of fingerprinting matching tasks by offloading tasks into the cloud, by including a flexible system with both edge and far cloud support.

Author Contributions: Conceptualization, P.J.F., J.S. and A.F.S.; methodology, P.J.F., J.S. and A.F.S.; software, P.J.F.; validation, P.J.F. and J.S.; formal analysis, P.J.F. and J.S.; investigation, P.J.F. and J.S.; resources, J.S. and A.F.S.; data curation, P.J.F.; writing—original draft preparation, P.J.F.; writing—review and editing, P.J.F. and J.S.; visualization, P.J.F., J.S. and A.F.S.; supervision, J.S. and A.F.S.; project administration, J.S. and A.F.S.; funding acquisition, J.S. and A.F.S. All authors have read and agreed to the published version of the manuscript.

Funding: This work has been supported by the Spanish Ministry of Science, Innovation and Universities, under the Ramon y Cajal Program (Grant No. RYC-2017-23823) and the project PERSEIDES (Grant No. TIN2017-86885-R); and by the European Commission, under the 5G-MOBIX (Grant No. 825496) project.

Acknowledgments: The authors would like to thank University of Murcia for letting us use the Faculty of Computer Science building and its facilities during the execution of experiments.

Conflicts of Interest: The authors declare no conflict of interest.

References

1. Macagnano, D.; Destino, G.; Abreu, G. Indoor positioning: A key enabling technology for IoT applications. In Proceedings of the 2014 IEEE World Forum on Internet of Things (WF-IoT), Seoul, Korea, 6–8 March 2014; pp. 117–118.
2. Khalajmehrabadi, A.; Gatsis, N.; Akopian, D. Modern WLAN Fingerprinting Indoor Positioning Methods and Deployment Challenges. *IEEE Commun. Surv. Tutor.* **2017**, *19*, 1974–2002. [CrossRef]
3. Bekkelien, A.; Deriaz, M. Hybrid positioning framework for mobile devices. In Proceedings of the 2012 Ubiquitous Positioning, Indoor Navigation, and Location Based Service (UPINLBS), Helsinki, Finland, 3–4 October 2012; pp. 1–7. [CrossRef]
4. Lilis, G.; Hoffet, A.; Kayal, M. GeoAware: A hybrid indoor and outdoor localization agent for smart buildings. In Proceedings of the 2016 18th Mediterranean Electrotechnical Conference (MELECON), Lemesos, Cyprus, 18–20 April 2016; pp. 1–6. [CrossRef]
5. Ficco, M.; Palmieri, F.; Castiglione, A. Hybrid indoor and outdoor location services for new generation mobile terminals. *Pers. Ubiquitous Comput.* **2014**, *18*, 271–285. [CrossRef]
6. Rashidian, M.; Karimi, H.A.; Rajabi, M.A.; Esmaeili, A. Developing a Hybrid GPS/Wi-Fi Navigation System for User Guidance. *Adv. Comput. Sci. Int. J.* **2016**, *5*, 12–20.
7. Angelis, G.D.; Angelis, A.D.; Pasku, V.; Moschitta, A.; Carbone, P. A hybrid outdoor/indoor Positioning System for IoT applications. In Proceedings of the 2015 IEEE International Symposium on Systems Engineering (ISSE), Rome, Italy, 28–30 September 2015; pp. 1–6. [CrossRef]
8. Lemelson, H.; Kjærgaard, M.B.; Hansen, R.; King, T. Error Estimation for Indoor 802.11 Location Fingerprinting. In *International Symposium on Location-and Context-Awareness*; Choudhury, T., Quigley, A., Strang, T., Suginuma, K., Eds.; Springer: Berlin/Heidelberg, Germany, 2009; pp. 138–155.
9. Wang, X.; Jiang, M.; Guo, Z.; Hu, N.; Sun, Z.; Liu, J. An Indoor Positioning Method for Smartphones Using Landmarks and PDR. *Sensors* **2016**, *16*, 2135. [CrossRef] [PubMed]
10. Li, X.; Wang, J.; Liu, C. A Bluetooth/PDR Integration Algorithm for an Indoor Positioning System. *Sensors* **2015**, *15*, 24862–24885. [CrossRef] [PubMed]
11. Tabata, K.; Konno, H.; Tsuno, K.; Morioka, W.; Nishino, A.; Nakajima, M.; Kohtake, N. The Design of Selective Hybrid Positioning by Utilizing Accuracy Information for Indoor-Outdoor Seamless Positioning and Verification in Tokyo Station. In Proceedings of the 2015 International Conference on Indoor Positioning and Indoor Navigation (IPIN), Calgary, AB, Canada, 13–16 October 2015; pp. 1–7.
12. Zeng, Q.; Wang, J.; Meng, Q.; Zhang, X.; Zeng, S. Seamless Pedestrian Navigation Methodology Optimized for Indoor/Outdoor Detection. *IEEE Sens. J.* **2018**, *18*, 363–374. . [CrossRef]
13. Want, R.; Hopper, A.; Falcao, V.; Gibbons, J. The Active Badge Location System. *ACM Trans. Inf. Syst.* **1992**, *40*, 91–102. [CrossRef]
14. Harter, A.; Hopper, A.; Steggles, P.; Ward, A.; Webster, P. The anatomy of a context-aware application. In Proceedings of the 5th Annual ACM/IEEE International Conference on Mobile Computing and Networking, (MOBICOM '99), Seattle, WA, USA, 15–19 August 1999; pp. 59–68.
15. Priyantha, N.; Chakraborty, A.; Balakrishnan, H. The Cricket Location-Support system. In Proceedings of the 6th Annual ACM/IEEE International Conference on Mobile Computing and Networking, (MOBICOM 2000), Boston, MA, USA, 6–11 August 2000; pp. 59–68.
16. Bahl, P.; Padmanabhan, V.N. RADAR: An In building RF-based User Location and Tracking System. In Proceedings of the Nineteenth Annual Joint Conference of the IEEE Computer and Communications Societies, INFOCOM 2000, Tel Aviv, Israel, 26–30 March 2000; Volume 2, pp. 775–784.
17. Takahashi, Y.; Honma, N.; Sato, J.; Murakami, T.; Murata, K. Accuracy Comparison of Wireless Indoor Positioning Using Single Anchor: TOF only Versus TOF-DOA Hybrid Method. In Proceedings of the 2019 IEEE Asia-Pacific Microwave Conference (APMC), Singapore, 10–13 December 2019; pp. 1679–1681.
18. Huh, J.H.; Seo, K. An Indoor Location-Based Control System Using Bluetooth Beacons for IoT Systems. *Sensors* **2017**, *17*, 2917. [CrossRef] [PubMed]
19. Machaj, J.; Brida, P. Performance Comparison of Similarity Measurements for Database Correlation Localization Method. In Proceedings of the ACIIDS 2011—Third Asian Conference on Intelligent Information and Database Systems, Daegu, Korea, 20–22 April 2011; pp. 452–461. [CrossRef]

20. Kang, W.; Han, Y. SmartPDR: Smartphone-Based Pedestrian Dead Reckoning for Indoor Localization. *IEEE Sens. J.* **2015**, *15*, 2906–2916. [CrossRef]
21. Piccinni, G.; Avitabile, G.; Coviello, G. Narrowband distance evaluation technique for indoor positioning systems based on Zadoff-Chu sequences. In Proceedings of the 2017 IEEE 13th International Conference on Wireless and Mobile Computing, Networking and Communications (WiMob), Rome, Italy, 9–11 October 2017; pp. 1–5.
22. Piccinni, G.; Avitabile, G.; Coviello, G.; Talarico, C. Analysis and Modeling of a Novel SDR-Based High-Precision Positioning System. In Proceedings of the 2018 15th International Conference on Synthesis, Modeling, Analysis and Simulation Methods and Applications to Circuit Design (SMACD), Prague, Czech Republic, 2–5 July 2018; pp. 13–16.
23. Vashist, A.; Bhanushali, D.R.; Relyea, R.; Hochgraf, C.; Ganguly, A.; Sai Manoj, P.D.; Ptucha, R.; Kwasinski, A.; Kuhl, M.E. Indoor Wireless Localization Using Consumer-Grade 60 GHz Equipment with Machine Learning for Intelligent Material Handling. In Proceedings of the 2020 IEEE International Conference on Consumer Electronics (ICCE), Las Vegas, NV, USA, 4–6 January 2020; pp. 1–6.
24. Morar, A.; Moldoveanu, A.; Mocanu, I.; Moldoveanu, F.; Radoi, I.E.; Asavei, V.; Gradinaru, A.; Butean, A. A Comprehensive Survey of Indoor Localization Methods Based on Computer Vision. *Sensors* **2020**, *20*, 2641. [CrossRef] [PubMed]
25. Machaj, J.; Brida, P.; Piché, R. Rank Based Fingerprinting Algorithm for Indoor Positioning. In Proceedings of the International Conference on Indoor Positioning and Indoor Navigation (IPIN), Guimaraes, Portugal, 21–23 September 2011; pp. 1–4. [CrossRef]
26. Moreno-Cano, M.; Zamora-Izquierdo, M.; Santa, J.; Skarmeta, A.F. An indoor localization system based on artificial neural networks and particle filters applied to intelligent buildings. *Neurocomputing* **2013**, *122*, 116–125. [CrossRef]
27. Santa, J.; Toledo-Moreo, R.; Zamora-Izquierdo, M.A.; Ubeda, B.; Gomez-Skarmeta, A.F. An analysis of communication and navigation issues in collision avoidance support systems. *Transp. Res. Part C Emerg. Technol.* **2010**, *18*, 351–366. [CrossRef]

Article

Improving Accuracy and Reliability of Bluetooth Low-Energy-Based Localization Systems Using Proximity Sensors

Marcin Kolakowski

Institute of Radioelectronics and Multimedia Technology, Warsaw University of Technology, Nowowiejska 15/19, 00-665 Warsaw, Poland; m.kolakowski@ire.pw.edu.pl; Tel.: +48-22-234-7635

Received: 29 August 2019; Accepted: 27 September 2019; Published: 30 September 2019

Abstract: One of the functionalities which are desired in Ambient and Assisted Living systems is accurate user localization at their living place. One of the best-suited solutions for this purpose from the cost and energy efficiency points of view are Bluetooth Low Energy (BLE)-based localization systems. Unfortunately, their localization accuracy is typically around several meters and might not be sufficient for detection of abnormal situations in elderly persons behavior. In this paper, a concept of a hybrid positioning system combining typical BLE-based infrastructure and proximity sensors is presented. The proximity sensors act a supporting role by additionally covering vital places, where higher localization accuracy is needed. The results from both parts are fused using two types of hybrid algorithms. The paper contains results of simulation and experimental studies. During the experiment, an exemplary proximity sensor VL53L1X has been tested and its basic properties modeled for use in the proposed algorithms. The results of the study have shown that employing proximity sensors can significantly improve localization accuracy in places of interest.

Keywords: localization; hybrid localization; Bluetooth Low Energy; extended kalman filter; internet of things; proximity sensors

1. Introduction

One of the biggest challenges that modern societies might face in the coming years is the aging of their populations. According to the European Commission report [1], the projected percentage of people aged 65 or higher in 2070 will be 42.1%, while in 2016 it was 25%. Additionally, the number of working-age persons for every person aged over 65 will fall from 3.3 to 2. It means that the traditional care system consisting in young, healthy people caring for the older ones will become ineffective. Therefore, there is an urgent need for solutions, which will support the elderly persons care and significantly reduce caregivers workload.

The above problem can be partially solved by prolonging elderly people's independent living by using modern Internet-of-Things technological solutions. Ambient and Assisted Living (AAL) programme supports their development by funding various projects aiming to improve older people's health. Among many AAL projects, there are solutions intended to support elderly people living alone at their homes. Their main goal is to reduce the burden of the caregivers and allow many older people to independently live at their homes for as long as possible [2].

The designed and implemented AAL platforms are usually comprised of a set of sensors and devices, which are used to monitor an older person's health state and behavior and help them in daily tasks. One of the functions allowing to monitor a user's activity and behavior is user localization, which is usually delivered by a localization system being a part of the platform [3]. The localization systems used in the AAL platforms should meet several requirements. First of all, their accuracy should be high enough to provide the caregivers with reliable data to detect abnormal behaviors [4] or life threatening situations. For example, the localization data should allow to detect presence near vital spots such as near a gas stove or in a shower, to properly react in case of the prolonged, unusual and possibly dangerous activities. Secondly, it should be energy efficient to avoid frequent charging of system devices. Unfortunately, meeting both of these requirements using conventional solutions is often hard.

1.1. Indoor Localization Systems

Conventional radio indoor localization systems can be divided into two groups based on the type of used signals: narrowband systems and ultra-wideband (UWB) systems. Narrowband systems, which usually use widely spread standards (e.g., WiFi [5] and Bluetooth [6]) typically allow to localize the user with accuracy from one to several meters. UWB-based solutions [7] are much more accurate. Their localization accuracy can be as high as a dozen centimeters, but since the employed technology is not as known as than the aforementioned, they are less popular.

In the UWB positioning systems [7], the users are usually localized based on the time of arrival measurements of signals transmitted from the localized tags to the system infrastructure. Due to the short duration of the UWB signals (less than 1 ns), the time measurements can be performed with a resolution as high as 15.65 ps [8], which under the right conditions, can allow for precise and accurate localization. The two typically used localization methods in the UWB based systems are Time of Arrival (ToA) and Time Difference of Arrival (TDoA).

In Time of Arrival-based UWB systems [9] users location is derived based on the signal propagation times between the localized tag and the anchors comprising the system infrastructure. Range to the anchor is usually measured using the Two-Way Ranging procedure (TWR) which consists of exchanging packets between the tag and the anchor, noting their transmission and reception times [9]. The main advantage of this method is that, since the distance between the tag and each of the anchors is measured independently, anchor synchronization is not needed. On the other hand, ranging with each of the anchors on its own takes time and in case of extensive packet exchange, the infrastructure's radio interface might become cluttered, which would limit the number of localized tags.

The Time Difference of Arrival method consists of localizing the user based on the difference of times, at which the anchors receive signals transmitted from the tag [10]. Typically, the only transmitting devices in this method are the tags, which helps to organize the transmission and allows to achieve higher system capacities when compared to ToA. The main problem with implementing TDoA is the requirement to achieve accurate and reliable anchor synchronization. In commercial systems, it is usually done by transmitting synchronization signals to the anchors with cables [11] but can also be achieved wirelessly by adding another reference anchor to the system infrastructure [10].

Both ToA and TDoA allow for precise localization in Line-of-Sight propagation conditions. The typically obtained accuraccies are at the level of a dozen centimeters [12]. When the propagation occurs in Non-Line-of-Sight Conditions (NLOS) the accuracy is usually lower.

Localization in narrowband systems is typically based on measurements of Received Signal Strength (RSS). In tracking and monitoring systems, where the transmitting tags are being localized, they are usually performed by the anchors. Those systems usually employ well known and widely spread standards such as WiFi or Bluetooth (more recently Bluetooth Low Energy BLE [6,13]). Most of the systems localize the users with one of two methods: fingerprinting or RSS ranging.

Fingerprinting is a method which consists of comparing the signal levels received by the anchors with a previously prepared radio map [14] containing information on signal levels received from particular places. Besides its easy implementation, the main advantage of the method is that it does not require any information on a system infrastructure placement or location. Unfortunately, achieving high accuracy localization demands preparing accurate high-density radio maps, which is usually done manually by placing the tag in points distributed in the area and storing received signal level values. Such a radio map is valid only for a specific configuration of a propagation environment and every change, such as moving a piece of furniture, may negatively impact a system's accuracy.

Radio map creation can be sped up using crowdsourcing [15] or interpolation [14], but since they still require performing some measurements, it would be hard to implement fingerprinting in a home environment.

Systems using RSS ranging are much easier to deploy. In that method, the user is localized by estimating their distance from the anchors based on a propagation model. The accuracy of the method depends on how accurately the assumed model reflects complex propagation conditions of the deployment area. Typically, the systems use simple models such as path loss exponentials [6] or multi-walls [16].

Narrowband systems average localization errors typically does not go below one meter. The accuracy highly depends on a place where the system is deployed and the number of anchors comprising the infrastructure. In apartments with extensive furniture, propagation models might not properly reflect the real conditions, and the accuraccy might suffer. Thus, the mean errors of the solutions found in the literature cover a broad range of 1–2 m [6] to 4–6 m [14].

A localization technique, which deserves attention is the Angle-of-Arrival (AoA). It can be used in both ultra wideband and narrowband systems and consists of estimating the directions from which the signals transmitted by localized object reach system anchor nodes. It can be done by rotating a directional antenna [17] or analyzing phase differences using antenna arrays [18]. The method does not require maintaing precise synchronization of the infrastructure [19] and allows to achieve results of accuracy comparable to RSS-based methods (average error of few meters) [20].

The localization accuracy that is possible to achieve with UWB-based systems is much higher than in case of narrowband WiFi or BLE solutions. Unfortunately, energy consumption of UWB devices is on a relatively high level. For the ToA-based system described in [9], it was estimated that the tag would work without recharging for 74 h if it were powered with 6000-mAh battery pack. In case of wearable devices, which usually use smaller batteries this time would be even shorter. BLE is more energy efficient. In case of BLE-based systems, in case of rare transmission tags, working times could reach 2 years powered by a coin battery. Moreover, due to the UWB technology being less popular and widespread, it is much more expensive and the overall cost of the localization system could be too high for the majority of older persons. That would make BLE a preferred choice for AAL solutions, but since its localization accuracy might not be enough for some services, it should be improved. It can be done by using hybrid localization methods combing measured BLE RSS with results obtained using different technologies.

1.2. Hybrid Localization Systems

In hybrid localization systems, user localization is computed based on different types of measurement data obtained using one or more various technologies.

The first group of hybrid localization systems are systems in which the user is localized using one technology based on various signal factors. Most notable examples combine typical measurements of RSS or TDOA with direction estimation AoA. In [21], a simple single device localization system is presented. The device uses RSS measurements to estimate a distance from the localized object and AoA to estimate direction, in which it is located. Such approach allows to minimize the number of reference anchor nodes to one. An interesting example of AoA usage in hybrid localization schemes is presented in [22], where a hybrid TDOA/AOA Extended Kalman Filter based algorithm for UWB localization system is described and tested. Another hybrid combination is RSS/TDOA, which can be used in WiFi-based systems [23].

In the second group of hybrid localization solutions, the users are localized using two or more different technologies. A very popular combination is fusing results obtained using the radio part with data from inertial measurement units. In [24], smartphone results from inertial sensors and magnetometer are used to support BLE-based fingerprinting. The presented solution allowed to localize the robot in places, where there was no coverage of BLE anchors, and quality of the radio map was poor. In [25], IMU senors were used for movement direction and gate analysis. The additional data allowed to significantly improve BLE positioning accuracy. A similar approach was taken in [26], where a single accelerometer was used for gait analysis, which allowed to improve localization accuracy in a UWB system.

Another large group is that of hybrid systems mixing two or more radio technologies. In [27], an asset tracking system using WiFi and Bluetooth is presented. It uses WiFi fingerprinting for early, rough asset localization and when the user gets closer, localizes it with greater accuracy using BLE trilateration. The same combination was used in a system deployed in a university library [28]. Here, the users were localized using BLE and WiFi fingerprinting. The system utilizes the fact that due to different anchor placement and propagation conditions, accuracy of the radio maps for BLE and WiFi varies in the same places. The localization errors for both maps are estimated during the map creation. During normal system operation the users are localized using both technologies concurrently and the more probable result is chosen. Another combination of technologies is BLE-UWB. In [29], UWB was used periodically to improve BLE-based localization. In [30], it was used in the process of radio map creation, which consisted in gathering points for the BLE radio map while localizing the user using UWB and then interpolating the missing parts based on the fitted exponential propagation model.

Solutions combining radio technologies with other types of sensors are much less common. In [31], a system concept, where the users are roughly localized using BLE and then tracked with vision systems is presented. Another example is [32], in which WiFi and ultrasound-based hybrid fingerprinting is presented.

An interesting type of sensors, which is not that popular and could be used in hybrid localization, are proximity and ranging sensors. They are usually used in robotics for SLAM (Simultaneous Localization and Mapping) but some examples of their use for localization can also be found. In [33], the sensors are used to form an array monitoring a small area of 600×300 mm dimensions. The sensors are used to track a round object moving around on a robot. The localization errors are very low and below 1 cm. Simulation of another example of a proximity sensor network is presented in [34].

The described proximity or ranging sensors network solutions would allow for easy device-free localization. Unfortunately, deploying such systems would require placing a large amount of sensors in fixed places, which makes them practically difficult to use in home environments. However covering with them selected spaces would not be problematic. Although, such sensors could be a good addition to radio-based systems, the author have not found any examples of employing them in hybrid localization schemes.

The paper presents a novel hybrid localization concept combining Bluetooth Low Energy with ranging sensors. The presented solution is the effect of the works started in [35], where a concept was briefly described and tested with very simple experiments. The proposed system concept is intended mainly for indoor user localization in AAL solutions. The user is localized using a BLE system, which is supported with a few singular proximity sensors placed in the areas, where BLE-based localization is less accurate or are important from older person's safety point of view (e.g., a shower, near a stove). The results from both parts are fused using two types of hybrid localization algorithms. The presented concept was tested with simulations and experiments conducted in a typical fully furnished flat. The paper also includes the results of experimental assessment of VL53L1X [36] proximity sensor properties.

2. Hybrid Localization Concept

The typical received signal strength-based positioning systems, due to difficult and complex multipath propagation environments typically present indoors, do not allow for very accurate and reliable localization. Localization errors are particularly high at the edges of the area covered by the system or in closed spaces such as toilets or bathrooms. It comes from the fact that the propagation between the localized devices and most parts of the system infrastructure occurs under OLOS (Obstructed Line-of-Sight) or NLOS (Non Line-of-Sight) conditions, which introduce significant signal attenuation. The concept proposed in the paper consist in additional coverage of those and other vital places with proximity sensors.

Proximity sensors are small and cheap devices, which allow to detect presence and measure distance from objects located at moderate distances up to a few meters. Although it is possible to use them for distance measurement, they should not be mistaken with much more advanced laser distance meters. Proximity sensor field of view typically has a shape of a conical beam covering an angle up to a few dozen degrees, whereas laser distance meter is much more directional.

An exemplary scenario of the proposed hybrid concept use is presented in Figure 1.

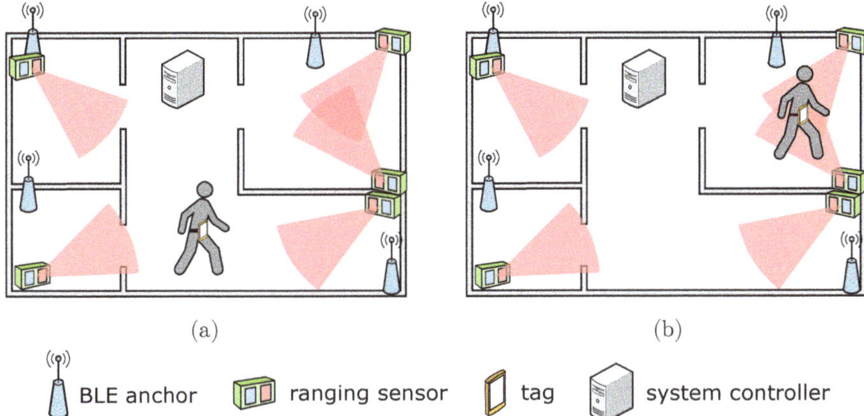

Figure 1. Hybrid localization use scenario. (**a**) User is localized using BLE only. (**b**) User is localized using BLE and proximity sensors.

In the proposed system concept, the localization system consists of three parts:

- a tag, which is worn by the user
- an infrastructure comprising a set of BLE anchors and proximity sensors
- a system controller

The system operates in the following way. The tag, which is worn by the user, repeatedly sends BLE packets in the advertisement channels. The anchors receive them and measure the RSS (Received Signal Strength). Meanwhile, the proximity sensors perform independent, periodical distance measurements. Results from the infrastructure are sent to the system controller, which is an application running on a regular PC or a dedicated server and is responsible for position determination. The results can be exploited by an external system delivering services to monitored persons and their caregivers. For example, in the IONIS project [37] cloud-based solution performs these functions. The system controller manages the localization system infrastructure. In case of an anchors or tags failures (that are reported by anchors or detected if the communication with devices is lost), the alert message is sent to the cloud.

For most of the time, a user's position is derived using BLE power measurement results only (Figure 1a). When the user steps into the area covered by one or more proximity sensors and their presence is detected, a hybrid algorithm combining BLE and ranging results is used. Unfortunately, since the proximity sensors have no way of discriminating between different users and the presented solution is intended for locating on individual only. That problem could be resolved by prior localization of the users based on BLE only and checking which user is closer to the sensor. However, because of low BLE accuracy, this approach would not be dependable when the users are close to each other. Since the system is intended for use by an older person living alone, it is not a major drawback.

In the presented system architecture, the BLE anchors and the proximity sensors are separate devices, but it is possible to integrate them. It would simplify communications with the system controller and greatly reduce number of power supplies needed by the infrastructure. However, such an approach is not the best for localization. A proximity sensor maximum range is usually limited to a few meters and it would be best to orient its beam parallel to the ground to maximize the covered area. It would also mean placing the sensors in lower positions, so the signal could easily reflect off the localized person. However, such placement might not be the best choice from a BLE signal propagation point of view, because of a large number of obstacles present at waist height.

3. Localization Algorithm

In the proposed concept, the user is localized based on two types of data coming from the system infrastructure: RSS measurements performed in BLE radio interface and ranging results from proximity sensors. The concept assumes use of different algorithms for the possible scenarios of user localization, which are:

- the user is not in the area covered by any of the proximity sensors and is localized using only BLE RSS measurement results (Extended Kalman Filter based algorithm),
- the user's tag is off or disabled and he is localized based solely on ranging data,
- the user wears the tag as normal and is present in at least one of the proximity sensors beams (hybrid algorithm in one of two versions: loosely or tightly coupled).

3.1. BLE Based Algorithm

The first localization method used in the proposed system is an Extended Kalman Filter (EKF)-based algorithm [38] utilizing solely BLE power measurement results. The main advantage of the EKF over traditional geometry-based methods such as trilateration is that the result of the EKF if a combination of the result calculated using performed measurements and the prognosed location based on the previous user position. It allows to mitigate abrupt changes of signal level, which often occur in the multipath environments. Additionally, the number of the results that can be used for localization is not limited. In EKF it is also possible to take into consideration higher variance of the results obtained by the anchors, which are far away from the tag. In the algorithm, the localized user is treated as a dynamic system and the state at the current moment k is described by a vector:

$$x_k = \begin{bmatrix} x & v_x & y & v_y \end{bmatrix} \quad (1)$$

containing the user's current coordinates: x, y and velocity components: v_x, v_y. The vector does not include information on the z coordinate (height, at which the localized tag is worn) and corresponding velocity. All of the algorithms presented in the paper are intended for 2D localization and z value is one of the algorithm parameters.

The EKF is an iterative algorithm consisting of two phases: time-update phase and measurement-update phase. In the time-update phase, the localization of the user at the current moment k is predicted based on the location calculated for the previous moment $(k-1)$ and the assumed movement model. The equations comprising this phase are:

$$\hat{x}_{k(-)} = F\hat{x}_{k-1(+)} \quad (2)$$

$$P_{k(-)} = FP_{k-1(+)}F^T + Q \quad (3)$$

where:

- $\hat{x}_{k(-)}$ is the predicted state vector value,
- $\hat{x}_{k-1(+)}$ is the state vector value obtained in the previous EKF iteration,
- $P_{k(-)}$ and $P_{k-1(+)}$ are the state covariance matrices of the above vectors,
- F is the state transition matrix containing the movement model,
- Q is the process noise covariance matrix.

The movement model used in the algorithm is the Discrete White Noise Acceleration (DWNA) [39] model, in which the movement between the analyzed moments is uniform and linear and acceleration is treated as process noise. For the DWNA, matrices F and Q have the following form:

$$F = \begin{bmatrix} 1 & \Delta t & 0 & 0 \\ 0 & 1 & 0 & 0 \\ 0 & 0 & 1 & \Delta t \\ 0 & 0 & 0 & 1 \end{bmatrix} \qquad Q = \begin{bmatrix} \Delta t^2 & \frac{1}{2}\Delta t^3 & 0 & 0 \\ \frac{1}{2}\Delta t^3 & \frac{1}{4}\Delta t^4 & 0 & 0 \\ 0 & 0 & \Delta t^2 & \frac{1}{2}\Delta t^3 \\ 0 & 0 & \frac{1}{2}\Delta t^3 & \frac{1}{4}\Delta t^4 \end{bmatrix} \sigma_a^2 \tag{4}$$

where Δt is algorithm update rate and σ_a^2 is acceleration variance, which is usually assumed to be in the range of 0.5–1 of a moving man's maximum acceleration.

The obtained predicted localization $\hat{x}_{k(-)}$ is updated with measurement results in the measurement-update phase. This phase consists of estimating measurement results, which would be obtained for the predicted localization based on assumed sensor model (BLE anchor model). The predicted measurement results are then compared with the actual values measured by the infrastructure. The set of equations making this phase is as follows:

$$K_k = P_{k(-)} H_k^T \left(H_k P_{k(-)} H_k^T + R_k \right)^{-1} \tag{5}$$

$$\hat{x}_{k(+)} = \hat{x}_{k(-)} + K_k \left(z_k - h_k(\hat{x}_{k(-)}) \right) \tag{6}$$

$$P_{k(+)} = \left(I - K_k H_k^T \right) P_{k(-)} \tag{7}$$

$$z_k = \begin{bmatrix} RSS_1 & \cdots & RSS_n \end{bmatrix} \tag{8}$$

$$h_k(x_k) = \begin{bmatrix} RSS_1(x_k) & \cdots & RSS_n(x_k) \end{bmatrix} \tag{9}$$

where:

- z_k is the measurement vector containing RSS (RSS_n) measurement results,
- $h_k(x_k)$ is the sensor model used to calculate measurement values which would be obtained for the predicted tag localization,
- H_k is a linearization of the sensor model,
- K_k is the Kalman gain ,
- R_k is the measurement covariance matrix.

In the proposed algorithm, the power levels received by the anchor (sensor model) were calculated using the exponential path loss model:

$$RSS_n(x_k) = RSS_0 - 10\gamma \log_{10} \frac{d(x_k)}{d_0} \tag{10}$$

where:

- RSS_n is the signal power received by the anchor n,
- d is the distance between the anchor and predicted tag localization,
- RSS_0 is the received power at the reference distance from the tag d_0,
- γ is the path-loss exponent.

The prediction updated with the measurement results is the final result of the algorithm iteration and is used as the input for the next EKF iteration.

3.2. Proximity Sensors Based Localization

The second proposed localization method consists of locating the user based on ranging results only. The basic idea behind the algorithm is illustrated with Figure 2.

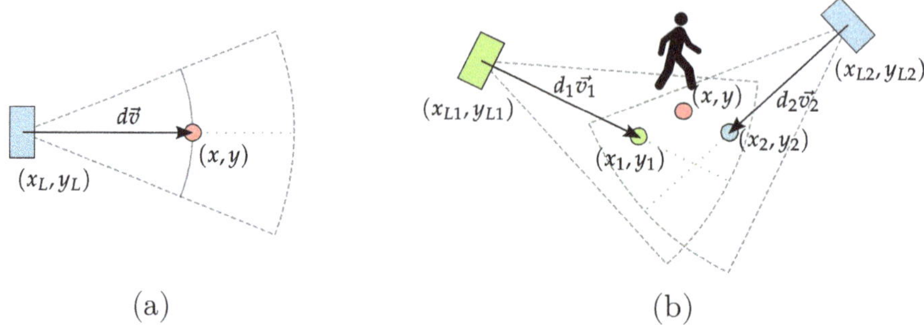

Figure 2. Localization using: (**a**) a single laser sensor, (**b**) multiple laser sensors.

When the result from only one sensor is available (Figure 2a) it is difficult to precisely localize the user—his location can only be expressed as an arc located at the measured distance. Since the system requires results as exact x-y coordinates, the user is roughly localized in the center of the sensor beam at the measured distance.

In this situation, when the user is in the range of two or more proximity sensors, his/her location can be inferred more precisely. One of the most popular choices for localization in case of ranging measurements is the trilateration algorithm [14]. It consists of finding an intersection of two arcs corresponding to distances measured by the sensors. Here, since the user, from whom the signal is reflected is not very small, a situation in which those arcs intersect would be extremely rare and the trilateration algorithm would suffer. Here, an alternative, simplified method is used. In this method, the user is located by each of the sensors separately at the sensors's main axes as presented in Figure 2b. Then, the obtained localizations are fused by calculating the weighted average:

$$x = \frac{\sum_i \frac{1}{\sigma_i^2} x_i}{\sum_i \frac{1}{\sigma_i^2}} \qquad y = \frac{\sum_i \frac{1}{\sigma_i^2} y_i}{\sum_i \frac{1}{\sigma_i^2}} \tag{11}$$

where σ_i^2 is the variance of the distance measurement performed by the sensor. Applying weights, which are inversely proportional to measurement variance allows to take into account the higher uncertainty of locations at bigger distances, which result from a lower level of the reflected signal received by the sensor and a longer arc due to a wide covered area. Assuming that all of the sensors are independent of each other, the variance corresponding to the computed localization can be approximated as:

$$\sigma^2 = \frac{1}{\sum_i \frac{1}{\sigma_i^2}} \tag{12}$$

The obtained value would be the final result in case of localizing a user without the tag. In case of a presented hybrid solution, it will be an input to loosely coupled version of the hybrid algorithm.

3.3. Loosely Coupled Hybrid Algorithm

The first version of the hybrid algorithm used in the system is loosely coupled. It means that the user localization is calculated separately using both technologies and then fused. Localizations based on BLE power measurements and ranging are calculated using algorithms presented in Sections 3.1 and 3.2 respectively. The obtained results are then fused using the following formula:

$$p = p_{BLE} \frac{\sigma_{RAN}^2}{\sigma_{BLE}^2 + \sigma_{RAN}^2} + p_{RAN} \frac{\sigma_{BLE}^2}{\sigma_{BLE}^2 + \sigma_{RAN}^2} \tag{13}$$

where:

- p_{BLE} is the localization calculated based on BLE power measurements (6) and σ_{BLE}^2 is the corresponding variance taken from covariance matrix P (7),
- p_{RAN} is the localization calculated based on ranging results (6) and σ_{RAN}^2 is its variance (12).

When the localization calculated using BLE is very inaccurate, it is possible that the result obtained using the hybrid algorithm will be positioned outside the used proximity sensor beam. Since the sensor detected the user, it is almost certain that he/she is in its field of view. In such case, to improve algorithms accuracy, the obtained result is pulled to the closest point on the arc corresponding to the measured distance.

3.4. Tightly Coupled Hybrid Algorithm

In the tightly coupled version of the hybrid algorithm, raw measurement results from the BLE anchors and ranging sensors are fused. For this purpose, the EKF is used. The assumed models and implementation of the time-update and measurement phase are almost identical to the BLE-based EKF presented in Section 3.1 (1)–(7). The major differences are the forms of the measurement vector z and the sensor model $h_k(x_k)$.

$$z_k = \begin{bmatrix} RSS_1 & \cdots & RSS_n & d_1 & \cdots & d_n \end{bmatrix} \tag{14}$$

$$h_k(x_k) = \begin{bmatrix} RSS_1(x_k) & \cdots & RSS_n(x_k) & d_1(x_k) & \cdots & d_n(x_k) \end{bmatrix} \tag{15}$$

The measurement vector contains both RSS measurement results and measured distances. In the sensor model $h_k(x_k)$ the RSS is estimated as in (10), whereas ranging results d_n are estimated as the distance of the sensor from the predicted location.

The other difference is verification and correction of the predicted state vector value. As in the loosely coupled version, if any of the proximity sensors detect the user and the predicted location is outside the sensors beam, it is pulled to the closest point on the arc corresponding to the measured distance.

4. Simulations

4.1. Simulation Environment

The proposed hybrid localization concept and algorithms were tested with simulations. Simulations were performed using a simple simulator implemented in Python. The simulator allowed to specify test paths, which could be covered by a user during his daily activities and simulate measurements performed by BLE anchors and ranging sensors comprising the system infrastructure.

The Received Signal Strength (RSS) of the BLE signals measured by the anchors was modeled using the multi wall propagation model [16]. The model takes into account the topography of the area, where the system is located and introduces additional attenuation due to signal propagation through multiple walls. The model is described by the following dependence:

$$P_R = P_{R0} - 10\gamma log_{10}\left(\frac{d}{d_0}\right) + \sum_{i=1}^{I}\sum_{k=1}^{K_i} L_{wik} + X_\sigma^2 \tag{16}$$

where:

- P_R is the power received by the anchor from the tag
- P_{R0} is the power received at the reference distance d_0 (1 m) (assumed −52 dB)
- γ is the path loss exponent (assumed 2.4)
- d is the distance between the tag and the anchor
- I is the number of different wall types in the simulated area
- K_i is the number of walls of type i
- L_{wik} is the attenuation due to k^{th} traversed wall of type i

- X_σ^2 is the log-normal random component present due to the shadowing effects and the receiver noise (3 dB standard deviation)

In the simulations, it was assumed that all of the walls are the same and introduce the same signal attenuation (2 dB).

The measured distance was simulated as:

$$d_m = d + n(d) \quad (17)$$

where:

- d is the distance between the localized person and the sensor,
- $n(d)$ is the measurement white noise of standard deviation calculated based on standard deviation model. The assumed deviation model was: $\sigma(d) = d^2/60$. According to the model, standard deviation rises with distance, which reflects less accurate measurements due to a lower level of reflected signals.

During the simulations, the sensors performed ranging only when the user was located in their beams. The areas covered by the sensors were modeled as conical beams of 30° width, and the maximum range of the sensor was set to 3.5 m.

The user was localized based on the generated results using three algorithms: EKF using RSS measurements and two hybrid algorithms proposed in Section 3.

4.2. Performed Simulations

The proposed system was simulated in an apartment consisting of few rooms. The simulations consisted in generating a set of simulated measurements for points distributed along two different test paths. It was assumed that the person was walking with a speed of 1.2 m/s, which is typical for the adults. The BLE RSS values and distances were measured five times per second. The layout of the apartment, the placement of the system infrastructure and both of the test paths are presented in Figure 3.

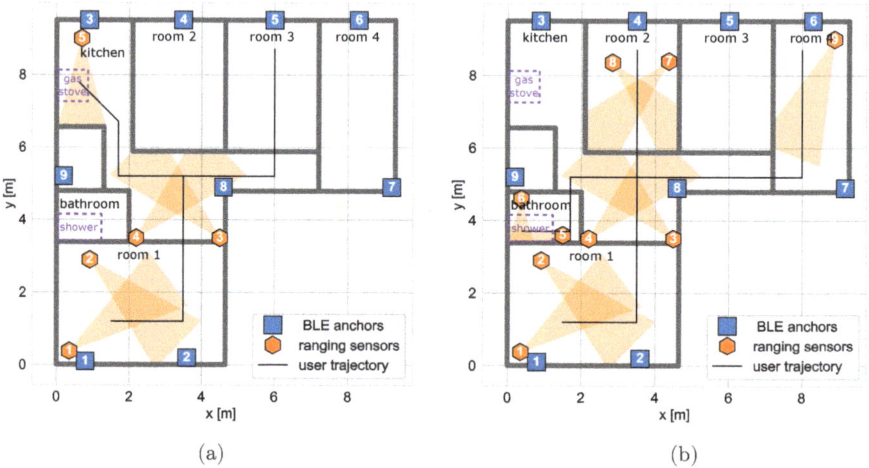

Figure 3. System layout and test paths used during the performed simulations. (**a**) scenario 1 (**b**) scenario 2.

The BLE system infrastructure consisted of nine anchors distributed in the apartment. They were placed in such a manner that in each of the rooms, at least one anchor was present. The anchors were fixed to the outer walls, which allowed to evenly cover the whole apartment. The number of used proximity sensors was different in both scenarios.

In the first simulated scenario, the user walked from room 1 to room 3 and then visited the kitchen, where he stood in front of the gas stove for 10 s, then he came back to room 1. In the scenario, five ranging sensors were used, two were placed in room 1, two in the corridor and one in the kitchen covering the area in front of the stove.

The second path included other rooms. The user started from room 1, then visited rooms 2 and 4 and went to the bathroom, where he stood in the shower cabin for 10 s. After that, he returned to room 1. In this experiment, nine proximity sensors were used, the first four were located in room 1 and the corridor as in the previous scenario. Additionally, two were placed in both bathroom and room 2 and one in room 4.

The localization results obtained for the first and second scenarios are presented in Figures 4 and 5, respectively.

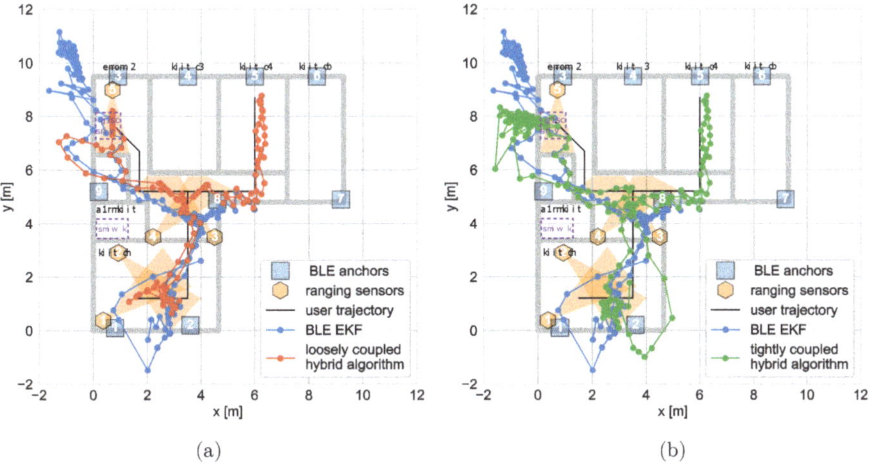

Figure 4. Scenario 1 results obtained with: (**a**) loosely coupled hybrid algorithm, (**b**) tightly coupled hybrid algorithm with BLE-based localization results as reference.

The accuracy of localizations calculated based on BLE RSS results only varies in the apartment. It is highest in the middle parts of the flat, whereas in the border rooms, the errors are significantly higher. It results from the fact that the signals transmitted from the tag located in those rooms propagate through multiple walls, which introduce significant attenuation. The obtained results are accurate enough to determine a room where the person is located, but nothing more. Using ranging sensors allows to improve localization accuracy in the border rooms, as well as in the corridor. As it can be seen, in case of a loosely coupled algorithm (Figure 4b) the system detects the prolonged presence in front of the gas stove. Results obtained with the tightly coupled version are less accurate (especially in the situation, where the results from only one proximity sensor are available).

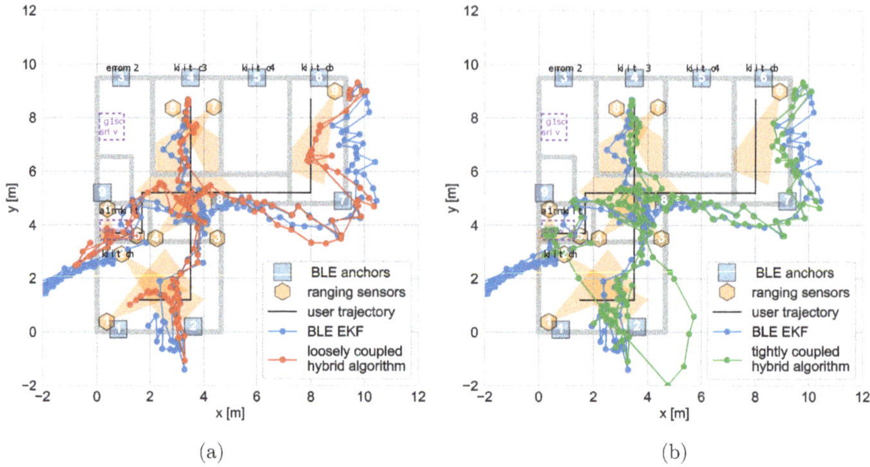

Figure 5. Scenario 2 results obtained with: (**a**) loosely coupled hybrid algorithm, (**b**) tightly coupled hybrid algorithm with BLE-based localization results as reference.

In the second scenario, similar results were obtained. BLE-based localization accuracy is low in places, from which the tag transmitted signal propagates through multiple walls. In case of localization in the bathroom, where the propagation to all of the anchors occurs under NLOS conditions, it is difficult to determine the room where the user is located. As in the previous example, using ranging sensors allows to significantly improve localization accuracy. The results from both algorithms allow us to say that the user spent some time in the shower cabin. For tightly coupled algorithms, it was also observed that fusing BLE RSS with only one distance measurement might not be enough. Localization accuracy in room 4 is much lower than in case of the bathroom or the corridor, which are covered by two sensors.

The localization methods can also be compared by analyzing localization errors defined as the Euclidean distance between the calculated localization and the point, where the simulated user was present. A convenient way to do that is to use the Empirical Cumulative Distribution Function (ECDF). The ECDF curves for both of the scenarios are presented in Figure 6.

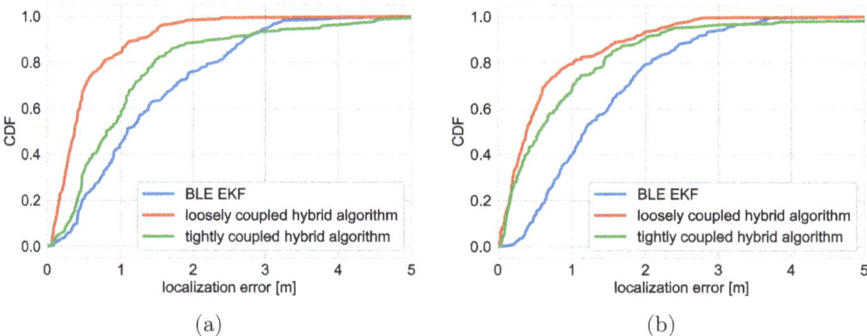

Figure 6. Empirical Cumulative Distribution Functions of localization error for: (**a**) scenario 1, (**b**) scenario 2.

In both cases, using hybrid positioning allowed to improve localization accuracy. Localizing using only BLE RSS resulted in median erros of about 1.1 m in both scenarios. The best accuracy was achieved using a loosely coupled version of the hybrid algorithm—median error was 0.4 m for both scenarios. In case of the tightly coupled algorithm, localization accuracy was lower and median error was about 0.8 m and 0.55 m for scenarios 1 and 2, respectively.

5. Experiments

The proposed concepts and algorithms were verified with experiments performed in one of the Warsaw University of Technology laboratories and a typical fully furnished flat. The experiment consisted of two phases. First, an exemplary ranging sensor available on the market—VL53L1X by ST Microelectronics was tested to verify whether its properties allow for use in the proposed hybrid localization scheme. This part was conducted in laboratory rooms. The second part of the experiment consisted of using the aforementioned sensors and a BLE-based localization system developed within the IONIS project [37] to verify the proposed concept in an apartment.

5.1. Properties of VL53L1X Proximity Sensor

An example of a proximity sensor, which could be easily adapted for use in a the presented hybrid localization system concept is the VL53L1X chip produced by ST Microelectronics [36]. It is a small device, which consists of a 940 nm invisible Class 1 laser emitter, SPAD (Single-photon avalanche diode) receiving array and additional optical filters. The main advantage of the sensor is the fact that unlike typical infrared sensors, it employs Time-of-Flight (ToF) method for ranging, which according to the producer is supposed to ensure accurate distance measurement regardless of target's color and surface reflectance [40].

The sensor's properties, which are most important from the localization point of view are listed in Table 1 (the values come from VL53L1X datasheet [36]).

Table 1. Selected VL53L1X parameters.

Parameter	Value
maximum range	400 cm
ranging resolution	1 mm
field of view	15–27°
maximum measurement rate	50 Hz
package size	4.9 × 2.5 × 1.56 mm

The sensor has a range of 4 m, which would be sufficient for indoor localization in typical apartments. Its field of view (FoV) is a conical-shaped beam, the width of which can be programmed in the range of 15–27°. Such property allows to tune the sensor and increase its selectivity. The measurement rate can also be programmed up to 50 Hz, which makes it possible to precisely match the localization rate of a radio subsystem. Additionally, the sensor is small and can be easily incorporated into the radio system's anchor nodes.

It is worth noting that the values stored in Table 1 were obtained in particular conditions and for a specific target. As stated in the application note, the measurements were taken for charts of different reflectance, which covered the full sensor FoV. In case of measuring distance from a person, those values might be different. The proposed hybrid algorithms utilize some of the above values (maximum range, field of view). Therefore, to avoid errors resulting from assuming sensor parameters from data sheets, it is necessary to estimate those parameters for reflection from a human body.

The performed sensor tests consisted of evaluating its basic properties such as ranging bias, ranging standard deviation and sensor FoV. In the experiment P-NUCLEO-53L1A1 [41] modules consisting of two evaluation boards: STM32F401RE (nucleo microcontroller board) [42] and X-NUCLEO-53L1A1 (nucleo board expansion with up to three VL53L1X sensors) [43] were used. The module was programmed with the exemplary SimpleRanging application [44]. The application measured distance and sent it to the attached PC over USB. The sensor working mode was set to 'long', ranging rate to 10 Hz and the FoV to the default 27° width.

The first part of the experiment consisted of ranging from a person located at different distances and dressed in different clothes. The sensor was placed at waist level. The measurements were conducted on a sunny day in an office space under two different lighting conditions (blinds opened and closed). The obtained results were compared to the distance measured with an accurate laser distance meter. The distance measurements were conducted for the following persons:

- person in a black shirt (blinds closed)
- bare-chested person (blinds closed)
- person in a white shirt (blinds closed)
- person in a black shirt (blinds opened)

For each of the measurements ranging bias and standard deviation were evaluated. The gathered results were used to model bias and standard deviation with polynomial functions, which would be used in the hybrid localization algorithm. The distances measured with VL53L1X and estimated ranging bias values are shown in Figure 7. Ranging standard deviation is presented in Figure 8.

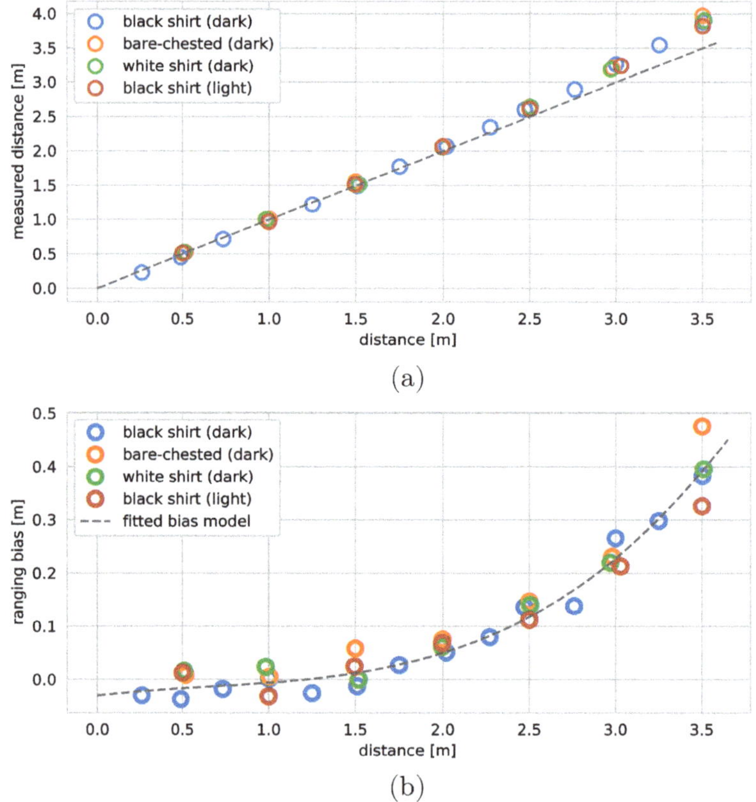

Figure 7. Ranging accuracy test results. (**a**) Distance measured from a person clad in different clothes. The grey dashed line is the actual distance drawn for reference. (**b**) Ranging bias. The coefficients of the fitted model are stored in Table 2.

During the tests, the maximum range, for which it was possible to properly measure distance from a person was about 3.5 m. At close ranges, up to 2 m, the measures are very accurate and the bias does not exceed 0.1 m. For larger distances, it rises and for 3.5 m is higher than 0.3 m. The measures show no clear dependence on the reflective surface (clothing color) or lightning conditions. Therefore, it makes sense to model the bias and introduce the correction in the localization algorithm. Ranging bias was modeled with a third degree polynomial and the resulting curve was plotted in Figure 7b. The polynomial coefficients are stored in Table 2.

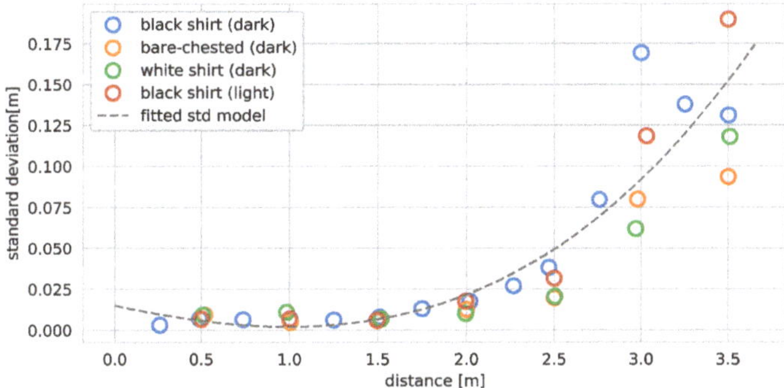

Figure 8. Ranging standard deviation for reflections from a person clad in different clothes. The coefficients of the fitted model are stored in Table 2.

Ranging standard deviation also depends on the measured distance. For ranging from close objects, it is relatively small (lower than 5 cm for distances up to 2.5 m). At the edge of the sensor range, measurement noise is significantly higher—for 3.5 m, in case of the black clad man in a bright room, standard deviation approaches 0.2 m. Based on the measurement results, standard deviation was modeled as a third degree polynomial. The coefficients are stored in Table 2. Large ranging errors caused by high measurement noise can be clearly seen on a ranging results boxplot presented in Figure 9.

Table 2. The bias and standard deviation models coefficients.

Model	a_0	a_1	a_2	a_3
bias	−0.0303	0.0385	−0.0293	0.0151
standard deviation	0.0151	−0.0229	0.0068	0.0031

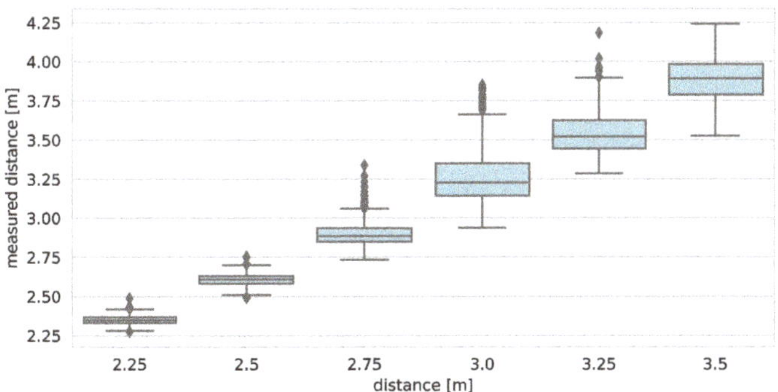

Figure 9. Boxplot of ranging results for the person dressed in a black shirt and located at larger distances.

Large standard deviations at long distances might pose a problem in terms of accurate localization. As it can be seen, for distances over 2.5 m, ranging errors sometimes have few dozen centimeter magnitudes. Such high errors can easily lead to very inaccurate localization. Therefore, the proposed localization algorithms (proximity sensors based localization and hybrid algorithms) will utilize both the above fitted models to mitigate ranging bias and lower the impact of low accuracy measurements on the final localization result.

The second phase of the VL53L1X properties testing was measurement of the sensor's FoV. The experiment was conducted in a bright, large (6×8 m) laboratory room. The test consisted of placing an A3 paper sheet attached to the utility cart in front of the sensor and moving it sideways till the sensor lost visibility of the object. The border positions at both sides were noted and the sensor's actual FoV was evaluated.

For the conditions in the laboratory room and the tested object, the measured FoV was smaller than claimed in the application note. For distances up to 50 cm, they were very similar, whereas at 3 m, the difference in the width of the covered area was almost 80 cm.

Based on the measurement results, a model of FoV was created. Sensor FoV was modeled as the conical beam of 20.3° width. The measured and modeled FoV is shown in Figure 10.

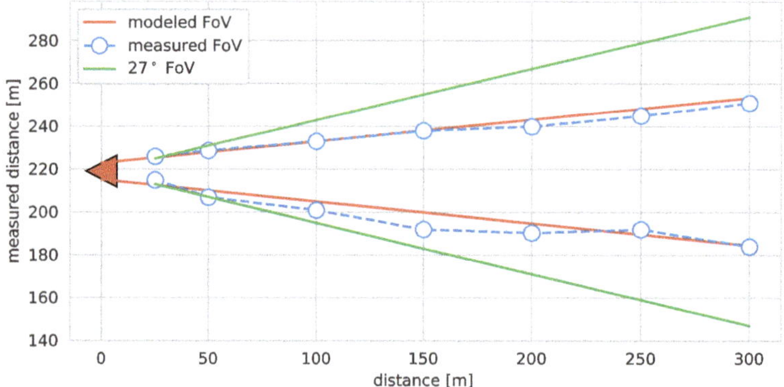

Figure 10. Measured FoV of VL53L1X.

The above results show that the VL53L1X meets the requirements to be used in the proposed hybrid localization scheme. The ranging results are accurate for close objects, and since the bias for large distances does not depend on lightning conditions and reflecting surface, it is easy to correct. The sensor FoV is wide enough to cover a significant area of typical rooms.

5.2. Hybrid Localization Concept Verification

The main part of the experiments was verification of the proposed concept using a BLE-based localization system and laser ranging sensors. The test was conducted in a typical, fully furnished flat, which was the flat modeled in the simulations. As in simulations, the experiment consisted of walking along two test paths and localizing the user using BLE-based and hybrid algorithms. Localization results were compared with the results obtained using a TDOA-based UWB positioning system developed at WUT during one of the earlier projects [10]. The layout of the apartment and places where the infrastructure elements were placed is presented in Figure 11.

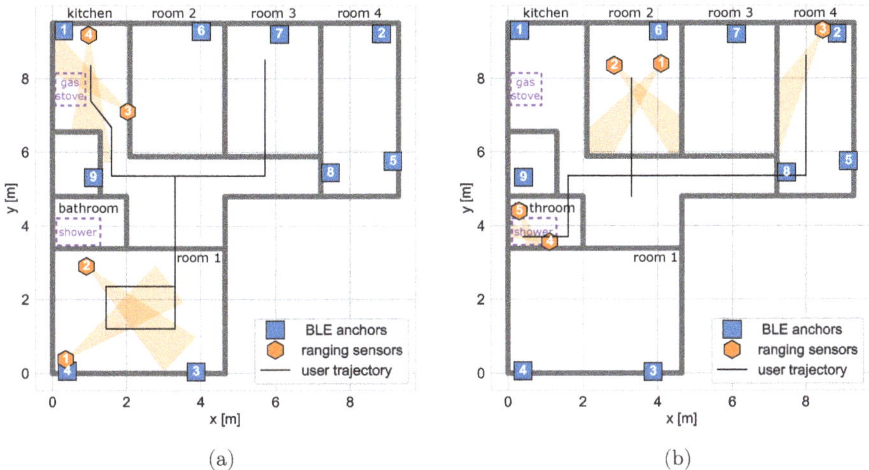

Figure 11. Layout of BLE system infrastructure and laser sensors used in the experiment for: (**a**) test path 1, (**b**) test path 2.

The infrastructure used in the experiment comprised two parts: BLE positioning system and ranging sensors.

The positioning system consisted of nine BLE anchors fixed to the walls and one tag worn by the user. All of the devices were equipped with Laird BLE652 modules, which were programmed to broadcast five BLE packets per second. The received results were averaged per second to mitigate signal power fluctuations. The tag was worn on a lanyard.

As the proximity sensors P-NUCLEO-53L1A1 modules including VL53L1X sensors were used. The modules were placed at the height of 1.4 m, so that the beam was directed at person's corpus. Ranging rate was set to 10 Hz, and the obtained distances were averaged per second to match BLE part localization rate.

The UWB-based localization system used as reference consisted of nine anchors, which were placed in the same places as the BLE-based infrastructure. In the system, the user was localized three times per second using an EKF-based algorithm utilizing TDOA measurement results [10].

The experiment consisted of walking along two test paths. In the first experiment, the user visited room 1, room 3 and the kitchen, where he stood in front of the stove for a few seconds. The second path included rooms 2 and 3 and the bathroom, where the user spent a few seconds in the shower cabin.

The collected measurement results were used to localize the user with the algorithms described in Section 3. The algorithms used were: EKF using only BLE RSS results, loosely coupled hybrid algorithm and tightly coupled hybrid algorithm. The results for both paths are presented in Figures 12 and 13.

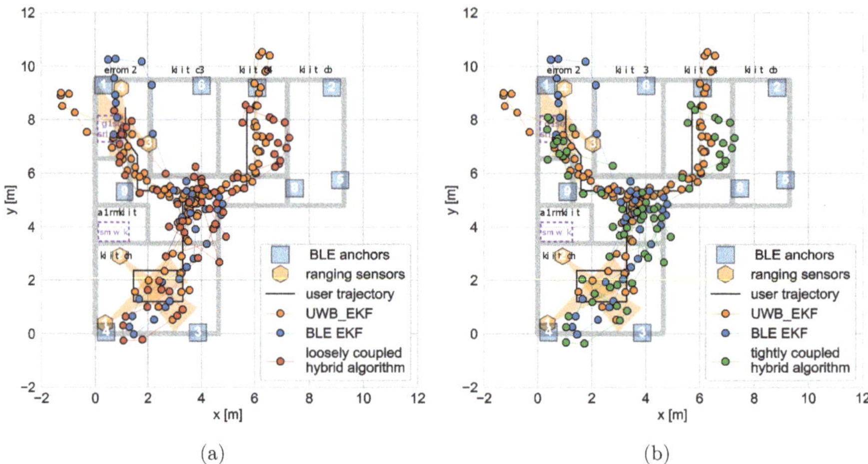

Figure 12. Experiment results obtained for test path 1 using: (**a**) loosely coupled hybrid algorithm, (**b**) tightly coupled hybrid algorithm with BLE-based localization results as reference.

The obtained localization results are similar to those calculated during the simulations. In case of all tested solutions, the system's accuracy is significantly lower in the border rooms, where most of the transmission occurs in NLOS conditions. It can especially be seen in the kitchen, where the user location estimated with both BLE and UWB based systems is outside the flat. Using ranging sensors in the kitchen allowed to localize the user in front of the stove. It did not have much effect in the living room, where the visualized path was still away from the reference. There were no significant differences in the results obtained using both algorithms.

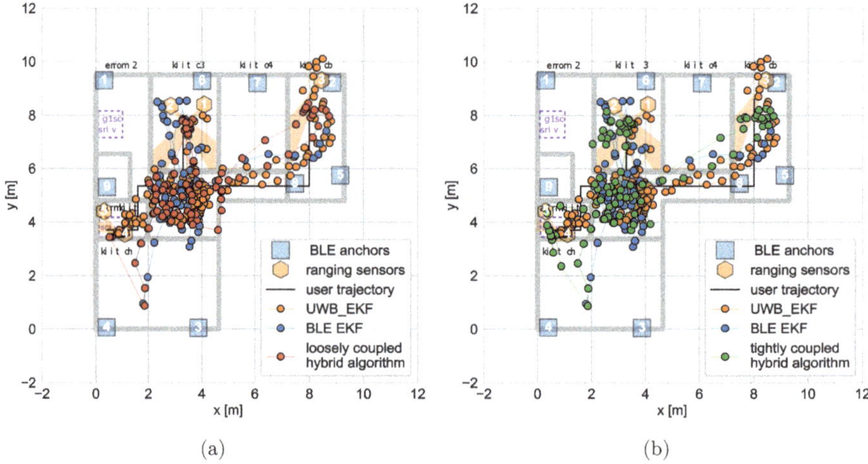

Figure 13. Experiment results obtained for test path 2 using: (**a**) loosely coupled hybrid algorithm, (**b**) tightly coupled hybrid algorithm with BLE-based localization results as reference.

For the second path, similar observations can be made. Using ranging sensors, visibly improved positioning accuracy in the bathroom and room number 2. The user is localized directly in the shower cabin, which was difficult using solely BLE (but was achieved using UWB). Additionally, the results are aligned close to the reference trajectory in room 2. Covering room 4 with one ranging sensor did not work very well. Since it was a single sensor and the covered area was small, there was no significant improvement in the accuracy.

The results obtained using both algorithms differ for localization under the shower. In case of the loosely coupled version, when the user entered the shower, his estimated locations are grouped in one point. For the tightly coupled version, due to high variance of RSS measurements, some of the results are located outside the bathroom.

The comparison of algorithm accuracy can also be performed based on Empirical Cumulative Distribution Function (ECDF) for localization errors. Since, during the performed experiment, the moving user was localized, it was difficult to precisely determine his true location for reference and calculate localization errors. Therefore, for accuracy comparison purposes, a trajectory error defined as the smallest distance between the localized point and the reference trajectory was used. The ECDF curves for calculated trajectory errors are presented in Figure 14.

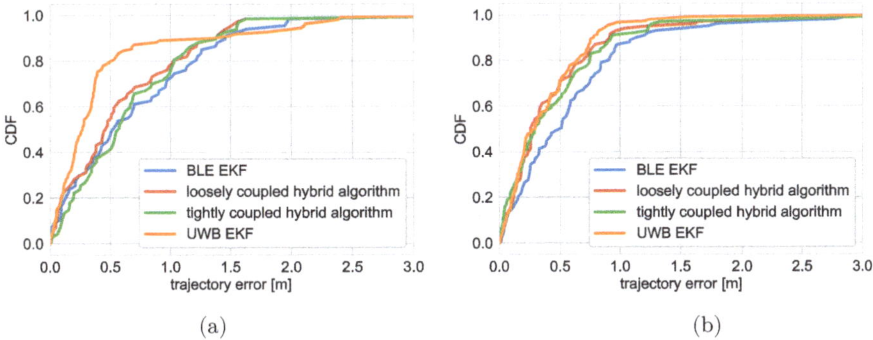

Figure 14. Empirical Cumulative Distribution Function of localization trajectory error (smallest distance from reference trajectory) for: (**a**) path 1, (**b**) path 2.

For both of the paths, the most accurate results were obtained with the UWB-based localization system. In case of the first path, the difference was significant (median trajectory errors equaled 0.27 m for UWB equaled, 0.45 m for loosely coupled hybrid algorithm and 0.51 for BLE-based EKF). In some cases, UWB system did not allow for accurate localization. For about 10 percent of the results the trajectory error exceeded one meter. It might have resulted from multipath propagation (in the apartment there were three large wardrobes with mirror covered doors). For the second path, the difference between UWB and hybrid algorithms was negligible. In both cases, the obtained trajectory errors were smaller when using additional ranging sensors than in case of using BLE only. The CDF curves show that a loosely coupled version of the hybrid algorithm was more accurate than a tightly coupled one but the difference was not significant.

To conclude, the performed experiments have shown that using proximity sensors as an addition to BLE-based localization system allows to improve localization accuracy. The overall accuracy of the solution is lower than of the UWB based positioning systems, but given the higher cost and energy demands of the UWB based systems, the proposed concept might be a good alternative for use in some AAL solutions.

6. Conclusions

In this article, a concept of a novel hybrid localization system intended for AAL solutions is presented. The system combines Bluetooth Low Energy technology with ranging sensors. Ranging sensors have a supporting role and are used to additionally cover places, where the BLE-based localization is not accurate enough. In case when the system is used by the person living alone, it can be useful in some cases even if the person does not wear the tag. In such a case, the user would be localized only in places covered by the sensors, but even such information can be valuable from a caregiver's point of view.

In the system, three types of localization algorithms can be used: an Extended Kalman Filter using only BLE results and two versions (loosely and tightly coupled) of the hybrid algorithm integrating results from both parts.

The proposed concept has been tested with simulations and experiments. Both have shown that the use of ranging sensors can improve localization accuracy in the covered areas. There were no problems with localizing the user directly in front of the stove or in the shower cabin, which was difficult using Bluetooth only. The results have shown that the best in terms of accuracy is the loosely coupled hybrid algorithm.

To conclude, the performed research has shown that the presented concept shows promise and could be efficiently implemented in the AAL platforms. The accuracy of the proposed solution could be improved by using ranging sensors, which would cover a larger area; for example, an array consisting of few VL53L1X sensors facing slightly different directions.

Funding: This work was financed by the National Centre for Research and Development, Poland under Grant AAL/Call2016/3/2017 (IONIS project).

Acknowledgments: I would like to thank the IONIS project team for making the localization system available for the performed test. I would like to also express my gratitude to Robert Kolakowski for his help in organizing and performing the measurement campaign.

Conflicts of Interest: The author declares no conflict of interest. The funders had no role in the design of the study; in the collection, analyses, or interpretation of data; in the writing of the manuscript, or in the decision to publish the results.

Abbreviations

The following abbreviations are used in this manuscript:

AAL	Ambient and Assisted Living
UWB	ultra-wideband
ToA	Time of Arrival
TDoA	Time Difference of Arrival
BLE	Bluetooth Low Energy
EKF	Extended Kalman Filter
ToF	Time of Flight
FoV	field of view
SPAD	Single-photon avalanche diode
LOS	Line of Sight
OLOS	Obstructed Line of Sight
NLOS	Non Line of Sight
RSS	Received Signal Strength

References

1. European Commission; Directorate-General for Economic and Financial Affairs; Economic Policy Committee of the European Communities. *The 2018 Ageing Report: Underlying Assumptions and Projection Methodologies*; OCLC: 1013458008; Publications Office of the European Union: Luxembourg, 2017.
2. Dawson, A.; Bowes, A.; Kelly, F.; Velzke, K.; Ward, R. Evidence of What Works to Support and Sustain Care at Home for People with Dementia: A Literature Review with a Systematic Approach. *BMC Geriatr.* **2015**, *15*, 59. [CrossRef] [PubMed]
3. Barsocchi, P.; Potortì, F.; Furfari, F.; Gil, A.M.M. Comparing AAL Indoor Localization Systems. In *Evaluating AAL Systems Through Competitive Benchmarking. Indoor Localization and Tracking*; Springer: Berlin/Heidelberg, Germany, 2011; pp. 1–13. [CrossRef]
4. Lin, Q.; Zhang, D.; Chen, L.; Ni, H.; Zhou, X. Managing Elders' Wandering Behavior Using Sensors-Based Solutions: A Survey. *Int. J. Gerontol.* **2014**, *8*, 49–55. [CrossRef]
5. Yang, C.; Shao, H.R. WiFi-Based Indoor Positioning. *IEEE Commun. Mag.* **2015**, *53*, 150–157. [CrossRef]
6. Cantón Paterna, V.; Calveras Augé, A.; Paradells Aspas, J.; Pérez Bullones, M.A. A Bluetooth Low Energy Indoor Positioning System with Channel Diversity, Weighted Trilateration and Kalman Filtering. *Sensors* **2017**, *17*, 2927. [CrossRef] [PubMed]
7. Alarifi, A.; Al-Salman, A.; Alsaleh, M.; Alnafessah, A.; Al-Hadhrami, S.; Al-Ammar, M.A.; Al-Khalifa, H.S. Ultra Wideband Indoor Positioning Technologies: Analysis and Recent Advances. *Sensors* **2016**, *16*, 707. [CrossRef] [PubMed]
8. DecaWave Ltd. *DW1000 User Manual*; Decawave Ltd.: Dublin, Ireland, 2015 .
9. Minne, K.; Macoir, N.; Rossey, J.; Van den Brande, Q.; Lemey, S.; Hoebeke, J.; De Poorter, E. Experimental Evaluation of UWB Indoor Positioning for Indoor Track Cycling. *Sensors* **2019**, *19*, 2041. [CrossRef] [PubMed]
10. Kolakowski, J.; Djaja-Josko, V.; Kolakowski, M. UWB Monitoring System for AAL Applications. *Sensors* **2017**, *17*, 2092. [CrossRef] [PubMed]
11. Dimension4 | Ubisense. Available online: https://www.ubisense.net/product/dimension4 (accessed on 30 September 2019).
12. Yassin, A.; Nasser, Y.; Awad, M.; Al-Dubai, A.; Liu, R.; Yuen, C.; Raulefs, R.; Aboutanios, E. Recent Advances in Indoor Localization: A Survey on Theoretical Approaches and Applications. *IEEE Commun. Surv. Tutorials* **2016**, *19*, 1327–1346. [CrossRef]
13. Mussina, A.; Aubakirov, S. RSSI Based Bluetooth Low Energy Indoor Positioning. In Proceedings of the 2018 IEEE 12th International Conference on Application of Information and Communication Technologies (AICT), Almaty, Kazakhstan, 17–19 October 2018; pp. 1–4. [CrossRef]
14. Yiu, S.; Dashti, M.; Claussen, H.; Perez-Cruz, F. Wireless RSSI Fingerprinting Localization. *Signal Process.* **2017**, *131*, 235–244. [CrossRef]
15. Wang, B.; Chen, Q.; Yang, L.T.; Chao, H.C. Indoor Smartphone Localization via Fingerprint Crowdsourcing: Challenges and Approaches. *IEEE Wirel. Commun.* **2016**, *23*, 82–89. [CrossRef]
16. Lott, M.; Forkel, I. A Multi-Wall-and-Floor Model for Indoor Radio Propagation. In Proceedings of the IEEE VTS 53rd Vehicular Technology Conference, Spring 2001. Proceedings (Cat. No.01CH37202), Rhodes, Greece, 6–9 May 2001; Volume 1, pp. 464–468. [CrossRef]
17. Kim, M..; Chong, N.Y. Direction Sensing RFID Reader for Mobile Robot Navigation. *IEEE Trans. Autom. Sci. Eng.* **2009**, *6*, 44–54. [CrossRef]
18. Chen, H.X.; Hu, B.J.; Zheng, L.L.; Wei, Z.H. An Accurate AoA Estimation Approach for Indoor Localization Using Commodity Wi-Fi Devices. In Proceedings of the 2018 IEEE International Conference on Signal Processing, Communications and Computing (ICSPCC), Qingdao, China, 14–16 September 2018; IEEE: Qingdao, China, 2018; pp. 1–5. [CrossRef]
19. Zhang, R.; Liu, J.; Du, X.; Li, B.; Guizani, M. AOA-Based Three-Dimensional Multi-Target Localization in Industrial WSNs for LOS Conditions. *Sensors* **2018**, *18*, 2727. [CrossRef] [PubMed]
20. Hou, Y.; Yang, X.; Abbasi, Q.H. Efficient AoA-Based Wireless Indoor Localization for Hospital Outpatients Using Mobile Devices. *Sensors* **2018**, *18*, 3698. [CrossRef] [PubMed]
21. Baik, K.J.; Lee, S.; Jang, B.J. Hybrid RSSI-AoA Positioning System with Single Time-Modulated Array Receiver for LoRa IoT. In Proceedings of the 2018 48th European Microwave Conference (EuMC), Madrid, Spain, 23–27 September 2018; IEEE: Madrid, Spain, 2018; pp. 1133–1136. [CrossRef]

22. Wann, C.-D.; Yeh, Y.-J.; Hsueh, C.-S. Hybrid TDOA/AOA Indoor Positioning and Tracking Using Extended Kalman Filters. In Proceedings of the 2006 IEEE 63rd Vehicular Technology Conference, Melbourne, VIC, Australia, 7–10 May 2006; IEEE: Melbourne, Australia, 2006; Volume 3, pp. 1058–1062. [CrossRef]
23. Kumarasiri, R.; Alshamaileh, K.; Tran, N.H.; Devabhaktuni, V. An Improved Hybrid RSS/TDOA Wireless Sensors Localization Technique Utilizing Wi-Fi Networks. *Mob. Netw. Appl.* **2016**, *21*, 286–295. [CrossRef]
24. Jadidi, M.G.; Patel, M.; Miro, J.V.; Dissanayake, G.; Biehl, J.; Girgensohn, A. A Radio-Inertial Localization and Tracking System with BLE Beacons Prior Maps. In Proceedings of the 2018 International Conference on Indoor Positioning and Indoor Navigation (IPIN), Nantes, France, 24–27 September 2018; IEEE: Nantes, France, 2018; pp. 206–212. [CrossRef]
25. Huang, K.; He, K.; Du, X. A Hybrid Method to Improve the BLE-Based Indoor Positioning in a Dense Bluetooth Environment. *Sensors* **2019**, *19*, 424. [CrossRef] [PubMed]
26. Kolakowski, M. Utilizing Acceleration Measurements to Improve TDOA Based Localization. In Proceedings of the 2017 Signal Processing Symposium (SPSympo), Jachranka, Poland, 12–14 September 2017; IEEE: Jachranka Village, Poland, 2017; pp. 1–4. [CrossRef]
27. Kao, C.H.; Hsiao, R.S.; Chen, T.X.; Chen, P.S.; Pan, M.J. A Hybrid Indoor Positioning for Asset Tracking Using Bluetooth Low Energy and Wi-Fi. In Proceedings of the 2017 IEEE International Conference on Consumer Electronics— Taiwan (ICCE-TW), Taipei, Taiwan, 12–14 June 2017; IEEE: Taipei, Taiwan, 2017; pp. 63–64. [CrossRef]
28. Antevski, K.; Redondi, A.E.C.; Pitic, R. A Hybrid BLE and Wi-Fi Localization System for the Creation of Study Groups in Smart Libraries. In Proceedings of the 2016 9th IFIP Wireless and Mobile Networking Conference (WMNC), Colmar, France, 11–13 July 2016; IEEE: Colmar, France, 2016; pp. 41–48. [CrossRef]
29. Kolakowski, M. Kalman Filter Based Localization in Hybrid BLE-UWB Positioning System. In Proceedings of the 2017 IEEE International Conference on RFID Technology & Application (RFID-TA), Warsaw, Poland, 20–22 September 2017; IEEE: Warsaw, Poland, 2017; pp. 290–293. [CrossRef]
30. Kolakowski, M. A Hybrid BLE/UWB Localization Technique with Automatic Radio Map Creation. In Proceedings of the 2019 13th European Conference on Antennas and Propagation (EuCAP), Krakow, Poland, 31 March–5 April 2019; pp. 1–4.
31. Kherani, A.A.; Bhogi, S.K.; Shin, B. Hybrid Location Tracking in BLE Beacon Systems with In-Network Coordination. In Proceedings of the 2016 13th IEEE Annual Consumer Communications & Networking Conference (CCNC), Las Vegas, NV, USA, 9–12 January 2016; IEEE: Las Vegas, NV, USA, 2016; pp. 814–815. [CrossRef]
32. Sosa-Sesma, S.; Perez-Navarro, A. Fusion System Based on WiFi and Ultrasounds for In-Home Positioning Systems: The UTOPIA Experiment. In Proceedings of the 2016 International Conference on Indoor Positioning and Indoor Navigation (IPIN), Alcala de Henares, Spain, 4–7 October 2016; IEEE: Alcala de Henares, Spain, 2016; pp. 1–8. [CrossRef]
33. Petryk, G.; Buehler, M. Dynamic Object Localization via a Proximity Sensor Network. In Proceedings of the 1996 IEEE/SICE/RSJ International Conference on Multisensor Fusion and Integration for Intelligent Systems (Cat. No.96TH8242), Washington, DC, USA, 8–11 December 1996; pp. 337–341. [CrossRef]
34. Qiang, L.; Kaplan, L.M. Target Localization Using Proximity Binary Sensors. In Proceedings of the 2010 IEEE Aerospace Conference, Big Sky, MT, USA, 6–13 March 2010; IEEE: Big Sky, MT, USA, 2010; pp. 1–8. [CrossRef]
35. Kolakowski, M. Improving BLE Based Localization Accuracy Using Proximity Sensors. In Proceedings of the 2018 26th Telecommunications Forum (TELFOR), Belgrade, Serbia, 20–21 November 2018; pp. 1–4. [CrossRef]
36. STMicroelectronics. *VL53L1X Datasheet*; Decawave Ltd: Dublin, Ireland, 2018.
37. IONIS | European Commission Programme. Available online: https://ionis.eclexys.com/ (accessed on 30 September 2019).
38. Grewal, M.S. *Kalman Filtering: Theory and Practice Using MATLAB*, 4th ed.; John Wiley & Sons: Hoboken, NJ, USA, 2015.
39. Bar-Shalom, Y. *Estimation with Applications to Tracking and Navigation*; John Wiley & Sons: New York, NY, USA, 2001.

40. VL53L1X—Long Distance Ranging Time-of-Flight Sensor Based on ST FlightSense Technology—STMicroelectronics. Available online: https://www.st.com/en/imaging-and-photonics-solutions/vl53l1x.html (accessed on 30 September 2019).
41. P-NUCLEO-53L1A1. Available online: https://www.st.com/en/ecosystems/p-nucleo-53l1a1.html (accessed on 30 September 2019).
42. STM32F401RE—STM32 Dynamic Efficiency MCU, ARM Cortex-M4 Core with DSP and FPU, up to 512 Kbytes Flash, 84 MHz CPU, Art Accelerator—STMicroelectronics. Available online: https://www.st.com/en/microcontrollers-microprocessors/stm32f401re.html (accessed on 30 September 2019).
43. X-NUCLEO-53L1A1—Long-Distance Ranging Sensor Expansion Board Based on VL53L1X for STM32 Nucleo—STMicroelectronics. Available online: https://www.st.com/en/ecosystems/x-nucleo-53l1a1.html (accessed on 30 September 2019).
44. X-CUBE-53L1A1—Long Distance Ranging Sensor Software Expansion for STM32Cube—STMicroelectronics. Available online: https://www.st.com/en/ecosystems/x-cube-53l1a1.html (accessed on 30 September 2019).

Article

A Novel Indoor Localization Method Based on Image Retrieval and Dead Reckoning

Jiuchao Qian *, Yuhao Cheng, Rendong Ying and Peilin Liu

School of Electronic Information and Electrical Engineering, Shanghai Jiao Tong University, Shanghai 200240, China; cyh958859352@sjtu.edu.cn (Y.C.); rdying@sjtu.edu.cn (R.Y.); liupeilin@sjtu.edu.cn (P.L.)
* Correspondence: jcqian@sjtu.edu.cn

Received: 27 April 2020; Accepted: 26 May 2020; Published: 29 May 2020

Abstract: Indoor pedestrian localization measurement is a hot topic and is widely used in indoor navigation and unmanned devices. PDR (Pedestrian Dead Reckoning) is a low-cost and independent indoor localization method, estimating position of pedestrians independently and continuously. PDR fuses the accelerometer, gyroscope and magnetometer to calculate relative distance from starting point, which is mainly composed of three modules: step detection, stride length estimation and heading calculation. However, PDR is affected by cumulative error and can only work in two-dimensional planes, which makes it limited in practical applications. In this paper, a novel localization method V-PDR is presented, which combines VPR (Visual Place Recognition) and PDR in a loosely coupled way. When there is error between the localization result of PDR and VPR, the algorithm will correct the localization of PDR, which significantly reduces the cumulative error. In addition, VPR recognizes scenes on different floors to correct floor localization due to vertical movement, which extends application scene of PDR from two-dimensional planes to three-dimensional spaces. Extensive experiments were conducted in our laboratory building to verify the performance of the proposed method. The results demonstrate that the proposed method outperforms general PDR method in accuracy and can work in three-dimensional space.

Keywords: indoor localization; PDR; VPR; fusion navigation

1. Introduction

Location information and location-based services are always hot topics. At present, GNSS (Global Navigation Satellite System)-based positioning methods are still dominant for outdoor environments. However, in complex indoor environments, the satellite signals received by users is so weak that the GNSS-based positioning method will fail, which makes people turn to other effective techniques to obtain accurate the information of indoor location.

Commonly, indoor positioning methods include RFID (Radio Frequency Identification), Bluetooth, IR (Infrared Ray), geomagnetism, WLAN (Wireless Local Area Network) and visual recognition [1]. In principle, indoor localization methods can be divided into five categories: triangulation, sensing, recognition, fingerprint and DR (dead reckoning) [2]. In these methods, RFID, Bluetooth, IR and WLAN require deploying additional equipment, resulting in an increased cost. These methods require special equipment to be laid and have special signal receiving equipment, which has a higher requirement on the scene. Meanwhile, magnetic-based positioning method is vulnerable to interference, which affects the positioning accuracy. Compared with other methods, DR does not rely on external environments and additional equipment, which can gain high accuracy in a short time. It achieves low-cost and independent localization through processing outputs of inertial sensors. With the development of MEMS (Micro Electro Mechanical System) technology, we pay more attention to DR. However, limited by cumulative error of gyroscopes, the positioning error will gradually increase when working continuously for long periods. Furthermore, DR can only work in a two-dimensional plane without additional information, which means that DR has trouble in dealing with pedestrians going up- and downstairs.

To reduce the cumulative error and extend application scenarios, we need to find an algorithm that can help PDR correct its cumulative error to increase its accuracy. There are many studies focused on the problem of cumulative error in PDR. A smartPDR using mobile phones has been proposed, which uses accelerometers, magnetometers and gyroscopes in mobile phones, proposing an indoor positioning model that can detect downstairs, headings, etc. for localization [3]. Another algorithm combines GNSS and PDR for positioning and navigation and uses GNSS for positioning correction [4]. In [5], a combination of GPS and dead reckoning is used to explore the trajectory of animal movement to help understand ecology. Thong [6] combined WI-FI and PDR and used the Gaussian process to estimate and construct the RSS fingerprints by K-nearest weight. In [7], RFID, PDR and magnetic matching are combined for positioning, where RSS technology and a floor map are used. In addition, some research focuses on PDR working in the three-dimensional environment. The air pressure is combined with the PDR system to solve the upstairs and downstairs problems. The air pressure of the sensor on the hand will change regularly with the swing of the hand. PDR can be realized by processing the air pressure signal [8]. Wang et al. [9] fused the magnetic signals and the PDR by Kalman filter and the PF scheme. The locations are obtained by the weighted mean of particles. It reduces the effect of heavy magnetic distortions. These works have achieved great results in the indoor location.

These researches provide us with a wide range of ways for improving PDR. Following their research, we aim to combine a more determined method with PDR, which might work better. Visual information is easy to obtain with low cost sensors and is easily identified by humans. The visual information includes common RGB pictures, including depth RGBD pictures, binocular camera pictures and other pictures. RGB images might be the simplest and lowest cost visual information. In addition, computer vision now delivers amazing results in a variety of areas, including image recognition and image classification, which are beneficial for indoor location. We believe that the combination of visual signal and PDR will be one of the solutions for indoor positioning. Therefore, it is meaningful to combine computer vision with PDR, which may better solve the problem of PDR position correction and work in a three-dimensional environment.

Visual position recognition is one of the visual methods that may be used to correct PDR. VPR is the task of identifying a place through a visual image of it. VPR is suitable to fill the gaps of positioning errors because of its high robustness and convenience of requiring no additional deployment. The principle of VPR is matching feature descriptors of the query image and other labeled images to find the most similar one. Then, from the label of the most similar image, we can get the location of the query image. Traditional visual place recognition algorithms use local or global feature method, such as SIFT [10], SURF [11], CenSurE [12], Gist [13] and ADMM-RS [14] to get the descriptors of images. However, each local and global descriptor loses some information of the image [15], which may result in a wrong match. Beside, the calculation of these traditional methods will cost a lot of computing resources and are too slow to use, which do not suit the real-time location. With the development of deep learning technology, CNNs (Convolutional Neural Networks) are used to extract feature descriptors from images and have made some improvements in recognition tasks [16–18]. CNNs can obtain the deep information in the images, so that they can focus on the deeper differences in the images, and thus can achieve better visual positioning effect. In VPR, CNN performs better than the traditional method and costs less time. In our method, CNN is used to build a real-time visual position recognition module.

In this paper, a novel and improved indoor localization method is proposed, which can effectively reduce accumulative errors derived from PDR and works well in a three-dimensional space by fusing vision information. Especially, we train a CNN architecture for indoor visual place recognition tasks and use it to aid PDR to gain better performance in indoor positioning. We use the VPR system to get a starting point for PDR, while the place coordinates are obtained from VPR to modify the localization results from PDR. The improved fusion indoor localization method has much less computing complexity than VPR system and makes PDR more flexible and accurate.

The main contributions of this article are: (1) A new method called V-PDR is proposed for indoor localization which combines the VPR and PDR. With the system, we can obtain the absolute position rather than the relative position. (2) V-PDR improves the localization accuracy and solves the problem that PDR cannot work properly in three-dimensional space by combining vision and inertial sensors.

The rest parts of this paper is arranged as follows. Section 2 is the related works for pedestrian dead reckoning. In Section 3, our PDR system is illustrated. Our design for the fusion indoor localization method is introduced in Section 4, which include our VPR network design and fusion method. Details of experiments implemented in our lab for testing our method and the results are listed in Section 5. Finally, the conclusions are drawn in the last section.

2. Related Works

2.1. PDR Improvements with Visual Information

Recently, many works have been done to improve PDR with visual information. These visual signals include 2D images, depth images, binocular cameras, etc. Some use a camera and PDR combined for localization and tracking, designing a new visual detection and tracking system [19]. For wearable devices, Roggen et al. [20] used optical flow and PDR to help get the trajectory of movement. They also used the previous vision to build a visual circuit, which can be used on maps without priors. They focused on an algorithm to combine two localization result to get the right result, which increases the confidence in positioning results [21]. In addition, Yan et al. [22] used target detection to get the 3D positions of pedestrians, which assists PDR to achieve good positioning. Some works [23] use the matching between 3D and 2D for position correcting, where visibility and co-visibility information is adopted. The visual position corrects the PDR positioning error, which achieves good results. In [24], the authors proposed helping indoor positioning by combining sign recognition and DR positioning They focused on correcting PDR movements by visual information, thereby reducing cumulative errors. Meanwhile, we directly use visual position recognition to obtain the exact positioning results to assist the correction of PDR.

2.2. Visual Position Recognition

In recent years, with the development of deep learning technology, position recognition based on computer vision has made great breakthroughs. The visual place recognition technology is not limited by application scenarios, and can intuitively provide users their locations. Chen et al. [25] used pre-trained CNNs that are trained for objection detection to realize a position recognition task. Krizhevsky et al. [26] obtained better performance using ImageNet to extract features from images than methods of SIFT features in recognizing locations. Arandjelovic et al. [27] designed a new CNN architecture to train a place recognition task in an end-to-end manner; their method outperforms existing feature descriptors. Chen et al. [28] considered place recognition as a classification task and built a massive specific dataset for VPR. Most of the currently used VPRs are applied to outdoor scenes; however, indoor scenes have many feature points, and the features are more homogeneous. Therefore, we need to make special designs for VPR in this regard to adapt to indoor positioning.

3. Pedestrian Dead Reckoning

Pedestrian dead reckoning (PDR) is an independent positioning system that estimates locations of pedestrians using accelerometer, magnetometer and gyroscope. It consists of four modules: step detection, stride length estimation, heading calculation and location update. The workflow of PDR is shown in Figure 1.

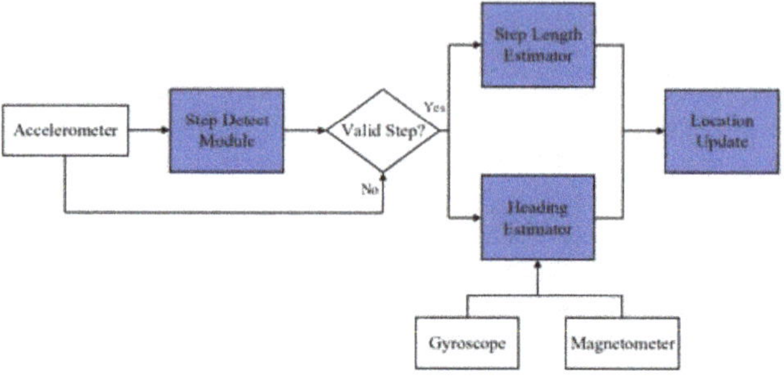

Figure 1. The workflow of PDR.

After filtered to reduce noise and aligned over time, the sensor data are assigned to different modules. The step detection module makes a judgement of whether a step occurs according to the acceleration values. If a step does occur, the system runs the stride length estimation module and heading calculation module to estimate the step length and heading angle at this step. Then, a new location update process is performed. The specific location update method is demonstrated as the following equations:

$$\begin{cases} x_{k+1} = x_k + s\_l_k \times \cos\varphi_k \\ y_{k+1} = y_k + s\_l_k \times \sin\varphi_k \end{cases} \quad (1)$$

where x and y are the coordinates of pedestrians, s_l_k is the stride length and φ is the heading. The subscript denotes step k. We can derive that, under the condition that the initial position is known, locations can be estimated continuously according to Equation (1). The principles of step detection, stride length estimation and heading calculation are illustrated in the next sections.

3.1. Step Detection

Step detection is one of the three basic modules of PDR system. Only when determining the occurrence of one step does the system use the stride length estimation and heading calculation modules. Therefore, the step detection module plays an important role in PDR system. Usually, we determine whether a step occurs or not by detecting the periodicity of two-peaks oscillation of accelerations in the process of walking [29]. There are some common step detection methods: zero-crossing detection [30], peak detection [31] and autocorrelation [32]. To gain high robustness, we choose improved zero-crossing detection, which uses the amplitude of accelerations to make the judgement, as our step detection method. The amplitude of accelerations can be calculated by triaxial outputs of the accelerometer:

$$a_m = \sqrt{{a_x}^2 + {a_y}^2 + {a_z}^2} \quad (2)$$

Then, we decide whether the step occurs according to the values of the interval between acceleration amplitude and the threshold and the time interval of two undetermined steps. The specific criteria are obtained by analyzing the waveform of accelerations, as shown in Figure 2.

- The peak value of acceleration amplitude is greater than the threshold, and the valley value is less than the threshold during the period of step k.
- The difference between the peak value of acceleration amplitude and the threshold in step k is within the restricted range. In addition, the difference between the threshold and the valley value of acceleration amplitude in step k is within the restricted range.
- The duration of step k is within the restricted range, which can be determined by the walking frequency.

With the restricted conditions above, the step detect module can eliminate the error steps introduced by perturbation of acceleration and gain a high accuracy in step counter.

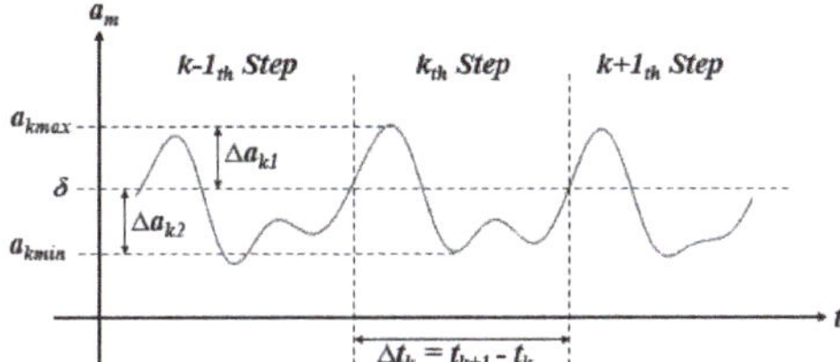

Figure 2. The waveform of accelerations.

3.2. Stride Length Estimation

In PDR system, the stride length estimation module takes the responsibility of estimating the step length of each step during the walking process. In a real scenario, the stride length of pedestrians varies from person to person and is also affected by stride frequency. Studies on stride length estimation mainly include empirical models related to pedestrian's height [33], linear models related to walking frequency [34,35] and so on. Considering the influence of walking speed on step length, we establish a linear model related to step frequency and acceleration to estimate the step length. The linear model is shown in Equation (3).

$$s\_l_k = \alpha + \beta \times f_k + \gamma \times {\sigma_{a_k}}^2 + \varepsilon \quad (3)$$

where s_l_k represents the step length, f is the walking frequency and σ is the variance of accelerations. α, β, and γ are coefficients associated with pedestrians. ε is random noise. The parameters in this model are determined by fitting and analyzing the data collected offline. We measured the interrelationship between the stride frequency and stride length of different people, and fit the data using the data of a large number of people. Note that the fitting is a nearly linear experimental value and can be applied to most people.

3.3. Heading Calculation

The heading calculation module is the most important part of PDR. The inherent defect of PDR algorithm is the cumulative error, which is largely caused by the cumulative error of integrating the angular velocity measured by the gyroscope. In the actual walking process, accompanied by swings, the postures of the sensors do not always represent the forward direction of the pedestrian, which, if not corrected, would introduce additional error into the heading calculation. To solve this problem, PDR algorithm integrates the outputs of gyroscope, accelerometer and magnetometer to determine the posture of sensors and to correct the heading of pedestrians. Kalman filtering [36] and particle filtering [37] are often used in the PDR system as the fused heading calculation algorithms. In this paper, we propose a recursively weighted fusion heading calculation algorithm with small computation. Firstly, the gyroscope is used to get three postures, which are named as angle of pitch, angle of yaw and angle of roll. Then, the acceleration is used to correct the pitch angle and the roll angle, and the magnetometer outputs are used to correct the yaw angle, which is the heading of the pedestrian. The values of correction parameters in this algorithm are directly related to the accuracy of the sensors.

Our algorithm uses quaternions to settle the attitude of AHRS (Attitude and Heading Reference System) and get the attitude information of the sensors. According to the mechanism of PDR, the angular rate and the attitude quaternion at step $k-1$ are known, thus the updated attitude quaternion at step k can be obtained by solving the gradient of the attitude quaternion (Equation (4)).

$$\begin{cases} q_{\omega,k} = q_{\omega,k-1} + \dot{q}_{\omega,k}\Delta t \\ \dot{q}_{\omega,k} = \frac{1}{2} q_{\omega,k-1} \otimes \omega_k \end{cases} \tag{4}$$

The attitude angles can be calculated using Equation (5).

$$\begin{cases} \varphi = \arctan\left[(2q_1q_2 - 2q_0q_3)\ /\ (2q_0{}^2 + 2q_2{}^2 - 1)\right] \\ \theta = \arcsin\left(2q_2q_3 + 2q_0q_1\right) \\ \gamma = -\arctan\left[(2q_1q_3 - 2q_0q_2)\ /\ (2q_0{}^2 + 2q_3{}^2 - 1)\right] \end{cases} \tag{5}$$

where φ is the yaw angle, θ is the pitch angle and γ is the roll angle.

In the navigation coordinate system, the gravity is a constant value, and, if the attitude matrix calculated by angular rate is accurate, it should be equal to the acceleration in the carrier coordinate system when projected onto it. Therefore, the inaccurate calculation of attitude matrix would inevitably lead to inconsistency of the outputs in the two coordinate systems. Thus, the gradient descent method can be adopted to minimize this inconsistency. When the loss function is zero, the error of the rotation matrix has been eliminated and an accurate attitude quaternion is obtained. The loss function of the attitude correction process is defined in Equation (6).

$$f(q,a) = \|q_a \times g - a_m\|^2 \tag{6}$$

Thus, to solve this problem, we get Equation (7)

$$\nabla f(q,a) = \mathcal{J}(q,a) \cdot e(q,a) \tag{7}$$

where $\mathcal{J}(q,a)$ is the Jacobian matrix of quaternions and $e(q,a)$ is the error function.

$$\begin{cases} \mathcal{J}(q,a) = \partial f / \partial q \\ e(q,a) = (q_a \cdot g - a_m)^T \end{cases} \tag{8}$$

The quaternion iteration equation represented by acceleration is shown in Equation (9).

$$\begin{cases} q_{a,k+1} = q_{a,k} - \mu_{a,k} \frac{\nabla f(q,a)}{\|\nabla f(q,a)\|} \\ \mu_{a,k} = \alpha_k \left\|\dot{q}_\omega\right\| \Delta t \end{cases} \tag{9}$$

Similar to the attitude correction by accelerations, we also adopt gradient descent method to achieve the purpose of correcting yaw angle of the attitude by minimizing the error between the yaw angle calculated by magnetic values and the yaw angle calculated by angular rates. Likewise, the loss function is defined in Equation (10).

$$f(q,m) = \|q_m \cdot b - m\|^2 \tag{10}$$

The quaternion iteration equation represented by magnetic values is shown in Equation (11).

$$q_{m,k+1} = q_{m,k} - \mu_{m,k} \frac{\nabla f(q,m)}{\|\nabla f(q,m)\|} \tag{11}$$

After correction of the fused attitude algorithm, the heading calculation module calculates a relatively accurate attitude of sensors and outputs accurate heading information to the location update module. All the necessary data for location update are obtained, thus the location update module can locate pedestrians continuously using Equation (1).

3.4. The Uncertainty of PDR

We test edour PDR system to estimate the performance of the PDR system. In the experiment, we tested the basic pedometer and positioning performance of the algorithm by holding a smartphone and walking clockwise along the pedestrian walkway back to the location near the starting point. The test results are shown in Figure 3.

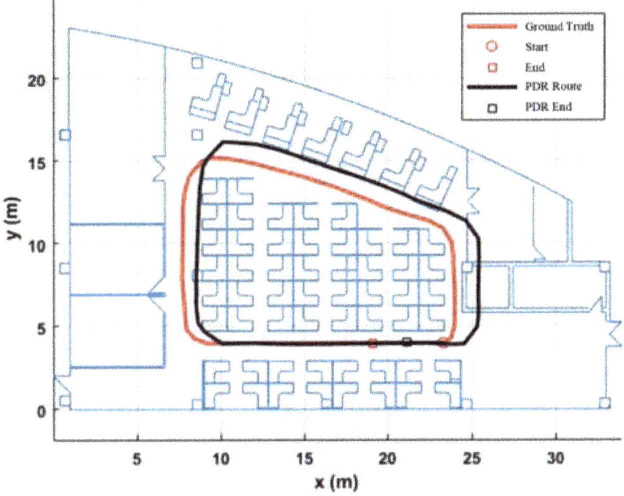

Figure 3. The test result of PDR.

In Figure 3, the red curve represents the actual route of pedestrian walking, while the black curve represents the pedestrian walking trajectory calculated by the PDR algorithm. The pedestrian walked for totally 55 m with 82 steps while the result of PDR is 81 steps. The distance between the end point of pedestrian position estimated by PDR and the end point of actual pedestrian position is 1.24 m, thus the positioning error is about 2.3% of the travel distance. The detailed results are shown in Table 1.

Table 1. The result of PDR.

Localization Method	Maximum Error (m)	Mean Error (m)	95% Situation Error (m)	Step Counting Accuracy
PDR	1.52	0.91	1.43	98.8%

Due to the short walking distance of pedestrians and the accumulated error of inertial sensors during walking, it is foreseeable that, as the walking distance continues to increase, the positioning error of pedestrian dead reckoning will also gradually increase. In a short distance (about 50 m), the error is within 3%. This means that, in short-distance operations, we can regard the PDR algorithm results as correct. We need to find a way not to modify the PDR all the time, but only to modify the location after the PDR has been running for a long time.

4. Vision-Aided PDR System Design

In this section, we introduce our vision-aided PDR (V-PDR) system. Our system is dividid into two parts: one is the visual place recognition model and the other is focusing VPR with PDR. The architecture of the proposed V-PDR system is exhibited in Figure 4.

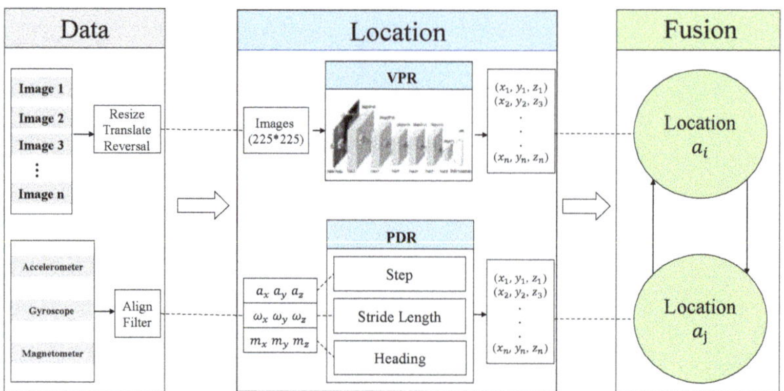

Figure 4. The system architecture of vision-aided PDR (V-PDR).

4.1. Visual Place Recognition

The main task of VPR is to find a similar image in the image dataset search by comparing the images with the stored database to find out where the image is located. The algorithm diagram is shown in Figure 5. The main steps of visual place recognition are:

- Extract features from the location image in the original map library, and establish a feature vector dataset.
- Extract the feature vector of the image to be predicted in the same way.
- Compare the feature vectors in the dataset and the feature of the image to be predicted by the method and select the location corresponding to the most similar feature vector as the predicted value.

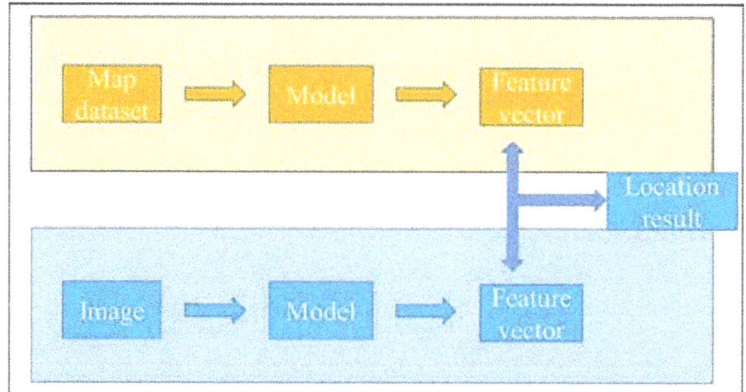

Figure 5. Visual Place Recognition Algorithm Diagram.

To assist PDR system to get better positioning performance, in this paper, we train a VPR network for indoor scenes. The following subsections present the structure of our VPR network, training process, ranking search algorithm and the results of place recognition.

4.1.1. Design of VPR Network and Training

Considering the real-time positioning and the simplicity of the device, it is not possible to use too complicated neural networks for the VPR here. The construction of a VPR network consists of two parts: design of structure and training of parameters. For the purpose of learning deep visual features to make exact matching, six convolution layers are used to learn enough deep features, and these features are combined by the fully connected layer. The activation function ReLu follows the convolution layer, and the max-pooling is selected. The structure of our CNN architecture is shown in Figure 6.

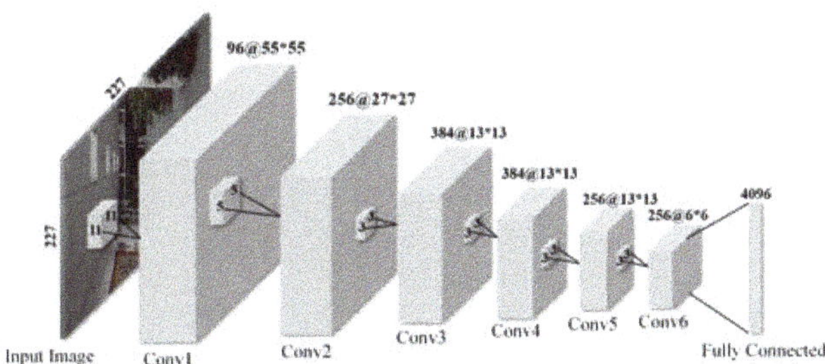

Figure 6. Visual place recognition network structure.

A classification model is not used here because classification models need to fine tune to the different datasets in other places, which is impossible in practical use. Thus, we use metric learning as our VPR backbone network to train our VPR model. In the training process, RGB images with three-dimensional location labels of any size are resized i to 256 × 256. Small blocks with size of 225 × 225 are extracted from the resized images and then flipped horizontally to increase total data amounts, which helps to overcome overfitting. The parameters are trained by hard triplet loss gradient descent method and back propagation mechanism. The triplet loss function can be expressed as Equation (12).

$$\sum \left[\|f(I_a) - f(I_p)\|_2^{\,2} - \|f(I_a) - f(I_n)\|_2^{\,2} + \alpha \right]_+ \tag{12}$$

where the first item is the Euclidean distance between the positive and the anchor and the second item is the Euclidean distance between the negative and the anchor. α serves as a margin parameter. $[x]_+ = \max(0, x)$ means the max distance between negative and positive queries. The purpose is to make the feature vectors of the same places more similar and make the distance between feature vectors in different places as large as possible.

Note that, because this is not a classification model but an encoder that encodes the image into the feature vectors, it can be used directly without modification for other scenes. In any scenario, this neural network can encode images into 4096-dimensional vectors. Vectors encoded by the same neural network will have the same feature extraction method, thus the encoding of similar images will be similar, and the encoding difference of different images will also be very large. Thus, when this network is used in other scenarios, no secondary training is required.

4.1.2. Ranking Search

To find the most similar location, because the same location can be viewed from different perspectives, our database needs to store images of different perspectives for the same location for searching [38]. Here, photos are stored every 20 degrees for a location, with a total of 18 images in one location. The image acquisition method is shown in Figure 7.

Every time a photos is searched, its encoding o is compared with the images in the database. For the same place, the similarity in a certain direction is the greatest, and the side angle is similar to the picture to be compared, but it is lower than the opposite angle. In this way, if there are multiple pictures of the same location and different angles in the search list in Recall@10 (R@10 denotes the top ten database candidates), then it can increase the certainty of the exact location and angle in this way, which reduces accidental location errors. According to this method, the positioning recognition rate of the image can be enhanced.

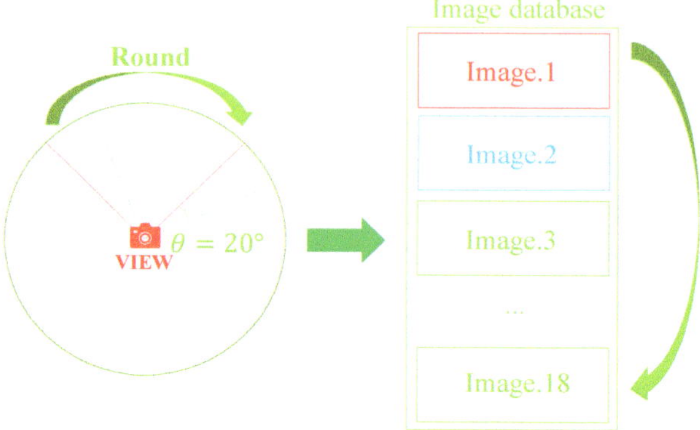

Figure 7. Image Acquisition Method.

4.1.3. Location of VPR

From the input image with size of 256 × 256, images are extracted with size of 224 × 224, flipped horizontally and then inputted into the trained network to get 4096-dimensional feature vectors. Before VPR is used, it is necessary to build an image vector dataset. When obtaining the images used for testing V-PDR, we extracted these images into the feature vector dataset. The vector dataset consists of the feature vectors of images and their labels. The labels are built for where the image is obtained. The images with the highest similarity are found by comparing the Euclidean distances between feature vectors and vector dataset. The location of the image is obtained by weighing the recognition results of these inputs. If the location of Recall@1 is not searched in the image from other angles in Recall@10, such locations are ignored. The most probable location is the VPR result. As for heading direction, the VPR direction is the most probable heading direction.

4.2. Vision-Aided PDR Fusing

In the case where the PDR positioning result is accurate in a short period and there is a cumulative error in long-term use, we can trust the PDR positioning information and only correct the position by VPR after the person has walked for a while. Since PDR and VPR are two independent systems, the purpose of fusion is to use VPR positioning results to provide absolute position reference information for PDR, without involving the exchange of internal information between the two systems. Therefore, loose coupling is adopted to achieve the fusion positioning of PDR and VPR.

The architecture of the proposed V-PDR system is exhibited in Figure 4. The system consists of three parts: data obtain and process, location and fusion. The first part is a layer of raw data acquisition and processing. In this part, input images shot by cameras on the smartphone cannot be used directly. Since these images are too large, it would take a long time to process them in neural networks directly. Besides, the images would be affected by the ambient light, thus other preprocessing is needed. These images are sent to the VPR system after some operations: resizing, translating and reversing. In this way, the images are suitable for neural network processing and some environmental impacts are reduced. The data of sensors acquired by accelerometer, gyroscope and magnetometer in the smartphone are passed to PDR system after being filtered. The second part is called location, which means using the input data to finish positioning. Obviously, it is the most important part of V-PDR. The algorithm for location by PDR and VPR is shown in Sections 3 and 4.1, respectively. Fusion is the third part and optimization and improvement of positioning results are achieved in this stage.

When positioning, the VPR system and the PDR system independently produce positioning results. The PDR position result is $p_j = [x_j \quad y_j \quad z_j]^T$, while the VPR position result is $p_v = [x_v \quad y_v \quad z_v]^T$. Due to the PDR making no correction to changes in floors, the weighted average of the positions obtained is:

$$p_f = \omega_j p_j + \omega_v p_v = [\omega_{j1} \quad \omega_{j2} \quad 0] \begin{bmatrix} x_j \\ y_j \\ z_j \end{bmatrix} + [\omega_{v1} \quad \omega_{v2} \quad 1] \begin{bmatrix} x_v \\ y_v \\ z_v \end{bmatrix} \tag{13}$$

where ω_j and ω_v obey the rules:

$$\begin{cases} \omega_{j1} + \omega_{v1} = 1 \\ \omega_{j2} + \omega_{v2} = 1 \end{cases} \tag{14}$$

The weight selection method related to the positioning errors of the PDR and VPR systems determines the weighting factors in the weighted average fusion algorithm by analyzing the positioning errors of the PDR and VPR systems. Subsystems with larger positioning errors have lower reliability and have less weight in the fusion system; subsystems with smaller positioning errors have higher reliability and have greater weight in the fusion system. In the indoor environment, the positioning error of PDR is about 3%. The positioning error of VPR is related to the construction of the local dataset. The distance between each location is about 1 m, and the positioning error of VPR is within 0.5 m. Therefore, the value of the weight in the weighted average algorithm is determined by the positioning performance of PDR and VPR.

$$\begin{cases} \omega_j = \begin{bmatrix} \frac{|p_{jx}^k - p_{0x}'| \cdot 0.03}{|p_{jx}^k - p_{0x}'| \cdot 0.03 + 1} & \frac{|p_{jy}^k - p_{0y}'| \cdot 0.03}{|p_{jy}^k - p_{0y}'| \cdot 0.03 + 1} & 0 \end{bmatrix} \\ \omega_v = \begin{bmatrix} \frac{|p_{jx}^k - p_{0x}'| \cdot 0.03}{|p_{jx}^k - p_{0x}'| \cdot 0.03 + 1} & \frac{|p_{jy}^k - p_{0y}'| \cdot 0.03}{|p_{jy}^k - p_{0y}'| \cdot 0.03 + 1} & 1 \end{bmatrix} \end{cases} \tag{15}$$

where p_j^k is the PDR position result at K and p_0' is the start or the last correction location. It corrects the location only when the distance between the estimated result of PDR at time k and the last correction point is between the thresholds ξ_1 and ξ_2 or the positioning result of VPR at time k changes in height compared to the positioning result at time $k-1$.

$$\begin{cases} z_v^k \neq z_v^{k-1} \\ \xi_1 < |p_j^k - p_0'| < \xi_2 \end{cases} \tag{16}$$

where the thresholds ξ_1 and ξ_2 are related to the positioning accuracy corrected by experiment.

As mentioned above, when working for long periods, the accumulative error leads to a decrease in positioning accuracy of PDR system. It is difficult for PDR to make reasonable estimates when spatial position changes. Focusing on these problems, we propose a solution by means of using VPR network to correct the localization results of PDR. VPR is called every 5 m or after going around a corner. The detailed fusion algorithm is shown in Algorithm 1.

Algorithm 1 Measure the location L_t at time t

Require: Picture x_i in the facing direction
  $i = 0$
  $L_0^v, p = VPR(x_0)$
  $L_0^p = L_0^v$
  while $p \leq \lambda$ **do**
    Turn right for 20 degrees and get new picture x_0
    $L_0^v, p = VPR(x_0)$
    $L_0^p = L_0^v$
  end while
  while $i \leq t$ **do**
    Get a new picture x_i in the facing direction
    $L_i^v, p = VPR(x_i)$
    if $\xi_2 \geq \left\| L_i^v - L_i^p \right\| \geq \xi_1$ **then**
      $L_i^p = Fusion(L_i^p, L_i^v)$
    end if
    $i = i + 1$
  end while
  return L_t^p

where L_t means the predicted location at time t, L_t^v means the location and direction predicted by VPR and the location predicted by PDR is L_t^p. λ is the setting threshold, obtained by correcting the median of Recall@1 in the test dataset. p is the probability of the top VPR location and ξ_1 and ξ_2 are the setting location values. The values of ξ_1 and ξ_2 were obtained through experiments. The function $Fusion()$ is shown above.

First, the starting point of PDR is determined by VPR. If the probability $p \geq \lambda$, the starting point is L_t^v. If not, then turn right for 20 degrees until $p \geq \lambda$. When the VPR performs well, it locates the position while turning around. If it does not perform well, it moves 0.5 m in a random direction for new positioning and infer the positioning at this step based on the relationship between the previous positioning and the movement calculated by PDR. Then, it repeats the steps above until the initial positioning is obtained. Then, PDR enters working state and starts to continuously update locations of pedestrians.

In the process, the pedestrian pauses to take images of a place, and VPR processes these images to get reasonable location coordinate to correct the location estimated by PDR. When the confidence of the positioning given by VPR is higher than the threshold σ, and the positioning result is within the distance between the previous correct positioning result and the PDR detection, we confirm that the positioning of VPR is accurate. Then, we judge the difference between PDR and VPR positioning. If the distance between results of VPR and PDR is between thresholds ξ_1 and ξ_2, the PDR positioning result is chosen to be the location. If not, the location is corrected to the VPR location, as the PDR has a large offset. When the confidence of VPR positioning is less than the threshold σ, or the distance between new positioning result and previous correct positioning result is more than the PDR detection distance, we regard the VPR result as incorrect and then locate the new place only by PDR. In the V-PDR, we determined the value of thresholds ξ_1 and ξ_2 through extensive experiments. The threshold σ was set as 0.6. After the correction, more accurate locations are estimated by V-PDR. There is no special requirement for selecting correction points, but VPR should be used after pedestrians crossing floors. The whole localization process of V-PDR is shown in Figure 8.

For location identification across floors, a new decision mechanism needs to be added here. The VPR location labels include the location in one floor and which floor it is. Besides, areas where floor transitions may occur are labeled. Note that the floor change here involves the floor change of stairs and elevators. A simple PDR system cannot judge the change of floors, thus the VPR positioning result is used for floor prediction. In addition, the scenes of the stairwell and the elevator doorway on each floor are probably similar; thus, the VPR cannot be identified only once. When entering the area where the floor may change, a new judgment mechanism is added. VPR judgment is performed here multiple times. Only when three consecutive floors are on the same floor does it judged that the floor has changed. With the above algorithm, the V-PDR can be used to solve the accumulative error due to PDR not being able work in a three-dimensional environment.

When there is a large error in PDR or VPR, it can be corrected by posterior. Only when $\xi_2 \geq \left\| L_i^v - L_i^p \right\| \geq \xi_1$ is the PDR corrected. This is already a correction process for outliers to ensure that the system positioning is not affected when the error is too large. Especially when abnormal conditions occur, to obtain reliable positioning results, the posterior needs to be processed in the subsequent positioning. Three consecutive VPR positioning distances and PDR measured distances are judged to determine whether there is an error.

$$\left\| \frac{|L_i^v - L_{i-1}^v|}{|L_i^p - L_{i-1}^p|} - 1 \right\| \leq 0.03 \tag{17}$$

Through Equation (17), whether outliers appear can be judged. If a positioning error occurs, the VPR is called multiple times in a short period during the operation. If Equation (17) holds, the positioning is corrected to the current positioning. With this method, rare errors are corrected. Note that all this is based on the fact that our PDR and VPR accuracy is high enough.

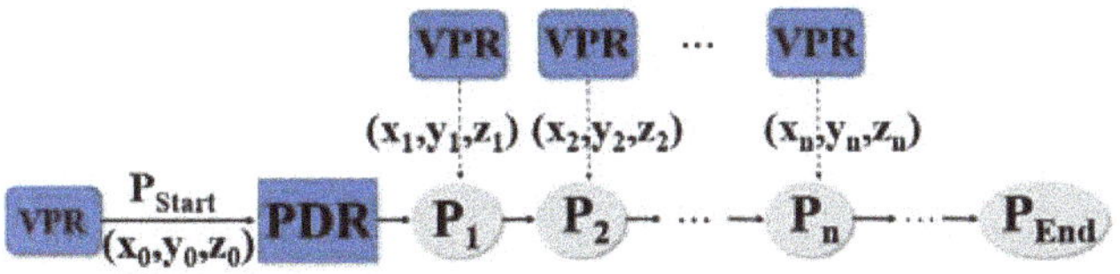

Figure 8. Fusion flow chart of V-PDR system.

5. Experiments and Results

The proposed fusion positioning system was tested in typical indoor scenarios. To verify the advantages of the proposed V-PDR algorithm compared with the traditional PDR algorithm in indoor positioning and to verify the feasibility of the V-PDR algorithm in three-dimensional space, the experiment was divided into two parts. The first was to verify the effectiveness of our proposed VPR model in indoor positioning. The second was to verify the advantage of the proposed V-PDR model compared to the traditional PDR model, which included comparing the positioning results between the V-PDR and PDR algorithms during driving and verifying that our V-PDR algorithm works properly in a cross-floor environment.

5.1. VPR Test Result

Our network training process was performed on an GTX1080Ti GPU. The margin parameter α was set to 0.2. We trained the model for 40 epochs, the learning rate was 0.0002 for the first 15 epochs, and was reduced by 90% every 15 epochs. The training was carried out on the Indoor Scene Recognition dataset [39], which contains 67 indoor categories and a total of 15,620 images. All images are in jpg format and there are at least 100 images per category in the dataset. It took about 8 h for training.

After the network was trained, it was necessary to test the model to see if the trained model is accurate enough to adapt to the V-PDR algorithm. We made two tests, respectively, on a subset of the training dataset and a dataset of our laboratory. In the experiment, we set the doorway as the origin, and east and north as the x-axis and y-axis. Note that, in actual use, since the location does not need to know the actual height but only the actual floor information, the 3D location information can be directly used as the label. In addition, we set the floor as the z-axis. Then, every location was set to a point with its location coding (x, y, z). Every 1 m we set a point. In every point, we obtained 18 images every 20 degrees. Note that, each image of one degree in different point was in the same direction. Overall, we obtained a dataset consisting of 36 locations and 648 images. In this way, the vector dataset was composed of the vectors and their labels, which consisted of location and direction. Sample images of the Indoor Scene Recognition and our laboratory datasets are shown in Figures 9 and 10, respectively.

Figure 9. Sample images in Indoor Scene Recognition dataset.

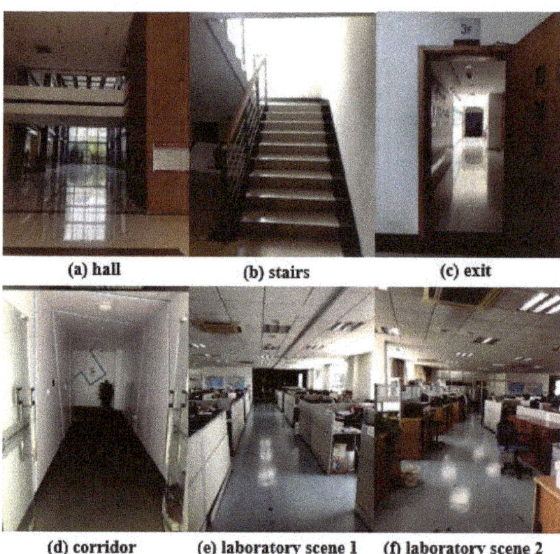

Figure 10. Sample images of our Laboratory.

Finally, we performed VPR model test experiments on indoor datasets and photos collected in our laboratory. First, images were extracted into the vector. After that, we compared the vector similarity in the vector dataset to check whether the predictions are correct. We used the Recall@1 to show the accuracy. As for indoor datasets, we used the test dataset for testing. For our laboratory dataset, a subset of the images we obtained was used for testing.

The VPR model obtains a result of a Recall@1 (@1 denotes one of top database candidates) of 80% on the Indoor Scene Recognition dataset and a recall@1 of near 92% on our laboratory. Note that the scenes in the Indoor Scene Recognition dataset are more complex than those in our laboratory. Some scenes are very similar, thus it performs worse than it in our laboratory. Performing VPR positioning once takes about 1.3 s. Since the VPR is only called occasionally, the speed is acceptable.

Analyzing the results in our laboratory dataset, we can find that good positioning results can be achieved for most images. However, for a few locations with incorrect positioning, it can be found that most of their positioning errors are due to the low similarity of all the images, even as low as 20%, resulting in inaccurate positioning. Such positioning errors can be repaired in subsequent V-PDR without affecting the overall positioning. Through testing, we found that our VPR model can identify the location well. This VPR is effective enough for our V-PDR algorithm. It is used as an intermittent auxiliary positioning and can be well adapted to actual use.

In addition, we performed an experiment to show the error between VPR route and the ground truth to show the VPR problems. We walked around our laboratory. For VPR positioning, in the experiment, the VPR worked for about 3 m walking, and the location is where the image is most similar. The positioning line is regarded as a walking path. The results are shown in Figure 11.

In Figure 11, the ground truth is shown in red while the VPR route is in black. The red triangle is where we used the VPR to locate while the black triangle is the location predicted by the VPR. We can observe that the VPR works well as the location predicted is not far from the ground truth. The average VPR location error is about 0.45. Since the images dataset is discrete, there are some errors in the VPR location. Therefore, VPR can only obtain approximate positioning results but not completely accurate positioning results. In addition, VPR predicts a point, and a continuous walking path cannot be obtained. It cannot be correctly positioned in the place without the images. In summary, it is difficult for VPR to perform independent and accurate positioning, and other algorithms are required to correct the VPR.

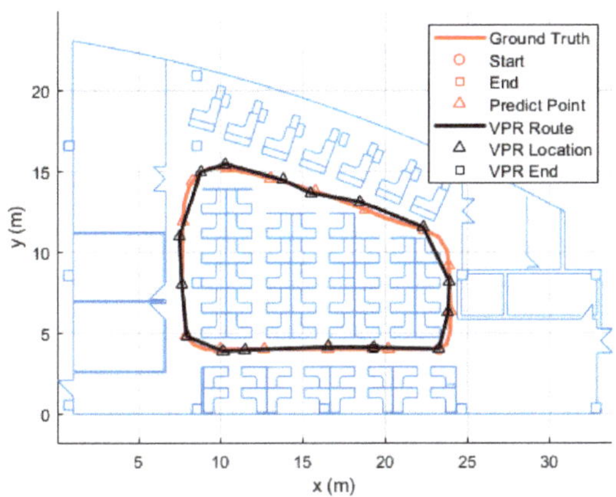

Figure 11. The test result of VPR.

5.2. V-PDR Testing Result

We performed our V-PDR experiments, which were divided into two parts. One was to show the ability of V-PDR to correct PDR cumulative errors, and the other was to test the ability of V-PDR to work in a three-dimensional environment. Experiments were conducted in our laboratory building and the ground truth was obtained by laboratory engineering drawings and high-precision measurement tools. The accuracy comparison test was conducted on the third floor and the total length of the route was about 64 m. The three-dimensional localization ability test of V-PDR was executed from the first floor to the third floor and the total length of this test route was about 120 m, not including the climbing process. The experiments were performed on an Android phone. We wrote a software application on a mobile phone for VPR and PDR. Every step, we took a photo with the mobile phone, the photo was transported to the cloud server for VPR localization and the software processed the data of PDR and VPR our algorithm.

The first experiment was performed on the third floor of our laboratory. We tested PDR and V-PPR with the same phone. We randomly specified a location as the starting point, and then randomly moved. After walking for a period of time, we obtained the relative motion path of the PDR. The V-PDR was tested through the same path to obtain the absolute position path of the V-PDR at the same time. Then, the paths were compared with the high-precision measurement path to analyze the results of our algorithm. The comparison results include maximum error, mean error, and 95% situation error. Maximum is the maximum location error distance in the experiment and the mean error is the average error for the path. The 95% situation error is the least 95% error in the experiment, which shows the effect for most locations.

The results of the first experiment conducted in the third floor are shown in Figure 12. The red curve is our ground truth. The blue and green curves represent the PDR trajectory and the V-PDR trajectory, respectively. The pedestrian departs from the start point and circles clockwise along the corridor to the end point. Obviously, the trajectory of PDR is quite far away from the ground truth, especially when the pedestrian turns corners. This bias denotes the accumulative localization errors of PDR, which is mainly caused by three reasons: step detection errors, stride length estimation errors and heading calculation errors. By contrast, in the V-PDR test, the results of PDR are successfully corrected by VPR at the turns, which makes the estimated pedestrian trajectory closer to the ground truth. The improvement of localization performance can be proved more intuitively from the estimated end positions of PDR and V-PDR.

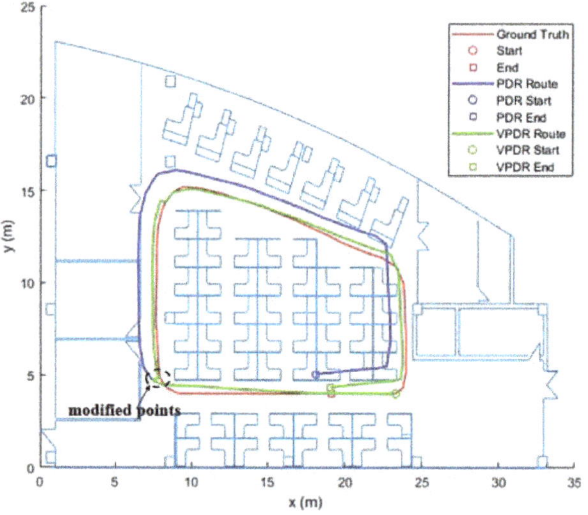

Figure 12. Estimated trajectories of PDR and V-PDR.

The end point of PDR is 1.43 m away from the actual end point, while the distance in V-PDR method is 0.35 m. The statistics of calculated errors of PDR and V-PDR are listed in Table 2. Due to the small scale of the neural network, it is fast to generate a vector and to find the similarity through matrix operations.

Table 2. Localization errors of PDR and V-PDR on the third floor.

Localization Method	Maximum Error (m)	Mean Error (m)	95% Situation Error (m)
PDR	1.74	1.04	1.66
V-PDR	0.91	0.44	0.81

From the mean error column, we know that a 51.2% improvement in localization accuracy was achieved by V-PDR compared with PDR. Through this experiment, it can be seen that our method is effective in reducing cumulative errors of PDR and improving localization accuracy.

The second experiment was carried out on the first floor to the third floor. The experimental setting was similar to the first experiment. The difference was that there were up and down movements in the elevator in the second experiment. The other settings were the same.

The results of the cross-floor experiment are shown in Figure 13. Again, the blue curve represents the trajectory of PDR, the green curve represents the trajectory of V-PDR and the red curve is the ground truth. The pedestrian walks along the ground truth shown in Figure 9 and goes upstairs from the first floor to the third floor. Our method makes some corrections at the stairwell exit and around the corner on the first and third floors. The results show that, after walking upstairs, the trajectory of PDR is still on the first floor, while the trajectory of V-PDR is on the correct (third) floor. The end point of PDR is 42.64 m away from the actual end point, while the distance in V-PDR system is merely 0.79 m. These comparisons demonstrate that our method outperforms PDR in a three-dimensional space. More detailed error information is shown in Table 3.

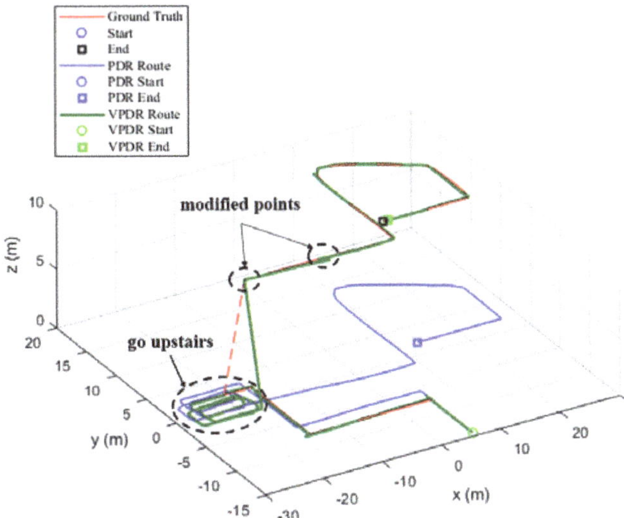

Figure 13. Estimated trajectories of PDR and V-PDR in cross-floor experiment.

Table 3. Localization Errors of PDR and V-PDR in Cross-floor Experiment.

Localization Method	Maximum Error (m)	Mean Error (m)	95% Situation Error (m)
PDR	14.96	4.29	10.30
V-PDR	1.68	0.46	0.90

The localization errors listed in Table 3 were calculated by projecting the trajectories into the horizontal plane. Comparing the average localization errors, V-PDR gains an 89.2% improvement in localization accuracy under the test containing stairs.

For the cost of the setup period, due to the accuracy of the VPR, it can be successfully located within three-dimension VPR. In most special cases, the local identification cannot be accurately performed, and the location may need to be changed to successfully locate. In this case, it may take a long time for the setup. However, the cost is acceptable for accurate positioning. A simple PDR algorithm can only obtain the relative position between two positions but not the exact position. With the assistance of VPR, the exact position can be obtained, which improves the positioning of the PDR. In the future, more improvements can be added to the setup, such as the use of panoramic images for positioning. This can obtain richer visual information and more accurate positioning.

Although our system uses a mobile phone to take pictures, which is complicated for personal use, this algorithm is to explore the fusion of PDR and vision. After wearable cameras and other devices are put into use, or used on a humanoid robot, users will no longer need to take pictures, which makes a lot of sense in actual use.

It can be indicated from above experiments that our indoor localization method successfully improves the localization accuracy by reducing accumulative errors of PDR and makes good performance in the cross-floor test.

There are many advantages of the proposed V-PDR compared with the standalone VPR system. Since VPDR can obtain a continuous walking path, VPR can only obtain discrete positioning points. The two method cannot be directly compared by experiments. However, we can analyze it from the experimental results of V-PDR. First, V-PDR can obtain continuous paths, and the amount of information obtained is greater than VPR. In addition, PDR has a correction effect on the inaccurate value of VPR, and V-PDR positioning is more accurate. Finally, V-PDR does not need to call the VPR network in real time, where it takes less calculation and time than VPR.

6. Conclusions

In this paper, we propose a novel indoor localization method to improve the localization accuracy and solve the problem that PDR cannot work properly in three-dimensional space by combining vision and inertial sensors. It was proved by experiments that our V-PDR method successfully reduces accumulated errors of PDR and improves the localization accuracy. In addition, it extends application scene of PDR to a three-dimensional space, which can still provide accurate localization service when pedestrians are moving across floors.

Author Contributions: J.Q. conceived and designed the experiments. J.Q. and Y.C. performed the experiments and analyzed the data. R.Y. contributed the experimental environment and analysis tools. P.L. provided critical comments to the paper. J.Q. wrote the paper. All authors have read and agreed to the published version of the manuscript.

Funding: This work was supported in part by the National Natural Science Foundation of China under Grants 61903246.

Conflicts of Interest: Jiuchao Qian mainly focuses on indoor localization technologies, video analysis, signal processing, deep learning, ubiquitous computing, and artificial intelligence.

Abbreviations

The following abbreviations are used in this manuscript:

PDR	Pedestrian Dead Reckoning
VPR	Visual Place Recognition
GNSS	Global Navigation Satellite System
RFID	Radio Frequency Identification
IR	Infrared Ray
WLAN	Wireless Local Area Network

MEMS Micro Electro Mechanical System
CNN Convolutional Neural Network

References

1. Liu, P.; Liu, D.; Qian, J. A Survey of Indoor Positioning Technology and Application. *Navig. Position. Timing* **2017**, *4*. [CrossRef]
2. Fallah, N.; Apostolopoulos, I.; Bekris, K. Indoor Human Navigation Systems: A Survey. *Interact. Comput.* **2013**, *25*, 21–33.
3. Kang, W.; Han, Y. Smartpdr: Smartphone-based pedestrian dead reckoning for indoor localization. *IEEE Sens. J.* **2015**, *5*, 2906–2916. [CrossRef]
4. Hsu, L.T.; Gu, Y.; Huang, Y.; Kamijo, S. Urban pedestrian navigation using smartphone-based dead reckoning and 3-d map-aided gnss. *Sens. J. IEEE* **2016**, *5*, 1281–1293. [CrossRef]
5. Bidder, O.R.; Walker, J.; Jones, M.W.; Holton, M.D.; Urge, P.; Scantlebury, D.; Wilson, R.P. Step by step: Reconstruction of terrestrial animal movement paths by dead-reckoning. *Mov. Ecol.* **2015**, *3*, 23–23. [CrossRef] [PubMed]
6. Thong, H.; Filippo, S.; Vinh, T. A Hybrid Algorithm Based on WiFi for Robust and Effective Indoor Positioning. In Proceedings of the 19th International Symposium on Communications and Information Technologies (ISCIT), Ho Chi Minh City, Vietnam, 25–27 September 2019; pp. 416–421.
7. Wu, J.; Zhu, M.; Xiao, B.; Qiu, Y. The Improved Fingerprint-Based Indoor Localization with RFID/PDR/MM Technologies. In Proceedings of the 2018 IEEE 24th International Conference on Parallel and Distributed Systems (ICPADS), Singapore, 11–13 December 2018; pp. 878–885. [CrossRef]
8. Katsuhiko, K.; Keisuke, I.; Tsubasa, T. Step Recognition Method Using Air Pressure Sensor. In Proceedings of the International Conference on Indoor Positioning and Indoor Navigation (IPIN), Pisa, Italy, 30 September–3 October 2019; pp. 1–8.
9. Wang, G.; Wang, X.; Nie, J.; Lin, L. Magnetic-Based Indoor Localization Using Smartphone via a Fusion Algorithm. *IEEE Sens. J.* **2019**, *19*, 6477–6485. [CrossRef]
10. Lowe, D.G. Object Recognition from Local Scale-Invariant Features. In Proceedings of the Seventh IEEE International Conference on Computer Vision, Kerkyra, Greece, 20–27 September 1999; Volume 2, pp. 1150–1157.
11. Bay, H. Surf: Speeded up robust features. In *European Conference on Computer Vision*; Springer: Berlin/Heidelberg, Germany, 2006; Volume 110, pp. 404–417.
12. Agrawal, M.; Konolige, K.; Blas, M.R. CenSurE: Center Surround Extremas for Realtime Feature Detection and Matching. In *European Conference on Computer Vision*; Proceedings, Part IV. DBLP; Springer: Berlin/Heidelberg, Germany, 2008.
13. Oliva, A. Building the gist of a scene: The role of global image features in recognition. *Prog. Brain Res.* **2006**, *155*, 23–36. [PubMed]
14. Wen, F.; Ying, R.; Gong, Z.; Liu, P. Efficient Algorithms for Maximum Consensus Robust Fitting. *IEEE Trans. Robot.* **2020**, *36*, 92–106. [CrossRef]
15. Lowry, S.; Sunderhauf, N.; Newman, P.; Leonard, J.J.; Milford, M.J. Visual place recognition: A survey. *IEEE Trans. Robot.* **2016**, *32*, 1–19. [CrossRef]
16. Zhou, B.; Lapedriza, A.; Xiao, J.; Torralba, A.; Oliva, A. Learning deep features for scene recognition using places database. *Adv. Neural Inf. Process. Syst.* **2014**, *1*, 487–495.
17. Xiao, J.; Hays, J.; Ehinger, K.A.; Oliva, A.; Torralba, A. SUN database: Large-scale scene recognition from abbey to zoo. In Proceedings of the Twenty-Third IEEE Conference on Computer Vision and Pattern Recognition, CVPR 2010, San Francisco, CA, USA, 13–18 June 2010; pp. 3485–3492. [CrossRef]
18. Razavian, A.S.; Azizpour, H.; Sullivan, J.; Carlsson, S. CNN Features Off-the-Shelf: An Astounding Baseline for Recognition. In Proceedings of the 2014 IEEE Conference on Computer Vision and Pattern Recognition Workshops, Columbus, OH, USA, 23–28 June 2014; pp. 512–519.
19. Li, D.; Lu, Y.; Xu, J.; Ma, Q.; Liu, Z. iPAC: Integrate Pedestrian Dead Reckoning and Computer Vision for Indoor Localization and Tracking. *IEEE Access* **2019**, 183514–183523. [CrossRef]

20. Roggen, D.; Jenny, R.; Hamette, P.D.L.; Tröster, G. Mapping by Seeing—Wearable Vision-Based Dead-Reckoning, and Closing the Loop. In *Smart Sensing and Context, Second European Conference, Eurossc, Kendal, England, October*; Springer: Berlin/Heidelberg, Germany, 2007.
21. Fu, W.; Peng, A.; Tang, B.; Zheng, L. Inertial sensor aided visual indoor positioning. In Proceedings of the International Conference on Electronics Technology (ICET), Chengdu, China, 23–27 May 2018; pp. 106–110.
22. Yan, J.; He, G.; Basiri, A.; Hancock, C. 3D Passive-Vision-Aided Pedestrian Dead Reckoning for Indoor Positioning. *IEEE Trans. Instrum. Meas.* **2020**, 1370–1386. [CrossRef]
23. Yan, Z.; Xianwei, Z.; Ruizhi, C.; Hanjiang, X.; Sheng, G. Image-based localization aided indoor pedestrian trajectory estimation using smartphones. *Sensors* **2018**, *18*, 258.
24. Fusco, G.; Coughlan, J.M. Indoor localization using computer vision and visual-inertial odometry. In *International Conference on Computers Helping People with Special Needs*; Springer: Cham, Switzerland, 2018.
25. Chen, Z.; Lam, O.; Jacobson, A.; Milford, M. Convolutional Neural Network-based Place Recognition. *arXiv* **2014**, arXiv:1411.1509.
26. Krizhevsky, A.; Sutskever, I.; Hinton, G.E. ImageNet Classification with Deep Convolutional Neural Networks. neural information processing systems. In Proceedings of the Advances in Neural Information Processing Systems 25, Lake Tahoe, NV, USA, 3–6 December 2012; pp. 1106–1114.
27. Arandjelovic, R.; Gronat, P.; Torii, A.; Pajdla, T.; Sivic, J. NetVLAD: CNN Architecture for Weakly Supervised Place Recognition. Computer vision and pattern recognition. *IEEE Trans. Pattern Anal. Mach. Intell.* **2018**, *40*, 1437–1451. [CrossRef] [PubMed]
28. Chen, Z.; Jacobson, A.; Sunderhauf, N.; Upcroft, B.; Liu, L.; Shen, C. Deep learning features at scale for visual place recognition. In Proceedings of the 2017 IEEE International Conference on Robotics and Automation (ICRA), Singapore, 29 May–3 June 2017; pp. 3223–3230.
29. Harle, R. A Survey of Indoor Inertial Positioning Systems for Pedestrians. *IEEE Commun. Surv. Tutor.* **2013**, *15*, 1281–1293. [CrossRef]
30. Beauregard, S. A Helmet-Mounted Pedestrian Dead Reckoning System. In Proceedings of the 3rd International Forum on Applied Wearable Computing 2006, Bremen, Germany, 15–16 March 2006; pp. 1–11.
31. Tom, J. A Personal Dead Reckoning Module. In Proceedings of the 10th International Technical Meeting of the Satellite Division of The Institute of Navigation (ION GPS 1997), Kansas City, MO, USA, 16–19 September 1997; pp. 47–51
32. Ying, H.; Silex, C.; Schnitzer, A.; Leonhardt, S.; Schiek, M. Automatic Step Detection in the Accelerometer Signal. In *Wearable and Implantable Body Sensor Networks*; Springer: Berlin/Heidelberg, Germany, 2007.
33. Nozawa, M.; Hagiwara, Y.; Choi, Y. Indoor human navigation system on smartphones using view-based navigation. In Proceedings of the 2012 12th International Conference on Control, Automation and Systems, JeJu Island, Korea, 17–21 October 2012; pp. 1916–1919.
34. Robert, R.; Thomas, J. Dead Reckoning Navigational System Using Accelerometer to Measure Foot Impacts. U.S. Patent 5,583,776, 10 December 1996.
35. Qian, J.; Pei, L.; Zou, D.; Liu, P. Optical Flow-Based Gait Modeling Algorithm for Pedestrian Navigation Using Smartphone Sensors. *IEEE Sens. J.* **2015**, *12*, 6797–6804. [CrossRef]
36. Kim, J.W.; Jang, H.J.; Hwang, D.; Park, C. A Step, Stride and Heading Determination for the Pedestrian Navigation System. *J. Glob. Position. Syst.* **2004**, *3*, 273–279. [CrossRef]
37. Qian, J.; Pei, L.; Ma, J.; Ying, R.; Liu, P. Vector graph assisted pedestrian dead reckoning using an unconstrained smartphone. *Sensors* **2015**, *3*, 5032–5057. [CrossRef] [PubMed]
38. Wang, T.; Huang, H.; Lin, J.; Hu, C.; Zeng, K.; Sun, M. Omnidirectional CNN for Visual Place Recognition and Navigation. In Proceedings of the 2018 IEEE International Conference on Robotics and Automation (ICRA), Brisbane, Australia, 21–25 May 2018; pp. 2341–2348. [CrossRef]
39. Quattoni, A.; Torralba, A. Recognizing indoor scenes. computer vision and pattern recognition. In Proceedings of the IEEE Conference on Computer Vision and Pattern Recognition, Miami, FL, USA, 20–25 June 2009; pp. 413–420.

Article

A Closed-Form Localization Algorithm and GDOP Analysis for Multiple TDOAs and Single TOA Based Hybrid Positioning

Zhongliang Deng, Hanhua Wang *, Xinyu Zheng, Xiao Fu, Lu Yin, Shihao Tang and Fuxing Yang

School of Electronic Engineering, Beijing University of Posts and Telecommunications, Beijing 100876, China; dengzhl@bupt.edu.cn (Z.D.); buptzxy@bupt.edu.cn (X.Z.); xiaofu@bupt.edu.cn (X.F.); inlu_mail@bupt.edu.cn (L.Y.); tsh_5529@foxmail.com (S.T.); yangfx@bupt.edu.cn (F.Y.)

* Correspondence: whh0710@bupt.edu.cn; Tel.: +86-010-6119-8509

Received: 20 September 2019; Accepted: 13 November 2019; Published: 16 November 2019

Featured Application: User terminal positioning based on the 5G system or other ground-based wireless systems.

Abstract: Cellular communication systems support mobile phones positioning function for Enhanced-911 (E-911) location requirements, but the positioning accuracy is poor. The fifth-generation (5G) cellular communication system can use indoor distribution systems to provide accurate multiple time-difference-of-arrival (TDOA) and single time-of-arrival (TOA) measurements, which could significantly improve the indoor positioning ability. Unlike iterative localization algorithms for TDOA or TOA, the existing closed-form algorithms, such as the Chan-Ho algorithm, do not have convergence problems, but can only estimate position based on one kind of measurement. This paper proposes a closed-form localization algorithm for multiple TDOAs and single TOA measurements. The proposed algorithm estimates the final position result using three-step weighted least squares (WLSs). The first WLS provides an initial position for the last two steps. Then the algorithm uses two WLSs to estimate position based on heteroscedastic TDOA and TOA measurements. In addition, the geometric dilution of precision (GDOP) of the proposed hybrid TDOA and TOA positioning has been derived. The analysis of GDOP shows that the proposed hybrid positioning has lower GDOP than TDOA-only positioning, which means the proposed hybrid positioning has a higher accuracy limitation than TDOA-only positioning. The simulation shows that the proposed localization algorithm could have better performance than closed-form TDOA-only positioning methods, and the positioning accuracy could approximate Cramer-Rao lower bound (CRLB) when the TDOA measurement errors are small.

Keywords: indoor positioning; 5G system; hybrid positioning; geometric dilution of precision; closed-form solution; Cramer-Rao lower bound

---

1. Introduction

Indoor positioning is important for location-based service, internet of things and health care [1,2]. Many wireless systems, such as Ultra-Wideband, Bluetooth and cellular communication systems, have been used for indoor positioning. As one of the most widely used wireless systems, the cellular communication system supports mobile phone positioning from 2G to 5G [3]. In recent years, with the development of the location service market, the cellular communication system based positioning can be used in many new fields such as vehicle positioning [4,5], location-aware communications [6] and location-based applications [7]. The 3rd Generation Partnership Project (3GPP) also puts higher requirements on the positioning accuracy of the 5G system [8,9].

The TDOA-only positioning technology based on cellular communication system is relatively mature. For a long time, the cellular communication system mainly realizes the user equipment (UE) positioning through the TDOA positioning method. Long Term Evolution (LTE) system provides a TDOA estimation technology called Observed Time Difference of Arrival (OTDOA) [10,11]. In general, UE can only receive downlink signals from the nearest base station. Using OTDOA technology, base stations broadcast positioning reference signal (PRS) to UE. PRS has a unique mechanism named power boosting to make sure that the UE could receive PRSs from multiple sources so that multiple time-difference-of-arrival (TDOA) measurements can be obtained [12]. UE estimates the TDOA measurements by finding the peak of correlation of the received PRS with a shifted and conjugated version of PRS [13]. When the signal sources' locations are known, a nonlinear equation can be constructed using sources' locations and TDOA measurements to calculate UE's position. There are many existing localization algorithms for the solution of the TDOA measurement equations [14]. These algorithms can be divided into iterative methods and closed-form solutions [15]. The iterative methods usually use the Taylor-series to linearize the nonlinear equations and use Gauss-Newton method to iterate location estimation errors. Those methods can provide accurate position estimation when the initial position is appropriate, otherwise, it cannot guarantee convergence [16]. Closed-form solutions could avoid the convergence problem. The most used closed-form algorithm is a two-step Weighted Least squares (WLS) estimators developed by Chan and Ho [17]. This solution can reach the Cramer-Rao lower bound (CRLB) when TDOA error is small enough. However, as the error increases, its performance would degrade. Ho also developed an improved version of the algorithm [18]. The improved algorithm has a better performance when the variance of the TDOA measurement errors is known. The constraint equation could improve the accuracy of the closed-form algorithm. An algorithm using earth constraint can improve accuracy [19], but this constraint is not suitable for indoor positioning. The above algorithms are often used for UE position estimation after obtaining TDOA measurements through OTDOA technology.

There are still many problems in the exiting system that make the positioning accuracy not high enough. Since the outdoor base station cannot meet the indoor signal coverage requirements, cellular communication system uses the indoor distribution system for indoor communication services [20]. For the indoor distribution system, all TDOA-only positioning methods have the same shortcoming. In the indoor environment, deployment of communication indoor distribution system leads to a higher geometric dilution of precision (GDOP), which is not suitable for TDOA-only based positioning. The GDOP is defined as the ratio of the accuracy limitation of a position fix to the accuracy of measurements and the concept is widely used in positioning performance analysis [21]. By calculating the GDOP, it can be found that the TDOA-only based positioning has a higher GDOP value when the receiver is not surrounded by signal sources [21,22]. However, in order to achieve better signal coverage and lower deployment costs, the indoor distribution antenna (ANT) of the mobile communication system is typically placed beneath the ceiling of the central area of the room. ANTs are never deployed in the indoor areas, which are near the outer wall. In those areas, UE can only receive signals from one side, resulting in a high GDOP value and low positioning accuracy.

The 5G system has the potential to solve this problem. The 5G system would also provide TDOA measurement based on OTDOA technology [23,24], same as the LTE system. Different from the LTE system, the 5G system could provide high-precision TOA measurement over a large bandwidth and continuous uplink and downlink signals [25–27]. In both LTE and 5G, due to network optimization, the UE can only transmit and receive signals with the nearest signal source. This means that UE could only estimate one signal's TOA measurement from the nearest source. TOA-only localization algorithms require at least three TOAs to perform two-dimensional position estimation. Therefore, the 5G system cannot support TOA-only based positioning. However, the 5G system could support new hybrid positioning based on multiple TDOA measurements and single TOA.

The GDOP of the TOA-only positioning is smaller than the TDOA-only positioning when the receiver is not surrounded by signal sources [28]. Therefore, the hybrid positioning based on multiple

TDOAs and single TOA will probably achieve better positioning performance than the TDOA-only positioning. TDOA-only closed-form solutions, such as Chan-Ho, are based on the assumption that measurement errors match Gauss-Markov theory. However, the random errors of TDOA and TOA measurements are not homoscedastic. It means those algorithms cannot be used for the hybrid TDOA and TOA positioning directly.

This paper aims to improve the positioning accuracy of the indoor distribution system by using multiple TDOAs and single TOA based hybrid positioning. In this paper, we establish a three-step WLS closed-form localization algorithm for heteroscedastic TDOA and TOA measurements. The first WLS provides an initial position for the last two steps. Then the algorithm uses two-step weighted least square (WLS) to estimate the final position result. We also derived the GDOP of the proposed hybrid positioning. We performed simulations in a typical indoor hall using an indoor distribution system. As shown in the simulation result, the GDOP of the hybrid positioning is lower than the GDOP of TDOA-only positioning. The simulations also show that the proposed method could approximate the CRLB when TDOA measurements errors are small enough and have the potential to achieve a better performance than two widely used closed-form TDOA-only positioning methods.

This paper is organized as follows: Section 2 describes the system model of hybrid TDOA and TOA positioning. In Section 3, a new closed-form localization algorithm is developed for the nonlinear and heteroscedastic hybrid TDOA and TOA positioning equations. Section 4 analyzes the GDOP of the hybrid positioning based on multiple TDOAs and single TOA. Simulation results and analysis are presented in Section 5. Finally, we draw our conclusions in Section 6.

2. System Model

The ANTs are deployed at the ceiling with same height, high vertical dilution of precision makes the system cannot provide reliable vertical positioning results [29], which is usually obtained by other sensors [30].

Therefore, for the system model, we assumed that ANTs are deployed at the ceiling and the height of UE is already known by other sensors, and only two-dimensional horizontal positioning is considered. The system model of the indoor distribution system based UE positioning is shown in Figure 1. We consider a UE position estimation problem using an indoor distribution system containing M ANTs. For the system shown in Figure 1, M is equal to 4. For each positioning process, in order to simplify mathematical representations, ANTs are sorted by the distance to the UE. The nearest ANT is numbered 1, and the farthest ANT is numbered M. UE could obtain $M - 1$ TDOA measurements and one TOA measurement from the nearest ANT. In Figure 1, the nearest ANT to UE is the ANT 1 and the farthest is the ANT 4.

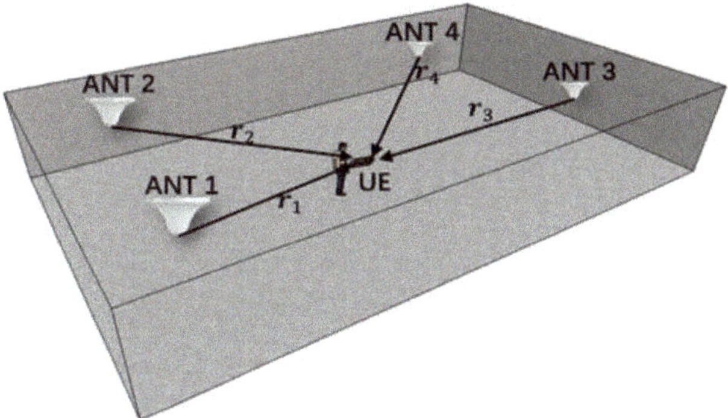

Figure 1. The system model of user equipment (UE) indoor positioning based on the indoor distribution system.

In the following description, the position of UE is denoted by $u = (x, y)^T$. The system contains M ANTs locating at $a_i = (x_i, y_i)^T$, $i = 1, 2, 3, \ldots, M$.

The distance from ANT i to UE is expressed as

$$r_i = \| a_i - u \| = \sqrt{(x_i - x)^2 + (y_i - y)^2}, \ i = 1, 2, 3, \ldots, M. \tag{1}$$

The distance difference from ANT 1 to UE and ANT i to UE is expressed as:

$$r_{i,1} = [\| a_1 - u \| - \| a_i - u \|] = r_1 - r_i. \tag{2}$$

Considering the TDOA ranging error $e_{i,1}$, TDOA measurement from ANT 1 and ANT i is represented as:

$$ct_{i,1} = d_{i,1} = r_{i,1} + e_{i,1}, \ i = 2, 3, 4, \ldots, M \tag{3}$$

where $t_{i,1}$ is TDOA measurement and c is the speed of light, $d_{i,1}$ is the distance difference calculated by TDOA measurement, $r_{i,1}$ is the real distance difference and $e_{i,1}$ is the ranging error caused by TDOA measurement error.

TOA measurement of ANT 1 to UE is expressed as:

$$ct_1 = d_1 = r_1 + e_1 \tag{4}$$

where t_1 is TOA measurement, d_1 is the distance calculated by TOA measurement, r_1 is the real distance from ANT 1 to UE and e_1 is the ranging error caused by TOA measurement error.

For brevity, we denote e_1 as the TDOA ranging error and e_i for $i = 2, 3, \ldots, M$ as TDOA ranging errors between the signal from ANT i and signal from ANT 1, which is denoted as $e_{i,1}$ in the above formula.

In a system with M ANTs, an equation system containing TDOA and TOA measurements is obtained as:

$$\begin{cases} d_1 = r_1 + e_1 \\ d_{2,1} = r_1 - r_2 + e_2 \\ d_{3,1} = r_1 - r_3 + e_3 \\ \vdots \\ d_{M,1} = r_1 - r_M + e_M \end{cases}. \tag{5}$$

3. Three-step WLS Localization Algorithm for Hybrid Positioning

In this section, a three steps WLS localization algorithm is proposed for positioning based on TDOA and TOA measurements shown in Section 2. The algorithm is an improvement of the two-step WLS TDOA-only localization algorithm, called the Chan-Ho algorithm, in Reference [17]. Those two algorithms use the same linearization method, and the third step WLS of the proposed algorithm is the same as the second step WLS of the Chan-Ho algorithm.

Using the following method, a linear equation about TDOA measurement and x, y coordination could be obtained. Moving the r_1 from the right side of Equation (2) to the left side and then squaring both sides, we obtain:

$$r_i^2 = (r_1 - r_{i,1})^2. \tag{6}$$

Expanding (6) we obtain:

$$x_{i,1}x + y_{i,1}y + r_{i,1}r_1 = \frac{1}{2}\left(K_1 - K_i + r_{i,1}^2\right) \tag{7}$$

where:

$$\begin{array}{ll} x_{i,1} = (x_1 - x_i), & i = 2,3,\ldots,M \\ y_{i,1} = (y_1 - y_i), & i = 2,3,\ldots,M \\ K_i = x_i^2 + y_i^2, & i = 1,2,3,\ldots,M. \end{array} \tag{8}$$

Different to the Chan-Ho algorithm, the proposed algorithm needs to consider the TOA measurement and uses a new solution as follows. Using Equation (8), write Equation (5) in matrix form as:

$$Gw + \varphi = h \tag{9}$$

where:

$$G = \begin{bmatrix} 0 & 0 & 1 \\ x_{2,1} & y_{2,1} & d_{2,1} \\ x_{3,1} & y_{3,1} & d_{3,1} \\ \vdots & \vdots & \vdots \\ x_{M,1} & y_{M,1} & d_{M,1} \end{bmatrix}, \; w = \begin{bmatrix} x \\ y \\ r_1 \end{bmatrix}, \quad h = \begin{bmatrix} d_1 \\ \frac{1}{2}\left[K_1 - K_2 + {d_{2,1}}^2\right] \\ \frac{1}{2}\left[K_1 - K_3 + {d_{3,1}}^2\right] \\ \vdots \\ \frac{1}{2}\left[K_1 - K_M + {d_{M,1}}^2\right] \end{bmatrix}, \; \varphi = \begin{bmatrix} \varphi_1 \\ \varphi_2 \\ \varphi_3 \\ \vdots \\ \varphi_M \end{bmatrix}. \tag{10}$$

The vector φ is the error between h and Gw. It can be obtained by:

$$\varphi = h - Gw \tag{11}$$

In accordance with Equations (3) and (4), the items in vector φ can be written as the following expressions:

$$\varphi_1 = d_1 - r_1 = e_1 \tag{12}$$

$$\begin{aligned} \varphi_i = \tfrac{1}{2}\left[K_1 - K_i + (d_{i,1} - e_i)^2\right] - x_{i,1}x - y_{i,1}y - (d_{i,1} - e_i)r_1 &= \tfrac{1}{2}e_i^2 - d_{i,1}e_i - r_1 e_i \\ &= \tfrac{1}{2}e_i^2 - r_i e_i \approx -r_i e_i\ ,\ i = 2,\ldots,M. \end{aligned} \tag{13}$$

where the assumption $r_i \gg e_i$ has been used. To minimize $\varphi^T\varphi$, the hybrid positioning problem is cast as the following quadratic optimization

$$\min_{w}(h - Gw)^T(h - Gw) \tag{14}$$

Ranging errors of TDOAs can be assumed as an independent, identical and zero-mean Gaussian distribution [31]. The ranging error of TOA can also be assumed to be a zero-mean Gaussian distribution [32]. Because TDOA and TOA measurements are estimated using different techniques, TOA and TDOA ranging errors can be assumed to be independent and have different variances. The parameter k is used to represent the ratio of TOA ranging error's standard deviation to TDOA ranging error's standard deviation. The ranging errors have the following statistical properties:

$$\begin{array}{l} E(e_i) = 0,\ \ i = 1,2,3,\ldots,M \\ Cov\left(e_i e_j\right) = 0,\ \ i \neq j \; and \; i,j = 1,2,3,\ldots,M \\ Var(e_i) = \sigma_{tdoa}^2,\ \ i = 2,3,\ldots,M \\ Var(e_1) = \sigma_{toa}^2 = k^2\sigma_{tdoa}^2,\ \ k = \frac{\sigma_{toa}}{\sigma_{tdoa}}. \end{array} \tag{15}$$

The errors in φ do not meet the Gauss-Markov assumption, ordinary least squares estimator is not the best linear unbiased estimator (BLUE) for this optimization problem. WLS could make it BLUE when each weight is equal to the reciprocal of the variance of the error [33]. The WLS solution of w that minimizes $\varphi^T W \varphi$ is:

$$w = \left(G^T W G\right)^{-1} G^T W h \tag{16}$$

where W is the weighting matrix which could be given by the covariance matrix of φ.

The covariance matrix of φ is:

$$\begin{aligned} W = Cov(\varphi)^{-1} = E\left(\varphi\varphi^T\right)^{-1} &= \begin{bmatrix} \sigma_{toa}^2 & 0 & 0 & \cdots & 0 \\ 0 & {r_2}^2\sigma_{tdoa}^2 & 0 & \cdots & 0 \\ 0 & 0 & {r_3}^2\sigma_{tdoa}^2 & \cdots & 0 \\ \vdots & \vdots & \vdots & \ddots & \vdots \\ 0 & 0 & 0 & \cdots & {r_M}^2\sigma_{tdoa}^2 \end{bmatrix}^{-1} \\ &= \frac{1}{\sigma_{tdoa}^2} \begin{bmatrix} k^2 & 0 & 0 & \cdots & 0 \\ 0 & {r_2}^2 & 0 & \cdots & 0 \\ 0 & 0 & {r_3}^2 & \cdots & 0 \\ \vdots & \vdots & \vdots & \ddots & \vdots \\ 0 & 0 & 0 & \cdots & {r_M}^2 \end{bmatrix}^{-1}. \end{aligned} \tag{17}$$

where $Cov(\varphi)^{-1}$ is the inverse of the covariance matrix of φ, k is obtained by prior probability statistics of the ranging errors after the system deployed.

Scaling of W does not affect the result of WLS [17]. Therefore, the weighting matrix W can be written as:

$$W = \begin{bmatrix} \frac{1}{k^2} & 0 & 0 & \cdots & 0 \\ 0 & \frac{1}{{r_2}^2} & 0 & \cdots & 0 \\ 0 & 0 & \frac{1}{{r_3}^2} & \cdots & 0 \\ \vdots & \vdots & \vdots & \ddots & \vdots \\ 0 & 0 & 0 & \cdots & \frac{1}{{r_M}^2} \end{bmatrix}. \tag{18}$$

However, the distances from ANTs and UE is unknown, to approximated the matrix W, an initial guess of UE position is need to be obtained in first-step WLS.

3.1. First-Step WLS

To calculate the initial position, we assume that the distances from ANTs to UE are equal to the ratio k. Rewriting the weighting matrix W as W':

$$W' = \begin{bmatrix} 1 & 0 & 0 & \cdots & 0 \\ 0 & 1 & 0 & \cdots & 0 \\ 0 & 0 & 1 & \cdots & 0 \\ \vdots & \vdots & \vdots & \ddots & \vdots \\ 0 & 0 & 0 & \cdots & 1 \end{bmatrix}. \tag{19}$$

We could obtain $w' = (x', y', {r_1}')^T$ using Equation (16) as

$$w' = \left(G^T W' G\right)^{-1} G^T W' h. \tag{20}$$

It is an ordinary least squares. When the UE positions are different, the deviation of the weighting matrix and the initial positioning error would change. The second-step WLS could mitigate this error and can be repeated again to give an even better estimate.

3.2. Second-Step WLS

After achieving the initial position $u' = (x', y')$, the distance r_i' from ANT i to u' can be calculated. Inserting u' and r_i' into Equation (18), the weighting matrix W could be rewritten as W'':

$$W'' = \begin{bmatrix} \frac{1}{k^2} & 0 & 0 & \cdots & 0 \\ 0 & \frac{1}{r_2'^2} & 0 & \cdots & 0 \\ 0 & 0 & \frac{1}{r_3'^2} & \cdots & 0 \\ \vdots & \vdots & \vdots & \ddots & \vdots \\ 0 & 0 & 0 & \cdots & \frac{1}{r_M'^2} \end{bmatrix}. \quad (21)$$

Inserting matrix W'' into Equation (16) in the same way as Equation (20), the solution $w'' = (x'', y'', r_1'')^T$ can be obtained.

3.3. Third-Step WLS

The former two steps assume no relationship between the UE location $(x,\ y)^T$ and r_1, but in fact they are related. Chan-Ho algorithm has the same problem in its first stage minimization and use a WLS to refine the result [17]. In the third-step WLS of this proposed algorithm, the same method as the Chan-Ho algorithm is used to refine the result of former two steps.

The former two steps' results location $(x'', y'')^T$ and distance r_1'' have a relationship shown in Equation (22). Therefore, the third-step WLS is to minimize the sum of the squared errors ε in:

$$\begin{cases} \varepsilon_1 = (x_1 - x'')^2 - (x_1 - x)^2 \\ \varepsilon_2 = (y_1 - y'')^2 - (y_1 - y)^2 \\ \varepsilon_3 = r_1''^2 - (x_1 - x)^2 + (y_1 - y)^2 \end{cases}. \quad (22)$$

Equation (22) could be rewritten in matrix form as:

$$\varepsilon = h' - G'\Delta u \quad (23)$$

where:

$$\varepsilon = \begin{bmatrix} \varepsilon_1 \\ \varepsilon_2 \\ \varepsilon_3 \end{bmatrix},\ h' = \begin{bmatrix} (x_1 - x'')^2 \\ (y_1 - y'')^2 \\ r_1''^2 \end{bmatrix} = \begin{bmatrix} (x_1 - x + e_x)^2 \\ \left(y_1 - y + e_y\right)^2 \\ (r_1 + e_r)^2 \end{bmatrix},\ G' = \begin{bmatrix} 1 & 0 \\ 0 & 1 \\ 1 & 1 \end{bmatrix},\ \Delta u = \begin{bmatrix} (x_1 - x)^2 \\ (y_1 - y)^2 \end{bmatrix}. \quad (24)$$

Because elements in vector ε have different variances, we need to use WLS to estimate the solution Δu as:

$$\Delta u = \left(G'^T W''' G'\right)^{-1} G'^T W''' h' \quad (25)$$

where weighting matrix W''' could be obtained by calculating the covariance matrix of ε. Assuming that the error of the solution w'' in second-step WLS are $e = \left(e_x, e_y, e_r\right)^T$, vector ε could be obtained as:

$$\begin{aligned} \varepsilon &= \begin{bmatrix} \varepsilon_1 \\ \varepsilon_2 \\ \varepsilon_3 \end{bmatrix} = \begin{bmatrix} (x_1 - x'')^2 \\ (y_1 - y'')^2 \\ r_1''^2 \end{bmatrix} - \begin{bmatrix} (x_1 - x)^2 \\ (y_1 - y)^2 \\ r_1^2 \end{bmatrix} = \begin{bmatrix} (x_1 - x + e_x)^2 \\ \left(y_1 - y + e_y\right)^2 \\ (r_1 + e_r)^2 \end{bmatrix} - \begin{bmatrix} (x_1 - x)^2 \\ (y_1 - y)^2 \\ r_1^2 \end{bmatrix} \\ &= \begin{bmatrix} 2(x_1 - x)e_x + e_x^2 \\ 2(y_1 - y)e_y + e_y^2 \\ 2r_1 e_r + e_r^2 \end{bmatrix} \approx \begin{bmatrix} 2(x_1 - x)e_x \\ 2(y_1 - y)e_y \\ 2r_1 e_r \end{bmatrix}. \end{aligned} \quad (26)$$

where the second-order error terms are ignored.

$$
\begin{aligned}
W'''^{-1} &= Cov(\varepsilon) = E\left(\varepsilon\varepsilon^T\right) = 4\begin{bmatrix} x_1 - x & 0 & 0 \\ 0 & y_1 - y & 0 \\ 0 & 0 & r_1 \end{bmatrix} Cov(e) \begin{bmatrix} x_1 - x & 0 & 0 \\ 0 & y_1 - y & 0 \\ 0 & 0 & r_1 \end{bmatrix} \\
&= 4BCov(e)B.
\end{aligned} \tag{27}
$$

The matrix B could be approximated by using (x'', y'', r_1'') as (x, y, r_1) and $Cov(e)$ could be approximated as $\left(G^T W' G\right)^{-1}$ [17].

Therefore, result Δu could be calculated by

$$
\Delta u = \left(G'^T \frac{1}{4} B^{-1} G^T W' G B^{-1} G'\right)^{-1} G'^T \frac{1}{4} B^{-1} G^T W' G B^{-1} h'. \tag{28}
$$

The Δu is the square of the difference between final positioning result u and ANT 1 on the x-axis and the y-axis. As with the Chan-Ho algorithm, u is given by the square root of Δu and ANT 1's location $(x_1,\ y_1)^T$.

4. GDOP of the Proposed Hybrid Positioning

GDOP is the ratio of the accuracy limitation of a position fix to the accuracy of measurements [21]. A positioning system has a small GDOP means that the positioning is accurate. In order to compare with GDOP of TDOA-only positioning, the GDOP of the proposed hybrid positioning is defined as the ratio of the positioning accuracy to the TDOA ranging accuracy.

In this section, the GDOP of the hybrid positioning based on multiple TDOAs and single TOA is derived.

To achieve the GDOP of the hybrid TDOA and TOA positioning of the system described in Section 2, the difference between TDOA measurements and real distances difference when UE is at the location (x, y) can be expressed as:

$$
f_i(x, y) = d_{i,1} - r_1 + r_i,\ i = 2, 3, \ldots, M. \tag{29}
$$

Difference between TOA measurements and real distances when UE is at the location (x, y) is expressed as

$$
f_1(x, y) = d_1 - r_1. \tag{30}
$$

Using a first order Taylor series to approximate Equations (6) and (7), the approximation of $f_i(x, y)$ at location $\left(x_p, y_p\right)$ can be obtained as

$$
f_i(x, y) \approx f_i\left(x_p, y_p\right) + \left(x - x_p\right)\frac{\partial f_i\left(x_p, y_p\right)}{\partial x} + \left(y - y_p\right)\frac{\partial f_i\left(x_p, y_p\right)}{\partial y},\ i = 1, 2, 3, \ldots, M. \tag{31}
$$

When there are no ranging errors, (9) can be obtained from (8).

$$
0 = f_i\left(x_p, y_p\right) + \left(x^0 - x_p\right)\frac{\partial f_i\left(x_p, y_p\right)}{\partial x} + \left(y^0 - y_p\right)\frac{\partial f_i\left(x_p, y_p\right)}{\partial y} \tag{32}
$$

where $\left(x^0, y^0\right)$ is UE's real location.

To simplify the mathematical derivations, we use e_1 to represent the TOA ranging error. Moreover, $e_i, i = 2, \ldots, M$, is used to represent TDOA ranging errors between the signal from ANT i and signal from ANT 0, which is represented as $e_{i,1}$ in Section 2. As the measurements have ranging errors e_i, Equation (33) can be obtained from (31):

$$
e_i = f_i\left(x_p, y_p\right) + \left(x\prime - x_p\right)\frac{\partial f_i\left(x_p, y_p\right)}{\partial x} + \left(y\prime - y_p\right)\frac{\partial f_i\left(x_p, y_p\right)}{\partial y} \tag{33}
$$

where $(x\prime, y\prime)$ is the position estimation result.

Let position estimation error vector $\Delta u = \left(e_x, e_y\right)^T$, $e_x = x\prime - x^0$, $e_y = y\prime - y^0$. Subtracting Equation (33) from Equation (32), the equation of e_i and $\left(e_x, e_y\right)^T$ can be obtained as:

$$e_i = \frac{\partial f_i\left(x_p, y_p\right)}{\partial x} e_x + \frac{\partial f_i\left(x_p, y_p\right)}{\partial y} e_y. \tag{34}$$

For TOA ranging error e_0, it is easy to obtain partial differential equations:

$$\alpha_1 = \frac{\partial f_1\left(x_p, y_p\right)}{\partial x} = \frac{x_1 - x_p}{r_1} \beta_1 = \frac{\partial f_1\left(x_p, y_p\right)}{\partial y} = \frac{y_1 - y_p}{r_1}. \tag{35}$$

For TDOA ranging error e_i, partial differential equations is:

$$\begin{gathered} \alpha_i = \frac{\partial f_i\left(x_p, y_p\right)}{\partial x} = \frac{x_i - x_p}{r_i} - \frac{x_1 - x_p}{r_1} \\ \beta_i = \frac{\partial f_i\left(x_p, y_p\right)}{\partial y} = \frac{y_i - y_p}{r_i} - \frac{y_1 - y_p}{r_1} \\ i = 2, 3, \ldots, M. \end{gathered} \tag{36}$$

Therefore, Equation (34) can be formulated as the following matrix form:

$$A\Delta u = e \tag{37}$$

where:

$$A = \begin{bmatrix} \alpha_1 & \beta_1 \\ \alpha_2 & \beta_2 \\ \alpha_3 & \beta_3 \\ \vdots & \vdots \\ \alpha_M & \beta_M \end{bmatrix} \Delta u = \begin{bmatrix} e_x \\ e_y \end{bmatrix} e = \begin{bmatrix} e_1 \\ e_2 \\ e_3 \\ \vdots \\ e_M \end{bmatrix}. \tag{38}$$

Different from the GDOP calculation process in [21], TOA and TDOA measurements ranging errors have different variances. Therefore, in order to make an unbiased and efficient estimator, the vector Δu cannot be calculated based on the ordinary least square (OLS) but weighted least square (WLS) as the following expression:

$$\Delta u = \left(A^T W A\right)^{-1} A^T W e. \tag{39}$$

where the weighting matrix W could be given by covariance matrix of error vector e.

Therefore, we could get a covariance matrix of error vector E as:

$$Cov(E) = E\left(\begin{bmatrix} e_1 \\ e_2 \\ e_3 \\ \vdots \\ e_M \end{bmatrix} \begin{bmatrix} e_1 & e_2 & e_3 & \cdots & e_M \end{bmatrix} \right) = \sigma_{tdoa}^2 \begin{bmatrix} k^2 & 0 & 0 & \cdots & 0 \\ 0 & 1 & 0 & \cdots & 0 \\ 0 & 0 & 1 & \cdots & 0 \\ \vdots & \vdots & \vdots & \ddots & \vdots \\ 0 & 0 & 0 & \cdots & 1 \end{bmatrix} = \sigma_{tdoa}^2 \Sigma \tag{40}$$

Therefore, the weighting matrix W in Equation (39) could be given as

$$W = \Sigma^{-1} = \begin{bmatrix} \frac{1}{k^2} & 0 & 0 & \ldots & 0 \\ 0 & 1 & 0 & \ldots & 0 \\ 0 & 0 & 1 & \ldots & 0 \\ \vdots & \vdots & \vdots & \ddots & \vdots \\ 0 & 0 & 0 & \ldots & 1 \end{bmatrix} \tag{41}$$

Expressing matrix $A^T W$ as

$$\begin{aligned} A^T W &= \begin{bmatrix} \alpha_1 & \alpha_2 & \alpha_3 & \ldots & \alpha_M \\ \beta_1 & \beta_2 & \beta_3 & \ldots & \beta_M \end{bmatrix} \begin{bmatrix} \frac{1}{k^2} & 0 & 0 & \ldots & 0 \\ 0 & 1 & 0 & \ldots & 0 \\ 0 & 0 & 1 & \ldots & 0 \\ \vdots & \vdots & \vdots & \ddots & \vdots \\ 0 & 0 & 0 & \ldots & 1 \end{bmatrix} = \begin{bmatrix} \frac{\alpha_1}{k^2} & \alpha_3 & \alpha_3 & \ldots & \alpha_M \\ \frac{\beta_1}{k^2} & \beta_2 & \beta_3 & \ldots & \beta_M \end{bmatrix} \\ &= \begin{bmatrix} a_1 & a_2 & a_3 & \ldots & a_M \\ b_1 & b_2 & b_3 & \ldots & b_M \end{bmatrix} \end{aligned} \tag{42}$$

From Equation (39), e_x can be calculated by:

$$e_x = \left[\left(A^T W A\right)^{-1}\right]_{1,1} \sum_{i=1}^{M} a_i e_i + \left[\left(A^T W A\right)^{-1}\right]_{1,2} \sum_{i=1}^{M} b_i e_i = \frac{s_1 e_1}{k^2} + \sum_{i=2}^{M} s_i e_i \tag{43}$$

where

$$s_i = \left[\left(A^T W A\right)^{-1}\right]_{1,1} a_i + \left[\left(A^T W A\right)^{-1}\right]_{1,2} b_i. \tag{44}$$

The mean of e_x could be calculated by:

$$E(e_x) = E\left(s_1 e_1 + \sum_{i=2}^{M} s_i e_i\right) = E(s_1 e_1) + \sum_{i=2}^{M} E(s_i e_i) = s_1 E(e_1) + \sum_{i=2}^{M} s_i E(e_i) = 0 \tag{45}$$

The variance of e_x could be calculated by:

$$\begin{aligned} \sigma_x^2 = Var(e_x) = \; & E\left(e_x^2\right) - E(e_x)^2 = E\left(e_x^2\right) = E\left(\left(s_1 e_1 + \sum_{i=2}^{M} s_i e_i\right)^2\right) \\ & = E\left((s_1 e_1)^2 + \sum_{i=2}^{M} (s_i e_i)^2\right) = E\left((s_1 e_1)^2\right) + \sum_{i=2}^{M} E\left((s_i e_i)^2\right) \\ & = \sigma_{toa}^2 s_1{}^2 + \sigma_{tdoa}^2 \sum_{i=2}^{M} s_i{}^2 = k^2 \sigma_{tdoa}^2 s_1{}^2 + \sigma_{tdoa}^2 \sum_{i=2}^{M} s_i{}^2. \end{aligned} \tag{46}$$

It is easy to obtain the mean and variance of e_y using the above equations:

$$E\left(e_y\right) = 0,\ \sigma_y^2 = Var\left(e_y\right) = k^2 \sigma_{tdoa}^2 g_1{}^2 + \sigma_{tdoa}^2 \sum_{i=2}^{M} g_i{}^2 \tag{47}$$

where:

$$g_i = \left[\left(A^T W A\right)^{-1}\right]_{2,1} a_i + \left[\left(A^T W A\right)^{-1}\right]_{2,2} b_i \tag{48}$$

GDOP was been definite as the ratio of the accuracy of a position fix to the accuracy of measurements [34]. In order to directly compare the GDOP of the proposed hybrid positioning and the GDOP of TDOA-only positioning, we redefine the GDOP of the hybrid positioning using multiple TDOAs and single TOA as the ratio of the accuracy of a position fix to the accuracy of TDOA measurements. The GDOP could be calculated as:

$$GDOP = \sqrt{\frac{\sigma_x^2 + \sigma_y^2}{\sigma_{tdoa}^2}} = \sqrt{k^2({s_1}^2 + {g_1}^2) + \sum_{i=2}^{M}({s_i}^2 + {g_i}^2)}. \tag{49}$$

5. Simulation Results and Analysis

To analyze the GDOP of multiple TDOAs and single TOA based hybrid positioning and the positioning accuracy of the proposed closed-form localization algorithm, we have built a simulation scenario according to the real scene of the zone 1, underground parking lot, Beijing University of Posts and Telecommunications. The real scene is shown in Figure 2 and the top view of the developed simulation scenario is shown in Figure 3.

Figure 2. The real scene of the underground parking lot. ANT: antenna.

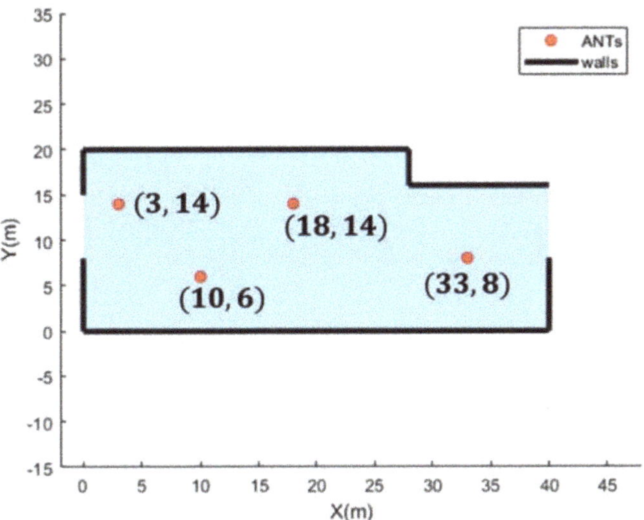

Figure 3. Simulation scenario. Dark area is the simulation area.

In Figure 3, a Cartesian coordinate system is constructed in meters. The ANTs of the indoor distribution system are referred to the real antennas' location in the underground parking lot.

In the following parts of this section, we compared the GDOP of the proposed hybrid positioning and the GDOP of TDOA-only positioning. Then, we analyzed the positioning accuracy of the proposed localization algorithm.

5.1. Analysis and Comparison of GDOP

To compare the GDOP of the TDOA-only positioning and the GDOP of the hybrid positioning using multiple TDOAs and single TOA in the same indoor scenario. 2889 points is selected every 0.5 m, as shown in Figure 4. We divide the selected points into two groups. One group contains the points within the quadrilateral surrounded by connections between ANTs, which is the green quadrilateral in Figure 4. The other group contains the points outside the quadrilateral. 2292 points are outside the green quadrilateral and 597 points are inside the green quadrilateral.

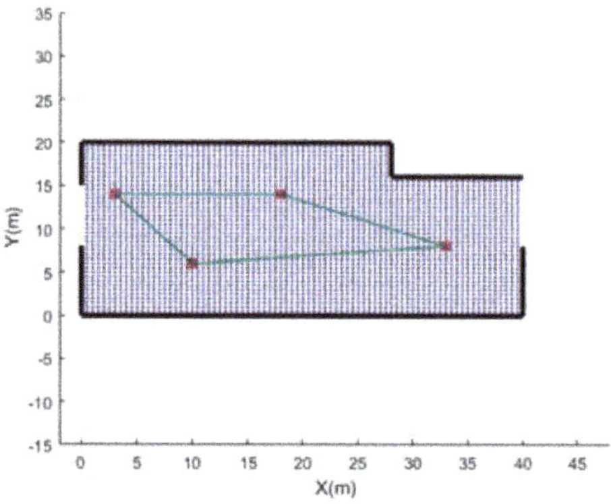

Figure 4. Selected points (blue points) in the developed indoor two-dimensional (2D) simulation scenario. The red dots are ANTs, the green lines are connections of ANTs.

The GDOPs of the hybrid positioning method based on three TDOAs and one TOA with different ratio k values are calculated at each points. The k is the ratio of TOA ranging error standard deviation to TDOA ranging error standard deviation definite in Section 4. We use four different k values and the result is shown in Figure 5.

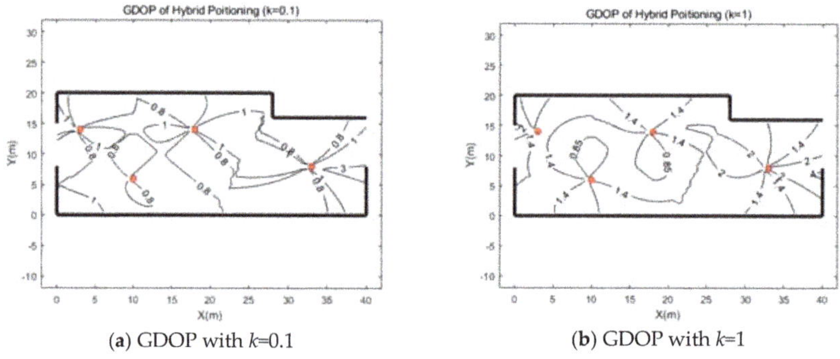

(**a**) GDOP with k=0.1 (**b**) GDOP with k=1

Figure 5. *Cont.*

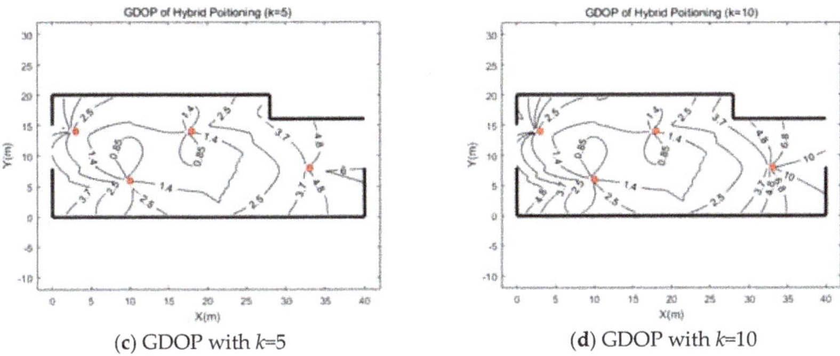

(**c**) GDOP with k=5 (**d**) GDOP with k=10

Figure 5. Geometric dilution of precision (GDOP) of the proposed hybrid time-difference-of-arrival (TDOA) and time-of-arrival (TOA) positioning in the developed indoor simulation scenario. The red dots are ANTs.

Figure 6 shows the GDOP of TDOA-only positioning in the same scenario. The TDOA-only positioning is based on three TDOA measurements which are the same as the proposed hybrid positioning method used above. The GDOP calculation method for TDOA-only positioning is similar to that GDOP calculation method for hybrid positioning described in Section 5 but does not contain the TOA measurement terms. The calculation method is the same as the method described in the definition of GDOP [21].

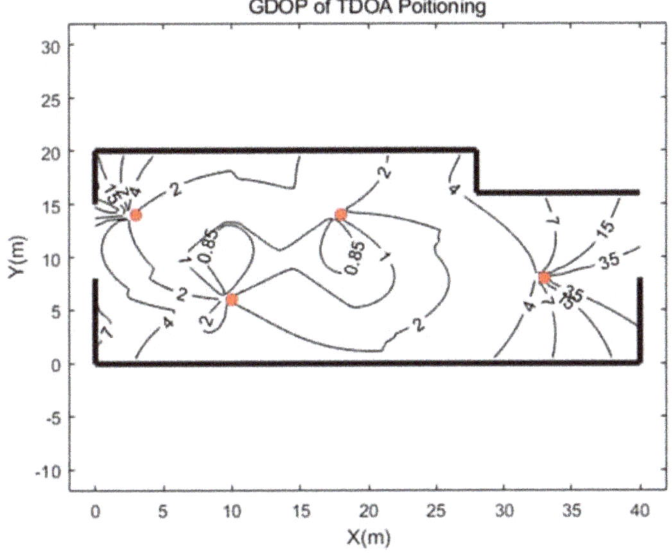

Figure 6. GDOP of TDOA positioning in the developed indoor 2D simulation scenario. The red dots are ANTs.

It can be easily seen from the comparison between Figures 5 and 6 that the GDOP of the TDOA-only positioning is not much different from the hybrid positioning in the center region (the quadrilateral area surrounded by the four ANTs). In other regions, the GDOP of hybrid positioning is significantly smaller than the TDOA-only positioning. However, the area of the center region only accounts for one-fifth of the total area. Assuming that the distribution of UEs is uniform, in most cases, using the hybrid positioning method will increase the positioning accuracy limitation.

The cumulative distribution functions (CDFs) of GDOPs is obtained and shown in Figure 7.

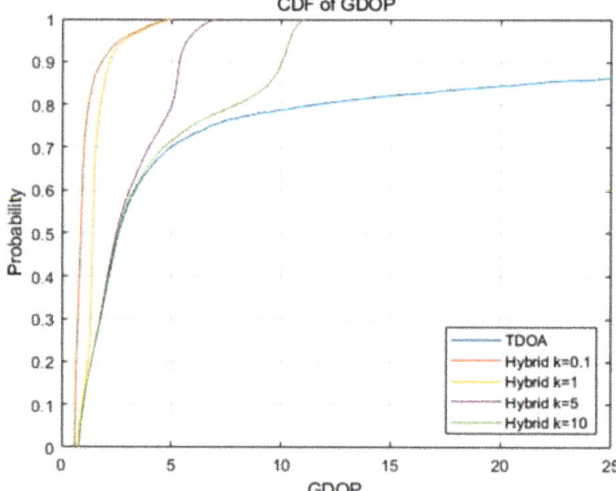

Figure 7. The cumulative distribution function (CDF) of different positioning methods GDOPs at all selected points.

From Figure 7, it is obvious that the GDOP values of the proposed hybrid positioning are significantly lower than the TDOA-only positioning. It can be seen in (16) that the greater the ratio k is, the smaller the weight of the TOA ranging error item in the weighting matrix (18) is, and the closer the GDOP calculated is to the GDOP of TDOA-only positioning. The result shows that it is possible to effectively improve the positioning performance by adding a TOA measurement.

To compare the GDOPs of the two kinds of positioning in the two different regions as shown in Figure 4, we separately counted the GDOPs of two kinds of positioning at points in the two groups. The CDFs of the GDOPs are shown in Figure 8.

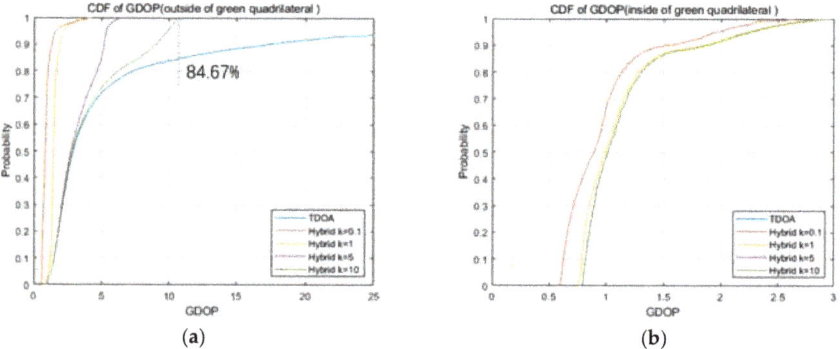

Figure 8. The GDOP CDF of different positioning methods; (**a**) the CDF of GDOPs at points outside the green quadrilateral; (**b**) the CDF of GDOPs at points inside the green quadrilateral, the curves for TDOA and Hybrid k = 10 are identical.

As shown in Figure 8a, at the points outside the green quadrilateral, GDOPs of the proposed hybrid positioning are significantly lower than the TDOA-only positioning. Even when the standard deviation of TOA ranging error is 10 times the TDOA ranging error, the 24.85% points still have a higher GDOP of the TDOA-only positioning than the highest GDOP of the proposed hybrid positioning.

However, as shown in Figure 8b, at the points inside the green quadrilateral, GDOPs of the proposed hybrid positioning are not much different than GDOP of the TDOA-only positioning. Even when the ratio k is equal to 0.1, the minimum GDOP of the proposed hybrid positioning is 0.59. And the minimum GDOP of the TDOA-only positioning is just 0.79, which is only 0.20 higher. When the ratio k is equal to 1, 5 and 10, the GDOPs of the proposed hybrid positioning are approximate to the GDOP of the TDOA-only positioning.

Analysis result shows that adding a TOA measurement to the TDOA measurements to achieve hybrid positioning have potential to greatly improve the positioning ability of the system in the edge area of the network. At the same time, in the core area of network coverage, although hybrid positioning has only a limited improvement in positioning ability, it is also superior to the TDOA-only positioning.

5.2. Analysis and Comparison of the Proposed Localization Algorithm

CRLB is the theoretical lower bound on the variance of unbiased estimator [35]. In order to analyze the performance of the proposed three-step WLS closed-form localization algorithm, we compare the algorithm with CRLB of the proposed hybrid positioning, CRLB of TDOA-only positioning and two widely used TDOA-only closed-form localization algorithms, Chan-Ho [17] and BiasRed [18]. CRLB is the trace of the inverse matrix of the Fisher information matrix [19].

$$\mathrm{CRLB} = \mathrm{tr}\left(J^{-1}\right) = \mathrm{tr}\left\{\left(A^T Q A\right)^{-1}\right\} \tag{50}$$

where matrix A and W are defined in Section 3.1, Q is a $M \times M$ matrix as

$$Q = \begin{bmatrix} \sigma_{toa}^2 & 0 & 0 & \cdots & 0 \\ 0 & \sigma_{tdoa}^2 & 0 & \cdots & 0 \\ 0 & 0 & \sigma_{tdoa}^2 & \cdots & 0 \\ \vdots & \vdots & \vdots & \ddots & \vdots \\ 0 & 0 & 0 & \cdots & \sigma_{tdoa}^2 \end{bmatrix}. \tag{51}$$

5.2.1. Simulations at Selected Point

According to the GDOP analysis in Section 5.1, we selected three test points at the right edge, the center, and the left edge of the scenario. Scenario shown in Figure 3 is used, and three test points are selected for the simulation, which is shown in Figure 9.

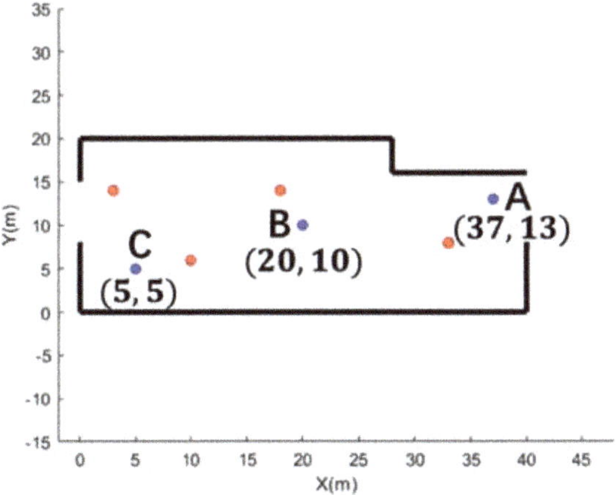

Figure 9. Selected Test Points: the blue dots are selected three test points; the red dots are the location of ANTs.

The simulation was performed 10,000 times at each test point. TDOA and TOA ranging errors were the zero-mean Gaussian distribution but with different standard deviations.

5G system could provide a TOA ranging accuracy with standard deviation $\sigma_{toa} = 0.3\ m$ or $\sigma_{toa} = 1.5\ m$ [25] and provide a TDOA ranging accuracy with a best $\sigma_{tdoa} = 0.03\ m$ [13].

In each simulation, we calculated the Mean Square Error (MSE) of the positioning errors for each test points as

$$\text{MSE} = \frac{\sum_{i=1}^{10,000}\left((x_i - x_t)^2 + (y_i - y_t)^2\right)}{10,000} \tag{52}$$

where x_t and y_t are the coordinate of the test point, x_i and y_i are the positioning result of the ith calculation.

Figure 10 shows the MSE of the positioning results and the calculated CRLBs at test point A. The result shows that the positioning using multiple TDOAs and single TOA has a lower CRLB than TDOA-only positioning. Regardless of whether $\sigma_{toa} = 1.5$ or $\sigma_{toa} = 0.3$ m, the MSE of proposed algorithm approaches CRLB when TDOA ranging error is small. When the TDOA ranging error is high, the positioning error of the proposed localization algorithm is higher than the CRLB but is still significantly lower than the other two TDOA-only closed-form algorithms. For the proposed algorithm, the positioning results with $\sigma_{toa} = 0.3$ m are slightly better than the positioning results with $\sigma_{toa} = 1.5$ m when the standard deviation of TDOA ranging error is around 0.5 m.

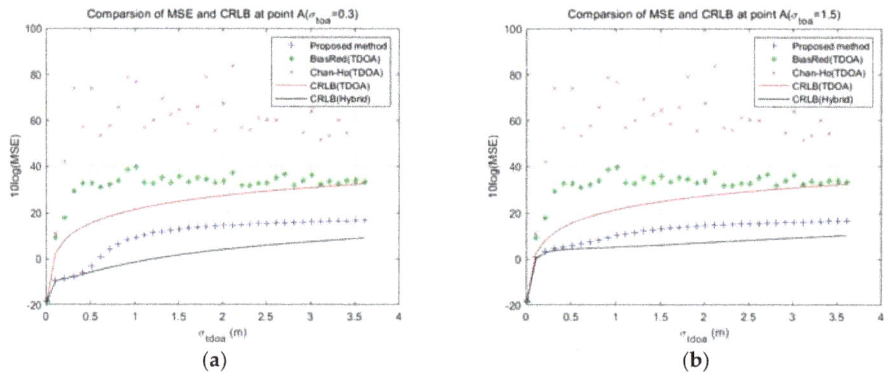

Figure 10. Accuracy of different positioning method at point A; (**a**) positioning error mean square error (MSE) when the standard deviation of TOA ranging error is 0.3 m; (**b**) positioning error MSE when the standard deviation of TOA ranging error is 1.5 m.

Figure 11 shows the MSE of positioning results and CRLBs at test point B. The result shows that the CRLB of the proposed hybrid positioning is close to the CRLB of TDOA-only positioning at test point B. Regardless of whether $\sigma_{toa} = 1.5$ or $\sigma_{toa} = 0.3$ m, the MSE of proposed algorithm approaches CRLB when TDOA ranging error is small. Even when the TDOA ranging error is higher, the positioning error of the proposed algorithm is only slightly higher than the CRLB. For the proposed algorithm, the positioning results with $\sigma_{toa} = 0.3$ m are slightly better than the positioning results with $\sigma_{toa} = 1.5$ m. However, regardless of the different performance caused by TOA ranging error, the positioning error of the proposed algorithm is lower than the other two TDOA-only positioning methods.

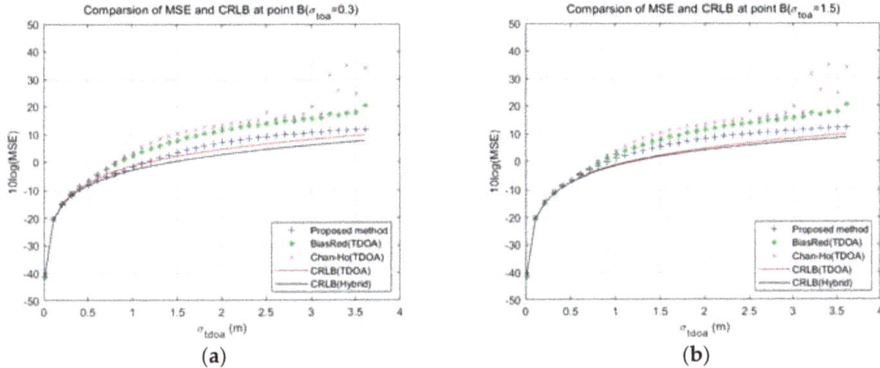

Figure 11. Accuracy of different positioning method at point B; (**a**) positioning error MSE when the standard deviation of TOA ranging error is 0.3 m; (**b**) positioning error MSE when the standard deviation of TOA ranging error is 1.5 m.

Figure 12 shows the MSE of positioning results and CRLBs at test point C. The result shows that the proposed hybrid positioning has a lower CRLB than TDOA-only positioning at test point C. Regardless of whether $\sigma_{toa} = 1.5$ or $\sigma_{toa} = 0.3$ m, the MSE of proposed algorithm approaches CRLB when TDOA ranging error is small. Even when the TDOA ranging error is higher, the positioning error of the proposed algorithm is very slightly higher than the CRLB. For the different results in (a) and (b), the proposed algorithm is more accurate while $\sigma_{toa} = 0.3$ m than $\sigma_{toa} = 1.5$ m, when the standard deviation of TDOA ranging error is around 0.5 m. However, regardless of the different performance caused by TOA ranging error, the positioning error of the proposed algorithm is significantly lower than the other two TDOA-only positioning methods.

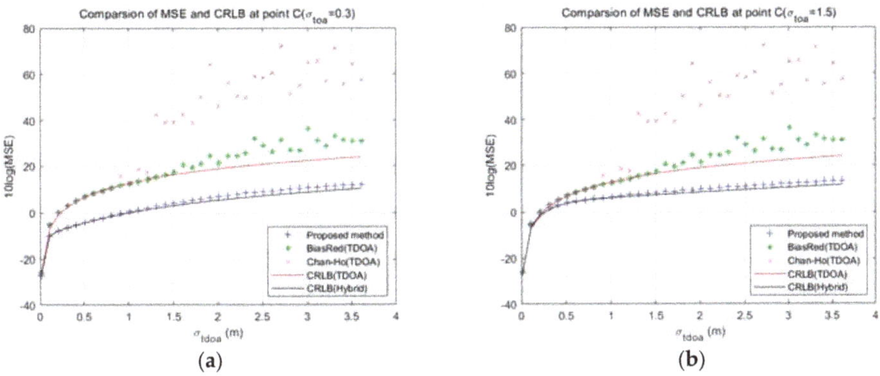

Figure 12. Accuracy of different positioning method at point C; (**a**) positioning error MSE when the standard deviation of TOA ranging error is 0.3 m; (**b**) positioning error MSE when the standard deviation of TOA ranging error is 1.5 m.

It can be seen from those three figures that the positioning error of Chan-Ho and BiasRed would increases irregularly with the increase of the error after the error becomes larger. This may be caused by the error of the weighting matrix of those algorithms. Those two algorithms use WLS in the calculation process and ignore some items about TDOA ranging errors to build the weighting matrix. When TDOA ranging errors are small, the weighting matrix is closed to be right. However, when the ranging errors are large, the error of the weighting matrix would increase. The positioning error, caused by weighted matrix, is irregular. The proposed algorithm also has this problem, but the item about TOA measurement in weighting matrix is not related with ranging error. This may inhibit the irregularity of the positioning error when TOA ranging error is low. The positioning error of the proposed algorithm would also increases irregularly when TOA ranging error is high.

5.2.2. Simulation Results in Different Regions

To compare the positioning accuracy of the three localization algorithms in different regions, the two groups of selected points in Figure 4 were used for simulation.

The positioning simulation is performed 10,000 times at each test point. For each test point group, the MSE of the positioning errors for each group test points was calculated and denoted as

$$\text{MSE}_{all} = \frac{\sum_{p=1}^{N} \sum_{i=1}^{10,000} \left(\left(x_{p,i} - x_p \right)^2 + \left(y_{p,i} - y_p \right)^2 \right)}{N \times 10,000} \tag{53}$$

where x_p and y_p are coordination of the pth test point in the group, $x_{p,i}$ and $y_{p,i}$ are the positioning result of the ith calculation at the pth test point. The MSE_{all} of the two groups' positioning results is counted separately. The MSEs and CRLBs of positioning results with different TDOA ranging error of the two groups are shown in Figures 13 and 14.

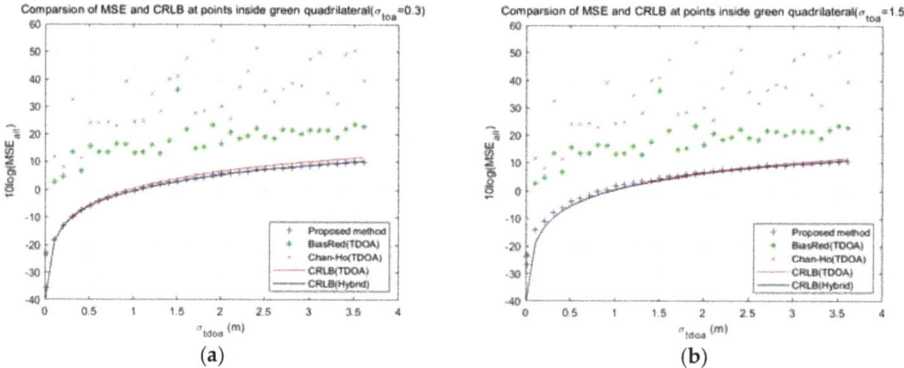

Figure 13. Accuracy of different positioning method at points inside green quadrilateral; (**a**) positioning error MSE when the standard deviation of TOA ranging error is 0.3 m; (**b**) positioning error MSE when the standard deviation of TOA ranging error is 1.5 m.

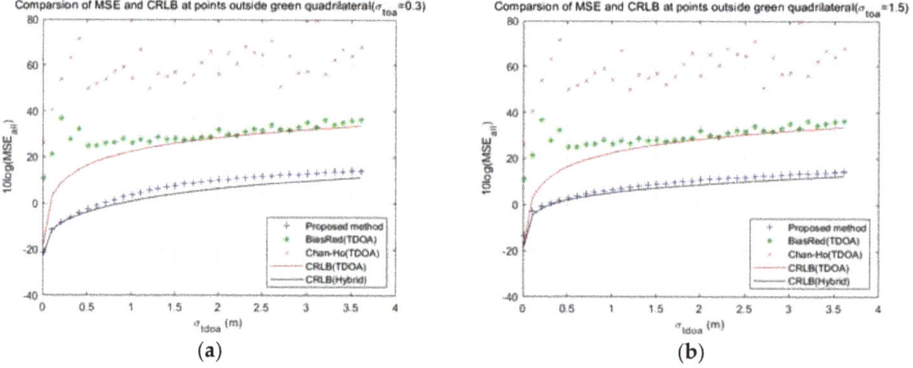

Figure 14. Accuracy of different positioning method at points outside green quadrilateral; (**a**) positioning error MSE when the standard deviation of TOA ranging error is 0.3 m; (**b**) positioning error MSE when the standard deviation of TOA ranging error is 1.5 m.

Figure 13 shows the MSE_{all} of the positioning results and CRLBs at test points inside the green square. The result shows that the CRLB of the proposed hybrid positioning is close to the CRLB of TDOA-only positioning. When $\sigma_{toa} = 0.3$ m, the MSE_{all} of proposed algorithm approach CRLB. When $\sigma_{toa} = 1.5$ m, the accuracy of the proposed algorithm is slightly higher than the CRLB while the standard deviation of TDOA ranging error is relatively small but has not yet approached zero. For the proposed algorithm with different TOA ranging error, the positioning results with $\sigma_{toa} = 0.3$ m are very slightly better than the positioning results with $\sigma_{toa} = 1.5$ m. However, regardless of the different performance caused by the TOA ranging error, the positioning error of the proposed hybrid positioning method is significantly lower than the other two TDOA-only positioning methods.

Figure 14 shows the MSE_{all} of positioning results and CRLBs at test points outside the green square. The result shows that the hybrid positioning has a lower CRLB than TDOA-only positioning. The MSE_{all} of the proposed algorithm approaches CRLB when TDOA ranging error is small. Even when the TDOA ranging error is higher, the positioning error of the proposed algorithm is also slightly higher than the CRLB and is significantly lower than the other two TDOA-only positioning methods. The positioning results of the proposed algorithm with $\sigma_{toa} = 0.3$ m are slightly better than it with $\sigma_{toa} = 1.5$ m when the standard deviation of TDOA ranging error is around 0.5 m.

The algorithms performance at each selected points in Figure 4 is also simulated as in Section 5.2.1. Different from Section 5.2.1, the square roots of MSE, called the root mean square error (RMSE), for different algorithms at every selected point are calculated. The CDF of the RMSE for the Chan-Ho, BiasRed, proposed algorithm with $\sigma_{toa} = 1.5$ m, proposed algorithm with $\sigma_{toa} = 0.3$ m and CRLBs are shown in Figure 15, when $\sigma_{tdoa} = 1$ m.

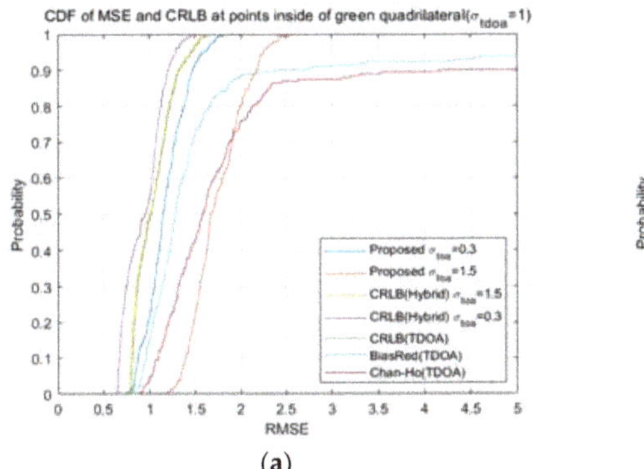

(a)

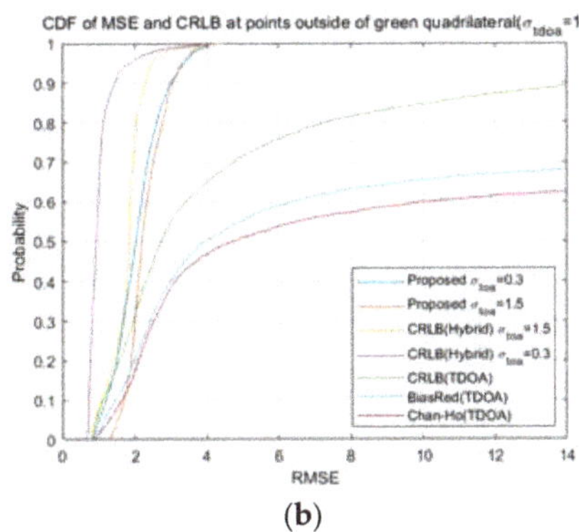

(b)

Figure 15. CDF of root mean square error (RMSE) different positioning method when $\sigma_{tdoa} = 1$ m; (**a**) CDF of RMSE of points inside green quadrilateral; (**b**) CDF of RMSE of points outside green quadrilateral.

As shown in Figure 15b, it is obviously that, regardless of $\sigma_{toa} = 0.3$ or 1.5 m, performance of proposed algorithm is better than Chan-Ho and BiasRed algorithms. Only at 16% points, BiasRed algorithm is more accurate than the proposed algorithm when $\sigma_{toa} = 1.5$ m. Moreover, the Chan-Ho algorithm is only more accurate at 13% points compared to the proposed algorithm when $\sigma_{toa} = 1.5$ m. When $\sigma_{toa} = 0.3$ m, the proposed algorithm has a better performance than the two algorithms at all points outside the green quadrilateral. In Figure 15a, the proposed algorithm still has a better performance than the two algorithms at all points inside the green quadrilateral when $\sigma_{toa} = 0.3$ m. In the comparison, the two algorithms and the proposed algorithm with $\sigma_{toa} = 1.5$ have their own advantages. BiasRed is more accurate than the proposed algorithm with $\sigma_{toa} = 1.5$ at 89% points and Chan-Ho is more accuracy than the proposed algorithm with $\sigma_{toa} = 1.5$ at 72% points. However, the proposed algorithm with $\sigma_{toa} = 1.5$ is more reliable than the two algorithm, since the maximum RMSE is only 2.613 and is much lower than the maximum RMSE of the two algorithms.

Based on the simulation results at the two groups' points and the three selected test points, the proposed closed-form localization algorithm using multiple TDOAs and single TOA could provide a better positioning accuracy than the two closed-form TDOA-only positioning methods, Chan-Ho and BiasRed, in the indoor distribution system application scenario.

6. Conclusions

The 5G system could provide accurate multiple TDOA measurements and single TOA measurement, which could improve UE positioning accuracy. However, existing closed-form localization algorithms are designed for measurements having the same variance, and not suitable for heteroscedastic TDOA and TOA measurements.

In this paper, a three-step WLS closed-form localization algorithm is developed for the nonlinear equation set of those hybrid heteroscedastic measurements. The algorithm is an extension of Chan-Ho algorithm [17]. The first WLS provides an initial position for the last two steps. Then the algorithm uses two WLSs to estimate position based on heteroscedastic TDOA and TOA measurements. Moreover, after the GDOP of this hybrid positioning is derived, it was proven that this hybrid positioning has a higher theoretical higher bound of accuracy than TDOA-only positioning.

The simulation scenario is built according to the real situation. By comparing with two TDOA-only closed-formed localization algorithm, Chan-Ho [17] and BiasRed [18], and CRLB [35] of TDOA-only positioning and the proposed hybrid positioning in the same simulation scenario, the results show that the proposed localization algorithm has better performance than the two closed-form TDOA-only positioning methods, and the positioning accuracy approaches CRLB when the TDOA measurement errors are small. In different areas of the scenario, the proposed positioning algorithm has a different improvement in positioning accuracy. In the area surrounded by the ANTs, the proposed algorithm shows only a slight improvement effect. However, in the rest of the area, the accuracy of the proposed algorithm is even higher than the CRLB of TDOA-only positioning. The simulations indicates that using the hybrid TDOA and TOA measurements as well as the localization algorithm proposed in this paper have the potential to effectively improve the positioning accuracy of the indoor distribution system.

The proposed localization algorithm and the two compared algorithms, Chan-Ho and BiasRed, were constructed and simulated according to the ranging errors under the line-of-sight wireless propagation channel. In real scenes, non-line-of-sight channels are ubiquitous, which may cause the positioning accuracy of these algorithms to decrease. In future work, the analysis and improvement of the proposed algorithm in the non-line-of-sight environment can be studied. Meanwhile, the proposed algorithm is only confirmed by simulation for a specific environment. The actual algorithm performance should be investigated in future work.

Author Contributions: Conceptualization, H.W., L.Y. and S.T.; Data curation, H.W.; Formal analysis, H.W.; Funding acquisition, Z.D. and F.Y.; Investigation, H.W. and X.Z.; Methodology, H.W.; Project administration, Z.D. and F.Y.; Resources, H.W. and X.Z.; Software, H.W. and X.Z.; Supervision, Z.D.; Validation, H.W. and X.F.; Visualization, H.W.; Writing—Original draft, H.W.; Writing—Review & editing, H.W., X.F. and S.T.

Funding: This work was financially supported by the National Key Research & Development Program under Grant No. 2018YFC0809702 and No. 2016YFB0502001.

Acknowledgments: The authors would like to express their thanks to Jiazhi Han for help in the simulation and Jun Mo for the help in the paper's revisions.

Conflicts of Interest: The authors declare no conflict of interest.

References

1. Zafari, F.; Gkelias, A.; Leung, K. A Survey of Indoor Localization Systems and Technologies. *IEEE Commun. Surv. Tutor.* **2017**, *21*, 2568–2599. [CrossRef]
2. Witrisal, K.; Hinteregger, S.; Kulmer, J.; Leitinger, E.; Meissner, P. High-accuracy positioning for indoor applications: RFID, UWB, 5G, and beyond. In Proceedings of the 2016 IEEE International Conference on RFID, Orlando, FL, USA, 3–5 May 2016; pp. 1–7.

3. Del Peral-Rosado, J.A.; Raulefs, R.; López-Salcedo, J.A.; Seco-Granados, G. Survey of Cellular Mobile Radio Localization Methods: From 1G to 5G. *IEEE Commun. Surv. Tutor.* **2018**, *20*, 1124–1148. [CrossRef]
4. Wymeersch, H.; Garcia, N.; Kim, H.; Seco-Granados, G.; Kim, S.; Wen, F.; Fröhle, M. 5G mmWave Downlink Vehicular Positioning. In Proceedings of the IEEE Global Communications Conference (GLOBECOM), Abu Dhabi, UAE, 9–13 December 2018; pp. 206–212.
5. Cui, X.; Gulliver, T.A.; Member, S. Vehicle Positioning Using 5G Millimeter-Wave Systems. *IEEE Access* **2016**, *4*, 6964–6973. [CrossRef]
6. Di Taranto, R.; Muppirisetty, S.; Raulefs, R.; Slock, D.; Svensson, T.; Wymeersch, H. Location-aware communications for 5G networks: How location information can improve scalability, latency, and robustness of 5G. *IEEE Signal Process. Mag.* **2014**, *31*, 102–112. [CrossRef]
7. Zhang, P.; Lu, J.; Wang, Y.; Wang, Q. Cooperative localization in 5G networks: A survey. *ICT Express* **2017**, *3*, 27–32. [CrossRef]
8. 3GPP TS 22.261 v16.7.0, Service Requirements for the 5G system; Stage 1. Available online: https://portal.3gpp.org/desktopmodules/Specifications/SpecificationDetails.aspx?specificationId=3107 (accessed on 24 May 2019).
9. 3GPP TR 22.872 v16.1.0, Study on Positioning Use Cases; Stage 1, 2018–09. Available online: https://portal.3gpp.org/desktopmodules/Specifications/SpecificationDetails.aspx?specificationId=3280 (accessed on 7 December 2018).
10. 3GPP TS 36.211 v15.0.0, Evolved Universal Terrestrial Radio Access (E-UTRA); Physical Channels and Modulation. Available online: https://portal.3gpp.org/desktopmodules/Specifications/SpecificationDetails.aspx?specificationId=2425 (accessed on 7 July 2018).
11. 3GPP TS 36.305 v15.3.0, Evolved Universal Terrestrial Radio Access Network (E-UTRAN); Stage 2 Functional Specification of User Equipment (UE) Positioning in E-UTRAN. Available online: https://portal.3gpp.org/desktopmodules/Specifications/SpecificationDetails.aspx?specificationId=2433 (accessed on 24 May 2019).
12. Fischer, S. Observed Time Difference Of Arrival (OTDOA) Positioning in 3GPP LTE. *Qualcomm White Pap.* **2014**, *1*, 1–62.
13. Del Paral-Rosado, J.A.; Lopez-Salcedo, J.A.; Seco-Granados, G.; Zanier, F.; Crisci, M. Preliminary Analysis of the Positioning Capabilities of the Positioning Reference Signal of 3GPP LTE. In Proceedings of the 5th European Workshop on GNSS Signals and Signal Processing, Toulouse, France, 8–9 December 2011; p. 9.
14. Zou, Y.; Liu, H.; Xie, W.; Wan, Q. Semidefinite Programming Methods for Alleviating Sensor Position Error in TDOA Localization. *IEEE Access* **2017**, *5*, 23111–23120. [CrossRef]
15. Jin, B.; Xu, X.; Zhang, T. Robust Time-Difference-of-Arrival (TDOA) localization using weighted least squares with cone tangent plane constraint. *Sensors* **2018**, *18*, 778.
16. Foy, W.H. Position-Location Solutions by Taylor-Series Estimation. *IEEE Trans. Aerosp. Electron. Syst.* **1976**, *AES-12*, 187–194. [CrossRef]
17. Chan, Y.T.; Ho, K.C. An efficient closed-form localization solution from time difference of arrival measurements. In Proceedings of the IEEE International Conference on Acoustics, Adelaide, Australia, 19–22 April 1994.
18. Ho, K.C. Bias reduction for an explicit solution of source localization using TDOA. *IEEE Trans. Signal Process.* **2012**, *60*, 2101–2114. [CrossRef]
19. Shu, F.; Yang, S.; Qin, Y.; Li, J.U.N. Approximate Analytic Quadratic-Optimization Solution for TDOA-Based Passive Multi-Satellite Localization With Earth Constraint. *IEEE Access* **2017**, *4*, 9283–9292. [CrossRef]
20. Yang, T.; Ouyang, Z.; Liu, J.; Luan, T.H. Design of TD-LTE based signal indoor distribution system. In Proceedings of the 2017 IEEE 86th Vehicular Technology Conference (VTC-Fall), Toronto, ON, Canada, 24–27 September 2017; pp. 1–7.
21. Sharp, I.; Yu, K.; Guo, Y.J. GDOP analysis for positioning system design. *IEEE Trans. Veh. Technol.* **2009**, *58*, 3371–3382. [CrossRef]
22. Sun, S.; Wang, Z.; Wang, Z. Study on Optimal Station Distribution Based on TDOA Measurements Shun Sun 1. In Proceedings of the 2016 International Conference on Computer Engineering, Information Science & Application Technology (ICCIA 2016), Guilin, China, 24–25 September 2016; pp. 31–36.
23. 3GPP TR 38.305 v15.4.0, NG Radio Access Network (NG-RAN); Stage 2 Functional Specification of User Equipment (UE) Positioning in NG-RAN, v15.4.0, 2019-06. Available online: https://portal.3gpp.org/desktopmodules/Specifications/SpecificationDetails.aspx?specificationId=3310 (accessed on 5 July 2019).

24. 3GPP TR 38.855 v16.0.0, Study on NR Positioning Support. Available online: https://portal.3gpp.org/desktopmodules/Specifications/SpecificationDetails.aspx?specificationId=3501 (accessed on 24 May 2019).
25. Koivisto, M.; Costa, M.; Hakkarainen, A.; Leppänen, K.; Valkama, M. Joint 3D positioning and network synchronization in 5G ultra-dense networks using UKF and EKF. In Proceedings of the 2016 IEEE Globecom Workshops (GC Wkshps), Washington, DC, USA, 4–8 December 2016.
26. Koivisto, M.; Costa, M.; Werner, J.; Heiska, K.; Talvitie, J.; Leppänen, K.; Koivunen, V.; Valkama, M.; Member, S. Joint Device Positioning and Clock Synchronization in 5G Ultra-Dense Networks. *IEEE Trans. Wirel. Commun.* **2017**, *16*, 2866–2881. [CrossRef]
27. Wymeersch, H.; Seco-Granados, G.; Destino, G.; Dardari, D.; Tufvesson, F. 5G mmWave positioning for vehicular networks. *IEEE Wirel. Commun.* **2017**, *24*, 80–86. [CrossRef]
28. Sharp, I.; Yu, K. GDOP Analysis for Positioning Design. In *Wireless Positioning: Principles and Practice*; Springer: Singapore, 2019; pp. 443–473. ISBN 9789811087912.
29. Yu, L.; Shijun, T. Vertical positioning technologies and its application of pseudolites augmentation. In Proceedings of the Int. Conf. Wirel. Commun. Netw. Mob. Comput., Dalian, China, 12–14 October 2008; pp. 1–3.
30. Del Peral-Rosado, J.A.; Bavaro, M.; López-Salcedo, J.A.; Seco-Granados, G.; Chawdhry, P.; Fortuny-Guasch, J.; Crosta, P.; Zanier, F.; Crisci, M. Floor detection with indoor vertical positioning in LTE femtocell networks. In Proceedings of the 2015 IEEE Globecom Workshops (GC Wkshps), San Diego, CA, USA, 6–10 December 2015.
31. Li, W.; Wei, P.; Xiao, X. A robust TDOA-based location method and its performance analysis. *Sci. China Ser. F Inf. Sci.* **2009**, *52*, 876–882. [CrossRef]
32. Li, Y.; Qi, G.; Sheng, A. General analytical formula of GDOP for TOA target localisation. *Electron. Lett.* **2018**, *54*, 6–7. [CrossRef]
33. Van Trees, H.L. *Detection, Estimation, and Modulation Theory, Part 1-Detection, Estimation, and Linear Modulation Theory*; John Wiley & Sons: Hoboken, NJ, USA, 1968; Volume 6, ISBN 9780471095170.
34. Lee, H.B. A Novel Procedure for Assessing the Accuracy of Hyperbolic Multilateration Systems. *IEEE Trans. Aerosp. Electron. Syst.* **1975**, *AES-11*, 2–15. [CrossRef]
35. Vankayalapati, N.; Kay, S.; Ding, Q. TDOA based direct positioning maximum likelihood estimator and the cramer-rao bound. *IEEE Trans. Aerosp. Electron. Syst.* **2014**, *50*, 1616–1635. [CrossRef]

Review

Automated Data Acquisition in Construction with Remote Sensing Technologies

Osama Moselhi [1], Hassan Bardareh [2,*] and Zhenhua Zhu [3]

1 Centre for Innovation in Construction and Infrastructure Engineering and Management (CICIEM), Department of Building, Civil and Environmental Engineering, Concordia University, 1455 de Maisonneuve Blvd, West, Montreal, QC H3G 1M8, Canada; moselhi@encs.concordia.ca

2 Department of Building, Civil and Environmental Engineering, Concordia University, Sir George Williams Campus, 1455 De Maisonneuve Blvd. W., Montréal, QC H3G 1M8, Canada

3 Department of Civil and Environmental Engineering, University of Wisconsin-Madison, 2258 Engineering Hall, 1415 Engine Drive, Madison, WI 53706, USA; zzhu286@wisc.edu

* Correspondence: hassan.bardareh@mail.concordia.ca; Tel.: +1-514-850-7731

Received: 7 April 2020; Accepted: 16 April 2020; Published: 20 April 2020

Featured Application: This review article aims to highlight the capabilities and limitations of a wide range of remote sensing (RS) technologies for their possible use for automated data acquisition on construction job sites. More attention is given to integrated RS technologies for the purpose of tracking, localization, and 3D modeling in construction projects. This article is expected to facilitate the use of RS technologies for automating progress reporting, improving safety and productivity in construction.

Abstract: Near real-time tracking of construction operations and timely progress reporting are essential for effective management of construction projects. This does not only mitigate potential negative impact of schedule delays and cost overruns but also helps to improve safety on site. Such timely tracking circumvents the drawbacks of conventional methods for data acquisition, which are manual, labor-intensive, and not reliable enough for various construction purposes. To address these issues, a wide range of automated site data acquisition, including remote sensing (RS) technologies, has been introduced. This review article describes the capabilities and limitations of various scenarios employing RS enabling technologies for localization, with a focus on multi-sensor data fusion models. In particular, we have considered integration of real-time location systems (RTLSs) including GPS and UWB with other sensing technologies such as RFID, WSN, and digital imaging for their use in construction. This integrated use of technologies, along with information models (e.g., BIM models) is expected to enhance the efficiency of automated site data acquisition. It is also hoped that this review will prompt researchers to investigate fusion-based data capturing and processing.

Keywords: automated data acquisition; remote sensing technologies; automated progress reporting; data fusion; tracking resources

---

1. Introduction

Having timely access to accurate and reliable onsite information is vital for the efficient management of construction operations. On the contrary, late detection of onsite operations can be problematic, leaving insufficient time for members of the project team to react which can result in negative impacts on project cost and schedule [1]. Real-time information can provide feedback to support tracking project operations [1–5], safety management [5], productivity analysis, and progress reporting [1,6]. This is especially critical for tracking activities with highly dynamic nature such as lifting activities [7].

Using the conventional manual progress reporting methods is time-consuming, labor-intensive and may not be accurate, even for small projects. Besides, the integration of manual data collection and available electronic interfaces can be a challenging task [1]. To overcome the drawbacks stated above, one can make use of automated site data acquisition for schedule updating and progress reporting. Such automation can provide project management teams with needed timely information on the status of onsite construction operations with sufficient accuracy [8]. The use of various remote sensing (RS) technologies for onsite automated data-capturing can result in more robust progress reporting. The captured data can be compared to related as-planned data to perform progress reporting [6,9].

This article encompasses four main areas: an overview of research on automated site data acquisition, standalone RS technologies, integrated use of RS technologies and a general discussion. It is based on a carefully selected set of articles from 20 diverse scientific journals and a couple of conference proceedings in the domain of construction engineering and management. In this search, keywords such as 'remote sensing', 'automated data acquisition', and 'progress tracking' were used. A total of 56 selected studies are used in this review, among which over 50% were published within last five years. Table 1 presents a list of selected journals and conference proceedings.

Table 1. Source journals, conferences, and the publications used in this review

Journal/Conference	Quantity
Automation in Construction	8
Advanced Engineering Informatics	4
Journal of Computing in Civil Engineering	2
Applied Sciences	2
Advances in Informatics and Computing in Civil and Construction Engineering	1
Computing in Civil Engineering	1
Construction Management and Economics	1
EURASIP Journal on Embedded Systems	1
Journal of Construction Engineering and Management	1
Journal of Construction Engineering and Project Management	1
Journal of Information Technology in Construction (ITcon)	1
Journal of IET (The Institution of Engineering and Technology)	1
Measurement	1
Remote Sensing	1
Robotica	1
Sensors	1
Visualization in Engineering	1
IEEE Transactions on Computers	1
IEEE Transactions on instrumentation and Measurement	1
IEEE Transactions on Intelligent Transportation Systems	1
Conference and congress (ISARC, CSCE, ICRA, etc.)	12
Thesis	6
Online archives	6
Total	56

2. Overview of Research on Automated Site Data Acquisition

Most previous attempts in this field were focused on single-sensor models which are not usually applicable for the entire duration of a project [10,11]. The literature reveals that the use of a single source of sensory data does not provide sufficient information about the status of onsite construction operations. For example, the point cloud data captured by laser scanners require a direct line-of-sight and they will become less effective as the project progresses and the site becomes obstructed due to increased congestion. In this way, the usage of another data acquisition technology could help to alleviate the drawbacks associated with individual usage of each RS technology in each of the project execution phases. That integration provides more timely and accurate information due to the complementarity and possible fusion of the captured data [3]. Further incorporation of captured data with building information modeling (BIM) can lead to more robust progress reporting [8,9].

Despite current progress, obstacles still exist for achieving a practical and comprehensive automated solution for onsite progress tracking [6].

Indoor tracking of the construction operations has some difficulties associated with a wide range of materials and structural objects that need to be recognized [6]. Activity tracking was another area of interest in recent years. In fact, an automated progress reporting system based on both activity and material (or objects) tracking has some advantages over systems which are solely based on object tracking [8].

In this article, the latest efforts on automated data acquisition are reviewed. More specifically, methods that integrate various RS technologies are investigated. For instance, the integration of the radio frequency identification (RFID) technology with GPS-based technologies are reviewed to enhance outdoor tracking of resources [12,13]. Data fusion of imaging technologies (e.g., laser scanning and photogrammetry) with technologies such as RFID, UWB, and GPS have also been investigated recently [14,15].

3. Standalone Remote Sensing (RS) Technologies

There are a wide range of RS technologies for data acquisition on site. Each of these technologies has its capabilities and drawbacks based on the domain of applications. Selection of one particular technology can vary depending on the type of information required, the accuracy and the environment in which the data will be acquired [16]. One group of these RS technologies includes real-time location systems (RTLSs) such as GPS, robotic total station (RTS), and ultra-wideband (UWB). These technologies generally function based on the time-of-arrival (TOA) principle, where the propagation speed and the propagation time of a signal directly lead to a corresponding distance [10]. Recently, there are some efforts to localize and track people and objects using Wi-Fi-/Bluetooth/BLE technologies for the purpose of occupancy measurement. In fact, by installing some receivers in a target area, the location and number of detected items are estimated using held and/or worn smart phones, Bluetooth headsets, and smart watches. It is done by tracking the changes in the received electromagnetic wave patterns and compare them with calibrated models to count the number of workers or vehicles on site, and to provide rough information about their location and behavioral algorithms. However, due to some legal restrictions for using smart phones on site and not fully accurate results of this technology its use in construction industry is not common yet [17–19]. In a different context, barcode and radio frequency identification devices (RFID) proven to be useful technologies for material identification with applications in tracking resources for a supply chain management system [20,21]. Vision-based technologies such as photo/videogrammetry and laser scanning have also found applications in construction industry for making 3D as-built models of site which provide sufficient information for an automated progress reporting [1]. Table 2 provides a brief overview of these technologies along with their capabilities, limitations, and accuracy.

Table 2. Capabilities and limitations of standalone remote sensing (RS) technologies

RS Technology	Capabilities	Limitations	Accuracy	Refs.
GPS, RTKGPS [1], and GNSS [2]	(1) Global access in an outdoor environment (2) RTK GPS provides centimeter-level positioning accuracy in real time over long distances (3) Flexibly and reacting quickly based on the needs in a construction site	(1) The acquired data are limited to position and time of objects, and not the identity or type of activities (2) Delays in data processing and transferring (an important issue specially for real-time operations such as equipment tracking or kinematic surveying) (3) Multipath errors in the congested environment (4) Signal blockage and building obstructiveness in an indoor environment	Almost one meter for GPS and few centimeters for RTKGPS	[5,22,23]
Robotic total station (RTS)	(1) Enhanced tracking and automation capabilities in the positioning of objects (2) High positioning update rate (3) Higher vertical positioning (elevation) accuracy than unaided GNSS and GPS-based technology	(1) The high investment (2) Limited capability to track only one point at a particular time (3) Positioning errors will occur if the angles and distances are not captured simultaneously	Down to few millimeters and arc seconds	[7,24]
Barcode	(1) Reasonable price (2) Straight forward usage with standard implementation protocols (3) Speeding up the computer data entry (4) Portability	(1) Very limited reading range (2) Sensitivity to the harsh environment [3] (e.g., tags are not readable if covered by snow) (3) Low data storage capacity	N.A	[20,21,25]
RFID	(1) Longer range than barcodes (up to 100 m for Ultra-high frequencies) (2) Non-line-of-sight (3) Providing cost efficient location information (4) Tags are light and easy to attach (5) Unlike barcodes, RFID tags are durable in construction environments (6) Batch readability of tags making identification process much more efficient	(1) Lack of accuracy and difficulties for 3D positioning (2) Calibration difficulties (e.g., the need for a path loss model) (3) Affected by multipath effect (4) Problems associated with simultaneous identification of many tags (5) Active tags are fairly expensive. (6) Plus, relatively wide range may cause obstruction object detection (7) The need for battery replacement in active tags (8) Influenced by metal and high humidity specially in high frequencies	Down to few meters for 2D localization	[4,5,11–13,21,25–30]
UWB	(1) A longer range (up to 1000 m), higher measurement rate and positioning accuracy (less than 1 m) than typical RFID systems (2) Less influenced by metal and high humidity (3) Suitable for both indoor and outdoor environment (4) Not affected easily by other RF systems (5) Relatively immunity to multipath fading (6) Reliable 3D localization even in harsh environment	(1) Violation of line-of-sight can lead to the performance degradation specially in congested areas (2) Limited update rate (3) Multipath and radio noise effect in the case of metal occlusion (4) Some calibration difficulties (e.g., geometry and number of sensors) (5) Tagging issues (e.g., battery replacement) (6) High cost (7) High missing data (8) Degraded range measurement as distance increases	Down to a few decimeter	[3,4,6,16,26,28,31,32]
Infrared	(1) High noise immunity (2) Low power requirement (3) Low circuitry costs (4) Higher security (5) Portability	(1) Dependency on line-of-sight due to its inability to penetrate materials (2) Sensitive to light and weather condition (3) Low data transmission rate (4) Short range (5) Not passing walls unlike other RF wireless links (6) Limitation associated with scalability	Down to a few centimeters	[26]
Photo/ Videogrammetry	(1) Straightforward configuration (2) Cost-effective field data collection (3) Portability (4) Provides information about the material, texture, and color of the target object (5) High update rate (6) Well known internal geometry (7) Good interpretability	(1) Calibration difficulties (e.g., sensitive to the surrounding light condition, visual occlusions, and moving background) (2) Mirror effect caused by reflective surfaces (3) Less accurate than laser scanners in generating point clouds (4) Problems associated with computing depth in 3D modeling (5) Difficulty in identification of objects with unclear geometric configurations (6) Computational complexity of photogrammetric surveying (7) The need for another technology (e.g., RTKGPS or RTS) to provide the geo-referenced inputs for photo-based 3D modeling	Around 1% error in volumetric measurement	[1,5,7,25,28,33–35]

Table 2. *Cont.*

RS Technology	Capabilities	Limitations	Accuracy	Refs.
Laser scanning	(1) High accuracy in generating point clouds (2) Simple and well-defined internal coordinate system (3) Homogeneous spatial distribution of range points (4) Ability to scan actively in darkness and shaded areas (5) Ability to measure areas without texture (6) Ability to scan a large area	(1) High cost of laser scanners (2) Scanning is a time consuming process (3) Occlusion problem and the need for a clear line-of-sight (4) More post-processing effort is required if the device moves (5) Limitations associated with modeling of edges and linear features (6) High storage capacity is needed (7) Not suitable for modeling moving objects (8) Not providing information about material type, texture, and color of the scanned objects (9) Eye-safety distance concerns	Around 2% error in volumetric measurement and few centimeters in ranging measurement	[28,34,36–39]

[1] Real-time kinematic GPS, [2] global navigation satellite system, [3] extreme cold or hot weather in an outdoor environment.

3.1. GNSS-Based Technologies

The RS technologies which operate based on signals received from satellites have various applications in civil engineering at global, regional, and local levels. Here we focus on the local applications of satellite systems in construction engineering and management, with emphasis on the global positioning system (GPS) as a real-time location tool. The most recent GPS-based systems with applications in outdoor positioning and tracking of resources are global navigation satellite system (GNSS), real-time kinematic (RTK), GNSS, and Russian GLONASS technology. These systems benefit from widespread access to the free global satellite system. However, since an unobstructed line-of-sight to the air is needed, the application of these systems for indoor tracking of resources is challenging in confined areas. Plus, relatively high cost of the accurate GPS receivers is another limitation of these systems for usage in a large-scale level. Below is a brief description of these tracking technologies along with their limitations and capabilities.

3.1.1. GNSS and Differential GNSS Techniques

The GNSS is part of a satellite application, providing signals from space that transmit positioning and timing data to GNSS receivers installed on the construction equipment. These receivers then use this data to determine the equipment location [5]. A typical 3D GNSS system, employed in a construction equipment consists of a roving receiver and a reference station [23]. The receiver located on the body of a construction vehicle receives signals from GNSS satellites to determine its position and direction. Besides, in a differential GNSS system the use of a reference GNSS station can enhance the location accuracy by sending corrections to the construction vehicle using a radio modem. These corrections can also be sent to a central unit over the internet through a GPRS modem connected to the levelling system. Furthermore, in a 3D GNSS levelling system, some highly resistant sensors are stuck on the body of a vehicle which make it possible to control any specific action demanded from the equipment in real-time. For applications related to the road pavement which need high accuracy in surface leveling, a 3D control unit processes the signals received from GNSS satellites and the reference station compares them with the stored designed values for quality measurement. A reference station is able to control any number of construction vehicles via onsite radio connections. The accuracy of such a system is around 2–3 cm [23].

3.1.2. RTK GNSS

One of the most recent innovations in using the GPS signals is the RTK GPS or RTK GNSS. It is a technology that provides position accuracy close to that achievable with conventional carrier-phase positioning, but in real-time. The RTK GPS employs a method of carrier-phase differential GPS positioning with centimeter-level positioning accuracy. It consists of a reference station and a roving station with a single or dual frequency GPS receiver. The GPS reference station sends carrier-phase and pseudorange data over a radio link to the roving station. Both single and dual frequency GPS receivers can be used, however, dual-frequency systems usually have faster ambiguity resolution and higher positioning accuracy over longer distances. To achieve the best results, the reference station antenna should be mounted in a location with less barrier and the radio link antennas should be as high as possible in order to maximize the link coverage [22].

Some RTK installations use combined GPS receivers at the reference and roving stations. Besides, a faster ambiguity resolution and higher positioning accuracies can be achieved by using GLONASS data in addition to GPS. However, the data collected at the reference station reaches the rover after some delays which causes some troubles for a real-time operation. This is mostly because of the need for the data to be formatted, packetized, transmitted over the link, decoded, and passed on to the roving receiver's software. This cannot all occur simultaneously and so there are some delays (latency) which depending on the link data rate might be up to two seconds. This delay might be acceptable for some static-point surveying applications, but it is not appropriate for kinematic surveys or for equipment navigation. To address this issue, the rover can extrapolate reference station measurements based on its own current measurements with an appropriate filter using an intelligent algorithm. This method can reduce solution latency to less than a quarter of a second, but accuracies are still limited to a few centimeters [22].

3.2. Robotic Total Station (RTS)

A RTS system adds tracking and automation capabilities to enhance localization and surveying applications on site. It consists of an electronic transit theodolite which is integrated with an electronic distance measurement (EDM) unit in order to measure both slope distance and angle from a prism attached to a target object. An on-board computer does the tasks associated with data collection and calculations pertinent to triangulation. The distance is measured by counting the integer number of wavelengths in a modulated infrared carrier signal, generated by a small solid-state emitter within the instrument's optical path. The angle is measured by electro-optical scanning of extremely precise digital barcodes etched on the rotating glass discs within the instrument. These localization devices have the angular accuracy of around one arc second, and distance measurement accuracy of one millimeter in static mode and few millimeters in dynamic mode with a high positioning update rate around 20 times per second. In fact, these systems are available in both statistic and dynamic mode for tracking and measurement of a steady and moving object respectively. For the latter one, positioning errors will occur if the angles and distances are not captured simultaneously depending on the vector and velocity of the moving target. To overcome this inherent error, manufacturers have developed total stations in which the angle and distance information are synchronized or they are with low latency [24].

Comparing RTS with technologies such as RTK GPS that we discussed earlier in this chapter, RTK GPS offers an alternative technology to RTS, but its reliability and accuracy in positioning a specific moving object are questionable. Furthermore, the RTS is one of the most accurate on-site survey instruments in the market with accuracy in millimeter level [7]. High vertical (elevation) accuracy is another factor that makes this system a better option than an un-aided GNSS system. These make RTS a very good choice for fine and finish grade applications. Moreover, a total station does not require line-of-sight to the satellites. It can be replaced with GNSS technology in case of obstructions or doing underground operations. Manufacturers have made it easy to switch between GPS and total station sensors for added flexibility. Having said that, total stations are constrained by the required line-of-sight between the total station device and the target prism which is installed on the machine or other resources. Moreover, their limited capability to track only one point at a particular time is another challenging issue [7]. Besides, their limited range can also become an issue. Some manufacturers have overcome this constraint by stringing total stations together in order to control an operation over greater distances without any interruption [24]. High investment required for implementing the devices of this type with acceptable performance specifications is another disadvantage of using these devices for localization purposes.

3.3. Barcode

Barcode is one of the most used technologies in industry for data entry with much less error rate and higher speed than those fully dependent on a human operator. A system using barcode technology consists of four main modules including barcode labels, scanner, decoder, and a database for gathering and analyzing information [20]. The barcode labels are available in one-and two-dimensional formats. Generally, the capacity of two-dimensional barcode systems is about 100 times larger than one dimensional system. However, one-dimensional barcodes are more common in industry by having a more accepted governing standard. The scanners are usually available in three main categories such as wand scanners, charged couple devices (CCD) scanners and laser scanners. They are categorized based on maximum length of barcode to be read, maximum distance away from barcode label, ability to read barcode labels attached to irregular objects and scanning reliability. Although wand scanners can read longer barcode labels, laser ones are considered superior with regard to all other criteria mentioned. Accordingly, laser scanners were used as data collection devices in the developed system. Decoders, on the other hand, receive the digitized barcode signals from scanners and then decode them to the correct data. Decoders are available in both fixed and portable versions; however, the latter one is more common in construction projects because of the distributed locations of activities in a construction project. Finally, the data are transmitted to the hub computer over a wired or wireless network [20,21].

A practical application of a barcode system in the construction industry is documentation of engineering data. In this way, barcode labels are produced and, then get assigned to each document. After that, coded information is scanned and transmitted to the database. For retrieving the information, developed queries are utilized and results will be shared with industry practitioners. Besides, the performance of this system can be improved by using feedbacks from industry practitioners. A system which utilize this technology is expected to improve productivity and to save time and cost of various tasks such as procurement and document control [20,25].

3.4. RFID

A typical RFID system consists of two components: readers and tags which are operating at a specific frequency, ranging from low frequencies to super high frequencies (e.g., from 125 KHz to over 5.6 GHz). RFID tags contain a microchip and an internal antenna with a specific ID and other data associated with an item, to send them to a reader upon its request. Based on their power source, tags can be distinguished as passive, active, and hybrid. Passive tags need to be activated by the electromagnetic energy that a reader emits. Therefore, they have shorter reading ranges and smaller data storage capacities. Active tags, on the other hand, rely on internal batteries for power supply, which increases their reading ranges significantly and enables additional on-board memory and local sensing and processing capacities. Hybrid tags also need to be turned on by a signal such as a satellite wave to transmit the data [1,11]. However, the cost of active tags is higher than the cost of passive tags and a battery replacement plan is also required [21]. Tags also can be categorized into two groups: read-only tags, on which data cannot be updated and, read/write tags with a memory that can be updated with new data. RFID readers, on the other hand, consist of a transceiver and an antenna which read data from tags (also write data on tags), and transfers the collected data to a host computer or server for future retrieval and analysis. The readers are available in both statistic format and hand-held roving devices based on the application required on site. With an antenna, the communication between the transponder and the transceiver is possible; however, its shape and dimension influence its performance characteristics such as frequency range. Available frequencies include low frequency (LF), high frequency (HF), and ultra-high frequency (UHF). Super-high frequency (SHF) or microwave is also used in some cases. Presently, UHF is the most widely used, because UHF passive tags offer simple and inexpensive solutions and most active tags operate on UHF [11].

Many researchers have investigated RFID technology with great capability in automatic identification and tracking of tagged objects on site. Moreover, it is applicable for the built environment due to its non-line-of-sight capability, wireless communication, and on-board data storage capacities [4,11]. Besides identification capability for tracking purposes on site, an RFID system can also be used for 2D localization of tagged objects. Studies have investigated the application of the RFID system for 3D localization of materials on site, however, only using RFID sensors faced some difficulties since a high density of reference tags in elevation are needed [30]. There are three main methods to localize an object using RFID sensors including triangulation, proximity, and scene analysis [4,26]. Triangulation determines the position of an object by measuring its distance from several reference positions (e.g., known position of some reference tags). Then, by using techniques such as trilateration, the position of other tags can be determined. The proximity method includes the measurement of the nearness of a set of neighboring reference points with fixed and known locations which are close to the target. In fact, in this method, the distance of the target object from reference points is not actually measured, but its presence within a certain range is only determined. Therefore, it has a rough localization accuracy. The scene analysis method estimates the location of a source of the signal (target object) by using a pre-observed data set of the monitoring scene with defined features in an electromagnetic signal characteristics map. Having said that, employment of this technique is not practical for construction sites with dynamic environment since it requires extra information and data storage [4]. Many researchers have used RFID for automated data acquisition from the site. For instance, in [4] a two-step process for localization of objects using a hand-held RFID reader and passive tags was used. In fact, they employed some reference tags for localization of RFID reader and then the location of the tags determined by using a path loss regression model and trilateration technique. Unlike barcodes, RFID tags can withstand harsh conditions with a much longer reading range which facilitates the automated material tracking on site. Having said that, its performance is degraded in vicinity of metal and liquid at high frequencies (e.g., UHF and microwave) [21].

3.5. UWB

Ultra-wideband (UWB) technology is a real-time location system (RTLS) which was initially developed for military use in the 1960s [16]. Its performance is almost similar to the active RFID system; however, it uses very narrow pulses of radio frequency (RF) energy which are occupied in a wide bandwidth for communication between tags and receivers. That is especially suitable for environments in which the multipath effect may happen. Like RFID technology, here we have some tags which are attached to the target objects, some receivers to read the tags data and a central hub with software to receive and analyze all these data for localization. However, the decision on choosing an appropriate type of tags really depends on the asset type (e.g., equipment, material, or people), its velocity and its operational environment. Various types of UWB sensors use different positioning techniques to estimate the tag's position. These techniques include the time of arrival (TOA) or time of flight, angle of arrival (AOA), time difference of arrival (TDOA), and received signal strength (RSS). AOA has some advantages over TDOA as it does not require synchronization of the sensors nor an accurate timing reference. While TDOA is less sensitive to changes in setup calibration but it still requires more cabling to have an accurate timing reference [3]. TOA, on the other hand, is more accurate when the two devices are in a clear line-of-sight, however, RSS is less influenced by changes in placement of objects in the environment.

A UWB system has two operational modes for data communication including wired and wireless modes. The wired mode in which all sensors are connected with timing cables can localize the tags using both AOA and TDOA techniques. While in the wireless mode, TOF and AOA are usually used for localization. It has been investigated that removing the timing cables will decrease the accuracy, but still a location accuracy of less than one meter is provided by the UWB system. However, in terms of spatial disruptions on the monitored area, the wireless mode is preferred since it imposes minimal spatial disruption due to less required cabling [16]. A number of companies [40–42] have provided newer versions of these sensors with better accuracy and reliability for various applications in construction. For example, Trek1000 Evaluation Kit is a product of an Irish company [42], which utilizes an atomic timer embedded in their PCB board that provides a high positioning accuracy in the range of a few decimeter for these wireless sensors. The product that is manufactured by this company is almost eight times less expensive than the available commercialized sensors, however, it is still not a final product with a protected enclosure. In [33], a set of experiments showed better performance of this product over other sensors by a mean ranging error of around 12 cm for this type of sensors. Plus, removal of the outlier data and using some data filtration techniques can improve both ranging and localization accuracy. Figure 1 illustrates a schematic view of these UWB sensors along with other RS technologies for automated onsite data acquisition.

(a)

Figure 1. *Cont.*

(b)

(c)

Figure 1. *Cont.*

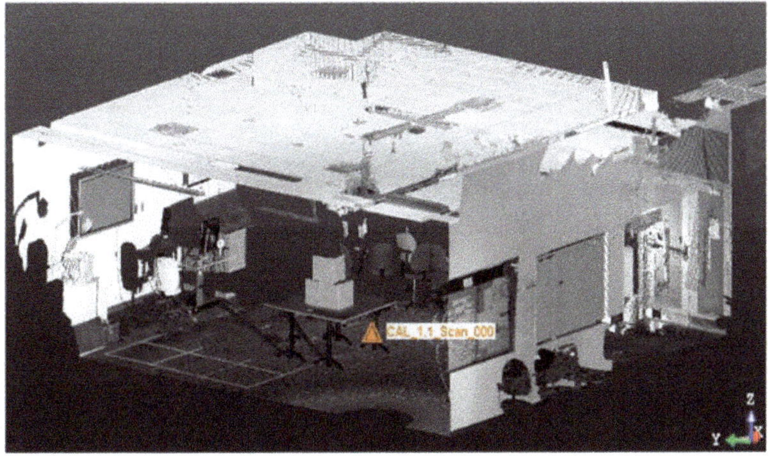

(d)

Figure 1. Various remote sensing technologies. (**a**) Automated data acquisition on site; (**b**) RFID system; (**c**) UWB system; (**d**) point cloud data for 3D modeling.

Several researchers have investigated the possibility of using this technology for identification, localization, and tracking of construction operations. They focused mostly on evaluating real-time tracking of workers, equipment, and materials in indoor and outdoor environments [3,16], and further on improving the quality of data received by the UWB system. It was shown in [16] that placement of static reference tags in the field can improve the positioning measurement of UWB sensors. Furthermore, the geometric configuration of the reference point and the receivers could improve the accuracy of the range measurements. Applying a suitable data enhancement method such as Kalman filter (KF) can also improve the localization accuracy. In fact, it mitigates the data uncertainties through prediction and update steps which would result in an optimal estimation of the desired parameter by using probability density functions and noise factors. In [36], the authors employed a particle filter to improve the localization accuracy of three different UWB devices by using a Bayesian filtering approach. In their experiment, they used the parameters in line-of-sight (LOS) condition to improve estimated locations in non-line-of-sight (NLOS) condition by using a combined Gaussian and Gamma distribution. In [43], the integration of extended Kalman filter (EKF) and extended unbiased finite impulse response (EFIR) filter was used in order to have more optimized and robust results. The integration of these two filters provided both robustness and accuracy required for localization with UWB sensors in a noisy environment like a construction site. The fusion of these two filters was based on defining a time-varying weighting factor for each of them. In fact, before reaching a specific time in the EFIR, the EKF was the active filter to find the optimized solution. Subsequently, the integration of these two filters comes to play. Therefore, the roving tag position is calculated by the summation of these two weighted factors in each estimated position. Researchers in reference [3] also showed that by attaching more than one tag to each object, and then averaging the positioning data acquired by these tags, the localization accuracy will be improved in comparison with the use of one tag. A higher number of receivers on site also increases the accuracy of this system [16].

The UWB sensors are less influenced by metals and high humidity (a limitation of ordinary RFID systems) and require only a few sensors. Comparing with other technologies such as RFID or ultrasound, UWB has unique advantages including longer range, higher measurement rate, better measurement accuracy, and relative immunity to multipath fading [11,16]. Besides, UWB offers a wide spectrum and fine time resolution of the signal which results in a more reliable positioning in a harsh environment (e.g., good performance in vicinity of moisture or snow). However, violation of the line-of-sight can lead to a decrement in the signal strength, so an accurate design and sensor placement strategy is needed. Unfortunately, the cost of UWB sensors is still high in comparison with other sensors [31].

3.6. Bluetooth, Infrared, and Ultrasound

Bluetooth technology has recently become more popular with the introduction of Bluetooth low energy (BLE) technology in 2010. In contrast to Wi-Fi, the battery usage on the client side is very low for BLE technology which can enable it to work up to three to four years with only a single small battery. However, the fluctuation in the strength of the received signal is a challenging issue when using these sensors. Saying that, due to the inherent nature of the system, it is highly recommended for accurate positioning. Besides, it is incorporating well with other techniques such as filtering techniques and motion sensors [31]. These technologies can also be used to prepare a congestion map which shows the areas onsite with more accident risk or activity conflicts, by information acquired from workers and vehicles behavioral patterns [19].

Infrared is electromagnetic radiation at a wavelength longer than red light. Infrared wireless technology can be used for short and medium range communications and control purposes, however, unlike radio-frequency (RF) wireless links, it cannot pass walls. Ultrasound, on the other hand, is usually implemented by using special microphones in a room. In fact, items that need to be tracked have a tag ID with a speaker attached. While the sound is stopped by the door, wall, or window but it is not stopped by other physical objects in the room such as people. These sensory systems need hardware infrastructure for signal communication and they have scalability limitations. Actually, they are rather focused on high accuracy than ease of deployment. Having said that, it requires a line-of-sight due to its limitation to penetrate in materials [31].

3.7. Vision-Based (Imaging) Technologies

Many pieces of research on automated progress reporting have focused on creating 3D models or point cloud models for both indoor and outdoor environments. These point cloud data are generated by using image frames or laser scanning data of the site. Generally, laser scanners can provide more accurate point cloud data but with more processing effort [1]. However, in recent years due to the availability of high-resolution cameras, more powerful computers, and new advancements in image processing techniques more accurate and real-time point clouds are generated by only using image frames.

3.7.1. Photo/Videogrammetry

Photogrammetric surveying has been widely used in construction engineering and manufacturing for 3D modeling, localization, and map production. Virtual reality (VR) and augmented reality (AR) research have referred photos as the most straightforward and cost-effective inputs for field data collection in construction management. The photogrammetric technique advantages over other traditional surveying methods include: (1) non-contact measurements, (2) relatively short turnaround cycle of data processing and modeling, and (3) measuring multiple points at one time [7]. Furthermore, researchers have extended photogrammetry into videogrammetry. Since video frames are sequential, pixels available in each frame are progressively reconstructed based on the previous frame(s). This characteristic of videogrammetry has made it possible to reconstruct various elements more quickly in civil infrastructures [1]. However, an extra video post-processing effort is necessary to match the time-stamped videos of each camera together. Generally, photogrammetry includes putting some cameras with a standard distance and angle of view from each other in a jobsite. Saying that, by advancement in simultaneous localization and Mapping (SLAM) techniques, the point clouds can also be generated by a moving camera. For registration of photo frames together, there is a need to have some stationary site control points (usually more than three) in each photo frame for 3D modeling. So, images taken from different angles are used to build the 3D model, by pairing and matching some common points in the two or more neighbor frames. It should be noted that the time differences between the photos and the control point survey data must not exceed the preset value (e.g., one second), otherwise, the photos and the survey data are discarded for ensuing 3D modeling. The 3D model coordinate system is set up by selecting an arbitrary origin and scale. The image and object coordinates are then evaluated based on defining specific parameters in related photogrammetry equations. The scale of the model is determined by using coordinates of some points which are determined by a terrestrial station such as robotic total station (RTS) unit. Reducing the modeling time and cost by generating the point cloud from both photos and structure from motion (SfM) method is also an acceptable approach. In fact, the point cloud-based application is aimed to reduce the effort in producing 3D as-built models. However, the model object must be stationary during the laser scanning process while removing the noise data (redundant or irrelevant information) requires considerable time and expertise. Thus, point cloud-based technologies are not suitable for modeling a particular moving object on a near real-time basis on site. In order to solve some of the limitations mentioned above, developing an alternative system, which can directly track and survey target points by use of synchronized RTS units would be useful. Such a system can automatically provide the geo-referenced inputs needed for photo-based 3D modeling [7]. This platform is reviewed in the section for integrated photogrammetry with RTS technologies.

Recent studies on this field have been focused on videogrammetry to create point cloud data through SLAM technique. In [35], a platform was designed for a real-time progress reporting in which a client connects to Wi-Fi and takes a video record of the site. Then this video (or image frames) is sent for mapping of the construction site by producing the site point cloud data. Then, through the reconstruction step, a dense point cloud with enough details is generated in which a mapping of the visual objects available on site is achieved. Then, for any new query image, the location/camera pose of the input image would be retrieved by using a feature-based matching with the base 3D point cloud map generated earlier. Finally, by aligning the parameters of the 3D point cloud with a geo-referenced BIM model, global localization of the frames can be estimated.

3.7.2. Laser Scanning

Using 3D point clouds produced by laser scanners has already been used as a standard practice for generating as-built information in construction, rehabilitation and facilities maintenance in areas ranging from process plants to historical preservation and structural health monitoring [9]. These devices use near-infrared (0.7–0.9 μm) wavelength in order to create 3D point clouds of the objects available in an environment. There are some important factors which determine a laser scanner performance such as angular resolution, scanning speed and the range required [37]. Clearly, the quality of the point clouds usually depends on the number of times doing the measurement for each point. As the number of measurements per point increases, the quality of the point cloud increases. Terrestrial laser scanning (TLS) is one of the best available technologies to capture 3D information on a project with high accuracy and speed. It has made it possible for construction managers to track progress, control quality, monitor health, and last but not least to create as-built 3D models of the site. Many studies have been focused on the interface between laser scanned data and a 3D model such as CAD or BIM models. For instance, in [9] a 3D CAD model was used as a prior source of information for detecting objects from the point clouds. However, these approaches are unable to identify objects which are not in their exact locations pre-described by as-designed model or located in the distance greater than a tolerance value specified in the object recognition software. Furthermore, they are incapable of providing sufficient information about non-structural activities such as the welding, inspection, piping system, etc.; however, BIM systems may help to facilitate these operations in future [6]. In fact, with the advent of 3D BIM, many newer approaches actively use the 3D information contained in BIM models to develop supervised object detection and recognition algorithms that can process the point cloud data more effectively. Nonetheless, the process of recognition, detection, and identification of objects in 3D TLS data remains an open challenge with marketed software while it is needed to be more studied in order to enhance the compatibility of these technologies with available software.

The objective of the process of 3D laser scanning is to scan a physical space or site to create an accurate digital representation of it. The scan to BIM is a term that refers to the process of scanning an existing building, area or site and creating a 3D model from this data. In fact, this process is proved to be more exact than conventional inspectional data that are usually accomplished using a measuring apparatus. The scan to BIM process can be divided into two stages. The first stage is the fieldwork where the surveying team goes to the site and collects data using their 3D laser scanners. The second stage is the office work where the raw data is calibrated and could data is used to create a 3D model [9].

4. Integration of Various RS Technologies

Due to the diversity of the construction resources, harsh environment, and complex nature of construction projects, the data acquired by each RS technology is missing or erroneous. Furthermore, each RS technology has its capabilities and drawbacks that limit its application. In fact, depending solely on a single technology to acquire data on-site in an accurate and timely manner is unreliable. In this way, the integration of various RS technologies together can improve the data acquisition process. Plus, it can help to overcome the limitations of an individual technology and makes it possible to provide the required information even if one of these data sources is not fully available [16,25]. This integration could happen in different scenarios. In a typical scenario, the integration includes the separate employment of various RS technologies in different stages of construction activity. Examples of this type of integration are available in [25], in which integrated barcode-RFID and photogrammetry-laser scanning systems are used for resource tracking and quantity measurement on site. In another scenario, the data acquired by each technology could enhance the data acquired by other technologies which result in having more accurate and robust information. An example of the latter integration approach is the data fusion of various RS technologies together or a multi sensor data fusion (MSDF) approach. In fact, data fusion refers to the combining of information from multiple sources in order to improve the quality of information obtained separately from each source. It is utilized to make inference decisions about the state of a construction project based on the data that are acquired from different sources [16]. Therefore, informed decisions and objective assessments of the activities progress on site are not only based on a single data source. In fact, a number of data sources are combined together in order to provide sufficient details about various activities on site. It should be noted that the level of data fusion is an important factor in order to have a more robust and accurate approach [28]. Recent studies in this field are focused on the integration of various RS technologies in order to benefit from identification, localization, and visualization capability of each of these technologies with applications in tracking objects and activities on site. Table 3 shows some of these integrated systems along with their capabilities over standalone usage of each technology and their drawbacks for implementation on site.

Table 3. Capabilities and limitations of integrated RS systems

Type	Integrated RS System	Capabilities	Limitations	Refs.
Positioning systems integration with other sensory data	GPS + Barcode	(1) Relatively low cost (mainly goes for GPS receiver); barcode: Label ~ $0.1, Reader ~ $100–500 and GPS: Receiver ~ $200 Satellite signal free (2) High level of standardization and reliability (3) More scalable for projects of varying sizes (4) Straight forward implementation	(1) Need for free access to space for GPS system which makes it unsuitable for interior environment (2) Limitation of barcode tags in differentiating between items of the same kind (3) Not fully automated approach	[5,22,23]
	GPS + RFID	(1) Easier material identification due to non-line-of sight capability of the RFID tags (2) Providing both identification and localization data simultaneously	(1) The need for free access to space for GPS system (2) A large number of RFID readers are required which increases the cost; RFID Tag ~ $1–50 and RFID Reader ~ $1k–5k (3) Boundary constraint limitations in cluttered environments	[7,24]
	Sensor-aided GPS (SA-GPS)	(1) Stability to be used in various construction operations (2) Real-time tracking and reporting capabilities (3) Not sensitive to the ambient environment (4) Having both location/action recognition capabilities (5) Providing continuous update of the location estimates	(1) Obstacles associated with data fusion, coordination, processing, and reduction of data to produce meaningful conclusions (2) Requiring relatively more time for post processing (3) Drift inherent to sensors (4) Initialization and calibration difficulties	[26,44–47]
RFID integration with other sensory data	RFID + WSN	(1) Make it possible for tags to communicate with each other (2) Facilitating the negotiation of RFID readers together (3) Increased positioning accuracy (4) Decreased energy consumption in comparison with an individual WSN [1] system	(1) Sensor does not provide any power until tag is not in the radio frequency field to communicate with reader (2) Reading range decreases as the system starts using energy	[26,29,48,49]
	RFID + LS	(1) Faster and more robust recognition and positioning of objects in 3D modeling (2) Alleviating the difficulties associated with the laser scanners object recognition specially in a crowded and full of furniture environment	(1) Poor positioning performance when the objects are in the corner of a room (2) Laser scanners are relatively expensive (TLS could cost more than $60k) (3) Need for a facility management database of the available objects in advance	[14,39]
Integration of the vision-based technologies together and with other RTLS	Photogrammetry + laser scanning	(1) Less number and time for scanning are required (2) Less number of digital photos are required which results in less computational effort (3) Better object recognition capabilities by adding depth information from laser data to the available digital photo planes (4) Convenient for modeling objects with unclear geometrical properties (e.g., excavation activities) (5) More accuracy in the localization of edge points (6) Providing more details about material, color, and texture of the objects which improves the geometry and visual quality of the 3D modeling (7) Self-calibrating capabilities for cameras	(1) Expensive laser scanners (2) Problems associated with calibration and orientation of data acquired from two sensory sources, e.g., geo-referencing issues (3) Less effective when the site is crowded or covered by enclosures	[25,28,33,34,38,50]
	Photogrammetry + RTS (robotic total station)	(1) Enhanced scaling and 3D modeling capabilities (2) Better measurement accuracy (down to few decimeter) rather than using individual photogrammetry techniques	(1) High cost of RTS devices (2) Synchronization of digital cameras with RTS devices (3) Limited capability for tacking multiple moving objects	[7,24]

Table 3. *Cont.*

Type	Integrated RS System	Capabilities	Limitations	Refs.
	Photo/ videogrammetry + UWB	(1) More reliability in positioning in comparison with individual use of each technology by replacing the data of one sensor with another in case of any failure in each sensor (2) More accurate and timely information are provided by proving more automation and decreasing computational effort in photogrammetry (3) Portability	(1) Calibration difficulties specially for large scale sites (2) Error propagation phenomenon due to the intrinsic inaccuracy in each technology (3) Less accuracy in comparison with other geo-referenced photogrammetry techniques (e.g., using a RTS or GNSS system)	[3,32,51]
	Laser scanning + UWB	(1) Overcoming challenges associated with the presence of multipath, NLOS [2] propagations and low visibility (e.g., occlusion, dust, fog, etc.) (2) Less accumulated error in comparison with an individual usage of laser scanning technology (3) Ease of need for prior knowledge or control inputs	(1) Discrepancy between the accuracy of the UWB mapping and that of LiDAR mapping (2) High implementation cost	[6,15,28]

[1] Wireless sensor network, [2] none line-of-sight.

4.1. GPS-Based Systems Integration with Other Sensory Data

The localization capability of GPS sensors could be integrated with the identification capability of technologies such as barcode and RFID for more accurate tracking of materials on site. Plus, integration of the GPS sensors with other aided sensors could provide more accurate and robust information about the site condition and the resources available in it.

4.1.1. GPS Integration with Barcode

It includes integration of the barcode and GPS technology for material identification and localization respectively. In this technique, various items are tagged by barcodes and their real-time locations are sent to a central database through Wi-Fi as scanned by a barcode reader equipped with GPS. For instance, in [52] an automated system for material tracking based on the integrated barcode and GPS technology was developed in which the location of materials is determined with three-meter accuracy. In fact, automated and real-time identification and localization of materials make the retrieval of these elements much easier which results in a better material tracking and management system.

The advantage of this tracking technology over other technologies includes relatively low cost, high level of standardization, and scalability for employment in various projects. Barcode labels have been widely used for many applications in industry with defined protocols to be attached to various materials for many years with relatively low cost (see Table 3). However, the reader should be very close to the barcodes which is not practical on many occasions for tracking items on site. Besides, the barcode system cannot differentiate between the items from the same type which decreases the level of detail provided for the identification process [52].

4.1.2. GPS Integration with RFID

Integrating GPS with RFID would provide a reasonable solution to benefit the localization and identification capability of both technologies. There are various scenarios to implement such a system in order to have a better outdoor resource tracking and management. In one scenario, the GPS receiver is attached to a roving RFID reader in order to provide information about the position of the RFID reader. By knowing the real-time location of the roving RFID reader, the 2D location of the tagged items are identified by using trilateration technique. Having said that, this technology is not accurate enough for localization of objects in elevation since GPS-based technologies cannot provide accurate information about height. In [12], a system consists of spatially distributed mobile RFID readers equipped with GPS technology was developed which could detect mobile RFID tags. The readers transfer the data to a central server via a communication network in which the server filters the data using a boundary condition-based algorithm to estimate the continuous and real-time 3D locations of tagged objects. In fact, boundary constraints were introduced to facilitate RFID-based localization applications to overcome the challenge associated with unknown tag-reader distance [12,13]. In another experiment [2], researchers tried to improve this approach by integrating the RFID sensors with a real time kinematic (RTK) GPS device. In their system, the location of the reader was recorded as a boundary point, which was acquired by the RTK GPS receiver roving together with the readers. Then, the tag's location was estimated using bilateral method in which by creating four equal-radius spheres centered at four boundary points (i.e., four readers that form the boundary condition with the target tag) and then checking whether an overlapped common volume exists. Besides, they identified two important factors to improve the accuracy of estimations including: (1) the reading angle between a tag's orientation and a reader's shooting direction, and (2) the geometric configuration of the reader locations in each time step. The geometric configuration of the readers was assessed based on a parameter called Spatial dilution of precision (SDOP) which was formulated to improve the localization accuracy [2].

In another scenario, a separated distribution of RFID readers and GPS receivers are employed to benefit the positioning and identification capability of these two technologies. In [5], a system based on the integration of RFID and GPS technology was developed in order to enhance the safety of workers on site by improving blind lifting and loading crane operations. The system architecture includes a GPS receiver and an RFID tag attached to the crane hook, and active RFID tags attached to the safety helmet of the workers. During the crane operations, the location of the workers and crane hooks are thus read by the RFID readers installed on the site and then imported to the BIM environment. In fact, their system obtains and compares the real-time position of tower cranes and construction workers and, if any unauthorized worker is detected in a pre-defined risk zone, an order by central hub would be sent to cut off the crane power. In a case study done for this approach, an accuracy of approximately ± 30 cm was recorded [5].

As stated, accuracy is an important factor when deploying a system for data acquisition in a construction site. Therefore, the integration of RFID with other positioning technologies like GPS-based systems is a solution to have higher accuracy in tracking tagged items. Having said that, GPS solutions present some drawbacks in certain environments especially when it comes to the interior of buildings. In order to avoid the limitations of RFID and GPS-based systems, the combination of ultrasound technologies and RFID can improve the positioning system [27,53].

4.1.3. SA-GPS

A sensor-aided GPS (SA-GPS) prototype consists of a microcontroller equipped with GPS module, and a number of sensors such as accelerometer, gyroscope, barometric pressure, strain gauges, etc. This prototype consists of some sensors that are installed on various resources on site and using a gateway to communicate together and with a central server. In this way, it can provide more accurate information about site conditions and working hours of the target resources [26,44]. Several studies have investigated the effectiveness of this system for progress estimation and tracking of various objects and activities. In [26], a system based on SA-GPS was designed for tracking and progress assessment of the earthmoving operations. They concluded that there is a considerable decrease in average absolute percentage error in the project progress estimation in comparison with a system working only based on GPS from 12% to almost 3%. In a similar study in [47], a smartphone with sensor technologies—such as GPS, accelerometer, and gyroscope—was put inside construction equipment. Then, the time-stamped data for the equipment position, acceleration and angular velocities were collected by using some commercial data logger applications available on the smart phone device. After a feature extraction process in which some features were defined based on the level of details required for an activity recognition process, a subset of originally extracted features were employed to train the supervised machine learning algorithms (e.g., artificial neural networks (ANN), K-nearest neighbor (K-NN), etc.)

for activity classification. Finally, through an activity recognition step the classified actions were defined as real actions in the site and these data got used as input for a simulation model. These estimations can also be used for a more accurate and enhanced earned value analysis (EVA) and productivity assessment [26,44]. Furthermore, integration of the GPS with inertial navigation system (consists of an accelerometer, a gyroscope sensor and a magnetic compass) can be used to estimate both the position and velocity of a moving object such as construction equipment. However, both GPS and INS estimations have some limitations and errors. In fact, the GPS error is caused by the time difference between the satellite and a moving device, geometrical difference among satellites and multi-paths of the GPS signal. INS, on the other hand, has good capability for precise location-information in short term but the accumulative error caused by environmental disturbances could be too big if it is used for a long time. Therefore, it would be very difficult to properly handle the noises of INS sensors for the precise localization of a moving object since the motion of the mobile object is nonlinear and the noises are not Gaussian. In this way, data fusion of these two sensory sources by appropriate filtering of the velocity-dependent noises in the INS signals would result in a more accurate and reliable data for the purpose of the outdoor tracking of the equipment [45]. In another study [46], other onboard aided sensors (e.g., inertial sensors) in addition to the GPS sensors were employed to investigate the effect of road slope in positioning and tracking of the equipment. They concluded that this approach can improve the accuracy and reliability of the equipment position estimations in comparison with approaches based on planar position estimation algorithms. In fact, depending on the planar models for positioning of the equipment in the sites with terrain condition would be more erroneous as the slope and curvature of the road vary. Therefore, it would be possible to improve the accuracy and reliability of pose estimations by compensating for the error due to the road slope. Finally, the integration of the GPS with other aided sensors provides capabilities with applications in the advanced driver assistance system (ADAS) and autonomous vehicles. This can provide more automation in the construction activities which can enhance the productivity and safety of the activities as a long-term objective in the construction industry [46].

Integrating GPS data with aided sensory data can compensate for the limitations in utilizing the individual GPS information. In fact, it would make it possible to capture the big picture of the construction operations by accurately distinguishing between productive, idle and out of service times during construction activity. Furthermore, the idea for using integrated sensors on a mobile phone device can mitigate the difficulties associated with mounting sensors on equipment body (e.g., attachment and detachment of various sensors for every data collection session), dust in a construction site, data storage issues, and unstructured arrangement of resources which makes calibrating of sensors difficult. Furthermore, it can improve inaccurate and unrealistic simulation models which are usually working by static input data and are built upon historical information during the pre-construction phase [47]. Having said that, the process of fusing the data acquired from various sources is an obstacle for achieving a real-time and automated procedure [26,44].

4.2. RFID Integration with Other Sensory Data

Only a few technology sets are available for indoor positioning. The reasons for this poor diffusion include high costs compared to the value achieved and technical limitations such as precision, reliability, and performance of such systems. In this way, some researchers have investigated the integration of a wireless sensor network (WSN) with the RFID. This approach provides new capabilities in order to have a better estimation by facilitating the identification of materials located in the scene and estimating their position at the same time [26,27]. The integration of the RFID technology with other RS technologies can also improve onsite information for having efficient management of construction operations during a project lifecycle [27].

4.2.1. RFID Integration with a Wireless Sensor Network (WSN)

In a system based on the integration of the RFID and WSN, smart nodes of sensors are employed to identify objects that their sensed information is needed, while the smart node readers detect tagged objects to recall them at a convenient time [26]. In fact, by integrating a large number of sensor nodes with RFID tags and readers, communication of the RFID tags with each other and also the negotiation of the readers together will be facilitated which can increase the efficiency of a tracking system. There are different scenarios for this integration including: (1) integrating tags with nodes in which sensors collect sensed information and RFID identify the objects, (2) integrating tags with sensor nodes and wireless devices in which tags can communicate with wireless devices and other tags and imitate a multiple hop network, (3) integrating RFID readers with the sensor nodes and wireless devices in which readers are able to communicate more efficiently with each other in this network and sending information to a host computer, and finally (4) integrating a separate combination of the RFID and the sensor nodes in which an integrated nodes system is not required since WSN and RFID data are processed in a software layer [48].

Various authors have investigated this integrated system for various tracking purposes by emphasizing on the indoor environment. For example, in [48] this system was used for the purpose of supply chain management. In their system, the RFID technology was employed for identification and WSN was applied for sensing the environmental factors such as humidity, temperature, and air quality since some products needed to be kept in a certain condition. In this way, tags were embedded in objects and readers were integrated with nodes for the purpose of identification and monitoring of the environment [27]. In another application, researchers in [26] employed a WSN for a better data transfer between RFID nodes in which they confirmed the positive contribution of WSN on communication and network flexibility. In another case study by them, they used a similar system for tracking construction operations by using tags as mobile units, which were carried by or mounted on the tracked resources. Besides, in their system, the third dimension (elevation) read from a pressure sensor [26]. In [49], also a hybrid WSN-RFID system was investigated for tracking people and objects in indoor scenarios. Through their case study, they showed that this hybridization provides an improvement of 1.6 m in RMSE and of 4% in data availability in a sample tracking experiment. The application of this integrated system for enhancement of safety management in a site has also been investigated by some authors. For instance, in [29] a system based on the integrated WSN and RFID technologies was employed to provide a preventive solution for enhancing the site safety. To do that, they proposed a real-time tracking system to provide pro-active information for crucial and controllable objects by attaching some RFID tags to those objects on site. A sensor node, on the other hand, transmitted sensing information to a router while the RFID tags transmit identification and positioning information to the active RFID readers. Finally, all information was sent to a central server for application rendering and alarm generation to make an early warning while a worker is too close to the hazard area [29].

The WSN offers good radio coverage but with low positioning accuracy due to high noise on received signal strength index (RSSI) measurements. On the other hand, RFID technology provides positioning information but with limited coverage and temporal discontinuity. Therefore, integration of these two technologies makes it possible to enhance the accuracy and robustness of the indoor positioning and tracking by providing better communication between RFID tags, RFID readers, and tag-readers. Having said that, some experiments have shown degraded performance of the RFID systems in object localization as active RFID tags and readers starts to get their required electrical power from WSN nodes [26,29,48].

4.2.2. RFID Integration with Laser Scanning (LS)

The integration of laser scanning with RFID technology could be performed in different scenarios. In a typical scenario, both RFID and LS technologies are employed separately based on the type of information required from each object or activity [14,39]. For example, in [25] a model to track the state of materials on site was proposed to integrate the RFID technology with a terrestrial laser scanning

(TLS). However, the two technologies were employed in two different stages through a construction supply chain system. Another scenario is an effective integration of these two technologies in which the identification and positioning capability of the RFID system help to enhance the accuracy of the point cloud data that are generated by a laser scanner [14]. Despite various techniques for object recognition from images on site, recently the process of directly extracting different types of objects from point cloud data through 3D reconstruction of building interiors is becoming more popular [54]. Besides, there are a number of advanced software solutions for 3D modeling in CAD/BIM models by using the TLS point cloud data. However, they still do not address the need for modeling all types of objects available in a room. In fact, in order to have robust point could data of the elements in a room, it is required that the room be as empty as possible when scanned. In this way, in [14,39] a methodology was proposed based on the integration of laser scanner technology with an identification technology such as RFID in order to enhance the process of object recognition out of the point cloud data. In their work, the process of extracting objects from point cloud data were not necessarily limited to structural elements (e.g., floor, ceiling, wall, windows, etc.) but also small objects and components on site were recognized, localized, and modeled in a 3D model (e.g., BIM or CAD). So, by attaching RFID tags to the objects (i.e., furniture, facilities, etc.) available on site, it was possible to identify, localize and model these components in a final 3D model too. In this scenario, the structural elements were modeled by using the TLS data. Then, the other objects were identified with RFID tags attached to them and sensed by a reader which was installed next to the laser scanner. To do that, discriminatory geometric information was used to recognize the target objects in a point cloud data. It was done by measuring dimensional parameters of an object (e.g., length, width, and height) or the pattern of different elements in the object. Since each object had a specific ID, it was possible to retrieve its geometric information from a facility management database and to model them more precisely in the 3D model. Saying that, in order to enhance the accuracy of the positioning algorithm, the 3D model of each piece of objects was stored in the facility management (FM) database with a mesh resolution which is automatically adopted from the point density of the acquired point cloud. This data fusion approach provided a more comprehensive and accurate 3D modeling of the as-built status of the site for the purpose of the progress reporting and productivity analysis of activities varying from structural to facility installations, etc. [14].

This integration provides an acceptable measurement accuracy by localization and angular error in the range of few centimeters and degrees respectively. Furthermore, it can facilitate the process of object recognition from the point cloud data acquired by a laser scanner. Having said that, the approach has some drawbacks such as degraded localization accuracy when the objects are in the corner of a room. Besides, it is needed to create a facility management database including all objects available in the site prior to the modeling process [14].

4.3. Integration of the Vision-Based Technologies Together and with Other RTLS

Using vision-based technologies such as photogrammetry, laser scanning, and their integration are appropriate choices to provide enough point cloud data in order to create 3D models of a site. More specifically, their integration enhances the point cloud data that are required for 3D reconstruction of the objects available on site. However, all these approaches have their limitations associated with their employment in a congested area and their drawbacks in modeling non-stationary and small objects such as furniture, equipment, etc. In this way, by integrating these vision-based technologies with a real-time locating systems (RTLS) technology such as robotic total station (RTS) and UWB, it is possible to overcome some of these problems. Table 4 provides a comparison between these integrated systems over the individual usage of photogrammetry or laser scanning. They are different in terms of the level of automation in data acquisition, cost, reliability, and scalability to be used in various projects.

Table 4. Capabilities of individual and integrated vision-based technologies

Method	Data Acquisition Effort	Processing Time	Affordability	Data Accuracy and Reliability	Scalability
Photogrammetry	√	√√	√√√	√	√
Laser scanning	√√√	√√√	√	√√	√√
Photogrammetry + laser scanning	√√	√√	√√	√√√	√√
Photogrammetry + RTS	√√	√	√√	√√	√√
Photogrammetry + UWB	√	√	√√	√	√√√
Laser scanning + RFID	√√	√√	√	√√	√√
Laser scanning + UWB	√√	√√	√	√√√	√√√

√: Low; √√: Medium; √√√: High.

4.3.1. Photogrammetry Integration with Laser Scanning

Due to the computational complexity of photogrammetric surveying, construction management researchers have attempted to decrease the number of required photos by imposing geometric constraints and automating the modeling process based on pattern recognition and feature detection [7]. Moreover, a solution based on the integration of the photogrammetry data with the data acquired by a 3D laser scanner is also investigated aiming to reduce the effort required for 3D modeling [25]. In fact, due to the geometric stability of digital images, they are very suitable references for inspection of the data acquired by laser scanners for 3D modeling. There are two approaches to integrating these two technologies. In one approach, the photos and scanned data are taken from the same positions, for example, the camera is installed on the laser scanner. In this approach, the extra process for coordination of different photos with scanned point cloud data is not needed, however, the photos are taken from the same distance and angle of view as scanning data which would decrease the details in photos. In another approach, photos are taken separately from scanned data but in a closer range and from different angles. Saying that in this configuration, an extra step for the orientation of the digital images with the scanned data is required [33]. This orientation can be done by choosing some common points or features in both data formats and then trying to merge them together, or by measuring the coordinates of several common points (tie points) available in both data formats in different positions [33,50].

According to the literature, the new capabilities provided by this integration have brought new applications in construction management. For instance, through a case study in [50] the quantity of an excavation work accomplished was rapidly tracked for automated progress reporting. By using this integration, they proposed a more robust and timely data acquisition procedure in which fewer images and less scanning time is required to produce acceptable results during the 3D modeling process. They could also overcome limitations in photogrammetry technique associated with modeling objects with unclear geometrical shapes (e.g., excavation work) [25,50]. Another application for this integration is the crack detection in an object or structure. In fact, this integration helps to improve the geometry and visual quality of the final 3D model. During the data collection phase, the information about the edges and linear features in the surface such as cracks are achieved by analysis of the images, while the information about the object geometry is provided by the scanning data. In this way, a more complete 3D model of the scene can be generated with sufficient and clear details about the color, texture, and material of various objects [33,34].

The integration of a laser scanning system with photogrammetry would provide a timely, cost effective, and accurate 3D model of a structure which can decrease the human exposure costs and risk exposures caused by the lack of information during construction [38]. In fact, more accurate visual information results in less amount of rework and change order which would improve the productivity indexes [50]. Moreover, these 3D models are used for making accurate rehabilitation, maintenance, and renovation plans during the structure operation. However, the registration and alignment of these two sensory data matters. On one hand, both digital images and scanning point cloud data have internal errors that affect the registration process. On the other hand, factors such as the required accuracy, resolution and accessibility of the object may affect the chosen of the optimal method [28,33,34].

4.3.2. Photogrammetry Integration with RTS

It is a time-dependent dynamic modeling technique in which the assets are identified by using digital imaging capabilities, while RTS units automatically provide the geo-referenced localization information for photo-based 3D modeling. In fact, each RTS unit automatically locks on a reflective glass prism which has been installed on the equipment or target object and tracks its location. Saying that, the cameras and RTS units should be placed on site with a good distance and angle of view from each other in order to enhance the quality of photogrammetry modeling and to reduce the residual error due to human factors. Besides, all cameras timer and RTS units should be time-stamped and synchronized together and being initialized in a predefined object coordinate system [7].

This integrated system makes it possible to have a dynamic visualization and tracking system for a moving object on site. In a case study [7], the sling lengths of a moving rigging system engineered for lifting heavy modules in industrial construction was tracked to ensure quality and safety. The absolute location error was estimated to be in the order of five centimeters for the measurement of the moving sling with a length of about 20 m.

Some newer versions of the total stations have spatial imaging capabilities with longer ranges. Moreover, by improving the machine control application, in future more integrated and synchronized actions will be done by these systems [24]. This integration enhances the 3D modeling of the moving objects by providing more accurate and reliable scaling information for the image processing-based model. However, the high cost of the RTS units and their limited capability to tack multiple moving objects are challenges for expanding their application in tracking a large number of moving objects [7].

4.3.3. Photo/Videogrammetry Integration with UWB

Some researchers have investigated the integration of the image-based data with UWB data in order to address problems associated with individual employment of an UWB system in a harsh and complex construction site with missing or erroneous data [3]. In fact, depending on only a single technology such as UWB to acquire required information from the site in an accurate and timely manner becomes unreliable. In this case, the employment of multiple sensors increases the reliability of the system by providing redundant and timely information as a result of either the speed of operations done by each sensor, or the parallel processing that may be achieved as part of the integration process [16]. In this integration, two important factors are scaling factor and transition factor which are used to register the point cloud data with the UWB data. Besides, there are some ways to improve the accuracy of such integration, for example by using optimization techniques, using UWB estimated positions for photogrammetric adjustment and time synchronization of the UWB and camera devices during the data acquisition process. Once the initial estimation for transformation from the photo coordinate system to UWB coordinate system is achieved, then the time-dependent camera position is corrected in order to re-estimate the transformation parameters with updated coordinates until convergence to an acceptable error [32,51].

Some potential applications of this integrated system on site include tracking of equipment and creating as-built 3D models for the purpose of activity progress reporting. This is done by improving the quality of the acquired data and providing more robust reconstruction information. Besides, it would be possible to localize the equipment even if one of these data sources is not fully available [16]. For the first application, the structure of such a system is based on the employment of the two sensory data sources in which each equipment is tagged with a UWB tag. Meanwhile, a camera is recording the operations in the monitored area. The information provided by the UWB system includes the position of the tag(s), tag update rate, the ID and type of the equipment. Imaging data, on the other hand, provides information about the position of the tracked equipment by using image processing techniques [3]. Then, the data acquired from the two sources should be integrated together. In this step, the UWB and image pixel positioning data should be aligned by transforming them from their local coordinate systems to a global coordinate system (GCS). After data association, in which the allowable error in UWB and associated image pixel positions is determined, the estimated position is calculated by averaging the position information of the UWB sensors and image processing modules. In cases, the integrated algorithm does not contain the locations of all equipment, the equipment position is estimated using the available data (e.g., UWB only or image processing only) [3,16,51]. In [3], a case study was implemented to measure the accuracy of this approach for resource tracking. They concluded that the UWB positioning data is better in comparison with the image processing data. However, applying image processing for more frames per second and using more advanced image processing techniques can improve the positioning accuracy of the video data.

For the second application, a UWB system is used for positioning of the camera in order to achieve information about the metric reconstruction of the site on a local coordinate system. Besides, using a 3D positioning technology such as UWB, a self-positioning system is achieved in which the data are directly geo-referencing (DG). It also helps to identify the parameters of the camera in a self-calibrated way which would reduce the error and computational complexity of the traditional photogrammetry techniques [32,51]. In [32], a case study was done for reconstruction of a building in which a camera and a Pozyx UWB device were mounted on an unmanned aerial vehicle (UAV) in a very close range from each other (less than the UWB accuracy range). In this way, first, the positioning of the receivers was done by both geo-referencing approaches included DG such as using RTS or GNSS devices, and self-positioning approach using the UWB system. The first approach was more accurate since the exact positioning of the receivers has an important role in the exactness of a proper transformation (i.e., scaling factor, translation, and rotation). After that, they achieved the position of other receivers in the same coordinate system by using trilateration and extended Kalman filter (EKF) techniques. The computed roving tag positions in the UWB coordinate system were used to estimate the transformation parameters (i.e., scaling and transition factors) for mapping the photogrammetry point clouds from the photo to the UWB frame. They concluded that introducing different weighting factors for various reading ranges, might improve the scale factor estimation. Besides, from these positioning data which will be used as reference positions, the camera external parameters can also be extracted [32,51].

The integration of these two sensing technologies provides accurate and timely positioning and visual data for various applications, especially in areas with limitation to use a positioning technology such as GNSS (e.g., in the tunnels or area covered with tall buildings) or using laser scanners due to limitation associated with scanning positions. Furthermore, there is no need for any extra devices for geo-referencing of images. In fact, in the case of direct geo-referencing using the UWB self-positioning technique, there is no need to use control points (CPs) anymore. However, the data acquired by UWB tag (s) need to get converted to the camera referenced frame and the photo coordinate system which may cause an error propagation phenomenon due to the intrinsic inaccuracy in each technology [32,51].

4.3.4. Point Cloud Data Fusion with UWB Data

While employment of the GPS-based systems for tracking resources is an appropriate choice for an unconfined environment and outdoor tracking, the use of UWB, RFID, and wireless local area network (WLAN) seems more appropriate for indoor tracking. However, on some occasions (e.g., the blockage, congestion, metal influence, etc.) the performance of the UWB based systems could be degraded. In this case, fusion of the UWB data with a laser scanning point cloud data can help to overcome these limitations. Plus, laser scanners provide a robust solution for object recognition, visualization, and 3D modeling of a site [6,28].

Integration of these two technologies, besides employment of other sources of information (e.g., 3D models, BIM, reports, etc.) have found very useful applications in construction industry. In fact, by having a better visualization model of a site and real-time information for progress reporting and schedule updating, it is possible to provide more robust and timely information for an earned value analysis (EVM) [6,15,28]. For instance, researchers in [15] integrated a 2D laser scanner with a UWB system in order to improve their simultaneous localization and mapping (SLAM) algorithm for localization of a moving robot. They demonstrated that this integration provides better visualization of the site due to its ability to accurately measure the range to nearby objects. Besides, this data fusion approach provided a real-time and drift-free SLAM algorithm based on ranging measurement information of the UWB system. It also helped to improve the UWB-only localization accuracy and overcoming the drawback associated with accumulated errors in scan matching needed in mapping with LiDAR. In fact, the pose estimation of a robot acquired by the UWB ranging measurements improved by subjecting it to a scan matching procedure. In [32], the results of their case study were validated for an integrated photogrammetry and UWB system with point cloud data acquired by a laser scanner to investigate the accuracy of their system for reconstruction of a building facade. In [6], a platform was proposed in which the data from this data fusion approach are integrated with other non-sensory data (e.g., reports, payments, 3D CAD, and BIM). Their platform provided more capabilities in which not only objects but also activities were tracked for real-time progress reporting on site. In fact, previous studies in this field were mainly focused on tracking objects on site and they have limitations to track non-structural activities which are not based on the tracking of a moving object, examples of such activities are welding, inspection, etc. They developed some algorithms for data fusion in high levels for object and activity recognition and tracking. In fact, they believed that data fusion in low levels especially in construction sites with complex working nature could be noisy, inaccurate, and incomplete due to the difference in dimensions and properties of the data acquired from various sources. In this way, trying to fuse data at higher levels decreases the computational complexity and reregistration efforts. In this way, they achieved the data for material and activity tracking by the UWB sensors and an algorithm called 3D marking, while the object recognition was done by using point cloud data acquired by a laser scanner (or photogrammetry). Besides, 3D CAD or BIM models were propped as prior information for having better object recognition. Finally, by adding administrative documents (e.g., schedule, progress reports, etc.) to the acquired information through project control applications, the EVM estimated and updated [6,28].

This integration makes it possible to detect and measure deviations from as-planned models by integrating the point cloud-based object recognition information with the UWB data for object and activity tracking on site. Besides, the UWB sensors can solve the problem of the 3D scanning in case of blockage or invisibility. However, the discrepancy between the accuracy of the UWB data and LiDAR mapping data need to be addressed [6,28].

5. Discussion on Current Solutions for Automated Data Acquisition

Considerable studies were reported on automated data acquisition on construction sites to overcome problems associated with manual data collection. However, most of these studies were based on single sensory data acquisition [11,16,20,23]. Examples of these individual sensing technologies include usage of the RTLSs such as GPS, RTS, and UWB technologies for real-time tracking of resources on site. Moreover, imaging technologies such as photogrammetry and laser scanning have been widely investigated by researchers [8,35]. In this way, recent studies in this field are focused on integrated data acquisition technologies. However, fusing the data acquired from sensing technologies is challenging in terms of the need for manual post-processing and variation in type of data that are acquired. Therefore, having an effective data management system has an important role in technologies working based on sensory data fusion for being employed in large-scale and complex construction environment [28].

Integrated systems based on fusion of the UWB data with point cloud data (acquired by using laser scanning, photogrammetry, or their integration) make it possible to track objects and activities on site by more accuracy and reliability. In fact, it provides more reliable as-built information for construction managers for the purpose of a real-time progress reporting, earned value analysis and automatic schedule updating. Besides, using BIM models during the construction phase has found applications with the advancement in updating BIM models with as-built models [1]. In fact, the integration of these vision-based technologies with the BIM model would make it possible to modify and complete the BIM models that were created at the design level. Then, by communicating back this information to the BIM model, it would be possible to make future maintenance and management plans [8,9,31]. However, it is still challenging specially for facilities in which the as-built conditions differs from the as-planned models. The two vision-based technologies, photo/videogrammetry and laser scanning, have made it possible to acquire 3D as-built data for 3D modeling in a point cloud format. Saying that, the use of camera-based technologies are beneficial in terms of cost and flexibility to be used in various site condition. However, the point clouds generated in this way are not as accurate as the ones acquired by the laser scanner since the quality of photogrammetry-based point clouds is very sensitive to constraints such as camera specification, position and calibration. Plus, the accuracy of the image frames registration in photogrammetry depends on finding common points between two or more frames, so if one constraint is not respected, then the overall image quality will be affected. A laser scanner with high resolution can generate over 700 million points in one scanning session, while the point cloud density generated by photogrammetry is much less than this value [55,56].

Besides, in order to make a 3D model more robust and to measure the deviation from as-planned models (e.g., placement deviation), the use of other supplementary technologies such as RFID, RTS, and UWB systems have been investigated by researchers. The identification and localization information provided by RFID or UWB systems are resulted in a better object reconstruction in an indoor environment. To do that, the UWB or RFID tags are attached to the subjected elements need to be tracked. In this way, it can contribute the process of object identification and extraction from the acquired point cloud data [14]. This information is used to create a better 3D as-built model of the elements associated with a specific construction activity by comparing them with as-planned models (e.g., BIM, CAD, etc.) for real-time progress reporting. Integration of the photogrammetry with the UWB technology also helps to have a direct geo-referencing system in an indoor environment with limitation to use GPS-based technologies [3,32]. Plus, by having a time-stamped location information of the digital camera, it would be possible to know from what perspective (or position) each photo frame has been taken in order to compare with the same position in the 3D models (e.g., BIM) which has applications for a more automated and real-time progress reporting [8].

Regarding the studies reviewed in this paper, improving the data collection process to acquire a complete set of as-built data with more efficiency as a project progresses and gets more complex would be an area of interest in the future. This can help to move from manual as-built modeling to an automated 3D modeling. However, investigating the use of the various RS technologies over manual techniques for automated data collection needs more attention in terms of providing economic justification for different applications. As a solution, the integration of the various RS technologies was investigated in this paper in order to address part of these issues. Suggestions for future research in this field include: (1) a more comprehensive data acquisition system in which a larger number of activities and objects are addressed, (2) improving the strategies for better fusion and registration of the sensing data acquired on site to benefit more efficiently from them. It can be done by deciding on the level of data fusion required (low to high) based on each application and the accuracy demanded, (3) deviation estimation from as-planned models to identify changes in a project more accurately and timely, (4) enhancing the compatibility and connection between these automated data acquisition technologies with the available software (e.g., Revit, Navisworks, etc.).

6. Summery and Concluding Remarks

Cost overrun and schedule delays are common problems, frequently leading to disputes and costly claims in the delivery of construction projects. Real-time progress reporting could address these issues by providing more timely and accurate information about activities on construction jobsites. Such timely reporting would enable decision makers and project managers to mitigate issues that give rise to the cost and schedule problems referred to above by making the schedule updating, progress reporting, and productivity analyses more frequent and accurate over project delivery period. Conventional systems for progress tracking, however, are usually manual, labor-intensive, time-consuming, and not accurate enough. In this way, automated data acquisition on site has been proposed by many researchers in order to have more robust and timely information to address the issues mentioned above. To do so, various studies have proposed the employment of different remote sensing technologies and their integrations in order to have a more automated data acquisition system. The selection of the technologies varies based on various factors such as the application, the required accuracy and the scale of a project. This study has investigated the advantageous and limitations of using these technologies by considering the applications required in construction industry.

Indoor progress tracking has also been investigated recently by some researchers. However, indoor tracking of resources and activities have some difficulties associated with a wide range of materials and structural objects that needs to be recognized. Besides, available remote sensing technologies with applications in indoor progress reporting (e.g., RFID, UWB, laser scanning, photogrammetry, etc.) have some limitations that may affect their performance in a confined area. In fact, vision-based technologies such as laser scanning and photogrammetry or their integration are good choices to generate point cloud data. However, they need post processing steps for 3D modeling which is manual and time consuming. Besides, the geo-referencing of the system during data collection really matters for having a more accurate data registration. In this way, the use of tracking technologies such as RTS and UWB with good localization accuracy can help to facilitate these problems and to have a more automated system for 3D modeling.

The fusion of the data acquired from various RS technologies and with other technologies has also been investigated as a solution to overcome the limitations of each technology. The activity-based approaches for progress reporting, is another area of interest in recent studies which enable us to not only track activities related to the moving objects but also to track activities which do not have such traceable objects. Finally, the registration of the data acquired by various RS technologies with available building models (e.g., 3D CAD, BIM, etc.) also needs more investigation in future.

Funding: This research received no external funding.

Conflicts of Interest: The authors declare no conflict of interest.

References

1. Omar, T.; Moncef, L.N. Data acquisition technologies for construction progress tracking. *Autom. Constr.* **2016**, *70*, 143–155. [CrossRef]
2. Su, X.; Li, S. Enhanced boundary condition–based approach for construction location sensing using RFID and RTK GPS. *J. Constr. Eng. Manag.* **2014**, *140*, 04014048. [CrossRef]
3. Siddiqui, H. UWB RTLS for Construction Equipment Localization: Experimental Performance Analysis and Fusion with Video Data. Master's Thesis, Dept. of Information Systems Engineering, Concordia University, Montréal, QC, Canada, 2014.
4. Montaser, A.; Moselhi, O. RFID indoor location identification for construction projects. *Autom. Constr.* **2014**, *39*, 167–179. [CrossRef]
5. Li, H.; Chan, G. Integrating real time positioning systems to improve blind lifting and loading crane operations. *Constr. Manag. Econ.* **2013**, *31*, 596–605. [CrossRef]

6. Shahi, A.; Safa, M. Data fusion process management for automated construction progress estimation. *J. Comput. Civ. Eng.* **2014**, *29*, 4014098. [CrossRef]
7. Siu, F.; Lu, M. Combining photogrammetry and robotic total stations to obtain dimensional measurements of temporary facilities in construction field. *Vis. Eng.* **2013**, *1*, 4. [CrossRef]
8. Kropp, C.; Koch, C. Interior construction state recognition with 4D BIM registered image sequences. *Autom. Constr.* **2018**, *86*, 11–32. [CrossRef]
9. Bosché, F.; Ahmed, M. The value of integrating Scan-to-BIM and Scan-vs-BIM techniques for construction monitoring using laser scanning and BIM: The case of cylindrical MEP components. *Autom. Constr.* **2015**, *49*, 201–213. [CrossRef]
10. Cheng, C.F.; Rashidi, A. Activity analysis of construction equipment using audio signals and support vector machines. *Autom. Constr.* **2017**, *81*, 240–253. [CrossRef]
11. Li, N.; Gerber, B.B. Performance-based evaluation of RFID-based indoor location sensing solutions for the built environment. *Adv. Eng. Inform.* **2011**, *25*, 535–546. [CrossRef]
12. Cai, H.; Andoh, A.R. A boundary condition based algorithm for locating construction site objects using RFID and GPS. *Adv. Eng. Inform.* **2014**, *28*, 455–468. [CrossRef]
13. Andoh, A.R.; Su, X.; Cai, H. A framework of RFID and GPS for tracking construction site dynamics. In *Proceedings of Construction Research Congress*; ASCE: West Lafayette, Indiana, 2012.
14. Valero, E.; Adán, A.; Bosché, F. Semantic 3D reconstruction of furnished interiors using laser scanning and RFID technology. *J. Comput. Civ. Eng.* **2015**, *30*, 04015053. [CrossRef]
15. Song, Y.; Guan, M.; Tay, W.P. UWB/LiDAR Fusion For Cooperative Range-Only SLAM. In Proceedings of the International Conference on Robotics and Automation (ICRA), Brisbane, Australia, 21–25 May 2018.
16. Cheng, T.; Venugopal, M. Performance evaluation of ultra wideband technology for construction resource location tracking in harsh environments. *Autom. Constr.* **2011**, *20*, 1173–1184. [CrossRef]
17. Yoo, J.; Park, J. Indoor Localization Based on Wi-Fi Received Signal Strength Indicators: Feature Extraction, Mobile Fingerprinting, and Trajectory Learning. *Appl. Sci.* **2019**, *9*, 3930. [CrossRef]
18. Kalikova, J.; Krcal, J. People counting by means of wi-fi. In Proceedings of the Smart City Symposium, Prague (SCSP), Prague, Czech Republic, 25–26 May 2017.
19. Ta, V.C. Smartphone-based indoor positioning using Wi-Fi, inertial sensors and Bluetooth. In *Machine Learning*; Université Grenoble: Alpes, France, 2017.
20. Shehab, T.; Moselhi, O. An automated barcode system for tracking and control of engineering deliverables. In *Construction Research Congress*; ASCE: San Diego, CA, USA, 2005.
21. Ergen, E.; Akinci, B. Life-cycle data management of engineered-to-order components using radio frequency identification. *Adv. Eng. Inform.* **2007**, *21*, 356–366. [CrossRef]
22. Langley, R.B. RTK GPS. GPS World 9, 9, 70-76. Available online: https://www.gpsworld.com (accessed on 1 September 1998).
23. Labant, S.; Gergel'ová, M.; Weiss, G. Analysis of the Use of GNSS Systems in Road Construction. In Proceedings of the IEEE Geodetic Congress, Gdansk, Poland, 22–25 June 2017.
24. Hahn, P. Total Stations The Other Machine Control Sensor and More Machine Control Magazine. 2(2). Available online: http://www.machinecontrolonline.com (accessed on 1 January 2012).
25. El-Omari, S.; Moselhi, O. Integrating automated data acquisition technologies for progress reporting of construction projects. *Autom. Constr.* **2011**, *20*, 699–705. [CrossRef]
26. Ibrahim, M. Models for Efficient Automated Site Data Acquisition. Ph.D. Thesis, Dept. of Building, Civil and Environmental Engineering, Concordia University, Montréal, QC, Canada, 2015.
27. Valero, E.; Adán, A. Integration of RFID with other technologies in construction. *Measurement* **2016**, *94*, 614–620. [CrossRef]
28. Shahi, A. Activity-Based Data Fusion for Automated Progress Tracking of Construction Projects. Ph.D. Thesis, Dept. of Civil Engineering, Waterloo, ON, Canada, 2012.
29. Wu, W.; Yang, H. An integrated information management model for proactive prevention of struck-by-falling-object accidents on construction sites. *Autom. Constr.* **2013**, *34*, 67–74. [CrossRef]

30. Maneesilp, J.; Wang, C. RFID support for accurate 3D localization. *IEEE Trans. Comput.* **2012**, *62*, 1447–1459. [CrossRef]
31. Park, J.; Cho, Y.K. A BIM and UWB integrated mobile robot navigation system for indoor position tracking applications. *J. Constr. Eng. Proj. Manag.* **2016**, *6*, 30–39. [CrossRef]
32. Masiero, A.; Fissore, F. A low cost UWB based solution for direct georeferencing UAV photogrammetry. *Remote Sens.* **2017**, *9*, 414. [CrossRef]
33. Rönnholm, P.; Honkavaara, E. Integration of laser scanning and photogrammetry. *Int. Arch. Photogramm. Remote Sens. Spat. Inf. Sci.* **2007**, *36.3/W52*, 355–362.
34. Alshawabkeh, Y. Integration of Laser Scanning and Photogrammetry for Heritage Documentation. Ph.D. Thesis, Institut für Photogrammetrie, Universität Stuttgart, Stutgart, Germany, 2006.
35. Feng, Y.; Golparvar-Fard, M. Image-based localization for facilitating construction field reporting on mobile devices. *Adv. Inf. Comput. Civ. Construct. Eng.* **2019**, 585–592.
36. Ruiz, A.; Jiménez, R. Comparing Ubisense, Bespoon, and Decawave uwb location systems: Indoor performance analysis. *IEEE Trans. Instrum. Meas.* **2017**, *66*, 2106–2117. [CrossRef]
37. Omari, S.; Moselhi, O. Integrating automated data acquisition technologies for progress reporting of construction projects. In Proceedings of the 26th International Symposium on Automation and Robotics in Construction, Austin, TX, USA, 24–27 June 2009.
38. Musa, R. BIM for Existing Buildings: A Study of Terrestrial Laser Scanning and Conventional Measurement Technique. Master's Thesis, Joint Study Programme of Metropolia UAS and HTW, Berlin, Germany, 2017.
39. Valero, E.; Adan, A. Automatic construction of 3D basic-semantic models of inhabited interiors using laser scanners and RFID sensors. *Sensors* **2012**, *12*, 5705–5724. [CrossRef]
40. Ubisense Series 7000. Available online: https://ubisense.com (accessed on 1 January 2009).
41. Bespoon SpoonPhone. Available online: http://bespoon.com (accessed on 1 January 2015).
42. Decawave TREK1000 Evaluation Kit. Available online: https://www.decawave.com/product/trek1000-evaluation-kit (accessed on 1 January 2016).
43. Xu, Y.; Shmaliy, Y.S. Robust and accurate UWB-based indoor robot localization using integrated EKF/EFIR filtering. *IET Radar Sonar Navig.* **2018**, *12*, 750–756. [CrossRef]
44. Magdy, I.; Moselhi, O. Automated productivity assessment of earthmoving operations. *J. Inf. Technol. Construct. (ITcon)* **2014**, *19*, 169–184.
45. Seo, W.; Hwang, S. Precise outdoor localization with a GPS–INS integration system. *Robotica* **2013**, *31*, 371–379. [CrossRef]
46. Jo, K.; Lee, M. Road slope aided vehicle position estimation system based on sensor fusion of GPS and automotive onboard sensors. *IEEE Trans. Intell. Trans. Syst.* **2016**, *17*, 250–263. [CrossRef]
47. Akhavian, R.; Behzadan, A.H. Construction equipment activity recognition for simulation input modeling using mobile sensors and machine learning classifiers. *Adv. Eng. Inform.* **2015**, *29*, 867–877. [CrossRef]
48. Mirshahi, S.; Uysal, S. Integration of RFID and WSN for supply chain intelligence system. In Proceedings of the International Conference on Electronics, Computers and Artificial Intelligence, Pitesti, Arkansas, Romania, 27–29 June 2013.
49. Xiong, Z.; Song, Z. Hybrid WSN and RFID indoor positioning and tracking system. *Eurasip J. Embed. Syst.* **2013**, *1*, 6. [CrossRef]
50. El-Omari, S. Automated Data Acquisition for Tracking and Control of Construction Projects. Ph.D. Thesis, Dept. of Building, Civil and Environmental Engineering, Concordia University, Montréal, QC, Canada, 2008.
51. Masiero, A.; Fissore, F. Performance Evaluation of Two Indoor Mapping Systems: Low-Cost UWB-Aided Photogrammetry and Backpack Laser Scanning. *Appl. Sci.* **2018**, *8*, 416. [CrossRef]
52. Song, L.; Tanvir, M. A cost effective material tracking and locating solution for material laydown yard. *Procedia Eng.* **2015**, *123*, 538–545. [CrossRef]
53. Skibniewski, M.J.; Jang, W.S. Localization technique for automated tracking of construction materials utilizing combined RF and ultrasound sensor interfaces. *Comput. Civ. Eng.* **2007**, 657–664.
54. Nasrollahi, M.; Bolourian, N.; Hammad, A. Concrete surface defect detection using deep neural network based on lidar scanning. In Proceedings of the CSCE Annual Conference, Laval, Greater Montreal, QC, Canada, 12–15 June 2019.

55. Ahmed, M.; Guillemet, A.; Shahi, A.; Haas, C. Comparison of point-cloud acquisition from laser-scanning and photogrammetry based on field experimentation. In Proceedings of the CSCE 3rd International/9th Construction Specialty Conference, Ottawa, ON, Canada, 14–17 June 2011.
56. James, D.; Eckermann, J.; Belblidia, F. Point cloud data from Photogrammetry techniques to generate 3D Geometry. In Proceedings of the 23rd UK Conference of the Association for Computational Mechanics in Engineering, Swansea University, Swansea, UK, 8–10 April 2015.

Article

RSS-Fingerprint Dimensionality Reduction for Multiple Service Set Identifier-Based Indoor Positioning Systems

Ahmed Abed [1,*] and Ikhlas Abdel-Qader [2]

1 Department of Electrical & Electronic Engineering-College of Engineering, Thi-Qar University, Thi-Qar 64001, Iraq
2 Department of Electrical & Computer Engineering-College of Engineering & Applied Sciences, Western Michigan University, Kalamazoo, MI 49008, USA
* Correspondence: ahmedkareemab.abed@wmich.edu; Tel.: +1-269-312-9305

Received: 30 June 2019; Accepted: 29 July 2019; Published: 2 August 2019

Abstract: Indoor positioning systems (IPS) have been recently adopted by many researchers for their broad applications in various Internet of Things (IoT) fields such as logistics, health, construction industries, and security. Received Signal Strength (RSS)-based fingerprinting approaches have been widely used for positioning inside buildings because they have a distinct advantage of low cost over other indoor positioning techniques. The signal power RSS is a function of the distance between the Mobile System (MS) and Access Point (AP), which varies due to the multipath propagation phenomenon and human body blockage. Furthermore, fingerprinting approaches have several disadvantages such as labor cost, diversity (in signals and environment), and computational cost. Eliminating redundancy by ruling out non-informative APs not only reduces the computation time, but also improves the performance of IPS. In this article, we propose a dimensionality reduction technique in a multiple service set identifier-based indoor positioning system with Multiple Service Set Identifiers (MSSIDs), which means that each AP can be configured to transmit N signals instead of one signal, to serve different kinds of clients simultaneously. Therefore, we investigated various kinds of approaches for the selection of informative APs such as spatial variance, strongest APs, and random selection. These approaches were tested using two clustering techniques including K-means and Fuzzy C-means. Performance evaluation was focused on two elements, the number of informative APs versus the accuracy of the proposed system. To assess the proposed system, real data was acquired from within the College of Engineering and Applied Sciences (CEAS) at the Western Michigan University (WMU) building. The results exhibit the superiority of fused Multiple Service Set Identifiers (MSSID) performance over the single SSID. Moreover, the results report that the proposed system achieves a positioning accuracy <0.85 m over 3000 m^2, with an accumulative density function (CDF) of 88% with a distance error of 2 m.

Keywords: Indoor Positioning System; WLAN; C-Means; K-Means; Access Point Selection; RSS-fingerprint

---

1. Introduction

A precise real-time indoor positioning system has recently attracted considerable attention from researchers due to location-aware services, which include patient location monitoring, mobile advertising, tag tracking, and robot guiding [1–5]. Unlike Global Positioning Systems (GPS), that require line of sight (LOS) transmission paths, indoor positioning faces many challenges due to changeable radio propagation environments inside the buildings [6]. Due to walls, celling, movement of people or furniture, and obstacles inside the buildings, indoor radio propagation is affected by multipath fading, shadowing, and delay distortion [7]. In addition to high-accuracy requirements, indoor position systems should also estimate the position of an object quickly with a light algorithm and low computational cost.

Many indoor positioning techniques have been promoted such as Wireless local Area Network (WLAN), Ultrasound, Lighting, Radio Frequency Identification (RFID), and Bluetooth. WLAN-based indoor positioning systems have been widely employed for IPS without additional cost because WLAN infrastructures are necessary in large buildings like airports, hospitals, universities, and museums. Positioning based on RSS can be divided into two models: signal propagation models, and fingerprint-based location models. The requirements of the first model for indoor propagation loss is relatively high, so this article focuses on the second model. Although the second model has been relatively mature, the WLAN-based fingerprinting technique can be considered as an important technique used to estimate the position of a target.

In general, fingerprinting-based positioning involves two phases: an offline and an online phase [8]. In the offline phase, the area of interest is divided into grid points, which are called reference points (RPs). Each RP_i is labeled by Cartesian coordinates (x_i,y_i). The MS at each RP_i collects the RSS readings from various APs, which are deployed in the Area of Interest (AOI). In the online phase, the MS records the RSS vector at an unknown location and pre-matches with the stored RM to determine the closest RPs to the MS by using Euclidian distance, which is the simplest method to estimate the position of MS.

Fortunately, Wi-Fi users are able to measure the RSS from various APs, which are deployed in the AOI. Therefore, location fingerprints are typically vectors of RSS of different APs, and the length of such vectors can be grown to include all the detectable APs in the AOI. Unfortunately, not all these APs can contribute positively to positioning, the majority of them are redundant. Therefore, including all these APs in the RSS vector results in superfluous computational costs, and in some cases, causes a deterioration in the accuracy [9]. Recently, most APs that are deployed in large buildings, use Multiple Service Set Identifier configurations, which means that each AP is configured to transmit N signals instead of one signal. MSSID is used in most buildings to advertise different WLANs where each WLAN has a different class of clients, and each class requires different keys, firewalls, privacy, and speeds [10]. This technique leads to increasing redundancy in the calculation process and a decrease in the accuracy of systems when each SSID is considered as a single AP [11].

In addition to the redundancy of APs, the accuracy of fingerprinting-based techniques is severely affected by the multipath issue and radio interference. Therefore, it is very important for any IPS designer to know the statistical properties of RSSI signals. Fluctuation in RSS signals due to the multipath phenomenon is an inherent issue in fingerprinting-based techniques. There are many solutions to mitigate the effect of a multipath issue in offline and online phases including the use of Kalman filter, averaging, and histograms [12–14]. The efforts in the online phase required many RSS time samples to be measured in the same position to reduce the effect of the multipath phenomenon. This kind of solution causes IPS to be slow and as such, is not appropriate for the tracking process.

The main contributions of this article include: (1) Study the characteristics of Multiple SSID signals in 2.4 GHz, which can be configured on the same AP. (2) Propose a new algorithm based on MSSIDs to reduce the effect of multi-path issues by fusing MSSIDs signals, and computational costs by selecting informative APs. (3) Study the effect of informative AP numbers in K-means and Fuzzy C-means clustering technique. (4) Compare the proposed system performance with state-of-the-art methods.

The rest of this article is arranged as follows: the related works are reviewed in Section 2. The fingerprinting technique is illustrated in Section 3. MSSID and the characteristics of MSSID signals are described in Section 4. The proposed system and measurement setup technique are illustrated in Sections 5 and 6, respectively. Section 7 shows the results and the discussion. In Section 8, the conclusion and future work are discussed.

2. Related Work

In the literature of indoor positioning, there are various taxonomies. However, there is a common classification, which falls into two separate categories: Radio Frequency (RF)-based techniques, and non-RF-based techniques, which use another kind of sensors. WLAN, RFID, and Bluetooth are among the RF-based techniques. Audio, visual, ultra-sonic, Infra-Red (IR), laser sensors, and magnetic field are considered as non-RF-based methods [15–18]. In this article, we mainly focus on RF-based techniques. Table 1 reports the main RF techniques that are utilized in indoor positioning.

Recently, many types of research have mainly focused on Wi-Fi fingerprint-based IPS challenges, such as collecting fingerprinting signals and how these signals are processed through IPS techniques [19]. Dimensionality reduction is also an important challenge and has been studied through different techniques.

Table 1. RF Position Techniques [19].

Technique	Advantage	Disadvantage
Cell of origin	Base stations are available and never move	Highly inaccurate
AoA	High accuracy	Requires additional hardware
ADoA	High accuracy	Requires additional hardware
ToA	High accuracy	Requires additional hardware synchronization issue
TDoA	High accuracy	Requires additional hardware synchronization issues
Fingerprinting- based location	High accuracy	Computational and labor costs

Generally, there is a tradeoff between dimensionality reduction and the accuracy of classifiers [20]. In the online phase, the reduction of computational costs for Wi-Fi fingerprint-based IPS is very crucial in a real-time tracking system because large data require a long time to be processed. The roaming of a MS inside the building makes IPS unable to track the MS's location inside the building. Therefore, working on designing a light algorithm with low computational cost makes IPS more feasible. In this section, we review key techniques for positioning, which are used in dimensionality reduction.

Fang et al. [21] proposed a new approach for a Wi-Fi-based location fingerprinting system. The proposed method transforms RSS signals into Principal Components (PCs) intelligently, such that the effect of all APs is considered. Instead of the selection of APs, the proposed technique replaces the RSS of APs by a subset of PCs to improve accuracy and reduce the computational cost as well. To evaluate the proposed system, a realistic WLAN environment was chosen to conduct the experimental work. The results exhibit that the mean distance error was eliminated by 33.75% and the complexity by 40%, as compared to traditional approaches.

Li et al. [22] proposed a novel technique for a WLAN-based location fingerprinting system. In the offline phase, the AOI was clustered into sub regions by fuzzy C-means and then the useful APs were selected to minimize the dimension of the fingerprint. In the online phase, the proposed system used a NN (Nearest Neighbor) method to choose subareas and to compute the coordinate of the target location by using relative distance fuzzy localization (RDFL) algorithms. The results showed the ability of the proposed system to reduce calculation time and improve positioning accuracy.

Jiang et al. [23] implemented a WLAN- fingerprinting-based localization method, which is based on selecting the important APs. The APs with the strongest RSS power were selected as the informative APs. The results showed that the proposed system exhibits superior performance as compared with traditional techniques.

Abusara et al. [24] modified a method, called Fast Orthogonal Search (FOS) method, to determine the singularity of RSS values, which are measured from various APs at each location point. In order to evaluate the modified FOS method, two important factors were taken into consideration, the amount of reduction, and the accuracy of IPS. The results with real RSS values measurement show that the modified FOS performance is better than the traditional FOS implementation.

Chen et al. [25] presented a novel algorithm, which is called a clustering and Decision Tree-based method (CaDet) for intelligent selection of APs in indoor positioning. CaDet is a union of information theory, clustering analysis, and a decision tree algorithm. By using machine-learning techniques, the results showed that CaDet is capable of selecting an informative small combination of APs to detect the user's location with high accuracy, as compared with traditional techniques.

Kanaris et al. [26] proposed a novel technique for indoor localization based on the cooperation between IEEE802.11 infrastructure and Bluetooth Low Energy (BLE). BLE is used to confine the location of the user within a specific subarea that means coarse localization. A new K-Nearest Neighbors (K-NN) algorithm is utilized to estimate the final user's location. Due to the use of a fragment of the initial fingerprint dataset, the results exhibit that the proposed system achieves fast positioning estimation and improves the accuracy of the proposed system as well.

Feng et al. [27] implemented a novel approach based on compressive sensing (CS) theory and selection APs to obtain high accuracy. The theory of CS is applicable for accurate IPS using RSS-based fingerprinting due to the location sparse nature on the map. There are two stages for finding the location of a user. The first stage is achieved by an Affinity propagation clustering method to reduce the computational cost and increase accuracy as well. In the second stage, fine positioning is applied by using the compressive sensing theory. Random and strongest APs selection were used in this article. The results show that the proposed system leads to robust enhancement in positioning accuracy and complexity over traditional fingerprinting techniques.

Sanchez et al. [28] proposed an indoor localization methodology based on data fusion and feature transformation. The methodology is formed by four phases: dataset building, feature fusion, feature transformation, and classification. Dataset consists of RSS from Wi-Fi AP, orientation from a compass, and simulated RSS from simulated Light Emitting Diode (LED) lamps. Principal Component is used to reduce dataset dimensionality and improve computational performance of system. The results exhibit that the proposed system considerably reduces overall computational cost and provides an acceptable location accuracy.

Lopez et al. [29] proposed a refinement cycle for an indoor positioning system based on dimensionality reduction techniques. Several dimensionality reduction techniques have been utilized to different datasets such as Principal Component Analysis (PCA), Linear Discrimination Analysis (LDA), and t-Stochastic Neighbor Embedding (t-SNE). Two different visualization methods are defined to obtain graphical information about the quality of the radio map in terms of overlapping areas and outliers. The proposed system is evaluated by two kinds of data: dataset, which is measured by the research group, and standard data. In general, the results show that there are some useful configurations which can be used to perform the refinement process.

3. Fingerprinting-Based Technique

Because of the unavailability of appropriate radio signal propagation models for indoor environments, fingerprinting-based location techniques became popular with RADAR [30]. RADAR is the first fingerprinting-based positioning system, which was developed by Microsoft. Fingerprinting is a scene analysis method in which location in the environment is associated with a unique signal parameter. In this article, RSS of different APs, which are deployed for network services, is used as feature location. In general, the fingerprinting-based technique is divided into two methods: probabilistic and deterministic methods [31]. In this article, the deterministic method is utilized because it does not require many RSS measurements [32]. This technique is performed into two phases:

- Offline phase: During this phase, the area of interest (AOI) is partitioned into grid points which are called reference points (RP). Each RP is labeled on the map by Cartesian coordinates (x_i, y_i). The RSS of various APs are recorded for a specific time period (KT), where K is an integer number which represents the number of samples that are taken at each RP, and T is a time period to record the value of RSS at RP. The fingerprint, $\overline{\varphi}_{i,j}$, of RP_j is generated by taking the average of the RSS signal values from APi for KT period where, $\overline{\psi}_{i,j} = \frac{1}{K}\sum_{a=1}^{K} \psi_{i,j}(aT) \mid a = 1, 2, \ldots\ldots, K$. Therefore, the radio map (RM) of the AOI is created by cardinality of the fingerprints, as shown in Equation (1):

$$\psi = \begin{bmatrix} \psi_{1,1} & \psi_{1,2} & \cdots & \psi_{1,\,M} \\ \psi_{2,1,} & \psi_{2,2} & & \psi_{2,M} \\ \vdots & & \ddots & \vdots \\ \psi_{L,1} & \psi_{L,2} & \cdots & \psi_{L,\,M} \end{bmatrix} \tag{1}$$

where, L and M represent a number of APs, and a number of RPs in the AOI, respectively.

Although the fingerprinting-based technique is an easy method to implement, this method has many challeges that should be considered when it is done practically. Establishing a site survey RM is normally a labor-intensive task. However, there are many methods for enhancing the efficiency of the offline phase. These methods are based on Compressive Sensing (CS) theory [27,33,34] to reconstruct the RM from a small number of RSS measuremnets instead of survying the entire AOI. Also, a divesity of MS arises an an important challenge for IPS designers. Most of the existing research assumes that the collecting of fingerprints of a radio map and online measurements are achieved in the same device. Practically, this assumption is not true because the training device, which is used to build the radio map during the offline phase, may be different from the positioning device used during the online phase. These result in different signal strength patterns across the devices, which could degrade the accuracy of the positioning system [35]. Hence, the problem of signal pattern variation across diverse devices should be considered when designing positioning systems.

- Online phase: In this phase, the MS at unknown location records $V = (RSS_1, RSS_2, \ldots RSS_L)$ from different APs in the AOI. This vector V is pre-matched with a RM which is created in the offline phase to determine the MS's location. Figure 1 shows the two phases which are used to estimate MS's location. The Euclidean distance between V and all RP vectors in the RM utilized to determine the estimated position, as shown in Equation (2):

$$Dj = ||V - \varphi_j|| \tag{2}$$

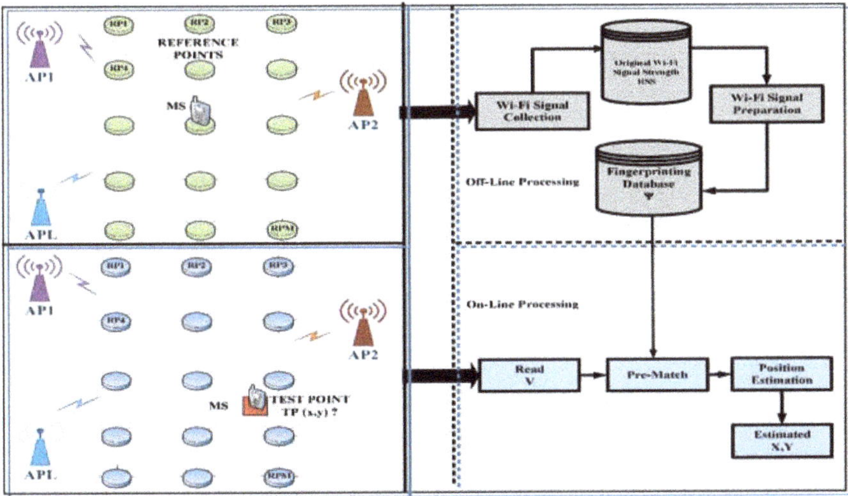

Figure 1. The fingerprinting-based method.

The signal distance vector, D_j, is sorted in ascending order. The first K RPs, which have the smallest signal distance, are selected to determine the position of MS, $\hat{p} = \frac{1}{K}\sum_{i=1}^{K} p_i$, where p_i represents the positions of K RPs. The reciprocal of distance, D_j, can be utilized as weights to determine the MS's location. This method is called weighted-KNN, where the estimated position, $\hat{p}$, is computed as shown in Equation (3):

$$\hat{p} = \frac{\sum_{i=1}^{K} \frac{1}{Di} p_i}{\sum_{i=1}^{K} \frac{1}{Di}} \tag{3}$$

4. Multiple Services Set Identifier (MSSIDs) Technique

In the past, a number of separated APs were required to present a multiple services network in a specific area to deploy for Wi-Fi users. Each AP advertises single SSID on a specific frequency channel. The raising number of APs in a particular area leads to a channel congestion issue. As a result, the Wi-Fi service will be inefficient.

Recently, most of the APs are configured to work with MSSID and feature simultaneously, as shown in Figure 2. MSSID services advertise various SSIDs within one AP instead of a group of separated APs [10]. This configuration is commonly used to advertise various networks for various kinds of users, where each kind of user requires different settings to each SSID, such as firewall, key, bandwidth, privacy, and security. MSSIDs' technique uses the same frequency channel. In order to avoid a collision or frame loss, MSSIDs utilizes a Clear Channel Assessment (CCA) protocol [36]. This protocol is used to investigate the channel state and whether it is used by another SSID. The channel is utilized by another SSID when the channel is clear. Therefore, the probability that all SSID signals are affected by the fading issue will be low [11]. In this article, we use this benefit to mitigate the variation of RSS signals by using a fusion approach for MSSIDs' signals.

Figure 2. Multiple Services Set Identifiers (MSSIDs) technique.

4.1. Mathematical Model of MSSID

This research aims to reduce the calculation cost of the MSSID-based indoor positioning system and increase the performance of IPS as well. Accordingly, N RMs (Ψ_1, Ψ_2, ... , and Ψ_N) are generated as in Equations (4)–(6):

$$\Psi_1 = \begin{bmatrix} \text{SSID}_{1,1,1} & \text{SSID}_{1,2,1} & \cdots & \text{SSID}_{1,M,1} \\ \text{SSID}_{2,1,1} & \text{SSID}_{2,2,1} & & \text{SSID}_{2,M,1} \\ & \vdots & \ddots & \vdots \\ \text{SSID}_{L,1,1} & \text{SSID}_{L,2,1} & \cdots & \text{SSID}_{L,M,1} \end{bmatrix} \tag{4}$$

$$\Psi_2 = \begin{bmatrix} \mathrm{SSID}_{1,1,2} & \mathrm{SSID}_{1,2,2} & \cdots & \mathrm{SSID}_{1,M,2} \\ \mathrm{SSID}_{2,1,2} & \mathrm{SSID}_{2,2,2} & & \mathrm{SSID}_{2,M,2} \\ \vdots & & \ddots & \vdots \\ \mathrm{SSID}_{L,1,2} & \mathrm{SSID}_{L,2,2} & \cdots & \mathrm{SSID}_{L,M,2} \end{bmatrix} \tag{5}$$

$$\Psi_N = \begin{bmatrix} \mathrm{SSID}_{1,1,N} & \mathrm{SSID}_{1,2,N} & \cdots & \mathrm{SSID}_{1,M,N} \\ \mathrm{SSID}_{2,1,N} & \mathrm{SSID}_{2,2,N} & & \mathrm{SSID}_{2,M,N} \\ \vdots & & \ddots & \vdots \\ \mathrm{SSID}_{L,1,N} & \mathrm{SSID}_{L,2,N} & \cdots & \mathrm{SSID}_{L,M,N} \end{bmatrix} \tag{6}$$

where, L and M represent a number of deployed APs, and a number of RPs in the AOI, respectively.

4.2. Time Samples of MSSID

RSS-based fingerprinting approaches have been widely used for indoor positioning. The behavior of RSS signals plays an important role in determining the characteristics and nature of location fingerprints, which are recorded in RM. The RSS is a function of the distance between AP and the user who receives this signal, and it is mainly affected by the multi-path issue [37]. 5000 RSS samples were taken at a specific location from AP with 5 SSIDs under the line-of-sight condition. The behavior of MSSID RSS at a specific location is shown in Figure 3, where 5 SSIDs (SSID_1, SSID_2, … , SSID_5) are configured on the same AP. Figure 3 clearly shows the effect of the multi-path issue on these signals.

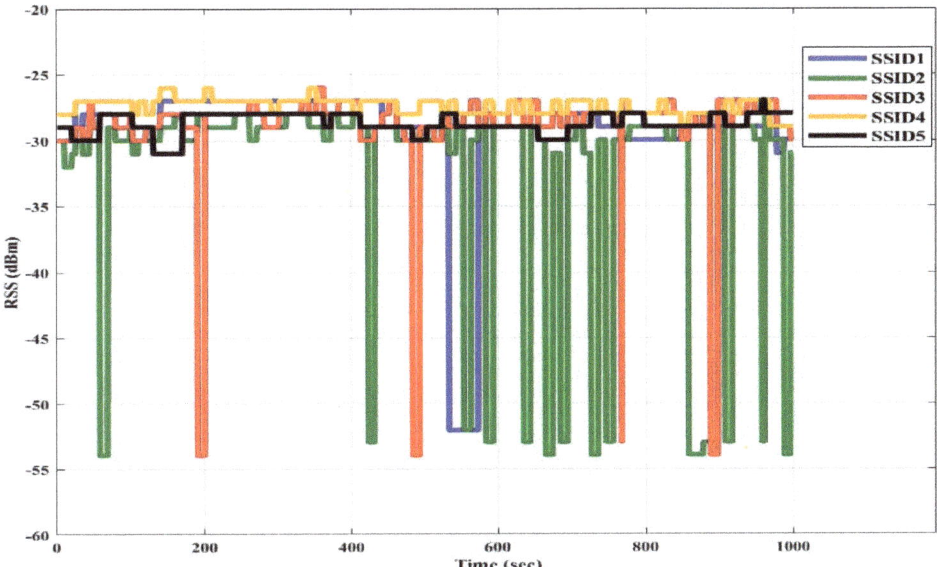

Figure 3. The MSSIDs' signals time behavior.

4.3. Time and Spatial Correlation Coefficients of MSSIDs' Signals

The behavior of N SSIDs' signals on the same physical AP is nearly identical to the deployment of N individual APs. In order to investigate the independency of these signals and the similarity over the AOI, Tables 2 and 3 illustrate the time correlation coefficients of SSIDs' signals at a specific location, and the spatial correlation coefficients over the AOI, respectively. In Table 2, the time correlation coefficients are $|Cij| < 0.05$, meaning all these signals are independent and its behaviors are similar to the behavior of 5 separated APs.

Table 2. The time correlation coefficient between MSSIDs' signals at a certain point.

SSID	SSID1	SSID2	SSID3	SSID4
SSID2	0.021	-	-	-
SSID3	0.000	0.013	-	-
SSID4	0.012	−0.01	0.045	-
SSID5	−0.012	0.007	0.029	0.012

Table 3. The spatial Correlation coefficient between MSSIDs' signals over AOI.

SSID	SSID1	SSID2	SSID3	SSID4
SSID2	0.98	-	-	-
SSID3	0.95	0.98	-	-
SSID4	0.94	0.98	0.95	-
SSID5	0.97	0.96	0.97	0.94

Table 3 shows the spatial correlation coefficient over the AOI. We can see that these values over the AOI are close to 1, meaning that all SSIDs' signals which are deployed on the same AP have a similar spatial power distribution. Furthermore, Figure 4 shows the heat map for two APs, each with three SSIDs in different locations on the AOI. From Figure 4 it can be seen that MSSIDs' signals have the same fingerprinting on the map. Therefore, the use of MSSID signals leads to an increase in the redundancy of these signals and a degradation of the IPS's performance.

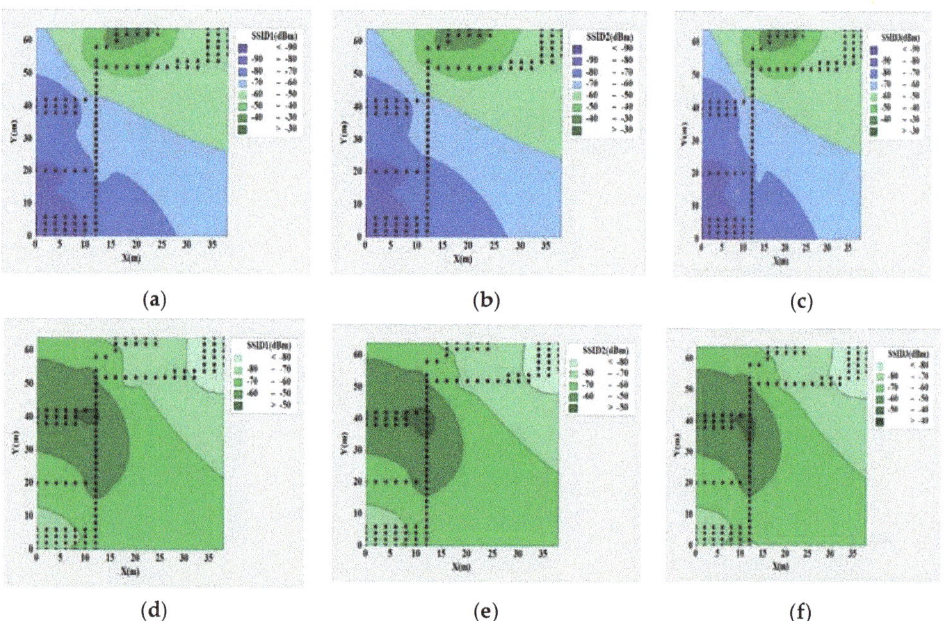

Figure 4. Spatial power distribution of two APs with three SSIDs. (**a**) $SSID_1$ of AP_1; (**b**) $SSID_2$ of AP_1; (**c**) $SSID_3$ of AP_1; (**d**) $SSID_1$ of AP_2; (**e**) $SSID_2$ of AP_2; (**f**) $SSID_3$ of AP_2.

4.4. Path Loss of MSSIDs RSS Signals

The RSS signal value decreases with distance, d, according to the function of the logarithm-distance path loss model [38]. Figure 5 shows the behavior of MSSIDs' signals and the mean of these signals versus distance d. The behavior of the mean of MSSIDs is close to the path loss model with path loss exponent (n = 3.5). The fusion of MSSIDs' signals by averaging them contributes to mitigating the effect of the multipath issue. Table 4 illustrates that the standard deviation of the RSS signal decreases with distance d. In addition, the standard deviation of the mean of MSSIDs is less than individual signals of MSSIDs.

Table 4. The Standard Deviation of MSSID signals versus distance (m).

No. SSID	L_1 = 2 m	L_2 = 10 m	L_3 = 20 m
SSID1	5.1	3	2.2
SSID2	5.5	2.8	2.4
SSID3	4.6	2.7	2.1
MSSIDs	2.6	1.5	1.4

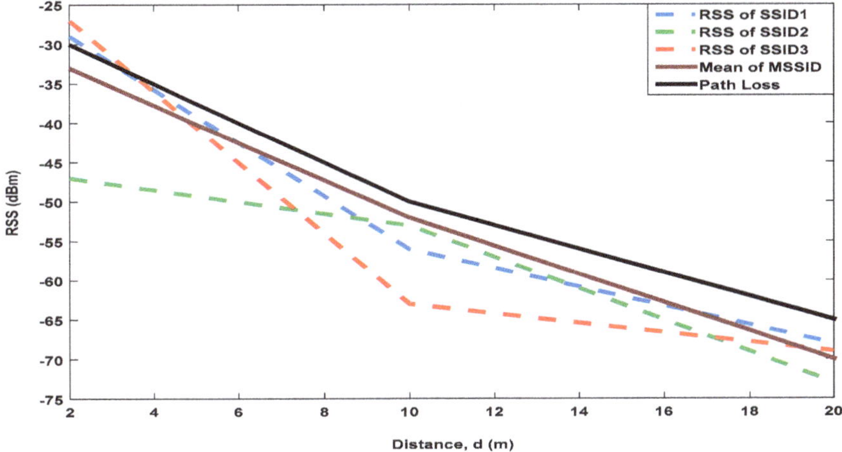

Figure 5. The change of Received Signal Strength (RSS) power of MSSIDs versus distance d (m).

Most researchers consider each RSS signal to be transmitted from a certain single AP with a unique Media Access Control (MAC) address. MSSID signals have a unique MAC address for each one but they are transmitted from the same AP. This article fills in this gap by investigating the behavior of MSSID signals and designing a classifier based on these signals to improve the performance of IPS.

5. Computational Cost Reduction Methods

In an indoor location system, eliminating redundant APs not only reduces the computation time of fingerprinting processing, but also enhances the accuracy of IPS. The WLAN-based fingerprinting technique has built RM for the AOI, where each vector in the RM comes from measuring the RSS of multiple APs, which are deployed in the AOI. To provide wireless network services for large buildings, a large number of APs are equipped. Nevertheless, all these APs contribute positively to the positioning accuracy where the majority are redundant. Therefore, including all detected APs in fingerprinting vectors leads to a confusing IPS [22]. In addition, the resources, which are equipped on MS, limit even the IPS use of the server's assistance to find the user's location. There are many techniques to reduce the computational costs for an IPS.

5.1. AP Selection Methods

To ensure high Wi-Fi quality in most large buildings, a large number of APs are deployed. This large number, L, of APs is greater than what is required to perfectly achieve indoor positioning estimation. The use of all APs leads to excessive redundancy and possible biased estimation. Thus, choosing the informative APs as a subset of L is efficient for reducing computation time and increasing the accuracy of IPSs. In this article, various kinds of AP selections, such as strongest, random, and stable APs, are investigated.

5.1.1. Strongest APs

The benefit of selecting the highest power APs is to provide a high probability of coverage over time. In the online phase, the strongest technique is achieved by selecting K subset of L APs with the highest power readings from the online RSS vectors, where K < L. The K subset APs with the highest power readings is created by sorting the online RSS vectors in descending order, then choosing the top L subset APs [39]. Accordingly, the indices of fingerprints on the RM are sorted to determine the location of MS.

5.1.2. Fisher Criterion

In this technique, the statistical properties of RM fingerprints are exploited during the offline phase to choose the informative APs for positioning [40–42]. The time stabilization and spatial discrimination of each AP across RPs are computed and sorted in descending order. A score is assigned for each AP separately as Equation (7):

$$\zeta^i = \frac{\sum_{j=1}^{M} \left(\psi_j^i - \overline{\psi}^i\right)^2}{\frac{1}{T-1}\sum_{l=1}^{T}\sum_{j=1}^{M}\left(r_j^i(t_l) - \psi_j^i\right)^2},\quad i = 1, 2, \ldots\ldots, L \tag{7}$$

where, $\overline{\psi}^i = \frac{1}{M}\sum_{j=1}^{M}\psi_j^i$, r represents an instantaneous vector at each RP, M represents a number of RPs in the RM, and T represents a number of time samples at each RP.

5.1.3. Random Combination

The above two methods select informative APs according to different criteria, but in this method, the selection of informative AP does not consider the behavior of APs in time or special domains. The random scheme selects K APs randomly of L APs, and RM is changed to indices of the RSS measured vector accordingly [43]. As a result, this method needs less computation complexity during the online phase, and it does not need a large number of RSS time samples.

5.1.4. Stable AP Method

This metric is based on the behavior of APs in a time domain. The APs with the highest time variance over the AOI are excluded. In the offline phase, K APs with low time variance over the AOI are selected from L APs. In the online phase, the measured vector, V, is sorted accordingly to determine the MS's position. The stability of APs ζ^i can be calculated as shown in Equation (8):

$$\zeta^i = \frac{1}{T-1}\sum_{l=1}^{T}\sum_{j=1}^{M}\left(r_j^i(t_l) - \psi_j^i\right)^2 \tag{8}$$

In the literature, we can find other techniques for the selection of Aps, such as Bhattacharyya distance, Information Potential (IP), Information Gain (Info-Gain), Entropy Maximization, and Group Discrimination (GD) [22].

5.2. Clustering Techniques

The computational cost for determining a position is directly proportional to the number of RPs. Therefore, a coarse positioning is introduced in the online phase to confine the positioning process within a small area. The clustering process is achieved in an offline phase where the RPs of the AOI are classified into small groups which are named clusters. Each cluster has a representative which is called an exemplar to represent RPs subset of cluster in an online phase. In this article, K-means and fuzzy C-means are discussed in detail.

5.2.1. K-Means Clustering

In general, the K-means clustering method divides the AOI into subareas to confine the location process within a small area instead of an entire area [44]. Therefore, the computational cost is divided by K, where K represents the number of clusters. In this technique, the center point of a cluster is determined by minimizing the signal distance between exemplar C_j and the members of the same clusters. Given a set of input patterns, $X = \left(x_1, \ldots, x_j, \ldots, x_M\right)$, where M represents number of FP, and $x_j = \left(x_{j1}, \ldots\ldots, x_{jL}\right) \in R^L$, we have $C = (C_1, \ldots \ldots ., C_k)$ $K \leq M$ exemplars for AOI. By minimizing an objective function, J, in this case a squared error function, the objective function is given in Equation (9):

$$J = \sum_{j=1}^{K} \sum_{i=1}^{M} ||r_i^j - C_j||^2 \tag{9}$$

where, $||r_i^j - C_j||$ represents a measured distance between data points r_i^j and exemplar of cluster C_j, it is an indicator of the distance of the n data points from their respective cluster centers. Figure 6 shows the distribution of FPs on K clusters $C = (C_1, \ldots \ldots ., C_k)$. The steps of K-means are summarized in Algorithm 1.

Algorithm 1. K-means clustering

Input:
Reference Points of Area of Interest (RM), **K** (number of clusters)

Output:
Exemplars vectors $C = (C_1, \ldots \ldots ., C_k)$, each RP is designated to a certain cluster

Mechanism of K-means

1- Initialize K centroid points
2- Compute the distances between RPs and centroids.
3- Assign each RP to a cluster.
4- Compute the cluster centroid again when all RPs are assigned,
5- Repeat steps 2, 3, and 4 until there is no change for each cluster.

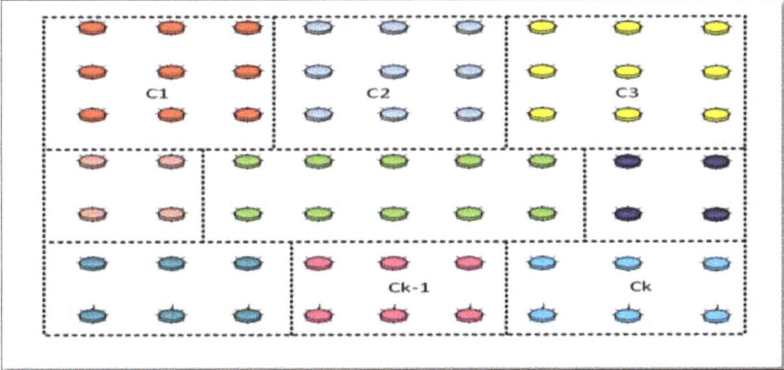

Figure 6. K-means clustering Technique.

5.2.2. Fuzzy C-Means Clustering

Unlike the K-means clustering technique, the membership of each RP is not crisp in the fuzzy C-means clustering approach [45]. Fuzzy C-means clustering can be divided using AOI into K subsets as shown in Figure 7 with clustering centers $C = (C_1, \ldots, C_k) \in R^M$ that verify the minimum cost function, which is a sum of the squared error between the RP's fingerprint and the clustering center. Its formula is given in Equation (10) as follows:

$$J_m = \sum_{i=1}^{M} \sum_{j=1}^{K} u_{ij}^{m} d_{ij}^{2} \quad , 1 \leq m < \infty \tag{10}$$

where, m is greater than 1, and u_{ij} is the degree of membership of RP_i in the cluster j, and $\sum_{i=1}^{K} u_{ij} = 1$. The $d_{ij} = ||x_i - c_j||^2$ represents the Euclidean distance between the PR_i fingerprint and the vector of cluster center C_j. The value of u_{ij} is randomly chosen between (0–1). Fuzzy partition is achieved through an iterative optimization shown in Equation (10). In order to find u_{ij} and C_j, the Equation (11) and Equation (12) should be applied:

$$u_{ij} = \frac{1}{\sum_{k=1}^{K} \left(\frac{||x_i - c_j||}{||x_i - c_k||} \right)^{\frac{2}{m-1}}} \tag{11}$$

$$c_j = \frac{\sum_{i=1}^{M} u_{ij}^{m} . x_i}{\sum_{i=1}^{M} u_{ij}^{m}} \tag{12}$$

The iteration will be stopped when $max_{ij} = \left(\left| u_{ij}^{(k+1)} - u_{ij}^{(k)} \right| \right) < \varepsilon$, where ε is a criterion value which falls in between (0–1), and k is the iteration steps.

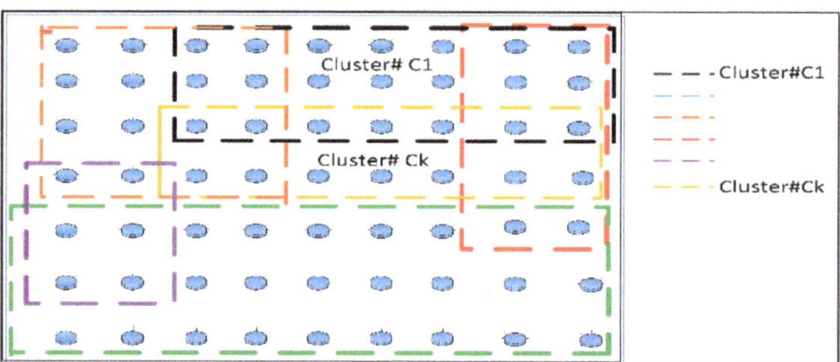

Figure 7. The distribution of RPs on the fuzzy C-means clusters.

6. The Multiple SSID-Based Proposed System

In this article, we propose a dimensionality reduction technique in a multiple service set identifier-based indoor positioning system. The main contribution of this article is to reduce the effect of the multipath phenomenon by fusing the MSSID-based fingerprinting and computational cost as well. The proposed IPS framework is presented in the block diagram, as shown in Figure 8. We have two phases: the offline phase and the online phase. We can describe the steps on each phase as summarized in Algorithm 2.

In the offline phase, fusion is the first step in the proposed system where the vectors of MSSID are fused to reduce the effect of the multipath problem and computational costs as well. In the second step, AOI is clustered into small subareas to confine the location of users by specific region. A maximum spatial variance AP selection is applied on each cluster to find the best joint combination of APs at each cluster. Therefore, there is a certain combination of APs at each cluster, *Pi* for *Ci* is used in the online phase to verify minimum computational costs as well as distance error.

In the online phase, the clustering technique reduces the positioning process time where the fused vector V is pre-matched with q exemplars of clusters *Ci* to find the closed cluster. The suitable pattern of APs is chosen correspondingly with assigned cluster C_j. The final estimated position is calculated in the last stage by using the NN approach.

The robustness of the proposed MSSID-based IPS comes from utilizing a fingerprinting approach where most of the indoor positioning systems based on RSS signals used fingerprinting-based systems due to unpredictable behavior of the RSS in an indoor environment. Unlike traditional fingerprinting systems, which use a single SSID at each AP, the proposed system uses MSSIDs at each AP to mitigate the effect of the multipath issue, which is noticed in an indoor environment. Therefore, designing the proposed system with low dimensionality due to fusing MSSID RM and AP selection makes the proposed system more applicable on limited MS resources.

Algorithm 2. The algorithm of the proposed system.

Stage No. 1 (Fusion):

$\Psi = \frac{1}{N}\sum_{i=1}^{N} \Psi_i$

Stage No. 2 (Clustering):

Clustering Ψ into small subset (Fuzzy C-Means, K-Means)

Stage No. 3 (AP Selection):

Selecting APs pattern for each cluster (Random, Strongest, Spatial Variance)

The proposed system Algorithm (ONLINE PHASE)

Stage No. 1 (Fusion):

$$V = \frac{1}{N}\sum_{i=1}^{N} V_i$$

Stage No. 2 (Clustering):

Apply coarse Positioning by assigning closed cluster

Stage No. 3 (AP selection):

Choose appropriate AP pattern at assigned cluster.

Stage No. 4 (position Estimation):

Find final estimated position (x_o,y_o).

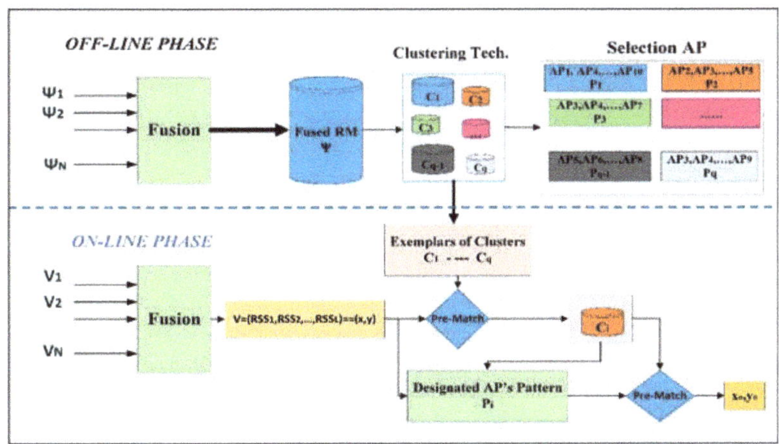

Figure 8. The block diagram of proposed system.

7. The Experiment

To validate the proposed system merits, we used real RSS data. This data was collected from CEAS and Waldo library at Western Michigan University to investigate the behavior of MSSID signals and validate the proposed system. All APs which are used in the AOI have worked with MSSIDs with an Industrial, Scientific, and Medical (ISM) band (2.4 GHz). The designated AOI at CEAS is a good environment to validate the proposed system because it has long corridors and classrooms with large study lounges. The AOI was divided into 117 RPs with a spacing of (2 m × 2 m) and a total area of 3000 m^2, as shown in Figure 9. Each RP was labeled by a Cartesian coordinate (x_i, y_i). The wireless Net View program was installed on a Toshiba Satellite laptop, which has an Intel wireless-N7260 adapter to record the vector of RSS values from various APs with a 1 sec period. Each RP was created from 200 RSS time samples from various directions (0°, 90°, 180°, and 270°) to mitigate the body presence. 100 different locations were chosen randomly on the AOI to evaluate the proposed system. At each selected location, 200 samples (20,000 TPs) were collected with different orientations to verify the validation of the proposed system.

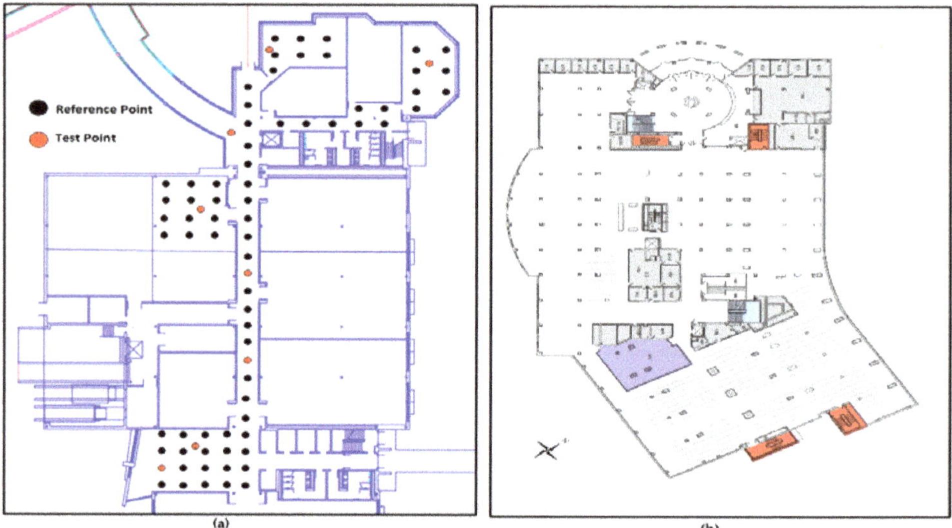

Figure 9. The map of the area of interest at Western Michigan University. (**a**) CEAS building; (**b**) waldo Library building.

8. Results and Discussion

In this section, the proposed system is investigated with two types of clustering techniques: K-means, and Fuzzy C-means.

8.1. K-Means Clustering Results

Figure 10 shows the performance of the proposed system when K-means clustering is applied for single SSID and multiple SSIDs. The outliers issue is clearly seen in the single SSID. Although the FPs come from averaging 200 RSS samples, the clustering of FPs with MSSIDs is more stable, and it is similar to the RPs' spatial distribution. Figure 11a shows the CDF of distance error for MSSIDs as compared to a single SSID at q = 10 (number of clusters). It is clear that the performance of MSSIDs outperforms the behavior of a single SSID. Figure 11b shows the average error of MSSID and single SSID for various values of q. The mean of the distance error is increased slightly with the increasing number of clusters. The number of APs is 34 for both experiments, and all results are considered without APs reduction.

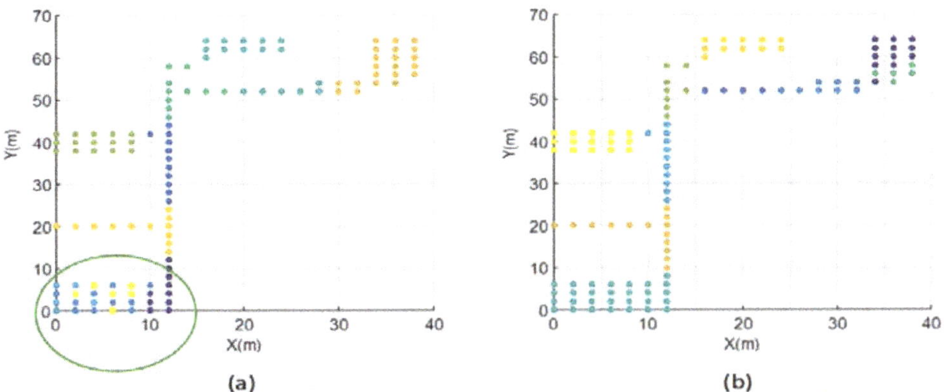

Figure 10. K-means clustering. (**a**) With a single SSID; (**b**) with multiple SSIDs.

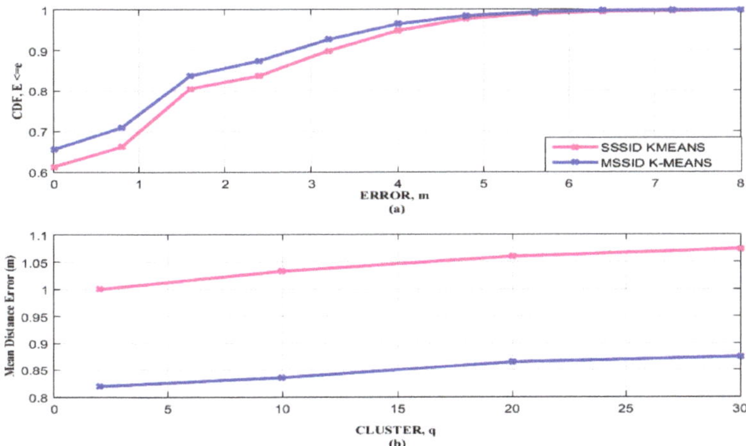

Figure 11. K-means clustering. (**a**) Cumulative Distribution Function (CDF) of Error; (**b**) mean error versus number of clusters.

8.2. Selection of APs with K-Means

Three kinds of approaches to the selection of APs were conducted on each cluster at the same time: random, strongest, and maximum spatial variance. Figure 12 shows the CDF of error for the selected approaches with q = 20, and AP reduction is 50% (number of APs = 17). The maximum spatial variance exhibits superior performance as compared with the other selected APs approaches.

Furthermore, the performance of the maximum spatial variance with 17 APs (50% reduction) is similar to the performance of all APs (34 APs). This confirms previous work [11]. We found the redundancy of MSSID occurred when SSIDs were exploited in the same fingerprinting vectors.

Table 5 summarizes the accuracy of the proposed system with various kinds of selection APs. The selection of informative APs leads to a reduction in computational cost of 50%, as well as the reduction that comes from the clustering technique. The performance of the K-means-based proposed system with various values of q (number of clusters) is illustrated in Figure 13. As we can see in the figure, the average error is decreased when the number of informative APs is increased. The maximum spatial variance selection AP exhibits a superior performance when compared to the strongest and random APs selection with various values of q (q = 10, and q = 20).

Table 5. The mean distance error of the K-means-based proposed system versus a single SSID.

Selection Technique	Strongest APs 50% Reduction	Random APs 50% Reduction	Max. Spatial Var. 50% Reduction	Whole APs
Single SSID	1.33 m	1.6 m	1.14 m	1.08 m
Multiple SSIDs	1.11 m	1.36 m	0.9 m	0.86 m

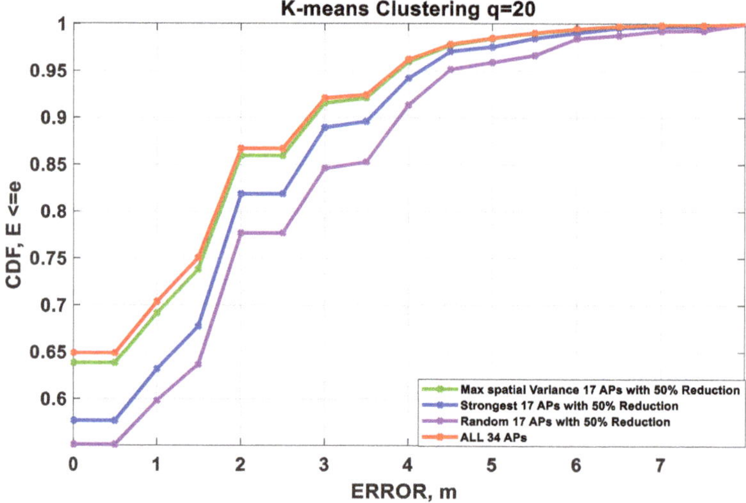

Figure 12. The proposed technique with various kind of AP selection methods.

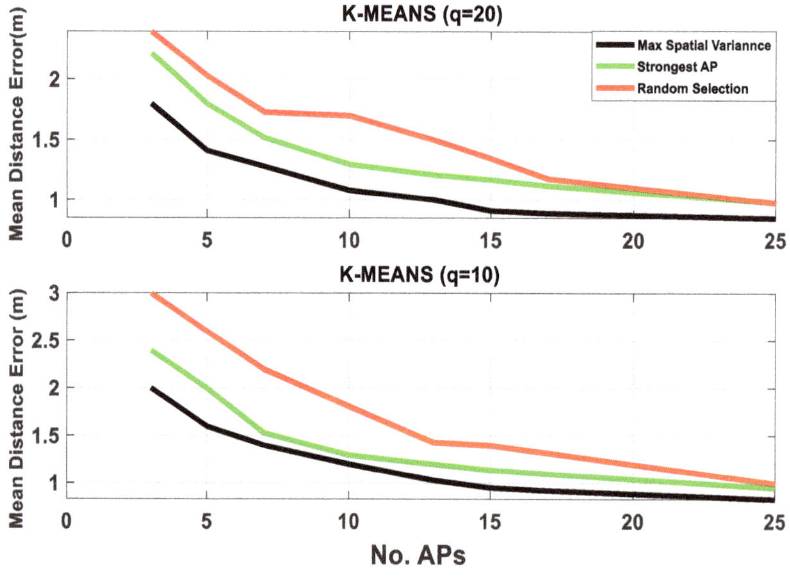

Figure 13. K-means-based proposed system with different values of clusters.

8.3. Fuzzy C-Means Results

The same methodology that is used with K-means is applied with the Fuzzy C-means method. Figure 14 shows the performance of Fuzzy C-means clustering, where Figure 14a shows the CDF of distance error for MSSIDs as compared to a single SSID at q = 10 (number of clusters), and the value of membership, which is issued by C-means, is (m = 1.8). It is clear that the performance of MSSIDs outperforms the behavior of a single SSID. Figure 14b shows the average distance error of MSSIDs and a single SSID for various values of q. From Figure 14b, we can see that the mean of error is increased slightly with the increasing number of clusters. The number of APs is also 34 APs for both experiments.

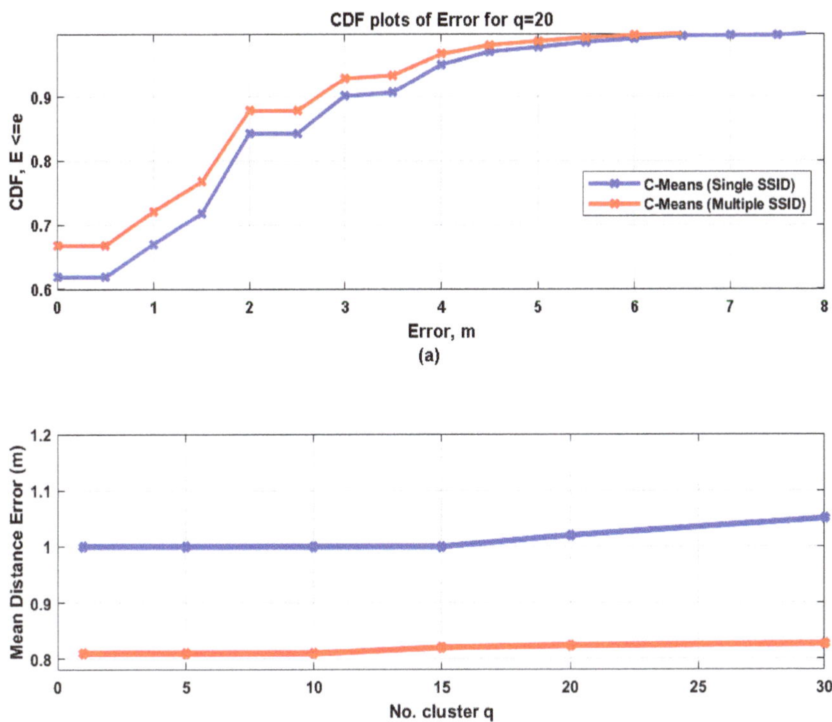

Figure 14. Fuzzy C-means clustering. (**a**) CDF of Error; (**b**) mean error versus number of clusters.

8.4. Selection of APs with Fuzzy C-Means

In this experiment, three types of selection APs approaches were also conducted on each cluster at the same time: random, strongest, and maximum spatial variance. Figure 15 shows the CDF of distance error for the selection approaches with q = 20, and AP reduction is 50% (number of APs = 17). The maximum spatial variance exhibits superior performance as compared with the other selected APs approaches. In addition, the performance of maximum spatial variance with 17 APs (50% reduction) is similar to the performance of all APs (34 APs). Table 6 summarizes the accuracy of the proposed C-means-based system with various kinds of selection APs.

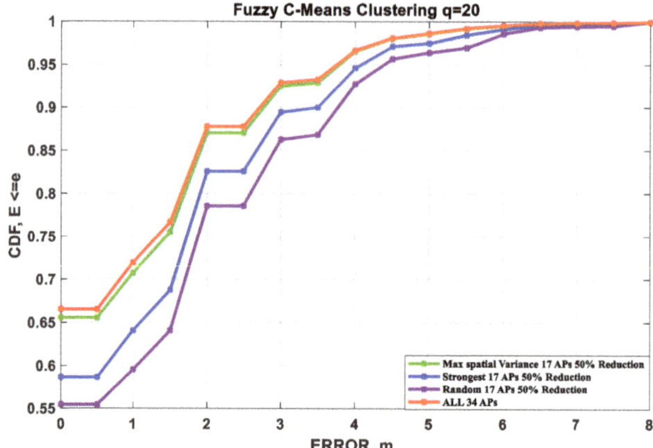

Figure 15. The proposed technique with the Fuzzy C-mean approach.

Table 6. The distance error of the C-means-based proposed system versus a single SSID.

Selection Technique	Strongest APs 50% Reduction	Random APs 50% Reduction	Max. Spatial Var. 50% Reduction	Whole APs
Single SSID	1.32	1.51	1.06	1.01
Multiple SSIDs	1.1	1.25	0.84	0.82

Figure 16 shows the comparison between C-means and the proposed K-means-based system with maximum spatial variance selection AP. The C-means clustering-based system performance exhibits a slight improvement in accuracy. The reason is the average number of FPs at each cluster in C-means is more than the average number of FPs in K-means.

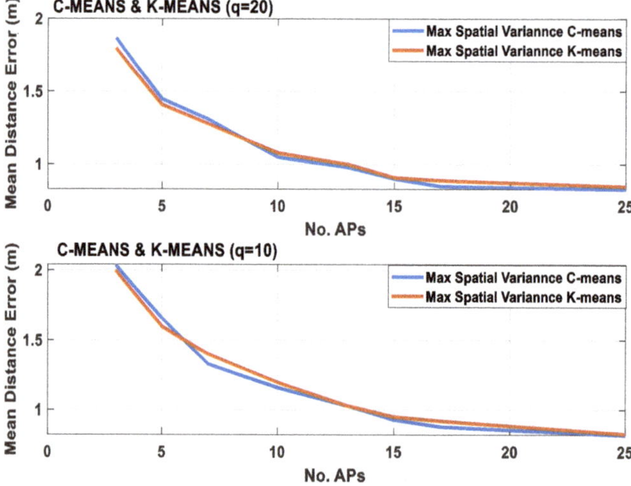

Figure 16. The proposed technique performance with the Fuzzy C-mean and K-mean approaches versus number of APs.

Figure 17 shows the compression between the proposed system and traditional technique that considers each SSID as a single AP. The superior performance of the proposed system with 17 SSID selection versus the traditional technique performance is clear. The MSSID-based technique selects 17 SSIDs (17 distinct APs), while the traditional technique selects 17 SSIDs, which means approximately 6 distinct APs (each AP transmits 3 SSIDs). The shortcoming of the traditional technique comes from exploiting the similar SSIDs in the same fingerprinting vector.

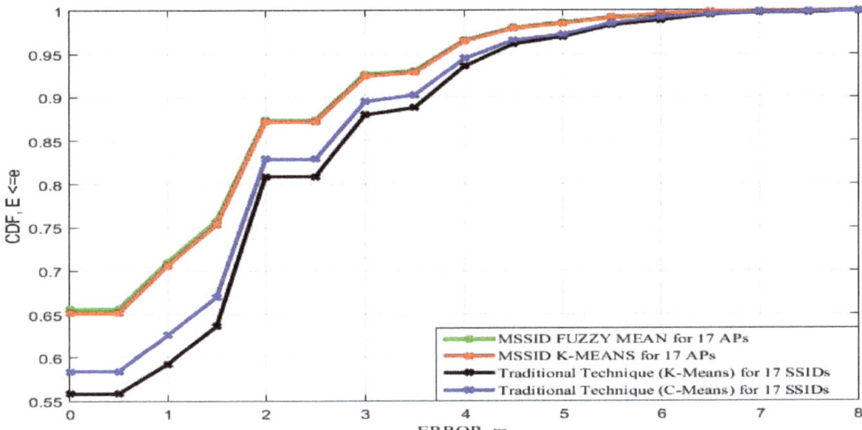

Figure 17. The proposed technique performance with the Fuzzy C-mean and K-mean approaches.

Table 7 summarizes the percentage of calculation reduction versus the number of SSIDs when q = 20. Choosing informative APs leads to a reduction in computational costs as well as the reduction that comes from fusing and the clustering technique.

Table 7. The Calculation reduction of positioning process versus number of SSIDs.

No. SSIDs	Distance Error (m) Traditional (K-Means)	Distance Error (m) Proposed (C-Means)	Calculation Reduction %
10	1.1	1.07	17%
17	0.9	0.84	12%
22	0.87	0.84	10%
28	0.85	0.83	5%
34	0.86	0.82	0%

9. Conclusions and Future Work

Indoor localization is an active and noble research area for its numerous applications in various areas of our daily lives. Within indoor environments, tracking people and localizing objects have become a necessity and thus motivates many researchers to tackle the challenges of IPS. In this article, the dimensionality reduction of MSSID's RM has been proposed. Reducing the computational cost of IPS is very important for real-time system applications because large data require a long time to be processed. The reduction comprises three parts: first, clustering AOI into subareas to confine the location process into a subset of RPs. Second, fusing N RM of MSSIDs to create a delegate RM, which is used to present N RM of MSSID in the positioning process. Third, the use of a maximum spatial variance to select the most informative APs. The proposed system is designed according to the behavior of MSSID signals in time and spatial domain. Fusing RM of MSSID mitigates the outlier's issue, which is created due to the multi-path phenomenon. The results show that the fusion of N RM of MSSID in both phases mitigates the effect of the fading issue and minimizes the length of fingerprinting vectors by using the maximum spatial variance selection APs. In addition, the results exhibit the ability of maximum spatial variance selection APs technique to enhance the precision of IPS and reduce the computational costs as well. Light algorithm approaches such as the proposed system are feasible for real-time tracking because the positioning process requires only a few calculations in the online phase to find the location of MS.

Future work will include several tasks to extend the current proposed system. First, the behavior of the mean of MSSIDs signals is close to path loss, as shown in Figure 5, because averaging the MSSIDs' signals contributes to mitigating the effect of the multipath issue. Accordingly, it might be possible to use the fused MSSIDs' RSS-based path loss model to determine the user's location. In this case, knowing the coordinates of APs in the AOI is sufficient to determine the user's location without creating a radio map, which requires more time and labor costs.

Second, it can be very fruitful to investigate the use of the 2.4 GHz and 5 GHz bands simultaneously with MSSID that are deployed on the same AP to enhance the WLAN-based IPS. The coverage area of 5 GHz is smaller than 2.4 GHz [46]. Therefore, we can use 5GHz for coarse positioning and the 2.4 GHz for fine positioning.

Author Contributions: A.A. and I.A.-Q. conceived the presented idea and developed the closed-form of each algorithm. Simulations and results presentation were conducted by A.A. All authors verified the analytical methods and discussed the results contributed to the final manuscript. Visualization and Supervision, I.A.-Q. Reviewing and Editing, A.A. and I.A.-Q.

Funding: This research received no external funding.

Conflicts of Interest: The authors declare no conflict of interest.

References

1. Zakavat, S.A.; Buehrer, R.M. *Handbook of Position Location: Theory, Practice, and Advances*; John, W.S., Ed.; IEEE Press: Hoboken, NJ, USA, 2011.
2. Gu, Y.; Lo, A.; Niemegeers, I. A survey of indoor positioning systems for wireless personal networks. *IEEE Commun. Surv. Tuts.* **2009**, *11*, 13–32. [CrossRef]
3. Roos, T.; Myllymaki, P.; Tirri, H.; Misikangas, P.; Sievanen, J. A probabilistic approach to WLAN user location estimation. *Int. J. Wirel. Inf. Netw.* **2002**, *9*, 155–163. [CrossRef]
4. Arzullaev, A.; Park, W.P.; Hoyoul, J. Accurate signal strength prediction based positioning for indoor WLAN systems. In Proceedings of the IEEE/ION Position Location Navigation Symposium, Monterey, CA, USA, 5–8 May 2008; pp. 685–688.
5. Shrestha, S.; Talvitie, J.; Lohan, E.S. Deconvolution-based indoor positioning with WLAN signals and unknown access point locations. In Proceedings of the International Conference on Localization and GNSS, Turin, Italy, 25–27 June 2013; pp. 1–6.
6. Kaplan, E.; Hegarty, C. *Understanding GPS Principles and Applications*, 2nd ed.; Artech House: Norwood, MA, USA, 2005.
7. Moghtadaiee, V.; Dempster, A.G. WiFi fingerprinting signal strength error modeling for short distances. In Proceedings of the 2012 International Conference on Indoor Positioning and Indoor Navigation (IPIN), Sydney, NSW, Australia, 13–15 November 2012; pp. 1–6.
8. Zegeye, W.K.; Amsalu, S.B.; Astatke, Y.; Moazzami, F. WiFi RSS fingerprinting indoor localization for mobile devices. In Proceedings of the 2016 IEEE 7th Annual Ubiquitous Computing, Electronics & Mobile Communication Conference (UEMCON), New York, NY, USA, 20–22 October 2016; pp. 1–6.
9. Duda, R.O.; Hart, P.E.; Stork, D.G. *Pattern classification*; Wiley-Interscience: New York, NY, USA, 2001.
10. Cisco. "Configure Multiple SSIDs on a Network". Available online: https://www.cisco.com/c/en/us/support/docs/smb/routers/cisco-rv-series-small-business-routers/smb5652-configure-multiple-ssids-on-a-network.pdf. (accessed on 12 December 2018).
11. Abed, A.; Al-Moukhles, H.; Abdel-Qader, I. Indoor positioning system using multiple services set identifiers-based fingerprints. In Proceedings of the 2017 IEEE International Conference on Electro Information Technology (EIT), Lincoln, NE, USA, 14–17 May 2017; pp. 295–300.
12. Sun, Y.; Sun, Q.; Chang, K. The application of indoor localization systems based on the improved Kalman filtering algorithm. In Proceedings of the 2017 4th International Conference on Systems and Informatics (ICSAI), Hangzhou, China, 11–13 November 2017; pp. 768–772.

13. Li, Y.; Pan, Y.Q.; Bai, D.F.; Liu, H.; Yang, B. A histogram-based cascade detector for radio tomographic localization. In Proceedings of the 2013 16th International Symposium on Wireless Personal Multimedia Communications (WPMC), Atlantic City, NJ, USA, 24–27 June 2013; pp. 1–5.
14. Abdullah, O. Convex Optimization via Symmetrical Hölder Divergence for a WLAN Indoor Positioning System. *Entropy* **2018**, *20*, 639. [CrossRef]
15. Tran, H.Q.; Ha, C. Improved Visible Light-Based Indoor Positioning System Using Machine Learning Classification and Regression. *Appl. Sci.* **2019**, *9*, 1048. [CrossRef]
16. Ashraf, I.; Hur, S.; Shafiq, M.; Park, Y. Floor identification using magnetic field data with smartphone sensors. *Sensors* **2019**, *19*, 2538. [CrossRef] [PubMed]
17. Bensky, A. *Wireless Positioning Technologies and Applications*; Artech House: Norwood, MA, USA, 2007.
18. Ashraf, I.; Hur, S.; Park, Y. Application of Deep Convolutional Neural Networks and Smartphone Sensors for Indoor Localization. *Appl. Sci.* **2019**, *9*, 2337. [CrossRef]
19. Gökhan, K.; Tansel, Ö.; Bülent, T. IEEE 802.11 WLAN based Real Time Indoor Positioning: Literature Survey and Experimental investigations. *Procedia Comput. Sci.* **2014**, *34*, 157–164.
20. Plastria, F.; De Bruyne, S.; Carrizosa, E. Advanced Data Mining and Applications. ADMA 2008. Lecture Notes in Computer Science. In *Dimensionality Reduction for Classification*; Tang, C., Ling, C.X., Zhou, X., Cercone, N.J., Li, X., Eds.; Springer: Berlin, Germany, 2008; Volume 5139.
21. Fang, S.H.; Lin, T. Principal Component Localization in Indoor WLAN Environments. *IEEE Trans. Mob. Comput.* **2012**, *11*, 100–110. [CrossRef]
22. Li, J.H.; Tian, J.B.; Fei, F.; Wang, Z.X.; Wang, H.J. Indoor localization based on subarea division with fuzzy C-means. *Int. J. Distrib. Sens. Netw.* **2016**, *12*, 1–16. [CrossRef]
23. Jiang, P.; Zhang, Y.Z.; Fu, W.Y.; Liu, H.Y.; Su, X.L. Indoor Mobile Localization Based on Wi-Fi Fingerprint's Important Access Point. *Int. J. Distrib. Sens. Netw.* **2015**, *11*, 429104. [CrossRef]
24. Abusara, A.; Hassan, M.S.; Ismail, M.H. RSS fingerprints dimensionality reduction in WLAN-based indoor positioning. In Proceedings of the 2016 Wireless TelecommunicationsSymposium(WTS), London, UK, 18–20 April 2016; pp. 1–6.
25. Chen, Y.Q.; Yang, Q.; Yin, J.; Chai, X.Y. Power-efficient access-point selection for indoor location estimation. *IEEE Trans. Knowl. Data Eng.* **2006**, *18*, 877–888. [CrossRef]
26. Kanaris, L.; Kokkinis, A.; Liotta, A.; Stavrou, S. Fusing bluetooth beacon data with Wi-Fi radiomaps for improved indoor localization. *Sensors* **2017**, *17*, 812. [CrossRef] [PubMed]
27. Feng, C.; Au, W.S.A.; Valaee, S.; Tan, Z. Received-Signal-Strength-Based Indoor Positioning Using Compressive Sensing. *IEEE Trans. Mob. Comput.* **2012**, *11*, 1983–1993. [CrossRef]
28. Sánchez-Rodríguez, D.; Alonso-González, I.; Ley-Bosch, C.; Quintana-Suárez, M. A Simple indoor localization methodology for fast building classification models based on fingerprints. *Electronics* **2019**, *8*, 103. [CrossRef]
29. Lopez-de-Teruel, P.; Oscar, C.; Felix, J.G. Using dimensionality reduction techniques for refining passive indoor positioning systems based on radio fingerprinting. *Sensors* **2017**, *17*, 871. [CrossRef] [PubMed]
30. Bahl, P.; Padmanabhan, V.N. RADAR: An in-building RF-based user location and tracking system. In Proceedings of the IEEE INFOCOM 2000. Conference on Computer Communications. Nineteenth Annual Joint Conference of the IEEE Computer and Communications Societies (Cat. No.00CH37064), Tel Aviv, Israel, 26–30 March 2000; Volume 2, pp. 775–784.
31. Khalajmehrabadi, A.; Gatsis, N.; Akopian, D. Modern WLAN Fingerprinting Indoor Positioning Methods and Deployment Challenges. *IEEE Commun. Surv. Tutor.* **2017**, *19*, 1974–2002. [CrossRef]
32. Myllymäki, P.; Roos, T.; Tirri, H.; Misikangas, P.; Sievänen, J. A Probabilistic Approach to WLAN User Location Estimation. In Proceedings of the 3rd IEEE Workshop on Wireless Local Areas Networks, Boston, MA, USA, 27–28 September 2001.
33. He, S.; Shin, K.G. Steering Crowdsourced Signal Map Construction via Bayesian Compressive Sensing. In Proceedings of the IEEE INFOCOM 2018—IEEE Conference on Computer Communications, Honolulu, HI, USA, 15–19 Apirl 2018; pp. 1016–1024.
34. Tan, J.; Fan, X.; Wang, S.; Ren, Y. Optimization-Based Wi-Fi Radio Map Construction for Indoor Positioning Using Only Smart Phones. *Sensors* **2018**, *18*, 3095. [CrossRef] [PubMed]
35. Hossain, A.K.M.M.; Jin, Y.; Soh, W.; Van, H.N. SSD: A Robust RF Location Fingerprint Addressing Mobile Devices Heterogeneity. *IEEE Trans. Mob. Comput.* **2013**, *12*, 65–67. [CrossRef]

36. Florwick, J.; Whiteaker, J.; Amrod, A.C.; Woodhams, J. Wireless LAN Design Guide for High Density Client Environments in Higher Education. Available online: https://www.cisco.com/c/en/us/products/collateral/wireless/aironet-1250-series/design_guide_c07-693245.pdf (accessed on 22 July 2017).
37. Rappaport, T.S. Characterization of UHF multipath radio channels in factory buildings. *IEEE Trans. Antennas Propagat.* **1989**, *37*, 1058–1069. [CrossRef]
38. Li, G.; Geng, E.; Ye, Z.; Xu, Y.; Lin, J.; Pang, Y. Indoor positioning algorithm based on the improved RSSI distance model. *Sensors* **2018**, *18*, 2820. [CrossRef]
39. Youssef, M.; Agrawala, A.; Shankar, A. WLAN Location Determination via Clustering and Probability Distributions. In Proceedings of the First IEEE International Conference on Pervasive Computing and Communications, Fort Worth, TX, USA, 23–26 March 2003; pp. 143–155.
40. Kushki, A. A Cognitive Radio Tracking System for Indoor Environments. Ph.D. Thesis, University of Toronto, Toronto, ON, Canada, 2008.
41. Shawe-Taylor, J.; Cristianini, N. *Kernel Methods for Pattern Analysis*; Cambridge University Press: Cambridge, UK, 2004.
42. Au, A.W.S. Rss-based wlan indoor positioning and tracking system using compressive sensing and its implementation on mobile devices. Master's Thesis, University of Toronto, Toronto, ON, Canada, 2010.
43. Luo, J.; Fu, L. A smartphone indoor localization algorithm based on WLAN location fingerprinting with feature extraction and clustering. *Sensors* **2017**, *17*, 1339.
44. Rui, X.; Wunsch, D. Survey of clustering algorithms. *IEEE Trans. Neural Netw.* **2005**, *16*, 645–678.
45. Zhou, H.; Van, N.N. Indoor Fingerprint Localization Based on Fuzzy C-Means Clustering. In Proceedings of the 2014 Sixth International Conference on Measuring Technology and Mechatronics Automation, Zhangjiajie, China, 10–11 January 2014; pp. 337–340.
46. Talvitie, J.; Renfors, N.; Lohan, E.S. A Comparison of Received Signal Strength Statistics between 2.4 GHz and 5 GHz Bands for WLAN-Based Indoor Positioning. In Proceedings of the 2015 IEEE Globecom Workshops (GC Wkshps), San Diego, CA, USA, 6–10 December 2015; pp. 1–6.

Article

Indoor Localization Based on Wi-Fi Received Signal Strength Indicators: Feature Extraction, Mobile Fingerprinting, and Trajectory Learning

Jaehyun Yoo [1,*] and Jongho Park [2]

1 Department of Electrical, Electronic and Control Engineering, Hankyong National University, Anseoung 17579, Korea
2 Department of Military Digital Convergence, Ajou Univeristy, Suwon 16499, Korea; parkjo05@ajou.ac.kr
* Correspondence: jhyoo@hknu.ac.kr; Tel.: +82-3167-05412

Received: 6 August 2019; Accepted: 17 September 2019; Published: 19 September 2019

Abstract: This paper studies the indoor localization based on Wi-Fi received signal strength indicator (RSSI). In addition to position estimation, this study examines the expansion of applications using Wi-Fi RSSI data sets in three areas: (i) feature extraction, (ii) mobile fingerprinting, and (iii) mapless localization. First, the features of Wi-Fi RSSI observations are extracted with respect to different floor levels and designated landmarks. Second, the mobile fingerprinting method is proposed to allow a trainer to collect training data efficiently, which is faster and more efficient than the conventional static fingerprinting method. Third, in the case of the unknown-map situation, the trajectory learning method is suggested to learn map information using crowdsourced data. All of these parts are interconnected from the feature extraction and mobile fingerprinting to the map learning and the estimation. Based on the experimental results, we observed (i) clearly classified data points by the feature extraction method as regards the floors and landmarks, (ii) efficient mobile fingerprinting compared to conventional static fingerprinting, and (iii) improvement of the positioning accuracy owing to the trajectory learning.

Keywords: indoor localization; Wi-Fi received signal strength indicator (RSSI); semisupervised learning; feature extraction; mobile fingerprinting; trajectory learning

---

Wi-Fi RSSI based indoor localization [1–4] is one of the standard approaches for indoor localization. It is able to utilize the RSSI measurements received from a large number of access points (APs) that are already built in construction. However, the Wi-Fi RSSI as a function of distance between a receiver (smartphone) and a transmitter (wireless AP) is nonlinear and varying due to interference of the indoor environments such as the other radio signals, walls, and obstacles. To address this problem, many machine learning based localization methods [5–8] have been developed, which learn a pattern of the RSSI measurements corresponding locations across the interested positioning area. In addition, due to its unbiased estimation capability, it is likely to be combined with other kinds of localization, such as pedestrian localization using inertial measurement unit (IMU) [9,10], visual localization [11,12], and magnetic sensor-based localization [13,14].

In particular, semisupervised learning algorithms have been recently suggested for efficient indoor localization, which reduce the human effort necessary for collecting training data [15–20]. For example, for indoor localization, a large amount of unlabeled data can be easily collected by recording only Wi-Fi RSSI measurements without requiring position labels, which can save resources for collection and calibration. By contrast, labeled training data have to be created manually. Adding a large amount of unlabeled data in the semisupervised learning framework can prevent the decrement in the estimation accuracy that occurs when using only a small amount of labeled data.

Given the advantage of semisupervised learning, this study describes (i) feature extraction, (ii) mobile fingerprinting, and (iii) mapless localization for efficient Wi-Fi RSSI-based localization. Figure 1 describes the interconnections between the research parts and the flow of our localization system.

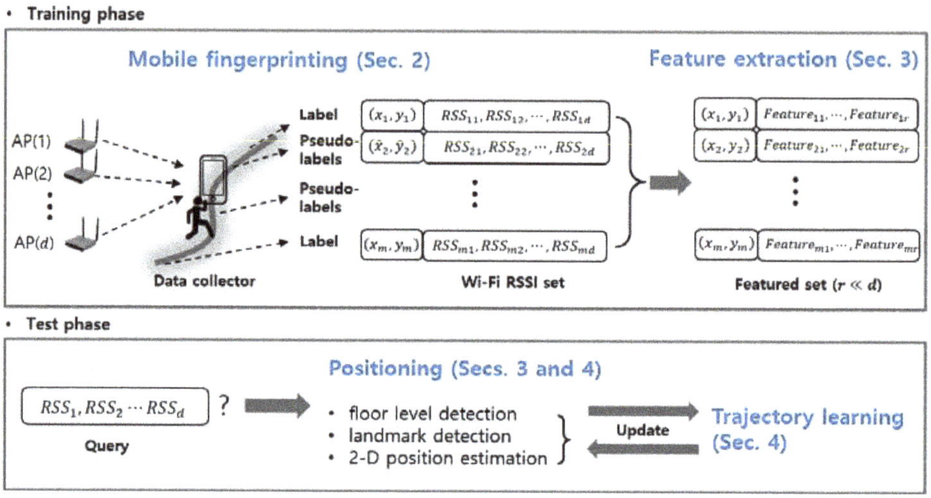

- Training phase

Mobile fingerprinting (Sec. 2) Feature extraction (Sec. 3)

AP(1) AP(2) AP(d) Data collector Label Pseudo-labels Pseudo-labels Label

(x_1, y_1) $RSS_{11}, RSS_{12}, \cdots, RSS_{1d}$

$(\tilde{x}_2, \tilde{y}_2)$ $RSS_{21}, RSS_{22}, \cdots, RSS_{2d}$

(x_m, y_m) $RSS_{m1}, RSS_{m2}, \cdots, RSS_{md}$

Wi-Fi RSSI set

(x_1, y_1) $Feature_{11}, \cdots, Feature_{1r}$

(x_2, y_2) $Feature_{21}, \cdots, Feature_{2r}$

(x_m, y_m) $Feature_{m1}, \cdots, Feature_{mr}$

Featured set ($r \ll d$)

- Test phase

Positioning (Secs. 3 and 4)

$RSS_1, RSS_2 \cdots RSS_d$? Query

- floor level detection
- landmark detection
- 2-D position estimation

Update

Trajectory learning (Sec. 4)

Figure 1. Wi-Fi received signal strength indicator (RSSI)-based indoor localization framework. In the training phase, the mobile fingerprinting for data collection and the feature extraction are conducted. The feature extraction models are applied to the floor level and landmark detections in the test phase. Also, the trajectory learning is proposed for improving position estimation accuracy. The positioning composes the 2D position estimation, floor classification, and landmark detections by using the same RSSI measurements.

First, the feature extraction considered in this paper aims to find a pattern from raw Wi-Fi RSSI and to reduce dimensionality. It allows not only recognizing different floor levels and different landmarks (e.g., toilet, room, and elevator), but also boosting calculation time. This study implements the multistory estimation including the classification of the floors and landmarks, by using the single Wi-Fi RSSI measurement set.

Second, most approaches to indoor localization have used the conventional static fingerprinting method in the training phase, where a trainer has to collect labeled training data stationarily at every position (or grid) while measuring and labeling Wi-Fi RSSI measurements. For more rapid and efficient collection, this study proposes the mobile fingerprinting method that allows a trainer to continue walking during the collection. Instead of the trainer's necessity to label the positions corresponding the measurements, the proposed algorithm automatically pseudolabels (or estimates position of) the unlabeled data with a small amount of labeled data. For accurate pseudolabeling, we design a semisupervised regression by considering both the spatial and temporal relationship of the Wi-Fi RSSI sets.

Third, in indoor localization, it is common to apply a filtering method, such as a particle filter, to estimate the position or to boost accuracy [21–23]. However, this implies the assumption that a localization service provider can give the floor plan of the area of interest, which causes a large volume of data being transmitted to users. In this paper, the trajectory learning algorithm is integrated with the floor and landmark classification, mobile fingerprinting, and positioning for the expanded multistory building experiment.

Last, a position estimation algorithm is created by combining a particle filter and Gaussian process (GP) [24], to exploit the learned trajectories as a prior function and to use probabilistic GP likelihood by modeling the relationship between Wi-Fi RSSI and position.

We evaluate the proposed algorithms in a multifloor building through the experience of a few users. The experimental results show that (i) clearly classified data points with respect to different floors and landmarks by the feature extraction, (ii) efficient mobile fingerprinting compared to the conventional fingerprinting method, and (iii) improvement of positioning accuracy up to the average 1.2 m in comparison to standard approach thanks to the trajectory learning.

The paper is organized as follows. Section 1 surveys some studies relevant to indoor localizations. In Section 2, details about the mobile fingerprinting are presented. Section 3 examines characteristics of Wi-Fi RSSI and describes a semisupervised discriminant analysis for extracting features from raw Wi-Fi RSSI observations, with a description of floor classification and landmark detection. Section 4 introduces the mapless localization and Section 5 summarizes the experimental results. Finally, Section 6 presents the conclusion.

1. Related Studies

This section provides a survey of studies relevant to indoor localization. Different semisupervised learning techniques are reviewed for feature extraction in Section 1.1 and for mobile fingerprinting in Section 1.2. Mapless localization, in Section 1.3, describes trajectory learning and position estimation.

1.1. Semisupervised Feature Extraction

A Wi-Fi RSSI dataset consists of RSSI values corresponding to a user's position, obtained from Wi-Fi access points (APs). The dimension of a raw Wi-Fi RSSI dataset is defined as the number of APs scanned in the entire area of interest in a building. Typically, the dimensionality is so high that it is impractical to perform real-time operations. Additionally, many of the elements in a raw RSSI database are usually empty because APs cannot cover a wide area. Therefore, reconstructing the data of a raw database is of paramount importance. Deep learning approaches [25–28] have been applied to the RSSI indoor localization. The high accuracy reported is due to its capability for feature extraction by the deep neural network structure, which learns complex pattern of the RSSI observations. The feature extraction in the standard deep learning except the autoencoder method [29] is computed among the hidden neural network layers, and then the hidden layers are connected to the latest localization layer. Thus, they cannot be used for our purpose to obtain the low-dimensional feature vector separately.

This paper combines two feature extraction methods: Fisher discriminant analysis (FDA) and principal component analysis (PCA). FDA is an a supervised dimensionality reduction method, and PCA is an unsupervised dimensionality reduction method. Supervised FDA tends to find embedded spaces overfitted to labeled samples. Therefore, it is effective to add unlabeled data, which can be collected easily and in large volumes. By using both labeled and unlabeled data, the combination of FDA and PCA can produce better accuracy compared to that achieved by solely using supervised FDA or unsupervised PCA. In this study, the role of semisupervised discriminant analysis is two-fold: dimension reduction and detection of floor level and landmarks.

1.2. Semisupervised Learning for Mobile Fingerprinting

In this section, the semisupervised learning methods for regression are presented, which are used to build the relationship between the Wi-Fi RSSI data sets and the locations. We then derive the contribution of the proposed mobile fingerprinting algorithm.

Utilization of the unlabeled data has been studied in the semisupervised learning to improve the efficiency and accuracy of the indoor localization. In those works, semisupervised deep learning methods [18,19,30] have been recently developed. The mobile fingerprinting requires the light computation and should be implemented as fast as the mobile device or the server deals with huge amount of the unlabeled data. Therefore, rather than the heavily computational deep learning methods, the support vector machine (SVM)-based [31] semisupervised learning algorithms are applied in this paper, which solves a convex optimization problem.

Semisupervised least square (SSL) [32] adds manifold regularization of unlabeled data into the standard Least Square SVR framework [33]. Because of its linearized setup, this algorithm is fast and useful for real-time application. Semisupervised colocalization (SSC) [34] builds an optimization framework consisting of a singular value decomposition, a manifold regularization, and a loss function. Because SSC estimates the locations of the APs as well as a target's location, it requires the known locations of the AP, whereas SVR-based semisupervised algorithms do not. Moreover, the large number of tuning parameters and the heavy computation may be a burden. Both SSL and SSC apply the unlabeled data only for making a manifold regularization, through the graph Laplacian. Further progressed utilization of unlabeled data occurs in transductive support vector machine (TSVM) [35] and Laplacian Embedded Regression Least Squares (LapERLS) [36]. TSVM attempts to find the labels of unlabeled data and obtain the decision function. Because finding a solution requires searching all candidate-labels of the unlabeled data, TSVM is not feasible in most applications.

LapERLS introduces an intermediate variable as a substitute for the original labeled data. During the optimization process via Karush–Kuhn–Tucker (KKT) conditions, pseudolabels and a transformed kernel matrix are generated. Then, the pseudolabels and transformed kernel matrix are used as a substitution for the original labeled data and the original kernel matrix, respectively. This algorithm is useful for large-scale problems, due to the light computation necessary for obtaining the transformed kernel matrix, because the standard kernel matrix and Laplacian matrix are decoupled. However, LapERLS becomes inaccurate when only a small number of labeled training data points are used.

We adopt the idea of pseudolabeling from LapERLS because the pseudolabels can compensate for the lack of labeled data. To improve the accuracy of the pseudolabels, we propose adding a temporal relation to unlabeled training data that are collected as time-series. A study [37] employing a Hodrick–Prescott (H–P) filter that captures a smoothed-curve representation for a time series from training data is helpful for assigning time-series pseudolabels. Consequently, our pseudolabeling is able to consider both the spatial and temporal aspect of the training data sets. Note that this pseudolabeling technique based on this semisupervised regression is used for the mobile fingerprinting, not for estimating the user's position.

1.3. Mapless Localization

Accommodating a mapless situation might be valuable for secure localization operations by keeping information private. The pedestrian localization based on IMU and camera sensors easily produces the integral error so the user should carefully hold the receiver such as smartphone stationary and should not rotate it. This limitation restricts its practical usage. This study eliminates the need to restrict the users' behavior and the need to assume on the accurate signal propagation model.

Crowd-sourcing has been a useful tool for indoor localization. Because the crowdsourced data are collected from a huge number of different users conveying various mobile devices, it has potential to help solving challenging problems such as heterogeneous hardware [38] and security issues [39]. In this study, we apply crowdsourced data to learn the hidden trajectory and extract the floor plan, to compensate for the absence of true map information. The trajectory learning algorithm originates in demonstration learning [40]. For the purpose of this study, trajectory learning is combined with semisupervised feature extraction in Section 1.1.

Finally, as the position estimator, the particle filter is employed for two reasons: First, this filter can use the learned trajectories as a prior distribution. Second, in the particle filter framework, the likelihood function can be defined as the function referring to the relationship between the RSSI measurement sets and the positions. In this study, the likelihood is defined as the probabilistic model by the Gaussian process.

2. Mobile Fingerprinting Based on Semisupervised Learning

Wi-Fi fingerprint localization estimates a location by matching the currently received Wi-Fi RSSI measurements to those in a training database. For creating this database, the conventional fingerprinting method requires a trainer to manually labels all the Wi-Fi RSSI measurements at every point of the grid. Instead of the time-consuming conventional method, this section introduces a new mobile fingerprinting data collection algorithm that allows the trainer to continue walking without the stationary calibration. In the training phase for data collection, it is common sense that data collecters recognize which floor they are located on. Thus, the proposed mobile fingerprint method aims to obtain 2D position of the unlabeled data.

Suppose $y = [y_1, \ldots, y_d]^{\mathrm{T}} \in R^d$ is the set of the Wi-Fi RSSI measurements received from d Wi-Fi APs. In 2-D space, the user's location is defined as (x_X, x_Y). The l number of the labeled training data points are defined as the set $\{(y_i, [x_{Xi}, x_{Yi}]^{\mathrm{T}})\}_{i=1}^{l}$ with $y_i \in X \subseteq \mathcal{R}^d$. The u number of unlabeled data set $\{y_i\}_{i=l+1}^{l+u}$ comprises only the RSSI measurements.

It is desired to find the separate mappings $f_X : X \to \mathcal{R}$ and $f_Y : X \to \mathcal{R}$, which denote the relationships between Wi-Fi signal strength and location of the smartphone, using the labeled training data $\{(y_i, x_{Xi})\}_{i=1}^{l}$ and $\{(y_i, x_{Yi})\}_{i=1}^{l}$, and the unlabeled data $\{y_i\}_{i=l+1}^{l+u}$. Because the models f_X and f_Y are learned independently, we omit the subscripts of f_X, f_Y, and of x_X, x_Y, for simplification.

In the SVM-based semisupervised learning framework, the optimization formulation is as follows,

$$f^* = \arg\min_{f \in \mathcal{H}_k} c \sum_i V(y_i, x_i, f) + \gamma_A \|f\|_A^2 + \gamma_I \|f\|_I^2, \tag{1}$$

where V is a loss function; $\|f\|_A^2$ is the norm of the function in the Reproducing Kernel Hilbert Space $\mathcal{H}_k$; $\|f\|_I^2$ is the norm of the function in the low dimensional manifold; and c, γ_A, and γ_I are the regularization weight parameters. To represent the manifold $\|f\|_I^2$, the method uses graph Laplacian, the so-called graph-based semisupervised learning, and it is also called a semisupervised support vector machine when we use an ϵ-insensitive loss function. Then, the solution is achieved by iterative quadratic programming. More details about the semisupervised algorithm can be found in [37].

2.1. Hodric–Prescott Filter

Let us describe a scenario of mobile fingerprinting where a trainer collects the training data during walking. The observations are naturally recorded in time-series. This section introduces capturing the temporal property from the mobile fingerprint data by exploiting the H–P filter [41]. By using the H–P filter, the optimization problem can be formulated as follows,

$$\min_f \sum_{i=1}^{K} (f(y_i) - x_i)^2 + \gamma_T \sum_{i=3}^{K} (f(y_i) + f(y_{i-2}) - 2f(y_{i-1}))^2 \tag{2}$$

where $\{(y_i, x_i)\}_{i=1}^{K}$ is the time-series labeled training data over discrete time horizon K. The second term renders the sequential functional values $f(y_i), f(y_{i-1}), f(y_{i-2})$ on a line in the embedded space. The solution of Equation (2) in matrix form is

$$f = \left(I + \gamma_T DD^{\mathrm{T}}\right)^{-1} x,$$

where $x = [x_1, \ldots, x_K]^{\mathrm{T}}$ and

$$D = \begin{bmatrix} 0 & 0 & \cdots & \cdots & \cdots & \cdots & 0 \\ 0 & 0 & 0 & \cdots & \cdots & \cdots & 0 \\ 1 & -2 & 1 & 0 & \cdots & \cdots & 0 \\ 0 & 1 & -2 & 1 & 0 & \cdots & 0 \\ \vdots & \vdots & \vdots & \vdots & \vdots & \vdots & \vdots \\ 0 & \cdots & \cdots & 0 & 1 & -2 & 1 \end{bmatrix}_{K \times K}. \tag{3}$$

In the following section, this H–P filter-based optimization is combined with the semisupervised learning framework to assign the temporal aspect to the unlabeled data.

2.2. Semisupervised Pseudolabeling

Here, the semisupervised optimization to create accurate pseudolabels for the unlabeled data by considering both the temporal and spatial aspect is presented.

Given the labeled and unlabeled training data $\{(y_i, x_i)\}_{i=1}^{K}$ arranged in chronological order, the optimization for generating pseudolabels $\tilde{X} \in R^K$ is

$$\begin{aligned} &\min_{\tilde{X} \in R^n} \frac{c}{2}E^{\mathrm{T}}E + \frac{1}{2}\mu_1 \tilde{X}^{\mathrm{T}} L \tilde{X} + \frac{1}{2}\mu_2 \tilde{X}^{\mathrm{T}} D D^{\mathrm{T}} \tilde{X} \\ &\text{subject to} : \Lambda \tilde{X} - X = E, \end{aligned} \tag{4}$$

where Λ is a diagonal matrix of trade-off parameters with $\Lambda_{ii} = 1$ if y_i is a labeled data point and $\Lambda_{ii} = 0$ if y_i is an unlabeled data point. μ_1 and μ_2 represent a trade-off relationship between spatial and temporal correlation, matrix D is defined in Equation (3), and $X \in R^K$ is given by

$$X = \left\{ \begin{array}{cc} x_i & \text{if } y_i \text{ labeled} \\ 0 & \text{else if } y_i \text{ unlabeled} \end{array} \right. ,$$

and graph Laplacian L is defined as $L = B^{-1/2}(B - C)B^{-1/2}$, with the adjacency matrix C and the diagonal matrix B given by $B_{ii} = \sum_{j=1}^{K} C_{ij}$. In general, the edge weights C_{ij} are defined as a Gaussian function of Euclidean distance, given by

$$C_{ij} = \exp\left(-\|x_i - x_j\|^2 / \sigma_w^2\right),$$

where σ_w^2 is the kernel width parameter.

With the introduction of a multiplier α_P, the Lagrangian of Equation (4) is given by

$$L = \frac{c}{2}E^{\mathrm{T}}E + \frac{1}{2}\mu_1 \tilde{X}^{\mathrm{T}} L \tilde{X} + \frac{1}{2}\mu_2 \tilde{X}^{\mathrm{T}} D D^{\mathrm{T}} \tilde{X} + \alpha_P \left(\Lambda \tilde{X} - X - E\right) \tag{5}$$

The derivatives of Equation (5) with respect to the variables $\tilde{X}$, α_P, and e set to zero are

$$\frac{\partial L}{\partial \tilde{X}} = \mu_1 L \tilde{X} + \mu_2 D D^{\mathrm{T}} \tilde{X} + \Lambda \alpha_P = 0 \tag{6}$$

$$\frac{\partial L}{\partial e} = cE - \alpha_P = 0 \tag{7}$$

$$\frac{\partial L}{\partial \alpha_P} = \Lambda \tilde{X} - X - E = 0 \tag{8}$$

Substituting Equation (7) into Equation (8) gives

$$\Lambda \tilde{X} - \frac{1}{c}\alpha_P = X. \tag{9}$$

Then, the linear algebraic equations satisfying the KKT conditions are defined as follows,

$$\Phi A = F, \tag{10}$$

where

$$\Phi = \left[\begin{array}{cc} \mu_1 L + \mu_2 D D^{\mathrm{T}} & \Lambda \\ \Lambda & -\frac{1}{c} \end{array} \right] \tag{11}$$

and,

$$A = \begin{bmatrix} \tilde{X} \\ \alpha_P \end{bmatrix}, \; F = \begin{bmatrix} 0 \\ X \end{bmatrix} \tag{12}$$

Therefore, the optimal pseudolabels are obtained by inverse calculation of the matrix Φ. The advantages of this optimization compared to traditional semisupervised learning are as follows.

- A closed-form solution can be obtained from the linear algebraic Equation (10), which is faster than for other semisupervised algorithms and requires an iterative quadratic programming (QP) solver.
- The solution for $\tilde{X}$ also inherits the sparsity characteristics of a support vector machine [42]. This is beneficial when we manage training data storage by saving reliable training data only, that is, large values of α_P i.e., the support vectors.
- In addition to spatial representation, a time-series representation is considered, where the relevant term, $\mu_2 DD^{\mathrm{T}}$, is inserted independently into the optimization problem. Without a substantial increase in computational time, it improves the accuracy of the pseudolabels. Its performance is analyzed in the next section.

In a summary, we address the role of the mobile fingerprinting and the connection to the next sections. The mobile fingerprinting is a method for database construction. Especially, it is useful when amount of the labeled data is not enough by its capability of accurately labelling the positions of the unlabeled data. As shown in Figure 1, the calibrated data from the mobile fingerprinting are used for the following feature extraction in Section 3 and the localization in Section 4.

3. Feature Extraction and Application to Floor Classification and Landmark Detection

Section 3.1 overviews the characteristics of Wi-Fi RSSIs and the need for feature extraction. Section 3.2 describes a semisupervised discriminant algorithm for the floor classification and landmark detection.

3.1. Wi-Fi RSSI Characteristics

This section examines characteristics of Wi-Fi RSSIs and discovers the necessity of the feature extraction for Wi-Fi RSSI-based indoor localization.

3.1.1. Nonlinearity and Uncertainty

According to a propagation model of Wi-Fi RSSIs [43–45], the RSSI value decreases exponentially when the distance between a transmitter and a receiver increases linearly. In reality, however, the interruption of the multipath generates uncertain Wi-Fi RSSIs because of the existence of a number of walls and the interference from other radio signals.

For validation, we record Wi-Fi RSSI values on an office floor shown in Figure 2a. Figure 2b shows the RSSI values according to the distance from Wi-Fi AP2; one graph shows the RSSI curve when a user moves along the corridor and the other is when the path is interrupted by walls. Because this study does not assume the that a map is available, it is impossible to predict the change in RSSI patterns with respect to the distance.

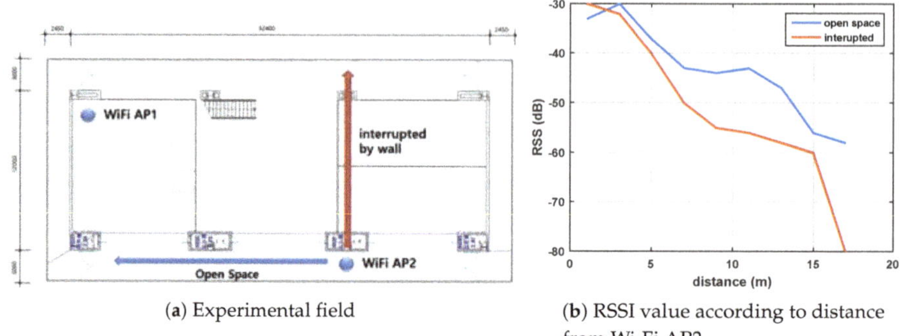

(**a**) Experimental field

(**b**) RSSI value according to distance from Wi-Fi AP2

Figure 2. (**a**) Experimental floor plan and (**b**) RSSI value of Wi-Fi AP2 according to the distance.

3.1.2. Sparsity

The other important characteristic is the sparsity of the raw Wi-Fi RSSI measurements. In a typical building, many Wi-Fi APs are installed, such as commercial APs, private internet-connected devices, and WLAN printers. For example, 193 APs were scanned in our experimental floors, and 531 APs are used in the the UJIIndoorLoc open dataset [46].

Not all the APs can be scanned at one point position because a single AP does not cover the entire positioning area. Thus, the RSSIs of APs located far from a user are recorded empty, as illustrated in Figure 3. The typical method to deal with the empty elements is to replace them by the possibly minimum RSSI value, e.g., −100 dBm. However, the sparsity still exists because the minimum RSSI values replace most of the elements, which may deteriorate localization performance. Therefore, we need to extract a feature. Further, substantially reduced dimensionality is helpful to reduce computational load.

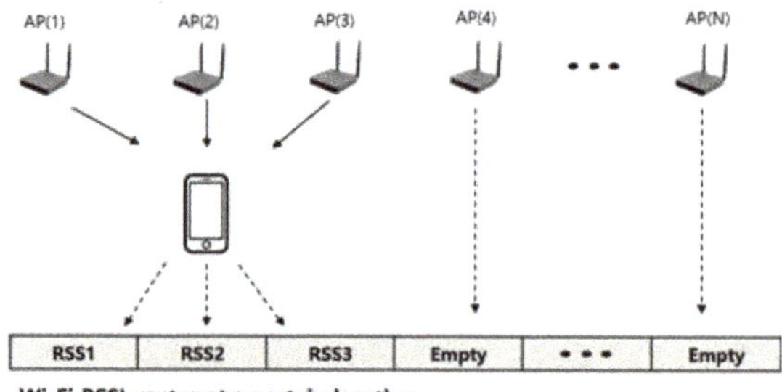

Figure 3. Sparsity of a raw Wi-Fi RSSI set.

3.2. Semisupervised Discriminant Analysis

Both principal component analysis (PCA) and Fisher discriminant analysis (FDA) have been widely used for feature extraction. These methods play the role of reducing dimensionality by highlighting meaningful elements from the original data vector. FDA is a supervised learning as it uses only labeled data, and PCA is an unsupervised learning because it uses only unlabeled data. The combination of FDA and PCA is categorized into the semisupervised learning.

3.2.1. Generalized Eigenvalue Problem

Remember that $y = [y_1, \ldots, y_d]^{\mathrm{T}} \in R^d$ is the set of the raw Wi-Fi RSSIs. The feature set is given $z \in R^r$ $(1 \leq r \leq d)$, which is the resultant low-dimensional set by a feature extraction implementation. Let $T \in R^{d \times r}$ be a transformation matrix; then, it has the following relationship,

$$z = T^{\mathrm{T}} y, \tag{13}$$

where $\cdot^{\mathrm{T}}$ represents the transpose operation. Both FDA and PCA solve the following optimization problem,

$$T^{\mathrm{opt}} = \mathrm{argmax}_T \left[\mathrm{tr} \left(T^{\mathrm{T}} B T (T^{\mathrm{T}} C T)^{-1} \right) \right], \tag{14}$$

where B is a quantity we want to increase, C is a quantity to be hopefully decreased, and $\mathrm{tr}(\cdot)$ represents the trace of a matrix.

Further, suppose that $\{\lambda_i\}_{i=1}^{d}$ is the set of generalized eigenvalues and $\{\varphi_i\}_{i=1}^{d}$ is the set of the corresponding generalized eigenvectors. Then, the generalized eigenvalue problem can be defined as

$$B\varphi = \lambda C \varphi. \tag{15}$$

The generalized eigenvectors are orthogonal to matrix C:

$$\varphi_i^{\mathrm{T}} C \varphi_j = 0, \tag{16}$$

for $i \neq j$, and the eigenvectors are normalized:

$$\varphi_i^{\mathrm{T}} C \varphi_i = 1 \quad \text{for } i = 1,\ 2,\ \cdots,\ d. \tag{17}$$

Additionally, assume that the eigenvalues are organized in descending order:

$$\lambda_1 \geq \lambda_2 \geq \cdots \geq \lambda_d. \tag{18}$$

Finally, the transformation matrix can be summarized as

$$T^{\mathrm{opt}} = \left(\sqrt{\lambda_1}\varphi_1 | \cdots | \sqrt{\lambda_r}\varphi_r | \cdots | \sqrt{\lambda_d}\varphi_d \right). \tag{19}$$

From Equation (19), it can be seen that the influence of the transformation becomes deemphasized as the dimension order increases, because the eigenvalues and eigenvectors are decreasing.

3.2.2. PCA

Let $\{y\}_{i=1}^{u}$ be a set of u unlabeled RSSI observations and S^p be the total scatter matrix, given by

$$S^p = \sum_{i=1}^{u} (y_i - \mu)(y_i - \mu)^{\mathrm{T}}, \tag{20}$$

where μ is the mean of all the observations:

$$\mu = \frac{1}{u} \sum_{i=1}^{u} y_i. \tag{21}$$

Equation (20) can be also expressed in a pairwise form as

$$
\begin{aligned}
S^p &= \sum_{i=1}^{u} y_i y_i^{\mathrm{T}} - n\mu\mu^{\mathrm{T}} \\
&= \frac{1}{2}\sum_{i,j=1}^{u} y_i y_i^{\mathrm{T}} - \frac{1}{2}\sum_{i,j=1}^{u} y_i y_j^{\mathrm{T}} \\
&= \frac{1}{2}\sum_{i,j=1}^{u} W_{ij}^{p}(y_i - y_j)(y_i - y_j)^{\mathrm{T}},
\end{aligned} \tag{22}
$$

with $W_{ij}^{p} = 1/u$. PCA finds the transformation matrix T^{pca} where

$$
T^{\mathrm{pca}} = \mathrm{argmax}_T \left[\mathrm{tr}\left(T^{\mathrm{T}} S^p T (T^{\mathrm{T}} T)^{-1} \right) \right]. \tag{23}
$$

Note that PCA is an unsupervised method that does not require label information such as floor level. Thus, PCA alone cannot be used for a supervised problem such as floor classification and landmark detection in this paper.

3.2.3. FDA

Let $\{(y_i, w_i)\}_{i=1}^{l}$ be a set of l labeled training data points, where $w_i \in \{1, 2, \ldots, C\}$ is the label of the RSSI vector y_i. For example, the label can be either floor level or a user-defined landmark. In FDA, the between-class covariance matrix S_b and the within-class covariance matrix S_w are defined as

$$
S^b = \sum_{c=1}^{C} l_c(\mu_c - \mu)(\mu_c - \mu)^{\mathrm{T}}, \tag{24}
$$

$$
S^w = \sum_{c=1}^{C} \sum_{i:w_i=c} (\mathrm{y}_i - \mu_c)(\mathrm{y}_i - \mu_c)^{\mathrm{T}}, \tag{25}
$$

where l_c is the number of the labeled data points in class $c \in \{1, 2, \ldots, C\}$ with $l = \sum_{c=1}^{C} l_c$, $\sum_{i:w_i=c}$ indicates summation over i such that $w_i = c$, and μ_c is the mean vector of the data points in class c, given by

$$
\mu_c = \frac{1}{l_c} \sum_{i:w_i=c} \mathrm{y}_i. \tag{26}
$$

The transformation matrix for FDA is given by

$$
T^{\mathrm{fda}} = \mathrm{argmax}_T \left[\mathrm{tr}\left(T^{\mathrm{T}} S^b T (T^{\mathrm{T}} S^w T)^{-1} \right) \right].
$$

Equations (24) and (25) can be expressed by the following weight matrices.

$$
W_{ij}^{b} = \begin{cases} \frac{1}{l} - \frac{1}{l_c} & \text{if } w_i = w_j \\ \frac{1}{l} & \text{if } w_i \neq w_j \end{cases} \tag{27}
$$

$$
W_{ij}^{w} = \begin{cases} \frac{1}{l_c} & \text{if } w_i = w_j \\ 0 & \text{if } w_i \neq w_j \end{cases} \tag{28}
$$

3.2.4. Semisupervised Combination of FDA and PCA

To extend to a semisupervised version, it modifies the original FDA in a way that utilizes the unlabeled data. Let $\{(y_i, w_i)\}_{i=1}^{l+u}$ be the set of both the labeled and unlabeled data points, where l is the number of the labeled data points and u is the number of the unlabeled data points. If $\{(y_i, w_i)\}$ is the labeled data, $w_i \in \{1, 2, \ldots, C\}$ denotes the class labels; for the unlabeled data, $w_j = 0$. Analogously to traditional semisupervised discriminant analysis, we define the new weight matrices, modified from Equations (27) and (28):

$$W_{ij}^{bS} = \begin{cases} \frac{1}{l} - \frac{1}{l_c} & \text{if } w_i = w_j = c \\ \frac{1}{l} & \text{if } w_i \neq w_j \\ \frac{1}{u} & \text{if } w_i = 0 \text{ or } w_j = 0 \end{cases} \tag{29}$$

$$W_{ij}^{wS} = \begin{cases} \frac{1}{l_c} & \text{if } w_i = w_j = c \\ 0 & \text{if } w_i \neq w_j \\ 0 & \text{if } w_i = 0 \text{ or } w_j = 0 \end{cases} \tag{30}$$

The corresponding matrices S^{bS} and S^{wS} are

$$S^{bS} = \frac{1}{2} \sum_{i,j=1}^{l+u} W_{ij}^{bS} (y_i - y_j)(y_i - y_j)^{\mathrm{T}}, \tag{31}$$

$$S^{wS} = \frac{1}{2} \sum_{i,j=1}^{l+u} W_{ij}^{wS} (y_i - y_j)(y_i - y_j)^{\mathrm{T}}. \tag{32}$$

The generalized eigenvalue problem for the semisupervised version is

$$T^{\text{semi}} = \operatorname{argmax}_T \left[\operatorname{tr} \left(T^{\mathrm{T}} S^{bs} T (T^{\mathrm{T}} S^{ws} T)^{-1} \right) \right], \tag{33}$$

$$S^{bs} = \alpha_{bal} S^{bS} + (1 - \alpha_{bal}) S^{p}, \tag{34}$$

$$S^{ws} = \alpha_{bal} S^{wS} + (1 - \alpha_{bal}) I, \tag{35}$$

where I is the identity matrix and α_{bal} is a parameter to adjust the balance between PCA and FDA.

3.3. Application to Floor Classification and Landmark Detection

3.3.1. Floor Classification

To apply the proposed semisupervised discriminant analysis algorithm in Section 3.2.4 to the floor classification, the label of the training data should be defined as the floor level. As a result, the transformation matrix plays a role as dividing the classes with respect to each floor level. In the test phase, when a test RSSI measurements set is arrived, it is first processed by the transformation matrix in Equation (13). Second, the floor level is estimated by k-NN method, which selects the averagely nearest class point in the training data to the test data.

3.3.2. Landmark Detection

Landmark detection intends to recognize whether a user is located at preliminarily defined points or not. Landmarks in an indoor environment can be toilets, elevators, and rooms, which create a particular pattern of Wi-Fi RSSI measurements such that their distinct features can be distinctly extracted. In this paper, the landmark detection is used for the trajectory learning, which will be introduced in Section 4.2.

The landmark detection implementation is as follows. First, similar to the floor classification, the label of the training data should be defined as the landmark index for the semisupervised discriminant analysis algorithm in the training phase, which is described in Section 3.2.4. In the test phase, when a test RSSI measurements set has come, the distance on the signal space is calculated between each landmark's feature set and the current RSSI feature vector. Let D_c be the distance given by

$$D_c = \|\bar{\mu}_c - z_*\|, \; c \in \{\text{landmark}_1, \text{landmark}_2, \cdots\}, \tag{36}$$

where $\bar{\mu}_c = T^{\text{semi}^{\text{T}}} \mu_c$ is the center of the training data points belong to the same landmark and $z_* = T^{\text{semi}^{\text{T}}} y_*$ is the test RSSI feature vector.

In sum, the semisupervised feature extraction is proposed to deal with the nonlinearity and sparsity of the raw Wi-Fi RSSI measurements. Two independent feature extraction models according to the different label type are applied to the floor classification and the landmark detection. In particular, the landmark detection can trim a trajectory sample by detecting two landmarks as the start and end points of a path, which is used for the trajectory learning in the next section.

4. Mapless Localization

In this section, we achieve localization that does not require true map information. Section 4.1 formalizes the position estimation based on a particle filter and Gaussian process. Section 4.2 introduces learning trajectories collected from a crowd for creating map information.

4.1. 2-D Position Estimator Based on Particle Filter and Gaussian Process

A particle filter involves obtaining a recursive estimate of the posterior distribution $x \in \mathcal{R}^2$ at current time k, given all the observations $z_{1:k} \in \mathcal{R}^r$. When we define $\{x_k^i, w_k^i\}_{i=1}^{N_p}$ as the set of N_p particles and corresponding weights, the posterior density function is

$$p\,(x_k|z_{1:k}) \approx \sum_{i=1}^{N_p} w_k^i \delta(x_k - x_k^i). \tag{37}$$

In Equation (37), $\delta(\cdot)$ is the Dirac delta function. The weights are normalized so that $\sum_{i=1}^{N_p} w_k^i = 1$. The estimate of the state x_k is given by

$$\hat{x}_k = E[x_k|z_{1:k}] \approx \sum_{i=1}^{N_p} w_k^i x_k^i, \tag{38}$$

and the weights are updated using the likelihood $p(z_k|x_k^i)$:

$$w_k^i = w_{k-1}^i p\left(z_k|x_k^i\right).$$

In this study, the likelihood is defined as a Gaussian process to achieve a nonlinear relationship between positions and RSSI observations, and is given by

$$p\,(z_k|x_k) \;=\; \prod_{j=1}^{r} p(z_k^j|x_k) = \prod_{j=1}^{r} N(\mu_{x_k}^j, \sigma_{x_k}^j), \tag{39}$$

where $N(\mu_{x_k}^j(\cdot), \sigma_{x_k}^j(\cdot))$ is a Gaussian distribution whose mean and variance are as follows,

$$\begin{aligned} \mu_{x_k}^j &= k_*^{\text{T}}(K(\tilde{X}, \tilde{X}) + \sigma_{GP}^2 I)^{-1} z_*^j \\ \sigma_{x_k}^{j2} &= k(x_k, x_k) - k_*^{\text{T}}(K(\tilde{X}, \tilde{X}) + \sigma_{GP}^2 I)^{-1} k_*. \end{aligned} \tag{40}$$

In Equation (40), training input $\tilde{X} \in R^{n\times 2}$ is defined as the pseudolabels of the x-y positions obtained in Section 2.2, and training input $z_*^j \in R^n$ is the Wi-Fi observation set corresponding to the j-th pseudolabels in $\tilde{X}$. Further, the kernel function and matrix, $k(\cdot,\cdot)$ and K, respectively, are defined by Gaussian kernel, and k_* is the $n \times 1$ vector of covariances between x_k and $\tilde{X}$. More details about the derivation of the Gaussian process and parameter selection can be found in the work by the authors of [47].

Here, z^j, for $j = 1, \cdots, r$ ($r = 10$ in this paper) are the PCA-driven observations from Section 3.2.2, that is, $z = T^{\text{pca}^{\text{T}}} y$, where y is the raw Wi-Fi RSSIs and T^{pca} is the transformation matrix in Equation (23). As $r = 10$, ten different Gaussian process models as in Equation (39) are used.

4.2. Trajectory Learning from a Crowd

In the particle filter framework, the sampling relies on the prior probability $p(x_t|x_{t-1})$. Under the unknown-map situation, the learned trajectory compensates for the absence of the true map. The prior function is defined as follows,

$$p(x_t|x_{t-1}) = P_{HF} \cdot P_{TL}, \tag{41}$$

where P_{TL} is a learned map and P_{HF} is for capturing a smoothed-curve representation of a time-series trajectory using the H–P filter introduced in Section 2.1:

$$P_{HF} = N\left(\| \ x_t - x_{t-2} - 2x_{t-1} \ \|^2; 0, \sigma_v\right), \tag{42}$$

In Equation (42), it does not require the estimation of velocity.

Now, we describe how to build P_{TL}. In indoor spaces, people trace similar trajectories to save their travel distance and time. The underlying idea for trajectory learning is that people tend to follow similar trajectories when they have the same departure point and destination. The departure and destination points are automatically obtained by the landmark detection algorithm described in Section 3.3.2. Suppose that we sample the trajectories X_j^k obtained from M different people $k = 0, \ldots, M-1$ and that the trajectories may have different trajectory lengths $j = 0, \ldots, N^{(k)} - 1$. Then, there might exist a hidden intended trajectory $h_{1:k}$, representative of all X_j^k. For example, $h_{1:k}$ can be the average trajectory of X_j^k.

The goal is to learn the hidden intended trajectory h_t. Dynamic time warping with Kalman smoothing [40] is applied to this problem. The hidden trajectory h_t has length O at $t = 0, \ldots, O-1$. The length O is initialized to an ample size, namely, $O = 2/M \sum_{k=1}^{M} N^{(k)}$. The trajectory learning method treats the samples X_j^k as observations of h_t such that

$$\begin{aligned} h_{t+1} &= f(h_t) + w_t^h, \ \ w^h \sim N(0, \Sigma^h) \\ X_j^k &= h_{\tau_j^k} + w_j^X, \ \ w^X \sim N(0, \Sigma^X). \end{aligned}$$

Here, the covariance Σ^h and Σ^X of the Gaussian noises w^h and w^X should be estimated. The subscript τ_j^k is the time index of h corresponding to X_j^k.

Log-likelihood maximization is used to estimate the hidden trajectory h_t and the time indices τ_j^k, as follows,

$$\max_{\tau, \Sigma^{(\cdot)}} \log p(h, \tau; \Sigma^z, \Sigma^X), \tag{43}$$

where $\Sigma^{(\cdot)}$ refers to Σ^h and Σ^X. In the work by the authors of [40], an iterative algorithm for solving Equation (43) is introduced. First, while keeping τ constant, it updates the covariance matrix $\Sigma^{(\cdot)}$ by separate E and M steps. In the E step, a Kalman smoother is applied to obtain the pairwise marginals over the latent variables $h_{1:t}$; the M step updates the covariance matrix $\Sigma^{(\cdot)}$. Then, dynamic time warping is applied to calculate $\hat{\tau}$ with the given $\Sigma^{(\cdot)}$ through the following optimization.

$$\begin{aligned} \hat{\tau} &= \text{argmax}_\tau \log p(h, \tau; \Sigma^{(\cdot)}) \\ &= \text{argmax}_\tau \sum_{k=0}^{M-1} \sum_{j=0}^{N^k-1} [\log p(X_j^k | h_{\tau_j^k|T}, \tau_j^k) \\ &\quad + \log p(\tau_j^k | \tau_{j-1}^k)]. \end{aligned} \tag{44}$$

The details of the dynamic programming for solving Equation (44) are described in the work by the authors of [40].

Now, we describe how to collect observations, that is, M trajectories X_j^k. By the landmark detection introduced in Section 3.3.2, we can estimate the edge of any pair of trajectories. Additionally, the elements of the trajectories are filled with the estimates obtained from the particle filter from Section 4.1.

To generalize map learning, suppose that we sample a set of n learned trajectories $h^{1:n} = \{h^{(1)}, h^{(2)}, \cdots, h^{(n)}\}$, where each $h^{(\cdot)}$ might have different start and end points and each $h^{1:n}$ is exploited to obtain P_{TL} in Equation (41). We assume that P_{TL} follows the Gaussian distribution given by

$$P_{TL} \sim N\left(x_k; \mu_{TL}, \Sigma_{TL}\right) \tag{45}$$

and

$$\mu_{TL} = \arg_h \min \|h^{1:n} - x_k\|,$$

where μ_{TL} is the trajectory among $h^{1:n}$ closest to x_k. The variance Σ_{TL} is set to the estimated covariance Σ^h of the learn trajectory, which is obtained from Kalman smoothing in Equation (43). The covariance Σ^X, which is also estimated in Equation (43), indicates how far the samples' trajectory is apart from the learned trajectory. Therefore, by inspecting Σ^X, we might detect an outlier that might arise when someone does not follow the common trajectory. In this paper, the outliers are filtered out by 95% confidence interval of the trajectory samples.

5. Experiment

The experimental field is a three-story building, where the area of each floor is 47 m $\times$ 36 m. The total number of scanned Wi-Fi APs is 193 and ten people are employed for training and testing.

Training data are divided into labeled data, composed of RSSI values and the corresponding labels and unlabeled data, which only consists of RSSI measurements. The labels of the labeled data are of three different categories, namely, floor level, landmark, and 2D position. The algorithm can be seen as a hierarchical structure; the floor level is first determined, and then the position is estimated. For the trajectory learning, the trajectory samples are generated whenever two landmarks indicating the start and end points (of the trajectory) are detected. Then, they are sent to a server to learn the hidden trajectories. Subsequently, the newly learned map information is used to update the particle filter algorithm.

5.1. Mobile Fingerprint

In this experiment, we compare all the benchmarks, that is, SSL, SSC, LapERLS, and the additional supervised k-NN algorithm. The parameters of the proposed algorithm are defined as $\mu_1 = 5$, $\mu_2 = 2$, $c = 1$ in Equation (5), and the other parameters of the compared methods are selected for by best performance. For the experimental study, we vary the number of the used labeled training data points out of the fixed total of the training data. When 283 number of additional test data points are used, and the results using 283 and 93 training data points are shown in Figure 4a,b, respectively. From the results shown in both parts of this figure, we observe that our algorithm outperforms the compared methods. In the cases of 100% and 75% labeled data in Figure 4b, SSC provides a slightly smaller error than the other methods. However, considering the advantage given to SSC, that is, as it knows the locations of the Wi-Fi APs, this result is not noteworthy. The major contribution of our algorithm can be seen in the case where a very small number of labeled data points are used. From Figure 4a,b, our algorithm shows a slightly increasing error as fewer labeled data points are used, whereas the others exhibit a substantially increasing error when comparing the case with 25% labeled data to that with 10% case.

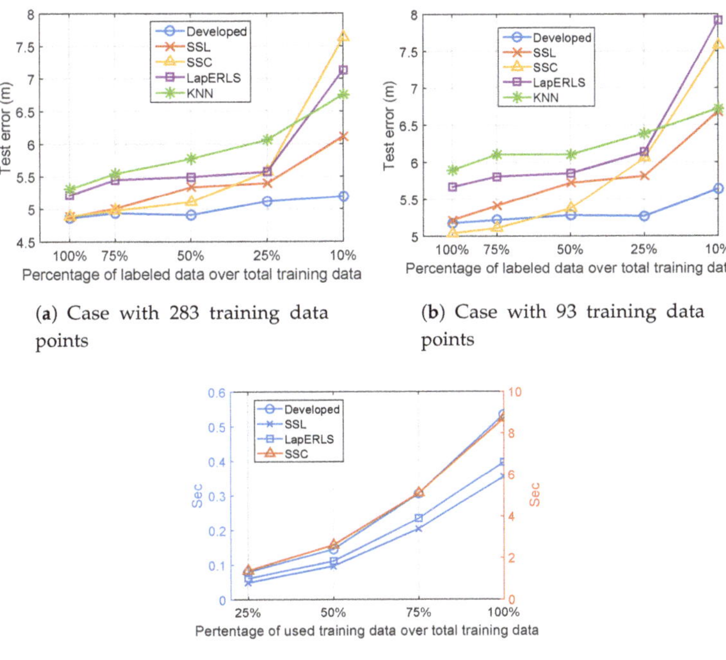

(**a**) Case with 283 training data points

(**b**) Case with 93 training data points

(**c**) Computational time

Figure 4. Localization results of the compared algorithms when we vary the percentage of the labeled training data: (**a**) case with 283 training data points, (**b**) case with 93 training data points, and (**c**) running time of the compared semisupervised algorithms for pseudolabeling with respect to the percentage of used training data.

Finally, Figure 4c shows the CPU running time of the compared algorithms with respect to the percentage of labeled data. SSC requires more computational time than the other methods. The proposed algorithm needs slightly more time than LapERLS and SSL (an additional 0.2 s at most), due to the additional time-series term to be solved in the optimization process.

5.2. Floor Classification and Landmark Detection

For the feature extraction, the dimensionality of the Wi-Fi RSSI set is reduced from 193 to 10, i.e., $d = 193$ and $r = 10$. All the RSSI values are scaled into the range [0, 1]. Figure 5a compares FDA (supervised) to the semisupervised discriminant analysis for floor level estimation. Note that because PCA is an unsupervised method, the PCA alone cannot be used for the supervised floor and landmark detections. To estimate the floor level, the k-NN method with $k = 1$ is employed. In Figure 5a, the ratio of the used labeled data varies from 10% to 100%. The developed algorithm provides better accuracy than FDA. The most noticeable result appears when the ratio of the labeled data is small. Although 10% of the data is labeled, semisupervised analysis achieves 95% accuracy whereas supervised analysis results in 60% accuracy.

Selecting a tuning the balancing parameter in Equation (35) depends on a cross validation. The optimal value is selected by minimizing the training error. In Figure 5b, the floor estimation accuracy is shown with respect to the variation of parameter values α_{bal} introduced in Equation (35) and the ratio of the labeled training data. Setting a relatively large α_{bal} value intends to focus more on FDA than on PCA in the semisupervised learning analysis. In this study, $\alpha_{bal} = 0.2$ is used in the rest of all experiments.

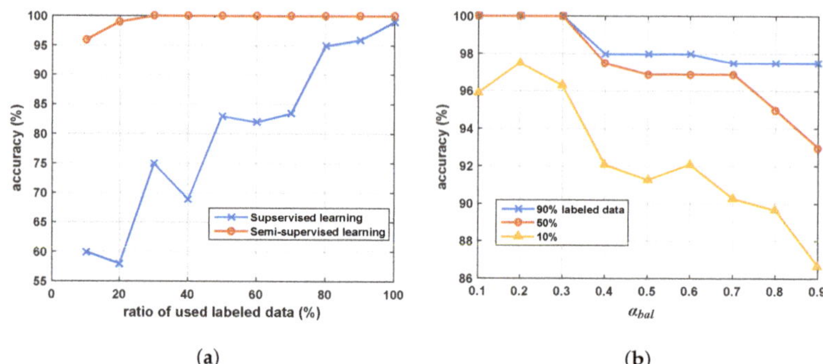

Figure 5. (**a**) Comparison of the supervised discriminant analysis and the semisupervised discriminant analysis for floor level estimation and (**b**) floor estimation accuracy with respect to the variation of α_{bal} parameter values introduced in Equation (35) and the ratio of the labeled training data.

Landmarks in an indoor environment, such as toilets, elevators, and rooms, create a particular pattern of Wi-Fi RSSI measurement sets so that their distinct features can be distinctly extracted. Figure 6a clarifies the extracted features at some landmarks using the semisupervised discriminant analysis algorithm. Figure 6b shows the distance on the signal space according to a user's path. The proposed algorithm detects the moment when the user arrives at each pre-defined landmark site with the threshold 0.3 in Figure 6b. Consequently, we can apply the landmark detection algorithm to calibrate any user's trajectory, by determining the start and end positions. Note that these trimmed trajectories are used for the trajectory learning in Section 4.

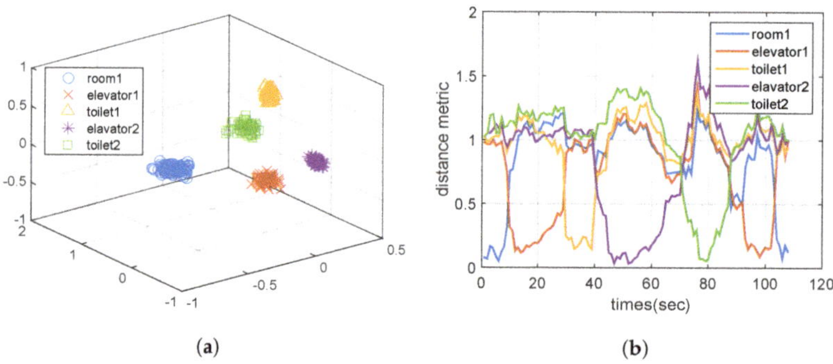

Figure 6. (**a**) Result of feature extraction from dataset on the landmarks. (**b**) Variation of the distance metric D_c in Equation (36), corresponding to a user's movement in the following order; $\text{room}_1 \rightarrow \text{elevator}_1 \rightarrow \text{toilet}_1 \rightarrow \text{elevator}_2 \rightarrow \text{toilet}_2 \rightarrow \text{elevator}_1 \rightarrow \text{room}_1$.

5.3. Trajectory Learning

Given the trajectory samples obtained from the landmark detection, Figure 7 shows the learned map on a floor with respect to the number of the participants. In Figure 7, seven different trajectories having different start and end landmarks are drawn with the different colors. We can observe that some learned trajectories are still inaccurate in Figure 7a, as they do not match the area on the true map when only two participants join. In Figure 7b, the more participants join in, the closer the learned map approaches the ground truth.

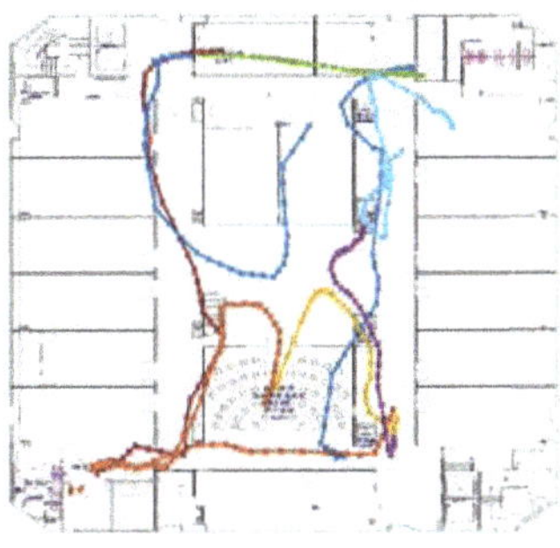

(a) Case with two participants

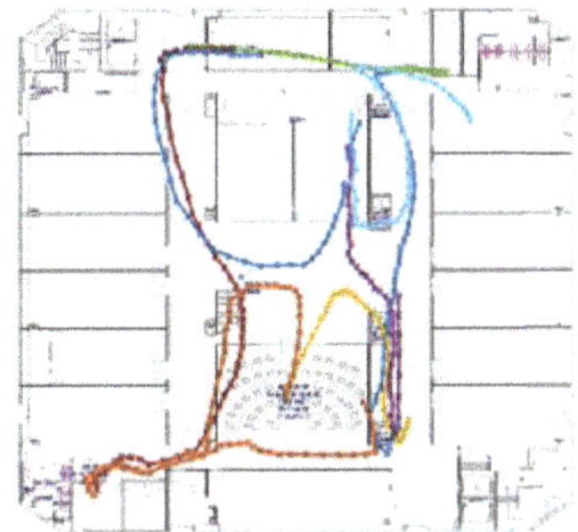

(b) Case with ten participants

Figure 7. Experimental result of trajectory learning.

5.4. Localization

To confirm the positioning error, two localization cases are compared with 50 number of particles setup for executing the particle filter. Figure 8a,b contains the localization results when using the learned trajectories as the prior for the particle filter. Figure 8c,d contains the localization results without using the learned trajectories. Because the user is moving, it is unable to measure the ground truths of the user's path all the time. Instead, some waypoints are designated to alarm the user to record its location with time stamps. Based on these waypoints, the average error is calculated. Ten positioning experiments by ten different users for each case are implemented. The average error of the developed algorithm is 2.2 m. By contrast, the average error for the localization without the learned trajectories is 3.4 m. In the worst case, the proposed algorithm has 2.8 m error and the benchmark one has 5.1 m. In the best case, each has 1.3 m and 2.5 m error, respectively. Thus, the map learning improves the localization accuracy by 1.2 m on average. It is noted that in Figure 8c the positioning is remarkably inaccurate because of the indoor environment composing majority of open space, whereas the localization on the aisles are relatively more precise. On the other hand, as shown in Figure 8a, the accurately learned map information supports to overcome this environmental restriction by enhancing the accuracy. The last remark of the experimental result analysis between Figure 8b,d is the positioning regarding the south room. Because the true map information is not given in this paper, it is unable for the localization in Figure 8d to recognize the wall between the room and aisle. On the other hand, due to aid of the trajectory learning, the proposed method recognizes the isolation by the wall, in which Figure 8b shows the clear–separate position estimations between the room and the aisle.

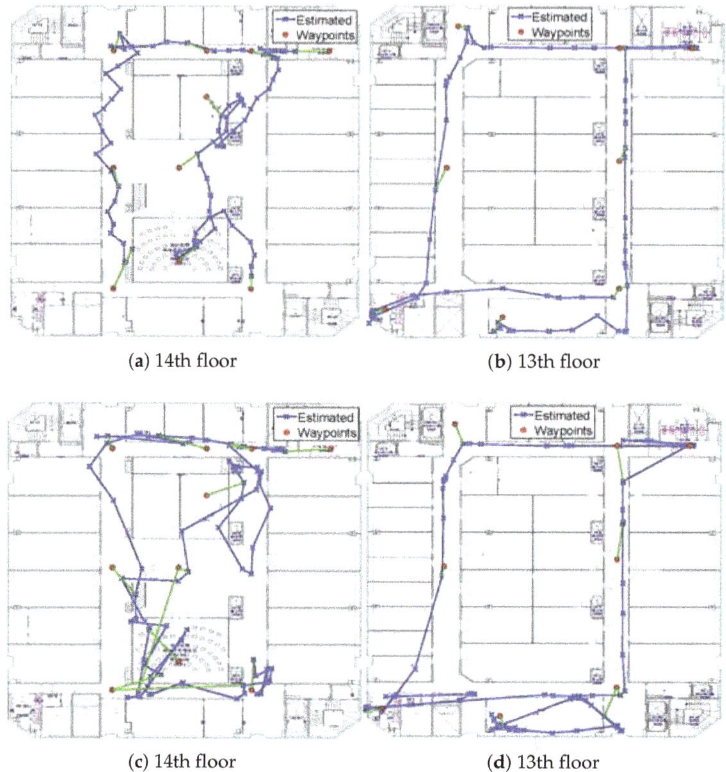

(**a**) 14th floor (**b**) 13th floor

(**c**) 14th floor (**d**) 13th floor

Figure 8. Positioning results with the trajectory learning (**a**,**b**) and without the trajectory learning (**c**,**d**).

6. Conclusions

In this study, we investigated indoor localization performing simultaneous floor classification, landmark detection, positioning, and map learning. The study was divided into three topics: (i) feature extraction from Wi-Fi RSSIs for floor classification and landmark detection, (ii) mobile fingerprinting and pseudolabeling for positioning, and (iii) mapless localization.

In the first part, characteristics of Wi-Fi RSSIs were determined by pattern recognition, using a semisupervised discriminant analysis. The proposed algorithm extracted the features from the noisy Wi-Fi signal data by reducing them from a high to a low dimension. During this process, worthless elements were removed to obtain clustered data points according to the different floors and different landmarks. At the same time, by investigating the distance between a test point and the training data on the reduced signal space, we successfully detected the floor and landmark changes.

The second part addressed the efficiency of fingerprinting. Compared to conventional static fingerprinting, our algorithm improved the efficiency for collecting the training data because a user could be mobile during the collection without manually labelling the position at every grid point in the area of interest. The proposed pseudolabeling algorithm based on the semisupervised regression aimed to obtain accurate pseudolabeled positions for the unlabeled training data. Considering both the spatial and temporal aspect, we formalized the optimization based on a graph Laplacian and H–P filter. Further, the optimization provided a closed-form solution, which enabled fast computation.

In the last part, we considered the situation where the true map information is not available. The key idea was crowdsourcing, from which we can obtain trajectory samples. From the experimental results, as more participants joined in, the learned map was updated more accurately. The experiments conducted in this study involved floors in a multifloor office building. Many people participated in validating our algorithm. The integration of all the parts led to accurate localization.

Author Contributions: J.Y. designed and developed the learning based localization algorithm, and implemented and validated the experiments; J.P. reviewed and edited the paper.

Funding: This work was supported by the National Research Foundation of Korea (NRF) grant funded by the Korean Government (MSIT) (No. 2019R1F1A1057516).

Conflicts of Interest: The authors declare no conflicts of interest.

References

1. Zhang, L.; Valaee, S.; Xu, Y.; Ma, L.; Vedadi, F. Graph-based semisupervised learning for indoor localization using crowdsourced data. *Appl. Sci.* **2017**, *7*, 467. [CrossRef]
2. Zhang, Z.; Tian, Z.; Zhou, M.; Li, Z.; Wu, Z.; Jin, Y. WIPP: Wi-Fi compass for indoor passive positioning with decimeter accuracy. *Appl. Sci.* **2016**, *6*, 108. [CrossRef]
3. Hernández, N.; Ocaña, M.; Alonso, J.; Kim, E. Continuous space estimation: Increasing WiFi-based indoor localization resolution without increasing the site-survey effort. *Sensors* **2017**, *17*, 147. [CrossRef] [PubMed]
4. Zheng, L.; Hu, B.; Chen, H. A high accuracy time-reversal based WiFi indoor localization approach with a single antenna. *Sensors* **2018**, *18*, 3437. [CrossRef] [PubMed]
5. Tran, H.Q.; Ha, C. Improved Visible Light-Based Indoor Positioning System Using Machine Learning Classification and Regression. *Appl. Sci.* **2019**, *9*, 1048. [CrossRef]
6. Wang, X.; Gao, L.; Mao, S. CSI phase fingerprinting for indoor localization with a deep learning approach. *IEEE Internet Things J.* **2016**, *3*, 1113–1123. [CrossRef]
7. Yan, J.; Zhao, L.; Tang, J.; Chen, Y.; Chen, R.; Chen, L. Hybrid kernel based machine learning using received signal strength measurements for indoor localization. *IEEE Trans. Veh. Technol.* **2017**, *67*, 2824–2829. [CrossRef]
8. Chen, Z.; Zou, H.; Yang, J.; Jiang, H.; Xie, L. WiFi Fingerprinting Indoor Localization Using Local Feature-Based Deep LSTM. *IEEE Syst. J.* **2019**. [CrossRef]
9. Ruiz, A.R.J.; Granja, F.S.; Honorato, J.C.P.; Rosas, J.I.G. Accurate pedestrian indoor navigation by tightly coupling foot-mounted IMU and RFID measurements. *IEEE Trans. Instrum. Meas.* **2011**, *61*, 178–189. [CrossRef]
10. Zhao, Y.; Liang, J.; Sha, X.; Yu, J.; Duan, H.; Shi, G.; Li, W.J. Estimation of pedestrian altitude inside a multi-story building using an integrated micro-IMU and barometer device. *IEEE Access* **2019**, *7*, 84680–84689. [CrossRef]
11. Brachmann, E.; Rother, C. Learning less is more-6d camera localization via 3d surface regression. In Proceedings of the IEEE Conference on Computer Vision and Pattern Recognition, Salt Lake City, UT, USA, 18–22 June 2018; pp. 4654–4662.
12. Elloumi, W.; Latoui, A.; Canals, R.; Chetouani, A.; Treuillet, S. Indoor pedestrian localization with a smartphone: A comparison of inertial and vision-based methods. *IEEE Sens. J.* **2016**, *16*, 5376–5388. [CrossRef]
13. Wang, G.; Wang, X.; Nie, J.; Lin, L. Magnetic-based Indoor Localization using Smartphone via a Fusion Algorithm. *IEEE Sens. J.* **2019**, *19*, 6477–6485. [CrossRef]
14. Shu, Y.; Bo, C.; Shen, G.; Zhao, C.; Li, L.; Zhao, F. Magicol: Indoor localization using pervasive magnetic field and opportunistic WiFi sensing. *IEEE J. Sel. Areas Commun.* **2015**, *33*, 1443–1457. [CrossRef]
15. Zhou, M.; Tang, Y.; Nie, W.; Xie, L.; Yang, X. GrassMA: Graph-based semisupervised manifold alignment for indoor WLAN localization. *IEEE Sens. J.* **2017**, *17*, 7086–7095. [CrossRef]
16. Zhou, M.; Tang, Y.; Tian, Z.; Xie, L.; Nie, W. Robust neighborhood graphing for semisupervised indoor localization with light-loaded location fingerprinting. *IEEE Internet Things J.* **2017**, *5*, 3378–3387. [CrossRef]
17. Yoo, J.; Kim, H. Target localization in wireless sensor networks using online semisupervised support vector regression. *Sensors* **2015**, *15*, 12539–12559. [CrossRef] [PubMed]
18. Mohammadi, M.; Al-Fuqaha, A.; Guizani, M.; Oh, J.S. Semisupervised deep reinforcement learning in support of IoT and smart city services. *IEEE Internet Things J.* **2017**, *5*, 624–635. [CrossRef]

19. Gu, Y.; Chen, Y.; Liu, J.; Jiang, X. Semi-supervised deep extreme learning machine for Wi-Fi based localization. *Neurocomputing* **2015**, *166*, 282–293. [CrossRef]
20. Xia, Y.; Ma, L.; Zhang, Z.; Wang, Y. Semisupervised Positioning Algorithm in Indoor WLAN Environment. In Proceedings of the 2015 IEEE 81st Vehicular Technology Conference (VTC Spring), Glasgow, UK, 11–14 May 2015.
21. Wu, Z.; Jedari, E.; Muscedere, R.; Rashidzadeh, R. Improved particle filter based on WLAN RSSI fingerprinting and smart sensors for indoor localization. *Comput. Commun.* **2016**, *83*, 64–71. [CrossRef]
22. Xie, H.; Gu, T.; Tao, X.; Ye, H.; Lu, J. A reliability-augmented particle filter for magnetic fingerprinting based indoor localization on smartphone. *IEEE Trans. Mob. Comput.* **2015**, *15*, 1877–1892. [CrossRef]
23. Pak, J.M.; Ahn, C.K.; Shmaliy, Y.S.; Lim, M.T. Improving reliability of particle filter-based localization in wireless sensor networks via hybrid particle/FIR filtering. *IEEE Trans. Ind. Inf.* **2015**, *11*, 1089–1098. [CrossRef]
24. Ko, J.; Fox, D. GP-BayesFilters: Bayesian filtering using Gaussian process prediction and observation models. *Auton. Robots* **2009**, *27*, 75–90. [CrossRef]
25. Belmonte-Hernández, A.; Hernández-Peñaloza, G.; Gutiérrez, D.M.; Álvarez, F. SWiBluX: Multi-Sensor Deep Learning Fingerprint for precise real-time indoor tracking. *IEEE Sens. J.* **2019**, *19*, 3473–3486. [CrossRef]
26. Adege, A.; Lin, H.P.; Tarekegn, G.; Jeng, S.S. Applying deep neural network (DNN) for robust indoor localization in multi-building environment. *Appl. Sci.* **2018**, *8*, 1062. [CrossRef]
27. Wang, X.; Gao, L.; Mao, S.; Pandey, S. DeepFi: Deep learning for indoor fingerprinting using channel state information. In Proceedings of the 2015 IEEE Wireless Communications and Networking Conference (WCNC), New Orleans, LA, USA, 9–12 March 2015; pp. 1666–1671.
28. Wang, X.; Gao, L.; Mao, S.; Pandey, S. A Deep Learning based Approach for Indoor Localization. *Technology* **2017**, *66*, 763–776.
29. Khatab, Z.E.; Hajihoseini, A.; Ghorashi, S.A. A fingerprint method for indoor localization using autoencoder based deep extreme learning machine. *IEEE Sens. Lett.* **2017**, *2*, 1–4. [CrossRef]
30. Jiang, X.; Chen, Y.; Liu, J.; Gu, Y.; Hu, L. FSELM: Fusion semisupervised extreme learning machine for indoor localization with Wi-Fi and Bluetooth fingerprints. *Soft Comput.* **2018**, *22*, 3621–3635. [CrossRef]
31. Burges, C.J. A tutorial on support vector machines for pattern recognition. *Data Min. Knowl. Discov.* **1998**, *2*, 121–167. [CrossRef]
32. Yoo, J.; Kim, H.J. Online Estimation using Semisupervised Least Square SVR. In Proceedings of the 2014 IEEE International Conference on Systems, Man, and Cybernetics (SMC), San Diego, CA, USA, 5–8 October 2014.
33. Suykens, J.A.; Vandewalle, J. Least squares support vector machine classifiers. *Neural Process. Lett.* **1999**, *9*, 293–300. [CrossRef]
34. Pan, J.J.; Pan, S.J.; Yin, J.; Ni, L.M.; Yang, Q. Tracking mobile users in wireless networks via semisupervised colocalization. *IEEE Trans. Pattern Anal. Mach. Intell.* **2012**, *34*, 587–600. [CrossRef] [PubMed]
35. Chapelle, O.; Vapnik, V.; Weston, J. *Transductive Inference for Estimating Values of Functions*; NIPS: Denver, CO, USA, 1999; Volume 12, pp. 421–427.
36. Chen, L.; Tsang, I.W.; Xu, D. Laplacian embedded regression for scalable manifold regularization. *IEEE Trans. Neural Networks Learn. Syst.* **2012**, *23*, 902–915. [CrossRef] [PubMed]
37. Yoo, J.; Johansson, K.H. Semi-supervised learning for mobile robot localization using wireless signal strengths. In Proceedings of the 2017 International Conference on Indoor Positioning and Indoor Navigation (IPIN), Sapporo, Japan, 18–21 September 2017.
38. Fang, S.H.; Wang, C.H. A novel fused positioning feature for handling heterogeneous hardware problem. *IEEE Trans. Commun.* **2015**, *63*, 2713–2723. [CrossRef]
39. Fang, S.H.; Chuang, C.C.; Wang, C. Attack-resistant wireless localization using an inclusive disjunction model. *IEEE Trans. Commun.* **2012**, *60*, 1209–1214. [CrossRef]
40. Abbeel, P.; Coates, A.; Ng, A.Y. Autonomous helicopter aerobatics through apprenticeship learning. *Int. J. Robot. Res.* **2010**. [CrossRef]
41. Ravn, M.O.; Uhlig, H. On adjusting the Hodrick–Prescott filter for the frequency of observations. *Rev. Econ. Stat.* **2002**, *84*, 371–376. [CrossRef]
42. Suykens, J.A.; Lukas, L.; Vandewalle, J. Sparse approximation using least squares support vector machines. In Proceedings of the 2000 IEEE International Symposium on Circuits and Systems (ISCAS), Geneva, Switzerland, 28–31 May 2000; Volume 2, pp. 757–760.

43. Zayets, A.; Steinbach, E. Robust WiFi-based indoor localization using multipath component analysis. In Proceedings of the 2017 International Conference on Indoor Positioning and Indoor Navigation (IPIN), Sapporo, Japan, 18–21 September 2017.
44. Wu, D.; Zhang, D.; Xu, C.; Wang, H.; Li, X. Device-free WiFi human sensing: From pattern-based to model-based approaches. *IEEE Commun. Mag.* **2017**, *55*, 91–97. [CrossRef]
45. Zhuang, Y.; Li, Y.; Lan, H.; Syed, Z.; El-Sheimy, N. Smartphone-based WiFi access point localisation and propagation parameter estimation using crowdsourcing. *Electron. Lett.* **2015**, *51*, 1380–1382. [CrossRef]
46. Torres-Sospedra, J.; Montoliu, R.; Martínez-Usó, A.; Avariento, J.P.; Arnau, T.J.; Benedito-Bordonau, M.; Huerta, J. UJIIndoorLoc: A new multi-building and multifloor database for WLAN fingerprint-based indoor localization problems. In Proceedings of the 2014 International Conference on Indoor Positioning and Indoor Navigation (IPIN), Busan, Korea, 27–30 October 2014; pp. 261–270.
47. Rasmussen, C.E. *Gaussian Processes for Machine Learning*; MIT Press: Cambridge, MA, USA, 2006.

Article

Robust and Accurate Wi-Fi Fingerprint Location Recognition Method Based on Deep Neural Network

Yifan Wang, Jingxiang Gao *, Zengke Li and Long Zhao

School of Environment Science and Spatial Informatics, China University of Mining and Technology, Xuzhou 221116, China; cumtwangyf@163.com (Y.W.); zengkeli@yeah.net (Z.L.); cehuizl@126.com (L.Z.)

* Correspondence: jxgao@cumt.edu.cn; Tel.: +86-516-8388-5785

Received: 4 December 2019; Accepted: 27 December 2019; Published: 1 January 2020

Abstract: Currently, indoor locations based on the received signal strength (*RSS*) of Wi-Fi are attracting more and more attention thanks to the technology's low cost, low power consumption and wide availability in mobile devices. However, the accuracy of Wi-Fi positioning is limited, due to the signal fluctuation and indoor multipath interference. In order to overcome this problem, this paper proposes a robust and accurate Wi-Fi fingerprint location recognition method based on a deep neural network (DNN). A stacked denoising auto-encoder (SDAE) is used to extract robust features from noisy *RSS* to construct a feature-weighted fingerprint database offline. We use the combination of the weights of posteriori probability and geometric relationship of fingerprint points to calculate the coordinates of unknown points online. In addition, we use constrained Kalman filtering and hidden Markov models (HMM) to smooth and optimize positioning results and overcome the influence of gross error on positioning results, combined with characteristics of user movement in buildings, both dynamic and static. The experiment shows that the DNN is feasible for position recognition, and the method proposed in this paper is more accurate and stable than the commonly used Wi-Fi positioning methods in different scenes.

Keywords: indoor location recognition; received signal strength (*RSS*); Wi-Fi fingerprint positioning; deep neural network (DNN); optimization methods; adaptive filter; hidden Markov models (HMM)

---

1. Introduction

With the popularization of wireless communication equipment and the improvement of location service technology, people are looking for more accurate location recognition technology in more extensive and complex applications [1,2]. Global navigation satellite systems (GNSS) play an important role in outdoor navigation due to their positioning ability, but their use is limited in indoor environments because the signal is blocked by houses and the inhabitants inside. In order to satisfy the needs of indoor location-based services (LBS), such as ultrasound [3], ultra-wideband [4], geomagnetic fields [5], images [6], light [7], inertia systems [8] and wireless signals [9] are used for indoor positioning. As infrastructure for widely deployed public networks, Wi-Fi access point (AP) networking hardware devices are now found in nearly every indoor area where people conduct their daily activities, which provides a huge application space for indoor location services. Therefore, Wi-Fi-based indoor positioning technology has become a preference for different kinds of indoor positioning technologies due to its low cost, low power consumption and wide availability in mobile devices [10].

As with other wireless technologies, time of arrival (TOA), time difference of arrival (TDOA), angle of arrival (AOA) and received signal strength (*RSS*) approaches are used to achieve Wi-Fi positioning. Unlike the other three methods of calculating user coordinates through geometric relationships, the Wi-Fi fingerprint location method based on *RSS* uses a matching algorithm to determine location with the character of signal strength. As a result, this method is less affected by the influence of environmental change, and there is no need to know the AP location, which explains why it has become mainstream Wi-Fi location technology [11]. However, wireless signal strength is not stable in complex indoor environments. Affected by multipath interference, *RSS* may fluctuate greatly, which will reduce its positioning accuracy.

At present, the advancement of micro-electro-mechanical-systems (MEMS) technology has led to more sensors in smartphones [12]. Researchers propose that combining a multi-sensor scheme with Wi-Fi data will profoundly improve the positioning accuracy [13,14]. However, equipment, such as accelerometers, gyroscopes and compasses, is prohibitively expensive for most customers, and the use of multiple sensors will inevitably increase power consumption. At the same time, users only need to know their outline coordinates and cooperate with the visual inspection to determine their location in daily life. Thus, it is more in accordance with users' requirements to obtain rapid and accurate positioning information by relying on fewer sensors in the public environment. Since Wi-Fi positioning has a huge advantage due to the wide deployment of its infrastructure, the focus is on how to improve its positioning performance in buildings. Machine learning has the ability to mine data features, which is a very promising technology for solving the problem of Wi-Fi signal mismatch, and has been the focus of much recent research.

In this paper, a robust and accurate Wi-Fi fingerprint location recognition method based on deep neural network is put forward, and its main contributions are the following:

(1) We use the stacked denoising autoencoders (SDAE) in a deep neural network (DNN) to perform Wi-Fi location recognition. Compared to the traditional Wi-Fi fingerprint location method, the proposed method can fully obtain feature representations from noisy *RSS*, reduce the dimension of the Wi-Fi signal and improve positioning accuracy and robustness. In addition, we use regularization techniques to avoid the model overfitting offline, and take into account the geometric distribution of fingerprint points in order to estimate the user's position and further ensure accuracy during the positioning stage.
(2) We construct a dynamic constrained Kalman filter (KF) based on the speed at which users inside a building move. This ensures the continuity of dynamic positioning results and improves the positioning accuracy by depending on identifying and smoothing out the gross errors caused by multipath interference. The hidden Markov model (HMM) is used to optimize multiple positioning results of the same unknown point to obtain the optimum coordinates of the unknown point based on the temporal correlation of multiple observations that are made when the users are not moving.
(3) We applied the UJIIndoorLoc indoor localization public dataset [15] to fully verify the feasibility of this method in floor recognition and location estimation, and analyzed the results at length. Moreover, we established an experimental site in an office building to test the positioning performance of constrained KF and HMM under dynamic and static conditions, respectively.

The rest of the paper is organized as follows. Section 2 introduces the related work. Section 3 provides the relevant methodology and more details about DNN-based Wi-Fi location recognition, constrained KF and HMM. Section 4 discusses the experimental results of the proposed method with different modes. Section 5 contains our conclusions, as well as suggestions for future research.

2. Related Works

In 2000, Microsoft developed RADAR, a radio frequency (RF)-based wireless local area network (WLAN) indoor positioning system for locating and tracking users inside buildings [16]. It used a k-nearest neighbors (KNN) approach, which used the European distance of signal space to match and locate, and became the gold standard of fingerprint positioning. Clustering-KNN [17] and improving weighted KNN [18] approached soon followed, but more time was needed to deal with the challenge of row data with the aim of improving accuracy. The traditional methods were unable to accomplish this because of the complex nature of the signal error mechanism [19]. In [20], the authors constructed a fingerprint system with the help of Wi-Fi channel State information (CSI), based on a neural network. The system achieved better accuracy, but because special equipment was required to obtain CSI, it was not considered a universal solution.

Wi-Fi signals are increasingly being received in public places. However, it is a challenge to use this big data fully and effectively, although the rapid development of machine learning theory makes it possible because it requires less parameter tuning and offers better scalability. Support vector machine [21] and random forest [22] have been used to capture the correlation between signal strength and physical location. Michael et al. [23] first used a DNN to realize floor classification based on Wi-Fi, but the solution could not estimate user location. In [24], the authors realized Wi-Fi fingerprint location based on deep learning by simulation, with location error less than 1 m. Zhang et al. [25] and Rizk et al. [26] proposed DNN-based solutions that extracted features from large noise signal samples for localization. They analyzed the impact of model parameters and data structures on positioning performance. With the gradual combination of crowdsourcing theory with wireless location, it is necessary to explore techniques for deep learning-based indoor positioning under the background of big data to obtain the better positioning accuracy and data utilization.

Multipath and non-line-of-sight (NLOS) propagation inevitably leads to gross errors in the Wi-Fi positioning results in complex indoor environments. However, many constraints exist when it comes to obtaining information about user navigation, and some of them can be used to improve the positioning accuracy. A smoothed constrained fingerprinting solution has also been adopted in a system that integrates a dual-filter MEMS sensor and Wi-Fi fingerprinting [27]. Poulose et al. [28] combined the Pedestrian Dead Reckoning localization algorithm with the linear Kalman filter to achieve Wi-Fi fusion positioning results. Li et al. [29] used fingerprinting accuracy indicators (FAIs) to predict wireless and magnetic fingerprinting accuracy, and then used the FAI-enhanced EKF to improve the stability of multi-sensor fusion. In [30], Bi et al. proposed an adaptive weighted KNN Wi-Fi location method, which improved the accuracy of single point positioning with the constraints of the user's orientation and the hybrid distance of the signal domain and position domain. In [26], the authors set up the CellinDeep system-based Wi-Fi signal, which used data augmenter techniques and mean filtering to address practical challenges caused by signal noise. Zhou et al. [31] used Kalman filtering to reduce drift and oscillation on a received signal strength indicator (RSSI) caused by the noise overlap-add in Bluetooth locations. Nevertheless, most of these methods rely on multiple sensors, and there has been relatively little research on a single Wi-Fi constraint location. Those Wi-Fi experiments that have taken place are based on discrete points. Inspired by the above research, we propose a robust and accurate Wi-Fi fingerprint location recognition method based on a DNN, along with optimization methods, which are adaptive constrained KF and HMM, to improve the accuracy and stability of a single Wi-Fi location on the basis of different movement modes of users.

3. Theoretical Frameworks

3.1. Overview

Figure 1 shows the architecture of Wi-Fi fingerprint location recognition method based on a DNN. The method has two phases: an offline training phase and online positioning phase. During the offline phase, the raw *RSS* of different Wi-Fi APs on the selected fingerprint points in the area of interest is recorded. We then trained the DNN and extracted features from the noisy *RSS* to construct an offline feature-weighted fingerprint database. During the online phase, the real-time AP information observed on unknown points is transferred to the feature fingerprint database trained in the offline phase to estimate the floor or location of the current user. We chose constrained Kalman filtering and hidden Markov model (HMM) to optimize the results of dynamic and static positioning, respectively, in order to reduce the number of outliers and obtain more accurate positioning results with regard to different motion modes of users.

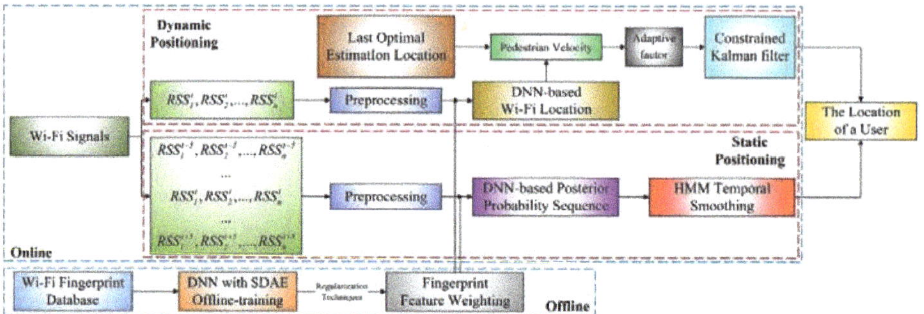

Figure 1. System framework.

3.2. DNN-Based Wi-Fi Location Recognition

3.2.1. Data Preprocessing

When every AP is observed, a pair information <MAC *RSS*> is recorded, in which MAC represents its unique ID and *RSS* its received signal strength. All AP information that can be observed in each scan is recorded as Wi-Fi training fingerprint or test data.

The preprocessing mechanism runs during both the offline and online stages. The first 15 APs with the largest *RSS* in every scan are retained to make the *RSS* set input to the depth model consistent, $o = \{rss_1, rss_2, \ldots, rss_{15}\}$, where each entry represents observation information in the test area. But it should be noted that there are only 15 currently observed elements in each recorded sample. Therefore, for the entire training set, all the elements in the sample are arranged uniformly in the AP list in the fingerprint database, and the *RSS* of the unobserved AP in each input is set at −100 dB.

We standardized the observed raw *RSS*, which ranges from −100 dB to 0 dB, to adjust its value to between 0 and 1 in order to reduce the distribution of *RSS* and facilitate the subsequent training. Formula (1) shows the standardization process. Formula (2) shows the fingerprint database built after standardizing the *RSS* of all fingerprint points.

$$RSS_i{}' = \frac{RSS_i + 100}{\max(RSS) + 100} \tag{1}$$

$$\textit{\textbf{Fingerprint Database}} \rightarrow \begin{bmatrix} RSS_1^{1\prime} & RSS_1^{2\prime} & RSS_1^{3\prime} & \cdots & RSS_1^{n\prime} \\ RSS_2^{1\prime} & RSS_2^{2\prime} & RSS_2^{3\prime} & \cdots & RSS_2^{n\prime} \\ \cdots & \cdots & \cdots & \ddots & \cdots \\ RSS_{z-1}^{1}{}' & RSS_{z-1}^{2}{}' & RSS_{z-1}^{3}{}' & \cdots & RSS_{z-1}^{n}{}' \\ RSS_z^{1\prime} & RSS_z^{2\prime} & RSS_z^{3\prime} & \cdots & RSS_z^{n\prime} \end{bmatrix} \tag{2}$$

where $RSS_i{}'$ represents the standardized value, $\max(RSS)$ is the maximum in the fingerprint database, n the number of APs observed in offline training, z the number of fingerprint points. Every row is an observation sample, and the *RSS* of the unobserved AP is set to −100 dB.

3.2.2. Offline Stage

The SDAE is used to capture nonlinear and implicit correlations between *RSS* and the spatial location through signal changes at different fingerprint points. The SDAE is able to represent input data in the form of hierarchically distributed data, using the observation information on fingerprint points and reducing the feature dimension of signal data [32].

An SDAE is stacked by multiple denoising autoencoders (DAEs) as hidden layers. Each DAE layer is used to train the pair of encoder-decoders to make the output value u equal to the input value

x as much as possible. If input data can be restored after encoding and decoding, then the feature representations trained in this way will contain relatively low noise levels. Figure 2 shows the process of a DAE. The encoder-decoder can be formulated as

$$\begin{cases} y(x) = \frac{1}{1+e^{-(wx+b)}} \\ u(y) = \frac{1}{1+e^{-(w'y+b')}} \end{cases} \tag{3}$$

where w and w' are the connection weights between the encoder and decoder, respectively; they are the transposition of each other, and b denotes the hidden units biases, while b' denotes the visible units biases.

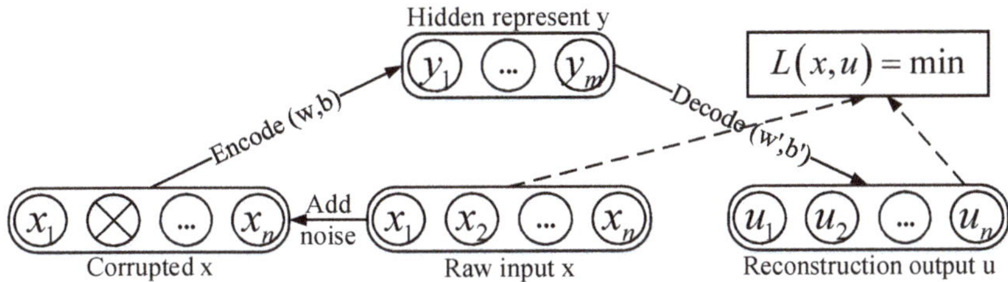

Figure 2. Training process of a denoising autoencoder (DAE).

We employed two techniques to suppress overfitting in training the DNN. First, we added a noise reduction mechanism during training [33]. Before the *RSS* in the fingerprint database is input into each DAE for unsupervised training, it will be "corrupted" by random setting to 0 based on a certain probability. The trained features are of higher robustness since corrupted input data are closer to actual test data. Second, we adopted dropout regularization, which has been shown to be useful for large networks [34]. This method randomly removed some neurons in different parts of the training stage, which means these neurons cannot participate in propagation. This prevented the interdependence of neurons because the DNN relied too much on several local features in the training process, and also enhanced the generalization of the model.

The softmax function is usually used for solving the finite discrete probability distribution in classification. After layer-by-layer unsupervised training of the DNN's three hidden layers, we treated the softmax function as the final output layer and prior probability of all fingerprint points as labeled information in consideration of classification. The feature weight obtained from the training was recorded to complete the offline training, but not before being fine-tuned from top to bottom.

The method constructs a four-layer DNN used for location recognition, as is shown in Figure 3. Boxes 1, 2 and 3 represent the hidden layers of DAE respectively, and the last box 4 represents the softmax output layer. After the standardized input *RSS* X' enters DNN, the training of each DAE hidden layer is completed according to the procedures shown in Figure 2 in turn, and the trained connection weight w and hidden units biases b are saved. The output y of the previous layer is taken as the input of the next layer to propagate layer by layer. When data reaches the last layer, the priority probability of all fingerprint points as labeled is used for global fine-tuning from top to bottom to realize offline training. N is the neuron dimension of input layer, and the number is equal to APs observed in offline training. Z is the output layer dimension and is equal to fingerprints when the model is used for positioning, and is equal to floors when the model is used for floor recognition. H_1, H_2 and H_3 represent the neuron dimension of each hidden layer. Their value is determined by the actual situation.

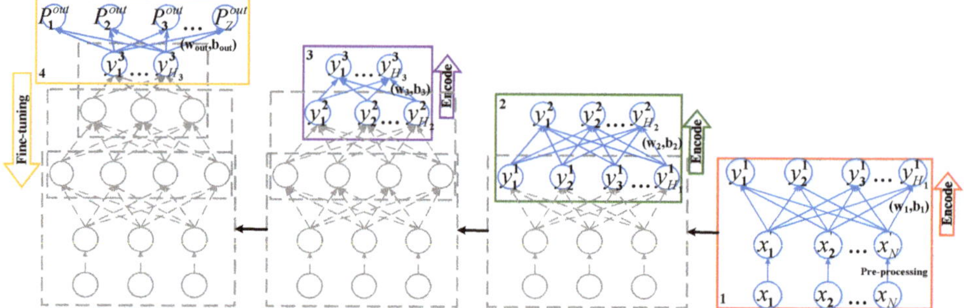

Figure 3. Deep neural network (DNN) structure used for location recognition.

3.2.3. Online Stage

In online positioning, we only need to propagate the preprocessed real-time observation $x^L = (RSS'_1, RSS'_2, \ldots, RSS'_n)$ from bottom to top into the trained DNN in the offline phase, without additional screening and clustering. The propagation process is shown as Formula (4), and y^1, y^2 and y^3 are the output of each hidden layer.

$$\begin{cases} y^1\left(x^L\right) = \frac{1}{1+e^{-(w_1 x^L + b_1)}} \\ y^2\left(y^1\right) = \frac{1}{1+e^{-(w_2 y^1 + b_2)}} \\ y^3\left(y^2\right) = \frac{1}{1+e^{-(w_3 y^2 + b_3)}} \end{cases} \tag{4}$$

The output feature y^3 is then passed into softmax output layer. According to Formula (5), we calculate the posterior probability $P(l_i|L, out), i = 1, 2, \ldots, z$ of the unknown point L being at each fingerprint point l_i:

$$P(l_i|L, out) = \frac{e^{\left(-w_i^{out} y^3 - b_i^{out}\right)}}{\sum\limits_{j=1}^{z} e^{\left(-w_j^{out} y^3 - b_j^{out}\right)}}, i = 1, 2, \ldots, z \tag{5}$$

During the process of floor recognition, the user's floor is the same as the floor with the maximum probability. During the estimate of positioning, the top K fingerprint points of maximum probability are selected as neighboring points to calculate the coordinates of the unknown point. In theory, neighboring points should be distributed around the unknown point. Taking the geometric relationship into consideration, we calculate the center of neighboring points and take the geometric distance between the center and each neighboring point as the weight factor. The larger the distance, the less significant the relationship between the neighboring point and its center, and the smaller the effect of the neighboring point on the unknown point. Neighboring points that are significantly far away from the center are removed. The coordinates of the unknown point are then calculated by combination weights of geometric distance and posterior probability of selected neighboring points. Figure 4 shows the process of combination weight determination, which is based on the following:

$$(X_{center}, Y_{center}) = \frac{\sum\limits_{i=1}^{K} \left(X_{l_i}, Y_{l_i}\right)}{K} \tag{6}$$

$$Distance_i = \left(\left(X_{l_i} - X_{center}\right)^2 + \left(Y_{l_i} - Y_{center}\right)^2\right)^{\frac{1}{2}}, i = 1, 2, \cdots, K \tag{7}$$

$$W_i = \frac{1/Distance_i + P(l_i|L, out)}{\sum\limits_{j=1}^{j=K}\left(1/Distance_j + P\left(l_j|L, out\right)\right)}, i = 1, 2, \ldots, K \tag{8}$$

$$(X_L, Y_L) = \sum_{i=1}^{K}\left(W_i\left(X_{l_i}, Y_{l_i}\right)\right) \tag{9}$$

where (X_{center}, Y_{center}) is the center coordinates of the neighboring points; W_i is the combination weight of the neighboring point l_i with respect to the unknown point L.

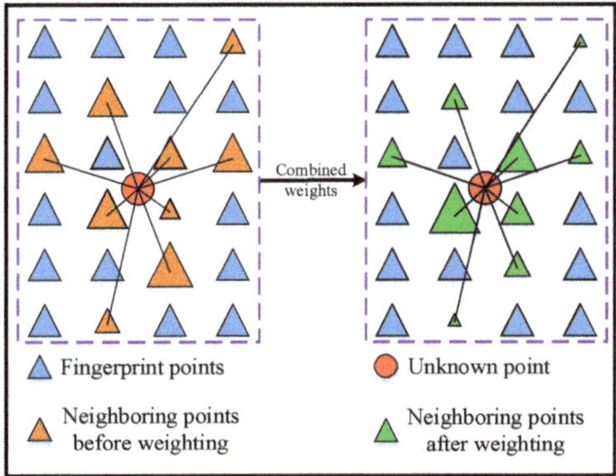

Figure 4. Process of the combination weighting method.

3.3. The Fine Localizer

The motion state is mainly divided into dynamic and static for the user in an indoor environment. According to the dynamic and static characteristics, we propose constrained KF and HMM, respectively, to further improve the accuracy of DNN-based Wi-Fi location estimation, under the condition of only using Wi-Fi.

3.3.1. Constrained Kalman Filter Smoothing

People usually move slowly in an indoor environment. We used KF to solve the smoothing problem in dynamic Wi-Fi location estimation to obtain the optimal estimation of a linear system. We took minimized mean square error as the criterion, estimated the current state by the previous optimal position, used the DNN-based Wi-Fi positioning results as the observation, then updated the information gain to obtain the optimal estimate of the current location.

In this paper, the state vector consists of user positions, velocities and accelerations, and is represented as

$$\boldsymbol{X} = [N, E, V_N, V_E, a_N, a_E] \tag{10}$$

In addition, the state estimation equation of the user at each time is expressed as:

$$\boldsymbol{X}_{t+1,t} = \boldsymbol{A}_{t+1}\boldsymbol{X}_{t,t} + \boldsymbol{Q} \Rightarrow \begin{cases} N_{t+1,t} = \hat{N}_{t,t} + \hat{V}_{t,t}^{N} * \Delta t + dw_N \\ E_{t+1,t} = \hat{E}_{t,t} + \hat{V}_{t,t}^{E} * \Delta t + dw_E \\ V_{t+1,t}^{N} = \hat{V}_{t,t}^{N} + \hat{a}_{t,t}^{N} * \Delta t + dw_V^N \\ V_{t+1,t}^{E} = \hat{V}_{t,t}^{E} + \hat{a}_{t,t}^{E} * \Delta t + dw_V^E \\ a_{t+1,t}^{N} = \hat{a}_{t,t}^{N} + dw_a^N \\ a_{t+1,t}^{E} = \hat{a}_{t,t}^{E} + dw_a^E \end{cases} \tag{11}$$

where dw_N, dw_E, dw_V^N, dw_V^E, dw_a^N and dw_a^E denote the process noises, have a Gaussian distribution. The standard deviations matrix of the process noises is written as $Q = diag[\delta w_N{}^2 \delta w_E{}^2\ \delta w_V^{N2}\ \delta w_V^{E2}\ \delta w_a^{N2}\ \delta w_a^{E2}]$, given $\delta w_N = \delta w_E = 3$, $\delta w_V^N = \delta w_V^E = 0.5$, $\delta w_a^N = \delta w_a^E = 0.1$.

The observation information consists of DNN-based Wi-Fi positioning results and the current user velocities. Therefore, the measurement equation of the system is:

$$Z_{t+1} = \begin{bmatrix} N_{t+1}^{Wi-Fi} \\ E_{t+1}^{Wi-Fi} \\ V_{t+1}^{N} \\ V_{t+1}^{E} \end{bmatrix} = HX_{t+1,t} + R \tag{12}$$

where $H = [diag[1\ 1\ 1\ 1]0_{4\times2}]_{4\times6}$, and $R = diag[\delta N^2\ \delta E^2\ \delta V_N^2\ \delta V_E^2]$ is measurement noise matrix, which has a Gaussian distribution, given $\delta N = \delta E = 12$, $\delta V_N = \delta V_E = 3$. The real-time velocities V_{t+1}^N and V_{t+1}^E are calculated by the current Wi-Fi positioning results and the last optimal location. The calculation formula is as follows:

$$\begin{cases} V_{t+1}^{N} = (N_{t+1}^{Wi-Fi} - \hat{N}_{t,t}) / \triangle t \\ V_{t+1}^{E} = (E_{t+1}^{Wi-Fi} - \hat{E}_{t,t}) / \triangle t \end{cases} \tag{13}$$

In the dynamic Wi-Fi positioning smoothing system, the Kalman filter is always adopted to update the state parameters by time and observation updates when Wi-Fi signals are available. In the prediction stage, we predict the state information and calculate its error covariance. The time-update method is expressed as:

$$\bar{X}_{t+1,t} = A_{t+1}\hat{X}_{t,t} \tag{14}$$

$$\bar{P}_{t+1,t} = A_{t+1}P_tA_{t+1}^{\mathrm{T}} + Q \tag{15}$$

We update the measurement to calculate KF gain in the correction stage. The estimation and the error covariance are updated according to real-time observation information. The Kalman filter observation-update equation is:

$$\bar{V}_{t+1} = Z_{t+1} - H\bar{X}_{t+1,t} \tag{16}$$

$$P_{\bar{V}_{t+1}} = H\bar{P}_{t+1,t}H^{\mathrm{T}} + R \tag{17}$$

$$G_{t+1} = \bar{P}_{t+1,t}H^{\mathrm{T}}P_{\bar{V}_{t+1}}^{-1} \tag{18}$$

$$\hat{X}_{t+1,t+1} = \bar{X}_{t+1,t} + G_{t+1}\bar{V}_{t+1} \tag{19}$$

$$P_{t+1} = (I - G_{t+1}H)\bar{P}_{t+1,t} \tag{20}$$

where P represent the covariance matrices of the state vector; G_{t+1} is the gain matrix of the Kalman filter; and subscript $t+1, t$ represents the state or covariance estimates forward from $t+1$ to t. Therefore, we can obtain the optimal location according the optimal estimation $\hat{X}_{t+1,t+1}$ of the system at time $t+1$.

Wi-Fi positioning belongs to absolute positioning, which has a significant impact on the final smoothing results. Therefore, the initial measurement noise R given in the KF smoothing system is relatively small. However, the interfered Wi-Fi signal will inevitably bring gross error to the positioning results in complex indoor environments. Further, the fixed measurement noise will affect the filter smoothing result. Accordingly, we add adaptive constraint on the basis of KF to automatically adjust the measurement noise in the system when detecting the abnormal observation information and thereby improve the positioning accuracy.

We regard the movement of low-speed users in a short time period as a uniform motion, and the velocity information as a constraint condition. When the gross error occurs, the two adjacent positioning results with low-speed motion will be far away from each other, which leads to an abnormal increase in the real-time velocities. We calculate the average of the past three velocities and use that as the standard to judge whether the current velocities are abnormal.

$$\bar{|\boldsymbol{V}_t|} = \frac{1}{3}\sum_{i=t-3}^{t-1} |\boldsymbol{V}_i| \tag{21}$$

When the real-time velocity is less than the standard $\bar{|V_t|}$, it is considered that the observation information is normal, and the measurement noise remains unchanged. Otherwise, it is considered that the Wi-Fi positioning results in gross error and measurement noise need to be enlarged to make the final positioning result of the system closer to the state estimation and reduce the gross error interference. We use the three-segment method to realize the adaptive calculation of constraint factor μ; setting $k_1 = 3\bar{|\boldsymbol{V}_t|}, k_0 = \bar{|\boldsymbol{V}_t|}$, the formula is:

$$\mu = \begin{cases} 1, \ |\boldsymbol{V}_t| < k_0 \\ \frac{k_1 - k_0}{k_1 - |\boldsymbol{V}_t|}, \ k_0 < |\boldsymbol{V}_t| < k_1 \\ 2, \ |\boldsymbol{V}_t| > k_1 \end{cases} \tag{22}$$

If the measurement noise before the constraint is $\boldsymbol{R} = diag[\delta N^2 \, \delta E^2 \, \delta V_N^2 \, \delta V_E^2]$, then the measurement noise is $\boldsymbol{R} = diag[\mu_N * \delta N^2 \mu_E * \delta E^2 \mu_N * \delta {V_N}^2 \mu_E * \delta {V_E}^2]$ after the constraint factor is added.

3.3.2. Hidden Markov Models (HMM) Optimization

The location of users in an indoor environment is uncertain; they may appear anywhere. At this point, users no longer need precise motion trajectory, but they do need accurate coordinates of their current position. However, as a form of absolute positioning, Wi-Fi positioning system cannot converge to a better accuracy as GPS does with the prolongation of observation time. Therefore, we construct HMM for optimization according to the temporal correlation of multiple observations at the same unknown point to improve the accuracy of static positioning and make effective use of the observation rather than simply calculating their mean location.

HMM is a probability model related to time series. We use the current RSS, o_t, with its five prior RSS, $o_{t-5}, \ldots o_{t-1}$, and five posterior RSS, $o_{t+1}, \ldots, o_{t+5}$, as observations at an unknown point. We take fingerprint points as hidden state and the posterior probability calculated by DNN as observation state to structure HMM $\lambda = (A, B, \pi)$.

- $A = (a_{ij}), i, j = 1, 2, \ldots, z$ is the hidden state transition matrix, which is made up of the transition probability among hidden states, and a_{ij} represents the probability of fingerprint point i transferring to j. We set the a_{ij} of non-adjacent fingerprint points to 0 and to 1 otherwise.
- $B = (b_{j\gamma}), j = 1, 2, \ldots z, \gamma = t-5, \ldots, t+5$ is the observation probability matrix, where $b_{j\gamma}$ denotes the probability of o_γ observed at the fingerprint point j. So, $b_{jr} = P(\gamma | l_j) = cP(l_j | \gamma, out)$, where $P(l_j | \gamma, out)$ is the posteriori probability of DNN output, and $c = P(\gamma)/P(l_j)$ is a constant that has nothing to do with final result.
- π is the prior state probability matrix. It represents the probability of the unknown point in each fingerprint point at the start time. In the system, we set the initial probability of each fingerprint point as equal, so $\pi = [1/z, \ldots, 1/z]$.

Therefore, the static optimal posterior probability $P(l_i | L, optimal\ out)$ of the unknown point L at each fingerprint point can be calculated by the forward and backward probability of the HMM when given the RSS observations $O = (o_{t-5}, \ldots o_{t-1}, o_t, o_{t+1}, \ldots, o_{t+5})$ and HMM model λ, as follows:

$$P(l_i|L, optimal\ out) = P_t(L_t = l_i, \lambda, O) = \frac{P(L_t = l_i, O|\lambda)}{P(O|\lambda)} = \frac{\mu_t(l_i)\rho_t(l_i)}{\sum_{j=1}^{z} \mu_t(l_j)\rho_t(l_j)}, i = 1, 2, 3, \ldots, z \tag{23}$$

$$\mu_\gamma(l_i) = P(O_{t-5}, \ldots, O_t, L_t = l_i|\lambda) = [\sum_{j=1}^{z} \mu_{\gamma-1}(j)a_{ji}]b_{i\gamma}, \gamma = t-5, \ldots, t-1, t \tag{24}$$

$$\rho_\gamma(l_i) = P(O_t, \ldots, O_{t+5}, L_t = l_i|\lambda) = \sum_{j=1}^{z} a_{ij}b_{j\gamma+1}\rho_{\gamma+1}(j), \gamma = t+5, \ldots, t+1, t \tag{25}$$

where $\mu_t(l_i)$ represents the forward probability of unknown point L at fingerprint point l_i with the observation sequence $o_{t-5}, \ldots o_{t-1}, o_t$, while $\rho_t(l_i)$ is the backward probability of the observation sequence $o_t, o_{t+1}, \ldots, o_{t+5}$ given unknown point L at fingerprint point l_i. Substituting Equation (23) into (8), we can calculate the optimal static coordinates of the unknown point.

4. Experience and Analysis

In order to evaluate the performance of the proposed DNN-based Wi-Fi fingerprint location recognition method, we carried out practical experiments with different scenarios in typical indoor environments. We first used the UJIIndoorLoc public dataset to prove the feasibility of the proposed method when compared against typical Wi-Fi positioning methods, and analyzed the effect of different model parameters on its performance. Next, we established the experiment site in order to perform further testing under the condition of avoiding the influence of heterogeneous equipment. We investigated user positioning performance in dynamic and static states and tested the effectiveness of optimization method on positioning results.

4.1. Experiments of UJIIndoorLoc Public Dataset

4.1.1. Experimental Setting

To evaluate the feasibility of the proposed method, we used the large Wi-Fi crowd-sourced UJIIndoorLoc dataset, which contains labeled positions and is publicly available. The dataset consists of 21,048 Wi-Fi *RSS* samples, of which 19,937 are training samples and 1111 are test samples. Each sample contains 520 *RSS* values from 520 APs. All the registered data were collected by 20 volunteers using 25 different Android devices on 13 floors of three buildings. (The reader is referred to Reference [15] for further details.)

Therefore, the established DNN contained 520 neurons in its input layer. We take every sample as a fingerprint to build the fingerprint database, and the *RSS* of the undetected AP that has positive value 100 in the sample is set to −100 dB. In an SDAE, we set the corrupted probability to 10%; the learning rate and scaling factor to 0.1 and 0.99, respectively; the non-sparse penalty to 0.05; and the dropout fraction to 0.2.

4.1.2. Results of Floor Recognition

Figure 5 shows the recognition results of several DNN architectures with different hidden layers and neurons. As the numbers of hidden layers and neurons increase, the recognition effect improves. However, if the network depth is increased optionally, not only is the running time increased, but the accuracy of recognition is also affected due to limited samples in the fingerprint database. The proposed four-layer DNN architectures obtained the highest recognition rate, as high as 94.60%. We next compared DNN with the standard neural network (NN), which was used for classification. Because the NN lacks the ability to effectively capture feature information, its recognition results were worse than those of the DNN under the same conditions. Compared with standard NN, the recognition rates of DNN under the one, two, three and four hidden layers architectures were increased 2.11%, 7.91%, 9.71% and 8.12%, respectively. In floor recognition, DNN obtains satisfactory recognition accuracy.

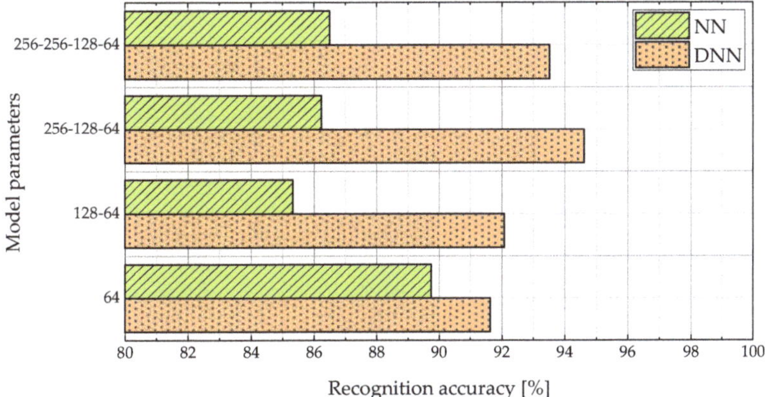

Figure 5. Recognition results for different architectures.

Figure 6 shows the effect of the dropout mechanism on recognition performance in the three-hidden-layer network. Dropout suppresses some neurons in the training process, reduces the model reliance on local features and enhances the generalization of the model. The figure shows that the recognition performance is highest when the dropout mechanism is at 0.2, and that excessive dropout reduces the accuracy of the model. This can be explained by noting that in a large sample model, dropout has a significant effect on restraining overfitting, while excessive dropout destroys the integrity of the model and makes it underfit with the training data.

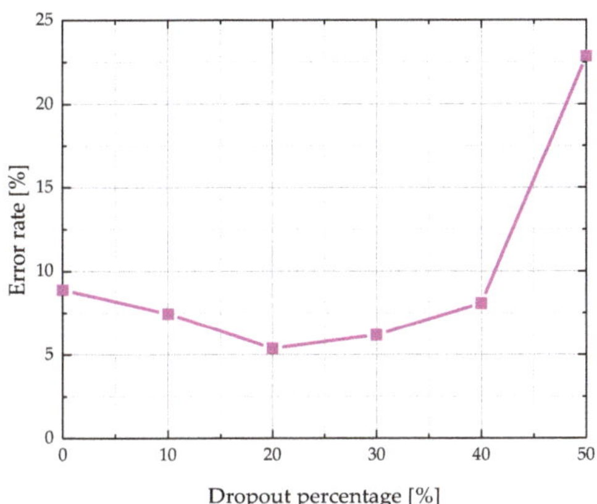

Figure 6. Effect of the dropout on recognition accuracy.

4.1.3. Results of Position Estimation

We chose all floors data to fully test the feasibility of DNN in position estimation and make it more effective. Tables 1–3 show the average error, root-mean-square error (RMSE), maximum error and minimum error of the DNN, as well as those of Microsoft's RADAR system. Figure 7 shows the positioning results of DNN-based Wi-Fi fingerprint location. The positioning error of RADAR is similar to that of Reference [15], while the DNN-based Wi-Fi fingerprint positioning results are better than those of the RADAR system. Compared to the RADAR system, the average errors of DNN on each floor were reduced as follows: (a) 7.90%, 5.20%, 9.13% and 3.50% for building 0 (Table 1); (b) 56.43%, 21.03%, 5.07% and 15.00% for building 1 (Table 2); (c) 3.66%, 15.98%, 8.64%, 27.03% and 1.32% for building 2 (Table 3). In general, the DNN-based method estimates the location of users better than the RADAR system. The localization results for different floors and different buildings show that the proposed location recognition method is highly robust in different scenes.

Table 1. Positioning errors of Building 0, m.

Floors	Methods	Average Error	Maximum Error	Minimum Error
Floor_0	RADAR	7.72	42.65	0.60
	DNN	7.11	42.79	0.22
Floor_1	RADAR	5.58	33.91	0.11
	DNN	5.29	29.87	0.14
Floor_2	RADAR	6.03	26.49	0.13
	DNN	5.48	23.88	0.33
Floor_3	RADAR	7.14	31.23	0.08
	DNN	6.89	32.75	0.22

Table 2. Positioning errors of Building 1, m.

Floors	Methods	Average Error	Maximum Error	Minimum Error
Floor_0	RADAR	19.14	75.99	0.46
	DNN	8.34	25.98	0.85
Floor_1	RADAR	13.17	54.90	0.19
	DNN	10.40	62.63	0.31
Floor_2	RADAR	13.21	54.91	0.26
	DNN	12.54	55.53	0.92
Floor_3	RADAR	10.27	58.12	0.37
	DNN	8.73	58.39	0.23

Table 3. Positioning errors of Building 2, m.

Floors	Methods	Average Error	Maximum Error	Minimum Error
Floor_0	RADAR	12.57	77.16	0.94
	DNN	12.11	74.53	0.41
Floor_1	RADAR	9.20	68.21	0.40
	DNN	7.73	55.06	0.47
Floor_2	RADAR	12.97	65.71	0.48
	DNN	11.85	52.12	0.79
Floor_3	RADAR	11.06	34.42	9.31
	DNN	8.07	33.83	0.41
Floor_4	RADAR	15.88	74.96	1.39
	DNN	15.67	59.99	0.05

By comparing the positioning results of the three buildings, we found that the positioning accuracy differed for each building, but the positioning results for each floor in the same building were similar. These findings mainly resulted from the deviation between the training data and the test data caused by different acquisition devices [35]. This deviation is a hot spot in the field of research on Wi-Fi signals. Many researchers have attempted to study its mechanism and solution. The relationship between the number of data acquisition devices and positioning errors is shown in Figure 8. It is apparent that the fewer devices used during the acquisition of training data, the higher the positioning accuracy. But this is not the focus of this paper, and therefore we do not carry out a detailed study. In addition, the test data were obtained four months after the original data gathering, which also contributed to increase of error. Therefore, although using the data for positioning is a challenge, DNN-based Wi-Fi fingerprint location recognition is more accurate than that of the widely used RADAR system, which proves the feasibility and robustness of this method.

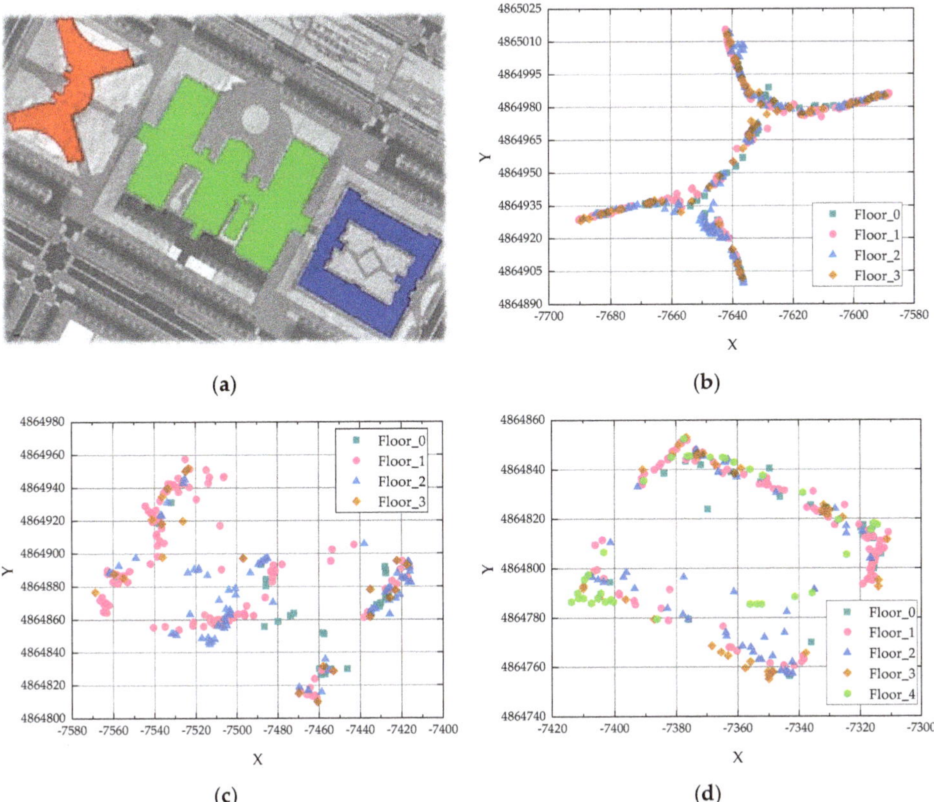

Figure 7. Positioning results of UJIIndoorLoc Public Dataset: (**a**) Building contours; (**b**) Building 0; (**c**) Building 1; (**d**) Building 2.

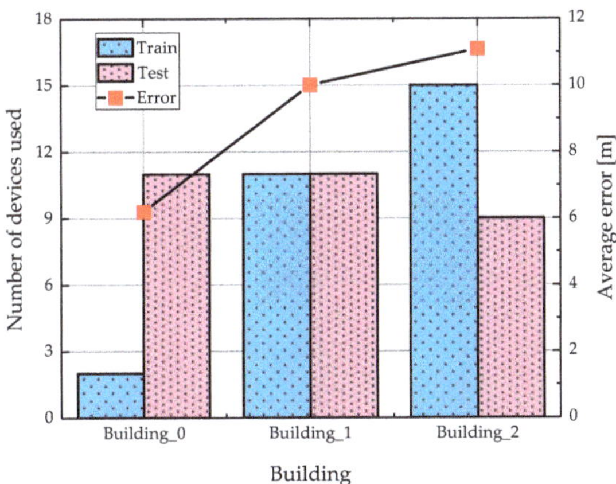

Figure 8. Relationship between the number of data acquisition devices and positioning errors.

With the rapid development of Wi-Fi protocols and hardware, the effect of the isomerism will be smaller and smaller. DNN-based Wi-Fi fingerprint location recognition shows vast potential.

4.2. Experiments of China University of Mining and Technology

4.2.1. Experimental Setting

In the UJIIndoorLoc public dataset, the test data are merely simple discrete points, which cannot be used to test the positioning accuracy of user motion. Therefore, a positioning experiment was set up on the fourth floor of the School of Environment Science and Spatial Informatics, China University of Mining and Technology, Xuzhou, China. This is a typical indoor environment with 800 square meters of the experimental area. We set up 330 fingerprint points in total (1.2 m × 1.2 m) and collected 60 Wi-Fi samples at each fingerprint point, for a total of 19,800 *RSS* training samples. During the offline stage, we observed 195 APs. The layout of the experimental site is shown in Figure 9.

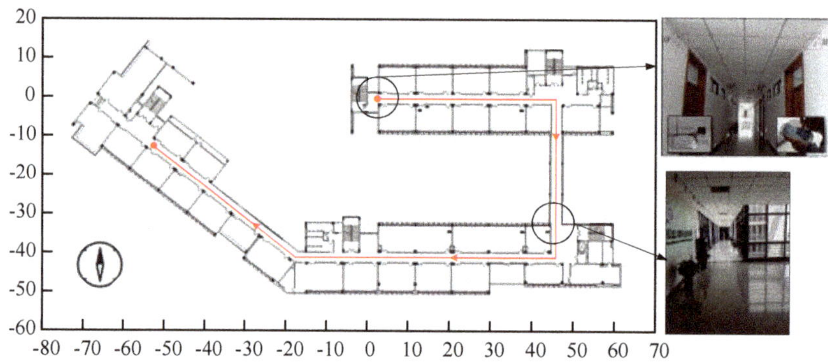

Figure 9. Layout of the experimental site.

We constructed a three-hidden-layer (400-300-200) DNN structure in line with the training data, which possessed 195 and 330 neurons in the input and output layers, respectively. The fingerprint database had 19,800 × 195 dimensions. The same smartphone models (Samsung Galaxy S5) were used as test devices to avoid the interference that heterogeneous devices can wreak on positioning results, and the sampling frequency of Wi-Fi was 0.35 Hz. The red line in Figure 9 represents the user's dynamic walking trajectory.

4.2.2. Analysis of Dynamic Positioning Results

Currently, KNN (RADAR) [16], weighted KNN (WKNN) [18] and Clustering-WKNN (CWKNN) (Horus) [17] are the gold standards of Wi-Fi positioning methods. Our evaluation of the DNN performance in dynamic Wi-Fi positioning can be seen in Figure 10, which illustrates the average error of the DNN as well as that of classic methods in dynamic motion. Further, the figure also shows the changes in the positioning accuracy of different methods two months after the establishment of the fingerprint database. Without the influence of device heterogeneity, the difference between the positioning results based on the DNN and traditional methods becomes smaller, but DNN still offers better performance. Compared with KNN, WKNN and CWKNN, the average errors of DNN are reduced as follows: 20.49%, 15.68% and 13.62%, respectively. As time goes by, the positioning errors of these methods increase to varying degrees, but the DNN still has the best positioning accuracy, with an error of 2.25 m. Figure 11 shows the average time consumption of these methods for each positioning. The DNN-based Wi-Fi positioning method has the least time consumption since there is no need to cluster and calculate the Euclidean distance of signal strength between unknown and all fingerprint points. Accordingly, experiments demonstrate that the DNN can not only offer better positioning accuracy and better real-time performance than the traditional methods, but also have a good time robustness.

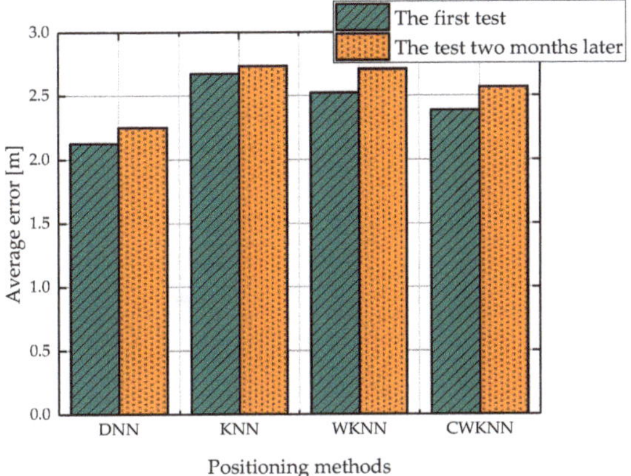

Figure 10. Positioning errors with different Wi-Fi positioning methods.

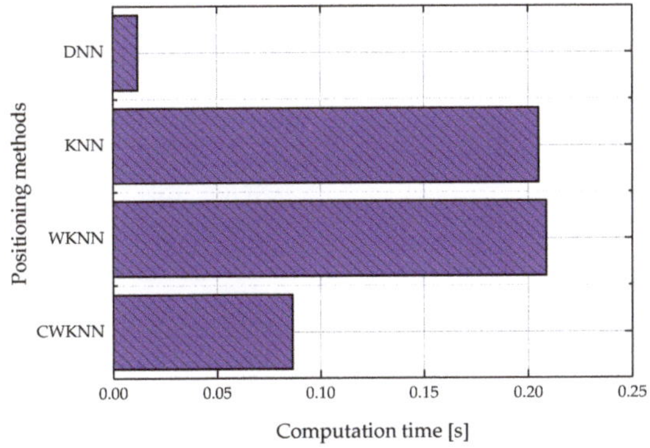

Figure 11. Average time consumption with different methods.

In order to test the effectiveness of constrained KF in dynamic positioning results, the following calculation schemes were conducted:

- Scheme 1: No use of Kalman filter.
- Scheme 2: Use of constrained Kalman filter.
- Scheme 3: Use of adaptive constrained Kalman filter.

Figure 12 shows the dynamic positioning results of the three schemes. It can be seen from the details of location trajectories that Scheme 1 without constraint inevitably suffers from typical overlapping and jumping due to the influence of multipath interference and *RSS* fluctuation. In Schemes 2 and 3, the accuracy of the positioning trajectory improved significantly with constrained KF. Further, Scheme 3 further adjusted the measurement noise adaptively to make trajectory smoother and reduce the occurrence of overlapping and jumping, so it estimated the location of the user very well. The trajectories show that the proposed adaptive constrained KF enhances the robustness of the localization results when gross errors of Wi-Fi appear.

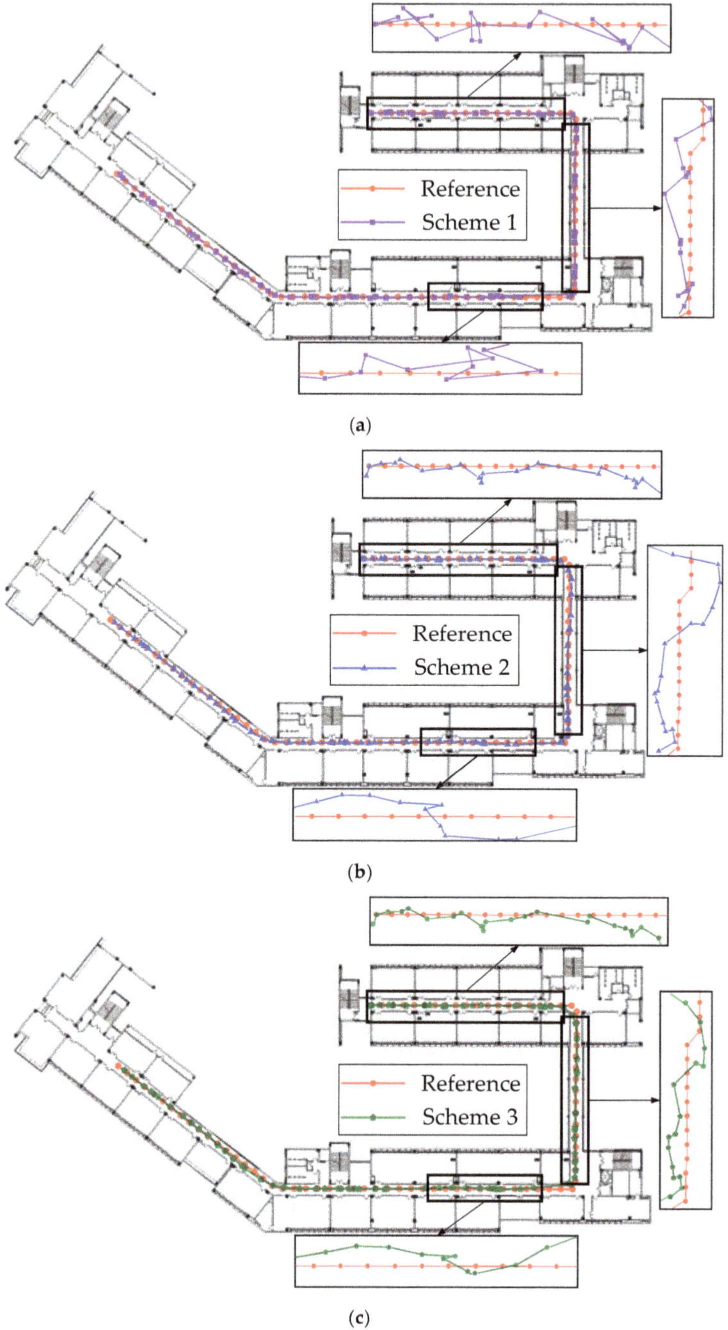

Figure 12. Positioning results of dynamic Wi-Fi positioning with different schemes: (**a**) Scheme 1; (**b**) Scheme 2; (**c**) Scheme 3.

Figures 13 and 14 show in detail the positioning errors and corresponding cumulative distribution functions (CDFs). The positioning result of Scheme 1 is not stable, and there is a large error. Schemes 2 and 3 can effectively suppress the error after using the constrained KF, and the maximum error is obviously decreased. Compared with Scheme 2, the main contribution of Scheme 3 is that the adaptive mechanism can recognize the gross error when it appears and reduce its influence on the positioning results by adjusting measurement noise. The probabilities of positioning errors of less than 3.5 m for the three schemes are 80.25%, 83.95% and 90.12% respectively.

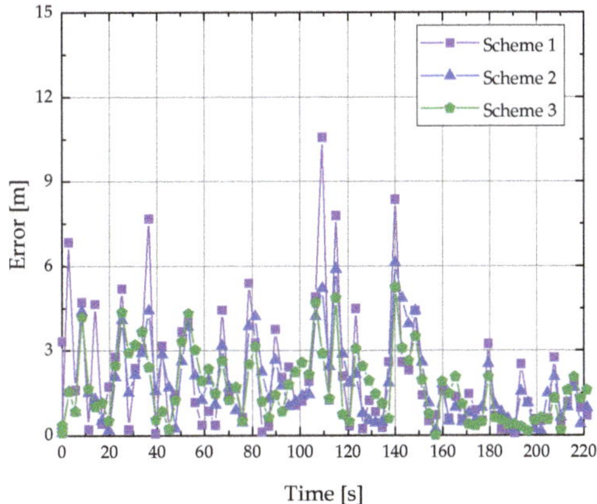

Figure 13. Positioning errors with different schemes.

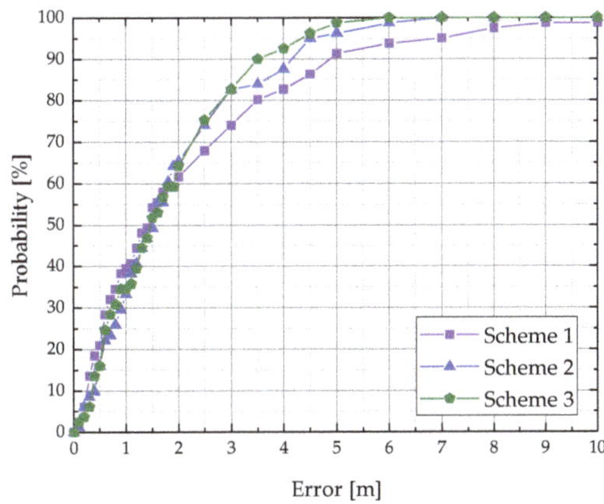

Figure 14. Cumulative distribution functions (CDFs) with different schemes.

Table 4 shows the average error, RMSE, circular-error-probability (CEP) and maximum error of the three schemes. Constrained KF improves the accuracy of dynamic positioning. Scheme 3, which uses the adaptive mechanism, offers the best performance. Compared with Schemes 1 and 2, Scheme 3 shows reduced errors as follows: (a) 19.25% and 7.03% for average error; (b) 29.47% and 8.97% for RMSE; (c) 39.07% and 5.86%for CEP; (d) 50.28% and 14.33%for maximum error. The positioning errors of Scheme 3 using adaptive constraint KF greatly reduce.

Table 4. Positioning errors for different schemes, m.

Schemes	Average Error	Room Mean Square Error (RMSE)	Circular-Error-Probability CEP (95%)	Maximum Error
Scheme 1	2.13	3.02	6.86	10.58
Scheme 2	1.85	2.34	4.44	6.14
Scheme 3	1.72	2.13	4.18	5.26

4.2.3. Analysis of Static Positioning Results

For a user who may be anywhere in an indoor environment, it is more important to determine the current position than to obtain an accurate trajectory. We utilized HMM to optimize our calculation of multiple posterior probability and obtain the optimal location estimation of the unknown point according to the temporal correlation of multiple observations on the same unknown point.

At the experimental site, we selected 39 test points at equal intervals of 5 m. Figure 15 shows the positioning situation of HMM optimal location and mean location of each test point, based on its five prior and five posterior *RSS* observations. Figure 16 and Table 5 show the error comparison between the two optimization methods. It can be seen that the results of the two methods are close when it comes to determining the real location, but HMM optimization has better performance. In addition, the positioning errors for HMM are fewer than that of using the mean. Compared with mean positions, the respective errors of HMM are reduced as follows: 0.36 m for average error, 0.46 m for RMSE, and 0.72 m for CEP. Accordingly, HMM optimization performs better than the mean when multiple observations are used to obtain accurate coordinates of the unknown point. However, it needs to perform multiple observations of the same point.

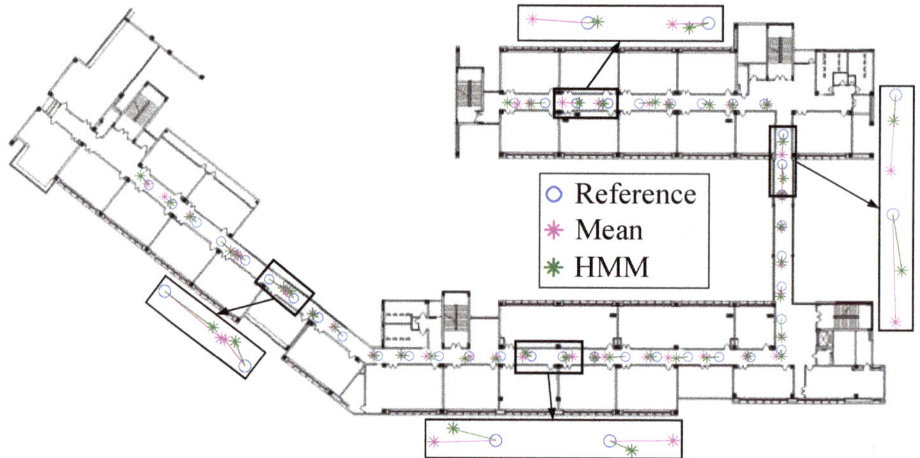

Figure 15. Positioning results calculated by mean and hidden Markov models (HMM).

Table 5. Positioning errors of mean location and HMM optimal location, m.

Algorithms	Average Error	RMSE	CEP (95%)	Maximum Error
Mean	1.58	1.89	3.18	4.52
HMM	1.22	1.43	2.46	2.84

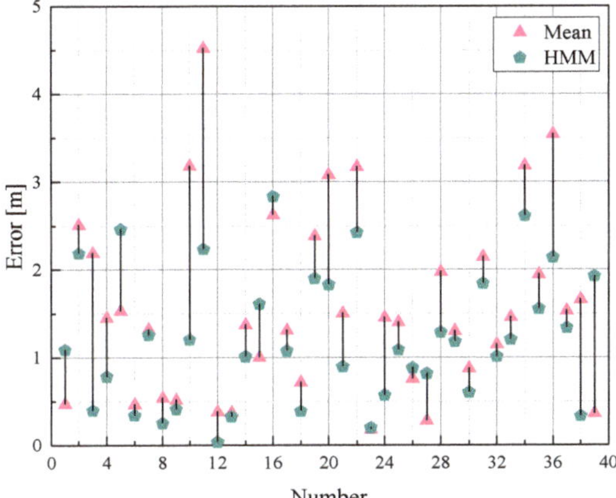

Figure 16. Positioning errors of mean and HMM locations compared with the reference.

5. Conclusions

In this paper, we propose a robust and accurate Wi-Fi fingerprint location recognition method based on a DNN. We use the adaptive constrained KF in the dynamic positioning and the HMM in the static positioning to overcome the gross error caused by the fluctuation in *RSS* and multipath interference and to improve positioning accuracy. We used a public dataset to fully verify the feasibility of DNN in Wi-Fi positioning. Experimental results show that the accuracy of DNN in floor recognition is 94.60%, and the error of position estimation is 5.29 m, which is superior to traditional methods. Further experiments in office buildings show that the positioning error of DNN is 2.13 m without interference caused using heterogeneous devices. Compared with classic methods, the DNN has higher positioning accuracy and better real-time performance. The dynamic positioning accuracy can reach 1.72 m after adaptive constrained KF smoothing. In static positioning, HMM is used to optimize the multi-observation results on the unknown points, which realizes 1.22 m positioning accuracy.

In the future, we will attempt to combine DNN with semi-supervised classification to reduce training time and update fingerprint database quickly and effectively when the environment changes, so as to ensure the accuracy of the positioning system. In the meantime, we will integrate DNN-based Wi-Fi positioning results with other sensors to build other positioning schemes based on real-world situations.

Author Contributions: Conceptualization, Y.W. and J.G.; methodology, Y.W.; software, Y.W. and Z.L.; validation, Y.W., Z.L. and L.Z.; data curation, Y.W.; writing—original draft preparation, Y.W. and J.G.; writing—review and editing, Z.L. and L.Z.; funding acquisition, J.G. and Z.L. All authors have read and agreed to the published version of the manuscript.

Funding: This work was supported by the National Natural Science Foundation of China (No. 41604006, No. 41874006, No. 41674008) and supported by Natural Science Foundation of Jiangsu Province (No. BK20160247).

Conflicts of Interest: The authors declare no conflicts of interest.

References

1. Dos Santos, I.L.; Pirmez, L.; Lemos, É.T.; Delicato, F.C.; Pinto, L.A.V.; de Souza, J.N.; Zomaya, A.Y. A localized algorithm for Structural Health Monitoring using wireless sensor networks. *Inf. Fusion* **2014**, *15*, 114–129. [CrossRef]
2. Khalajmehrabadi, A.; Gatsis, N.; Akopian, D. Modern WLAN fingerprinting indoor positioning methods and deployment challenges. *IEEE Commun. Surv. Tutor.* **2017**, *19*, 1974–2002. [CrossRef]
3. Bordoy, J.; Wendeberg, J.; Schindelhauer, C.; Reindl, L.M. Single transceiver device-free indoor localization using ultrasound body reflections and walls. In Proceedings of the International Conference on Indoor Positioning and Indoor Navigation (IPIN), Banff, AB, Canada, 13–16 October 2015; pp. 1–7.

4. Ruiz, A.R.J.; Granja, F.S. Comparing ubisense, bespoon, and decawave uwb location systems: Indoor performance analysis. *IEEE Trans. Instrum. Meas.* **2017**, *66*, 2106–2117. [CrossRef]
5. Daroogheha, S.; Lasky, T.A.; Ravani, B. Position Measurement Under Uncertainty Using Magnetic Field Sensing. *IEEE Trans. Magn.* **2018**, *54*, 1–8. [CrossRef]
6. Zheng, Y.; Luo, P.; Chen, S.; Hao, J.; Cheng, H. Visual search based indoor localization in low light via rgb-d camera. *World Acad. Sci. Eng. Technol. Int. J. Comput. Electr. Autom. Control Inf. Eng.* **2017**, *11*, 349–352.
7. Xie, B.; Tan, G.; He, T. Spinlight: A high accuracy and robust light positioning system for indoor applications. In Proceedings of the 13th ACM Conference on Embedded Networked Sensor Systems, Seoul, Korea, 1–4 November 2015; pp. 211–223.
8. Harle, R. A survey of indoor inertial positioning systems for pedestrians. *IEEE Commun. Surv. Tutor.* **2013**, *15*, 1281–1293. [CrossRef]
9. Li, Q.; Li, W.; Sun, W.; Li, J.; Liu, Z. Fingerprint and assistant nodes based Wi-Fi localization in complex indoor environment. *IEEE Access* **2016**, *4*, 2993–3004. [CrossRef]
10. Gansemer, S.; Pueschel, S.; Frackowiak, R.; Hakobyan, S.; Grossmann, U. Improved RSSI-based Euclidean distance positioning algorithm for large and dynamic WLAN environments. *Int. J. Comput.* **2010**, *9*, 37–44.
11. Talvitie, J.; Renfors, M.; Lohan, E.S. Distance-based interpolation and extrapolation methods for RSS-based localization with indoor wireless signals. *IEEE Trans. Veh. Technol.* **2015**, *64*, 1340–1353. [CrossRef]
12. Shaeffer, D.K. MEMS inertial sensors: A tutorial overview. *IEEE Commun. Mag.* **2013**, *51*, 100–109. [CrossRef]
13. Li, Y.; Zhuang, Y.; Zhang, P.; Lan, H.; Niu, X.; El-Sheimy, N. An improved inertial/wifi/magnetic fusion structure for indoor navigation. *Inf. Fusion* **2017**, *34*, 101–119. [CrossRef]
14. Chen, J.; Ou, G.; Peng, A.; Zheng, L.; Shi, J. An INS/WiFi indoor localization system based on the Weighted Least Squares. *Sensors* **2018**, *18*, 1458. [CrossRef] [PubMed]
15. Torres-Sospedra, J.; Montoliu, R.; Martínez-Usó, A.; Avariento, J.P.; Arnau, T.J.; Benedito-Bordonau, M.; Huerta, J. UJIIndoorLoc: A new multi-building and multi-floor database for WLAN fingerprint-based indoor localization problems. In Proceedings of the International Conference on Indoor Positioning and Indoor Navigation (IPIN), Busan, Korea, 27–30 October 2014; pp. 261–270.
16. Bahl, P.; Padmanabhan, V.N.; Bahl, V.; Padmanabhan, V. RADAR: An in-building RF-based user location and tracking system. In Proceedings of the IEEE INFOCOM 2000, Tel Aviv, Israel, 26–30 March 2000.
17. Youssef, M.A.; Agrawala, A.; Shankar, A.U. WLAN location determination via clustering and probability distributions. In Proceedings of the First IEEE International Conference on Pervasive Computing and Communications, Fort Worth, TX, USA, 26 March 2003; pp. 143–150.
18. Khodayari, S.; Maleki, M.; Hamedi, E. A RSS-based fingerprinting method for positioning based on historical data. In Proceedings of the International Symposium on Performance Evaluation of Computer and Telecommunication Systems, Ottawa, ON, Canada, 11–14 July 2010; pp. 306–310.
19. Wu, C.; Yang, Z.; Zhou, Z.; Liu, Y.; Liu, M. Mitigating large errors in WiFi-based indoor localization for smartphones. *IEEE Trans. Veh. Technol.* **2016**, *66*, 6246–6257. [CrossRef]
20. Wang, X.; Gao, L.; Mao, S.; Pandey, S. CSI-based fingerprinting for indoor localization: A deep learning approach. *IEEE Trans. Veh. Technol.* **2016**, *66*, 763–776. [CrossRef]
21. Figuera, C.; Rojo-Álvarez, J.L.; Wilby, M.; Mora-Jiménez, I.; Caamaño, A.J. Advanced support vector machines for 802.11 indoor location. *Signal Process.* **2012**, *92*, 2126–2136. [CrossRef]
22. Elbasiony, R.; Gomaa, W. WiFi localization for mobile robots based on random forests and GPLVM. In Proceedings of the 13th International Conference on Machine Learning and Applications, Detroit, MI, USA, 3–6 December 2014; pp. 225–230.
23. Nowicki, M.; Wietrzykowski, J. Low-effort place recognition with WiFi fingerprints using deep learning. In Proceedings of the International Conference Automation, Warsaw, Poland, 15–17 March 2017; pp. 575–584.
24. Félix, G.; Siller, M.; Alvarez, E.N. A fingerprinting indoor localization algorithm based deep learning. In Proceedings of the Eighth International Conference on Ubiquitous and Future Networks (ICUFN), Vienna, Austria, 5–8 July 2016; pp. 1006–1011.
25. Zhang, W.; Liu, K.; Zhang, W.; Zhang, Y.; Gu, J. Deep neural networks for wireless localization in indoor and outdoor environments. *Neurocomputing* **2016**, *194*, 279–287. [CrossRef]
26. Rizk, H.; Torki, M.; Youssef, M. CellinDeep: Robust and accurate cellular-based indoor localization via deep learning. *IEEE Sens. J.* **2018**, *19*, 2305–2312. [CrossRef]

27. Zhuang, Y.; Li, Y.; Qi, L.; Lan, H.; Yang, J.; El-Sheimy, N. A two-filter integration of MEMS sensors and WiFi fingerprinting for indoor positioning. *IEEE Sens. J.* **2016**, *16*, 5125–5126. [CrossRef]
28. Poulose, A.; Kim, J.; Han, D.S. A Sensor Fusion Framework for Indoor Localization Using Smartphone Sensors and Wi-Fi RSSI Measurements. *Appl. Sci.* **2019**, *9*, 4379. [CrossRef]
29. Li, Y.; He, Z.; Gao, Z.; Zhuang, Y.; Shi, C.; El-Sheimy, N. Toward Robust Crowdsourcing-Based Localization: A Fingerprinting Accuracy Indicator Enhanced Wireless/Magnetic/Inertial Integration Approach. *IEEE Internet Things J.* **2018**, *6*, 3585–3600. [CrossRef]
30. Bi, J.; Wang, Y.; Li, X.; Cao, H.; Qi, H.; Wang, Y. A novel method of adaptive weighted K-nearest neighbor fingerprint indoor positioning considering user's orientation. *Int. J. Distrib. Sens. Netw.* **2018**, *14*, 1–13. [CrossRef]
31. Zhou, C.; Yuan, J.; Liu, H.; Qiu, J. Bluetooth indoor positioning based on RSSI and Kalman filter. *Wirel. Pers. Commun.* **2017**, *96*, 4115–4130. [CrossRef]
32. Vincent, P.; Larochelle, H.; Lajoie, I.; Bengio, Y.; Manzagol, P.A. Stacked denoising autoencoders: Learning useful representations in a deep network with a local denoising criterion. *J. Mach. Learn. Res.* **2010**, *11*, 3371–3408.
33. Vincent, P.; Larochelle, H.; Bengio, Y.; Manzagol, P.A. Extracting and composing robust features with denoising autoencoders. In Proceedings of the 25th International Conference on Machine Learning, Helsinki, Finland, 5–9 July 2008; pp. 1096–1103.
34. Krizhevsky, A.; Sutskever, I.; Hinton, G.E. Imagenet classification with deep convolutional neural networks. In Proceedings of the Advances in Neural Information Processing Systems, Nevada, CA, USA, 3–6 December 2012; pp. 1097–1105.
35. Peng, Z.; Richter, P.; Leppäkoski, H.; Lohan, E.S. Analysis of crowdsensed wifi fingerprints for indoor localization. In Proceedings of the 21st Conference of Open Innovations Association (FRUCT), Helsinki, Finland, 6–10 November 2017; pp. 268–277.

Article

A Novel Indoor Ranging Algorithm Based on a Received Signal Strength Indicator and Channel State Information Using an Extended Kalman Filter

Jingjing Wang † and Joon Goo Park *

School of Electronics Engineering, Kyungpook National University, 80 Daehak-ro, Buk-gu, Daegu 41566, Korea; wjj0219@naver.com

* Correspondence: jgpark@knu.ac.kr; Tel.: +82-10-8560-6580

† Current address: Mobile Software and Navigation Laboratory, Room 721, IT-1, Kyungpook National University, 80, Daehak-ro, Bukgu, Deagu 41566, Korea.

Received: 20 April 2020; Accepted: 25 May 2020; Published: 26 May 2020

Abstract: With the increasing demand of location-based services, the indoor ranging method based on Wi-Fi has become an important technique due to its high accuracy and low hardware requirements. The complicated indoor environment makes it difficult for wireless indoor ranging systems to obtain accurate distance measurements. This paper presents an Extended Kalman filter-based approach for indoor ranging by utilizing transmission channel quality metrics, including Received Signal Strength Indicator (RSSI) and Channel State Information (CSI). The proposed ranging algorithm scheme is implemented and validated with experiments in two typical indoor environments. A real indoor experiment demonstrates that the ranging estimation accuracy of our algorithms can be significantly enhanced compared with the typical algorithms. The ranging estimation accuracy is defined as the cumulative distribution function of the distance error.

Keywords: indoor ranging algorithm; channel state information; received signal strength indicator; extended Kalman filter

---

1. Introduction

For positioning in outdoor environments, the Global Position System (GPS) [1,2] can provide very accurate positioning results. Since acquiring a satellite signal inside a building is not possible, the GPS can not be applied to the purpose of indoor positioning. Therefore, it is of great significance to study the indoor positioning technology not relying on the GPS. In recent years, Wireless Local Area Network (WLAN)-based [3] positioning technology has become a research hotspot because of its widespread deployment and ease of use WLAN. In the design of a positioning system, the accuracy of ranging is one of the factors that must be considered. The most important reason for affecting the positioning accuracy in the indoor positioning system is the complex multipath transmission environment and the limited bandwidth of the wireless signal, which cannot effectively distinguish the signals transmitted through multiple paths.

According to the position estimation method in wireless sensor networks, the ranging algorithm is divided into two categories: ranging-based algorithm and range-free algorithm. The former method includes Angle of Arrival(AOA) [4], Time of Arriva (TOA) [5], Time Difference of Arrival (TDOA) [6], Received Signal Strength (RSS) [7], Channel State Information (CSI) [8] etc., and calculate or estimate the distance between the node and the reference. The latter one uses the relationship of the geometric position information between the receiving node and the transmitting node to estimate the distance, including the fingerprint positioning method [9] and approximate estimation method [10]. Figure 1 shows the position estimation methods between range-based and range-free algorithms.

Due to the complex indoor environment, RSSI is often affected by multipath effects and noise signals, and the positioning performance is not stable. With the availability of channel state information from the physical layer, Wi-Fi-based indoor positioning schemes have gradually shifted from adopting RSSI indicators to higher resolution CSI indicators. In recent years, commercial Wi-Fi devices (such as the Intel 5300 wireless network card) have begun to support the acquisition of CSI at the physical layer. CSI can characterize signals with finer granularity. By analyzing the transmission of different sub-channel signals separately, CSI can avoid the multipath effects and noise as much as possible.

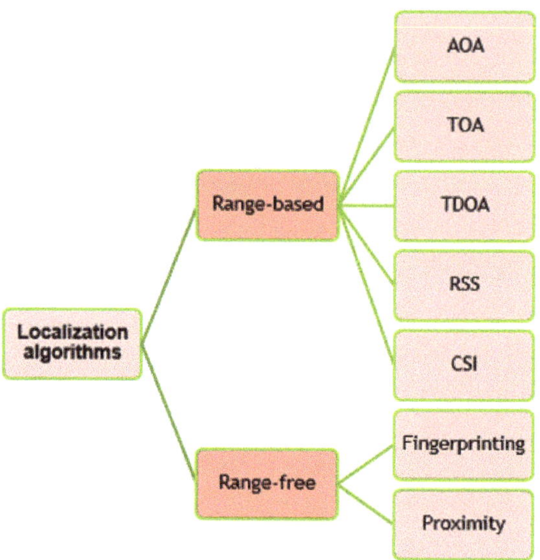

Figure 1. The position estimation methods between range-based and range-free algorithms.

Wu et al. [8] explore the frequency diversity of subcarriers in the Orthogonal Frequency Division Multiplexing (OFDM) system and propose a new method called fine-grained indoor localization (FILA), which uses CSI to establish the propagation model and fingerprint identification system at the receiver. FILA can achieve a median accuracy of 1.2 m in a corridor environment. In paper [11], Destiarti et al. proposed a mobile cooperative tracking method based on extended RSSI ranging. The error estimation range of this algorithm was between 0.22 m and 2.84 m. The fine-grained indoor fingerprinting system (FIFS) [12] explores CSI of the physical layer which specifies channel status on all subcarriers for location fingerprint identification in WLAN. The system uses the amplitude and phase of CSI on multiple propagation paths to display a single position. In experiments, for over 90% of data points, the error of FIFS was within the range of 1.3 meters. In [13], a new indoor fingerprint recognition system based on deep learning using CSI is introduced called deep learning for indoor fingerprinting (DeepFi). This method based on three hypotheses of CSI, the DeepFi system architecture includes an off-line training phase and an online localization phase. In the living room environment, the mean error is 0.9245 m. In the laboratory environment, the mean error distance is 1.8081 m. In paper [14], Chapre et al. proposed a novel Wi-Fi fingerprint recognition system CSI-MIMO, which uses a multiple-input-multiple-output (MIMO) system to utilize frequency diversity and spatial diversity. The system combines the spatial attributes of CSI to improve the FIFS system. The CSI-MIMO system uses the amplitude difference and the phase difference matrix of adjacent subcarriers as the unique fingerprints of the sample points to achieve more accurate positioning than FIFS. The paper [15] presents a deep learning-based method for indoor positioning by utilizing transmission channel quality metrics, including RSSI and CSI.

However, most of the current algorithms are range-free-based using RSI and CSI. These positioning methods are currently based on the similarity of the indoor environment during the positioning and training phases. These positioning methods do not consider the factors that change the indoor environment during the positioning phase, that is, the difference between the indoor environment during the positioning phase and the environment when the database is established. For example, a fingerprint database was established in an unmanned indoor environment, but location matching was performed when the number of people in the room increased. In the case of a large difference in indoor environment between the positioning and training phases, if fingerprint matching is still performed in the traditional way, the positioning error will inevitably increase, and the accuracy rate will also be greatly reduced.

Based on the reasons above, we propose the indoor ranging algorithm based on RSSI and CSI that are simultaneously processed by the extended Kalman filter (EKF) to address the complex indoor situation. In this paper, we analyze the different characteristics of RSSI and CSI and propose a novel indoor ranging algorithm. We use two different granularities of RSSI and CSI to realize regional ranging and precise positioning respectively and make the best use of the advantages of different granularity information as much as possible. EKF [16] is used to filter nonlinear attenuation ranging model based on RSSI and CSI. The proposed algorithm breaks the limitations of traditional RSSI or CSI-based fingerprint positioning and can be applied to online positioning in any indoor environment.

The main contributions of this paper are:

1. A cross-layer approach including MAC layer and physical layer that enable fine-grained indoor ranging in WLANs. Our proposed method includes two parts, RSSI-based ranging model and CSI-based ranging model.
2. Indoor ranging research based on RSSI and CSI in the environment of high-load Access Point (AP). This paper demonstrates the feasibility of this method in a high-load AP environment.
3. The method we propose is also the first one that uses extend Kalman filtering to combine RSSI-based signal attenuation model and CSI-based ranging model to perform distance estimation.
4. The experimental evaluation in two representative indoor environments to confirm the feasibility of our design and its effect on the ranging results. The experimental results show that the proposed algorithm outperforms existing algorithms.

The rest of this paper is organized as follows: Section 2 gives a preliminary of the basics of RSSI, CSI and EKF. In Section 3, the proposed method is described in detail; We use extended Kalman filter to estimate the distance through the nonlinear relationship between CSI, RSSI and distance. The experimental experiment and result are detailed in Section 4. Finally, conclusions and the future research are presented in Section 5.

2. Related Work

2.1. Characteristics of RSSI and CSI

The low-dimensional RSS information makes it possible to perform the regional level of indoor positioning. RSSI, as an easy to get a signal feature, is often used in an indoor positioning system. RSSI belongs to the Medium Access Control (MAC) layer and comes from each packet. Unlike RSS information, CSI information [17–20] considers the signal transmission between different subchannels and different antennas. In addition, its dimension and complexity provide information much richer than the RSS information.

This unique physical feature meets the following requirements:

(1) It has excellent resistance to interference in the 2.4 GHz band signal and has less fluctuation in a stable environment. It can also reflect the changes in the environment.
(2) The use of OFDM technology to distinguish signals of different paths as finely as possible.

CSI is a fine-grained attribute value of the physical layer that describes the amplitude and phase of the frequency domain corresponding to each subcarrier. The CSI can reflect the attenuation of the wireless signal as it travels between the transmitter and receiver. Table 1 demonstrates the differences between RSSI and CSI.

Table 1. The differences between Received Signal Strength Indicator (RSSI) and Channel State Information (CSI).

Category	RSSI	CSI
Time resolution	Packet	Multipath signal cluster
Frequency resolution	None	Subcarrier
Stability	Low	High
Dimension	One dimension	High dimension
Universality	All Wi-Fi devices	Some Wi-Fi devices

In this paper, the RSSI value and CSI value were collected at fixed positions of 1 m, 4 m and 7 m in the same indoor environment, and the results of their multipath effects were compared. Figure 2 compares the stability between the sampled RSSI value and the amplitude of the CSI sampled on channel 2.

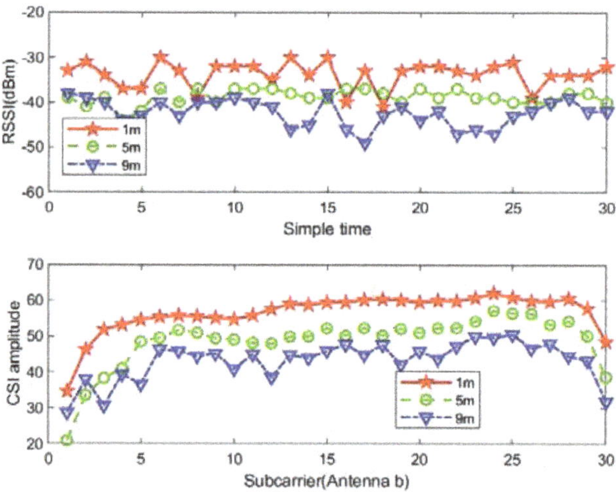

Figure 2. Comparison of the stability between the sampled RSSI value and the amplitude of the CSI sampled on channel 2.

Although the CSI value collected on a certain channel will change, compared with the collected RSSI value, the CSI value collected less varies in time and remains stable.

2.2. RSSI-Based Signal Attenuation Model for Indoor Ranging

Ranging information between two points is calculated using the received signal strength information. In free space, RSSI is inversely proportional to the square of the distance d between the receiving point and the transmitting point. Their relationship [21–24] can be expressed using the famous Friis formula:

$$P_r\left(d\right) = \frac{P_t G_t G_r \lambda^2}{\left(4\pi\right)^2 d^2 L} \tag{1}$$

In Equation (1), $P_r(d)$ is the received power, and its unit is watts. P_t is the transmit power in watts. G_t is the transmit antenna gain. G_r is the gain of the receiving antenna. λ is the wavelength of the transmitted signal; its unit is meter. L is a loss parameter which is irrelevant to the propagation environment. The system loss parameters represent the total loss of the actual system hardware. It includes transmission lines, filters, and antennas.

In general, L is greater than one. However, if we assume that the system hardware has no loss, we can let $L = 1$. From Equation (1), we can observe that the attenuation of the received power is exponential with the distance. So, the free-space path loss can be directly derived from Equation (1) without any system loss.

$$PL_F(d)\,[dB] = 10\log(\frac{P_t}{P_r}) = -10log(\frac{G_tG_r\lambda^2}{(4\pi)^2d^2}) \tag{2}$$

We can ignore the antenna gain and let $G_t = G_r = 1$. In a free-space model, the average received signal is in a logarithmic relationship with the distance d between the transmitter and receiver in all environments. Basically, a more general path loss model can be constructed using the environment-dependent signal attenuation factor to change the free-space path loss model. The mathematical expression of the signal attenuation log model is as follows:

$$RSSI = A - 10n\log(\frac{d}{d_0}) + X_0, \tag{3}$$

where, RSSI indicates the received signal strength indication value in dBm. A is the signal strength at 1m from the source. d represents the distance between the transmitting node and the receiving node in meters. d_0 is the unit distance and usually takes 1 m. $X_{(0)}$ is a Gaussian random number taking a mean of 0 and its standard deviation range between 4 and 10. When the n value is smaller, the signal attenuation in the transmission process is smaller, and the signal can spread farther away. The range is generally between 2 and 4.

2.3. CSI-Based Signal Attenuation Model for Indoor Ranging

OFDM systems can modulate signals into multiple subcarriers and transmit them simultaneously in 802.11 a/g/n networks [25–27]. The CSI information obtained from the physical layer can reflect the channel quality between the transmitter and receiver. The CSI describes that signals are affected by multiple paths during propagation. Therefore, the channel can be estimated by CSI [28] analyzing the channel characteristics of the communication link. Table 2 shows the available information about CSI.

Table 2. Channel state information.

Data Information	Properties
Bfee-count	Number of Bfee count beamforming sent to user space by drive record
Nrx	Number of Nrx receiving antennas (Intel5300 network card is usually 3)
Ntx	Number of Ntx transmit antennas
rssi-a, rssi-b, rssi-c	Received signal strength of each receiving antenna
rate	Rate the transmission rate of each packet
noise	noise
CSI	CSI data itself is a three-dimensional arrays of Nrx * Ntx * 30

In an OFDM system, the received signal in the indoor environment can be expressed by Equation (4):

$$Y = HX + N \tag{4}$$

where Y is the received signal vector. H is the channel matrix. X is the transmitted signal vector. H is the channel matrix, and N is the additive Gaussian white noise.

The estimated CSI of all subcarriers is:

$$\hat{H} = \frac{Y}{X} \tag{5}$$

where $\hat{H}$ represents the channel frequency response (CFR) in the frequency domain. A set of CSI values can be obtained from the received packets, which can be expressed by Equation (6):

$$H = [H_1, H_2, H_3, ..., H_k, ..., H_n], \tag{6}$$

where H_k is the k-th subcarrier of CSI.

By modifying the wireless network card, CFR samples with 30 subcarriers ($n = 30$) can be obtained by the Wi-Fi devices. Each group of CSI includes the amplitude and phase of subcarriers, as shown in Equation (7):

$$H_k = \|H_k\| \, e^{j\angle H_k} \tag{7}$$

where $\|H_k\|$ represents the amplitude. $\angle H_k$ represents the phase of k-th subcarriers.

An Intel 5300 wireless network card operating in a 20 MHz high-throughput mode (HT mode) under a MIMO system has p transmitting antennas and q receiving antennas. CSI can be expressed as a matrix of pxq dimension, as follows:

$$H(f_x) = \begin{bmatrix} h_{11} & h_{22} & \cdots & h_{1q} \\ h_{21} & h_{22} & \cdots & h_{2q} \\ \vdots & \vdots & \ddots & \vdots \\ h_{p1} & h_{p2} & \cdots & h_{pq} \end{bmatrix} \tag{8}$$

where H_{pq} is a complex number, which contains the amplitude and phase of the subcarrier on each antenna.

Figures 3 and 4 show the amplitude–frequency response and phase–frequency response of CFR under three receiving antennas. Analysis of Figure 4 shows that the phase information measured at the receiver is chaotic and we can not use it directly.

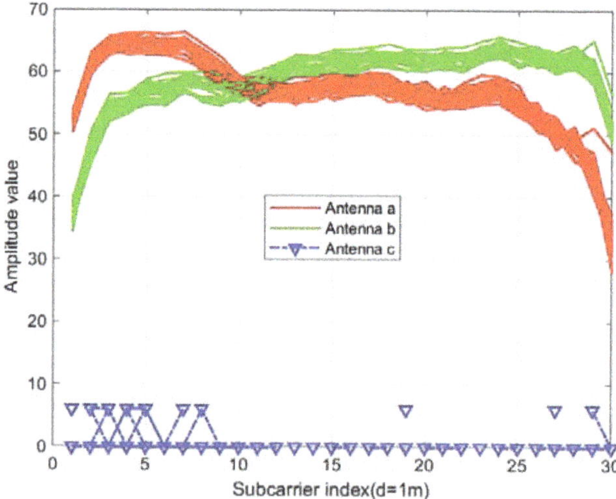

Figure 3. The amplitude frequency response of channel frequency response (CFR) under three receiving antennas.

Currently, WLAN protocols, especially 802.11n, use OFDM and MIMO as their standard technologies. The MIMO technology enables the diversity transmission and reception of signals. These two technologies play an important role in the formation of CSI data. FILA [8] utilizes effective CSI to propose an indoor ranging algorithm. The CSI-based ranging formula is as follows:

$$CSI_{eff} = \frac{1}{K}\sum_{k}\frac{f_k}{f_c} \times ||A||_k \tag{9}$$

where CSI_{eff} represents the effective CSI. K is the number of subcarriers. f_c represents the calculated center frequency, and $||A||_k$ represents the amplitude of the filtered CSI on the kth subcarrier. The propagation distance between the transmitter and receiver can be represented by effective CSI as follows:

$$d = \frac{1}{4\pi}[(\frac{c}{f_c \times |CSI_{eff}|})^2\sigma]^{\frac{1}{n}} \tag{10}$$

where d represents the distance between the transmitter and the receiver, c is the radio velocity, f_c is the central frequency of CSI, n is the path loss attenuation factor, and σ is the environmental factor.

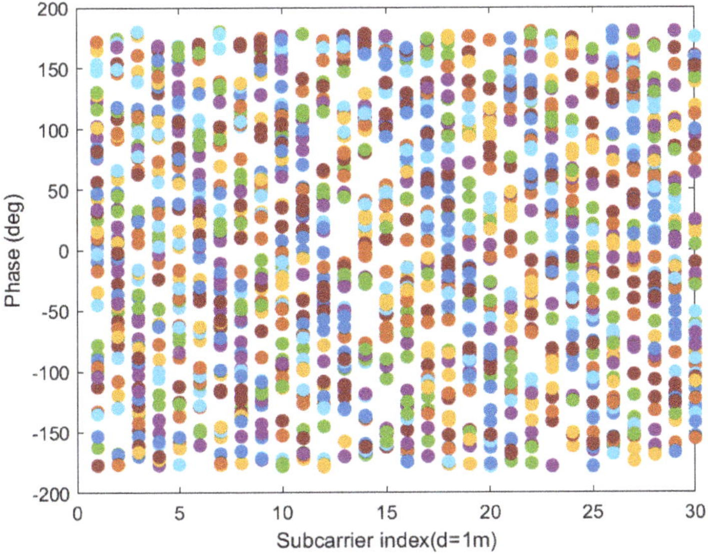

Figure 4. The phase frequency response of CFR under three receiving antennas.

The basic process is to collect the original CSI (CFR form existence) value of the ranging node at each reference node. According to the characteristics of the signal attenuation model, the specific relationship between CSI value and distance is obtained, and the mathematical model of CSI value and distance is established. Since CSI ranging is easily affected by the indoor environment, the indoor attenuation model based on the original CSI can not meet the requirements of the target object for ranging accuracy. At the same time, the mathematical model of CSI value and distance is not a linear equation. The function relationship between CSI and d can be obtained from Equation (11).

$$CSI_{eff} = \frac{c}{f_c}\sqrt{\frac{\sigma}{(4\pi d)^n}}. \tag{11}$$

2.4. The Extended Kalman Filtering Algorithm

The Kalman Filter (KF) is an optimal sequential state estimation algorithm for a linear state-space model in terms of the mean squared estimation error. When the state-space model is nonlinear, the KF is no longer applicable and nonlinear filters, especially the Extended Kalman Filter (EKF) [14–16], become necessary. For the EKF, the nonlinear Gaussian state-space model is linearized around the current best estimates of states and the KF algorithm is applied to the linearized models. The EKF algorithm is divided into two parts: time and observation updates. In the time update, the state estimation is predicted by the system model; for the observation update, the predicted state estimate is corrected using the observation based on the observation model. This corrected state estimation is called the filtered state estimate and it is then used for the next time update.

The nonlinear state-space model can be described as:

$$\begin{cases} X(k) = f[X(k-1)] + GV(k), \; V(k) \sim N(0, Q_k) \\ Z(k) = h[X(k)] + W(k), \; W(k) \sim N(0, R_k) \end{cases} \tag{12}$$

Here, $k = 1, 2, \cdots$ and $X(k)$ and $Z(k)$ are N and L dimensional random vectors called the state and the observation at k. $V(k)$ and $W(k)$ are independent zero-mean white Gaussian noises which covariance matrices are $Q(k)$ and $R(k)$, respectively. These noises are also assumed as independent from $X(0)$. G is the noise input matrix.

The predicted state estimate and the covariance matrix are:

$$\hat{X}(k|k-1) = f[\hat{X}(k-1)] \tag{13}$$

$$P(k|k-1) = A(k-1)P(k-1|k-1)A(k-1)^T + GQ_kG^T. \tag{14}$$

The filtered state estimate and the covariance matrix are:

$$K(k) = \frac{P(k|k-1)H(k)^T}{H(k)P(k|k-1)H(k)^T + R(k)} \tag{15}$$

$$\hat{X}(k) = \hat{X}(k|k-1) + K(k)[Z(k) - h[\hat{X}(k|k-1)]] \tag{16}$$

$$P(k) = [I - K(k)H(k)]P(k|k-1) \tag{17}$$

Among them, $A(k-1)$ and $H(k)$ are Jacobian matrices of non-linear systems, which are derived from partial derivatives of $f[X(k-1)]$ and $h[X(k)]$, respectively.

3. Indoor Localization Architecture and Methodology

In view of the disadvantages of traditional indoor positioning methods based on RSSI information or CSI information alone, we propose a novel positioning algorithm based on RSSI and CSI to make up for their shortcomings. This section focuses on the realization of the indoor ranging model. First, the structure of the model is introduced in its entirety. Then, the filtering fusion process of the extended Kalman filter is described in detail.

3.1. Indoor Localization Architecture

The basic flow of the algorithm starts from sending a ping command to the IP address of the AP at the receiving end to collect data packets. After collecting the data packets, the algorithm imports the data into MATLAB for processing and obtains the original RSSI value and original CSI value on each antenna of the MIMO system. The original RSSI value is averaged and the obtained CSI matrix is averaged with weights to obtain the effective CSI value. After that, the EKF is used to fuse the nonlinear RSSI ranging model and the nonlinear CSI ranging model to obtain the distance between the receiver and the transmitter. When using RSSI and CSI values to measure the access point or mobile terminal, the ranging model Equations (3) and (11) are usually used. The unknown parameters in the formula usually need to be trained to obtain. For different environments, the parameters will be different. In order to eliminate the influence of parameter changes on the ranking results, the propagation model formula is linearized. The architecture of the ranging algorithm is shown in Figure 5.

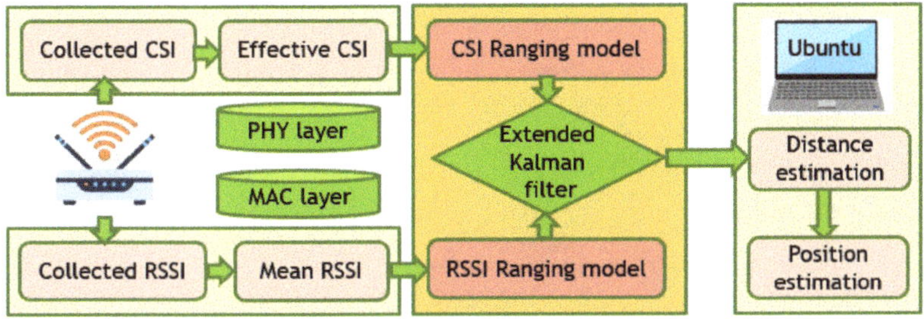

Figure 5. Indoor localization architecture.

3.2. The Extended Kalman Filtering Algorithm Model

This paper uses the EKF filtering [29,30] function to filter the obtained RSSI average value and effective CSI value, and then combines the distance-related mathematical ranging model to reduce the impact of noise on the system and improve the ranging accuracy. The observation equation comes from the RSSI-based ranging model and effective CSI-based ranging model, which is shown as follows.

$$h[X(k)] = \begin{bmatrix} RSSI_{antenna_1} - 10\sigma_1 log_{10}(dist(X_k, X_{ref_1})/d_0) \\ \vdots \\ RSSI_{antenna_L} - 10\sigma_L log_{10}(dist(X_k, X_{ref_L})/d_0) \\ CSI_{eff\,antenna_1} - \frac{c}{f_c}\sqrt{\frac{\sigma}{(4\pi dist(X_{(k)}, X_{(ref_1)}))^n}} \\ \vdots \\ CSI_{eff\,antenna_L} - \frac{c}{f_c}\sqrt{\frac{\sigma}{(4\pi dist(X_{(k)}, X_{(ref_L)}))^n}} \end{bmatrix} \tag{18}$$

where d is the distance between the transmitter and receiver in indoor environments. $RSSI_{antenna_L}$ and $CSI_{eff\,antenna_L}$ represent the RSSI value and the effective CSI at a distance $d_0 = 1$ m from the receiver reference point, respectively. $L = p * q$, where p and q represent the number of antennas at the transmitting node and the receiving node respectively. c is the radio velocity, f_c is the central frequency of CSI. n is the path loss attenuation factor, and σ is the environmental factor, and $dist()$ is the Euclidean distance function.

We can calculate the Jacobian matrix of the observation function:

$$H(k) = \begin{bmatrix} -\frac{10}{dist(X_k, X_{ref_1})} \\ \vdots \\ -\frac{10}{dist(X_k, X_{ref_L})} \\ -\frac{2\pi cn\sqrt{\sigma}}{f_c}(4\pi dist(X_{(k)}, X_{(ref_1)}))^{-\frac{n}{2}-1} \\ \vdots \\ -\frac{2\pi cn\sqrt{\sigma}}{f_c}(4\pi dist(X_{(k)}, X_{(ref_L)}))^{-\frac{n}{2}-1} \end{bmatrix} \tag{19}$$

The covariance matrix of the observation noise is:

$$R_k = \begin{bmatrix} R_{11} & 0 & \cdots & 0 \\ 0 & R_{22} & \cdots & 0 \\ \vdots & \vdots & \ddots & \vdots \\ 0 & 0 & 0 & R_{2L2L} \end{bmatrix} \tag{20}$$

The R_k value is affected by the indoor environment. When the indoor environment is more complicated, the value is larger. Since the system model is linear, $A(k) = I$ and $G = I$.

4. Experimental Environment and Results

4.1. Experimental Environment

This experiment was conducted at the corridor on the 3rd floor of the IT-1 building and the hall on the 2nd floor of the IT-1 building at Kyungpook National University (KNU) as shown in Figures 6 and 7. Figure 8 shows the schematic diagram of the experimental environment of the IT-1 building, which is a 2.5 m by 10 m corridor environment. Figure 9 shows the schematic diagram of the experimental environment of the IT-2 building, which is an 8 m by 10 m hall environment. In the collection of sample data, there are 10 reference points between AP and reference point (RP), each of which collects 1000 samples, and the interval between each reference point is 1 m.

Figure 6. Indoor experiment environment at the 3rd floor of IT-1 building.

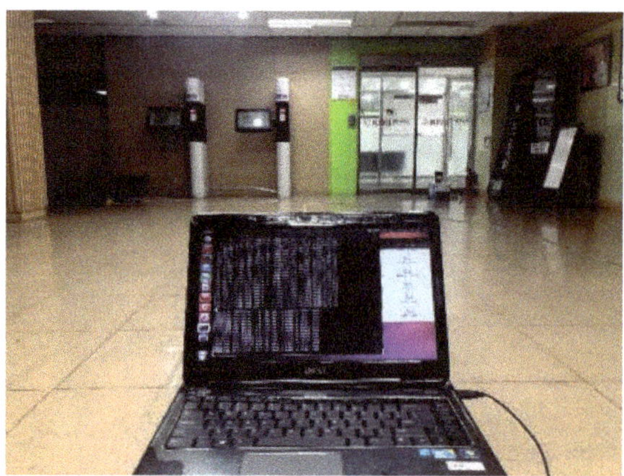

Figure 7. Indoor experiment environment at the 2rd floor of IT-2 building.

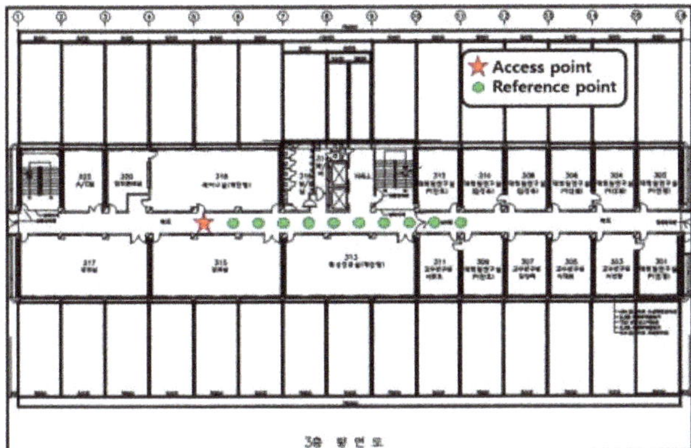

Figure 8. Schematic diagram of the indoor experiment environment at the 3rd floor of IT-1 building.

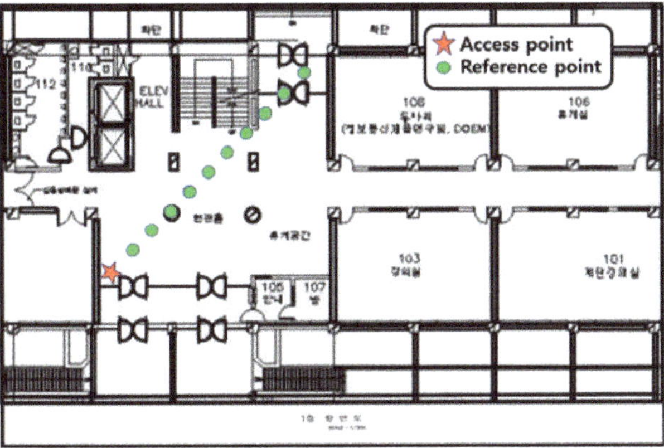

Figure 9. Schematic diagram of indoor experiment environment at the 2rd floor of IT-2 building.

In this paper, we use a laptop with a modified wireless network card and CSI tools [31,32] to obtain CSI data. The equipment required was: (1) a laptop with Intel 5300 for which the Ubuntu 14.04 operating system is installed with the customized kernel and wireless card drivers; (2) wireless AP based on 802.11n. We connect to the laptop through the Wi-Fi hotspot provided by the wireless AP. Since the 5GHz wireless signal has a strong ability to penetrate the wall, 5GHz was selected as the experimental frequency band. Currently, the commercial wireless network card supporting the IEEE 802.11n standard can provide the amplitude and phase difference information of different subcarriers in the form of the CSI matrix. The OFDM system of the IEEE 802.11n standard contains 56 subcarriers.

In this paper, the modified device can extract 30 subcarriers. Therefore, CSI is a numerical matrix of $3 \times 3 \times 30$. Figure 3 shows that the antenna c did not work as expected. Thus, we use a $2 \times 3 \times 30$ CSI numerical matrix for ranging. The collected CSI subcarrier information is an imaginary number that contains the amplitude and phase values of the subcarriers between different channels. Analysis of Figure 4 shows that the phase information measured at the receiving end is chaotic. There is no way to use it directly.

4.2. Experimental Results

4.2.1. Data Collection and Processing of RSSI and CSI

The data collection and processing module include the collection and processing of RSSI and CSI. In this paper, the transmitter AP was IPtime N3004 which had three transmitting antennas. In this paper, the transmit frequency of IPtime N3004 is 5 GHz. The receiver was a Dell Inspiron n4010 laptop. Each pair of transmitting and receiving antennas can obtain the RSSI value and CSI of 30 subcarriers. For each packet, the RSSI value on different antennas and CSI data matrix of 270-dimension can be obtained. The third antenna did not work normally and therefore we chose 180 ($2 \times 3 \times 30$) dimensional CSI data matrix for experiments. Figure 10 shows the changes in the RSSI values obtained on different antennas in two different indoor environments. Figure 11 shows the variation of the subcarrier values of CSI on antenna 2 in two different indoor environments.

According to Equation (9), we process the acquired subcarriers to obtain an effective CSI value, as shown in the Figure 12.

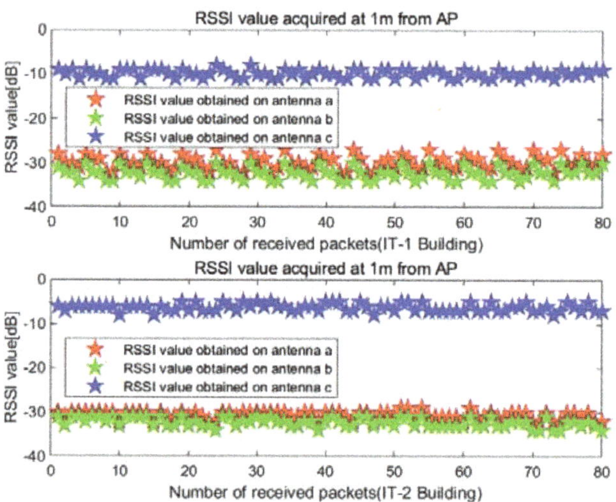

Figure 10. RSSI values obtained on different antennas in different indoor environments.

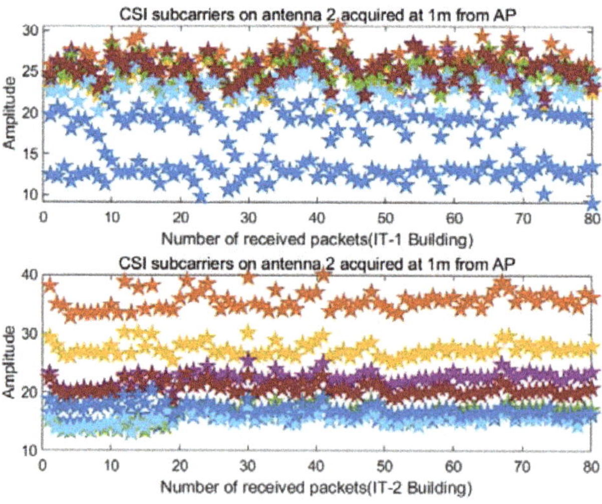

Figure 11. CSI subcarrier values obtained on antenna 2 in different indoor environments.

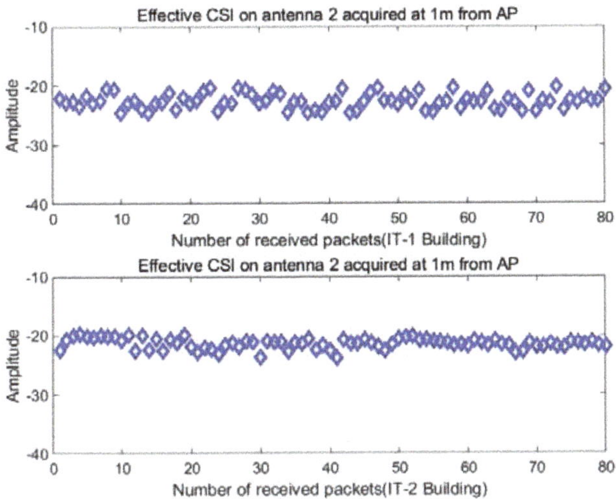

Figure 12. CSI effective values obtained on antenna 2 in different indoor environments.

4.2.2. Three-Dimensional Diagram of Nonlinear Ranging Model

The experiment was conducted in IT-1 building and IT-2 building of Kyungpook National University. Select the area covered by the Wi-Fi signal, select 10 test points in this area, and evenly distribute each test point for 100 times and 10 tests. Every reference point can get a total of 100,000 data values. We obtained the RSSI value and CSI value through MATLAB extraction and analysis. In the actual signal propagation environment, the RSSI value is susceptible to environmental influences. If the measured RSSI value is simply converted into a distance, the ranging error becomes larger. Therefore, when measuring distance, we should pre-process the RSSI value. We use the measured RSSI value multiple times to perform arithmetic average to improve the accuracy of the RSSI value. While CSI is relatively stable, we obtain the effective value of the CSI value by weighting the subcarriers.

The acquisition process of the ranging model can be divided into three stages; First, the RSSI and CSI of the transmitted signal of the transmitting node are obtained. Next, the receiving node uses the strength value of the received signal and the effective value of CSI to calculate the propagation loss of the ranging model and the parameters related to the indoor environment. Finally, the receiving node uses theoretical or empirical models to convert the signal propagation loss into the distance. The empirical model is a model formed by setting multiple reference points in the actual environment, and then measuring the relationship between signal strength and distance while at the same time combined with a mathematical fitting algorithm. The theoretical model selected in this paper is shown in Equations (3) and (11). Figure 13 shows the 3D graph of RSSI signal attenuation with distance in IT-1 building. Figure 14 shows the 3D graph of CSI effective value attenuation with distance in IT-1 building.

In practical applications, the complex and changeable indoor environment can cause multipath noise in RSSI and CSI measurements. It is assumed that this noise is additive white noise, which conforms to the Gaussian distribution. We need to choose the most suitable filter to eliminate the influence of noise on RSSI and CSI values, and obtain accurate distance measurement results.

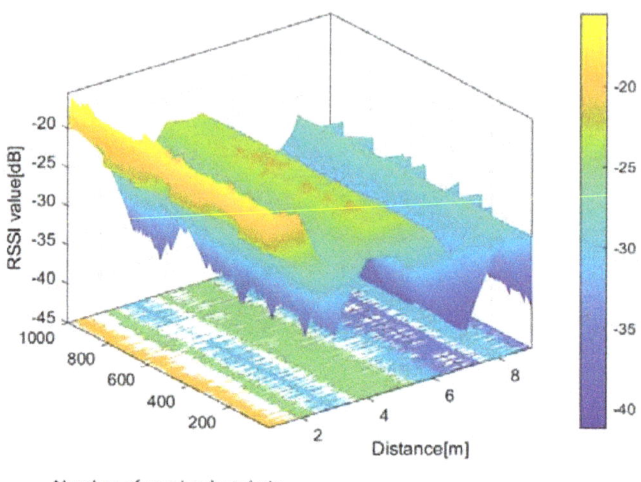

Figure 13. 3D graph of RSSI signal attenuation with distance in IT-1 building.

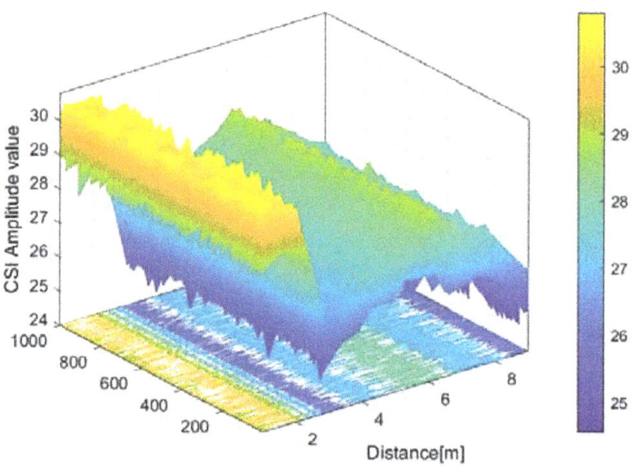

Figure 14. 3D graph of CSI effective value attenuation with distance in IT-1 building.

4.2.3. Distance Estimation Based on Extended Kalman Filtering

EKF has important theoretical significance and broad application prospects for the state estimation of nonlinear systems. Using the Extended Kalman Filter (EKF) method, a linearized standard Kalman filter model of the system can be established. The EKF algorithm has a simple structure and a certain accuracy, so it is widely used.

At each reference point, the EKF is used to filter the obtained RSSI value and CSI effective value. The processing procedure (the reference point distance is 1 meter) is shown in Figure 15. Figure 15 shows the comparison of the distance value and the real value after the extended Kalman filter.

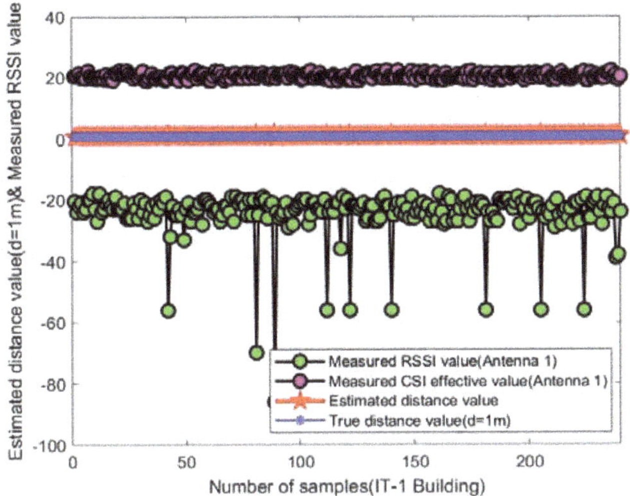

Figure 15. The accuracy of distance estimation (the proposed algorithm).

To analyze the improvement of the distance estimation accuracy of the proposed method more intuitively and clearly, we compare the obtained estimated distance with the real distance. Figure 16 shows the accuracy of the proposed algorithm for distance estimation. We used the proposed algorithm to filter all reference points within 10m in two indoor environments. We use MATLAB to plot the estimated distance error. Figure 17 shows a comparison of estimated distance error values between 1 m and 10 m.

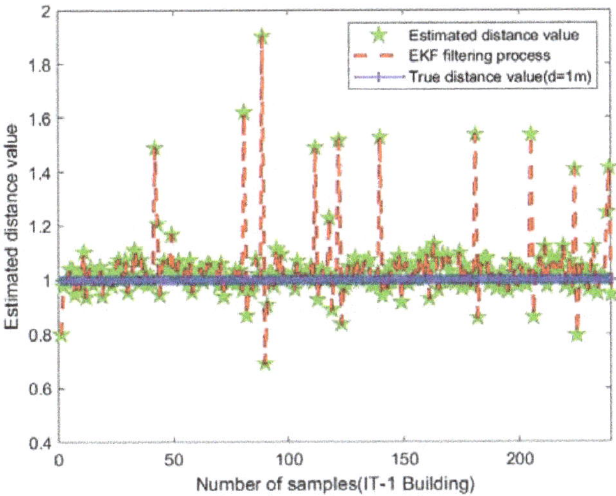

Figure 16. The comparison of the distance value and the real value after the extended Kalman filter (EKF).

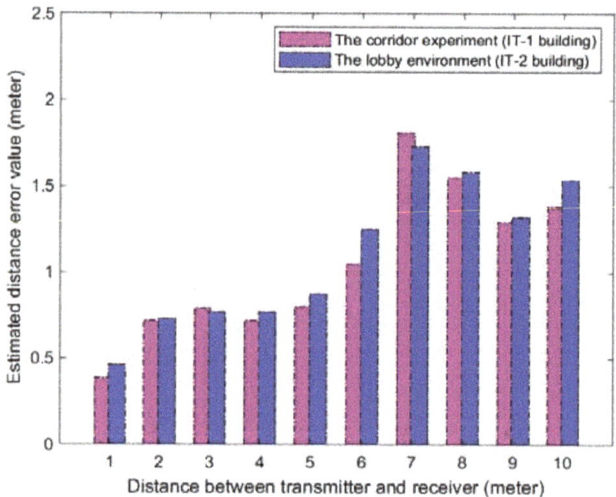

Figure 17. The comparison of the estimated distance error value between 1 m and 10 m.

4.2.4. Performance Evaluation

We evaluate the performance of our proposed ranging method under the two representative indoor environments. Figure 18 presents the Cumulative Distribution Function (CDF) of distance errors with the four ranging methods in the corridor experiment. In the IT-1 building experiment, the mean distance error is about 1.049 m and a standard deviation of 0.623 m for our proposed ranging method. In Figure 19, we plot the CDF of distance errors with the four ranging methods in the IT-2 building of KNU. In this propagation environment, our proposed method can achieve a 1.5 m distance error for over 70% of the reference points. The algorithm we proposed can obtain the best ranging results, and is the most accurate of the four ranging methods.

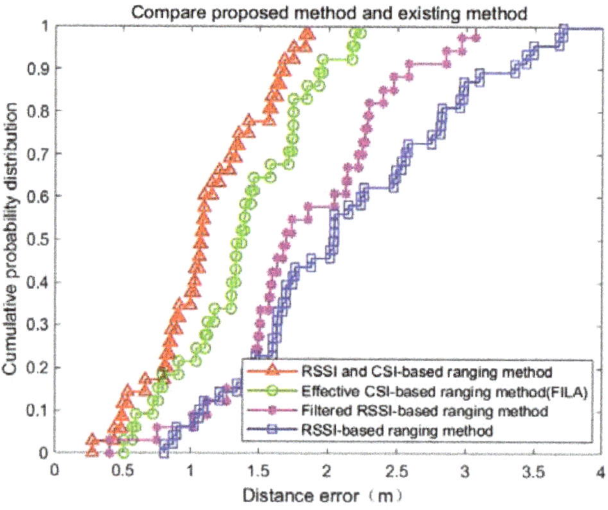

Figure 18. CDF of distance errors with the four ranging algorithms in IT-1 building.

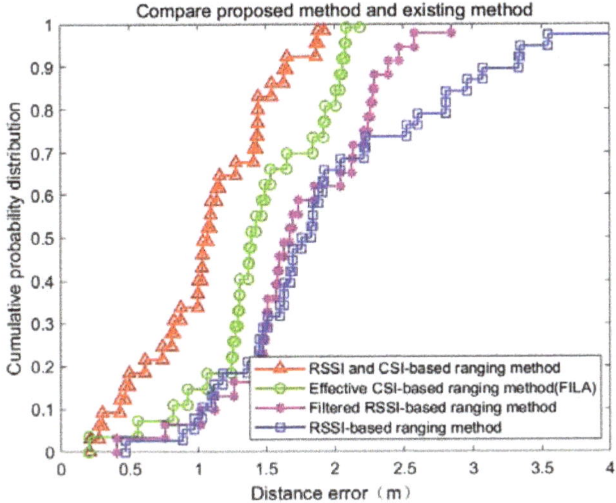

Figure 19. Cumulative Distribution Function (CDF) of distance errors with the four ranging algorithms in IT-2 building.

Figures 18 and 19 shows our proposed ranging method outperforms the existing ranging methods in both scenarios. The RSSI-based ranging method and the CSI-based ranging methods do not perform as well as our RSSI and CSI-based scheme. In the IT-1 building environment, the RSSI-based ranging method only manages to achieve 25% of the points that have an error under 1.5 m. The filtered RSSI-based ranging method only manages to achieve 36% of the points that have an error under 1.5 m. This is in contrast to that of the CSI-based ranging method which achieves 67% of the points that have an error under 1.5 m. These same methods achieved 30%, 37% and 60% of the points that have an error under 1.5 m in IT-2 building environment respectively.

We evaluate the performance of the proposed ranging algorithm based on RSSI and CSI using EKF through the statistical results of two indoor environments. Tables 3 and 4 list the average and standard deviation of the ranging error, respectively. Unlike FILA, the proposed algorithm utilizes RSSI and various CSI subcarriers acquired on each antenna. The algorithm can achieve higher accuracy. Two ranging methods using CSI (i.e., the proposed method and FILA) are superior to the two ranging methods based on RSSI.

Table 3. Comparison of four indoor ranging algorithms (IT-1 building).

Indoor Ranging Algorithm	Mean Ranging Error	Standard Deviation
RSSI-based algorithm	2.041 m	1.214 m
Filtered RSSI-based algorithm	1.696 m	0.908 m
CSI-based algorithm (FILA)	1.381 m	0.577 m
RSSI and CSI-based algorithm (Proposed algorithm)	1.049 m	0.623 m

Table 4. Comparison of four indoor ranging algorithms (IT-2 building).

Indoor Ranging Algorithm	Mean Ranging Error	Standard Deviation
RSSI-based algorithm	1.798 m	1.279 m
Filtered RSSI-based algorithm	1.669 m	1.243 m
CSI-based algorithm (FILA)	1.395 m	0.972 m
RSSI and CSI-based algorithm (Proposed algorithm)	1.103 m	0.612 m

Different from the RSSI, the filtered RSSI and the CSI-based ranging methods, our RSSI and CSI-based ranging method using EKF suppresses rich multipath effects in an indoor environment. It also achieves significantly higher accuracy. This result indicates that our proposed method is able to effectively estimate and compensate for the ranging result.

5. Conclusions and Future Work

The main contributions of this article are as follows: First, a novel ranging scheme based on CSI and RSSI using EKF is proposed. Second, an indoor ranging algorithm based on RSSI and CSI in the environment of high-load AP is proposed, which solves the ranging accuracy problem under high load AP. This paper demonstrates the feasibility of this method in a high-load AP environment and proves the accuracy of the algorithm through related experiments. We also compared it with several other related algorithms. Third, we use the DELL Inspiron n4010 laptop and commercial 802.11n wireless network card driver to build the experimental environment. Compared with existing other ranging methods, the proposed method does not require additional equipment and can be applied to a wider range of indoor situations. The experimental results show that the proposed ranging method effectively improves the accuracy of the indoor environment. Of course, the current system is still insufficient and we need to continue improving it in our future work. We could consider the participation of multiple APs in the experiment. Additionally, as the indoor environments are complex and vary with time due to people, we will analyze impacts brought about by these factors.

Author Contributions: J.W. proposed the idea and implementation methodology, performed all experiments and wrote the paper, verified the experiment process and results; J.G.P. reviewed and edited paper, and supervised the work. All authors have read and agreed to the published version of the manuscript.

Funding: This study is supported by the BK21 Plus project funded by the Ministry of Education, Korea (21A20131600011). This study is supported by Smart City R&D project of the Korea Agency for Infrastructure Technology Advancement (KAIA) grant funded by the Ministry of Land, Infrastructure and Transport (Grant 18NSPS-B149843-01).

Conflicts of Interest: The authors declare no conflict of interest.

References

1. Rycroft, M.J. Understanding GPS. Principles and applications.Second Edition *J. Atmos. Sol.-Terr. Phys.* **1997**, *59*, 598–599. [CrossRef]
2. El-rabbany, A. *Introduction to GPS: The Global Position System*; Artech House: New York, NY, USA, 2006.
3. Crow, B.P.; Widjaja, I.; Kim, J.G.; Sakai, P.T. IEEE 802.11 wireless local area networks. *IEEE Commun. Mag.* **1997**, *35*, 116–126. [CrossRef]
4. Chen, H.X.; Hu, B.J.; Zheng, L.L.; Wei, Z.H. An Accurate AoA Estimation Approach for Indoor Localization Using Commodity Wi-Fi Devices. In Proceedings of the 2018 IEEE International Conference on Signal Processing, Communications and Computing (ICSPCC 2018), Qingdao, China, 14–17 September 2018.
5. Güvenç, I.; Chong, C.C. A survey on TOA based wireless localization and NLOS mitigation techniques. *IEEE Commun. Surv. Tutor.* **2009**, *11*, 107–124. [CrossRef]
6. Makki, A.; Siddig, A.; Saad, M.; Cavallaro, J.R.; Bleakley, C.J. Indoor Localization Using 802.11 Time Differences of Arrival. *IEEE Trans. Instrum. Meas.* **2016**, *65*, 614–623. [CrossRef]
7. Xu, J.; Liu, W.; Lang, F.; Zhang, Y.; Wang, C. Distance Measurement Model Based on RSSI in WSN. *Wirel. Sens. Netw.* **2010**, *2*, 606.
8. Wu, K.; Xiao, J.; Yi, Y.; Gao, M.; Ni, L.M. FILA: Fine-grained indoor localization. In Proceedings of the IEEE INFOCOM, Orlando, FL, USA, 25–30 March 2012.
9. Wang, K.; Yu, X.; Xiong, Q.; Zhu, Q.; Lu, W.; Huang, Y.; Zhao, L. Learning to improve WLAN indoor positioning accuracy based on DBSCAN-KRF Algorithm from RSS Fingerprint Data. *IEEE Access* **2019**, *7*, 72308–72315. [CrossRef]

10. Xie, C.; Guan, W.; Wu, Y.; Fang, L.; Cai, Y. The LED-ID Detection and Recognition Method Based on Visible Light Positioning Using Proximity Method. *IEEE Photonics J.* **2018**, *10*, 1–16. [CrossRef]
11. Destiarti, A.R.; Kristalina, P.; Sudarsono, A. Mobile cooperative tracking with RSSI ranging in EKF algorithm for indoor wireless sensor network. In Proceedings of the 2016 International Conference on Knowledge Creation and Intelligent Computing (KCIC 2016), Manado, Indonesia, 15–17 November 2016.
12. Xiao, J.; Wu, K.; Yi, Y.; Ni, L.M. FIFS: Fine-grained indoor fingerprinting system. In Proceedings of the 2012 21st International Conference on Computer Communications and Networks (ICCCN 2012), Munich, Germany, 30 July–2 August 2012.
13. Wang, X.; Gao, L.; Mao, S.; Pandey, S. DeepFi: Deep learning for indoor fingerprinting using channel state information. In Proceedings of the 2015 IEEE Wireless Communications and Networking Conference (WCNC 2015), New Orleans, LA, USA, 9–12 March 2015.
14. Chapre, Y.; Ignjatovic, A.; Seneviratne, A.; Jha, S. CSI-MIMO: Indoor Wi-Fi fingerprinting system. In Proceedings of the Conference on Local Computer Networks (LCN), Edmonton, AB, Canada, 8–11 September 2014.
15. Hsieh, C.H.; Chen, J.Y.; Nien, B.H. Deep Learning-Based Indoor Localization Using Received Signal Strength and Channel State Information. *IEEE Access*, **2019**, *7*, 33256–33267. [CrossRef]
16. Simon, D. Kalman filtering with state constraints: A survey of linear and nonlinear algorithms. *IET Control Theory Appl.* **2010**, *4*, 1303–1318. [CrossRef]
17. Zhang, H.; Du, H.; Ye, Q.; Liu, C. Utilizing CSI and RSSI to Achieve High-Precision Outdoor Positioning: A Deep Learning Approach. In Proceedings of the IEEE International Conference on Communications, Shanghai, China, 20–24 May 2019.
18. Zhang, L.; Ding, E.; Hu, Y.; Liu, Y. A novel CSI-based fingerprinting for localization with a single AP. *EURASIP J. Wirel. Commun. Netw.* **2019**, *2019*, 51. [CrossRef]
19. Yang, Z.; Zhou, Z.; Liu, Y. From RSSI to CSI: Indoor localization via channel response. *ACM Comput. Surv.* **2013**, *46*, 1–32.
20. Wang, X.; Gao, L.; Mao, S.; Pandey, S. CSI-Based Fingerprinting for Indoor Localization: A Deep Learning Approach. *IEEE Trans. Veh. Technol.* **2017**, *66*, 763–776. [CrossRef]
21. Zhu, X.; Feng, Y. RSSI-based Algorithm for Indoor Localization. *Commun. Netw.* **2013**, *5*, 37. [CrossRef]
22. Pathak, O.; Palaskar, P.; Palkar, R.; Tawari, M. Wi-Fi Indoor Positioning System Based on RSSI Measurements from Wi-Fi Access Points—A Tri-lateration Approach. *Int. J. Sci. Eng. Res.* **2014**, *5*, 1234–1238.
23. Li, G.; Geng, E.; Ye, Z.; Xu, Y.; Lin, J.; Pang, Y. Indoor positioning algorithm based on the improved rssi distance model. *Sensors* **2018**, *18*, 2820. [CrossRef] [PubMed]
24. Zanca, G.; Zorzi, F.; Zanella, A.; Zorzi, M. Experimental comparison of RSSI-based localization algorithms for indoor wireless sensor networks. In *Proceedings of the Workshop on Real-World Wireless Sensor Networks (REALWSN 2008)*; ACM: Glasgow, Scotland, 2008.
25. Adachi, T. IEEE802.11n. *Kyokai Joho Imeji Zasshi/J. Inst. Image Inf. Telev. Eng.* **2011**, *65*, 950–953. [CrossRef]
26. Letor, N.; Torfs, W.; Blondia, C. Multimedia multicast performance analysis for 802.11n network cards. In Proceedings of the IFIP Wireless Days, Dublin, Ireland, 21–23 November 2012.
27. Lin, M.F.; Tzu, J.Y.; Lin, L.; Lee, H.M. The IEEE802.11n capability analysis model based on mobile networking architecture. In Proceedings of the IEEE International Conference on Systems, Man and Cybernetics, San Antonio, TX, USA, 11–14 October 2009.
28. Wang, X.; Gao, L.; Mao, S. CSI Phase Fingerprinting for Indoor Localization with a Deep Learning Approach. *IEEE Internet Things J.* **2016**, *3*, 1113–1123.
29. Lee, S.H.; Lim, I.K.; Lee, J.K. Method for Improving Indoor Positioning Accuracy Using Extended Kalman Filter. *Mob. Inf. Syst.* **2016**, *2016*. [CrossRef]
30. Ben Kilani, M.; Raymond, A.J.; Gagnon, F.; Gagnon, G.; Lavoie, P. RSSI-based indoor tracking using the extended Kalman filter and circularly polarized antennas. In Proceedings of the 2014 11th Workshop on Positioning, Navigation and Communication (WPNC 2014), Dresden, Germany, 12–13 March 2014.

31. Halperin, D.; Hu, W.; Sheth, A.; Wetherall, D. Tool release: Gathering 802.11n traces with channel state information. *Comput. Commun. Rev.* **2011**, *41*, 53. [CrossRef]
32. Liu, X.L.; Hu, W.; Pu, Q.; Wu, F.; Zhang, Y. ParCast: Soft video delivery in MIMO-OFDM WLANs. In Proceedings of the Annual International Conference on Mobile Computing and Networking (MOBICOM), Istanbul, Turkey, 22–26 August 2012.

Article

Finite Memory Structure Filtering and Smoothing for Target Tracking in Wireless Network Environments

Pyung Soo Kim

Department of Electronic Engineering, Korea Polytechnic University, Siheung-si, Gyeonggi-do 15073, Korea; pskim@kpu.ac.kr; Tel.: +82-31-8041-0489

Received: 26 June 2019; Accepted: 17 July 2019; Published: 18 July 2019

Abstract: In this paper, a state estimation problem is considered for a target tracking scheme in wireless network environments. Firstly, a unified algorithm of finite memory structure (FMS) filtering and smoothing is proposed for a discrete-time state-space model. As shown in the terminology *unified*, both FMS filter and smoother are derived by solving one optimization problem directly with incorporation of the unbiasedness constraint. Hence, the unified algorithm provides simultaneously the current state estimate as well as the lagged state estimate using only finite measurements and inputs on the most recent window. The proposed unified algorithm of FMS filtering and smoothing shows that there are some unique properties such as unbiasedness, deadbeat, time-invariance and intrinsic robustness, which cannot be obtained by the recursive infinite memory structure (IMS) filtering such as Kalman filter. The on-line computational complexity of the proposed unified algorithm is discussed. Secondly, as an application of the proposed unified algorithm, a target tracking scheme in wireless network environments is considered via computer simulations for moving target's accelerations of various shapes. The proposed unified algorithm-based target tracking scheme provides estimates for position as well as acceleration of moving target in real time, while eliminating unwanted noise effects and maintaining desired moving positions. Due to intrinsic robustness and deadbeat properties, the proposed unified algorithm-based scheme can outperform the existing IMS filtering-based scheme when acceleration suddenly changes.

Keywords: filter; finite memory structure; infinite memory structure; smoother; target tracking

---

1. Introduction

The Kalman filter and smoother [1–5] have been the most commonly fundamental tools for filtering and smoothing in statistical time series analysis. Thus, the Kalman filter and smoother have been a standard choice and a beautiful reference for the state estimation and applied successfully for diverse engineering problems. However, due to their recursive formulations and infinite memory structure (IMS), the Kalman filter and smoother may exhibit performance degradation and even divergence in severe cases for mismodeling and temporary uncertainties.

Therefore, as an alternative to the Kalman filter, the finite memory structure (FMS) filter has been designed for state estimation and shown inherently to have BIBO stability and be more robust against temporary uncertainties [6–9]. This FMS filter has been applied successfully in various engineering fields [10–12]. Likewise the recursive Kalman filter in [1–5], the FMS filter in [6–9] is a causal filter that provides state estimates at given times based only on the relative past. Hence, as a noncausal filter, the FMS smoother has been also designed for estimation problems when there is a fixed delay d between the original state and the estimated state[13–16]. The FMS smoother has been shown to be much less computationally complex and more robust against temporary uncertainties than the recursive Kalman smoother. Of course, noncausal filters such as the recursive Kalman smoother and the FMS smoother naturally yield more accurate estimates when there is a fixed delay d between the original state and the estimated state. However, the FMS smoother could not be better than the FMS filter when there is no fixed delay between the original state and the estimated state. That is, the FMS filter should be applied when there is no fixed delay between signal generation and signal estimation.

Meanwhile, a target tracking problem in indoor positioning systems for wireless network environments has been an interesting research problem and thus applied successfully for locating and tracking people within a building and products stored within a warehouse using wireless sensors and mobile devices [17–19]. Due to a variety of indoor wireless channel characteristics, an accurate tracking estimation of a moving target is required for indoor positioning systems. However, a couple of noises such as system noise and measurement noise can contaminate the estimated position and thus can cause deterioration of target tracking performance. Hence, these noises must be filtered for tracking a true path of a moving target. To provide the most accurate position and velocity of a moving target on the real-time indoor positioning in wireless network environments, some research adopted the Kalman filter to reduce the effect of noises [20–22]. Actually, past measurement data might have a little information about the target's current motion since the moving target changes its motion dynamics. In other words, the valid duration of the moving target's dynamics might be limited to the recently finite time interval. In addition, estimation filters are typically designed with the state-space model called the constant velocity motion model. The constant velocity motion model assumes that targets move with constant velocity, that is, zero-acceleration within a short sampling time. However, in real situations, moving targets maneuver and change velocity and thus move temporarily with nonzero acceleration. Although this can be a temporary uncertainty and thus effects typically occur over a short time interval, the state estimation filter should be robust to diminish the effects of the temporary uncertainty. Therefore, the FMS filter and smoother might be very appropriate for target tracking approaches. The FMS filter was applied successfully for the target tracking in wireless network environments [23,24] while the FMS smoother has not been addressed.

In this paper, a state estimation problem is considered for a target tracking scheme in wireless communication environments. Firstly, an alternative FMS filtering and smoothing algorithm is derived for both cases with delay and without delay between the original state and the estimated state. This algorithm is thus called the unified algorithm of FMS filter and smoother. As shown in the terminology *unified* which can also mean *united*, *combined*, *integrated*, both FMS filter and smoother are derived by solving one optimization problem directly with incorporation of the unbiasedness constraint using only finite measurements and inputs on the most recent window $[i-M,i]$. Thus, the proposed unified algorithm provides simultaneously state estimates at the current time i as well as at the lagged time $i-d$, given finite measurements and inputs on the most recent window $[i-M,i]$ where $i-M$ is the window initial time and i is the current time. The proposed unified algorithm of FMS filtering and smoothing shows that there are some unique properties such as unbiasedness, deadbeat, time-invariance and intrinsic robustness while the recursive IMS filter such as Kalman filter does not have these properties. The on-line computational complexity of the proposed unified algorithm is discussed and compared with the IMS filter. Secondly, a target tracking in wireless network environments is considered as an application of the proposed unified algorithm of FMS filtering and smoothing. Through extensive computer simulations for moving target's accelerations of various shapes, the proposed unified algorithm-based target tracking scheme is shown to provide estimates for position as well as acceleration of moving target in real time, while eliminating unwanted noise effects and maintaining desired moving positions. In particular, due to intrinsic robustness and deadbeat properties, the performance of the proposed unified algorithm-based scheme can be shown to be better than the existing IMS filtering-based scheme for suddenly changing acceleration.

This paper has the following structure. In Section 2, a unified algorithm of FMS filtering and smoothing is proposed. In Section 3, a target tracking scheme using the proposed unified algorithm is considered via extensive computer simulations. Then, concluding remarks are given in Section 4.

2. Unified Algorithm of Finite Memory Structure Filter and Smoother

Consider the following linear discrete-time state-space model:

$$\begin{aligned} x_{i+1} &= Ax_i + Bu_i + Gw_i, \\ z_i &= Cx_i + v_i, \end{aligned} \tag{1}$$

where $x_i \in \Re^n$ is the system state variable, $u_i \in \Re^p$ is the control input, $z_i \in \Re^q$ is the sensor measurement output. A couple of noises, the system noise $w_i \in \Re^p$ and the measurement noise $v_i \in \Re^q$, are zero-mean white Gaussian. These noises are mutually uncorrelated and their covariances are denoted by positive definite matrices Q and R, respectively.

At the current time i, finite measurements Z_i and inputs U_i on the most recent window $[i-M, i]$ can be expressed by the regression form with the current state x_i as follows:

$$Z_i - \Xi U_i = \Gamma x_i + \Lambda W_i + V_i, \tag{2}$$

where

$$Z_i \triangleq \begin{bmatrix} z_{i-M} \\ z_{i-M+1} \\ \vdots \\ z_{i-2} \\ z_{i-1} \end{bmatrix}, U_i \triangleq \begin{bmatrix} u_{i-M} \\ u_{i-M+1} \\ \vdots \\ u_{i-2} \\ u_{i-1} \end{bmatrix}, W_i \triangleq \begin{bmatrix} w_{i-M} \\ w_{i-M+1} \\ \vdots \\ w_{i-2} \\ w_{i-1} \end{bmatrix}, V_i \triangleq \begin{bmatrix} v_{i-M} \\ v_{i-M+1} \\ \vdots \\ v_{i-2} \\ v_{i-1} \end{bmatrix}, \tag{3}$$

and matrices Γ, Ξ, Λ are defined by

$$\begin{aligned} \Gamma &\triangleq \begin{bmatrix} CA^{-M} \\ CA^{-M+1} \\ \vdots \\ CA^{-2} \\ CA^{-1} \end{bmatrix}, \ \Xi \triangleq - \begin{bmatrix} CA^{-1}B & CA^{-2}B & \cdots & CA^{-M}B \\ 0 & CA^{-1}B & \cdots & CA^{-M+1}B \\ \vdots & \vdots & \cdots & \vdots \\ 0 & 0 & \cdots & CA^{-2}B \\ 0 & 0 & \cdots & CA^{-1}B \end{bmatrix}, \\ \Lambda &\triangleq - \begin{bmatrix} CA^{-1}G & CA^{-2}G & \cdots & CA^{-M}G \\ 0 & CA^{-1}G & \cdots & CA^{-M+1}G \\ \vdots & \vdots & \cdots & \vdots \\ 0 & 0 & \cdots & CA^{-2}G \\ 0 & 0 & \cdots & CA^{-1}G \end{bmatrix}. \end{aligned} \tag{4}$$

Meanwhile, at the lagged time $i-d$, the lagged state x_{i-d} can be represented by

$$x_{i-d} = A^{-d}x_i - \tilde{\Xi}U_i - \tilde{\Lambda}W_i, \tag{5}$$

where matrices $\tilde{\Xi}$ and $\tilde{\Lambda}$ is defined by

$$\tilde{\Xi} \triangleq -\Big[\overbrace{0\,0\,\cdots\,0}^{M-d}\; A^{-1}B\; A^{-2}B\; \cdots\; A^{-d}B\Big],$$
$$\tilde{\Lambda} \triangleq -\Big[\overbrace{0\,0\,\cdots\,0}^{M-d}\; A^{-1}G\; A^{-2}G\; \cdots\; A^{-d}G\Big].$$

From (2) and (5), the current state x_i and the lagged state x_{i-d} can be represented by the regression form with the augmented state X_i as follows:

$$\bar{Z}_i - \bar{\Xi}U_i = \bar{\Gamma}X_i + \bar{\Lambda}W_i + \bar{V}_i, \tag{6}$$

where

$$X_i \triangleq \begin{bmatrix} x_i \\ x_{i-d} \end{bmatrix}, \; \bar{Z}_i \triangleq \begin{bmatrix} Z_i \\ 0 \end{bmatrix}, \; \bar{V}_i \triangleq \begin{bmatrix} V_i \\ 0 \end{bmatrix},$$
$$\bar{\Gamma} \triangleq \begin{bmatrix} \Gamma & 0 \\ A^{-d} & -1 \end{bmatrix}, \; \bar{\Xi} \triangleq \begin{bmatrix} \Xi \\ \tilde{\Xi} \end{bmatrix} \; \bar{\Lambda} \triangleq \begin{bmatrix} \Lambda \\ \tilde{\Lambda} \end{bmatrix}.$$

The noise term $\bar{\Lambda}W_i + \bar{V}_i$ in (6) is zero-mean white Gaussian and its covariance given by the positive definite matrix Π as follows:

$$\Pi \triangleq \bar{\Lambda}\bar{Q}\bar{\Lambda}^T + \begin{bmatrix} \bar{R} & 0 \\ 0 & 0 \end{bmatrix} = \begin{bmatrix} \Lambda\bar{Q}\Lambda^T + \bar{R} & \Lambda\bar{Q}\tilde{\Lambda}^T \\ \tilde{\Lambda}\bar{Q}\Lambda^T & \tilde{\Lambda}\bar{Q}\tilde{\Lambda}^T \end{bmatrix}, \tag{7}$$

where

$$\bar{Q} \triangleq \Big[\mathrm{diag}(\overbrace{Q\,Q\,\cdots\,Q}^{M})\Big], \; \bar{R} \triangleq \Big[\mathrm{diag}(\overbrace{R\,R\,\cdots\,R}^{M})\Big]. \tag{8}$$

Based on the approach of *best linear unbiased estimation* in [25], the unified algorithm of FMS filter and smoother is derived using the following series of equations. The unified algorithm of FMS filter and smoother $\hat{X}_i$ is assumed to be obtained from a matrix form using only finite measurements $\bar{Z}_i$ and inputs U_i on the most recent window $[i-M, i]$ as follows:

$$\hat{X}_i = \begin{bmatrix} \hat{x}_i \\ \hat{x}_{i-d} \end{bmatrix} \triangleq \mathcal{H}\big(\bar{Z}_i - \bar{\Xi}U_i\big), \tag{9}$$

where $\mathcal{H}$ is the unified gain matrix. When taking the expectation both sides of (9) as follows:

$$\mathbf{E}\big[\hat{X}_i\big] = \mathbf{E}\begin{bmatrix} \hat{x}_i \\ \hat{x}_{i-d} \end{bmatrix} = \mathbf{E}\Big[\mathcal{H}\big(\bar{Z}_i - \bar{\Xi}U_i\big)\Big] = \mathcal{H}\bar{\Gamma}\mathbf{E}\big[X_i\big], \tag{10}$$

$\hat{X}_i$ is unbiased, i.e., $\mathbf{E}\big[\hat{X}_i\big] = \mathbf{E}\big[X_i\big]$, with the following constraint:

$$\mathcal{H}\bar{\Gamma} = I. \tag{11}$$

Thus, the constraint (11) can be called the *unbiasedness constraint* for the unified algorithm of FMS filter and smoother $\hat{X}_i$.

Subject to the unbiasedness constraint (11), the objective is now to obtain the gain matrix $\mathcal{H}_*$ in order that the error of $\hat{X}_i$ has a minimum variance as follows:

$$\mathcal{H}_* = \arg\min_{\mathcal{H}} \mathbf{E}\Big[(X_i - \hat{X}_i)^T (X_i - \hat{X}_i)\Big]. \tag{12}$$

Then, by solving the optimization problem directly with incorporation of the unbiasedness constraint, the unified algorithm of FMS filter and smoother $\hat{X}_i$ is obtained by the solution of (12) as follows:

$$\hat{X}_i = \begin{bmatrix} \hat{x}_i \\ \hat{x}_{i-d} \end{bmatrix} = \mathcal{H}_*\Big(\bar{Z}_i - \bar{\Xi} U_i\Big), \tag{13}$$

where

$$\mathcal{H}_* = (\bar{\Gamma}^T \Pi^{-1} \bar{\Gamma})^{-1} \bar{\Gamma}^T \Pi^{-1}. \tag{14}$$

The proposed unified algorithm (13) of FMS filtering and smoothing shows that there are some unique properties such as unbiasedness, deadbeat, time-invariance and intrinsic robustness. The gain matrix $(\bar{\Gamma}^T \Pi^{-1} \bar{\Gamma})^{-1} \bar{\Gamma}^T \Pi^{-1}$ in (14) of the unified algorithm requires only once computation on the interval $[0, M]$ and then is used for all windows, which means the time-invariance property. As shown in (10) and (11), the unified algorithm has the unbiasedness property *by design*. In addition, the proposed unified algorithm $\hat{X}_i$ in (13) for both the current state estimate $\hat{x}_i$ and the lagged state estimate $\hat{x}_{i-d}$ has the deadbeat property as following theorem.

Theorem 1. *When $M \geq n$, the proposed unified algorithm of FMS filter and smoother $\hat{X}_i$ on the most recent window $[i - M, i]$ is exact for noise-free systems.*

Proof of Theorem 1. When there are no noises on the most recent window $[i - M, i]$ for the discrete-time state-space model (1), $\bar{Z}_i - \bar{\Xi} U_i$ (6) is represented by

$$\bar{Z}_i - \bar{\Xi} U_i = \bar{\Gamma} X_i = \bar{\Gamma} \begin{bmatrix} x_i \\ x_{i-d} \end{bmatrix}.$$

Hence, the following is satisfied:

$$\begin{aligned} \hat{X}_i = \begin{bmatrix} \hat{x}_i \\ \hat{x}_{i-d} \end{bmatrix} &= (\bar{\Gamma}^T \Pi^{-1} \bar{\Gamma})^{-1} \bar{\Gamma}^T \Pi^{-1} \Big(\bar{Z}_i - \bar{\Xi} U_i\Big) \\ &= (\bar{\Gamma}^T \Pi^{-1} \bar{\Gamma})^{-1} \bar{\Gamma}^T \Pi^{-1} \bar{\Gamma} \begin{bmatrix} x_i \\ x_{i-d} \end{bmatrix} \\ &= \begin{bmatrix} x_i \\ x_{i-d} \end{bmatrix}. \end{aligned}$$

This completes the proof of the deadbeat property. □

The deadbeat property means that the unified algorithm of FMS filter and smoother $\hat{X}_i$ in (13) tracks exactly the current state x_i and the lagged state x_{i-d} at every time for noise-free systems although the proposed unified algorithm has been designed assuming a couple noises such as w_i and v_i in the discrete-time state-space model (1). Since the deadbeat property indicates finite convergence time and fast tracking ability, of the proposed unified algorithm of FMS filter and smoother. Thus, it can be expected that the proposed unified algorithm of FMS filter and smoother might be appropriate for a variety of applications requiring fast tracking capability such as fault detection and diagnosis for dynamic process systems as well as maneuver detection and target tracking for indoor positioning systems. Moreover, in contrast to the recursive IMS filter such as the Kalman filter, the proposed unified algorithm can have intrinsic robustness due to its finite memory structure. This means that the proposed unified algorithm can be more robust against round-off errors, mismodeling and temporary uncertainties.

The on-line computational complexity of the proposed unified algorithm is discussed. To simply compare the computational complexity of the proposed unified algorithm and the IMS filter, the measurement z_i is assumed as a scalar-valued (i.e., $q = 1$). In the case of the IMS filter such as the fixed-lag Kalman smoothing filter, the coefficient values for fixed-lag Kalman filtering and fixed-lag smoothing algorithms must be always computed for d iterations by on-line computing before inputting any data to the fixed-lag Kalman smoothing filter algorithm stage. Thus, the IMS filter such as the fixed-lag Kalman smoothing filter with nth dimension can be done with $4n^3 + 8n^2 + 6n + 1$ flops as shown in [26]. On the other hand, as mentioned before, the gain matrix (14) of the unified algorithm requires only once computation on the interval $[0, M]$ and then is used for all windows. Thus, FMS filter and smoother with nth dimension in the proposed unified algorithm can be done respectively with $n \times M$ flops as shown in Table 1. Thus, as the dimension grows, the on-line computational complexity of the proposed unified algorithm is shown to be much less than that of the fixed-lag Kalman smoothing filter. However, the proposed unified algorithm requires the memory with size of M to store finite measurements on the most recent window $[i - M, i]$ and the memory shift with $M - 1$ operations. On the other hand, the fixed-lag Kalman smoothing filter requires to store only intermediate estimates with size of d. Since $M > d$, the fixed-lag Kalman smoothing filter can be better than the proposed unified algorithm in terms of the memory management.

Table 1. Comparison of on-line computation and memory size.

	FMS Filter and Smoother	Fixed-Lag Kalman Smoothing Filter
On-line computation (flops)	$n \times M$ (respectively)	$4n^3 + 8n^2 + 6n + 1$
Memory shift	$M - 1$	*None*
Memory size	M	d

3. Application for Target Tracking Problem

As an application of the proposed unified algorithm of FMS filtering and smoothing, a target tracking problem is considered through extensive computer simulations. In these days, a target tracking problem in indoor positioning systems for wireless network environments has been an interesting research problem and thus been applied successfully for locating and tracking people within a building and products stored within a warehouse using wireless sensors and mobile devices. Due to a variety of indoor wireless channel characteristics, an accurate tracking estimation of a moving target is required for indoor positioning systems. However, a couple of noises such as system noise and measurement noise can contaminate the estimated position and thus can cause deterioration of target tracking performance. These noises can be caused by various factors such as environmental interference, turbulence affecting the target's movement, inaccuracies of sensor measurement data, and human's navigation error in a perfect straight line. Hence, these noises must be filtered for tracking a true path of a moving target. To provide the most accurate position and velocity of a moving target on the real-time indoor positioning in wireless network environments, some research adopted both IMS and FMS filters to reduce the effect of noises [20].

This paper considers only the one dimensional moving target for the X direction for the simplicity. The state estimation filter for the target tracking is typically designed with the state-space model called the constant velocity motion model. The constant velocity motion model assumes that targets move with constant velocity, that is, zero-acceleration within a short sampling time. However, in real situations, moving targets maneuver and change velocity and thus move temporarily with nonzero acceleration. Thus, the moving target's successive locations for the X direction can be represented by the 3rd-order discrete-time state-space model with sampling time T and acceleration term:

$$\begin{aligned} x_{i+1} &= Ax_i + Gw_i, \\ z_i &= Cx_i + v_i \end{aligned} \tag{15}$$

where

$$x_i \triangleq \begin{bmatrix} x_p : \text{position} \\ x_v : \text{velocity} \\ x_a : \text{acceleration} \end{bmatrix}, A = \begin{bmatrix} 1 & T & T^2/2 \\ 0 & 1 & T \\ 0 & 0 & 1 \end{bmatrix}, G = \begin{bmatrix} T/2 & 0 \\ 1 & 0 \\ 0 & 1 \end{bmatrix}, C = [1 \ 0 \ 0]. \tag{16}$$

The state vector x_i at time i consists of three state variables, x_p to represent the moving target's random *position*, x_v to represent the corresponding *velocity*, x_a to represent the corresponding *accelerations*. Typically, since the acceleration occurs over a short time interval, this can be considered as a temporary uncertainty, that is, an unknown input and thus considered as an unknown state term. Hence, since the target tracking problem in this paper does not need the control input term, the proposed unified algorithm (13) is applied with assuming that the control input matrix is zero, that is, $B = 0$. Because of random disturbances by fading and shadowing, the state vector x_i cannot be measured directly. Thus, the measurement vector z_i is modeled by only the position to take these effects into account filtering or smoothing estimates. As shown in [23], target tracking applications using the constant velocity motion model often adopt the state-space model where the state equation is linear and only the measurement equation is nonlinear. Since little research is going on in the case of the FMS smoother for nonlinear systems, the proposed unified algorithm is currently difficult to use this type of state space model. Thus, this paper consider only the linear measurement equation. The application of the nonlinear measurement equation can be a future work of the current paper.

Using (13) with (15) and (16), the proposed unified algorithm of a FMS filtering and smoothing-based target tracking scheme provides estimates for position as well as acceleration of moving target in real time, while eliminating unwanted noise effects and maintaining desired moving positions. Via extensive computer simulations, the performance of the proposed unified algorithm-based target tracking scheme is evaluated and compared with the existing IMS filtering-based scheme of [20–22]. Computer simulations are performed for moving target's accelerations of four kinds of shapes. The first and second scenarios consider that the target moves with suddenly changing acceleration such as step-type and ramp-type. The third case considers that the target moves with slowly changing acceleration such as triangle-type. The last case considers that the target moves with randomly changing acceleration. The sampling period is taken by $T = 2$. The covariance of the system noise is $Q = diag(0.173^2 \ \ 0.01^2)$ and the covariance of measurement noise is $R = 0.05^2$. The window length is taken by $M = 15$. The lagged length is taken by $d = 5$. Simulations of 30 runs are performed using different system and measurement noises to make the comparison clearer. Each single simulation run lasts 800.

Figures 1–4 show simulation results for moving target's accelerations of four kinds of shapes. The first plot of figures shows the root-mean-square(RMS) estimation error of the moving target's random position for 30 simulations. The second plot of all figures shows the estimation error of the moving target's random position for one of 30 simulations. The last plot of figures shows the estimate of unknown acceleration. In addition, time averaged values of RMS estimation errors are presented by Table 2. As shown in simulation results, the proposed unified algorithm-based target tracking scheme can outperform the IMS filtering-based target tracking scheme when the target moves with suddenly changing acceleration. On the interval where the acceleration varies suddenly, the estimation error of the proposed unified algorithm-based scheme is remarkably smaller than that of the IMS filtering-based scheme. In addition, when the acceleration varies constantly, the convergence of the estimation error for the proposed unified algorithm-based scheme is much faster than that of the IMS filtering-based scheme. These observations for simulation results might come from the fast convergent time and the fast tracking ability due to intrinsic robustness and deadbeat properties of the proposed unified algorithm-based scheme.

Table 2. Comparison of mean of RMS estimation errors.

	Step-Type	Ramp-Type	Triangle-Type	Randomly Changing
FMS Smoother	0.0062	0.0060	0.0059	0.0059
FMS Filter	0.0477	0.0441	0.0408	0.0445
Kalman Filter	0.0505	0.0456	0.0415	0.0481

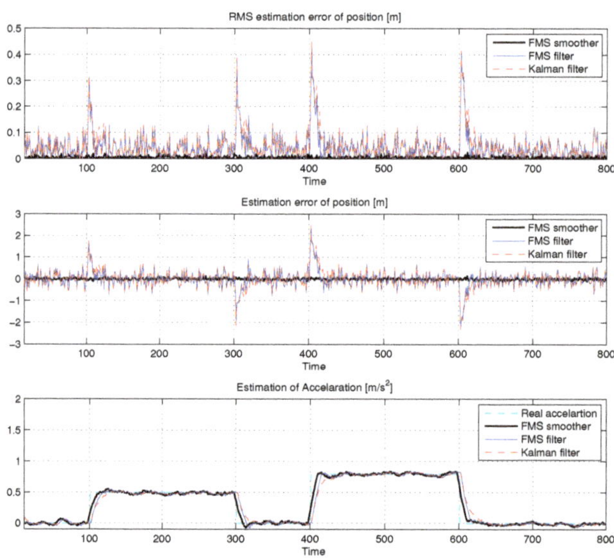

Figure 1. Simulation result for suddenly changing acceleration(step-type).

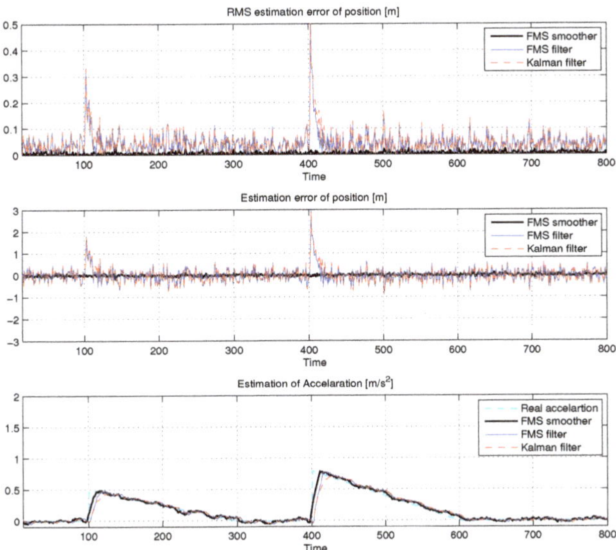

Figure 2. Simulation result for suddenly changing acceleration(ramp-type).

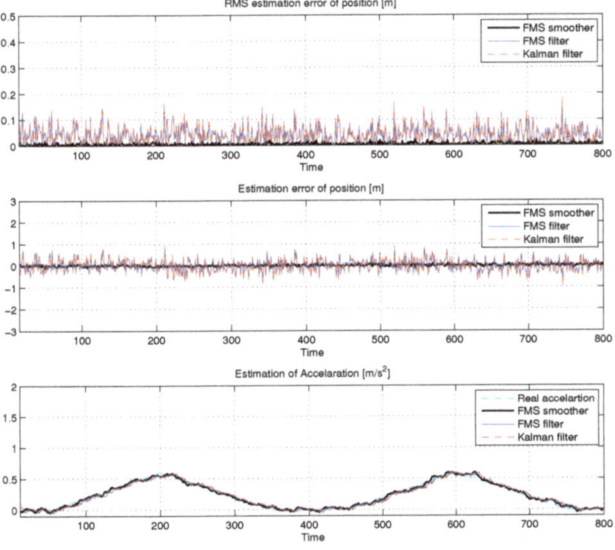

Figure 3. Simulation result for slowly changing acceleration(triangle-type).

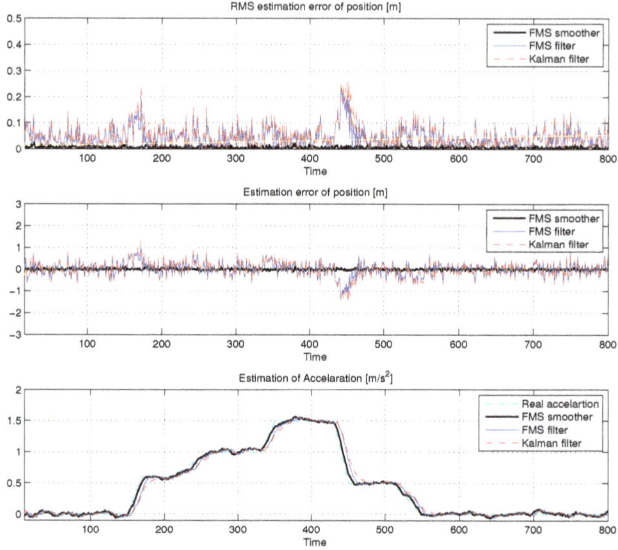

Figure 4. Simulation result for randomly changing acceleration.

4. Conclusions

This paper has dealt with the state estimation problem for the target tracking scheme in wireless communication environments. Firstly, the unified algorithm of FMS filtering and smoothing has been proposed. The unified algorithm has been derived by solving one optimization problem directly with incorporation of the unbiasedness constraint using only finite measurements and inputs on the most recent window. The proposed unified algorithm provides simultaneously the current state estimate as well as the lagged state estimate. The proposed unified algorithm has shown that there are some unique properties such as unbiasedness, deadbeat, time-invariance and intrinsic robustness while the recursive IMS filter such as Kalman filter does not have these properties. The on-line computational complexity of the proposed unified algorithm has been compared with the IMS filter such as the fixed-lag Kalman smoothing filter. Secondly, the target tracking in wireless network environment has been considered via extensive computer simulations as an application of the proposed unified algorithm of FMS filtering and smoothing. Through simulation results for moving targets' accelerations of various shapes, the proposed unified algorithm-based target tracking scheme has been shown to provide estimates for position as well as acceleration of moving target in real time, while eliminating unwanted noise effects and maintaining desired moving positions. It has been shown that the performance of the proposed unified algorithm-based target tracking scheme can be better than the IMS filtering-based scheme for suddenly changing acceleration.

Actually, research on the FMS filter and smoother in nonlinear systems is relatively inactive compared to research in linear systems. There is some research, such as an adoption of nonlinear measurement model, on the FMS filter for nonlinear systems, but little research is going on in the case of the FMS smoother for nonlinear systems. Thus, for nonlinear systems, research on the FMS smoother should be preceded and then a unified algorithm of FMS filter and smoother can be researched as a future work. In addition, the estimation of target's acceleration can be extended to identification and classification of moving targets, which could be another future work.

Funding: This research was supported by Basic Science Research Program through the National Research Foundation of Korea (NRF) funded by the Ministry of Education (NRF-2017R1D1A1B03033024).

Conflicts of Interest: The authors declare no conflict of interest.

References

1. Grewal, M. Applications of Kalman filtering in aerospace 1960 to the present. *IEEE Control Syst.* **2010**, *30*, 69–78. [CrossRef]
2. Auger, F.; Hilairet, M.; Guerrero, J.M.; Monmasson, E.; Orlowska-Kowalska, T.; Katsura, S. Industrial applications of the Kalman filter: A review. *IEEE Trans. Ind. Electron.* **2013**, *60*, 5458–5471. [CrossRef]
3. Rhudy, M.; Salguero, R.; Holappa, K. A Kalman filtering tutorial for undergraduate students. *Int. J. Comput. Sci. Eng. Surv.* **2017**, *8*, 1–18. [CrossRef]
4. Barrau, A.; Bonnabel, S. Invariant Kalman filtering. *Ann. Rev. Control Robot. Auton. Syst.* **2018**, *1*, 237–257. [CrossRef]
5. Aravkin, A.; Burke, J.V.; Ljung, L.; Lozano, A.; Pillonetto, G. Generalized Kalman smoothing: Modeling and algorithms. *Automatica* **2017**, *86*, 63–86. [CrossRef]
6. Kim, P.S. An alternative FIR filter for state estimation in discrete-time systems. *Digit. Signal Process.* **2010**, *20*, 935–943. [CrossRef]
7. Shmaliy, Y.S.; Simon, D. Iterative unbiased FIR state estimation: A review of algorithms. *EURASIP J. Adv. Signal Process.* **2013**, *2013*, 113. [CrossRef]
8. Shmaliy, Y.S.; Zhao, S.; Ahn, C.K. Unbiased finite impluse response filtering: An iterative alternative to Kalman filtering ignoring noise and initial conditions. *IEEE Control Syst.* **2017**, *37*, 70–89.
9. Shmaliy, Y.S.; Neuvo, Y.; Khan, S. Review of unbiased FIR filters, smoothers, and predictors for polynomial signals. *Front. Signal Process.* **2018**, *2*, 1–29.
10. Zhao, S.; Shmaliy, Y.S.; Liu, F. Fast Kalman-Like optimal unbiased FIR filtering with applications. *IEEE Trans. Signal Process.* **2016**, *64*, 2284–2297. [CrossRef]
11. Shmaliy, Y.; Khan, S.H.; Zhao, S.; Ibarra-Manzano, O. General unbiased FIR filter with applications to GPS-based steering of oscillator frequency. *IEEE Trans. Control Syst. Technol.* **2017**, *25*, 1141–1148. [CrossRef]
12. Kim, P.S. A design of finite memory residual generation filter for sensor fault detection. *Meas. Sci. Rev.* **2017**, *17*, 75–81. [CrossRef]
13. Shmaliy, Y.S.; Morales-Mendoza, L.J. FIR smoothing of discrete-time polynomial signals in state space. *IEEE Trans. Signal Process.* **2010**, *58*, 2544 – 2555. [CrossRef]
14. Kim, P.S. A computationally efficient fixed-lag smoother using recent finite measurements. *Measurement* **2013**, *46*, 846–850. [CrossRef]
15. Kwon, B.K.; Quan, Z.; Han, S.H. A robust fixed-lag receding horizon smoother for uncertain state space models. *Adapt. Control Signal Process.* **2015**, *29*, 1354–1366. [CrossRef]
16. Kim, P.S. A finite memory structure smoother with recursive form using forgetting factor. *Math. Probl. Eng.* **2017**, *2017*, 1–6. [CrossRef]
17. Ciuonzo, D.; Buonanno, A.; D'Urso, M.; Palmieri, F.A.N. Distributed classification of multiple moving targets with binary wireless sensor networks. In Proceedings of the 14th International Conference on Information Fusion, Chicago, IL, USA, 5–8 July, 2011; pp. 1–8.
18. Ciuonzo, D.; Willett, P.K.; Bar-Shalom, Y. Tracking the tracker from its passive sonar ML-PDA estimates. *IEEE Trans. Aerosp. Electron. Syst.* **2014**, *50*, 573–590. [CrossRef]
19. Ez-Zaidi, A.; Rakrak, S. A comparative study of target tracking approaches in wireless sensor networks. *J. Sens.* **2016**, *2016*, 1 – 11. [CrossRef]

20. Lee, S.; Cho, B.; Koo, B.; Ryu, S.; Choi, J.; Kim, S. Kalman filter-based indoor position tracking with self-calibration for RSS variation mitigation. *Int. J. Distrib. Sens. Netw.* **2015**, *2015*, 1–10. [CrossRef]
21. Lee, S.H.; Lim, I.K.; Lee, J.K. Method for improving indoor positioning accuracy using extended Kalman filter. *Mob. Inf. Syst.* **2016**, *2016*, 1–15. [CrossRef]
22. Fariz, N.; Jamil, N.; Din, M.M.; Rusli, M.E.; Sharudin, Z.; Mohamed, M.A. An improved indoor location technique using Kalman filter. *Int. J. Eng. Technol.* **2018**, *7*, 1–4. [CrossRef]
23. Pak, J.M.; Kim, P.S.; You, S.H.; Lee, S.S.; Song, M.K. Extended least square unbiased FIR filter for target tracking using the constant velocity motion model. *Int. J. Control Autom. Syst.* **2017**, *15*, 947–951. [CrossRef]
24. Kim, P.S.; Lee, E.H.; Jang, M.S.; Kang, S.Y. A finite memory structure filtering for indoor positioning in wireless sensor networks with measurement delay. *Int. J. Distrib. Sens. Netw.* **2017**, *13*, 1–8. [CrossRef]
25. Mendel, J. *Lessons in Estimation Theory for Signal Processing, Communications, and Control*; Prentice-Hall: Englewood Cliffs, NJ, USA, 1995.
26. Grewal, M.S.; Anderews, A.P. *Kalman Filtering—Theory and Practice*; Prentice-Hall: Englewood Cliffs, NJ, USA, 1993.

Article

A Novel Method of Adaptive Kalman Filter for Heading Estimation Based on an Autoregressive Model

Dashuai Chai [1], Guoliang Chen [1,*] and Shengli Wang [2]

[1] School of Environment Science and Spatial Informatics, China University of Mining and Technology, Xuzhou 221000, China

[2] Ocean Science and Engineering College, Shandong University of Science and Technology, Qingdao 266000, China

* Correspondence: chglcumt@163.com

Received: 31 July 2019; Accepted: 3 September 2019; Published: 6 September 2019

Abstract: With the popularity of smartphones and the development of microelectromechanical system (MEMS), the pedestrian dead reckoning (PDR) algorithm based on the built-in sensors of a smartphone has attracted much research. Heading estimation is the key to obtaining reliable position information. Hence, an adaptive Kalman filter (AKF) based on an autoregressive model (AR) is proposed to improve the accuracy of heading for pedestrian dead reckoning in a complex indoor environment. Our approach uses an autoregressive model to build a Kalman filter (KF), and the heading is calculated by the gyroscope, obtained by the magnetometer, and stored by previous estimates, then are fused to determine the measurement heading. An AKF based on the innovation sequence is used to adaptively adjust the state variance matrix to enhance the accuracy of the heading estimation. In order to suppress the drift of the gyroscope, the heading calculated by the AKF is used to correct the heading calculated by the outputs of the gyroscope if a quasi-static magnetic field is detected. Data were collected using two smartphones. These experiments showed that the average error of two-dimensional (2D) position estimation obtained by the proposed algorithm is reduced by 40.00% and 66.39%, and the root mean square (RMS) is improved by 43.87% and 66.79%, respectively, for these two smartphones.

Keywords: smartphone; pedestrian dead reckoning; heading estimation; autoregressive model; adaptive Kalman filter

1. Introduction

Location-based-services (LBSs) have developed rapidly in recent years, and their application brings great convenience to people's lives [1]. The core idea of LBSs is to determine a user's location quickly and effectively [2]. Outdoors, the global navigation satellite system (GNSS) provides accurate and reliable locations in open-sky environments, but it faces the challenges of signal blockage and invisibility in an indoor environment. With the development of indoor positioning technology, various positioning methods, based on WiFi [3], Bluetooth [4], ultra-wide band (UWB) [5], and inertial sensors [6], have been applied for personal navigation, emergency rescue, and tracking [7]. With the popularity of smartphones and the development of microelectromechanical system (MEMS), indoor positioning technology based on smartphones has found wide application in pedestrian navigation [8].

MEMS inertial sensors have become an appropriate choice for pedestrian navigation because they are small, low-cost, and completely autonomous [9]. Pedestrian navigation based on inertial measurement units (IMUs) can be divided into two categories [10]. One is the strap-down inertial navigation system (SINS) and the other is pedestrian dead reckoning (PDR) based on a human-motion model. SINS based on a three-axis accelerometer and three-axis gyroscope can provide high-frequency three-dimensional positions, velocities, and attitudes. Good results were obtained from the use of the error of velocity to establish an observation equation for personal navigation [11]. Gu et al. [12] applied SINS for a foot-mounted inertial pedestrian navigation system (PNS), with 0.3% accuracy of location estimation for walking distance. However, these methods require extra MEMS sensors, which are not suitable for LBSs based on a smartphone's built-in sensors. Zhuang et al. [13] applied SINS for PNS using the error of velocity from SINS and step estimation as the observation, and considering zero-velocity update technology (ZUPT) and zero-angular rate update (ZARU) as constraints. The experiment showed that the accuracy of navigation parameters was greatly improved. Kuang et al. [10] used the velocity, position, pseudo-velocity, and pseudo-position as observations, and a tri-axial accelerometer and tri-axial magnetometer of the quasi-state (QS) were introduced into observations for SINS. The accuracy of position was greatly improved compared to the previously mentioned experiment. However, the error of navigation parameters solved by SINS gradually accumulates because of the integral. Estimation of the user's position by PDR can be divided into four procedures: Step detection, step-length estimation, heading estimation, and position calculation [14]. Common methods of step detection include peak detection, cross-zero, autocorrelation algorithm and dynamic time warp [15,16]. The peak-detection algorithm is relatively optimal for step detection. Both linear and nonlinear models can be used for step-length estimation to estimate the moving distance of pedestrians [17,18]. In general, an accurate heading is the key to estimating the position of PDR, and some methods have been used to obtain the heading [19]. Afzal et al. [20] estimated the heading by magnetometer, and analyzed the influence of various environments on magnetic field strength. Zheng et al. [21] used the raw output of a gyroscope to update the heading by quaternion, and the position gradually deviated from the reference trajectory with time. Kang et al. [22] fused these headings calculated by the outputs of a magnetometer and gyroscope, and stored by the previous estimate, to determine the current heading, which was better than the heading calculated by a single sensor. Valenti et al. [23] used the quaternion of accelerations, magnetic field strength, and the outputs of a gyroscope to build a linear Kalman filter (KF) for heading estimation. Rennaudin et al. [24] applied the acceleration, magnetic field strength in QS, and the outputs of gyroscope to estimate the heading by extended Kalman filter (EKF). The linearization of the coefficient matrix by EKF was unstable, and the accuracy of linearization was perhaps not high enough. An unscented Kalman filter (UKF) does not require linearization of the coefficient matrix, and has at least second-order approximation accuracy. Therefore, UKF has a better result than EKF. Yuan et al. [25] used UKF for pedestrian navigation, and the mean heading error was less than 10°. Although UKF can achieve relatively high accuracy, it requires a certain amount of computation. The magnetic field is susceptible to contamination in complex indoor environments, and this can cause deviation of heading estimation [26]. Due to the accumulation of errors, a gyroscope is insufficient for long-term heading estimation [27]. Therefore, an accurate heading is still a challenge for pedestrian navigation with a smartphone.

Inaccurate prior knowledge of stochastic models and model function error will affect the accuracy of KF, and the adaptive Kalman filter (AKF) can address this problem. Mohamed et al. [28] used the innovation-based adaptive estimation (IAE) to adjust the state noise matrix or measurement noise matrix, showing that an AKF was better than KF. Chang et al. [29] applied the square of the Mahalanobis distance to detect model function error, and a fading factor was introduced to inflate the prior covariance to reduce the influence of inaccurate model functions. Jiang et al. [30] used two factors to adjust the measurement noise matrix, and a hypothesis test was introduced to avoid interference from abnormal states. Wu et al. [31] applied a two-stage function to obtain an adaptive factor, and a robust-adaptive KF was used to estimate the pedestrian heading. Li et al. [32] adaptively adjusted the measurement noise matrix according to the level of acceleration. Currently, AKF has been widely used in various fields of navigation, but is rarely applied in PDR.

This paper proposes a method of adaptive heading estimation. A novel heading prediction based on an autoregressive model (AR) prediction model is introduced to KF, and the current heading—determined by fusing these headings calculated by the gyroscope, obtained by the magnetometer and stored by the previous estimate—is used as the measurement of KF. Taking into account model function error and inaccurate prior noise, AKF is used to improve the accuracy of heading estimation. The change rate of the total magnetic field is used to detect its QS, and the heading calculated by gyroscope is corrected by the feedback of the estimated heading by AKF in QS. In order to intuitively and precisely express the accuracy of heading estimation, the estimated heading was used for the position estimation of PDR. Two kinematic experiments are used to verify this algorithm.

2. Heading Estimation Based on an AR Model

A smartphone contains a variety of sensors, and it is an effective device for pedestrian navigation. When a smartphone is used for navigation, it is held in the hand more than 87% of the time [26], so we assume that the heading of a pedestrian is consistent with the smartphone in this paper. That phone is held in the hand of a pedestrian. The heading is determined by the KF based on AR model in this paper.

2.1. KF Process

The state equation and measurement equation of a linear discrete KF system can be expressed as

$$\begin{cases} x_k = \Phi_{k,k-1}x_{k-1} + w_{k-1} \\ z_k = H_k x_k + v_k \end{cases} \tag{1}$$

where x_{k-1} and x_k represent the state vector at times $k-1$ and k, respectively; $\Phi_{k,k-1}$ is the state transition matrix; z_k is the measurement vector; H_k is the measurement matrix; and w_{k-1} and v_k are zero-mean white noise, whose variances can be expressed as Q_{k-1} and R_k, respectively.

2.2. AR Model

The AR model not only satisfies the polynomial constraint of state variables, but also can reduce the noise by the criterion of minimizing the mean-square error (MMSE) using redundant observations [33]. In order to enhance the accuracy of heading estimation, we use the AR model to build KF.

Here, we suppose that the current heading can be predicted by the previous S samples:

$$att_k = \sum_{s=1}^{S} h_s att_{k-s} \tag{2}$$

where att_k is the current heading and h_s is the coefficient matrix of the AR model.

Without loss of generality, the current heading can be fitted by L polynomials,

$$att_k = \sum_{l=1}^{L} a_l (t_k)^l \tag{3}$$

where a_l is the coefficient matrix of polynomials, $t_k = kT$, and T is the sample interval.

Substituting Equation (3) into Equation (2), we can obtain

$$(k+1)^l = \sum_{s=1}^{S} h_s (k+1-s)^l, l = 0, 1, \ldots, L \tag{4}$$

When $l = 0$, we can obtain the equation as follows

$$\sum_{s=1}^{S} h_s = 1 \tag{5}$$

Similarly, the recursive expression of Equation (4) can be obtained by

$$\sum_{s=1}^{S} h_s s^l = 0, l = 1, 2, ..., L \tag{6}$$

In summary, Equations (5) and (6) can be expressed in matrix form as follows.

$$A\mu_k = b \tag{7}$$

where $u_k = \begin{bmatrix} h_1 & h_2 & \cdots & h_S \end{bmatrix}^T, b = \begin{bmatrix} 1 & 0 & \cdots & 0 \end{bmatrix}^T,$

$$A = \begin{bmatrix} 1 & 1 & \cdots & 1 \\ 1 & 2 & \cdots & S \\ \vdots & \vdots & \ddots & \vdots \\ 1 & 2^L & \cdots & S^L \end{bmatrix}_{(L+1)\times S} \tag{8}$$

When $S = L + 1$, the matrix A is nonsingular, and Equation (7) has a unique solution. When $S > L + 1$, the AR model has redundant observations. Namely the AR model not only satisfies the polynomial constraint but also can reduce the noise by extra observations.

2.3. Derivation of AR Model under KF Frame

In this section, the AR model is introduced into KF. The state equation of KF based on the AR model can be expressed as follows,

$$x_k^{AR} = \Phi_{k,k-1}^{AR} x_{k-1}^{AR} + w_k^{AR} \tag{9}$$

where $x_{k-1}^{AR} = \begin{bmatrix} att_{k-1} & att_{k-2} & \cdots & att_{k-S} \end{bmatrix}^T$, w_k^{AR} is the state noise, its variance matrix is Q_k^{AR}, and the state transition matrix is

$$\Phi_{k,k-1}^{AR} = \begin{bmatrix} h_1 & h_2 & \cdots & h_{S-1} & h_S \\ 1 & 0 & \cdots & 0 & 0 \\ 0 & 1 & \cdots & 0 & 0 \\ \vdots & \vdots & \ddots & \vdots & \vdots \\ 0 & 0 & \cdots & 1 & 0 \end{bmatrix}_{S\times S} \tag{10}$$

Therefore, KF based on AR model prediction can be expressed as follows.

$$x_{k,k-1}(\mu_k) = \Phi_{k,k-1}(\mu_k) x_{k,k-1} \tag{11}$$

$$P_{k,k-1}(\mu_k) = \Phi_{k,k-1}(\mu_k) P_{k-1,k-1} \Phi_{k,k-1}^T(\mu_k) + Q_{k-1} \tag{12}$$

$$C_k(\mu_k) = HP_{k,k-1}H^T + R_k \tag{13}$$

$$K_k(\mu_k) = P_{k,k-1}(\mu_k) H^T C_k^{-1}(\mu_k) \tag{14}$$

$$x_{k,k}(\mu_k) = x_{k,k-1}(\mu_k) + K_k(\mu_k)\left[z_k - Hx_{k,k-1}[\mu_k]\right] \tag{15}$$

$$P_{k,k}(\mu_k) = [I - K_k[\mu_k]H] x_{k,k-1}(\mu_k) P_{k,k-1}(\mu_k) \tag{16}$$

where P is the state variance matrix, C_k is the variance matrix of the innovation sequence, and K is the gain matrix.

We know from Equations (11)–(16) that KF based on the AR model needs to know μ_k. A cost function based on MMSE can be constructed as follows

$$\begin{array}{c} \min \mu_k^T P_{k-1,k-1} \mu^T \\ subject\ to\ A\mu_k = b \end{array} \quad (17)$$

Equation (17) can be solved by the Lagrangian multiplication operator [34]

$$L(\mu_k, \lambda) = \mu_k^T P_{k-1,k-1} \mu_k - \lambda^T (A\mu_k - b) \quad (18)$$

where λ is a Lagrange multiplier.

Taking the partial derivatives with respect to μ_k and λ for Equation (18) and setting them to zero. We can get

$$\begin{cases} \frac{\partial L\{\mu_k,\lambda}{\partial \mu_k} = 0 \\ \frac{\partial L\{\mu_k,\lambda}{\partial \lambda} = 0 \end{cases} \quad (19)$$

We can obtain μ_k by solving Equation (19),

$$\mu_k = P_{k-1,k-1}^{-1} A^T \left(A P_{k-1,k-1}^{-1} A^T \right)^{-1} b \quad (20)$$

We can see from Equation (20) that the solution of μ_k uses the information polynomial and state variance. Thus, compared to traditional KF, KF based on the AR model can effectively reduce the noise.

2.4. The Adaptive Kalman Filter

KF based on the AR model can work well if the model function and noise statistic are both accurately known. However, the model function is an approximate expression. Moreover, a smartphone may be shaken due to human motion when the pedestrian is holding the smartphone for navigation, which makes the noise statistic unreliable. Therefore, it is necessary to adaptively adjust the noise for obtaining reliable filtering results.

To reduce the influence of inaccurate state noise, we use an adaptive method based on the innovation sequence to adjust the state covariance matrix [35].

Substituting Equation (14) into Equation (16), we can get

$$P_{k,k} = P_{k,k-1} - K_k C_k K_k^T \quad (21)$$

Substituting Equation (12) into Equation (21), the state noise covariance can be got as follows.

$$Q_k = P_{k,k} - F_{k,k-1} P_{k-1,k-1} F_{k,k-1}^T + K_k C_k K_k^T \quad (22)$$

We can get the approximate solution of the state noise covariance based on Equation (22).

$$\hat{Q}_k = K_k \hat{C}_k K_k^T \quad (23)$$

The covariance matrix of innovation can be obtained by averaging the innovation sequence over N windows.

$$\hat{C}_k = \frac{1}{N} \sum_{j=0}^{N-1} v_{k-j} v_{k-j}^T \quad (24)$$

where $v_k = z_k - H_k x_{k,k-1}$ is the innovation vector.

3. The Calculation of Measurement Vector for Heading Estimation

In this section, we discuss the calculation of measurement vector in KF for heading estimation. In general, two methods are used to calculate the heading. One is based on the outputs of the magnetometer, and the other is obtained from raw gyroscope data

3.1. Heading Estimation by Magnetometer

The accelerometer, that measure specific force, can be used to obtain roll and pitch, and the heading can be solved by the leveled magnetic field intensity that is obtained by using roll and pitch. Many references discuss the solution of roll and pitch using acceleration, so we mainly discuss the heading calculated by magnetometer in this section. In this paper, the smartphone axis is show in Figure 1.

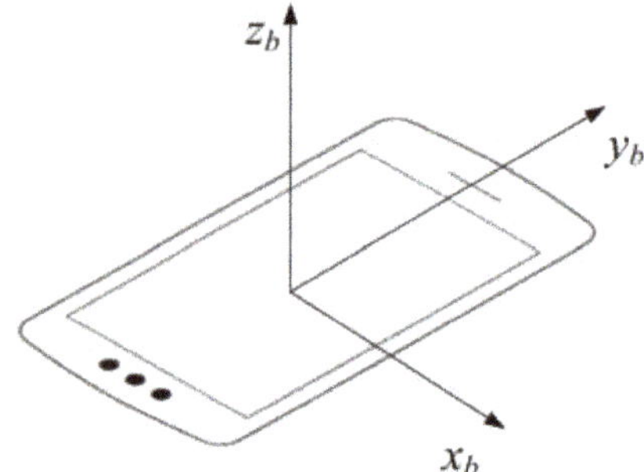

Figure 1. The body coordinate system of smartphone.

Due to the building structure and materials, the magnetic field is more susceptible to contamination by hard iron, soft iron, and scale factors. This causes bias in magnetometer outputs; hence it is necessary to calibrate the magnetometer. The commonly used calibration method is ellipsoid fitting [36,37].

After the magnetometers are calibrated, the heading can be solved from calibrated magnetic field intensities. Firstly, the magnetic field intensities can be leveled by roll and pitch.

$$\begin{bmatrix} M_x \\ M_y \\ M_z \end{bmatrix} = \begin{bmatrix} \cos\theta & \sin\gamma\sin\theta & \sin\gamma\cos\theta \\ 0 & \cos\theta & -\sin\theta \\ -\sin\gamma & \cos\gamma\sin\theta & -\sin\gamma\cos\theta \end{bmatrix}^T \begin{bmatrix} m_x \\ m_y \\ m_z \end{bmatrix} \tag{25}$$

where (m_x, m_y, m_z) and (M_x, M_y, M_z) are the geomagnetic field intensities in the body frame and navigation frame, respectively, and γ and θ are the roll and pitch, respectively, calculated from the raw acceleration data.

Then, taking into account the local magnetic declination, the heading can be calculated by:

$$\psi_{mag} = \arctan\left(\frac{m_x\cos\theta - m_z\sin\gamma}{m_x\sin\gamma\sin\theta + m_y\cos\theta + m_z\cos\gamma\sin\theta}\right) + D \tag{26}$$

where ψ_{mag} is the heading and D is the magnetic declination. γ and θ, calculated by accelerations, can be expressed as follows:

$$\gamma = \arctan\left(\frac{a_x}{\sqrt{a_y^2 + a_z^2}}\right) \tag{27}$$

$$\theta = -\arctan\left(\frac{a_y}{a_x * \sin(\gamma) + a_z * \cos(\gamma)}\right) \tag{28}$$

where $\left(a_x, a_y, a_z\right)$ are the accelerations in the body frame.

3.2. Heading Estimation Based on Gyroscope

The relative heading can be calculated using the raw outputs of a gyroscope, and the absolute heading can be obtained when the initial heading is known. The quaternion heading can be directly derived from the outputs of the gyroscope [25].

$$\overset{\bullet}{q} = \frac{1}{2}M(\omega)q = \frac{1}{2}\begin{bmatrix} 0 & -\omega_x & -\omega_y & -\omega_z \\ \omega_x & 0 & \omega_z & \omega_y \\ \omega_y & -\omega_z & 0 & \omega_x \\ \omega_z & \omega_y & -\omega_x & 0 \end{bmatrix}\begin{bmatrix} q_1 \\ q_2 \\ q_3 \\ q_4 \end{bmatrix} \tag{29}$$

where $\overset{\bullet}{q}$ is the differentiation of the quaternion, ω is the angular rate measured by gyroscope, and $M(\bullet)$ is the mathematical function of ω.

The quaternion-based update with the angular rate can be obtained from the discrete form of Equation (29),

$$q_k = \left[\cos\left[\frac{\eta}{2}\right]I + \frac{1}{2}\frac{\sin\left[\frac{\eta}{2}\right]}{\frac{\eta}{2}}M[\omega]dt\right]q_{k-1} \tag{30}$$

where $\eta = \sqrt{(\omega_x dt)^2 + \left(\omega_y dt\right)^2 + (\omega_z dt)^2}$, dt is the sampling interval, and I is unit vector.

In this paper, the interconversion between quaternion and the Euler angle is required. Their relationship can be expressed as follows.

$$\begin{bmatrix} \theta_{gyro} \\ \gamma_{gyro} \\ \psi_{gyro} \end{bmatrix} = \begin{bmatrix} \arcsin[2[q_3q_4 + q_1q_2]] \\ \arctan\left[-[2[q_2q_4 - q_1q_3]], q_1^2 - q_2^2 - q_3^2 + q_4^2\right] \\ \arctan\left[-[2[q_2q_3 - q_1q_4]], q_1^2 - q_2^2 + q_3^2 - q_4^2\right] \end{bmatrix} \tag{31}$$

3.3. Fused Heading Estimation

The heading estimated by single sensor is noisy due to sensor performance and external disturbance, and a reliable heading can be determined by fusing headings calculated by different sensors [22]. A heading estimation by fusing these headings that are calculated by a magnetometer, obtained by gyroscope and stored by previous estimate can be expressed as follows.

$$\psi_k = \begin{cases} \beta^{mgp}\left\{\beta^{mag}\psi_k^{mag} + \beta^{gyro}\psi_k^{gyro} + \beta^{prev}\psi_{k-1}^{prev}, \psi_\Delta^{cor} \le \psi_\delta^{cor}, \psi_\Delta^{mag} \le \psi_\delta^{mag}\right. \\ \beta^{mg}\left\{\beta^{mag}\psi_k^{mag} + \beta^{gyro}\psi_k^{gyro}, \psi_\Delta^{cor} \le \psi_\delta^{cor}, \psi_\Delta^{mag} > \psi_\delta^{mag}\right. \\ \psi_{k-1}^{prev}, \psi_\Delta^{cor} > \psi_\delta^{cor}, \psi_\Delta^{mag} \le \psi_\delta^{mag} \\ \beta^{gp}\left\{\beta^{gyro}\psi_k^{gyro} + \beta^{prev}\psi_{k-1}^{prev}, \psi_\Delta^{cor} > \psi_\delta^{cor}, \psi_\Delta^{mag} > \psi_\delta^{mag}\right. \end{cases} \tag{32}$$

where

$$\beta^{mgp} = \left(\beta^{mag} + \beta^{gyro} + \beta^{prev}\right)^{-1} \tag{33}$$

$$\beta^{mg} = \left(\beta^{mag} + \beta^{gyro}\right)^{-1} \tag{34}$$

$$\beta^{gp} = \left(\beta^{gyro} + \beta^{prev}\right)^{-1} \tag{35}$$

$$\varphi_\Delta^{cor} = \left|\psi_k^{mag} - \psi_k^{gyro}\right| \tag{36}$$

$$\varphi_\Delta^{mag} = \left|\psi_k^{mag} - \psi_{k-1}^{mag}\right| \tag{37}$$

where β^{prev}, β^{mag}, and β^{gyro} are the respective weights of the heading stored by the previous estimate, the heading calculated by the outputs of the current magnetometer and the heading obtained by the gyroscope—ψ_k^{mag} is the current heading calculated by Equation (26), ψ_k^{gyro} is the current heading calculated by Equation (31), ψ_{k-1}^{prev} is the heading of previous time updated by KF, ψ_δ^{cor} and ψ_δ^{mag}—are thresholds.

4. The Filter Design of KF Based on AR Model for Heading Estimation

In this paper, KF based on AR model is used to estimate the heading. In Section 2, the AR model is introduced to build KF for heading estimation, and the heading measurement is introduced in Section 3. In this section, we sum up the process of KF for heading estimation. In this paper, the current heading is predicted by the previous four samples in AR model constructed in Section 2, so S is set as four. The fused heading presented in Section 3.3 is used for the measurement of the KF update. The state and measurement equations of KF based on AR model for the proposed heading estimation can be expressed as follows,

The state model can be obtained by

$$x_k^{AR}(-) = \Phi_{k,k-1}^{AR} x_{k-1}^{AR}(+) + w_k^{AR} \tag{38}$$

where, $x_k^{AR}(-) = \begin{bmatrix} att_k^- & att_{k-1}^- & att_{k-2}^- & att_{k-3}^- \end{bmatrix}^T$, $x_{k-1}^{AR}(+) = \begin{bmatrix} att_{k-1}^+ & att_{k-2}^+ & att_{k-3}^+ & att_{k-4}^+ \end{bmatrix}^T$ and $\Phi_{k,k-1}^{AR}$ is the state transition matrix that can be obtained by Equation (10). att_{k-1}^+ is the estimated heading at time $k-1$, att_k^- is the predicted heading at time k.

The measurement model can be calculated by

$$Z_k^{AR} = H_k^{AR} x_k^{AR}(-) + V_k^{AR} \tag{39}$$

where, $Z_k^{AR} = \psi_k$ that can be calculated by Equation (32), $H_k^{AR} = \begin{bmatrix} 1 & 0 & 0 & 0 \end{bmatrix}_{1\times 4}$.

In this paper, Equations (38) and (39) were constructed to predict and update the KF for heading estimation.

5. The Detection of a QS Magnetic Field

The geomagnetic field suffers severe perturbations in indoor environments, causing the magnetic field magnitude to not be constant. However, locations are possible where these magnetic field magnitudes are constant, and these situations can be considered as QS [38]. The heading calculated by magnetic field intensity in QS can achieve an accurate result. In this paper, the change rate of the magnetic field intensity is used to detect QS.

The detection model based on the change rate of magnetic field magnitude is expressed as

$$y_k^m = o_k + v_k \tag{40}$$

where $o_k = \left\| \dot{m}_k \right\|$, $\left\| \dot{m}_k \right\|$ is the change rate of magnetic field magnitude and v_k is noise.

The magnetic field can be considered as QS if $o_k = 0$. Therefore, the detection of QS can be expressed as a binary hypothesis test. Here, the two hypotheses H_0 and H_1 are

$$\begin{array}{lll} H_0 : \exists k \in \Gamma_n & s.t. & o_k \neq 0 \\ H_1 : \forall \in \Gamma_n & then & o_k = 0 \end{array} \tag{41}$$

where $\Gamma_n = \{l \in N : n \leq l \leq n + N - 1\}$.

The probability density function (PDF) of Equation (40) based on hypothesis H_0 can be given by

$$f(y; o_k, H_0) = \prod_{k \in \Gamma_n} \frac{1}{\sqrt{2\pi\sigma_{o_k}^2}} \exp\left(\frac{-1}{2\pi\sigma_{o_k}^2} \left(y_k^m - o_k \right)^2 \right) \tag{42}$$

where $\sigma_{o_k}^2$ is the noise variance of o_k.

We assume that the maximum likelihood estimate (MLE) of o_k is $\hat{o}_k$, which can be obtained by the mean of samples as

$$\hat{o}_k = \frac{1}{N}\sum_{k\in\Gamma_n} y_k^m \tag{43}$$

Therefore, the PDF of H_0 can be changed to

$$f(y; o_k, H_0) = \prod_{k\in\Gamma_n} \frac{1}{\sqrt{2\pi\sigma_{o_k}^2}} \exp\left(\frac{-1}{2\pi\sigma_{o_k}^2}\left(y_k^m - \hat{o}_k\right)^2\right) \tag{44}$$

For the hypothesis H_1, $o_k = 0$, so the PDF of H_1 can be obtained by

$$f(y; H_1) = \prod_{k\in\Gamma_n} \frac{1}{\sqrt{2\pi\sigma_{o_k}^2}} \exp\left(\frac{-1}{2\pi\sigma_{o_k}^2}\left(y_k^m\right)^2\right) \tag{45}$$

The detection of QS based on the generalized likelihood ratio test (GLRT) can be expressed as

$$\wedge(y) = \frac{f(y; \|\hat{o}_k\|, H_0)}{f(y; H_1)} < \chi_m \tag{46}$$

Substituting Equations (44) and (45) into Equation (46), we can obtain

$$\wedge(y) = \prod_{k\in\Gamma_n} \exp\left(\frac{1}{2\sigma_{o_k}^2}\left(y_k^m\right)^2 - \frac{1}{2\sigma_{o_k}^2}\left(y_k^m - \hat{o}_k\right)^2\right) < \chi_m \tag{47}$$

Taking the natural logarithm on both sides of Equation (47), we get the test statistics as follows.

$$\left|\frac{1}{\sqrt{N}}\left|\sum_{k\in\Gamma_n} y_k^m\right|\right| < \chi_m' \tag{48}$$

where $\chi_m' = \sqrt{2\sigma_{o_k}^2 \ln(\chi_m)}$ is the threshold and N is the window size.

If Equation (48) is true, the magnetic field is considered as QS. Then the heading calculated by Equation (31) can be corrected with the heading updated by Equation (15). Similarly, the roll and pitch can also be corrected with the solution estimated by KF.

The following algorithm is tabulated for expression of the proposed heading estimation.

Algorithm 1: The heading estimation with KF based on AR model

Step 0. Given S, L with $S \geq L + 2$. We set $S = 4$, $L = 1$. Step 0.1 Initial state: $x_0 = \begin{bmatrix} z_0 & z_{-1} & z_{-2} & z_{-3} \end{bmatrix}^T$, where z is the measurement heading calculated by Equation (32). Step 0.2 Initial state variance matrix P_0, state noise covariance Q_0 and measurement noise covariance R_0.

Step 1. Calculation of coefficients matrix: Calculate the coefficient of an AR model by Equation (20), and calculate the state transition matrix by Equation (10).

Step 2. KF prediction: Performance the KF time update of Equations (11) and (12).

Step 3. KF update: Obtain the measurement vector of heading by Equation (32); Performance the KF measurement update of Equations (13)–(16); Update the state noise covariance by Equation (23).

Step 4. Correct the heading calculated by gyroscope: If Equation (48) is true, the heading calculated by Equation (31) is corrected by the updated heading of Step 3.

Step 5. Terminate the process, otherwise return to Step 1.

6. Position Estimation of PDR Algorithm

6.1. The Principle of PDR Algorithm

The PDR algorithm includes four procedures: Step detection, step-length estimation, heading estimation, and position calculation. In this paper, peak detection is used for step detection, and the time interval of two peaks is considered as the time of a step. A linear model is used to estimate the step-length [39].

$$SL = a * F + b + v \tag{49}$$

where SL is the step-length, F is the step frequency, v is noise, and a and b are coefficients dependent on human motion.

After step detection, step-length estimation, and heading estimation are obtained, the position can be calculated as follows.

$$\begin{cases} N_k = N_{k-1} + SL_{k-1} \cos \varphi_{k-1} \\ E_k = E_{k-1} + SL_{k-1} \sin \varphi_{k-1} \end{cases} \tag{50}$$

where N and E are the north and east direction positions, respectively.

6.2. Filter Design of UKF for Position Estimation

In this paper, the position estimation is solved based on the UKF that reference the method proposed in [38]. After the heading is updated and the step-length is estimated, these values can be used to estimate the position based on the UKF.

According to Equation (50), a system model of nonlinear formula can be built as follows:

$$X_k = \begin{pmatrix} N_k \\ E_k \\ SL_k \\ \varphi_k \\ \Delta\varphi_k \end{pmatrix} = \begin{pmatrix} N_{k-1} + SL_{k-1} \cos \varphi_{k-1} \\ E_{k-1} + SL_{k-1} \sin \varphi_{k-1} \\ SL_{k-1} \\ \varphi_{k-1} + \Delta\varphi_{k-1} \\ \Delta\varphi_{k-1} \end{pmatrix} + W_{k-1} \tag{51}$$

where, $\Delta\varphi_k$ is the increment of heading at time k, W_{k-1} is state noise.

The measurement equations can be obtained by

$$Z_k = \begin{pmatrix} SL_k \\ \varphi_k \\ \Delta\varphi_k \end{pmatrix} = \begin{pmatrix} SL_k \\ \varphi_k \\ \Delta\varphi_k \end{pmatrix} + V_k \tag{52}$$

where, V_k is measurement noise.

In this paper, Equations (51) and (52) are used to estimate the position at every step. And the calculated process of UKF can be found in [25,40].

7. Experimental Analysis

7.1. Static Experiment

A Huawei Honor V9 (HV9) and XiaoMi8 (XM8) smartphone were used to collect static data. The collection time was about 30 min, and the sampling frequency was 50 Hz. Three methods were used to calculate the heading using these data. The first was the heading calculated by the outputs of the magnetometer (the method discussed in Section 3.1), the second was the heading obtained by a gyroscope (the method introduced in Section 3.2), and the last was the proposed method by this paper.

Figure 2 shows the heading error of different methods. The solid line is the error of heading calculated by magnetometer (Mag), and we can see that the margin of error is about 5 degrees even though the data are static in Figure 2a, and the jump range of error is 1 degree for XM8 in Figure 2b. The dashed line is the error of heading obtained by gyroscope (Gyr). The error is low and smooth in the short term, but it gradually accumulates with time, and the maximum error reaches 50 degrees for HV9. The chain line expresses the error of heading estimated by the proposed method that the AKF based on AR is used to estimate heading (ProA). We can see from Figure 2 that the drift error of heading is effectively contained by the proposed algorithm compared with the method of Gyr. Compare with the method of Mag, the error is smooth and accurate for XM8, however, the result of Pro is worse than that of Mag for HV9 with time. We can see from Equation (32) that the accuracy of a heading calculated by Gyr will affect the result of ProA, so the accuracy of ProA is affected by the cumulative error of Gyr in Figure 2a. Comparing Figures 2a and 2b, we found that the heading calculated by XM8 is better than that estimated by HV9. That's because the sensors accuracy of XM8 are better than that of HV9.

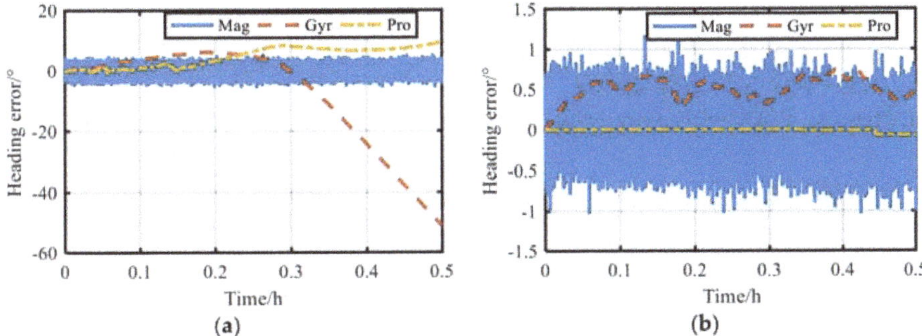

Figure 2. Heading error of static data for different methods: (**a**) The heading error of Huawei Honor V9 (HV9); (**b**) The heading error of XiaoMi8 (XM8).

In summary, the heading calculated by a single sensor embedded in a smartphone still has disadvantages even under great observation.

7.2. Kinematic Experiment

The kinematic data were used to further verify the effectiveness of the proposed algorithm. The site of data collection was on the third floor of the J9 office building of Shandong University of Science and Technology, and many iron materials were on site. The floor plan is shown in Figure 3. The tester held the HV9 and XM8 smartphones in their hand to collect data along the preplanned trajectory.

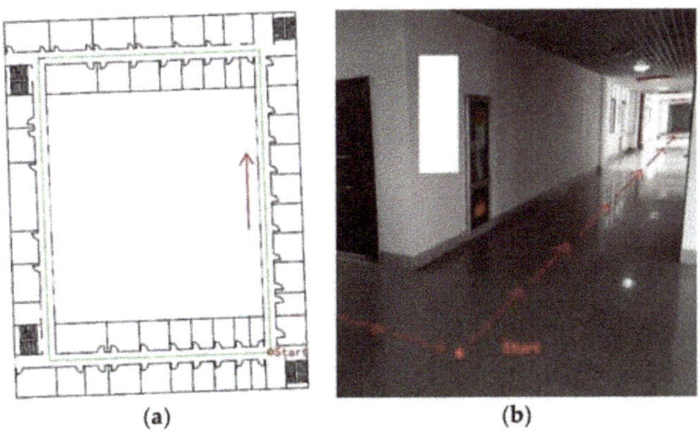

Figure 3. Trajectory of experiment: (**a**) Floor plan; (**b**) Experimental environment.

Figure 4 shows the calculated heading that used the data collected by the HV9 and XM8, and true north was used to largely avoid the heading jumps between 0 and 360. Five methods were used for heading estimation in these kinematic experiments. The first was Mag, the second was Gyr, the third was proposed in [22] that the current heading was determined by magnetometer, gyroscope and previous estimate (MGP), the fourth was Pro that is the proposed method in this paper without the adaptive filter, and the last was ProA that the adaptive filter is used in Pro. And the reference is the reference heading in Figure 4. Compared to the other four methods, the heading based on the magnetometer severely jumped for these smartphones, and the maximum jump achieved 30 degrees, although the heading calculated by magnetic field was close to the reference at some locations. Thus, it is not appropriate to obtain the heading just using the outputs of the magnetometer for pedestrian navigation in complex indoor environments.

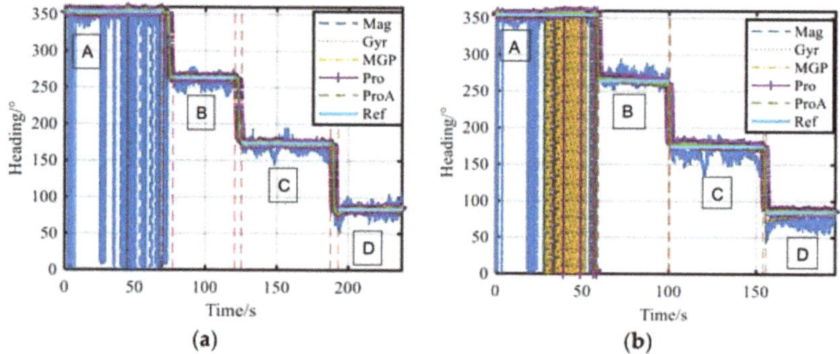

Figure 4. Heading estimation with different methods for different smartphones: (**a**) Heading of HV9; (**b**) Heading of XM8.

Figure 5 shows the heading errors of the local trajectory marked in A, B, C, and D in Figure 4a, and Figure 6 shows the heading errors of the local trajectory of Figure 4b. The heading errors of Mag for HV9 and XM8 are not shown in Figures 5 and 6 because these errors fluctuate widely. However, these statistical results, of heading errors of these five methods for HV9 and XM8, are all shown in Tables 1 and 2, respectively. In Tables 1 and 2, the mean RMSs of heading error for Mag are 8.62° and 10.88° for HV9 and XM8, respectively, and these accuracies are not appropriate for pedestrian navigation. In Figure 5a, most heading errors of Pro and ProA are within 4°, and these heading errors are smaller than those of Gyr and MGP. In Figure 5b–d, these heading errors of Pro and ProA are within 2°, and these accuracies of heading estimation are better than those of Gyr and MGP. In Table 1, the RMS of MGP is slightly superior to that of Gyr, and the mean RMS decreases from 2.38° to 2.19°. The accuracy of Pro is better than that of MGP in every section. The RMS decreases from 2.20° to 1.74° in the local trajectory of mark A, the RMS drops from 1.17° to 1.09° in the local trajectory of mark B, the RMS reduces from 3.01° to 1.26° in the local trajectory of mark C, and the RMS decreased from 2.38° to 1.55° in the local trajectory of mark D. And the mean accuracy increased by 35.62%. In Table 1, the accuracy of Pro is basically the same as that of ProA. We also can see from these lists of Mean in Table 1 that these results of heading estimation are bias. The first reason for this bias is that the tester cannot strictly walk in accordance with the preplanned trajectory, and the other is that the smartphone will inevitably shake with the movement of the tester. These conditions can cause the biased estimation. We can see from Table 1, the bias of Pro and ProA is smaller than that of the other methods. In summary, compared with the heading estimated by Mag, Gyr and MGP, the heading solved by the proposed algorithm is close to the reference for this experiment.

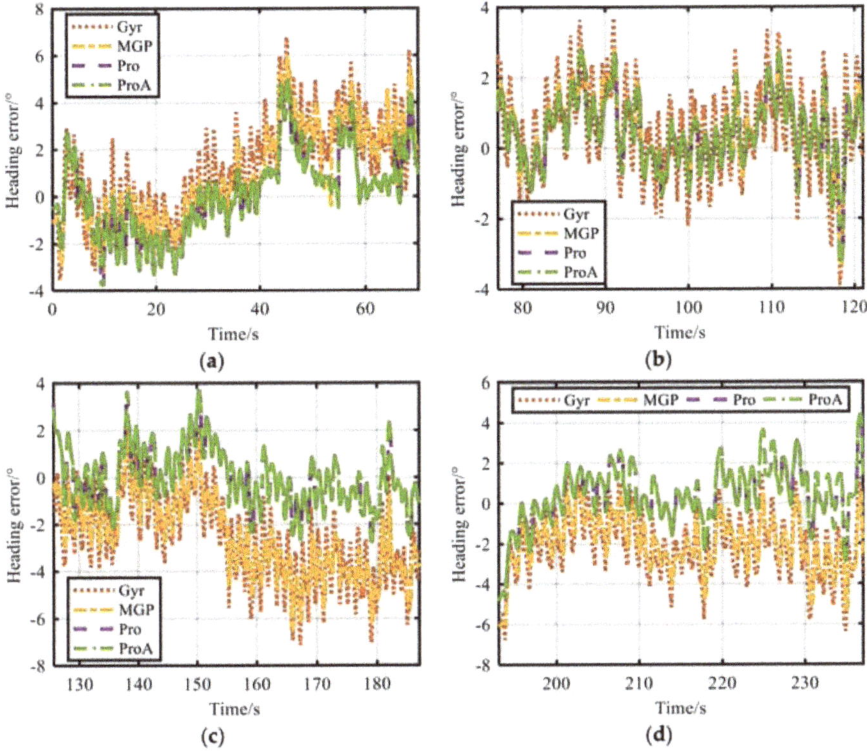

Figure 5. Heading error of local trajectory for HV9: (**a**) Heading error of mark A; (**b**) Heading error of mark B; (**c**) Heading error of mark C; (**d**) Heading error of mark D.

Figure 6 shows local heading error of Figure 4b. In Figure 6a, most errors of Pro and ProA are within 4°, and these errors are smaller than those of Gyr and MGP. In Figure 6b–d, these errors of Pro and ProA are principally less than 5°, and these results of Pro and ProA are superior to those of Gyr and MGP. In Table 2, The RMS of MGP is slightly smaller than that of Gyr, and the mean RMS is decreased by 1.97%. For these two kinematic experiments, the accuracy of heading estimation for MGP is better than that of Gyr, and this conclusion is consistent with that of the literature [22]. We can also see from Table 2 that these accuracies of Pro are better than those of MGP in these four sections. The RMS decrease from 3.45° to 2.25° in the local trajectory of mark A, the RMS drops from 3.66° to 3.28° in the local trajectory of mark B, the RMS reduces from 5.35° to 3.39° in the local trajectory of mark C, and the RMS decreases from 5.51° to 2.58° in the local trajectory of mark D. Compared with the mean accuracy of MGP, the mean accuracy of Pro increases by 35.86%. In Table 2, the accuracy of heading estimation is further improved by the adaptive algorithm. The RMS decreases from 2.25° to 1.70° in the local trajectory of mark A, the RMS drops from 3.28° to 2.69° in the local trajectory of mark B, the RMS reduces from 3.39° to 2.02° in the local trajectory of mark C, and the RMS decreases from 2.58° to 2.19° in the local trajectory of mark D. Also, the mean of RMS is improved by 25.35% for this kinematic data. Similarly, the biased heading estimation can be observed in Table 2, and the bias of mean for ProA is the smallest among these methods. Compared with these methods of Mag, Gyr and MGP, these methods of Pro and ProA can obtain the higher accuracy of heading estimation.

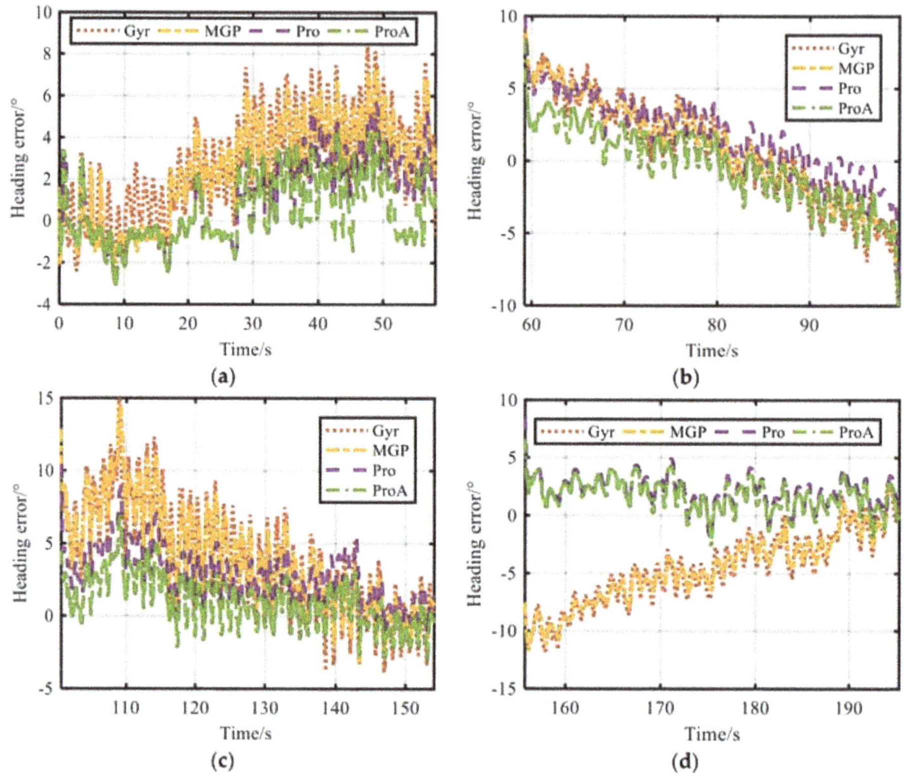

Figure 6. Heading error of local trajectory for XM8: (**a**) Heading error of mark A; (**b**) Heading error of mark B; (**c**) Heading error of mark C; (**d**) Heading error of mark D.

Table 1. The statistic of heading errors for HV9.

	Mag		Gyr		MGP		Pro		ProA	
	RMS/°	**Mean/°**	**RMS/°**	**Mean/°**	**RMS/°**	**Mean/°**	**RMS/°**	**Mean/°**	**RMS/°**	**Mean/°**
A	8.28	−0.84	2.40	1.10	2.20	1.00	1.74	0.09	1.74	0.12
B	8.73	−4.32	1.39	0.53	1.17	0.55	1.09	0.49	1.09	0.47
C	9.61	−5.74	3.13	−2.50	3.01	−2.47	1.26	0.05	1.27	0.04
D	7.85	−0.40	2.59	−2.06	2.38	−1.98	1.55	0.43	1.55	0.42
Mean/°	8.62	−2.83	2.38	−0.73	2.19	−0.73	1.41	0.27	1.41	0.26

Table 2. The statistic of heading errors for XM8.

	Mag		Gyr		MGP		Pro		ProA	
	RMS/°	**Mean/°**	**RMS/°**	**Mean/°**	**RMS/°**	**Mean/°**	**RMS/°**	**Mean/°**	**RMS/°**	**Mean/°**
A	3.55	−0.08	3.55	2.67	3.45	2.50	2.25	1.07	1.70	0.47
B	10.20	6.50	3.77	0.76	3.66	0.88	3.28	1.85	2.69	−0.36
C	14.76	−10.95	5.41	3.92	5.35	3.99	3.39	2.85	2.02	1.00
D	15.02	−13.00	5.57	−4.58	5.51	−4.58	2.58	2.24	2.19	1.76
Mean/°	10.88	−4.38	4.58	0.69	4.49	0.69	2.88	2.00	2.15	0.71

In order to intuitively and precisely express the accuracy of heading estimation, the estimated heading was used for the positioning estimation of PDR. The method of Mag was not used to estimate the position in this section because the heading accuracy is poor, and we designed four programs to obtain the position estimation of these two kinematic experiments.

(1) The Gyr is used to calculate heading, and UKF is used for position estimation (Gyr-UKF).
(2) The MGP is used to obtain heading, and position is estimated based on UKF (MGP-UKF).
(3) The Pro is applied to estimate heading, and position estimation is obtained by UKF (MGP-UKF).
(4) The ProA is applied to estimate heading, and UKF is used for position estimation (MGP-UKF).

We compare the position estimation of these four methods in this section.

Figure 7 shows the position trajectories calculated by the outputs of the HV9. We can see from Figure 7 that these position trajectories of Gyr-UKF and MGP-UKF deviate from the reference with time, and the results of position estimation for these two methods are almost identical. The trajectory is improved by the method of Pro-UKF, and the position trajectory of ProA-UKF is basically the same as that of Pro-UKF. Figure 8 shows these trajectories obtained by XM8. We can see that these trajectories of Gyr-UKF and MGP-UKF deviate from the reference with time, and the trajectory of MGP-UKF is slighter better than that of Gyr-UKF. The trajectory is improved by the method of Pro-UKF, and the trajectory is further improved when ProA is used to estimate heading. Therefore, the trajectory calculated by a ProA-UKF algorithm is more accurate than that obtained by the other three methods.

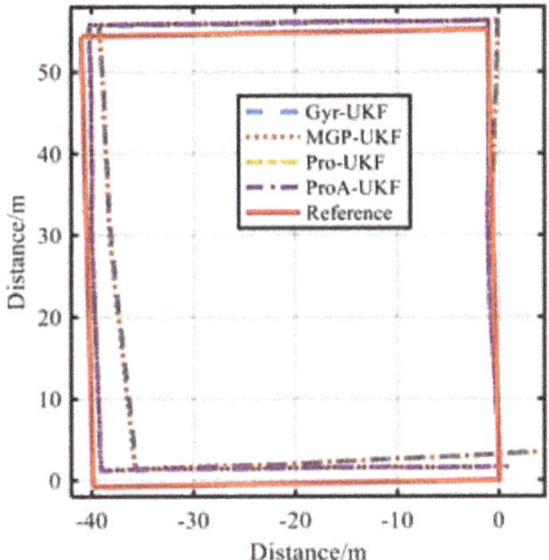

Figure 7. Trajectories of different heading estimations for HV9.

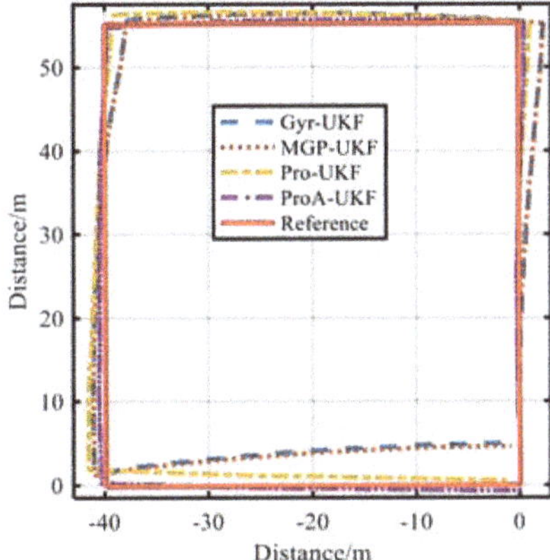

Figure 8. Trajectories of different heading estimations for XM8.

In order to provide a reference at each step, these timestamps were recorded when the tester passed corners, and the coordinates of these corners can be obtained by the reference trajectory and reference heading. The coordinates of each step between these two corners can be calculated by linear interpolation in time. In order to accurately analyze the heading, the step-length error is ignored here. Figure 9 shows the 2D position error of each step using HV9 and XM8. Figure 9a shows the position error of HV9. When the method of Gyr-UKF and MGP-UKF is used, the position error gradually accumulates with the walking distance, and the maximum error exceeds 5.5 m. The estimated position of MGP-UKF is slightly better than that of Gyr-UKF. The position error is improved by the Pro-UKF, and the maximum error is less than 2.8 m. We can also see that the position error of ProA-UKF is slightly smaller than that of Pro-UKF. Figure 9b is the position error of XM8 for different methods. The position error deviates from the reference with time when the Gyr-UKF is used, and the maximum error is over 5.0 m. The position estimation is slightly improved when MGP-UKF is used. The position errors are effectively constrained when the proposed algorithm of Pro-UKF is used, and the maximum error is less than 3.0 m. The position errors are further improved if the adaptive algorithm is used to estimate the heading, and the maximum error is less than 2.0 m. In Figure 9b, the positions accuracy of Pro-UKF is worse than those of Gyr-UKF and MGP-UKF at times. We can see from Figure 6c that the heading error is mostly positive for Pro, that will make the error accumulate, and the heading errors have positive and negative values for Gyr and MGP that will offset the error accumulation.

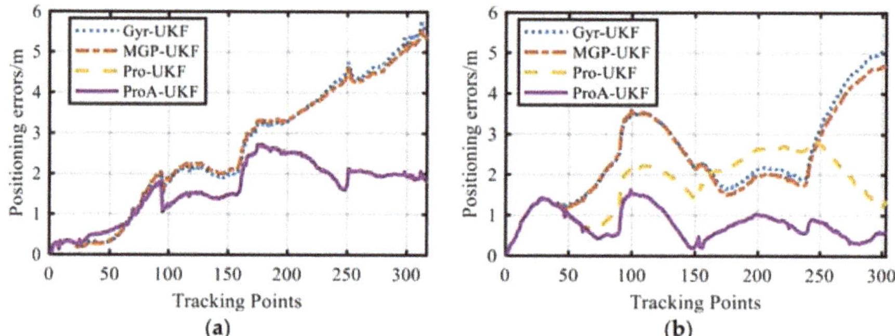

Figure 9. Position error of each step with different methods for different smartphones: (**a**) The position error of HV9; (**b**) The position error of XM8.

The corresponding position cumulative error estimation percentages (CEF) of Figure 9 is explained in Figure 10. In Figure 10a, the CEF within 1.5 m is 24.92% and 25.24% for Gyr-UKF and MGP-UKF, respectively, and the CEF is raised to 40.38% and 41.01% for Pro-UKF and ProA-UKF, respectively. The CEF within 2 m are 38.80% and 31.86% for Gyr-UKF and MGP-UKF, respectively. The CEF is increased to 68.45% for both Pro-UKF and ProA-UKF. When the CEF is about 75%, the position error is over 4 m for Gyr-UKF and MGP-UKF, and the position error is less than 2.25 m for Pro-UKF and ProA-UKF. For HV9, the proposed method can enhance the accuracy of position estimation. Figure 10b shows the CEF of XM8. In Figure 10b, the CEF within 1.5 m is just 20.79% for both Gyr-UKF and MGP-UKF, the CEF achieves 36.30% for Pro-UKF, and the CEF is, as expected, increased to 97.03% for ProA-UKF. When the position error is within 2 m, the CEF is 38.94% and 47.52% for Gyr-UKF and MGP-UKF, respectively, and the CEF is slight improved for Pro-UKF with 50.50%, however, the CEF is raised to 100% when the ProA-UKF is used. When the CEF is about 75%, the position errors are both over 3.25 m for Gyr-UKF and MGP-UKF, the error is within 2.50 m for Pro-UKF, and the error is less than 1.25 m for ProA-UKF. Therefore, we can obtain the conclusion that the proposed algorithm can improves the accuracy of heading estimation for smartphone.

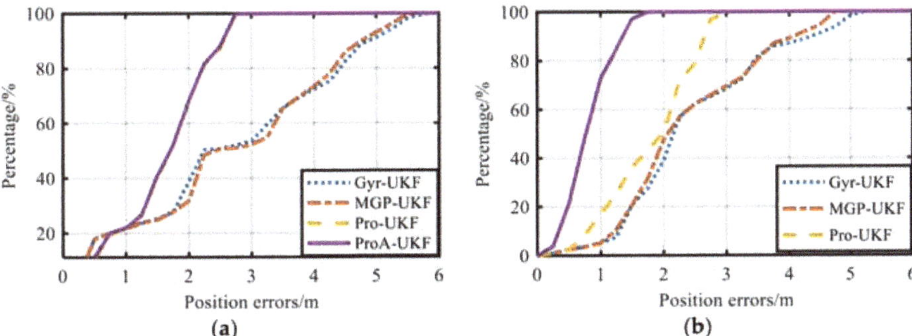

Figure 10. Position cumulative error percentages of different methods for different smartphones: (**a**) The error percentage for HV9; (**b**) The error percentage for XM8.

Table 3 shows the statistics of position errors. In Table 3, the mean errors of position estimation for PDR are 2.67 m, 2.65 m, 1.60 m and 1.59 m for these four respective schemes of position estimation, and the corresponding RMS values are 3.14 m, 3.10 m, 1.75 m and 1.74 m for the HV9 smartphone. For the XM8 smartphone, the mean errors are 2.47 m, 2.38 m, 1.81 m and 0.80 m, and the RMS values are 2.73 m, 2.62 m, 1.94 m and 0.87 m for these four respective schemes. The estimated position by ProA-UKF, where the adaptive filter is used to estimate heading, is better than that by Pro-UKF, the position error of Pro-UKF is smaller than that of the traditional method and the position error of MGP-UKF is slightly superior to that of Gyr-UKF. Compared with the method of MGP, the mean error of ProA-UKF is reduced by 40.00%, and RMS is improved by 43.87% for the HV9. And the mean error is reduced by 66.39%, and the RMS is improved by 66.79% for the XM8. Therefore, the method of heading estimation proposed in this paper can improve the accuracy of heading estimation, and it directly enhances the accuracy of position estimation for pedestrian navigation.

Table 3. Statistics of position error with different method for different smartphone.

		Gyr-UKF	MGP-UKF	Pro-UKF	ProA-UKF
HV9	Max/m	5.80	5.63	2.75	2.73
	Mean/m	2.67	2.65	1.60	1.59
	RMS/m	3.14	3.10	1.75	1.74
XM8	Max/m	5.05	4.69	2.89	1.63
	Mean/m	2.47	2.38	1.81	0.80
	RMS/m	2.73	2.62	1.94	0.87

8. Conclusions

A novel method of heading estimation based on smartphones is proposed in this paper. This method uses the AR model to build KF, and the current heading—determined by fusing these headings calculated by magnetometer, estimated by gyroscope and stored by the previous estimate—is used as measurement value of KF update. To effectively suppress the accumulated error of the heading estimated by the outputs of the gyroscope, the estimated heading by KF in QS is used to correct the heading calculated by gyroscope. To improve the accuracy of heading estimation, an AKF based on the innovation sequence is used to adaptively adjust the state noise covariance. HV9 and XM8 smartphones were used to collect data. In order to intuitively and precisely express the accuracy of heading estimation, the estimated heading is also used for the position estimation of PDR. The following conclusions were obtained,

(1) The heading calculated by a magnetometer is susceptible to disturbances in the complex indoor environment that can cause dozens of degrees bias. Also, it is possible that the heading calculated

by magnetometer is close to the reference at some locations. The heading estimated by gyroscope is affected by cumulative error. The heading obtained by the proposed method can not only reduce the noise but also constraint the cumulative error.

(2) Compared with the mean accuracy of traditional methods for heading estimation, the mean accuracy of the heading estimated by proposed algorithm is increased by 35.62% and 35.86%, respectively, for HV9 and XM8. And compared with KF, AKF can further improve the accuracy of heading estimation.

(3) The estimated heading is used for position calculation of PDR. Compared to traditional methods, the mean errors of position estimation for the proposed method decrease by 40.00% and 66.39%, and the corresponding RMS is improved by 43.87% and 66.79%, respectively, for HV9 and XM8 smartphones with different performance sensors.

Author Contributions: D.C. proposed the research ideas and technical lines. G.C. and S.W. provided research guidance. D.C. wrote the paper and G.C. corrected this paper.

Funding: This work was supported by the national Key Research and Development Program of China under grant 2016YFB0502105, a project Funded by the Science Foundation of Jiangsu Province under grant bk20161181 and the National Natural Science Foundation of China under grant 41371423.

Conflicts of Interest: The authors declare no conflict of interest.

References

1. Alletto, S.; Cucchiara, R.; Fiore, G.; Mainetti, L.; Mighali, V.; Patrono, L.; Serra, G. An indoor location-aware system for an IoT-based smart museum. *IEEE Internet Things J.* **2016**, *3*, 244–253. [CrossRef]
2. Chen, J.; Ou, G.; Peng, A.; Zheng, L.; Shi, J. An INS/WiFi indoor localization system based on the weight least squares. *Sensors* **2018**, *15*, 1458. [CrossRef] [PubMed]
3. Liu, H.; Yang, J.; Sidhom, S.; Wang, Y.; Chen, Y.; Ye, F. Accurate WiFi based localization for smartphone using peer assistance. *IEEE Trans. Mob. Comput.* **2014**, *13*, 2199–2213. [CrossRef]
4. Li, X.; Wang, J.; Liu, C. A Bluetooth/PDR integration algorithm for an indoor positioning system. *Sensors* **2015**, *15*, 24862–24885. [CrossRef]
5. Chen, P.; Kuang, Y.; Chen, X. A UWB/Improved PDR Integration Algorithm Applied to Dynamic Indoor Positioning for Pedestrians. *Sensors* **2017**, *17*, 2065. [CrossRef]
6. Harle, R. A survey of indoor inertial positioning systems for pedestrians. *IEEE Commun. Surv. Tutor.* **2013**, *15*, 1281–1293. [CrossRef]
7. Li, X.; Wang, J.; Liu, C.; Zhang, L.; Li, Z. Integrated WiFi/PDR/Smartphone using an adaptive system noise extended Kalman filter algorithm for indoor localization. *Int. J. Geo-Inf.* **2016**, *5*, 8. [CrossRef]
8. Chen, R.; Chen, L. Indoor positioning with smartphones: The state-of-art and challenges. *Acta Geod. Cartogr. Sin.* **2017**, *46*, 1316–1326.
9. Li, Y.; Zhuang, Y.; Lan, H.; Zhou, Q.; Niu, X.; El-Sheimy, N. A hybrid WiFi/Magnetic matching/PDR approach for indoor navigation with smartphone sensors. *IEEE Commun. Lett.* **2016**, *20*, 169–172. [CrossRef]
10. Kuang, J.; Niu, X.J.; Chen, X. Robust pedestrian dead reckoning based on MEMS-IMU for smartphone. *Sensors* **2018**, *18*, 1391. [CrossRef] [PubMed]
11. Davidson, P.; Takala, J. Algorithm for pedestrian navigation combining IMU measurements and gait models. *Gyroscope Navig.* **2013**, *4*, 79–84. [CrossRef]
12. Gu, Y.; Song, Q.; Li, Y.; Ma, M. Foot-mounted pedestrian navigation based on particle filter with an adaptive weight updating strategy. *J. Navig.* **2015**, *68*, 23–38. [CrossRef]
13. Zhuang, Y.; Lan, H.; Li, Y.; El-Sheimy, N. PDR/INS/WiFi integration based on handheld devices for indoor pedestrian navigation. *Sensors* **2015**, *6*, 793–812. [CrossRef]
14. Kim, Y.; Choi, S.; Kim, H.; Lee, J. Performance improvement and height estimation of pedestrian dead-reckoning system using a low cost MEMS sensor. In Proceedings of the 12th International Conference on Control, Automation and Systems, Jeju Island, Korea, 17–21 October 2012.

15. Kang, W.; Nam, S.; Lee, S. Improved heading estimation for smartphone-based indoor positioning systems. In Proceedings of the 2012 IEEE 23rd International Symposium on Personal, Indoor and Mobile Radio Communications, Sydney, Australia, 9–12 September 2012.
16. Brajdic, A.; Harle, R. Walk detection and step counting on unconstrained smartphones. In Proceedings of the 2013 ACM International Joint Conference on Pervasive and Ubiquitous Computing, Zurich, Switzerland, 8–12 September 2013.
17. Hasan, M.; Mishuk, M. MEMS IMU based pedestrian indoor navigation for smart glass. *Wirel. Pers. Commun.* **2018**, *101*, 287–303. [CrossRef]
18. Tian, Z.; Zhang, Y.; Zhou, M.; Liu, Y. Pedestrian dead reckoning for MARG navigation using a smartphone. *J. Adv. Signal Process.* **2014**, *65*, 65–73. [CrossRef]
19. Gade, K. The seven ways to find heading. *J. Navig.* **2016**, *69*, 955–970. [CrossRef]
20. Afzal, M.; Renaudin, V.; Lachapelle, G. Magnetic field based heading estimation for pedestrian navigation environments. In Proceedings of the 2011 International Conference on Indoor Positioning and Indoor Navigation, Guimaraes, Portugal, 21–23 September 2011.
21. Zheng, Z.; Hu, Y.; Yu, J.; Na, Z. Extended Kalman filter for real time indoor localization by fusing WiFi and smartphone inertial sensors. *Sensors* **2015**, *6*, 523–543.
22. Kang, W.; Han, Y.N. SmartPDR: Smartpone-based pedestrian dead reckoning for indoor localization. *IEEE Sens. J.* **2015**, *15*, 2906–2916. [CrossRef]
23. Valenti, R.; Dryanovski, I.; Xiao, J. A linear Kalman filter for MARG orientation estimation using the algebraic quaternion algorithm. *IEEE Trans. Instrum. Meas.* **2015**, *65*, 467–481. [CrossRef]
24. Renaudin, V.; Combettes, C.; Peyret, F. Quaternion based heading estimation with handheld MEMS in indoor environments. In Proceedings of the IEEE/ION Positioning, Location and Navigation Symposium, Monterey, CA, USA, 5–8 May 2014.
25. Yuan, X.; Yu, S.; Zhang, S. Quaternion-based unscented Kalman filter for accurate indoor heading estimation using wearable multi-sensor system. *Sensors* **2015**, *15*, 10872–10890. [CrossRef]
26. Afzal, M.; Renaudin, V.; Lachapelle, G. Use of earth's magnetic field for mitigating gyroscope errors regardless of magnetic perturbation. *Sensors* **2011**, *11*, 11390–11414. [CrossRef]
27. Ali, A.; El-Sheimy, N. Low-cost MEMS-based pedestrian navigation technique for GPS-denied areas. *J. Sens.* **2013**, *2013*, 197090. [CrossRef]
28. Mohamed, A.; Schwarz, K. Adaptive Kaman filtering for INS/GPS. *J. Geod.* **1999**, *73*, 193–203. [CrossRef]
29. Chang, G.; Liu, M. An adaptive fading Kalman filter based on Mahalanobis distance. *Proc. Inst. Mech. Eng. Part G J. Aerosp. Eng.* **2015**, *229*, 1114–1123. [CrossRef]
30. Jiang, C.; Zhang, S. A novel adaptively-robust strategy based on the Mahalanobis distance for GPS/INS integrated navigation systems. *Sensors* **2018**, *18*, 695. [CrossRef]
31. Wu, D.; Xia, L.; Geng, J. Heading estimation for pedestrian dead reckoning based on robust adaptive Kalman Filtering. *Sensors* **2018**, *18*, 1970. [CrossRef]
32. Li, W.; Wang, J. Effective adaptive Kalman filter for MEMS-IMU/Magnetometers integrated attitude and heading reference system. *J. Navig.* **2013**, *66*, 99–113. [CrossRef]
33. Jin, B.; Guo, J.; He, J.; Guo, W. Adaptive Kalman filter based on optimal autoregressive predictive model. *GPS Solut.* **2017**, *21*, 307–317. [CrossRef]
34. Jin, B.; Su, T.; Zhang, W.; Zhang, L. Joint optimization of predictive model and transmitted waveform for extended target tracking. In Proceedings of the 2014 12th International Conference on Signal Processing (ICSP), Hangzhou, China, 19–23 October 2014.
35. Ding, W.; Wang, J.; Rizos, C. Improving adaptive Kalman estimation in GPS/INS integration. *J. Navig.* **2007**, *60*, 517–529. [CrossRef]
36. Renaudin, V.; Afzal, M.; Lachapelle, G. Complete triaxis magnetometer calibration in the magnetic domain. *J. Sens.* **2010**, *2010*, 967245. [CrossRef]
37. Klingbeik, L.; Eling, C.; Zimmermann, F.; Kuhlmann, H. Magnetic field sensor calibration for attitude determination. *J. Appl. Geod.* **2014**, *8*, 97–108.
38. Ma, M.; Song, Q.; Gu, Y.; Zhou, Z. Use of magnetic field for mitigating gyroscope errors for indoor pedestrian positioning. *Sensors* **2018**, *18*, 2592. [CrossRef]

39. Li, F.; Zhao, C.; Ding, G.; Gong, J.; Liu, C.; Zhao, F. A reliable and accurate indoor location method using phone inertial sensors. In Proceedings of the 2012 ACM Conference on Ubiquitous Computing, New York, NY, USA, 5–8 September 2012.
40. Chen, G.; Meng, X.; Wang, Y.; Zhang, Y.; Tian, P.; Yang, H. Integrated WiFi/PDR/Smartphone using an unscented Kalman filter algorithm for 3D indoor localization. *Sensors* **2015**, *15*, 24595–24614. [CrossRef]

Article

Mobile Robot Path Planning Using a Laser Range Finder for Environments with Transparent Obstacles

Jin-Woo Jung *, Jung-Soo Park, Tae-Won Kang, Jin-Gu Kang and Hyun-Wook Kang

Department of Computer Science and Engineering, Dongguk University, Seoul 04620, Korea; hostkit@naver.com (J.-S.P.); ktw3388@dgu.ac.kr (T.-W.K.); kanggu12@dongguk.edu (J.-G.K.); kangtaepoong@naver.com (H.-W.K.)

* Correspondence: jwjung@dongguk.edu; Tel.: +82-2-2260-3812

Received: 10 March 2020; Accepted: 15 April 2020; Published: 17 April 2020

Abstract: Environment maps must first be generated to drive mobile robots automatically. Path planning is performed based on the information given in an environment map. Various types of sensors, such as ultrasonic and laser sensors, are used by mobile robots to acquire data on its surrounding environment. Among these, the laser sensor, which has the property of being able to go straight and high accuracy, is used most often. However, the beams from laser sensors are refracted and reflected when it meets a transparent obstacle, thus generating noise. Therefore, in this paper, a state-of-the-art algorithm was proposed to detect transparent obstacles by analyzing the pattern of the reflected noise generated when a laser meets a transparent obstacle. The experiment was carried out using the environment map generated by the aforementioned method and gave results demonstrating that the robot could avoid transparent obstacles while it was moving towards the destination.

Keywords: transparent obstacle recognition; reflection noise; laser range finder; path planning; mobile robot

---

1. Introduction

The goal of path planning is to provide a path that is safe from obstacle collisions and to guide any object through the most appropriate path that has the shortest distance [1–5]. To accomplish this goal, a robot must collect surrounding data to generate an environment map.

An environment map refers to a map that features data on the surroundings in which a mobile robot stands. Although a robot uses sensors to generate an environment map, the data acquired through these sensors are always affected by elements of the surrounding environment. For this reason, such data always contain inaccuracies. Therefore, these inaccuracies in the data must be considered when creating environment maps.

Sensors that are often used to detect distances from obstacles include ultrasound sensors [6], laser sensors [7,8], etc. Among these, ultrasound sensors use the elapsed time it takes for a sound wave to come back to the sensor after hitting the obstacle to detect distance. However, because ultrasound sensors use sound as a medium, the measurable distance is short, and detecting distance is difficult if the direction of the obstacle and the sound wave do not align vertically. Laser sensors measure an object's distance away from obstacles using lasers [9,10]. The properties of laser sensors include their ability to direct straight beams, they are capable of measuring long distances, and they are more accurate compared to other sensors [11]. However, due to these properties, a drawback of laser sensors is that they cannot accurately detect transparent obstacles. Therefore, a mobile robot misoperates in an environment that includes transparent obstacles [6,12].

When a laser beam meets a transparent obstacle, as shown in Figure 1, reflection, refraction, and penetration occur, and they cause the inaccurate detection of transparent obstacles.

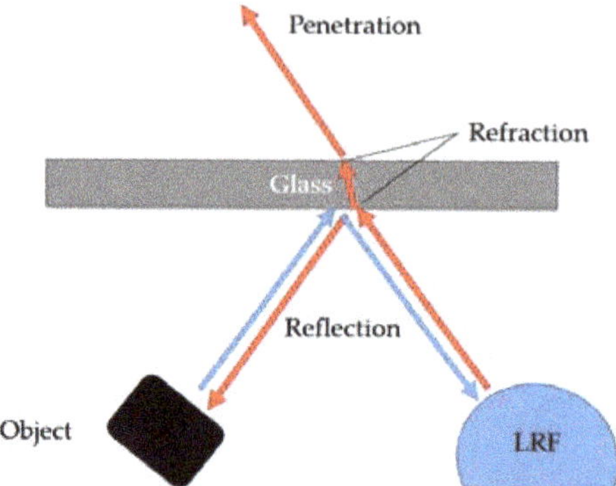

Figure 1. The phenomenon when measuring transparent object using a laser range finder (LRF).

The reflection noise that occurs at the opposite side of the transparent obstacle could be found from the results of measuring an object's distance away from the transparent obstacle using an actual laser range finder (LRF). Accordingly, in this paper, a state-of-the-art algorithm that can detect the boundaries of transparent obstacles from reflection data obtained solely by the LRF has been proposed.

2. Related Works

2.1. Environment Map

The first step to path planning for automatic mobile robots is to configure the environment map. Many different methods of generating environment maps according to different methods of expression exist [13,14]. For example, there are grid maps, feature-based maps, etc [15,16].

A grid map divides the map into cell units, and the occupancy state of each cell is denoted between 0 and 1 to represent probability with the inaccuracy of the cell being measured taken into consideration [17] (Figure 2). The benefit of a grid map is that it is easy to generate, applies quickly to the changing environment that surrounds an object, and provides good accessibility. However, the grids result in inaccuracies in the real environment, and the complexity of grid maps is high because all cells must hold an occupancy state [14,18]. Recently, there has been the paper on path planning using a grid-based potential field [19].

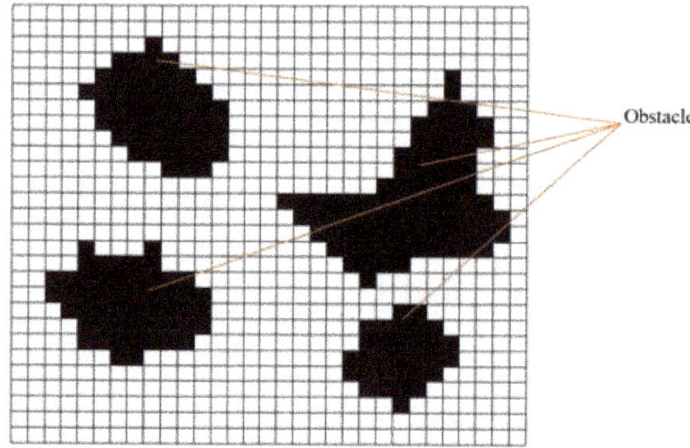

Figure 2. An example of an occupy grid map.

Feature-based maps (Figure 3) can better express the real environment compared to grid maps because they indicate obstacles with points and lines. However, when it comes to map generation, the accuracy of an environment map is increased only when sensors with high accuracy are used because feature-based maps are highly affected by the performance of the sensor [20].

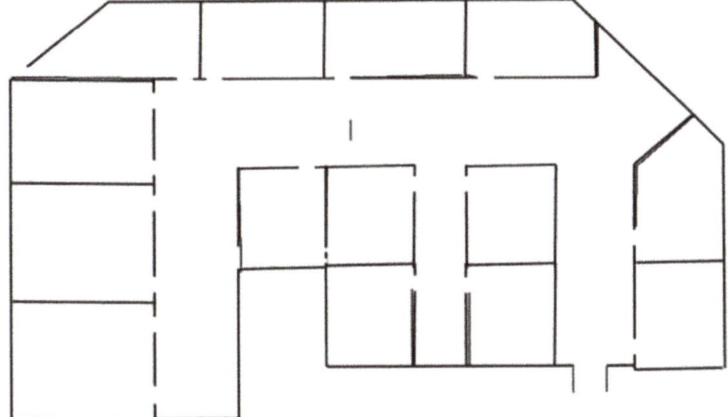

Figure 3. An example of a feature-based map.

2.2. Path Planning

Path planning is about creating the shortest path with safe guidance and obstacle avoidance to the destination from the starting point in which a mobile robot stands [21]. Previous studies on environment map generation and path planning have been carried out under the assumption that a mobile robot already knows of the surrounding environment information [1,3,4,13].

However, in real environments, obstacles that had not existed in the environment map previously happen to appear, and mobile robots operate in dynamic environments where obstacles are moving; it is hard to apply static environment maps to the real environment under the assumption that robots are already equipped with all the required information [22]. Therefore, as a solution to the aforementioned problem, there are algorithms, such as D* algorithm [23–25], Wavefront-propagation algorithm [1,4,26,27], etc., that can be applied to make path planning easier.

The D* algorithm allows robots to avoid obstacles by partially performing path planning in the vicinity of obstacles if these robots meet an obstacle while they are moving. The D* algorithm has the benefit of a fast reaction to obstacles compared to the A* algorithm that needs to recalculate the whole environment map to generate a new path whenever an object meets a new obstacle.

Also, the D* algorithm is applied as the D* Lite algorithm [13] and is used as a motion-planning algorithm to ensure collision-free movement with a transportation mobile robot [28].

In addition, the algorithms based on A* algorithm also have been widely studied recently. For example, the improved A* algorithm [29] has reduced the computing time by considering the parent nodes during path planning and the smoothed A* algorithm [30] is used for vessels path planning.

The Wavefront-propagation algorithm has the advantage of being able to quickly generate a path that is close to the optimized one within the occupancy grid map. In addition, it has the advantage of being able to quickly apply the newly detected obstacle information from a dynamic environment to the environment grid map [21].

2.3. Detection of Transparent Obstacles

The goal of this paper is to detect transparent obstacles using LRF. In recent previous studies, some efforts have been carried out to detect the presence of transparent obstacles, which use the red-green-blue (RGB) cameras and depth to recognize transparent obstacles [31], and using the difference between one image of a transparent object that is captured with lights and the other without lights [32].

However, these methods require the use of additional measurement equipment such as RGB-D camera for the RGB-D method [31] or lighting device for the capturing with lights method [32].

3. Mobile Robot System

3.1. Overall System Configuration

For a mobile robot to generate an environment map and perform path planning, the overall system configuration of the mobile robot should be conducted as shown in Figure 4. Regarding hardware, the mobile robot collects data through sensors, and based on these data, an environment map is generated through software. A mobile robot executes path planning based on environment map and move along the path using actuator.

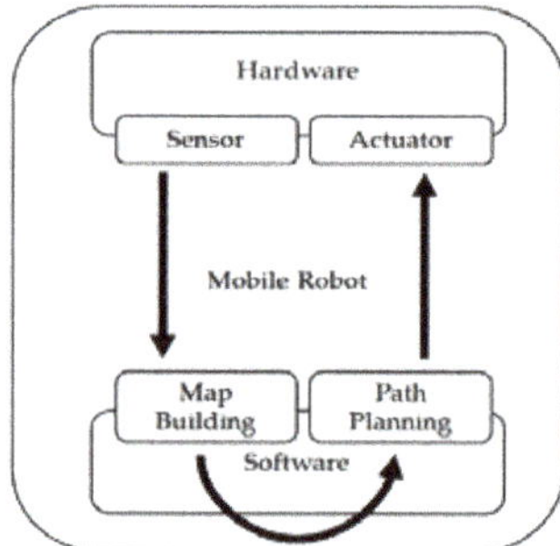

Figure 4. System overview.

3.2. Mobile Robot

Our experiment was carried out using the actual mobile robot. The Pioneer (Pioneer company, Japan) [33] P3DX (Figure 5) that is widely used for research purposes had been used in experiments related to mobile robots.

Figure 5. Pioneer P3-Dx.

3.3. Laser Range Finder (LRF)

The goal of this paper was to create an environment map that allows mobile robots to conduct automatic driving using only LRF within an environment where transparent obstacles are included. As such, the LMS100-10000 laser range finder (Figure 6) from SICK (Sick AG company, Germany) [34] was used.

Figure 6. SICK (Sick AG company, Germany) laser range finder (LMS100-10000).

The maximum measurable angle of the corresponding LRF is 270 degrees, the measurable distance varies from a minimum of 50 cm to a maximum of 20 m, the systemic error is + −30mm, and the stochastic error rate is + −12 mm. Additionally, it has an angular resolution of 0.25 degrees or 0.5 degrees [17]. In this study, an angular resolution of 0.25 degrees was used, and 1081 units of information can be collected within the range when the robot stands between −135 and 135 degrees using the 0.25 degree of angular resolution.

After collecting data on the obstacles surrounding the robot, the coordinate system requires additional transformation to apply this data to the environment map [16,26,34,35]. The data collected using LRF corresponds to the polar coordinate system and contains information of the distance away from the sensor and direction. This data must also be converted to a rectangular coordinate system that the robot uses.

The following equation can be derived from Figure 7.

$$x_{obstacle} = x_{robot} + d * \cos\left(\theta_{lrf} + \theta_{robot}\right),\ y_{obstacle} = y_{robot} + d * \sin\left(\theta_{lrf} + \theta_{robot}\right) \quad (1)$$

x_{robot} and y_{robot} correspond to the current coordinates of the robot in the rectangular coordinate system. θ_{robot} is the direction of the robot, and θ_{lrf} is the angle between the beam and the direction that the robot is moving towards. d indicates the distance to the obstacle from the robot's position. Using these methods, the coordinates of $x_{obstacle}$ and $y_{obstacle}$ can be found on the rectangular coordinate system.

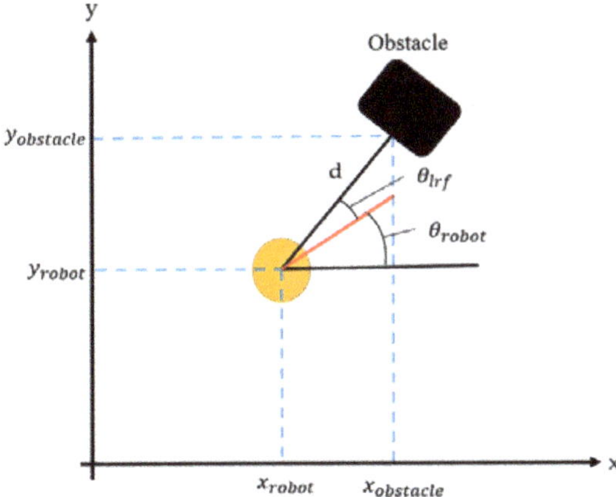

Figure 7. The relationship between the coordinates of detected obstacles and the coordinates of the robot.

4. Environment Map Generation Algorithm in a Transparent Obstacle Environment

The environment map generation algorithm using LRF requires the three following assumptions. *[Assumptions]*

1. Transparent obstacles are in the form of a straight line.
2. No opaque obstacle exists adjacent to the transparent obstacle that is on the opposite side of the LRF.
3. All obstacles are fixed in their positions, and they are not known.

The proposed environment map generation for a transparent obstacle environment processes the reflection noises that occur by using LRF. This method is used to extract reflection noise that is different from common noise, and the boundaries of the transparent obstacles are found based on the form and characteristics of the extracted reflection noises [27].

In this paper, the occupancy grid map is chosen in the creation of an environment map. Although feature-based maps better represent actual environments, they have the drawback of being weak to noises; however, occupancy grid maps are highly resistant to noise, and their measured data can be applied quickly to environment maps without additional operations [15]. Occupancy grid maps still require a lot of computing time [36]; however, this can be overcome through recent improvements in computing power [5,9,10].

The following table explains the terminology required for the algorithm.

Based on the terminology in Table 1, the algorithm used to generate an environment map within an environment that includes transparent obstacles is shown below.

Table 1. Terminology essential to algorithm explanation.

Terminology	Explanation
$\theta_n(t)$	The angle value of nth sensor beam measured at time t $\theta_n(t) = \theta_{min} + \theta_{step}*(n\text{-}1)$, n:1 ~ 1081, θ_{min} = -45(deg), θ_{step} = 0.25(deg)
$d_n(t)$	The distance towards the direction $\theta_n(t)$ measured at time t
D_{min}	The minimum distance that sensor can recognize: 50 cm
D_{max}	The maximum distance that sensor can recognize: 20 m
$P_R(t)$	The position (x, y) of the robot at time t
$\theta_R(t)$	The direction that robot is heading towards at time t
$f(d_n(t), \theta_n(t), P_R(t), \theta_R(t))$	The position of the cell located from $P_R(t)$ with the distance $d_n(t)$ and the heading angle $\theta_R(t) + \theta_n(t)$ (Represented as cell No.)
$cell_n(t)$	The position of the cell on the grid map calculated using $f(d_n(t), \theta_n(t), P_R(t), \theta_R(t))$ at time t (Represented as cell No.)
$cell_R(t)$	The position of the robot center point on the grid map at time t (represented as cell No.)
WDCMap(cell)	The temporary local map representing the number of detected sensor beams in a certain cell (weakly detected counter (WDC) map)
SDCMap(cell)	The global map representing the accumulated number of detected sensor beams in a certain cell to decide obstacle cell (strongly detected counter (SDC) map)
$CELL_0(t)$	The set of cells with *WDCMap*(cell) = 0 at time t
$CELL_{neighor}(cell_{n-1}(t-1))$	The set of cells pointed by n-1-rth, ⋯ , n-1-1th, n-1th, n-1+1th, ⋯ , n-1+rth beams at time t-1 (default value: r = 1)

Table 1. *Cont.*

Terminology	Explanation
$g(x)$	A function emphasizing the extent of increase of the *SDCMap*(cell) (default function: $g(x) = x$)
$TH_{non-glass}$	The threshold number of *WDC* to determine opaque obstacle (default value = 5, WDC = weakly detected counter)
TH_{glass}	The threshold number of WDC to determine the transparent obstacle candidate (default value = 1, $TH_{glass} < TH_{non-glass}$)
$TH_{glass-surface}$	The threshold number of WDC to determine the surface of transparent obstacle (default value = 3)

Algorithm 1. Pseudo code of the algorithm used to create environment maps for environments with transparent obstacles.

Input:
$d_n(t) \leftarrow$ Distance towards the direction $\theta_n(t)$ measured at time t
$\theta_n(t) \leftarrow$ Angle value of nth sensor beam measured at time t
$P_R(t) \leftarrow$ Position (x, y) of the robot at time t
$\theta_R(t) \leftarrow$ Direction that robot is heading towards at time t
Output:
SDCMap $\leftarrow$ Global map representing the accumulated number of detected sensor beams in a certain cell to decide obstacle cell
Begin ***Algorithm Creating Environment maps with Transparent Obstacle***
For $d_n(t)$ ***in*** all sensor beams n: $\{1 \sim 1081\}$ ***do***
  If $D_{min} \leq d_n(t) \leq D_{max}$ ***then***
    $cell_n(t) \leftarrow f(d_n(t), \theta_n(t), P_R(t), \theta_R(t))$
  Else
    $cell_n(t) \leftarrow -1$
  If $cell_n(t) == cell_{n-1}(t)$ ***then***
    $WDCMap(cell_{n-1}(t)) \mathrel{+}= 1$
  Else If $P_R(t-1) \neq P_R(t)$ ***then***
    If $WDCMap(cell_{n-1}(t)) \geq TH_{non-glass}$ ***then***
      $SDCMap(cell_{n-1}(t)) \mathrel{+}= g(WDCMap(cell_{n-1}(t)))$
    Else If $WDCMap(cell_{n-1}(t)) \geq TH_{glass}$ ***Or*** $cell_{n-1}(t) \in CELL_{neighor}(cell_{n-1}(t-1))$ ***then***
      $L \leftarrow$ ***Rasterization***$(cell_R(t), cell_{n-1}(t))$
// L: The list of cells included in the virtual line segment
// connecting $L[0]$: $cell_R(t)$ and $L[\text{NUM\_CELLS}]$: $cell_{n-1}(t)$
      $cell_{L1} \leftarrow$ ***Find_NearCell***$(cell_R(t), L)$ // Function finding the closest but not the
// same cell from $cell_R(t)$ in the list L
  loop:
      SDCMap $\leftarrow$ ***DetectObstacleCandidate***$(cell_{L1}, L, \text{NUM\_CELLS})$
      Goto *loop*
      If there is no $cell_{L1}$ in L s.t. $SDCMap\ (cell_{L1}) > 0$ ***then***
        $SDCMap(cell_{n-1}(t)) \mathrel{+}= 1$
    Else
      If $cell_{n-1}(t) \in CELL_0(t)$ ***then***
        $SDCMap(cell_{n-1}(t)) \mathrel{+}= 1$
      Else
        Append $cell_{n-1}(t)$ ***to*** $CELL_0(t)$
    $WDCMap(cell_n(t)) \leftarrow 0$
End ***Algorithm Creating Environment maps with Transparent Obstacle***

In Algorithm 1, whether $d_n(t)$ belongs to the range of $D_{min} \leq d_n(t) \leq D_{max}$ is checked first to delete all inaccurate distance information about $d_n(t)$ according to its $\theta_n(t)$ measured from LRF at time t. The distance is calculated using Equation (1) if it belongs to the range.

In this paper, LRF's angular resolution is 0.25 degrees, and the size of the cell is 15 cm in the grid map; therefore, two beams detect the same $cell_n(t)$ for a cell that has an opaque obstacle [28]. The angle between the *n-1*th beam and the *n*th beam is 0.25 degrees, more than 35 m distance is needed if two beams point to different cells (Figure 8). The value of D_{max} is 20 m. Therefore, a cell that has an opaque obstacle is always detected by more than two beams.

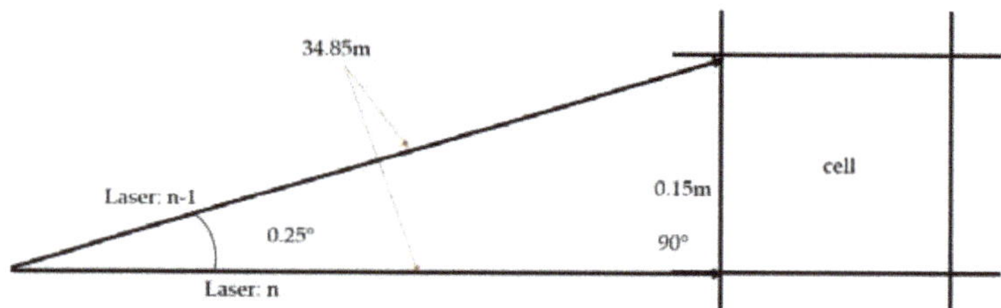

Figure 8. The relationship between the cell and the laser beam.

Based on the above information, if $cell_n(t)$ and $cell_{n-1}(t)$ point to the same cell, the probability of $cell_{n-1}(t)$ being an obstacle is increased by adding 1 to *WDCMap*($cell_{n-1}(t)$). If ($cell_n t$) and $cell_{n-1}(t)$ are different, the *SDCMap*($cell_{n-1}(t)$) is updated according to the value of *WDCMap*($cell_{n-1}(t)$). The algorithm proposed in this paper uses the common feature of reflection noises generated from transparent obstacles while the robot is moving; if the *SDCMap* is updated while the robot is not moving, redundant detections are accumulated. Therefore, the *SDCMap* may not be updated while the robot is not moving.

Updates to the *SDCMap* are provided through different methods according to measurements of *WDCMap*($cell_{n-1}(t)$). If the *WDCMap*($cell_{n-1}(t)$) > $TH_{non-glass}$, this denotes that many beams are indicating the same cell simultaneously. This denotes that the probability of ($cell_{n-1}t$) being an opaque cell is high; therefore, the value of the *SDCMap*($cell_{n-1}(t)$) should be increased significantly.

Algorithm 2. Pseudo code of the algorithm used to ***DetectObstacleCandidate***().

Input:
$cell_{L1}$ ← Result of function finding the closest but not the same cell from $cell_R(t)$ in the list L
L ← List of cells included in the virtual line segment
NUM_CELLS ← Last index number of list L
Output:
SDCMap ← The global map representing the accumulated number of detected sensor beams in a certain cell to decide obstacle cell
Begin ***Algorithm Detect Obstacle Candidate***
If *SDCMap*($cell_{L1}$) > 0 ***then***
    If *WDCMap*(L[NUM_CELLS]) ≥ $TH_{glass-surface}$ ***then*** // L[NUM_CELLS]: $cell_{n-1}(t)$
        SDCMap($cell_{L1}$) += g(*WDCMap*(L[NUM_CELLS]))
    Else
        SDCMap($cell_{L1}$) += 1
    INIT_CELL ← 1
    $cell_{L'}$ ← ***Find_NearCell***($cell_{L1}$, L[INIT_CELL:NUM_CELLS])
// Function finding the closest but not the same cell from $L[0] == cell_{L1}$ in the list L
loop:

```
    If WDCMap(cell_{n-1}(t)) ≥ TH_{glass-surface} then
        SDCMap(cell_{L'}) -= g(WDCMap(cell_{n-1}(t)))
    Else
        SDCMap(cell_{L'}) -= 1
    INIT_CELL += 1 // INIT_CELL: 1, 2, ... , NUM_CELLS
    cell_{L'} ← Find_NearCell(cell_{L'}, L[INIT_CELL:NUM_CELLS])
    Goto loop
  L ← L-{cell_{L1}}  // Update L With the reduced size
  cell_{L1} ← Find_NearCell(cell_{L1}, L[INIT_CELL:NUM_CELLS-1])
End Algorithm Detect Obstacle Candidate
```

If the value of the $WDCMap(cell_{n-1}(t))$ is not high, whether $cell_{n-1}(t)$ is detected by reflection noise or common noise should be distinguished. For this purpose, the following three conditions are provided.

[Conditions]

1. The cell containing transparent obstacles have higher *WDCMap* compared to a cell detected by irregular noise due to the regularity of the reflection noise generated by transparent obstacles. Although the value of the $WDCMap(cell_{n-1}(t))$ is not higher than $TH_{non-glass}$, if it is higher than the value of TH_{glass}, it is marked as a candidate of transparent obstacle.
2. While the robot is moving, the reflection noise occurs continuously, but common noises occur intermittently by opaque obstacles and the surrounding environment. Using this property, it is considered as a candidate for a transparent obstacle if the $cell_{n-1}(t)$ at time t correspond to any of $CELL_{neighbor}(cell_{n-1}(t-1))$ at time t.
3. The reflection noise generated by transparent obstacles may contain a cell that has low *WDCMap* because it is not detected continuously; however, if it is detected regularly, it is considered to be a candidate for a transparent obstacle.

The following steps progress when conditions 1 and 2 from the above are met. At time t, a list L of cells is generated (Figure 9), consisting of the cells included in the virtual line segment connecting $cell_R(t)$ and $cell_{n-1}(t)$.

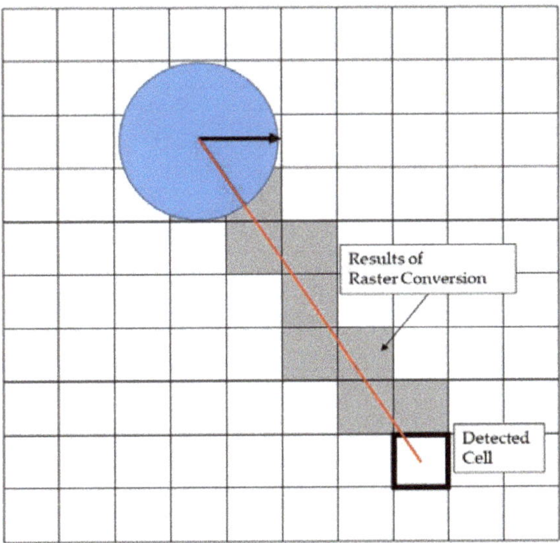

Figure 9. Example of raster conversion.

The *SDCMap* is checked from $cell_{L1}$, which is closest to the mobile robot, to all other elements in the list L if a cell identified as an obstacle candidate is found; (Algorithm 2) $cell_{n-1}(t)$ is checked whether it is detected by reflection noise by the boundaries of the transparent obstacles ($WDCMap(cell_{n-1}(t)) \geq TH_{glass-surface}$). If $cell_{n-1}(t)$ is detected by reflection noise, the *SDCMap* of $cell_{L1}$ is increased instead of $cell_{n-1}(t)$ with a large amount.

If $cell_{n-1}(t)$ is not detected by reflection noise, the *SDCMap* is increased with a small amount for later updates. Once the *SDCMap* of $cell_{L1}$ is increased by a large amount, the *SDCMap* of those elements that belong to the list L but $cell_{L1}$ are decreased by a large amount, and if *SDCMap* of $cell_{L1}$ is increased by a small amount, the *SDCMap* of those elements that belong to the list L but $cell_{L1}$ is decreased by a small amount.

If all the elements of the list L are not detected as candidates for obstacles, $cell_{n-1}(t)$ has a low probability of being an obstacle but would be considered for its probability of being a candidate for an obstacle; $SDCMap(cell_{n-1}(t))$ is then increased by a small amount.

If condition 3 is met, if $cell_{n-1}(t)$ belongs to $CELL_O(t)$, then it is a case of not being detected continuously but regularly; therefore, considering the probability of being a later candidate of an obstacle, $cell_{n-1}(t)$ is increased by a small amount, and if it does not belong to $CELL_O(t)$, then $cell_{n-1}(t)$ is added to $CELL_O(t)$.

5. Path Planning in an Environment with Transparent Obstacles

The following conditions are required when making an environment map in an environment with transparent obstacles.

[Conditions]

1. A path planning method that can be applied to the occupancy grid map must be used.
2. A new path must be generated quickly within the environment map being updated by the sensor data being collected in real time.
3. The generated path should be close to the optimized one.

Other than the above conditions, an environment map uses the *SDCMap*, and by using it, three types of cells are generated as not obstacle cell, obstacle cell, and obstacle candidate cell. Therefore, path planning is run according to the Wavefront-propagation algorithm (Table 2, Algorithm 3) with all these conditions considered [24].

Table 2. Terminology essential to understanding the Wavefront-propagation algorithm.

Terminology	Explanation
i	The cost required to propagate Wavefront
W_i	The set of cells for ith propagation
X_G	The set of cells in which the destination is stored (default: The number of destination cells is 1)
$\Phi(x)$	The function to store the optimized propagation cost of cells that belong to W_i
All unexplored neighbors of x	The non-obstacle cells not yet propagated among adjacent cells with four different directions

Once the grid map is completed using the Wavefront-propagation algorithm, the cost of all the cells to the destination is calculated as in Figure 10a. The spot with a cost of 0 is the destination, and path planning is run based on this. The next cell is chosen by looking at the eight adjacent cells from which the robot stands. The current position of the robot is colored blue in Figure 10b, and the eight adjacent cells are colored red. The box with slashes represents the path the robot travels.

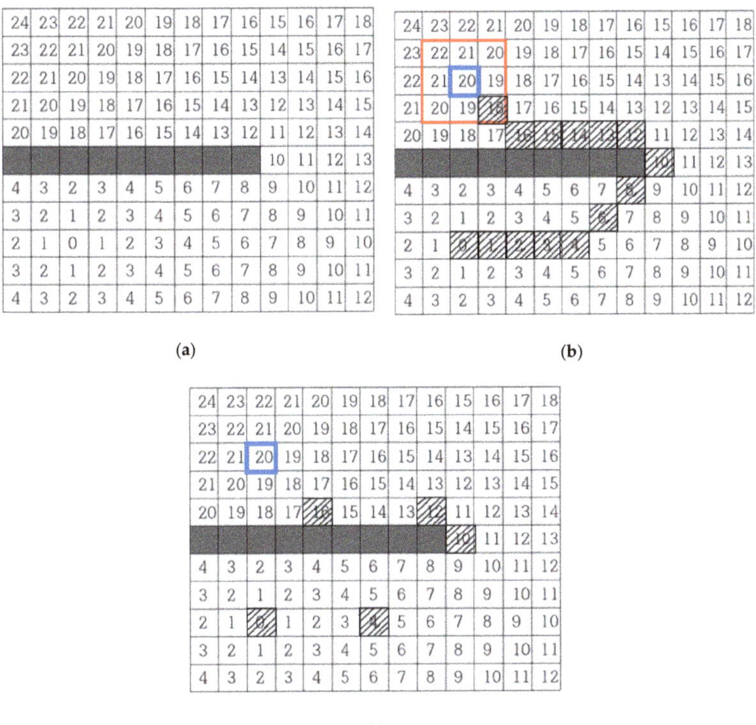

Figure 10. Path planning using the Wavefront-propagation algorithm: (**a**) environment map; (**b**) path planning results; and (**c**) optimization results.

As the robot travels towards the destination, the robot selects the box with the minimum cost. However, following all the boxes with minimum cost results in awkward movement. Therefore, boxes that are not needed to control the robot's movement are removed. Not only does this contribute to more natural movement, but computing time can also be significantly reduced. The optimized map is shown in Figure 10c.

However, in this paper, both the map in Figure 10b,c are used; they represent the original and optimized paths, respectively, for path planning in environments with transparent obstacles. The original path is used to detect the newly found obstacles while the robot is travelling in a dynamic environment. The optimized path is used to direct the robot's motion control.

Algorithm 3. Pseudo code for the Wavefront-propagation algorithm [24]. (*Citation mark*)

Begin *Algorithm Wavefront-propagation*
1 ***Initialize*** $W_0 \leftarrow X_G$, $i \leftarrow 0$
2 ***Initialize*** $W_{i+1} \leftarrow \varnothing$
3 ***For*** every $x \in W_i$, assign $\Phi(x) = i$ and insert All unexplored neighbors of x ***do***
4 ***If*** $W_{i+1} == \varnothing$ ***then***
5 ***Terminate***
6 ***Else***
7 $i \leftarrow i + 1$
8 ***Goto*** step 2
End *Algorithm Wavefront-propagation*

The time complexity of Wavefront-propagation is O(n).

6. Experimental Results

6.1. Experimental Environment

Three environments for the experiments were configured. All the experimental environments were 4 m × 3.53 m, and four sides were surrounded with opaque obstacles. Three transparent obstacles (objects 1–3) were used in all the experiment environments and as substitutes for window frames in a real building; an opaque rectangular obstacle (5 cm × 4 cm) was placed in the gap between each transparent obstacle.

The first experimental environment in Figure 11a and the obstacles (objects 1–3) were placed in line. In the second experiment represented by Figure 11b, object 3 was placed orthogonally to object 2 to configure a corner composed of transparent obstacles. Lastly, a diamond-shaped opaque obstacle was added to the first experimental environment, as shown by Figure 11c.

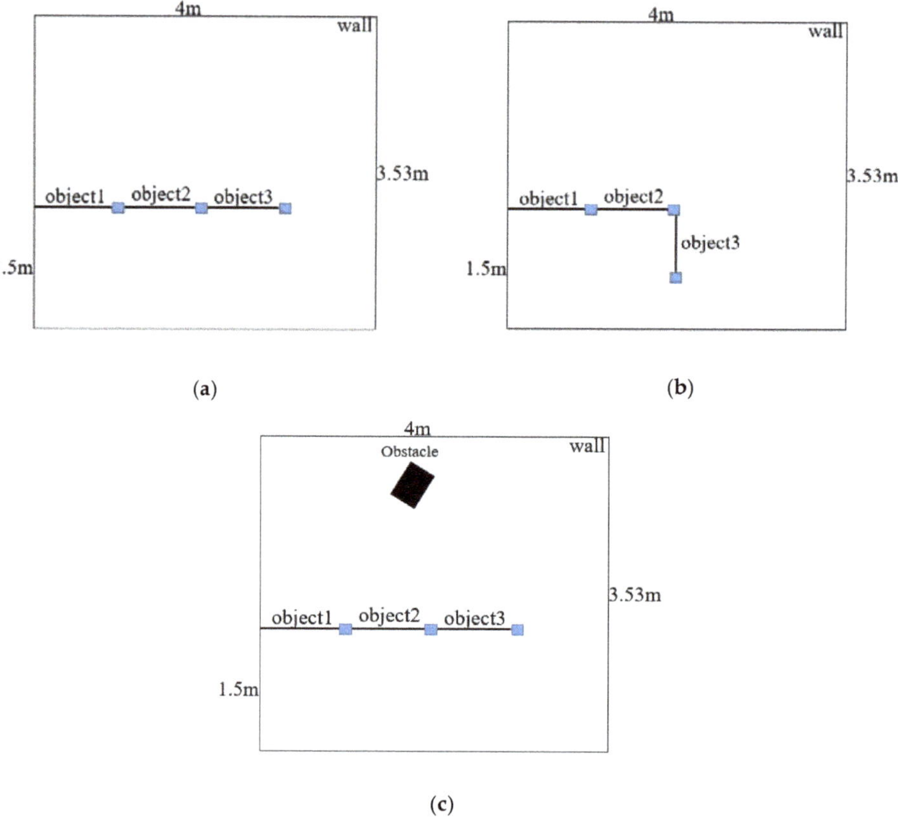

Figure 11. Experimental environment overview: (**a**) experimental environment with transparent obstacles arranged in a line; (**b**) experimental environment with transparent obstacles arranged vertically; (**c**) experimental environment with non-transparent obstacles added to (a).

Figure 12 shows the actual experimental environment. As can be seen on Figure 12, the transparent materials used are thick sheets of glass with 0.5 cm thickness.

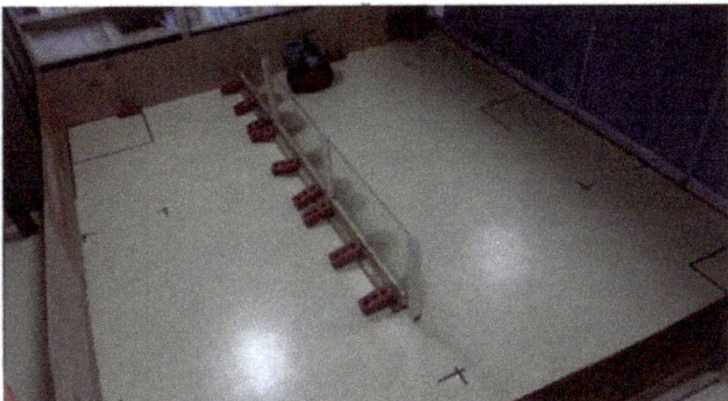

Figure 12. The figure of the actual experimental environment.

6.2. Environment Map Generation Experiment in an Environment with Transparent Obstacles

6.2.1. Experimental Method

The LRF was fixed to the robot to conduct experiments to create environments using data collected from the surrounding environment. Experiments were run in two different ways.

[Method]

1. The robot drove alongside the wall of transparent obstacles with a 30 cm distance away from it.
2. The robot drove towards the wall in a direction angled 45 degrees at a position far from the wall of the transparent obstacle, and the robot's position moved parallel to the wall once it reached 30 cm distance away from the wall.

Based on the collected data, two occupied grid map results were generated for comparison. The first generated map was the result of using only data collected from LRF in an environment with transparent obstacles, and the second generated map was the result of using LRF and the proposed algorithm for transparent obstacle recognition in the same environment.

6.2.2. Analyses of the Experimental Results

The black represents cells with a high probability of having obstacles; the closer the cells are to white, the lower the probability of the grids having obstacles. The size of each cell was fixed to 15 cm.

The figures include environment maps for experimental environment 1. Figure 13a is the baseline for the other environment maps. This environment map is generated if the transparent obstacles (objects 1–3) are transferred to opaque maps. Figure 13b is a result generated by using only collected data from LRF. A lot of cells with obstacle candidates were detected because of the reflection noises caused by transparent obstacles on the opposite side of the transparent obstacle. Figure 13c is a result generated by applying the environment map generation algorithm in an environment with transparent obstacles. The results show that the robot detecting the transparent obstacles exclude the place at the start of the robot. However, some of the cells being left as obstacle candidates can be seen on the opposite side of the transparent obstacles due to the algorithm's failure to remove all of the reflection noise.

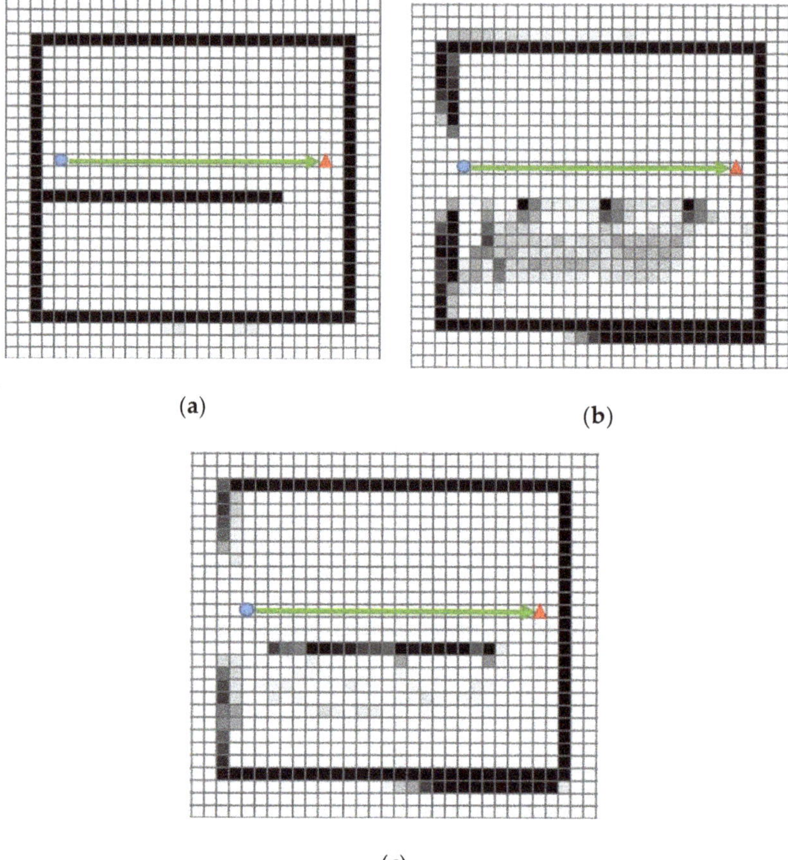

Figure 13. Environment maps corresponding to experimental environment 1 (the starting point, destination point, and the travelled paths of the robot are represented by the blue circle, the red triangle and the arrow, respectively): (a) map using the baseline; (b) map using the only collected data from LRF; (c) map using LRF and the algorithm.

Figure 14 corresponds to environment maps generated in response to experimental environment 2. Figure 14a is the baseline for experimental environment 2. Figure 14b is the result generated by using only collected data from LRF. As the results show, a lot of obstacle candidates are generated on the opposite side of transparent obstacles as the result of experimental environment 2. The transparent obstacles placed vertically look as though the transparent obstacles had been found, but this is a result of a lot of reflection noises distributed by the transparent obstacle. Figure 14c is the result generated using the environment map generation algorithm in an environment with transparent obstacles. The boundaries of the transparent obstacles placed horizontally show a low probability of being an obstacle candidate. This phenomenon results from the process of the rasterization of the reflection noise for transparent obstacles in the vertical line, which also decreases their probability of being an obstacle candidate for cells belonging to transparent obstacles on the horizontal line.

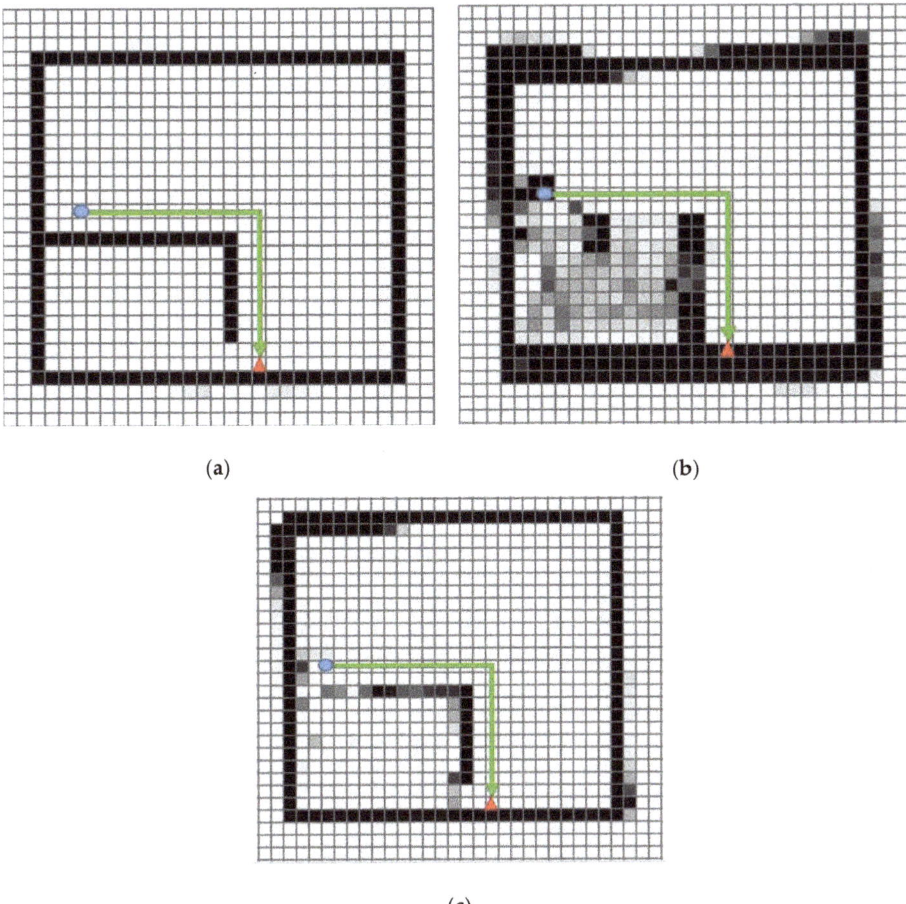

(a) (b) (c)

Figure 14. Environment map corresponding to experimental environment 2 (the starting point, destination point, and the travelled paths of the robot are represented by the blue circle, the red triangle and the arrow, respectively): (**a**) map representing the baseline; (**b**) Map using only collected data from LRF; (**c**) map using LRF and proposed algorithm.

Figure 15 represents the three environment maps generated in accordance with experimental environment 3. Figure 15a is the baseline for the environment map. Figure 15b is the result of using only collected data from LRF. Although it detected the diamond-shaped opaque obstacle, it could not detect the boundaries of transparent obstacles, and many obstacle cells were represented on the opposite side. On the other hand, Figure 15c represented the diamond-shaped opaque obstacle well, and the surface of the transparent obstacles were detected as well.

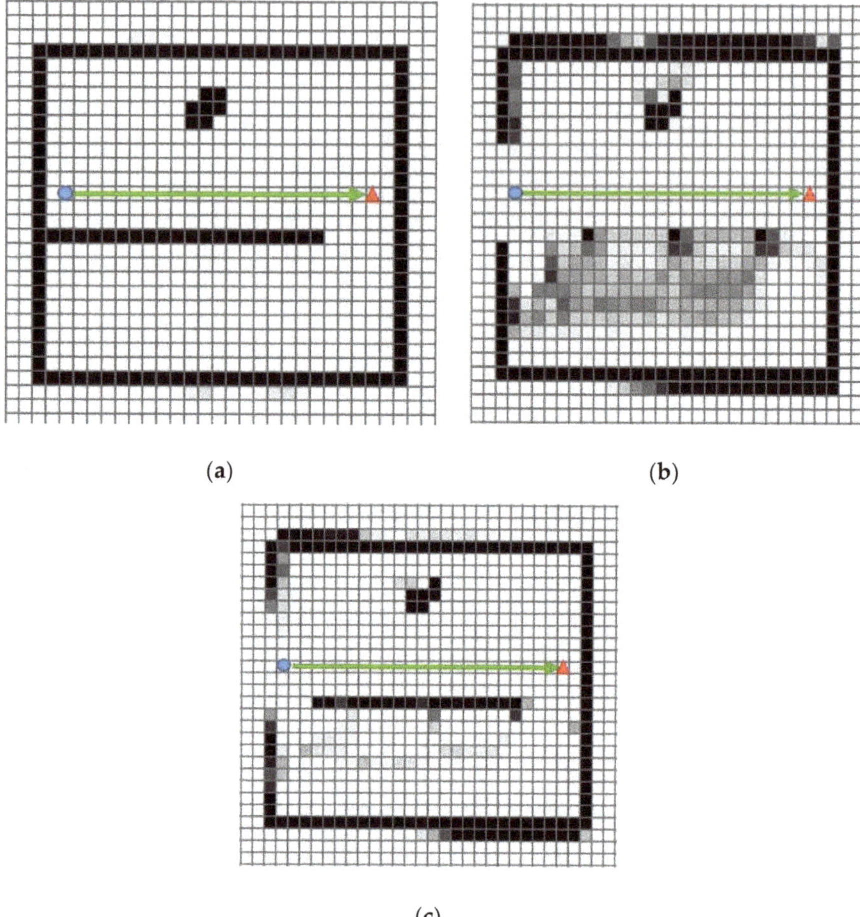

(a) (b) (c)

Figure 15. Environment map corresponding to experiment environment 3 (the starting point, destination point, and the travelled paths of the robot are represented by the blue circle, the red triangle and the arrow, respectively): (**a**) map representing the baseline; (**b**) map using only collected data from LRF; (**c**) map using LRF and proposed algorithm.

As the figures above demonstrate, the robot drives towards the wall of the transparent obstacle at an angle of 45 degrees, stops near the transparent obstacle, rotates, and resumes the movement alongside the wall of the transparent obstacle. In Figure 16a,c, the cells corresponding to the obstacle candidates show a high probability because the reflection noise keeps occurring at the fixed place while driving towards the transparent obstacle. In Figure 16b,d a high probability for being obstacle candidates is shown for the same reasons as Figure 16a,c; however, most of the obstacle candidate cells could be deleted while driving alongside the transparent obstacle.

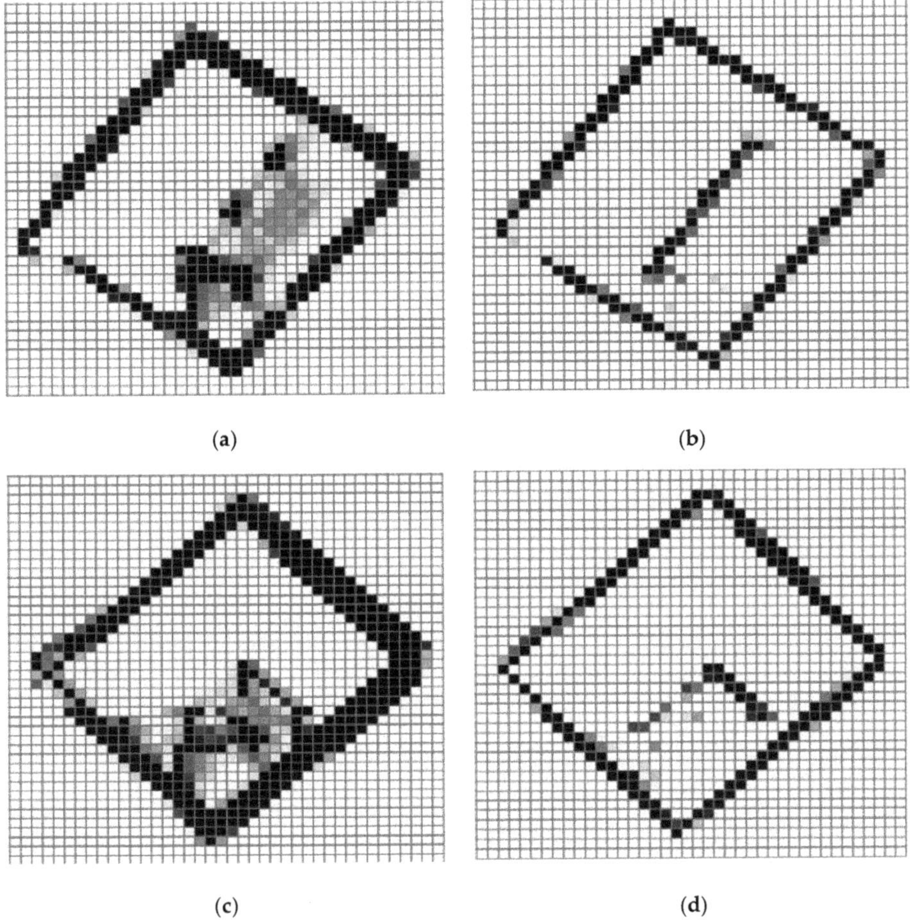

Figure 16. Environment map corresponding to experiment method 2: (a) environment map generated by using only collected data from LRF in experimental environment 1; (b) environment map generated using LRF and the proposed algorithm in experimental environment 1; (c) environment map generated by using only collected data from LRF in experimental environment 2; (d) environment map generated by using LRF and the proposed algorithm in experimental environment 2.

6.3. Mobile Robot Path Planning Experiment in an Environment with Transparent Obstacles

6.3.1. Experimental Method

As can be seen in Figure 17, the results can be checked by setting the starting point and destination at experimental environment 1 and by enabling the robot to do path planning and travel by itself with the environment map generated from the real algorithm. A transparent obstacle is placed between the start position and destination.

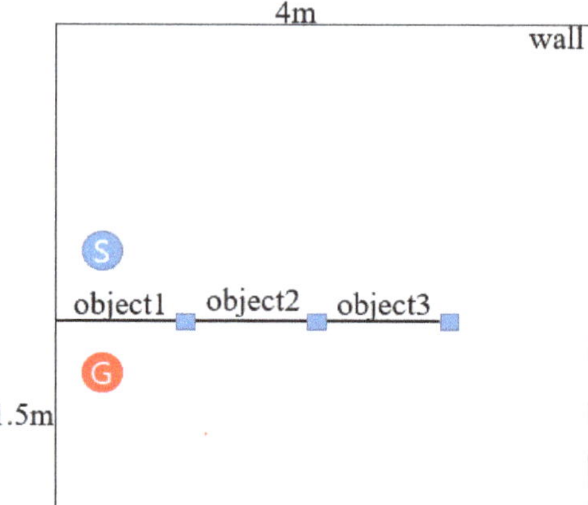

Figure 17. Starting point and the destination in experimental environment 1.

6.3.2. Experiment Result Analysis

Figure 18a shows the position in which the robot takes before it starts moving. As shown, the information from the sensors is not yet stored in the environment map that is held by the robot; it rotates to move towards the destination marked x. Figure 18b shows the intermediate destinations as it has acquired some of the information during its rotation. Figure 18c–e shows the process of modifying the paths based on the sensor data the robot acquires while it travels through the intermediate destinations. Additionally, the transparent obstacles looking clearer as the robot moves towards the wall can be seen. Figure 18f shows the robot arriving at the destination.

6.4. Error Rate Experiment

Experimental Method

In the experiment, a total three experiment environments were configured, and each one had its own baseline environment map. Therefore, in this experiment, the environment map generated by using only collected data from LRF and the environment map generated by using LRF and the algorithm proposed in this paper were compared to their baseline, respectively, and the error rates were measured.

The following equation was proposed for the sake of the above experiment:

$$\text{errorate}(\%) = \frac{\sum_{i=1}^{n}\sum_{j=1}^{m}|cell_{result}[i][j]-cell_{base}[i][j]|}{n*m} * 100\ (\,if\ cell_{result}[i][j] > \text{thrshold}\ then\ cell_{result}[i][j] = 1\ else\ cell_{result}[i][j] = 0) \quad (2)$$

The environment map for the baseline can be indicated as either 1 or 0; 1 represents the cell with obstacles, and 0 represents the cell with no obstacles. However, the environment map generated by the algorithm contains cells with obstacle candidates; therefore, they are encoded into binary numbers as either 1 or 0 by the threshold.

In Equation (2), each of n and m represent the maximum number of rows and columns when they are expressed by a vector represented with rows and columns. Each of i and j represent the index of the array. The meaning of $cell_{result}[i][j]$ is the rate of how long the cell had been occupied at row i and column j in the environment map, which is the result of the algorithm. However, the value of $cell_{result}[i][j]$ is transferred to binary numbers compared to the threshold. $cell_{base}[i][j]$ represents the occupied rate of the cell on the baseline environment map. The baseline environment map is the most optimized one; therefore, the gap of cells that are obstacles and not obstacles are clearly distinguished.

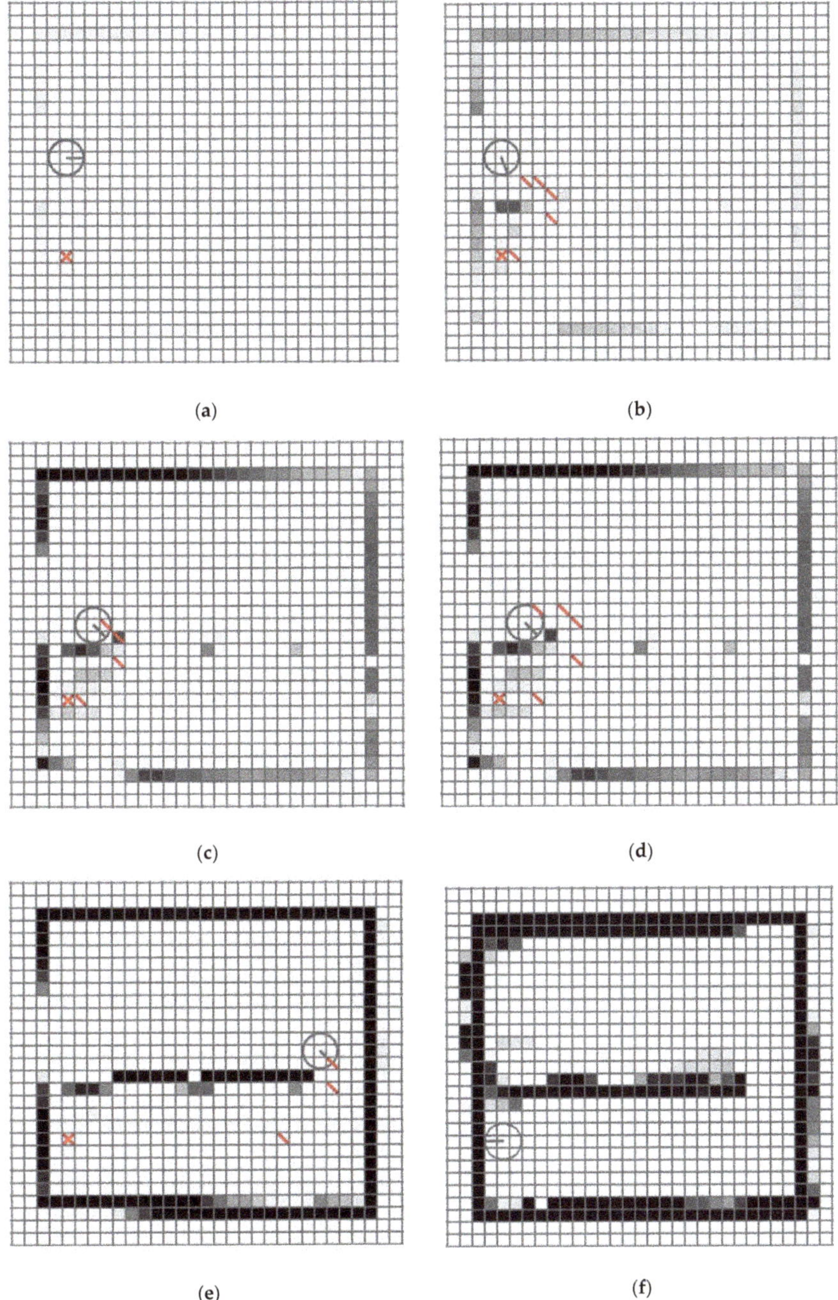

Figure 18. Results of path planning in an environment with transparent obstacles, the red diagonal line represent waypoint and the red cross line represent goal: (**a**) starting point; (**b–e**) steps in finding the transparent obstacle and modifying the planned path; (**f**) step in which the robot reaches the destination and stops.

Table 3 shows the proposed algorithm having a lower error rate in all experimental environments compared to using only data collected from LRF.

Table 3. Error rate compared to the environment maps (%).

	Environment 1	Environment 2	Environment 3
Using only data from LRF	12.26	19.57	21.40
Using LRF and the algorithm	6.45	7.20	4.73

Also, the case of using only data collected from LRF shows the average error rate was 17.74%. On the other hand, the case of using LRF and the proposed algorithm shows the average error was 6.12%. On average, the error rate of using LRF and the proposed algorithm is 11.62% lower than error rate of using only data from LRF.

7. Conclusions

In this paper, a map-generation algorithm using LRF was proposed to help path planning in environments with transparent obstacles. In this algorithm, the reflection noises that may occur in the vicinity of transparent obstacles are used to detect transparent obstacles. Additionally, the possibility of the robot's automatic driving was verified by checking whether the robot could detect the surface of the transparent obstacles and arrive at the destination while avoiding the obstacles by using the environment map generated by the algorithm. The performance for map building improved by 11.62% when comparing the environment map with only using data collected from LRF as the baseline. Therefore, the results show that the surface of transparent obstacles could be detected solely by LRF even in the environment with transparent obstacles.

Author Contributions: Idea and conceptualization: J.-W.J. and J.-S.P.; methodology: J.-W.J. and J.-S.P.; software: J.-S.P.; experiment: J.-S.P., T.-W.K. and J.-G.K.; validation: J.-W.J., J.-S.P. and J.-G.K.; investigation: J.-S.P. and T.-W.K.; resources: J.-W.J. and J.-S.P.; writing: J.-W.J., J.-S.P., T.-W.K., J.-G.K. and H.-W.K.; visualization: J.-S.P., T.-W.K. and J.-G.K.; project administration: J.-W.J. All authors have read and agreed to the published version of the manuscript.

Funding: This research was supported by the Ministry of Science and ICT, Republic of Korea, under the National Program for Excellence in Software supervised by the Institute for Information and Communications Technology Promotion (2016-0-00017), the KIAT (Korea Institute for Advancement of Technology) grant funded by the Korean government (MOTIE: Ministry of Trade Industry and Energy) (No. N0001884, HRD program for Embedded Software R&D), the Basic Science Research Program through the National Research Foundation of Korea (NRF) funded by the Ministry of Education, Science and Technology (2015R1D1A1A09061368), and the KIAT (Korea Institute for Advancement of Technology) grant funded by the Korea Government (MSS: Ministry of SMEs and Startups). (No. S2755615, HRD program for Enterprise linkages R&D).

Conflicts of Interest: The authors declare no conflicts of interest.

References

1. Tsubouch, T. Nowadays trend in map generation for mobile robots. In Proceedings of the IEEE International Conference on Intelligent Robot and System, Osaka, Japan, 8 November 1996; Volume 2, pp. 828–833.
2. Moravec, H.P.; Elfes, A.E. High resolution maps from wide angle sonar. In Proceedings of the IEEE International Conference on Robotics and Automation, St. Louis, MO, USA, 25–28 March 1985; Volume 2, pp. 116–121.
3. Leonard, J.J.; Durant-Whyte, H.F. Dynamic Map Building for an Autonomous Mobile Robot. *Int. J. Robot. Res.* **1992**, *11*, 286–298. [CrossRef]
4. Stateczny, A.; Kazimierski, W.; Gronaska-Siedz, D.; Motly, W. The Empirical Application of Automotive 3D Rader Sensor for Target Detection for an Autonomous Surface Vehicle's Navigation. *Remote Sens.* **2019**, *11*, 1156. [CrossRef]

5. Jung, J.-W.; So, B.-C.; Kang, J.-G.; Lim, D.-W.; Son, Y. Expanded Douglas–Peucker Polygonal Approximation and Opposite Angle-Based Exact Cell Decomposition for Path Planning with Curvilinear Obstacles. *Appl. Sci.* **2019**, *9*, 638. [CrossRef]
6. Feder, H.J.S.; Leonard, J.J.; Smith, C.M. Adaptive mobile robot navigation and mapping. *Int. J. Robot. Res.* **1999**, *18*, 650–668. [CrossRef]
7. Park, J.-S.; Jung, J.-W. Transparent Obstacle Detection Method based on Laser Range Finder. *J. Korean Inst. Intell. Syst.* **2014**, *24*, 111–116. [CrossRef]
8. Jensgelt, P. Approaches to Mobile Robot Localization in Indoor Environments. Ph.D. Thesis, KTH Royal Institute of Technology, Stockholm, Sweden, 2001.
9. Zhuang, Y.; Sun, Y.; Wang, W. Mobile Robot Hybrid Path Planning in an Obstacle-Cluttered Environment Based on Steering Control and Improved Distance Propagation. *Int. J. Innov. Comput. Inf. Control* **2012**, *8*, 4098–4109.
10. Sohn, H.-J. Vector-Based Localization and Map Building for Mobile Robots Using Laser Range Finder in an indoor Environments. Ph.D. Thesis, Korea Advanced Institute of Science and Technology (KAIST), Daejeon, Korea, 2008.
11. Jung, Y.-H. Development of a Navigation Control Algorithm for Mobile Robot Using D* Search Algorithm. Master's Thesis, Chungnam University, Daejeon, Korea, 2008.
12. Park, J.-S.; Jung, J.-W. A method to LRF noise Filtering for the transparent obstacle in mobile robot indoor navigation environment with straight glass wall. In Proceedings of the IEEE International conference on Ubiquitous Robots and Ambient Intelligence, Xi'an, China, 19–22 August 2016; pp. 194–197.
13. Koenig, S. Fast Replanning for Navigation in Unknown Terrain. *IEEE Trans. Robot.* **2005**, *21*, 354–363. [CrossRef]
14. Guo, J.; Liu, Q.; Qu, Y. An improvement of D* Algorithm for Mobile Robot Path Planning in Partial Unknown Environments. Intell. *Comput. Technol. Automob.* **2009**, *3*, 394–397.
15. Kuipers, B.; Byum, Y.T. A robot exploration and mapping strategy based on a semantic hierarchy of spatial representations. *Robot. Auton. Syst.* **1991**, *8*, 47–63. [CrossRef]
16. Choset, H.; Burdick, J. Sensor-Based exploration: The hierarchy generalized voronoi graph. *Int. Robot. Res.* **2000**, *19*, 96–125. [CrossRef]
17. Young, D.K.; Jin, S.L. A Stochastic Map Building Method for Mobile Robot using 2D Laser Range Finder. *Auton. Robot* **1999**, *7*, 187–200.
18. Rudanm, J.; Tuza, G.; Szederkenyi, G. Using LMS-100 Laser Range Finder for indoor Metric Map Building. In Proceedings of the 2010 IEEE International Symposium on Industrial Electronics, Bari, Italy, 4–7 July 2010; pp. 520–530.
19. Jung, J.-H.; Kim, D.-H. Local Path Planning of a Mobile Robot Using a Novel Grid-Based Potential Method. *Int. J. Fuzzy Log. Intell. Syst.* **2020**, *20*, 26–34. [CrossRef]
20. Siciliano, B.; Khatib, O. *Handbook of Robotics*; Springer: Berlin, Germany, 2008.
21. Latombe, J. *Robot Motion Planning*; Kluwer Academic Publisher: Cambridge, MA, USA, 1991.
22. Nohara, Y.; Hasegawa, T.; Murakami, K. Floor Sensing System using Laser Range Finder and Mirror for Localizing Daily Life Commodities. In Proceedings of the IEEE/RSJ International Conference on Intelligent Robots and Systems, Taipei, Taiwan, 18–22 October 2010; pp. 1030–1035.
23. Choset, H.; Lynch, K.M.; Hutchinson, S.; Kantor, G.; Burgard, W.; Kavaraki, L.E.; Thrum, S. *Principles of Robot Motion: Theory, Algorithm and Implementation*; MIT Press: Cambridge, MA, USA, 2005.
24. La Valle, S.M. *Planning Algorithms*; Cambridge University Press: Cambridge, MA, USA, 2006.
25. Stentz, A. Optimal and Efficient Path Planning for Partially-Known Environments. In Proceedings of the IEEE International Conference on Robotics and Automation, San Diego, CA, USA, 8–13 May 1994; Volume 4, pp. 3310–3317.
26. Siegwart, R.; Nourbakhsh, I.R. *Introduction to Autonomous Mobile Robot*; MIT Press: Cambridge, MA, USA, 2004.
27. Willms, A.R.; Yang, S.X. An Efficient Dynamic System for Real-Time Robot—Path Planning. *IEEE Trans. Syst. Cybern. Part B Cybern.* **2006**, *36*, 755–766. [CrossRef] [PubMed]
28. Nguyen Van, V.; Kim, Y.-T. An Advanced Motion Planning Algorithm for the Multi Transportation Mobile Robot. *Int. J. Fuzzy Log. Intell. Syst.* **2019**, *19*, 67–77.

29. Mingxiu, L.; Kai, Y.; Chengzhi, S.; Yutong, W. Path Planning of Mobile Robot Based on Improved A* Algorithm. In Proceedings of the IEEE 29th Chinese Control and Decision Conference (CCDC), Chongqing, China, 28–30 May 2017.
30. Rui, S.; Yuanchang, L.; Richard, B. Smoothed A* Algorithm for Practical Unmanned Surface Vehicle Path Planning. *Appl. Ocean Res.* **2019**, *83*, 9–20.
31. Guo-Hua, C.; Jun-Yi, W.; Ai-Jun, Z. Transparent object detection and location based on RGB-D Camera. *J. Phys. Conf. Ser.* **2019**, *1183*, 012011. [CrossRef]
32. Cui, C.; Graber, S.; Watson, J.J.; Wilcox, S.M. Detection of Transparent Elements Using Specular Reflection. U.S. Patent 10,281,916, 7 May 2019.
33. Pioneer Corporation. Available online: https://www.pioneerelectronics.com/ (accessed on 16 April 2020).
34. SICK Sensor Intelligence. Available online: https://www.sick.com/ (accessed on 16 April 2020).
35. Lumelsky, V.; Skewis, T. Incorporation Range Sensing in The Robot. *IEEE Trans. Syst.* **1990**, *20*, 1058–1068.
36. Moravec, H.P. Sensor fusion in certainty grids for mobiles robots. *Ai Mag.* **1988**, *9*, 61–74.

Article

GPS-Based Indoor/Outdoor Detection Scheme Using Machine Learning Techniques

Van Bui, Nam Tuan Le, Thanh Luan Vu, Van Hoa Nguyen and Yeong Min Jang *

Department of Electronics Engineering, Kookmin University, Seoul 02707, Korea; buivandut@gmail.com (V.B.); namtuanlnt@gmail.com (N.T.L.); vuthanhluan999@gmail.com (T.L.V.); vanhoahd95@gmail.com (V.H.N.)

* Correspondence: yjang@kookmin.ac.kr; Tel.: +82-2-910-5068

Received: 6 December 2019; Accepted: 6 January 2020; Published: 10 January 2020

Featured Application: Supporting commercial mobile applications, improving indoor channel estimation and mobile performance in indoor environment. Applied in Optical Camera Communication (OCC) applications to detect the outdoor environment.

Abstract: Recent advances in mobile communication require that indoor/outdoor environment information be available for both individual applications and wireless signal transmission in order to improve interference control and serve upper-layer applications. In this paper, we present a scheme to identify the indoor/outdoor environment using GPS signals combined with machine learning classification techniques. Compared to traditional schemes, which are based on received signal strength indicator (RSSI), the proposed scheme promises a robust approach with high accuracy, smooth operation when moving between indoor and outdoor environments, as well as easy implementation and training. The proposed scheme combined information from a certain number of GPS satellites, using the GPS sensor on mobile devices. Then, data are collected, preprocessed, and classified as indoor or outdoor environment using a machine learning model that is optimized for the best performance. The GPS input data were collected in the Kookmin University area and included 850 training samples and 170 test samples. The overall accuracy reached 97%.

Keywords: I/O detection; GPS signal; machine learning; positioning applications.

---

1. Introduction

Recent developments in Mobile and mobile communication means that an excessive amount of information about the working environment is required for both individual applications and wireless signal transmission. One type of information is indoor/outdoor detection (I/O detection). I/O detection refers to the estimation of the mobile user's working condition, i.e., whether the device is being operated in an outdoor or indoor environment. This information is beneficial in many mobile applications and wireless communication. However, few studies focus on I/O detection.

I/O detection can provide useful information about user behavior and the appropriate use of mobile network resources [1,2]. I/O detection can also provide essential information for the upper layer of mobile applications. For example, a Global Positioning System (GPS) location management system can use this information to turn a GPS off to save power. GPSs only work well in outdoor environments; therefore, turning the system off GPS in indoor environments can save energy for other applications [3]. Another example where I/O detection can improve energy consumption is Wi-Fi access point searching behavior. Typically, indoor environments provide more access points as well as stronger signals. I/O detection can reduce energy consumption of both GPSs and Wi-Fi connection applications, as well as mobile behaviors. I/O detection can help a mobile device adapt to an indoor environment. For example, I/O detection can facilitate adjusting the device volume, trigger reminders, changing the device's home screen, and adjusting screen brightness. I/O detection also plays an important role in indoor channel estimation. Indoor channels block some radio signal and have significant interference noise. I/O detection can improve channel estimation prediction as well as data transmission [4]. I/O detection can be used in Optical Camera Communication (OCC) applications since the outdoor environment decreases the OCC performance. Based on the I/O information, OCC applications will have more information to adapt and improve its performance.

Some GPS constellations consist of 31 satellites, of which 27 are used [5]. GPS satellite systems, produced and maintained by the United States of America, provide an efficient way to detect a mobile user's location, sometimes within centimeters. GPS signals are always available and provide a reliable working environment and low-energy consumption compared to other communication systems, such as Wi-Fi or 3G. However, GPS signals from a single satellite are unstable, vary from location to location, and can change when the user moves or when the signal is occluded. The number of available satellites from which a GPS sensor receives signals changes over time and cannot be determined without a complex calculation. In this paper, GPS signals will be synthesized to provide information for I/O detection applications. The proposed I/O detection method for mobile devices has low-power consumption, easy implementation, and flexible operation compared to other I/O detection methods.

As mentioned previously, raw GPS signals contain information about indoor/outdoor condition; however, indoor/outdoor condition information is difficult to synthesize and calculate. Currently, machine learning, which is attracting a great deal of attention as a popular classifier technique, provides a robust solution for I/O detection problems. The advantages of machine learning compared to other classification methods are its high accuracy and ability to handle all types of input data. The redundant GPS signal data is the best condition for machine learning to perform effectively, despite the complex structure of GPS input data. In this paper, we provide a supervised algorithm as the primary solution because it is more suitable for classifier tasks. The supervised machine learning family includes support vector machines (SVM), logistic regression technique, naive Bayes classifiers, k-nearest neighbor (KNN) technique, and neural networks. A machine learning approach provides a simple solution, low-power costs, and high accuracy for I/O detection problems.

The proposed scheme based on the GPS signal and machine learning methods require a very small volume of dataset, as only 850 training samples are enough for an accuracy classification. Furthermore, a mobile phone application can apply offline learning, which uses the trained model for the classification and updated if requested, so it does not require a training dataset. Therefore, the proposed scheme will be suitable for almost all mobile applications that require I/O environment information.

The remainder of this paper is organized as follows. Section 2 describes related work. We discuss GPS signal characteristics and how we can take advantage of them in Section 3. Section 4 provides the working scheme for an I/O detection scheme, including the data collection process, preprocessing, and the training and testing phases. The supervised evaluation results will be discussed in Section 5. The paper will be concluded in Section 6.

2. Related Work

Current research into I/O detection is very limited and many studies in this field are primarily based on image processing, which is not suitable for mobile applications due to the complex calculation capacity and large memory required for input data [6–8]. However, there are practical methods to deal with this problem. One such method is based on collecting data using various sensors that are available on mobile devices [1,9–12]. Another is based solely on collecting data using a GPS sensor [13,14].

In a previous study [9], the authors proposed an I/O detection method that used a set of mobile sensors, including a light sensor, a cellular module, and a magnetic sensor for I/O detection. That proposed method promised high accuracy and energy efficiency. This combination of sensors provided high accuracy (88%–90%) for both indoor and outdoor environments with average power consumption. Although the input data is available with mobile, this method requires complex calculations and a large amount of input data to work well. In another study, the author used even more sensors and data for I/O detection, including an activity recognition Application Programming Interface (API), barometric pressure, a light sensor, cloud coverage data, a timer, Global System for Mobile Communications (GSM) signal strength, an accelerometer, a magnetometer, and ambient noise [10]. These sets of input data increase I/O detection accuracy to more than 98%; however, calculation and data collection are complicated and consume significant power.

Harsh testing conditions also affect I/O detection accuracy. In [11], the authors used the same input dataset with previous study [10], including an accelerometer, cellular radio, light sensor, magnetometer, and proximity sensor data; however, the accuracy is only 80%. Power consumption is also greater compared to a previous study [9]. Primarily based on light sensor data and a Wi-Fi sub-detector with a machine learning approach, another [12] obtained more than 90% accuracy. However, the training model is very complicated and difficult to implement. Similarly, reference [1] uses a machine learning approach for data processing and obtained high accuracy. Again, data consumption and a complicated calculation are disadvantages of this approach.

Another approach is using only GPS signal because it is always available and provides significant information for I/O detection. In addition, this approach consumes less power compared to Wi-Fi and 3G signals. In a study about I/O detection based only on sparse GPS positioning information, Iwata et al. obtained 98% accuracy for I/O detection [14]. Although the test sample was very small (53 samples) and the approach required significant time and a complex calculation, this study promised highly accurate I/O detection based on GPS signals and a machine learning classifier. Based on a direct analysis of GPS signals, Chen et al. provided three GPS-based techniques for I/O detection [13]. The technique that provided the best performance was SatProbe, which searches the raw signals from all available GPS satellites to identify input data patterns. With this technique, using a sizable dataset, accuracy reached 86% and it worked much better than the traditional method. Although the accuracy is not too high, as well as the complex calculation, this study also emphasizes the importance of raw GPS signals in I/O detection problems.

A breakthrough approach for the indoor classification and localization is the combination of GPS and Wi-Fi signal, which is a flexible method to deal with emergency situations, where the GPS signal becomes unstable and weaker. In particular emergency situations, when the transmission of the GPS signal was obstructed, the localization process is impossible with GPS signal only, the Wi-Fi combination will be helpful in both indoor/outdoor detection and localization. Reference [15] provides a robust system for tracking GPS in such situation, that the Navigation System can be based on the Wi-Fi signal to improve the accuracy of uses. The same concept can be applied in the I/O detection scheme and improve its performance in the emergency situation, when the GPS signal becomes unstable and weaker.

3. Data Analysis

3.1. Global Positioning System

The GPS is a global navigation satellite system that geolocation information to GPS receivers [16]. GPS is a satellite-based radio navigation system; consequently, solid obstacles, such as mountain and buildings, weaken and can even block GPS signals. The GPS does not require any transmission data from users, and its operation is independent from telephonic or internet signal. Currently, the GPS is owned by the United States' government and operated by the United States Air Force. The United States' government is responsible for creating and maintaining the system and makes it freely accessible to any GPS receiver [17].

The GPS calculates the receiver's position using the time and position of the satellite system. The satellites use stable atomic clocks that are synchronized all satellites and with the ground receivers. Any offset from GPS time of ground receiver is corrected daily. Similarly, satellite locations are controlled with great accuracy. Each GPS satellite continuously transmits a radio signal to provide the current time and its position; therefore, GPS sensors can receive signals without issuing a request. The distance between a satellite and a GPS receiver is calculated based on the delay time of the transmitted and received signals due to the constant speed of radio waves. A GPS receiver updates the distance of multiple satellites and determines the accurate position of the receiver and its deviation from real time. At least four satellites must be used to compute four unknown quantities and solve the accuracy equation [5].

The GPS satellite system comprises 24–32 satellites in medium Earth orbit. Originally, the GPS had only 24 satellites. In that configuration, each set of eight satellites were in three approximately circular orbits [18]. However, the system was modified to six orbital planes with only four satellites each (Figure 1). The orbital period is one-half a day (11 h and 58 min) so that the satellites pass over the same locations every day. The satellite orbits are carefully arranged so that at least six satellites are in line of sight from anywhere on the Earth's surface [5].

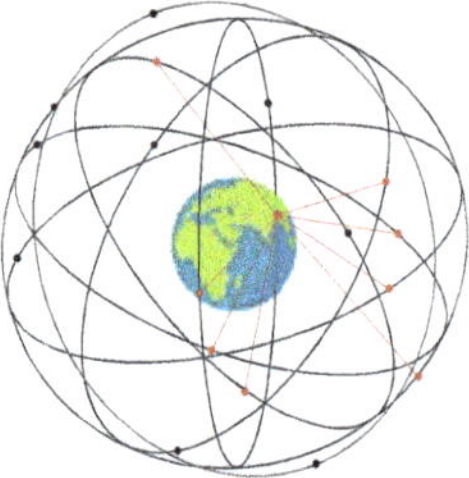

Figure 1. Illustration of a 24 GPS satellite constellation in motion relative to the Earth's rotation (Source: Wikipedia [5]).

As of February 2019, there were 32 GPS satellites in orbit, and 27 satellites are in use [19]. Increasing the number of satellites improves GPS sensor calculations by providing redundant measurements, which leads to better position estimation. With the addition of redundancy satellites, the constellation was changed to a non-uniform arrangement. Compared to a uniform arrangement, a non-uniform arrangement dramatically improves the reliability and availability of the GPS system if multiple satellites malfunction [19]. At any given time, at least nine satellites can be observed from a single point on the ground, which is significantly more than the minimum four satellites required for accurate positioning.

Other satellite navigation systems, including GLONASS (GLObal Navigation Satellite System) developed by Russia and GALILEO developed by Europe [20], can be used in the I/O detection applications as an alternative option besides the GPS. The combination of GPS and other navigation systems can also be a promising approach, since it increases the reliability in both navigation [21] and I/O detection. However, the requirement of I/O detection is different from navigation systems since it prefers the unification of satellites over the number of it. Using only GPS signal guarantees a united system, signal, and the number of satellites, that can be adopted in all mobile phones nowadays. The other reason is that I/O detection does not require as much information as the navigation system, and the redundancy of GPS satellites is enough for the application.

In this study, we use two characteristics of raw GPS signals for I/O detection. The first is GPS signal redundancy. Assuming no occluding obstacles, approximately nine satellites can be observed from any one point on the ground. Consequently, GPS signals can always be received with desirable signal strength. The second characteristic is the fact that the GPS satellite navigation system is radio-based. Therefore, solid obstacles, such as mountains and tall buildings, can weaken and even block GPS signals. We used this characteristic to identify if the sensor is in an indoor or outdoor environment.

3.2. Data Analysis

As mentioned previously, raw GPS signals vary relative to time and place, which makes the input data complex and difficult for analyzing. The machine learning technique requires simpler input data in order to extract unique features and find similar patterns. Therefore, in this study, the preprocessing phase is the most important part. We will investigate some features of raw GPS signals to identify the best way to deal with input data.

- GPS signals are always available.

Of the 31 GPS satellites in Earth orbit, 27 are in use. The minimum number of GPS satellites required for normal operation is 24, which means that at least seven satellites are redundant. In practice, in a defined position on the Earth's surface, at least nine satellites always available to send GPS signals to receivers. Due to this redundancy, receivers can obtain stable GPS signals anywhere on Earth, and at least two or three signals will be sufficiently strong for data collection and analysis. However, the satellites always moving; thus, tracking signals from a fixed GPS satellite is impractical. A large number of satellites is also a challenge. The transmission satellites ID changes frequently; therefore, GPS sensors cannot obtain a fixed set of satellite IDs at one time. Due to these problems, a practical approach is to measure all GPS satellites signals and ignore the individual IDs. This technique will provide at least two or three strong GPS signals in an outdoor environment and weak signals in an indoor environment in an unorganized order.

- GPS signal transmission is radio-based.

The GPS is a satellite-based radio navigation system; therefore, solid obstacles, such as mountains and tall buildings, can weaken and even block the signal. If the GPS sensor is in an indoor environment, the signal will be weakened and may even disconnect (Figure 2a). Based on this characteristic, indoor environments will be detected easily because the GPS signal is weakened dramatically. In the indoor environment, the GPS signal is weakening, so the navigation cannot be done with lacking information. However, that is not the case of proposed I/O detection scheme, since it based on the weakened signal of GPS satellites to identify the indoor environment. Detecting signals in an outdoor environment is more difficult because the signal may be blocked by tall buildings or mountains (Figure 2a). As shown in Figure 2b, the GPS signal is blocked by a tall building. Therefore, if we use signals from that satellite, we cannot determine if the device is indoors or outdoors. A similar situation is shown in Figure 3. Here the GPS sensor can receive signals from a satellite, even indoors. Considering these problems, we investigated another approach and determined that stronger GPS signals are more important than weak signals, and the number of GPS signals we collect is not significant.

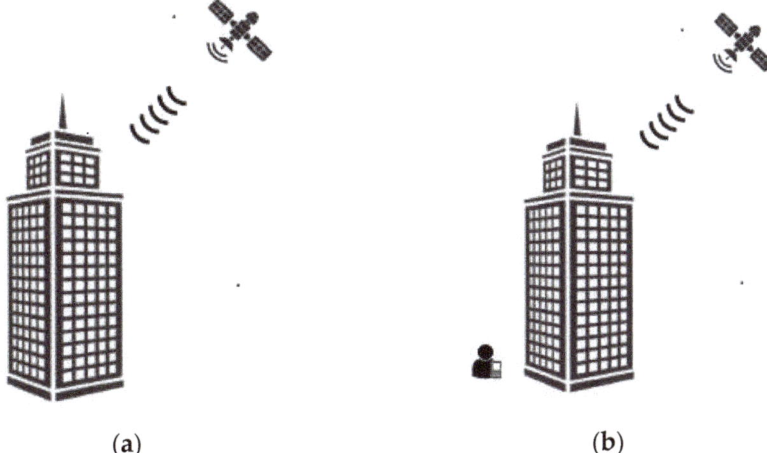

Figure 2. (**a**) GPS signal is blocked inside the building and (**b**) GPS signal is blocked by a tall building.

Figure 3. GPS sensors can receive signals when the receiver is close to an external door.

- Every mobile device has a GPS sensor.

Every mobile device has a GPS sensor that receives many active GPS positions constantly. Therefore, the input data for an I/O detection application is always redundant. In addition, GPS sensors are consistent among mobile devices; thus, we can apply offline learning for our application, leading to a simple and fast calculation model.

- Machine learning techniques are most suitable for I/O detection.

As mentioned previously, machine learning is a robust classifier technique to identify the input pattern on I/O detection applications. Machine learning has high accuracy, can use all types of input data, and is adaptable to complex input patterns. Furthermore, compared to traditional approaches, redundant GPS input data also increases the accuracy of machine learning techniques. Another advantage of ML is the diversity of ML techniques, which allows us to select the best technique to address our I/O detection problem. Finally, offline learning for a mobile application is easy to implement and low cost. In addition, calculations are performed quickly, and offline learning does not require a large amount of input data.

4. I/O Detection Scheme

The proposed scheme is shown in Figure 4: firstly, signal received from GPS satellites was collected through a specialized Mobile application. After collected, the raw input was reprocessed to filter out the errors and missing data. Input data was then analyzed and classified by various machine learning techniques to find the best solution for the I/O detection problem. For special cases of semi-indoor environments, we proposed another preprocessing process to increase the accuracy of proposed scheme. By using different types of machine learning techniques, the proposed scheme becomes flexible to deal with various problems in real life. Furthermore, the proposed preprocessing process will increase the overall accuracy in the semi-indoor environment, which is hard to deal with.

4.1. Data Collection

We used a mobile application to collect and preprocess the raw GPS signal information. This application collects information from GPS sensors, rearranges the information, and analyses it for better input data. The mobile application was selected for data collection and preprocessing rather than a GPS sensor module because it is simple, suitable for mobile devices, and provides stable input data. In Figure 5, the number and the magnitude of GPS signals in orbit is determined and preprocessed for easier access for data collection. The satellite ID is given below each GPS signal; however, the ID is ignored during the data collection phase. The magnitude of all GPS signals will be recorded and used as input data for the machine learning classifier.

In this study, we analyzed statistical differences by focusing on the difference between GPS signals received in indoor and outdoor environments. A large dataset comprising real signal values collected at multiple locations and in various environments was used. We investigated the impact of indoor, semi-indoor, and outdoor environments on the received GPS signal.

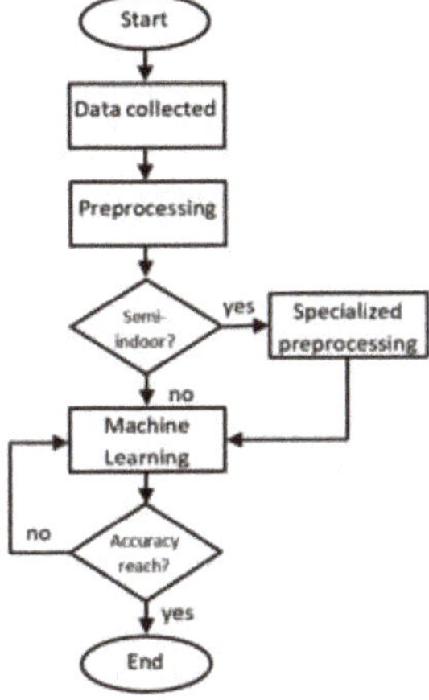

Figure 4. Proposed scheme.

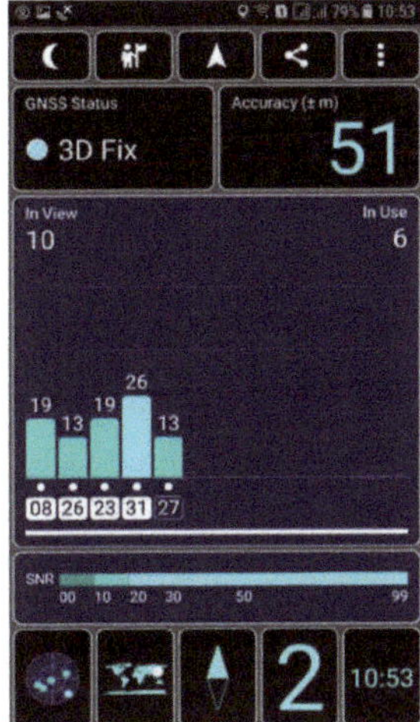

Figure 5. GPS signal determination application.

4.1.1. Data Collection

The input dataset consists of received GPS signals read directly from the GPS Test application after the preprocessing phase. For convenience, the signals were analyzed and transformed into numbers. Every received GPS signal has an ID number that corresponds to the sending GPS satellite. The magnitude of GPS signals varied from 6 to 99 to indicate signal strength, i.e., stronger signals have greater magnitude. Signals with magnitude less than 6 were not collected and forced to a default value of zero. This approach can reduce noise probability and decrease the number of satellite signals that cannot be used as an anchor. As described previously, we only needed a maximum of nine GPS signals per set. After collecting, each set of GPS signals was sorted from strongest to weakest. This data was checked again and used as input data for I/O detection problems.

The signals were collected in a three-week period from 2 to 23 May 2019. Each sample was obtained in a different location within a 10 km^2 area. The collection area included many different environments, such as buildings, restaurants, urban areas, hills, loose forests, and basements. The data collection area is a complex urban area with tall building and other obstacles, such as upper highways and mountains. These obstacles block GPS signal from different directions, which makes the dataset varied and difficult to predict. The buildings also have different architectures. Some architectures block GPS signal and other allow the signal to pass into or through the building. The collection period occurred during spring in Korea. Therefore, we were able to collect data under various types of weather conditions, such as foggy, sunny, windy, and rainy. The collection period occurred from 07:00 to 21:00. This time period reflects the changing position of GPS satellites moving in Earth orbit. These conditions were designed to reflect the complexity and variety of a real-world mobile user.

The set of indoor environment signals were collected in the Kookmin University's Campus and the surrounding area (Figures 6 and 7). The largest datasets were collected in the College of Engineering building (number 3), which, compared to other buildings, is large and isolated. The semi-indoor dataset was primarily collected mainly in Building 7 (number 14) and the College of Business Administration (number 12). These two buildings have a large number of windows and some places have glass walls, which GPS signals can pass through. The other dataset was collected in Bugak Hall (number 2), the College of Law (number 4), the College of Design (number 5), a gymnasium (number 6), the International Hall (number 11), the Student Union building (number 15), the Global Center (number 17), and a dormitory (number 21). Based on the various locations, we are confident that the data covers the overall indoor environment characteristics.

Figure 6. Survey area, includes Kookmin University Campus and the surrounding area.

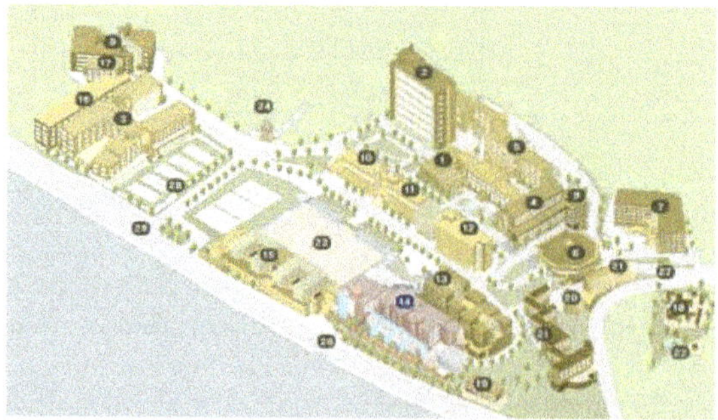

Figure 7. Kookmin University's Campus.

The set of outdoor environments were also collected in Kookmin University's Campus, as well as the surrounding urban area and highways. The data collection was large and included various types of environments, some of which had high obstacles that can block and weaken GPS signals (Figure 8).

Figure 8. Outdoor environment for data collection: (**a**) under a highway, (**b**) walking road, (**c**) high obstacles, (**d**) thin forest, (**e**) beside a mountain, (**f**) University campus, (**g**) street, (**h**) playground.

Another scenario that can be considered is the vehicular application when vehicles access urban canyons that could significantly impair signal reception from navigation satellites. The test set included a high building similar to the urban canyons and also the parking ground. In Figure 9, the GPS signals are different between the underground parking lot and the outside environment. According to this scenario, the I/O detection can be applied to the vehicular application. I/O detection can classify the indoor parking lot and the outside environment. Due to this characteristic, the proposed scheme can improve the quality of the Navigation System. I/O detection scheme can provide information of places where the navigation system required an enhancement method, such as the combination of Wi-Fi signal [15] or other GPS systems [20,21] to improve its accuracy.

4.1.2. Data Description

The collected data comprise GPS signals measured by the mobile phone application (GPS signal magnitude mode) and recorded. The dataset described in this subsection was collected using this mode.

Figure 9 also shows the GPS signal magnitude obtained with the dataset. The significant offset between indoor and outdoor values is the result of the substantial difference and attenuation variation in radio signal propagation due to wall or roof obstacles in the indoor environment. However, in some cases, this difference overlaps due to variation in the environment. Each data is denoted as vector X, with components $x_1, x_2 \ldots x_i$, where x_i is a set with nine integer numbers. The dataset includes 1020 vectors X and 1020 corresponding classes Y. Training and test sets were collected without simulation or adding noise. Note that some data were collected from a semi-indoor environment (this problem will be discussed later).

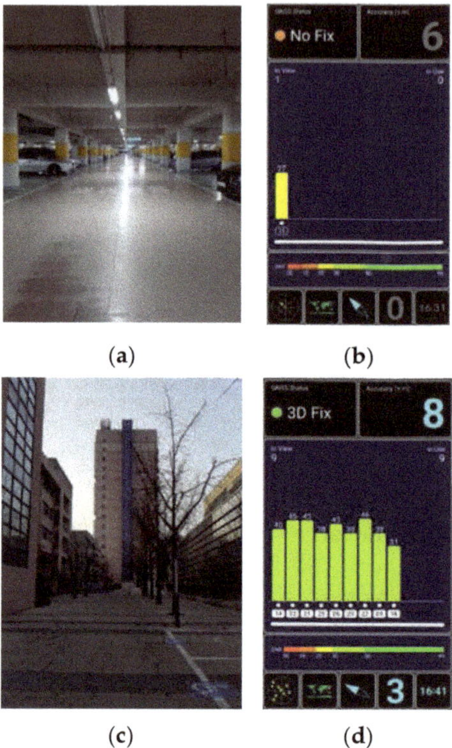

Figure 9. Data collected similar to a traffic environment: (**a**) underground parking, (**b**) corresponding GPS signal, (**c**) outdoor environment, (**d**) corresponding GPS signal.

4.2. Machine Learning Approaches

4.2.1. Logistic Regression

Logistic regression is a machine learning technique to estimate the parameters of a logistic model, where the logarithm of the equation with a value labeled as "1" is a linear combination of one or more independent variables. The probability of a value being labeled "1" can vary between 0 and 1, and the function that converts the equation to a probability is the logistic function.

Logistic regression is useful in various fields, such as data analysis, medical, economics, and social sciences. The advantages of logistic regression are simplicity, ease of use and calculation, applicability to any type of inputs, and identification of the probability of output values. However, logistic regression is unsuitable for complex problems. In addition, it is sensitive to noise and can easily suffer from overfitting.

Consider a model with two variables x_1, x_2 and one binary response Y, which have probability $p = P(Y = 1)$. Consider a linear relationship between the variables and the log-odds of p that $Y = 1$. This linear relationship can be expressed as follows:

$$l = \log_b \frac{p}{1-p} = \beta_0 + \beta_1 x_1 + \beta_2 x_2, \quad (1)$$

where ℓ is the log-odds, b is the base of the logarithm, and β_i represents the model's parameters.

Here, we can calculate that

$$p = \frac{1}{1 + b^{-(\beta_0+\beta_1 x_1+\beta_2 x_2)}},$$

is the probability of Y = 1.

The formula shows that once β_i are fixed, we can effortlessly compute the probability that gives Y = 1 for a given observation. The main purpose of a logistic model is to calculate observation (x_1, x_2) and estimate the probability that $P(Y = 1)$. Logistic regression, which is a general statistical model, was originally developed and introduced by Joseph Bergson [22].

4.2.2. Naive Bayes Classifier

Naive Bayes classifiers are based on "naive" independence assumptions between the features used to construct classifiers, i.e., models that assign class labels to problem instances, where the class labels are drawn from some finite set [23]. Maximum-likelihood training can be performed by evaluating the number of occurrences of the sample [24], which requires linear time, rather than expensive iterative approximation (as used for many other types of classifiers).

In our context, the goal of naive Bayes classification is to find the probability that data x belongs to indoor class I or outdoor class O based on Bayes' theorem.

$$P(A|B) = \frac{P(B|A)P(B)}{P(A)}. \tag{2}$$

The probability of x belonging to the indoor or outdoor class is given as follows.

$$P(y = I|x);\ P(y = O|x). \tag{3}$$

We can find the class of data x by selecting the class with the highest probability.

$$c = \underset{c\in\{I,O\}}{arcmax} p(c|x). \tag{4}$$

By applying naive Bayes to Equation (4), we can obtain c as follows.

$$c = \underset{c\in\{I,O\}}{arcmax} \frac{p(x|c)p(c)}{p(x)}, \tag{5}$$

$$c = \underset{c\in\{I,O\}}{arcmax} p(x|c)p(c). \tag{6}$$

In addition, by applying the assumption that all features in x are independent, we can calculate c as follows.

$$c = \underset{c\in\{I,O\}}{arcmax} p(x_1, x_2 \ldots, x_9|c)p(c)c = \underset{c\in\{I,O\}}{arcmax} \prod_{i=1}^{9} p(x_i|c)\, p(c). \tag{7}$$

In the test phase, the class of new data x is found as follows.

$$c = \underset{c\in\{I,O\}}{arcmax} \prod_{i=1}^{9} p(x_i|c)\, p(c). \tag{8}$$

Despite their naive design and apparently oversimplified assumptions, naive Bayes classifiers have worked quite well in many complex real-world situations. Due to the "naive" independent assumption between the features, the naive Bayes classifier is simple and provides fast coverage during the training phase. It can also solve pattern recognition problems with a small amount of training data to estimate the parameters required for classification. However, naive Bayes cannot work with data that never happened before, and the "naive" assumption is often impractical for many real-world problems.

4.2.3. Support Vector Machine

SVMs are supervised learning models that analyze data used for classification and regression analysis. An SVM training technique assigns all new examples to a single category, thereby making it a non-probabilistic binary linear classifier. An SVM model is a representation of the examples in space that maps the examples of separate categories divided by a clear gap calculated to be as wide as possible. All new examples are mapped into that same space and predicted to belong to a category based on which side of the gap they fall. In addition to linear classification, SVMs can classify nonlinear data efficiently using a kernel trick, i.e., mapping their inputs to high-dimensional feature spaces.

In the SVM technique, the training set for the I/O detection problem is $(x_1, y_1), (x_2, y_2), \ldots, (x_{850}, y_{850})$, where $x_i \in \mathrm{R}^9$ are the input data and y_i is the class of x_i. The class of the input data is selected based on the sign of y_i ($y_i = 1$ when x_i is outdoor, and $y_i = -1$ when x_i is indoor). In the training phase, the I/O detection program finds hyperplane that classifies the two classes. The hyperplane is expressed as follows.

$$\mathbf{w}^{\mathrm{T}}\mathbf{x} + b = w_1x_1 + w_2x_2 + \cdots + w_9x_9 + b = 0. \tag{9}$$

With a specific data point (x_i, y_i), the distance to the hyperplane is calculated as follows.

$$\frac{y_n(\mathbf{w}^{\mathrm{T}}\mathbf{x}_n + b)}{\|w\|_2}. \tag{10}$$

Since Equation (10) is positive, we can find the margin of a data point to hyperplane as follows.

$$\text{margin} = \min_n \frac{y_n(\mathbf{w}^{\mathrm{T}}\mathbf{x}_n + b)}{\|w\|_2}. \tag{11}$$

The training phase is finding the margin as large as possible.

$$(\mathbf{w}, b) = \operatorname*{argmax}_{w,b}\left\{\min_n \frac{y_n(\mathbf{w}^{\mathrm{T}}\mathbf{x}_n + b)}{\|w\|_2}\right\} = \operatorname*{argmax}_{w,b}\left\{\frac{1}{\|w\|_2}\min_n y_n(\mathbf{w}^{\mathrm{T}}\mathbf{x}_n + b)\right\}. \tag{12}$$

Equation (12) can be simplified as follows.

$$(\mathbf{w}, b) = \operatorname*{argmin}_{w,b} \frac{1}{2}\|w\|_2^2 \text{ subject to : } 1 - y_n(\mathbf{w}^{\mathrm{T}}\mathbf{x}_n + b) \leq 0, \forall n = 1, 2, \ldots, N. \tag{13}$$

We find the weigh matrix of Equation (13) as follows.

$$\mathbf{w} = \sum_{n=1}^{850} \lambda_n y_n \mathbf{x}_n \text{ where } S = \{n : \lambda_n \neq 0\}, \tag{14}$$

$$b = \frac{1}{N_S}\sum_{n \in S}(y_n - \mathbf{w}^{\mathrm{T}}\mathbf{x}_n) = \frac{1}{N_S}\sum_{n \in S}(y_n - \sum_{m \in S}\lambda_m y_m \mathbf{x}_m^T \mathbf{x}_m). \tag{15}$$

SVMs are one of the most popular classifier techniques and are highly effective when dealing with unstructured and multidimensional input data. SVMs are very flexible, and we can use an SVM even if we do not possess knowledge of the data structure. However, SVMs are very complicated and provide slow coverage. SVM models also are hard for hyperparameters tuning.

4.2.4. K-Nearest Neighbors Classifier

The KNNs technique is a nonparametric method used for both classification and regression [25], where the input consists of training examples in the feature space. The KNN technique is one of the simplest of all machine learning techniques, in which it only approximated locally, and all computations are calculated in the classification state. Here, an object is classified by the majority of its neighbors, where the object is assigned to the most common class among its nearest neighbors.

Due to the simplicity of the technique, KNN is very easy to implement; however, it can still handle unstructured data and simple valuation with good estimation of the test set. With lazy learning, the KNN calculation is performed in the test phase; thus, with a large dataset, the test phase requires more time than other machine learning techniques. This characteristic makes offline learning impossible for KNN. Another drawback of KNN is its sensitivity to noise, which makes it unattractive for complex problems.

4.2.5. Neural Network

Artificial neural networks (ANN) are computing systems that are including many different machine learning techniques to work together and process complex data inputs. Such systems "learn" to perform tasks using examples, and they increase performance gradually without being programmed according to any task-specific rules. Instead, they automatically generate identifying characteristics from the learning material they process. The original goal of the ANN approach was to solve problems in the same manner a human brain would. However, ANNs have been developed over time to perform specific tasks, leading to deviations from biology. ANNs have been used in a variety of fields, including computer science, security, medicine, social network science, board and video games, and military purposes.

In this study, the neural network model was designed as follows: nine for input layer (each receiving a GPS signal), 18 hidden neural, and two output neural (Figure 10).

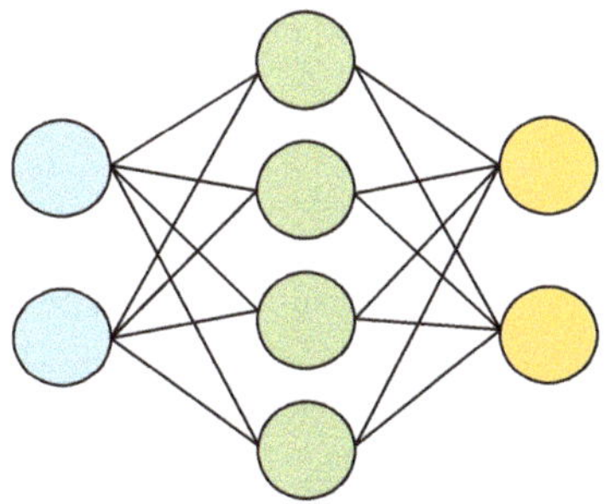

Figure 10. Neural network model.

4.3. *Training Phase*

Since the input data are well collected, the training phase is easier, which leads to high accuracy in the model. Errors will fall to the semi-indoor environment, where the GPS signal is still strong despite the indoors environment. This is the characteristic of the input data, yet the learning model can reduce the impacts to minimum.

We employed L2 norm for error calculation, where errors are calculated as the Euclidean distances between two points. Note that L2 norm is the best choice for multidimensional input vector data.

Based on the required accuracy and user devices, we can also apply offline training. Offline training is suitable for low-accuracy applications or devices that do not require permanent mobility because it does not require significant resources, has a fast calculation process, and does not require an update operation. Offline learning is also an advantage of this scheme because the final model is very simple, offers fast calculation, and requires only the current GPS signal information. In contrast, online training can increase the accuracy of the model; however, online training results in a complex model and requires extra calculations to update the model. Online training is more suitable for mobile applications, where the device's resources are available and applications requires high accuracy.

In this study, we focus on offline training, and we tested with many machine learning techniques to identify the best solution for the GPS signal classifier.

4.4. Testing Phase

Generally, the preprocessing phase improves the quality of the input data; thus, the machine learning models can work effectively with very high accuracy. First, we had a training set with 423 indoor samples, 427 outdoor samples, and a small number of semi-indoor samples that we set as indoor samples. Note that we selected this training set to simulate normal activities (most time is spent in indoor and outdoor environments rather than semi-indoor environments). With the test set with only indoor and outdoor samples, all machine learning models reach very high accuracy (minimum was 96% with the SVM technique, and the maximum was 98% with the KNN technique). To facilitate better evaluation of the data set, we increased the number semi-indoor environment samples as an individual test set to compare the accuracy of all models in a high noise environment. This test set included 170 samples (70 indoor samples, 50 outdoor samples, and 50 semi-indoor samples categorized as indoor samples). We also proposed a unique preprocessing method to improve the final result of this test set.

We employed an offline learning method for I/O detection. Offline learning uses the trained model for the actual application, and it updates the model based on the programmer's schedule. Offline learning provides a simple model with fast calculations. The application does require a large amount of training data, as well as learning phase calculation. Another advantage of offline training is the uniformity of all applications: note that the model will do the same classification for all GPS sensors.

5. Evaluation

5.1. Testing with Indoor/Outdoor Sample Test Set

Figure 11 shows the accuracy of the test set with indoor and outdoor samples. We also included I/O detection using a combined sensor [1,11] for comparison. The first test set included 70 indoor and 50 outdoor samples, with high differences in both the number and strength of the received GPS signal. The proposed scheme demonstrated an impressive result, with the highest overall rate of 98% and lowest of 96%. The results indicate that the strength of the GPS signal from the smartphone sensor can be used to extract features to be used in the I/O detection classifier, and the method can be used for other applications. Note that the sample data can be representative for the overall feature of indoor/outdoor areas because the GPS signal is essentially the same at other locations on Earth. With the indoor/outdoor test set, the I/O detection application can obtain impressive accuracy, even without performing a preprocessing phase. This high accuracy is due to the large bias of the indoor and outdoor environments, in which the GPS signal is very weak inside buildings. However, this test set is not included the semi-indoor environment, which is the most difficult environment for I/O detection; therefore, the application does not promise smooth transfer from an outdoor environment to an indoor environment (and vice versa). Thus, another test set for semi-indoor environments was added and evaluated with the indoor/outdoor environment test set. Here, we tested using the semi-indoor environment and set the samples to indoor samples.

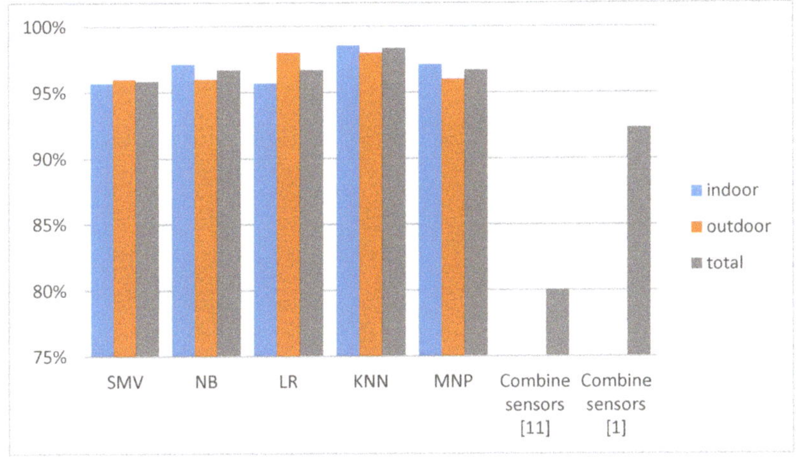

Figure 11. Indoor/outdoor (I/O) detection accuracy.

5.2. *Testing with Indoor/Outdoor and Semi-Indoor Sample Test Set*

This test set was included to improve the application's ability to identify an indoor environment even if the GPS signal was weaker due to the semi-indoor environment. This characteristic is extremely helpful when moving from an outdoor environment to an indoor environment (and vice versa). The second test set included 70 indoor, 50 outdoor, and 50 semi-indoor samples. For the comparison, we also included I/O detection using a combined sensor [9] for comparison. Figure 12 shows the accuracy for indoor and outdoor cases separately, in which the semi-indoor environment is categorized as indoor. The accuracy of indoor detection was reduced significantly, which indicates confusion between semi-indoor and outdoor environments in the classifier phase. The result of indoor environment detection was further analyzed to identify the root causes of the poor performance of several classifiers. In consideration of the lower accuracy compared to the outdoor environment, the indoor and semi-indoor samples were misclassified, which indicates that the GPS signal strength of semi-indoor cases is similar to outdoor cases, and it is very difficult to distinguish these cases. In many semi-indoor environments, there is very little change in the GPS signal compared to outdoors, which leads to a malfunction in the I/O detection application. The noise added to the GPS signal in an outdoor environment is also a reason for errors because this makes the GPS signal become weaker and confused with the semi-indoor signal, which decreases the accuracy of the learning model.

Compared to the combined sensor technique, the accuracy of the proposed scheme is still better (95% for the KNN technique compared to 88% for the combined sensor [9]). This emphasizes the effectiveness of the GPS-based I/O detection scheme. However, the result is insufficient, and we can improve it by employing a unique preprocessing process.

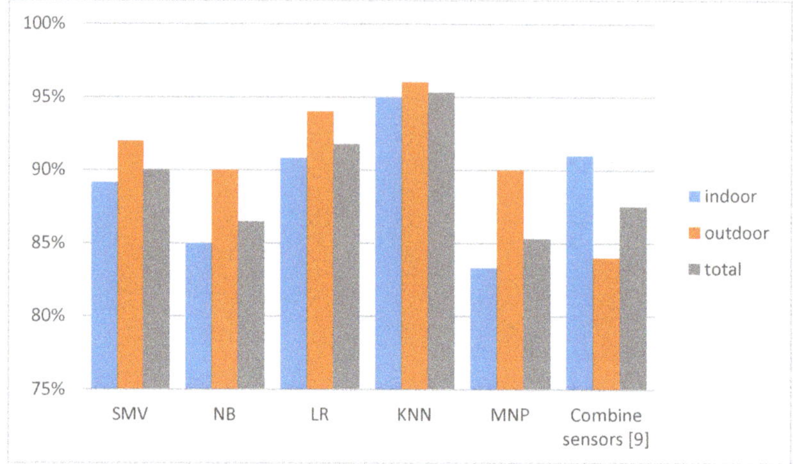

Figure 12. Accuracy of I/O detection with semi-indoor samples (classified as indoor samples).

5.3. Proposed Preprocessing Process

For the dataset including 1020 individual data, each of them was based on the strength of GPS signals corresponding to a specific place. All possible GPS signals were collected, and, using the mobile application, the analog input was transformed to digital data for further analysis. Here, the data included information about the GPS signal (zero to 10 features). We denoted the ith data as X_i, and the GPS signal from satellite j in X_i data as x_{ij}. We regularized the input data as follows.

Step 1 Reduce the number of GPS signals to nine.

The tenth GPS appears very infrequently, and, as the GPS signal is very small, this information is not useful for analysis.

Step 2 Fill the missing datum.

As mentioned previously, low-strength GPS signals are removed by the mobile application. Thus, if the GPS signal is too weak (in a building, to be blocked by obstacle), nine GPS signals cannot be obtained. The datum will be filled by zeroes; thus, these data will not affect the strongest signals, which contain the most information about I/O detection.

Step 3 Rearrange data for further analysis.

The strong signals contain the most information about I/O detection; however, the order of these signals is chaotic and independent from each other. Thus, we cannot find any pattern because the strong signals mix with the weak signal, which makes the entire dataset chaotic and unpredictable. Therefore, we propose a simple solution for this problem, i.e., we rearrange the GPS signals from strongest to smallest. As a result, strong GPS signals will take priority during the training and testing phases. Even if this does not work effectively, we can add a heavyweight to strong GPS signals before applying provided schemes.

The preprocessing sequence is shown in Figure 13.

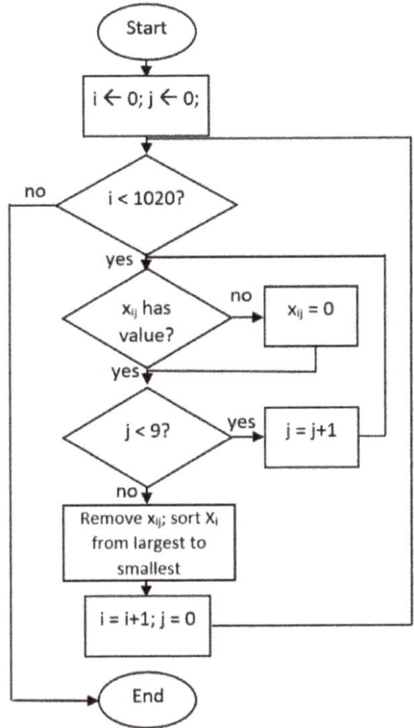

Figure 13. Preprocessing flow.

5.4. Testing with Indoor/Outdoor and Semi-Indoor Sample Test Set with Proposed Data Processing

Figure 14 shows the accuracy for indoor and outdoor cases obtained after applying the proposed preprocessing. The preprocessing phase increases the I/O detection accuracy significantly (approximately 4% for the worst case and 2% for the best case). The machine learning techniques clean the input data, as well as propose a better structure for the learning model.

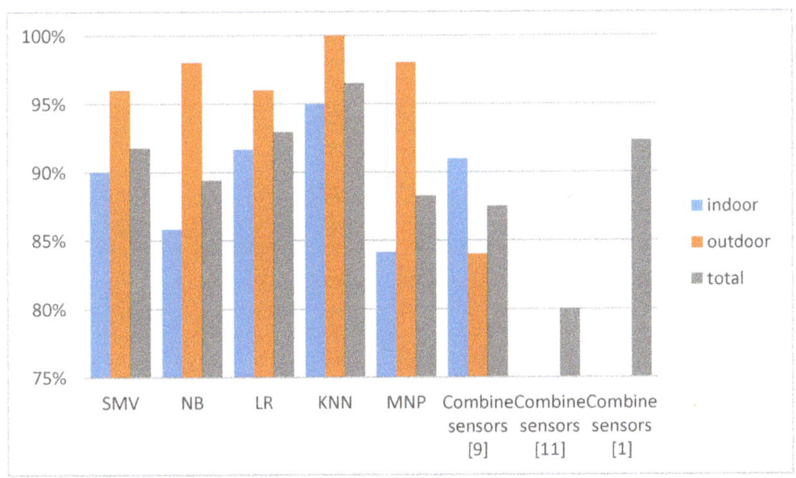

Figure 14. I/O detection accuracy obtained with proposed preprocessing.

The performance of the proposed approach was compared to traditional technologies commonly used for I/O detection. The first comparison was against I/O detection using GPS. The proposed approach was compared to three different studies [1,9,11]. The proposed method is clearly better because it uses more information, has better preprocessing, and applies suitable Machine Learning model.

Based on the overall scheme, other classifier methods, such as LSTM and Meta-classifier are applied. However, they are not suitable for the I/O detection scheme and the accuracy is also lower than KNN. In proposed scheme, I/O detection focused on the simple and effective methods, as it does not require mobile resources. Based on this approach, the machine learning methods are more suitable for the scheme. In contrast, both LSTM and Meta-classifier required large resources for both training and testing phases, so that they are not considered as promised methods for proposed scheme. Another reason is that the accuracy of LSTM and Meta-classifier is lower than KNN. LSTM reaches 91% overall accuracy and Meta-classifier is 94%, which cannot be considered in the proposed scheme.

The machine learning model will be evaluated based on the accuracy of test sets, as well as training coverage. We use offline training for the application; thus, training coverage may not be important. Nonetheless, it will remain a criterion used to evaluate the machine learning model.

6. Conclusions

In this paper, we proposed an indoor and outdoor environment detection scheme that classifies indoor and outdoor environments at high accuracy under a simple implementation. We also proposed a data preprocessing technique. Using the proposed data preprocessing process and suitable machine learning techniques, the proposed I/O detection scheme obtains quick and accurate detection results in various time and environments. We comprehensively tested I/O detection through a prototype implementation and evaluated the system based on different data samples collected in the Kookmin University area. The prototype systems obtained an accuracy of 97%. We particularly conducted a case study where we made use of I/O detection results to infer the GPS availability and accuracy under various indoor/outdoor environments.

Author Contributions: All authors contributed to this paper: Van Bui and Nam Tuan Le proposed the idea and implementation methodology, reviewed and edited paper; Van Bui performed all experiments and wrote the paper, verified the experiment process and results; Thanh Luan Vu collected data and created the first version of data augmentation software; Van Hoa Nguyen proposed and optimized machine learning techniques, performed parts of experiments; and Yeong Min Jang supervised the work and provided funding support. All authors have read and agreed to the published version of the manuscript.

Funding: This research was financially supported by the Ministry of Trade, Industry and Energy (MOTIE) and Korea Institute for Advancement of Technology (KIAT) through the International Cooperative R&D program.

Conflicts of Interest: The authors declare no conflict of interest.

References

1. Radu, V.; Katsikouli, P.; Sarkar, R.; Marina, M.K. A semi-supervised learning approach for robust indoor-outdoor detection with smartphones. In Proceedings of the 12th ACM Conference on Embedded Network Sensor Systems, Memphis, TN, USA, 3–6 November 2014.
2. Mekki, S.; Karagkioules, T.; Valentin, S. HTTP adaptive streaming with indoors-outdoors detection in mobile networks. *arXiv* **2017**, arXiv:1705.08809.
3. Carroll, A.; Heiser, G. An analysis of power consumption in a smartphone. In Proceedings of the 2010 USENIX Annual Technical Conference (USENIXATC'10), Boston, MA, USA, 23–25 June 2010.
4. Jiang, L.; Tan, S.Y. Geometrically Based Statistical Channel Models for Outdoor and Indoor Propagation Environments. *IEEE Trans. Veh. Technol.* **2007**, *56*, 3587–3593. [CrossRef]
5. Global Positioning System. Available online: https://en.wikipedia.org/wiki/Global_Positioning_System (accessed on 21 September 2019).
6. Pei, L.; Chen, R.; Chen, Y.; Leppäkoski, H.; Perttula, A. Indoor/Outdoor Seamless Postioning Technologies Integrated on Smart Phone. In Proceedings of the First International Conference on Advances in Satellite and Space Communications, Colmar, France, 20–25 July 2009.

7. Lipowezky, U.; Vol, I. Indoor-Outdoor Detector for Mobile Phone Cameras Using Gentle Boosting. In Proceedings of the 2010 IEEE Computer Society Conference on Computer Vision and Pattern Recognition, San Francisco, CA, USA, 13–18 June 2010.
8. AzizyanIonut, M.; Constandache, I.; Roy Choudhury, R. Surround-Sense: Mobile Phone Localization via Ambience Fingerprinting. In Proceedings of the 15th Annual International Conference on Mobile Computing and Networking, Beijing, China, 20–25 September 2009.
9. Zhou, P.; Zheng, Y.; Li, Z. IODetector: A generic service for indoor outdoor detection. In Proceedings of the 10th ACM Conference on Embedded Network Sensor Systems, 6–9 November 2012.
10. Anagnostopoulos, T.; Garcia, J.C.; Goncalves, J.; Ferreira, D.; Hosio, S.; Kostakos, V. Environmental exposure assessment using indoor/outdoor detection on smartphones. *Pers. Ubiquitous Comput.* **2017**, *21*, 761–773. [CrossRef]
11. Capurso, N.; Song, T.; Cheng, W.; Yu, J.; Cheng, X. An Android-Based Mechanism for Energy Efficient Localization Depending on Indoor/Outdoor Context. *IEEE Internet Things J.* **2017**, *4*, 299–307. [CrossRef]
12. Shengnan; Qin, L.; Song, H. A lightweight and aggregated system for indoor/outdoor detection using smart devices. *Future Gener. Comput. Syst.* **2017**. [CrossRef]
13. Chen, K.; Tan, G. SatProbe: Low-energy and fast indoor/outdoor detection based on raw GPS processing. In Proceedings of the IEEE INFOCOM 2017—IEEE Conference on Computer Communications, Atlanta, GA, USA, 1–4 May 2017.
14. Iwata, S.; Ishikawa, K.; Takayama, T.; Yanagisawa, M.; Togawa, N. A Robust Indoor/Outdoor Detection Method based on Sparse GPS Positioning Information. In Proceedings of the IEEE 8th International Conference on Consumer Electronics, Berlin, Germany, 2–5 September 2018.
15. Femminella, M.; Reali, G. A Zero-Configuration Tracking System for First Responders Networks. *IEEE Syst. J.* **2015**, *11*, 2917–2928. [CrossRef]
16. What Is GPS? Available online: https://www.loc.gov/rr/scitech/mysteries/global.html (accessed on 28 January 2018).
17. Available online: https://web.archive.org/web/20181231195431/https://www.youtube.com/watch?v=ozAPGnr-934. (accessed on 17 June 2017).
18. Daly, P. Navstar GPS and GLONASS: Global satellite navigation systems. *Acta Astronaut.* **1991**, *25*, 399–406. [CrossRef]
19. Space Segment GPS.GOV. Available online: https://web.archive.org/web/20190718190908/https://www.gps.gov/systems/gps/space/ (accessed on 14 September 2019).
20. Alkan, R.M.; Karaman, H.; Sahin, M. GPS, GALILEO and GLONASS satellite navigation systems & GPS modernization. In Proceedings of the 2nd International Conference on Recent Advances in Space Technologies (RAST 2005), Istanbul, Turkey, 9–11 June 2005.
21. Van Diggelen, F.; Abraham, C.; de Salas, J.; Silva, B. Insidegnss. BROADCOM CORPORATION. Available online: https://www.insidegnss.com/auto/marapr11-VanDiggelenREV.pdf (accessed on 7 December 2019).
22. Cramer, J.S. *The Origins of Logistic Regression*; Tinbergen Institute Working; Tinbergen Institute: Amsterdam, The Netherland, 2002; pp. 167–178.
23. Rish, I. An Empirical Study of the Naive Bayes Classifie. In Proceedings of the IJCAI Workshop on Empirical Methods in AI, Seattle, WA, USA, 4 August 2001.
24. Russell, S.; Norvig, P. *Artificial Intelligence: A Modern Approach*; Prentice Hall: Upper Saddle River, NJ, USA, 2010.
25. Altman, N.S. An introduction to kernel and nearest-neighbor nonparametric regression. *Am. Stat.* **1992**, *46*, 175–185.

Article

A Holistic Approach to Guarantee the Reliability of Positioning Based on Carrier Phase for Indoor Pseudolite

Yinzhi Zhao [1,2,3], Jiming Guo [1,*], Jingui Zou [1,*], Peng Zhang [1,2], Di Zhang [1], Xin Li [4], Gege Huang [5] and Fei Yang [1,3]

1 School of Geodesy and Geomatics, Wuhan University, Wuhan 430079, China; yzzhao_gnss@whu.edu.cn (Y.Z.); pzhang@sgg.whu.edu.cn (P.Z.); zhangdi@whu.edu.cn (D.Z.); coffeeyang@whu.edu.cn (F.Y.)
2 Research Center for High Accuracy Location Awareness, Wuhan University, Wuhan 430079, China
3 Guangxi Key Laboratory of Spatial Information and Geomatics, Guilin 541004, China
4 College of Geology Engineering and Geomatics, Chang'an University, Xi'an 710054, Shannxi, China; lixin2017@chd.edu.cn
5 State Key Laboratory of Information Engineering in Surveying, Mapping and Remote Sensing, Wuhan University, Wuhan 430079, China; geehuang@whu.edu.cn
* Correspondence: jmguo@sgg.whu.edu.cn (J.G.); jgzou@sgg.whu.edu.cn (J.Z.); Tel.: +86-1330-862-0798 (J.G.); +86-1338-757-1063 (J.Z.)

Received: 15 January 2020; Accepted: 6 February 2020; Published: 11 February 2020

Featured Application: This work is potentially applied to indoor high-precision engineering/industrial measurement, such as parts installation, tunnel engineering, dynamic rail inspection, path planning, etc. Especially in case of pseudolite faulty, the reliability and availability of positioning can be guaranteed.

Abstract: The integrity monitoring algorithm based on pseudorange observations has been widely used outdoors and plays an important role in ensuring the reliability of positioning. However, pseudorange observations are greatly affected by the error sources such as multipath, clock drift, and noise in indoor pseudolite system, thus the pseudorange observations cannot be applied to high-precision indoor positioning. In general, double differenced (DD) carrier phase observations are used to obtain a high-precision indoor positioning result. What's more, the carrier phase-based integrity monitoring (CRAIM) algorithm is applied to identify and exclude potential faults of the pseudolites. In this article, a holistic method is proposed to ensure the accuracy and reliability of positioning results. Firstly, if the reference pseudolite operates normally, extended Kalman filter is used for parameter estimation on the premise that the number of common pseudolites meets positioning requirements. Secondly, the innovation sequence in the Kalman filter is applied to construct test statistics and the corresponding threshold is determined from the Chi distribution with a given probability of false alert. The pseudolitehorizontal protection level (HPL) is calculated by the threshold and a prior probability of missed detection. Finally, compared the test statistics with the threshold to exclude the faultypseudolite for the reliability of positioning. The experiment results show that the proposed method improves the accuracy and stability of the results through faults detection and exclusion. This method ensures accuracies at the centimeter level for dynamic experiments and millimeter levels for static ones.

Keywords: carrier phase; differential pseudolite system; extended Kalman filter; reliability; integrity monitoring

1. Introduction

The global navigation satellite system (GNSS) has been widely applied in outdoor positioning. Nevertheless, it is difficult to obtain reliable GNSS signals in sheltered environments or indoors [1], where a pseudolite system can make up for the shortage of GNSS [2,3].

Relative to the GNSS, indoor pseudolite observations have no tropospheric and ionospheric errors. The PL/receiver clock and antenna phase error can also be eliminated by DD observations which greatly reduce the difficulty in processing error sources in indoor PL positioning [4]. However, the multipath effect, noise, and fixed satellite constellations are still the problems indoors. What is worse, the indoor pseudolite system has a great nonlinear effect due to the close distance between the receivers and pseudolites. If initial position estimation has poor accuracy, the nonlinear error caused by linearization will be severe. Therefore, large code measurement errors caused bythe above factors may result in low accuracy and reliability of positioning results.

At present, the research on the indoor pseudolite system mainly focuses on the design of pseudolite constellation or high-precision positioning based on the carrier phase, but few on the reliability of positioning. According to quality control methods of GNSS carrier phase positioning, they can be divided into two categories: priori and posteriori quality control methods. As to priori methods, there are some classical methods, such as cycle slip detection, DD method. Cycle slip detection can effectively eliminate such faults as cycle slip or signal interruption. The double differenced method will eliminate clock difference and weaken troposphere and ionosphere delays. In general, the posteriori quality control method does not distinguish which kind of error sources, such as RAIM, IGG III, and variance component estimation. These methods can effectively improve the positioning accuracy. However, the existing legacy approaches cannot eliminate all faults, such as multipath effect, noise and so on, which is also the current problem. In an indoor environment, the influence of multipath and noise is greater than outdoors. A holistic approach to guarantee the reliability of positioning for indoor pseudolite based on carrier phase is proposed and the feasibility of the method is verified by some experiments. The proposed method creatively considers the fault of indoor positioning source (pseudolite) and other error sources, including complete fault detection, exclusion procedure, andformula derivation of pseudolite protection level. Faults on reference pseudolite are also considered. Although we cannot deal with any specific error source, the data with poor quality can be effectively identified no matter which error source. By excluding the corresponding epochs, the positioning accuracy and reliability will be guaranteed.

The research on indoor pseudolite high-precision positioning algorithms has become a hotspot in recent years and the key is to use carrier phase observations and ambiguity resolution [4–6].Reference [4] has successfully introduced AFM (Ambiguity Function Method) into PL-AR (Pseudolite-Ambiguity Resolution). However, due to the multi-peak characteristic of AFM, the reliability of the solutions is sometimes difficult to be guaranteed compared with the LAMBDA method. In addition, the way to ensure the reliability of positioning has not been discussed. References [5,6] proposed a new OTF-AR method for ground-based positioning systems which solely relies on carrier phase measurements and applies to cases where no sufficiently accurate measurements were provided in advance; however, the gross error must be given careful consideration in practical applications and the experiments in the references were conducted under relatively good conditions. Actually, it is the poor conditions that lead to unreliable positioning results. Some research institutes used KPI when LAMBDA and ratio tests are applied to obtain the integer ambiguity resolution [7]. The most mature indoor pseudolite system is produced by LOCATA [8]. ProprietaryTimeLoc technology is utilized for tight time synchronization, allowing the clock difference in the carrier phase measurement equation to be ignored. The horizontal error of the non-differential positioning is less than 6 cm, and the precision of the altitude direction is about 15 cm [9]. The frequency stability and accuracy of the PL clock are much inferior to the GPS, and PL clock errors are difficult to obtain accurately [2]. Although LOCATA can ensure PL clock synchronization, many pseudolite systems lack this costly technology at present. Reference [10] achieved centimeter-level PPP (Precise point positioning) results with a new prototype pseudolite system without relying on accurate time synchronization. It is believed that PL would be more extensively applicable for integration with GNSS and INS (Inertial Navigation System). Reference [11] proposed a new concept of pseudolite-based navigation to overcome the receiver modification problem of systems. However, only meter-level positioning precision could be achieved in poor geometry tests. References [10,11] did not consider quality control under severe conditions as well. In reference [12], Xu et al. built a new pseudolite-based indoor positioning system and the clock could be synchronized with each other. Nevertheless, only simulated data wereused to obtain high-precision positioning results. References [13,14] believed that the convergence precision of initial coordinate values was important for positioning, but how to ensure centimeter-level positioning precision has not been considered with bad initial coordinate value. Reference [15] used carrier interferometry to conduct analyses on indoor pseudolite positioning. However, this method has great limitations in applications. Reference [16] proposed a new method for indoor pseudolite positioning based on AFM and DD pseudorange observations with good data. However, it is generally known that data quality has a great influence on the AFM.

Current high-precision PL positioning algorithms mainly are applied based on carrier phase observations. However, the indoor environment is more severe than outdoors. Whether the carrier phase observations or the positioning results are reliable remains to be discussed. To the authors' knowledge, there are few methods to guarantee the reliability of the results for indoor pseudolite positioning and no effective strategy to ensure the positioning accuracy whena fault happens. Therefore, the application of receiver autonomous integrity monitoring (RAIM) for pseudolite system is of great significance in practical positioning.

There are two types of RAIM algorithm, snapshot RAIM, and sequential RAIM. The snapshot RAIM includes the parity vector method, the least-square residual method [17–19], weighted RAIM algorithm, etc. Sequential RAIM is almost based on the Kalman filter which is an indispensable component of many navigation systems [20]. In view of the above, the carrier phase-based RAIM (CRAIM) algorithm [21] proposed by Feng.was applied tothe high-precision positioning of the indoor pseudolite system. Reference [22] proposed an INS-assisted RAIM algorithm to increase the level of reliability and integrity for an integrated navigation system. Reference [23] explored the application of the snapshot RAIM method for GNSS vessel receivers when space segment geometry was poor. However, the proposed algorithm did not discuss how to determine the protection level with a given probability of missed detection (MD) and false alarm (FA). Reference [24] showed a correlation between traditional Shapiro-Wilk test statistics when multiple faults occur. The paper proposed a criterion to overcome the correlation based on the reliability principle and designed a multi-fault processing method based on an extended Shapiro-Wilk test. Li et al. proposed a complete algorithm to ensure reliability in attitude measurement [25]. In this article, a new integrity monitoring method was proposed for the GNSS-based attitude determination and they constructed the test statistic by using the measurement residuals to detect the potential failure. Furthermore, they derived the attitude protection level based on the worst-case protection concept under the multiple faults hypothesis. Reference [26] indicated that carrier phase observations error does not subject to the standard normal distribution and a carrier phase-based RAIM algorithm using a Gaussian sum filter was proposed. Since pseudorange-based RAIM (PRAIM) cannot meet the requirement of PPP, References [27–29] applied CRAIM to GNSS PPP and RTK algorithm for airport surface movement. Similar to GNSS PPP and RTK, carrier phase observations must also be used in high-precision indoor pseudolite positioning. In references [30,31], a specific algorithm for multiple constellations was studied and a method of integrity monitoring-based ambiguity validation for triple-carrier ambiguity resolution was proposed. They considered that ambiguity validation methods include the ratio test, the quantitative evaluation based on the success rate and the equivalent failure rate of ambiguity resolution. However, the ambiguity correctness validation cannot necessarily guarantee the reliability of indoor pseudolite positioning. Sometimes, an incorrectly fixed ambiguity will be obtained because of the severe environment and the poor satellite geometry. Therefore, it is necessary to carry out the integrity monitoring for indoor pseudolite rather than solely using the ambiguity validation method.

The RAIM algorithm is mainly used in safety-critical fields such as aircraft navigation at present and is almost unmanned in indoor pseudolite. In fact, indoor pseudolite positioningis seriously influenced by many environmental factors, such as people's walking, multipath effects from wallsand so on. Sometimes it is difficult to obtain a high-precision position result. In addition, even if a specific positioning result can be obtained, the reliability may not be guaranteed. This paper focuses on the RAIM algorithm of indoor pseudolite system based on EKF, including PRAIM and CRAIM algorithm. Some experiments are operated to verify the feasibility of the indoor pseudolite RAIM algorithm and the reliability of the positioning results, which can expand the availability of pseudolite positioning. According to the actual characteristics of pseudolite, an algorithm to guarantee the accuracy and reliability of indoor positioning for pseudolite based on CRAIM is proposed, which is named PLRAIM (Pseudolite based RAIM). It is especially important to use this method in harsh indoor environments.

2. Algorithm

The algorithm includes two aspects, the calculation of the protection level and fault detection. All algorithms in this paper are implemented in horizontal positioning.

2.1. Double Differenced Positioning Algorithm in Pseudolite System

2.1.1. Observation Model in Pseudolite System

Due to the complex indoor environment and the poor stability of commercial pseudolite clocks, the pseudolite high-precision positioning algorithms are generally realized based on DD carrier-phase positioning mode. The error of clock synchronization and antenna phase can be eliminated or weakened by theDD model [32]. The DD observations do not include tropospheric and ionospheric delay because the signal from the pseudolite itself does not pass through the troposphere and ionosphere. Furthermore, since the pseudolite constellation is fixed, there are no ephemeris errors and relativistic effects. Therefore, the indoor single-frequency pseudolite DD observation equations can be formulated as [33–35],

$$\begin{gathered} \nabla\Delta P_{br}^{ks} = \nabla\Delta\rho_{br}^{ks} + \nabla\Delta e_P + \nabla\Delta M_P \\ \nabla\Delta L_{br}^{ks} = \lambda_1\nabla\Delta\varphi_{br}^{ks} = \nabla\Delta\rho_{br}^{ks} + \nabla\Delta\varepsilon_L + \nabla\Delta m_L + \lambda_1\nabla\Delta N_{br}^{ks} \end{gathered} \tag{1}$$

where $\nabla\Delta$ is a DD operator; r and b represent the rover and base stations, respectively. k and s represent pseudolites. The pseudorange and carrier phase observations are represented by P and φ. ρ represents the geometric distance between pseudolites and the receivers. e and ε denote the noise of the pseudorange and carrier phase observations, respectively. M and m represent the multipath effects of pseudorange and carrier phase observations. N indicates the integer ambiguities of carrier phase observations. λ indicates the wavelength.

2.1.2. Parameter Estimation

Extended Kalman filter (EKF) is applied to estimate the parameters ina high-precision positioning solution. EKF is mainly to solve the linearization problem in KF. It can be considered that the coordinates remain the same because the rover station is a state of low dynamic or static in indoor positioning. The transition matrix is I and the covariance matrix of the system process noise is O. The measurement noise should be considered. The observation and the state model are as follows,

$$\begin{gathered} X_k = \Phi_{k-1}X_{k-1} + \omega_{k-1} \\ Z_k = H_kX_k + v_k \end{gathered} \tag{2}$$

where k is the current epoch. X represents the state vector. Φ is the transition matrix I. ω is the dynamic process noise of the system. Z is a measurement vector after linearization. H is a design matrix $[I_r\ \lambda]$

and v indicates the measurement noise. If the dynamic process noise and the measurement noise of the system are Gaussian white noise, then the parameters have the following statistical characteristics,

$$\begin{gathered} E(\omega_k) = 0, Cov(\omega_k, \omega_k) = Q_k \\ E(v_k) = 0, Cov(v_k, v_k) = R_k \\ Cov(\omega_k, v_k) = 0 \end{gathered} \tag{3}$$

After the initial coordinate being obtained, the corresponding initial state parameter X_0 and its variance-covariance matrix P_0 can be given. The basic formula of EKF follows,

$$\begin{gathered} K_k = P_k^{(-)} H_k^T W_k \\ W_k = \left(H_k P_k^{(-)} H_k^T + R_k \right)^{-1} \\ P_k^{(-)} = \Phi_{k-1|k} P_{k-1}^{(+)} \Phi_{k-1|k}^T + Q_{k-1}, P_k^{(+)} = (I - K_k H_k) P_k^{(-)} \\ X_k^{(-)} = \Phi_{k-1} X_{k-1}^{(+)}, X_k^{(+)} = X_k^{(-)} + K_k V_k \\ V_k = Z_k - H_k X_k^{(-)} \end{gathered} \tag{4}$$

where subscript k denotes the current epoch. X is state vector; P is covariance matrix. K is Kalman gain matrix and V is innovation vector. W is weight matrix. R is measurement noise matrix. (+) and (−) denote predicted and updated values. When positioning, the iterative EKF is usually used to weaken the linearization error. In order to avoid the problem of divergence caused by the imprecise of a given dynamic model during filtering, the weight can be adjusted according to the residual matrix to achieve the robust Kalman filter [36].

2.2. The Procedure of PLRAIM

The reliability of the positioning algorithm to obtain the rover coordinates is greatly affected by multipath and ambiguity resolution. Therefore, this paper focuses on studying the reliability of positioning results. Integrity monitoring mainly includes two aspects: fault detection and alert and availability of the RAIM algorithm. The main idea is to detect and excludefaultsbased on the consistency between redundant observations [37,38]. The construction of test statistics is one of the key factors, consequently. Considering the actual situation of our indoor pseudolite positioning system, the steps of PLRAIM are as follows:

(a) Obtain data of rover and base station and ensure there are more than three common visible pseudolites in order for the horizontal positioning to be performed;

(b) Judge fault on reference pseudolite (rpl). After the reference pseudolite is determined by the elevation angle, the priori residuals of DD pseudorange observations can be calculated. If the residuals of all DD observations perform largely, it can be considered that the reference pseudolite is faulty and it should be replaced by another one;

(c) When the number of common pseudolites meets requirements and reference pseudolite is normal, parameter estimation and ambiguity resolution will be performed. The classical LAMBDA method is used for ambiguity resolution (AR) [39,40];

(d) Judge the availability of RAIM. Compare HPL calculated by the probability of missed detection (P_{md}) with HAL. We set $P_{md} = 1 \times 10^{-3}$ [21]. As to HAL, it is only used to set a maximum error range for comparison with HPL. Referring to GNSS positioning, the accuracy of high-precision positioning needs at least a centimeter-level. Therefore, in high-precision indoor pseudolite positioning, the positioning accuracy withcentimeter-level would be reliable and available. The HAL of indoor pseudolite high-precision positioning is set to 0.1m;

(e) Fault detection of pseudolite positioning system. Construct the test statistic T and compare it with threshold T_h determined from the probability of false alert (P_{fa}). We set $P_{fa} = 3.33 \times 10^{-9}$ [21]. If T is larger than the threshold, the pseudolite may be faulty;

(f) Fault exclusion. The most intuitive method is to remove pseudolites in turn and compare the test statistic. Then exclude this pseudolite or delete corresponding epochs when a fault occurs;

(g) Positioning. After the above steps, the reliability of results is effectively improved. The flow chart follows,

In Figure 1, HPL is calculated before fault detection. If HPL is greater than HAL, the current observation condition (pseudolite constellation) would not be good enough and the unavailable situation of the system shown in Figure 2 may appear. In Figure 2b, HMI means hazardously misleading information and MI means misleading information. There are two important concepts in fault detection, false alert (FA) and missed detection (MD). P_{fa} refers to the probability of the fault detected by the system and alerts on the condition that the pseudolite is not faulty or HPE is less than HPL. FA occurs when the navigation is normal in Figure 2. P_{md} is the probability that the system fails to detect it while in fact, the fault happens and HPE is greater than HPL. The MD shown as MI and HMI in Figure 2 mainly occurs when the navigation is not normal. HPL guarantees that P_{md} will not exceed demand under the current pseudolite constellation and the threshold ensures that there is no FA in fault detection.

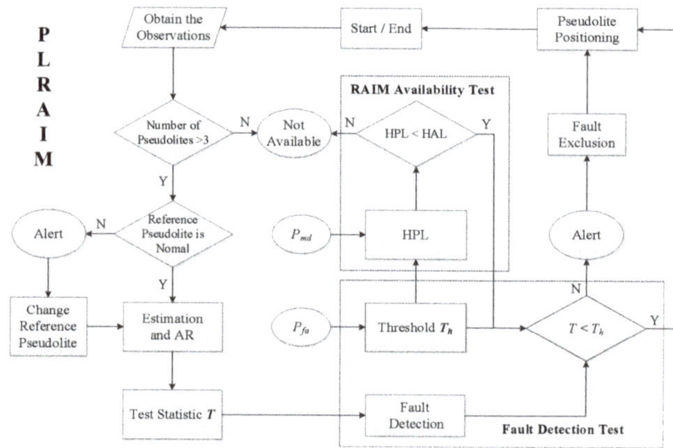

Figure 1. The Procedure of PLRAIM.

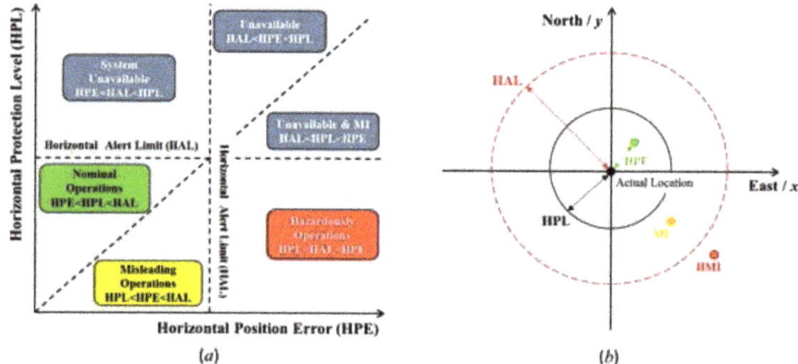

Figure 2. (**a**) Events according to the Stanford Diagram, (**b**) Sketch Map of MI and HMI.

The determination of HPL and test statistic threshold will be presented in the next section in conjunction with the pseudolite system. In summary, under the premise that RAIM availability is guaranteed (HPL<HAL), the following four situations in Figure 3 may occur during fault detection.

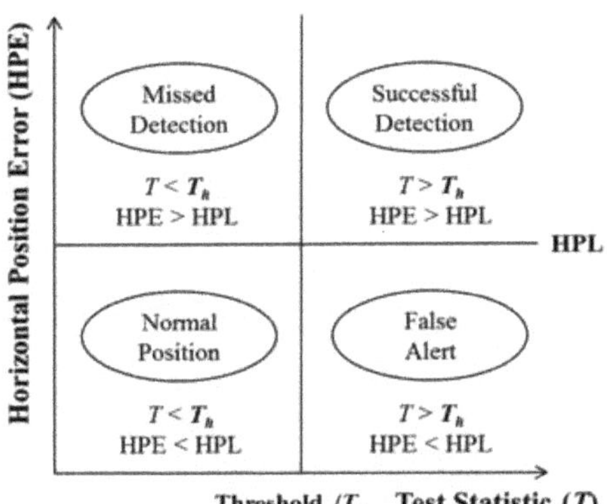

Figure 3. Conditions in FaultDetection.

2.2.1. Fault Detection

In fault detection, the most critical step is the construction of test statistic T and the calculation of threshold T_h. The test statistic is a value calculated from a set of sample data used to determine whether the null hypothesis should be rejected in the hypothesis test. The construction of T depends on the probability model (distribution) and the assumptions of the problem. As long as T is constructed, T_h can be determined from a probability model with a given confidence level.

According to (2), the residual vector of the DD observations is obtained as,

$$r_k = Z_k - H_k X_k^{(+)} = Z_k - H_k(X_k^{(-)} + K_k V_k) \tag{5}$$

$$r_k = Z_k - H_k X_k^{(-)} - H_k K_k V_k \tag{6}$$

$$V_k = Z_k - H_k X_k^{(-)} \tag{7}$$

The simultaneous (6) and (7) can be written as,

$$\begin{aligned} r_k &= V_k - H_k K_k V_k \\ &= (I - H_k K_k) V_k \end{aligned} \tag{8}$$

The residual squared SSE is a scalar,

$$SSE = r_k^T r_k = V_k^T (I - H_k K_k)^T (I - H_k K_k) V_k \tag{9}$$

The posterior unit weight mean error and test statistic can be constructed as,

$$T = \sqrt{V_k{}^T (I - H_k K_k)^T (I - H_k K_k) V_k \hat{\sigma}} = \sqrt{SSE/(n-t)} \tag{10}$$

where t represents the necessary number of observations. In experiments, parameters to be estimated are the horizontal coordinates x, y, and ambiguity, thereby t = 3. The innovation sequence V provides the corresponding source of information for PLRAIM. Nevertheless, when DD observations are applied, the noise is no longer independent. The dependence is a mathematical correlation rather than a physical correlation. To de-correlate the dependence, T can be constructed as [21],

$$T = \sqrt{r_k^T W_k r_k} \tag{11}$$

where residual vector $r = [r_{L1} r_{P1}]^T$, thus, the statistic of each observation can be expressed as,

$$T_{L1} = \sqrt{r_{L1}^T W_{L1} r_{L1}} T_{P1} = \sqrt{r_{P1}^T W_{P1} r_{P1}} \quad (12)$$

In (12), T_{L1} is a test statistic based on carrier phase *L1* observations and T_{P1} is constructed based on pseudorange observations *P1*. These two statistics form the core of the CRAIM and PRAIM algorithm respectively. Some faults can be detected when using PRAIM, for example, indoor multipath. However, there are also some faults that cannot be reflected from the test statistic based on pseudorange (e.g., cycle slips) and such faults cannot be ignored in high-precision indoor pseudolite positioning. Consequently, the proposed indoor PLRAIM can also achieve the goal of data pre-processing such as cycle slip detection and gross error detection. Moreover, since the results of indoor pseudolite positioning are greatly affected by data quality, the PLRAIM algorithm is also a guarantee for the availability and reliability of positioning.

The innovation sequence in a Kalman filter is assumed to subject to Gaussian distribution (Mehra 1970), $N(0, \sigma^2)$. According to the statistical distribution theory, test statistic T^2/σ^2 subjects the chi-square distribution with a degree of freedom of m. When there is no deviation in the observations, the system is under a normal state. If an alert is performed, it would be a false alert. The following probability equation can be established from a given P_{fa},

$$P_r(T^2/\sigma^2 < T_h^2) = \int_0^{T_h^2} f_{\chi^2(m)}(x)dx = 1 - P_{fa} \quad (13)$$

The threshold T_h can be determined by (13) and T/σ is used to compare with T_h. If T/σ is greater than T_h, it can be considered that the fault exists in observations that should be excluded. If T/σ is less than the threshold, the positioning result is reliable and the observations can be applied for positioning.

In addition, the selection principle of reference pseudolite is also based on the elevation angle in indoor pseudolite positioning. However, elevation angles of all pseudolites are relatively close owing to the small scene indoors. Currently, it seems not reliable to select a reference pseudolite according to the highest elevation angle. Therefore, the possibility of reference pseudolite fault is also considered in PLRAIM. The fault of reference pseudolite is easier to detect. Once the fault in reference pseudolite occurs, the prior residuals of DD pseudorange observations of all pseudolites will be large and the reference pseudolite should be replaced in time.

2.2.2. HPL and the Availability of PLRAIM

This article focuses on horizontal positioning, so only the horizontal protection level (HPL) is analyzed. HPL refers to the horizontal positioning error that can be guaranteed when a fault occurs in current pseudolite constellation before fault detection. If HPL exceeds a given HAL, the current pseudolite constellation is not available. In other words, HPL is to ensure that the maximum P_{md} under the current pseudolite constellation does not exceed the demand. There are usually two methods to calculate:

The first way is to use the geometric factor SLOPE between horizontal position error (HPE) and test statistic [41]. SLOPE represents the mapping of the test statistic to HPE, reflecting the relationship between them. If there is a deviation in the test statistic, it will be reflected in the SLOPE.

The horizontal position errors are the first two elements of $K_k V_k$ inan extended Kalman filter. Let $S = (I - H_k K_k)^T (I - H_k K_k)$, assuming that the deviation occurs only in the ith pseudolite and the other pseudolites are normal. $HSLOPE(i)$ can be used to indicate the sensitivity of horizontal position error to the deviation of the ith observations,

$$HSLOPE(i) = \sqrt{(K_{1i}^2 + K_{2i}^2)/S_{ii}} \quad (14)$$

where S_{ii} represents the ith element in diagonal matrix S. SLOPE is purely a geometric factor. Corresponding to the same test statistic, P_{md} is the highest one when a fault occurs on apseudolite with the largest SLOPE. Therefore, the larger the $HSLOPE$, the more difficult it is to detect. In other words, when SLOPE is the largest, its statistic is the smallest in the case of the same HPE, which means that its residuals are the smallest, too. At this time, the minimum detectable deviation bias (MDB) is set to $pbias$ and it can be calculated according to the threshold and a given P_{md}.

If a fault occurs, T/σ should theoretically be greater than T_h. If T/σ is smaller than T_h, it indicates that a missed detection happens. The following probability equation can be established for a given P_{md},

$$P_r(T^2/\sigma^2 < T_h^2) = \int_0^{T_h^2} f_{\chi^2(m,\ \lambda)}(x)dx = P_{md} \tag{15}$$

where λ is the noncentral parameter, $pbias^2 = \sigma^2\ \lambda/(n-t)$. $\boldsymbol{\sigma}$ represents the standard deviation of carrier phase measurements. The horizontal protection level HPL can be calculated as,

$$\begin{aligned} HPL_1 &= HSLOPE_{\max} \times pbias \\ &= HSLOPE_{\max} \times \sigma\ \sqrt{\lambda}/\ \sqrt{(n-t)} \end{aligned} \tag{16}$$

The other way is to use the position estimate uncertainty, in which the first three elements of covariance matrix (P) is applied,

$$HPL_2 = k_H\sigma_H \tag{17}$$

The positioning error can be regarded as Gaussian noise and $k_H(P_r)$ can be obtained based on a given confidence probability P_r (usually 1 $-P_{md}$). In (17), σ_H represents the horizontal position uncertainty,

$$\sigma_H = \sqrt{P_{11} + P_{22}} \tag{18}$$

where P represents covariance matrix in expression (4). According to (16) and (17), HPL can be expressed as,

$$HPL = \max(HPL_1,\ HPL_2) \tag{19}$$

3. Experiment and Analyses

Five pseudolites have been set up in the laboratory. The ranging signals with C/A codes are transmitted on the GPS L1 frequency. The RF front end of the receiver is a USRP and a full series of GNSS signals can be captured with a suitable daughter board. Taking a certain mark as the origin point, the two horizontal directions of the floor are X and Y axis, respectively. The Z-axis is perpendicular to the floor and a right hand indoor independent coordinate system is established. All the experiments are carried out on the floor. In addition, the scope of the laboratory is about 10m×7m×4m. The pseudolites constellation is as shown in Figure 4a,b. The HDOP (Horizontal Dilution of Precision) on the floor under this condition of pseudolite layout is shown in Figure 4c. HDOP is less than 6 in 92.8% of the laboratory area, and the mean value of HDOP is 3.1055.

Two sets of experiments were performed. It should be noted that this pseudolite system is in good condition in most cases. In most of the other experiments, high precision positioning results can be obtained in our laboratory. The faults in the following two groups of experiments are accidental. However, these kinds of accidental faults should also be considered in high-precision positioning.The first group of experiments with a zero baseline is named Experiment 1. The purpose is to verify the availability of an indoor positioning system by eliminating all the spatial correlation errors (for example, multipath).Through conducting a zero-baseline experiment, the receiver bias is estimated.A total of 20 epochs are sampled with a sampling interval of 1s. The second group is a short baseline dynamic experiment, including the static stage before moving, named Experiment 2.The sampling interval is set to 1 s, and a total of 123epochs are sampled. Firstly, the static measurement of the rover station at an initial position is maintained and then the rover station is controlled to move back and forth on the fixed rail for two periods. One period is divided into two datasets and a total of four datasets are obtained.

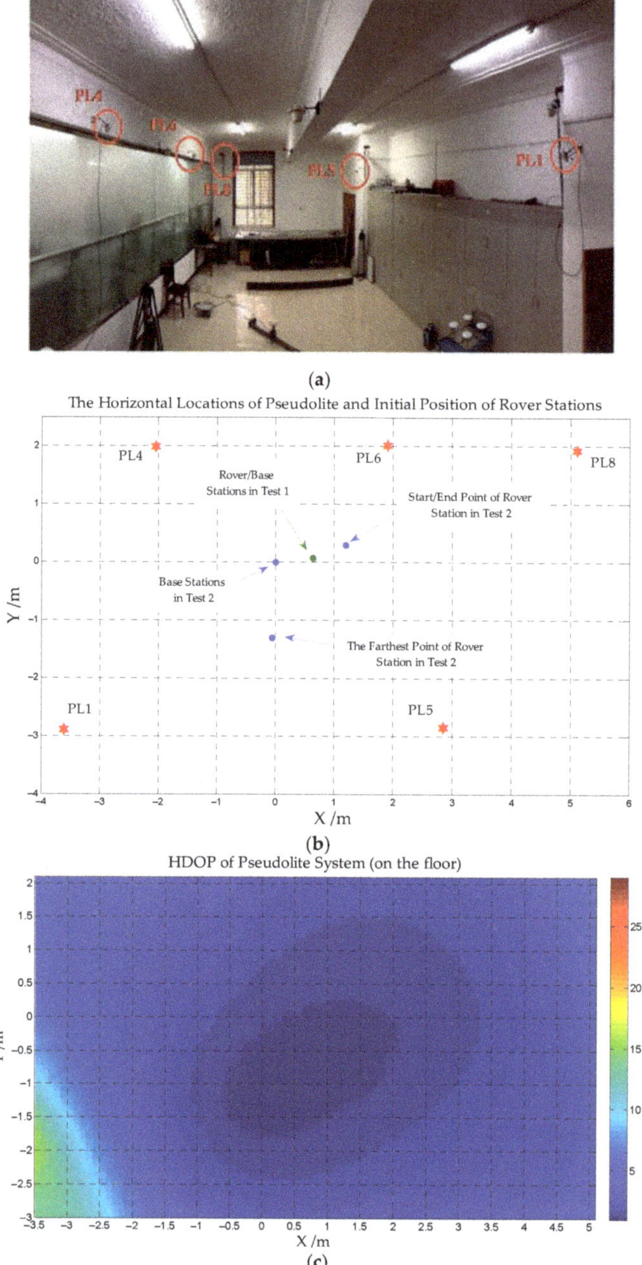

Figure 4. (**a**) Pseudolite Constellation in the Laboratory, (**b**) The Horizontal Locations of Pseudolite and Initial Position of Rover Stations, (**c**) HDOP of Pseudolite System(On the Floor).

3.1. Static Experiment with Zero-Baseline

3.1.1. Fault Identification

Since the antenna is shared between the rover and the base stations, the error source such as multipath has been eliminated by DD and the factor affecting positioning only remains noise. Therefore, the DD pseudorange residuals should be close to zero in theory. In Experiment 1, the receiver antenna is placed near (0.64, 0.08) and the positioning results based on carrier phase observations are shown in Figure 5. The color bar represents epochs.

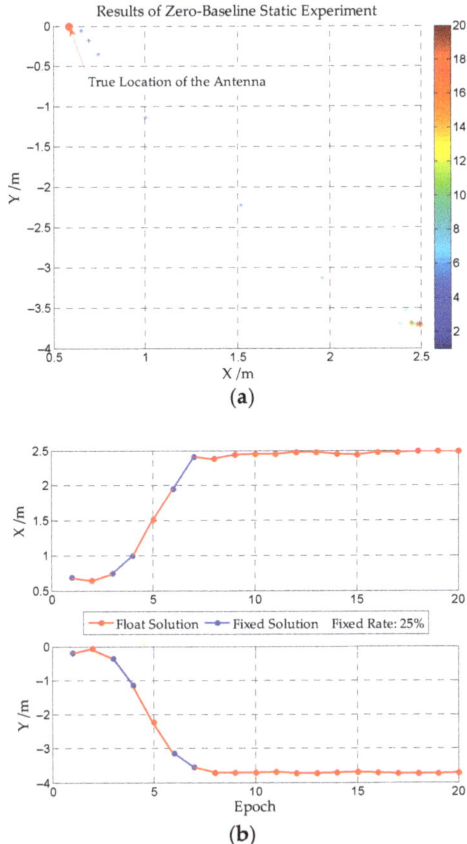

Figure 5. (**a**) Horizontal Positioning Results of Experiments 1, (**b**) Positioning Results on X, Y.

According to Figure 5, the ambiguity fixed rate is 25%. However, even if the ambiguities can be fixed in some epochs, it is fixed incorrectly. There is a large deviation in positioning results, and they are not reliable. In Experiments 1, the reference pseudolite is PL5 and DD pseudorange observations are shown in Figure 6.

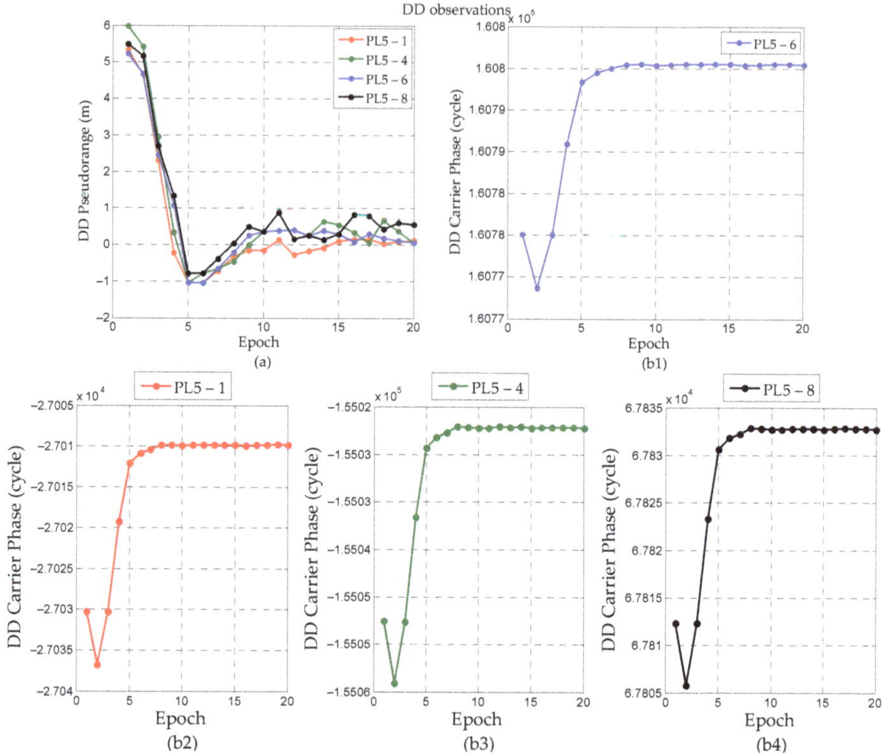

Figure 6. (**a**) Double Differenced (DD) Observations of Pseudorange, (**b1–b4**) DD Observations of the Carrier Phase.

In theory, DD pseudorange residuals in Figure 6 are close to 0, but it is obvious that there are abnormalities in DD observations of all pseudolites. The same circumstance is also found in DD carrier phase observations. It returned to normal around the 10th epoch. Therefore, it can be judged that the reference pseudolite PL5 is faulty. The reason for the bad pseudorange quality of PL5 may be an accidental fault from pseudolite hardware itself.

3.1.2. Fault Exclusion

Two strategies are considered for fault exclusion to ensure the reliability of positioning. The first one called Strategy 1 is to replace the reference pseudolite from PL5 to PL6 and retain the observations of original reference pseudolite. The other one named Strategy 2 is eliminating PL5 directly. In theory, Strategy 1 is more rigorous because subsequent observations return to normal. After the 10th epoch, the STD of PL6–5 for DD pseudorange observations is 0.27 m and 0.04 cycles forthe DD carrier phase. Therefore, after observations return to normal, the signals of PL5 should continue to be received. Figure 7 shows the DD pseudorange and carrier phase observations after using both two strategies. Figure 8 shows the test statistics after applying two strategies. The construction of test statistics and the determination of thresholdare shown in expression (12) and (13).

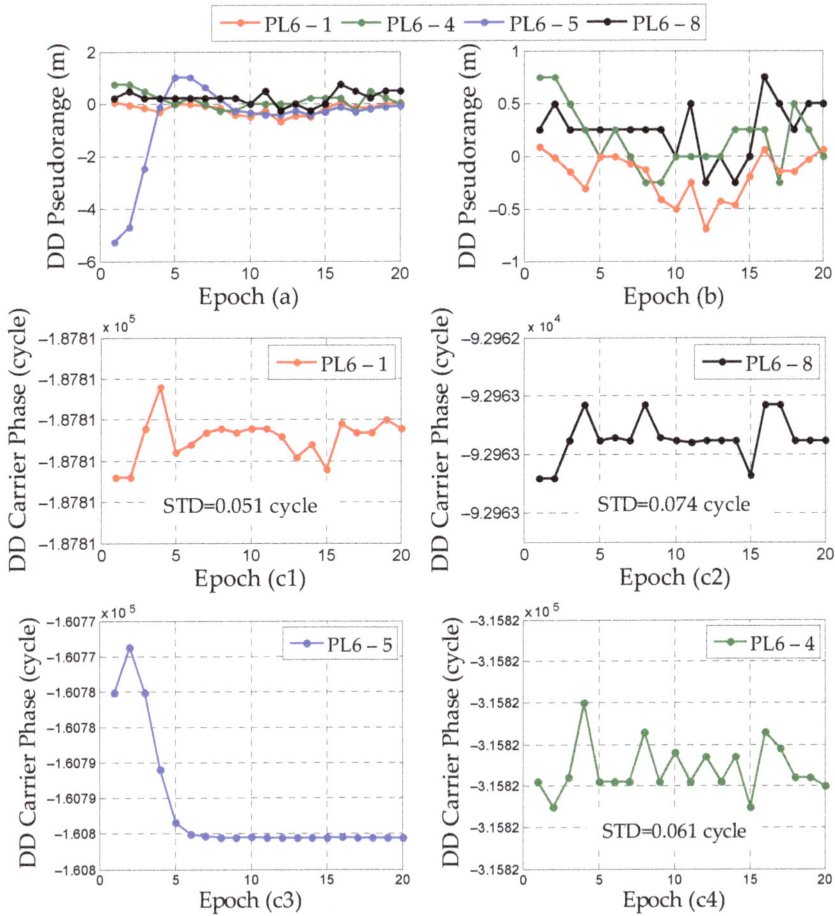

Figure 7. (**a**) DD Pseudorange Observations after Using Strategy 1, (**b**) DD Pseudorange Observations after Using Strategy 2, (**c1–c4**) DD Carrier Phase Observations after Using Strategy 1.

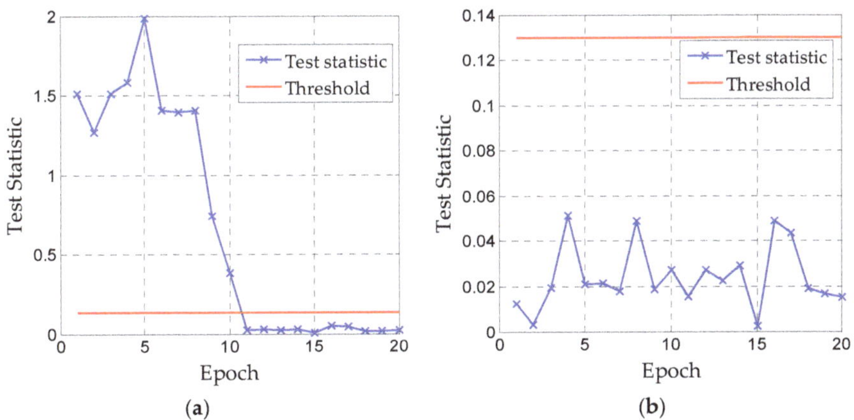

Figure 8. (**a**) CRAIM Test Statistic after Using Strategy 1, (**b**) CRAIM TestStatistic after Using Strategy 2.

According to Figure 7, the DD observations of all pseudolites return to normal after the reference pseudolite is replaced. When using Strategy 1, the test statistics exceed threshold during the previous 10 epochs because the data quality of PL5 is poor, but test statistics remain normal in the last 10 epochs. When using Strategy 2, test statistics are below the threshold since PL5 is removed. It can be considered that the fault has been excluded and high-precision positioning can be performed.

In theory, as long as PL5 returns to normal, the test statistics are less than the set threshold due to a large reduction in residuals of DD observations. In Experiment 1, the faulty pseudolite PL5 is the reference pseudolite, so the method of replacing reference PL is used in Strategy 1. However, even if the reference PL is replaced, PL5 is still in a faulty state in the first 10 epochs and test statistics are higher than the threshold. However, PL5 is eliminated in Strategy 2 and test statistics are less than the threshold, which can ensure that there is no obvious deviation in the whole positioning process. After PL5 returns to normal, more redundant observations will be received in Strategy 1 and the theoretical positioning precision will be higher than Strategy 2. Figure 9 shows the results of horizontal positioning with two strategies and the color bar represents epochs.

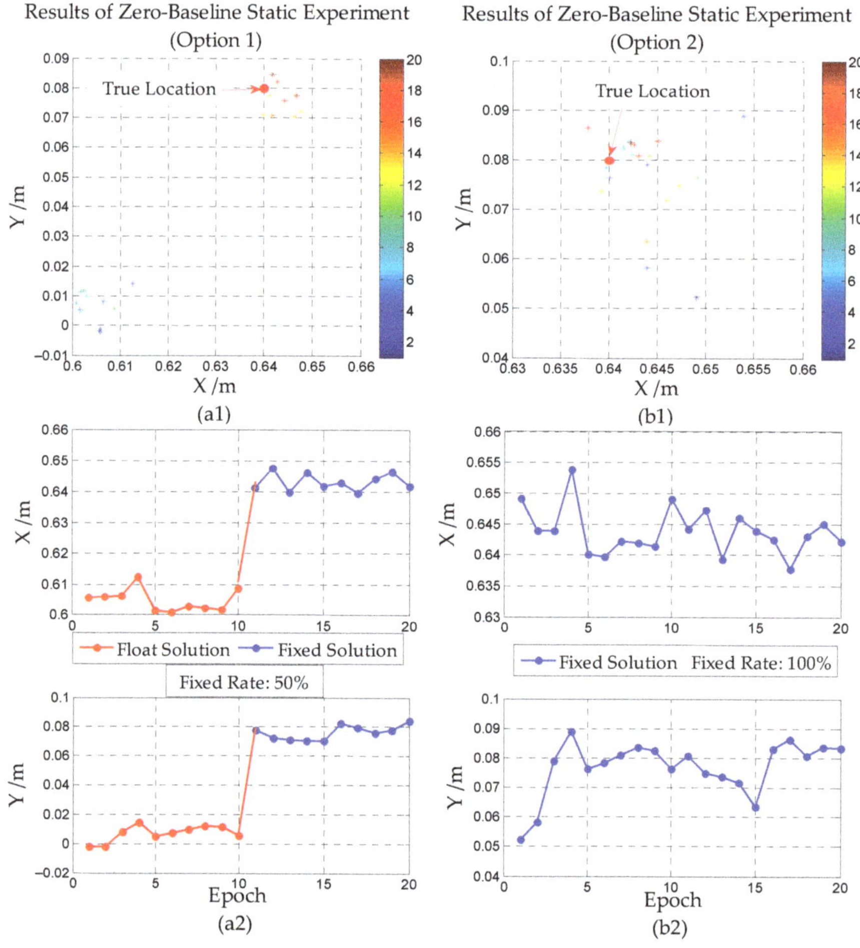

Figure 9. (**a1**,**a2**) Results of Experiment 1 when Using Strategy 1, (**b1**,**b2**) Results of Experiment 1 when Using Strategy 2.

According to Figure 9, it can be found that the fixed rate has been increased to 50% and the results of the fixed solution are close to true value when using Strategy 1. The STD of horizontal results is 5.4 mm (X direction is 2.7 mm and Y direction 4.7mm). When applying Strategy 2, the fixed rate has been increased to 100% and the STD of horizontal results is 9.8 mm (X direction is 3.7 mm and Y direction 9.1mm).

This phenomenon is also mutually confirmed with the previous analysis. In Strategy 1, when PL5 is faulty, the positioning result shows a large deviation and correct ambiguity fixed solutions cannot be obtained. At present, the positioning results of Strategy 2 are stabilized near true value. After PL5 returns to normal, the positioning result of Strategy 1 is also close to a true value and STD is better than Strategy 2. Nevertheless, we cannot decide which strategy is better without in combination with specific scenarios. The applicable scenarios of the two strategies and the best methods in the dynamic scenario will be discussed in this article.

Combined with Figure 6, the following conclusions can be obtained: if observations return to normal, PL5 can be resumed for positioning. Considering Strategies 1 and 2 comprehensively, Strategy 2 is better from the perspective of a fixed rate. However, considering the precision of the positioning results, Strategy 1 is better. No matter which strategy is adopted, it will be more accurate and reliable than before.

In practical applications, both strategies are acceptable for users according to the results (the precision can reach less than 1cm) and the positioning results are more reliable than before. The availability can be also improved. However, for dynamic positioning (or real-time navigation), Strategy 2 is more suitable than Strategy 1, because Strategy 2 can ensure that a correct positioning result is obtained in each epoch. The continuity and integrity of dynamic positioning will be guaranteed. From the perspective of redundant observations and final positioning precision, the operating state of PL5 (or faulty pseudolite) should be monitored in real-time. Once it returns to normal, the stations should continue to receive its signal, which will make the constellation configuration better. The number of redundant observations can also be added.

The following experiment is a dynamic one and the most reasonable strategy will be determined according to the real-time state of faulty pseudolite.

3.2. Dynamic Experiment

3.2.1. Fault Identification

A dynamic experiment is carried out in this chapter (named Experiment 2). Firstly, the static measurement of the rover station at the initial position is maintained and then the rover station is controlled to move back and forth on the fixed rail for two periods. Each period possesses two datasets. The horizontal coordinate of the start point is about (1.2, 0.3) and the endpoint coincides with the start point. The positioning results are shown in Figures 10 and 11. The color bar represents the epochs. The red line represents the reference trajectory.

According to Figure 10, it can be found that the positioning results are basically inconsistent with the reference line, which indicates a large deviation comparing with the actual situation. This conclusion can also be drawn from Figure 11. Although the ambiguity fixed rate is 31.7%, it is fixed incorrectly. High-precision positioning cannot be guaranteed, and the results are not reliable.

Similar to Experiment 1, judging whether reference pseudolite performs normally is the first step and reference pseudolite in Experiment 2 is PL6. Figure 12 presents the DD carrier-phase observations of four pseudolites groups. Since Experiment 2 takes "stationary—motion—stationary—motion—stationary" as a period, the observations are periodic and contain two cycles in theory.

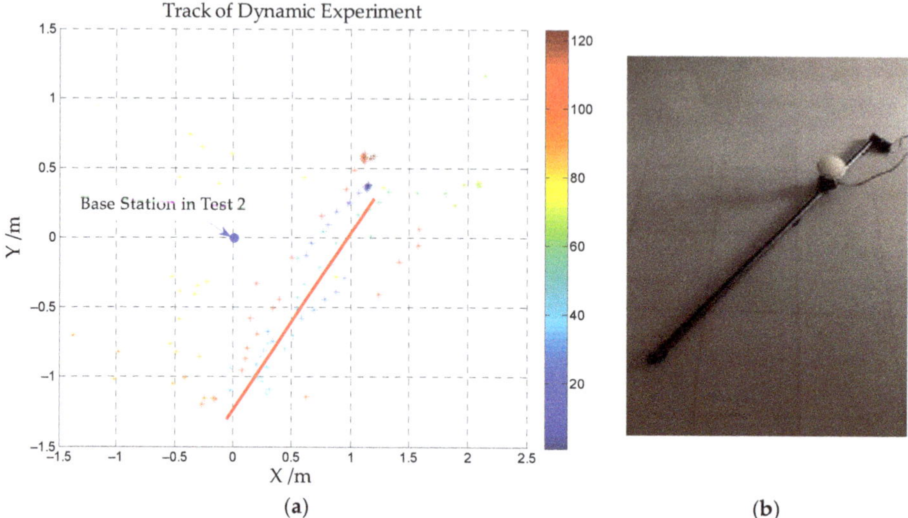

Figure 10. (**a**) The trajectory of Experiment 2, (**b**) The Equipment of Experiment 2.

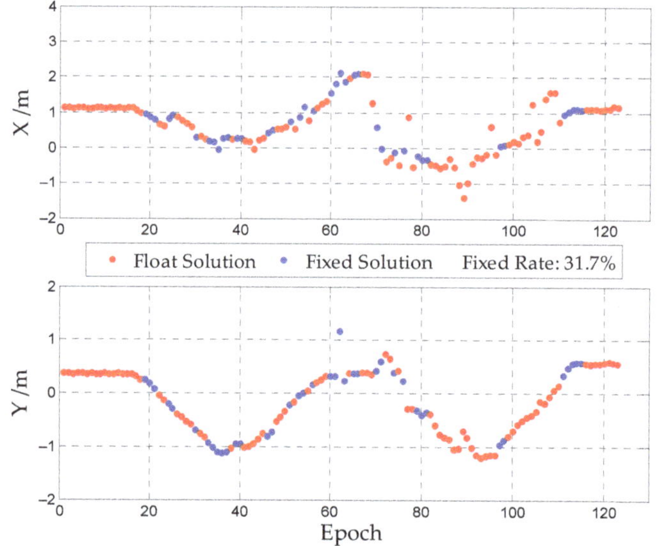

Figure 11. Results in X, Y Components of Experiment 2.

Figure 12 shows that DD carrier-phase observations of PL6–1, PL6–4, and PL6–5 remain normal, which confirms the regular pattern of two motion periods and observations are stable. The reason for such regular observations is that the pseudolite constellation is fixed. The observations will maintain this periodicity as long as the rover station moves back and forth along the fixed rail. Frequent cycle slips and data interruptions only occur in DD carrier-phase observations of PL6–8, indicating that the reference pseudolite PL6 is normal in Experiment 2 and the signals of PL8 may be faulty. It is possible that some environmental factors (such as multipath, shelters) lead to frequent signals interruption and cycle slips of PL8. However, further fault identification is required to determine faulty pseudolite or epochs theoretically. The test statistics and threshold for fault identification in Experiment 2 are shown in Figure 13. The construction of test statistics and the determination of thresholdare shown in (12) and (13).

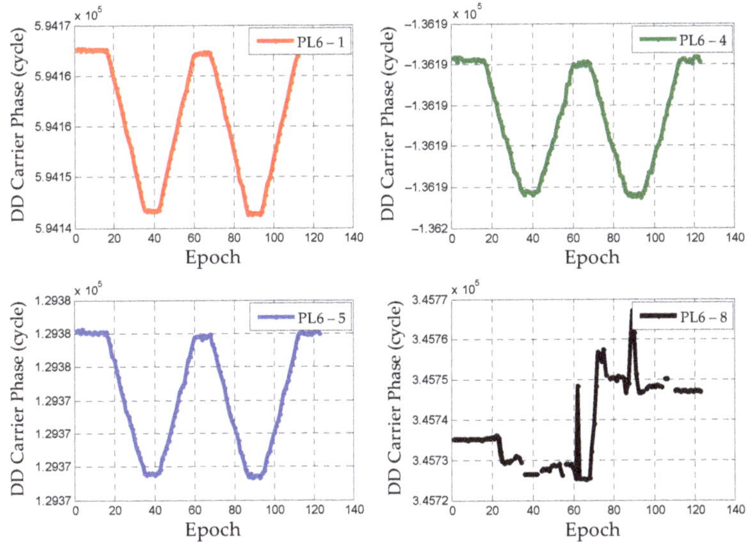

Figure 12. DD Carrier Phase Observations of Experiment 2.

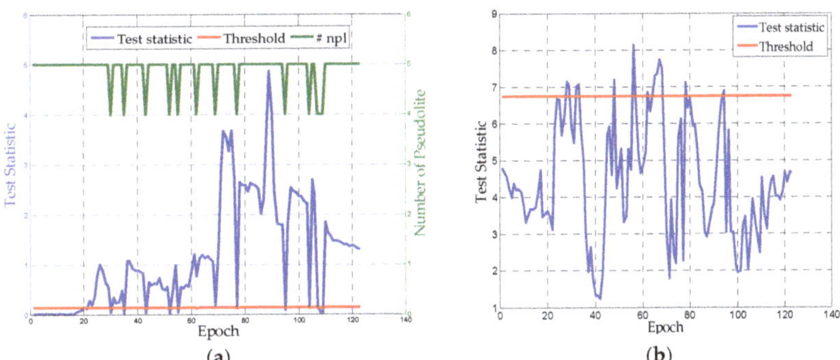

Figure 13. (**a**) The Test Statistics Based on Carrier Phase Observations, (**b**) The Test Statistics Based on Pseudorange Observations.

The CRAIM test statistics in Figure 13a obviously exceeds the threshold. It is less than or close to the threshold in some epochs, where the signal of PL8 is interrupted instead of returning to normal. The number of visible pseudolites (# npl) has also changed from five to four. When the number of visible pseudolites is four, the test statistics wereless than the threshold. At this moment, the receivers lose lock to PL8 signals.It indicates that other pseudolites' observations are normal except PL8. Besides, the incorrect ambiguity fixed solutions in Figures 10 and 11 are also verified, because a wrong ambiguity value can cause slow growth effects in both test statistics and position error. The PRAIM test statistics in Figure 11b also exceed the threshold in very few epochs and it is not as significant as test statistics of CRAIM. According to the introduction, pseudorange is greatly affected by indoor environments, such as multipath, clock, noise and so on. Sometimes the pseudorange-based test statistics cannot truly reflect the fault because test statistics may become very large due to other factors even if there is no fault. More importantly, the PRAIM test statistics can only reflect noise and multipath but cannot recognize cycle slips. Therefore, pseudorange-based RAIM is not applicable in high-precision indoor PL positioning, which illustrates the significance and purpose of this study again.

Since the cycle slip is the most threatening fault for the carrier phase-based positioning system and it can frequently occur in reality, the analysis under cycle slip fault cases is also included. The simulated cycle slips (3 cycles) on PL1-L1, PL4-L1, PL5-L1 and PL6-L1 during the last 20 epochs are added in the dynamic data collection for further analysis. In order to avoid the influence of PL8, PL8 is excluded first, and the simulated cycle slips are added respectively. The corresponding test statistics are shown in Figure 14.

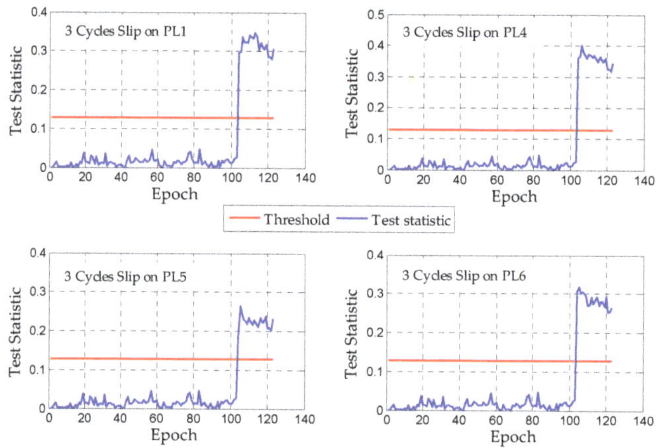

Figure 14. Test Statistics when simulated cycle slips are added.

In this case, the error source is only simulated cycle slips. According to Figure 14, the data quality of the other four pseudolites except PL8 is good. What's more, the test statistics based on carrier phase observations are sensitive to cycle slips and this kind of fault can be also effectively detected.

3.2.2. Fault Exclusion

In combination with Figures 12 and 13, in order to ensure the accuracy and reliability of positioning, the best way to exclude the fault is to eliminate PL8. Figure 15 shows the test statistics based on pseudorange and carrier phase after PL8 is eliminated.

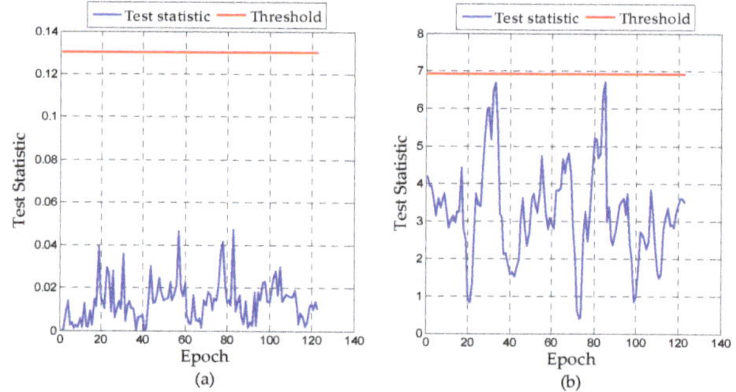

Figure 15. (**a**) The Test Statistics of CRAIM after PL8 is excluded, (**b**) The Test Statistics of PRAIM after PL8 is excluded.

According to Figure 15, it can be found that the test statistics based on pseudorange and carrier phase return to normal after PL8 areexcluded and the carrier phase-based statistics are much smaller than the threshold. Since the normal DD observations can be received after fault exclusion (PL8 is eliminated). The observations are only affected by general factors such as noise and the test statistics based on posterior residuals are less than the threshold. At the same time, the correctness of AR is verified. Although the pseudorange-based test statistics are below the threshold, they are relatively large, which further confirms the unavailability of pseudorange observations in high-precision positioning for indoor pseudolites.

In theory, the best way is to continue to receive the signals of PL8 when it returns to normal. However, it is revealed that PL8 has been in a faulty state during the dynamic process according to Figure 13. Therefore, the signals of PL8 are not received during the whole experiment. In addition, PL8 is normal in the first 18 epochs, but the rover station is in a static state, so whether the PL8 is used at this stage does not affect the positioning results. The result without PL8 is shown in Figure 16. The red line represents the reference trajectory.

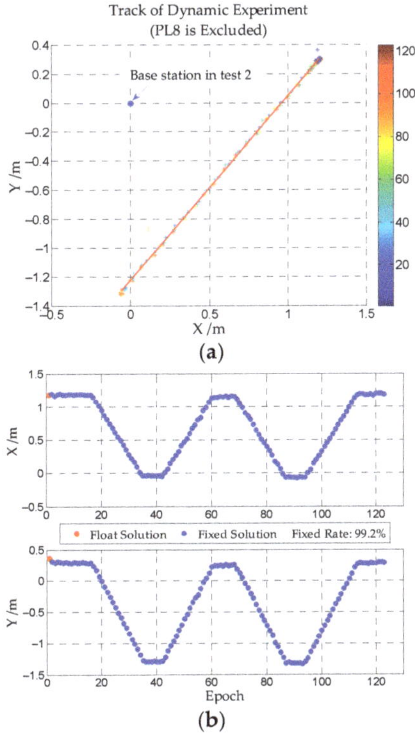

Figure 16. (**a**) The trajectory of Experiment2 after FaultExclusion, (**b**) Results in the X, Y Direction of Experiment 2 after Fault Exclusion.

According to Figure 16 a,b, the motion trajectory conforms to the actual situation basically that the carrier is controlled to move back and forth along a straight rail. The positioning precision is at centimeter-level in ambiguity fixed solution and the fixed-rate reaches 99.2%. Therefore, it can be judged that PL8 is faulty and the normal observations cannot be obtained during the whole experiment according to the test statistics and threshold. The signals of PL8 should not be received in the whole process. After fault exclusion, all DD observations with good quality can be obtained and indoor positioning results with high-precision can be achieved, which ensures the reliability and availability of indoor pseudolite positioning.

3.2.3. The Horizontal Protection Level

In Experiment 2, HPL is calculated according to position estimate uncertainty, which is obtained by the first two elements of the covariance matrix P. In practical applications, HPL being less than HAL is a prerequisite for RAIM. When the system is operating, the circle centered on the motion position with the HPL as the radius will contain the true position in this epoch. Figure 17 shows the HPL calculated in Experiment 2. The calculation of HPL is obtained by (19). The color bar represents the epochs.

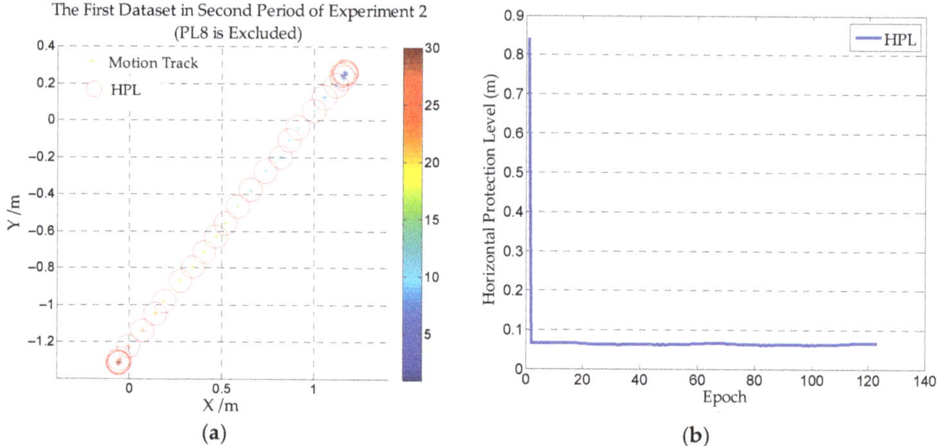

Figure 17. (**a**) The First Dataset in the Second Period of Experiment 2, (**b**) Horizontal protection level (HPL) of Experiment 2.

Figure 17a shows the motion trajectory of the first dataset in the second period and the corresponding HPL. In this dataset, all ambiguities are fixed, and the constellation is stable without the replacement of the reference pseudolite. Figure 17b shows the HPL of the full dataset in Experiment 2. The HPL changes within 5 cm and is relatively stable throughout the movement except for the first epoch. The reason is that the ambiguity has not been fixed and the variance is relatively large, which is why the peak appears in Figure 17b. In addition, according to the definition of HPL and the theory of RAIM, it can be considered that HPE is smaller than HPL, that is, the positioning precision of the dynamic experiment is below 5cm. Since there is no true value in positioning, this method also provides a reference index for the accuracy assessment of positioning. Therefore, according to Experiments 1 and 2, it can be considered that the proposed indoor PLRAIM algorithm has the abilitytoensuring the accuracy and integrity of pseudolite system. All test statistics and HPL in Figures 15 and 17 are relatively stable, which also proves the reliability of AR.

4. Conclusions and Discussion

Due to the influence of various error factors, pseudorange observations are not reliable and the accuracy and reliability of positioning cannot be guaranteed in indoor high-precision pseudolite positioning. Considering these problems, a holistic indoor pseudolite integrity monitoring algorithm PLRAIM is proposed and the specific steps are shown in Section 2.2. The feasibility of the whole algorithm is analyzed by static and dynamic experiments. The ambiguity fixed rate of both experiments is greatly improved when using PLRAIM and the reliability of AR is also guaranteed due to the stability of test statistics and HPL. In terms of positioning precision, the RMS of the zero-baseline static experiment reaches mm level when PLRAIM is applied and the maximum deviation in the dynamic experiment is within 5 cm, which ensures the positioning precision well.

The proposed method can be further improved. In fact, it is not reliable to choose the reference pseudolite of an indoor pseudolite system based on the elevation angle because of the fixed pseudolite constellation and the small scope of the indoor environment. Therefore, the elevation angles of all pseudolites are basically very close. Besides, whether the reference pseudolite is faulty is analyzed based on the stability of DD observations.A more suitable fault identification method for reference pseudolite is the next research focus. In addition, the threshold of test statistics is determined by a given P_{fa}, but it may not be the best way for an indoor pseudolite system. How to adjust the threshold dynamically according to the actual situation is also the next important issue.

Furthermore, this set of pseudolite hardware and software devices will be applied to some actual engineering situations or more complex indoor environments in the future. More actual data will be collected for a longer period. The available state of pseudolites will be monitored in real-time to verify the robustness of our method.

Author Contributions: P.Z. performed all data acquisition. Y.Z. and P.Z conceived the experiments and analyzed the data. Y.Z. and X.L. performed data processing and contributed to the manuscript. J.G. and J.Z. provided critical comments and contributed to the final revision of the paper. D.Z., F.Y. and G.H. contributed to the expression and the design of programs. All authors contributed to the manuscript and discussed the results. All authors together developed the idea that led to this paper. All authors have read and agreed to the published version of the manuscript.

Funding: This work was supported by "National Nature Science Foundation of China" (Grant Nos. 41604019, 41474004, 41871373) and Guangxi Key Laboratory of Spatial Information and Geomatics (Grant Nos. 19-050-11-10, 19-050-11-01).

Conflicts of Interest: The authors declare no conflict of interest.

Abbreviations

DD	Double differenced
GNSS	Global navigation satellite system
INS	Inertial Navigation System
PL	Pseudolite
AR	Ambiguity resolution
CRAIM	Carrier phase-based RAIM
PRAIM	Pseudorange-based RAIM
PLRAIM	Pseudolite based RAIM
HPL	Horizontal protection level
HAL	Horizontal alert limit
HPE	Horizontal positioning error
MI	Misleading information
HMI	Hazardously MI
MDB	Detectable deviation bias
RF	Radio frequency
FA	False alert
MD	Missed detection
KPI	Known point initialization
PPP	Precise point positioning
RTK	Real-time kinematic positioning
EKF	Extended Kalman filter
SSE	The sum of squares due to error
USRP	Universal software radio peripheral
rpl	Reference pseudolite
npl	Number of visible pseudolites
AFM	Ambiguity function method
STD	Standard deviation
RMS	Root mean square

References

1. Su, J.; Yao, Z.; Lu, M. An improved position determination algorithm based on nonlinear compensation for ground-based positioning systems. *IEEE Access* **2019**, *7*, 23675–23689. [CrossRef]
2. Wang, J. Pseudolite applications in positioning and navigation progress and problems. *J. Glob. Position. Syst.* **2002**, *1*, 48–56. [CrossRef]
3. Dai, L.; Wang, J.; Rizos, C.; Han, S. Pseudo-satellite applications in deformation monitoring. *GPS Solut.* **2002**, *5*, 80–87. [CrossRef]
4. Li, X.; Zhang, P.; Guo, J.; Wang, J.; Qiu, W. A new method for single-epoch ambiguity resolution with indoor pseudolite positioning. *Sensors* **2017**, *17*, 921. [CrossRef]
5. Wang, T.; Yao, Z.; Lu, M. On-the-fly ambiguity resolution based on double-differential square observation. *Sensors* **2018**, *18*, 2495. [CrossRef]
6. Wang, T.; Yao, Z.; Lu, M. On-the-fly ambiguity resolution involving only carrier phase measurements for stand-alone ground-based positioning systems. *GPS Solut.* **2019**, *23*, 36. [CrossRef]
7. Counselman, C.C.; Gourevitch, S.A. Miniature interferometer terminals for earth surveying: Ambiguity and multipath with Global Positioning System. *IEEE Trans. Geosci. Remote Sens.* **1981**, *GE-19*, 244–252. [CrossRef]
8. Bonenberg, L.K. Closely-Coupled Integration of Locata and GPS for Engineering Applications. Ph.D. Thesis, University of Nottingham, Nottingham, UK, 2014.
9. Rizos, C.; Roberts, G.; Barnes, J.; Gambale, N. Experimental results of Locata: A high accuracy indoor positioning system. In Proceedings of the 2010 International Conference on Indoor Positioning and Indoor Navigation, Zurich, Switzerland, 15–17 September 2010.
10. Guo, X.; Zhou, Y.; Wang, J.; Liu, K.; Liu, C. Precise point positioning for ground-based navigation systems without accurate time synchronization. *GPS Solut.* **2018**, *22*, 34. [CrossRef]
11. Kim, C.; So, H.; Lee, T.; Kee, C. A pseudolite-based positioning system for legacy GNSS receivers. *Sensors* **2014**, *14*, 6104–6123. [CrossRef] [PubMed]
12. Xu, R.; Chen, W.; Xu, Y.; Ji, S. A new indoor positioning system architecture using GPS signals. *Sensors* **2015**, *15*, 10074–10087. [CrossRef] [PubMed]
13. Fujii, K.; Sakamoto, Y.; Wang, W.; Arie, H.; Schmitz, A.; Sugano, S. Hyperbolic positioning with antenna arrays and multi-channel pseudolite for indoor localization. *Sensors* **2015**, *15*, 25157–25175. [CrossRef] [PubMed]
14. Wan, X.G.; Zhan, X.Q.; Du, G. Carrier phase method for indoor pseudolite positioning system. *Appl. Mech. Mater.* **2011**, *130–134*, 2064–2067. [CrossRef]
15. Zhang, P.; Sun, F.; Xu, Y.; Wang, J. Indoor pseudolite positioning based on interferometric measurement of carrier phase. In Proceedings of the China Satellite Navigation Conference (CSNC), Changsha, China, 18–20 May 2016; Sun, J., Liu, J., Fan, S., Wang, F., Eds.; Springer: Singapore, 2016.
16. Zhao, Y.; Zhang, P.; Guo, J.; Li, X.; Wang, J.; Yang, F.; Wang, X. A new method of high-precision positioning for an indoor pseudolite without using the known point initialization. *Sensors* **2018**, *18*, 1977. [CrossRef] [PubMed]
17. Bhattacharyya, S.; Gebre-Egziabher, D. Kalman filter–based RAIM for GNSS receivers. *IEEE Trans. Aerosp. Electron. Syst.* **2015**, *51*, 2444–2459. [CrossRef]
18. Sturza, M.A. Navigation system integrity monitoring using redundant measurements. *Navigation* **1988**, *35*, 483–501. [CrossRef]
19. Parkinson, B.W.; Axelrad, P. Autonomous GPS integrity monitoring using the pseudorange residual. *Navigation* **1988**, *35*, 255–274. [CrossRef]
20. Joerger, M.; Pervan, B. Kalman filter–based integrity monitoring against sensor faults. *J. Guid. Control Dyn.* **2013**, *36*, 349–361. [CrossRef]
21. Feng, S.; Ochieng, W.; Moore, T.; Hill, C.; Hide, C. Carrier phase-based integrity monitoring for high-accuracy positioning. *GPS Solut.* **2009**, *13*, 13–22. [CrossRef]
22. Abuhashim, T.S.; Abdel-Hafez, M.F.; Al-Jarrah, M.A. Building a robust integrity monitoring algorithm for a low cost GPS-aided-INS system. *Int. J. Control Autom. Syst.* **2010**, *8*, 1108–1122. [CrossRef]
23. Nowak, A. The proposal to snapshot raim method for GNSS vessel receivers working in poor space segment geometry. *Pol. Marit. Res.* **2015**, *22*, 3–8. [CrossRef]

24. Hewitson, S.; Wang, J. GNSS receiver autonomous integrity monitoring for multiple outlines. *J. Navig.* **2006**, *4*, 47–54.
25. Na, L.; Lin, Z.; Liang, L.; Jia, C. Integrity monitoring of high-accuracy GNSS-based attitude determination. *GPS Solut.* **2018**, *22*, 120.
26. Yun, H.; Yun, Y.; Kee, C. Carrier phase-based RAIM algorithm using a gaussian sum filter. *J. Navig.* **2011**, *64*, 75–90. [CrossRef]
27. Shi, X. Carrier-Phase Based Real-time Static and Kinematic Precise Point Positioning Using GPS and GALILEO. Ph.D. Thesis, Imperial College, London, London, UK, 2010.
28. Jokinen, A.; Feng, S.; Schuster, W.; Ochieng, W.; Hideb, C.; Mooreb, T.; Hill, C. Integrity monitoring of fixed ambiguity Precise Point Positioning (PPP) solutions. *Geo-Spat. Inf. Sci.* **2013**, *16*, 141–148. [CrossRef]
29. Schuster, W.; Bai, J.; Feng, S.J.; Ochieng, W. Integrity monitoring algorithms for airport surface movement. *GPS Solut.* **2012**, *16*, 65–75. [CrossRef]
30. Li, L.; Wang, H.; Jia, C.; Zhao, L.; Zhao, Y. Integrity and continuity allocation for the RAIM with multiple constellations. *GPS Solut.* **2017**, *21*, 1503–1513. [CrossRef]
31. Li, L.; Jia, C.; Zhao, L.; Yang, F.; Li, Z. Integrity monitoring-based ambiguity validation for triple-carrier ambiguity resolution. *GPS Solut.* **2017**, *21*, 797–810. [CrossRef]
32. Liu, Y.; Lian, B. Indoor pseudolite relative localization algorithm with kalman filter. *Acta Phys. Sin.* **2014**, *63*, 1–7.
33. Cellmer, S.; Wielgosz, P.; Rzepecka, Z. Modified ambiguity function approach for GPS carrier phase positioning. *J. Geod.* **2010**, *84*, 267–275. [CrossRef]
34. Alfred, L. *GPS Satellite Surveying*, 3rd ed.; Wiley: New York, NY, USA, 2004.
35. Teunissen, P.J.G.; Kleusberg, A. *GPS for Geodesy*; Springer: Berlin, Germany, 1998.
36. Koch, K.R.; Yang, Y. Robust Kalman filter for rank deficient observation models. *J. Geod.* **1998**, *72*, 436–444. [CrossRef]
37. Ochieng, W.; Sauer, K.; Walsh, D.; Brodin, G.; Griffin, S.; Denney, M. GPS integrity and potential impact on aviation safety. *J. Navig.* **2003**, *56*, 51–65. [CrossRef]
38. Weng, D.; Chen, W. SBAS enhancement using an independent monitor station in a local area. *GPS Solut.* **2019**, *23*, 12. [CrossRef]
39. Teunissen, P.J.G. The least-squares ambiguity decorrelation adjustment: A method for fast GPS integer ambiguity estimation. *J. Geod.* **1995**, *70*, 65–82. [CrossRef]
40. Li, X.; Zhang, P.; Huang, G.; Zhang, Q.; Guo, J.; Zhao, Y.; Zhao, Q. Performance analysis of indoor pseudolite positioning based on the unscented Kalman filter. *GPS Solut.* **2019**, *23*, 79. [CrossRef]
41. Song, J.; Hou, C.; Xue, G. GNSS receiver autonomous integrity monitoring based on EKF. *J. Tianjin Univ. Sci. Technol.* **2017**, *50*, 405–410.

Article

A Robust Tracking Algorithm Based on a Probability Data Association for a Wireless Sensor Network

Long Cheng *, Mingkun Xue, Ze Liu and Yong Wang

Department of Computer and Communication Engineering, Northeastern University, Qinhuangdao 066004, China; mingkunxue61@gmail.com (M.X.); liuze.mail.1234@gmail.com (Z.L.); 1143539166cjj@sina.com (Y.W.)

* Correspondence: chenglong@neuq.edu.cn; Tel.: +86-189-3134-7611

Received: 16 November 2019; Accepted: 15 December 2019; Published: 18 December 2019

Abstract: As one of the core technologies of the Internet of Things, wireless sensor network technology is widely used in indoor localization systems. Considering that sensors can be deployed to non-line-of-sight (NLOS) environments to collect information, wireless sensor network technology is used to locate positions in complex NLOS environments to meet the growing positioning needs of people. In this paper, we propose a novel time of arrival (TOA)-based localization scheme. We regard the line-of-sight (LOS) environment and non-line-of-sight environment in wireless positioning as a Markov process with two interactive models. In the NLOS model, we propose a modified probabilistic data association (MPDA) algorithm to reduce the NLOS errors in position estimation. After the NLOS recognition, if the number of correct positions is zero continuously, it will lead to inaccurate localization. In this paper, the NLOS tracer method is proposed to solve this problem to improve the robustness of the probabilistic data association algorithm. The simulation and experimental results show that the proposed algorithm can mitigate the influence of NLOS errors and achieve a higher localization accuracy when compared with the existing methods.

Keywords: wireless sensor network; indoor localization; time of arrival (TOA); NLOS; modified probabilistic data association (MPDA)

---

1. Introduction

Due to there being many obstacles, it is difficult to provide accurate localization indoors. Applying wireless sensor network technology to indoor localization can solve the problem of indoor localization. Wireless positioning systems usually use time of arrival (TOA), time difference of arrival (TDOA), angle of arrival (AOA), and received signal strength (RSS) to estimate the location information to realize the tracking and positioning of a mobile node. This paper uses the TOA method to obtain the Euclidean distance measurements between the beacon nodes and the mobile node. The principle of the TOA method is to calculate the arrival time from the mobile node to the beacon nodes, and then these measurements are multiplied by the speed of light in order to obtain the Euclidean distance measurements between the beacon nodes and the mobile node. Trilateration techniques can be used to estimate the position of the mobile node. If the channels between the mobile node and the beacon nodes have a direct path, the channels are considered to be in line-of-sight (LOS) and the TOA measurements obtained in LOS environments are perfect. The accurate position can be obtained using the extended Kalman filter to process the TOA measurements. However, the assumption of the LOS channel is ideal and impractical [1]. In practice, obstacles such as persons, furniture, or walls block the propagation path, causing diffraction and refraction. The propagation path of the signal becomes long, and the channel is in a non-line-of-sight (NLOS) environment. The NLOS factor results in a positive bias of the TOA measurements, leading to a decreased positioning accuracy. Thus, algorithms that reduce NLOS errors can achieve high-precision positioning.

Many positioning algorithms have emerged based on various positioning methods. A semidefinite programming algorithm was proposed in Su et al. [2], where it transforms a TDOA model into a TOA model and uses new constraints to mitigate NLOS errors. A bisection-based approach is proposed in Tomic and Beko [3], which can accurately solve the maximum likelihood estimation derived from the measurement model through the bisection procedure to achieve accurate positioning. Yang et al. [4] proposed a high-precision and low-complexity localization algorithm based on an imported vector machine (IVM), which employs the probability output of an input vector machine, and has a higher localization accuracy than the corresponding methods, such as a support vector machine (SVM) and correlation vector machine (RVM). Zhang et al. [5] proposed a novel distributed consensus-based adaptive Kalman estimation algorithm. In order to estimate the states of the target more precisely, an optimal Kalman gain is obtained by minimizing the mean-squared estimation error. An adaptive consensus factor is employed to adjust the optimal gain, as well as to acquire a better filtering performance. In the filter's information exchange, dynamic cluster selection and a two-stage hierarchical fusion structure are employed to get a more accurate estimation. When all the range measurements estimation is from the LOS environments, the position estimation obtained by the least-squares method is accurate; if there is at least a range measurement estimate from the NLOS condition, the position estimation of the mobile node will be inaccurate, which will deviate from the real position. The more NLOS data that is used, the less accurate the positioning is. According to the idea of data fusion, combining different positioning methods is also a common method to improving positioning accuracy. An improved positioning method was proposed to improve the accuracy by combining TOA and AOA positioning methods [6,7]. Vaghefi and Buehrer [8] used semidefinite programming to solve the problem of collaborative localization. In Vaghefi and Buehrer [9], a novel cooperative localization algorithm of source nodes is proposed. Based on the interrelation of multiple source nodes, the novel extended Kalman filter (EKF) integrated with semidefinite programming method is used for localization. It not only solves the problem of the cooperative localization using multiple source nodes, but also improves the localization performance compared with the classic EKF. In addition, it can also be used for traditional non-cooperative localization.

In the localization algorithms, many algorithms are studied to mitigate the interference of NLOS factors [10–17]. Chen [10] proposed a residual weighting (RWGH) algorithm, which can mitigate NLOS errors to a certain extent, but the computational complexity is high. The algorithm proposed by Park and Chang [11] also uses a residual weighting method.Jiao and coworkers [12,13] used a method that selects the smallest normalized residual combination using different methods, and then performs weighted summation, which reduces the complexity of the residual weighting algorithm. The introduction of iterative ideas in the NLOS algorithms and the optimization of the calculation results are also effective methods for reducing the NLOS errors. Li [15] proposed an iterative minimum residual algorithm, which iteratively selects the minimum residual combination as the final estimated position of the mobile station (MS) by iterating the residual size in each combination whose value is less than the predetermined threshold. Horiba and coworkers [16,17] used a TOA/AOA hybrid positioning method to improve the performance of the iterative minimum residual algorithm by selecting the appropriate iterative minimum residual criterion. In References [18–20], the NLOS mitigation algorithms proposed by the authors can reduce the NLOS errors without prior knowledge of the NLOS errors.

The NLOS identification methods can effectively eliminate the influence of NLOS errors on the positioning accuracy. Wylie and Holtzman [1] proposed a method for judging whether the range measurements contain NLOS errors based on the measurement variance (standard deviation). Location spoofing is an important factor for producing NLOS errors, and Liu et al. [21] eliminated NLOS errors by identifying location spoofing. Han proposed a probabilistic position selection algorithm [22], which is based on the received signal strength indication (RSSI) and pedestrian dead reckoning (PDR) in the mixed LOS environment and NLOS environment for low-complexity identification. In the case of an unknown LOS/NLOS propagation prior probability, NLOS propagation can be identified by examining whether the range measurements obey a Gaussian distribution [23–25]. In recent years, Kolmogorov-Smirnov (K-S), Aderson-Drling (A-D), chi-square, Gruss test, skewness test, and kurtosis test have appeared successfully [26–28]. If the LOS/NLOS propagation prior probability is known, the NLOS can be identified using a generalized likelihood ratio test based on the statistical distribution of the different probabilities of the error. Large outliers sometimes occur in the range measurements. These outliers can seriously interfere with the positioning result, resulting in a positioning failure, thus we need to discard the outliers.

In this paper, we propose a robust tracking algorithm based on an improved modified probabilistic data association (MPDA) and an interacting multiple model (IMM). In this paper, the improved probabilistic data association filter is used in the NLOS model. Through NLOS recognition, the position estimation with the NLOS error is discarded, and the correct position estimation is weighted with the corresponding correlation probability to obtain the final position estimation to reduce the NLOS error. The proposed algorithm in this paper has the following advantages:

(1) The traditional MPDA algorithm is used for the NLOS recognition, where there may be continuous incorrect position estimation, resulting in inaccurate positioning. Therefore, the NLOS tracer method is proposed in this paper to record the occurrence of an incorrect position estimation. When we use the NLOS tracer method to find that there are two or more consecutive occurrences of incorrect position estimation, the EKF is used for updating to reduce the adverse influence of this situation on positioning and improve the robustness of the algorithm.

(2) The improved probabilistic data association algorithm has a high positioning accuracy and robustness to NLOS errors, therefore it is used to reduce the NLOS errors.

(3) The simulation and experimental results show that the proposed algorithm in this paper can mitigate the influence of NLOS errors when the NLOS error obeys different distributions.

The paper is structured as follows. Section 2 introduces the signal model and provides an overview of existing techniques from the literature. The proposed algorithm is explained in Section 3, and Section 4 illustrates the simulation results. Conclusions are drawn in Section 5.

2. Problem Statement

2.1. Signal Model

The signal transmission channel between the mobile node and the beacon nodes changes between the LOS condition and NLOS condition, where this transformation is considered a switching mode system, as shown in the two-state Markov process of Figure 1. The state vector of the mobile node is $x(k) = \begin{bmatrix} x(k) & y(k) & \dot{x}(k) & \dot{y}(k) \end{bmatrix}^T$, where $(x(k), y(k))$ denotes the position of the mobile node and $(\dot{x}(k), \dot{y}(k))$ denotes the velocity of the mobile node.

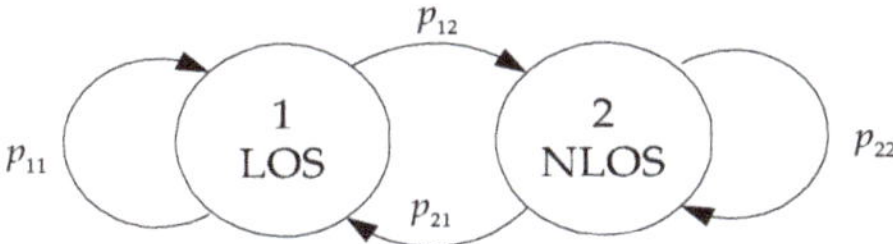

Figure 1. Markov switching model. LOS: line of sight, NLOS: non-LOS.

The state vector of the mobile node changes over time according to a force model:

$$x(k) = Ax(k-1) + C\omega(k-1);\ k = 1, \ldots, K, \tag{1}$$

where, K is the number of time steps, and:

$$A = \begin{bmatrix} 1 & 0 & \Delta t & 0 \\ 0 & 1 & 0 & \Delta t \\ 0 & 0 & 1 & 0 \\ 0 & 0 & 0 & 1 \end{bmatrix}, C = \begin{bmatrix} \Delta t^2/2 & 0 \\ 0 & \Delta t^2/2 \\ \Delta t & 0 \\ 0 & \Delta t \end{bmatrix},$$

where, A is the state transition matrix that describes the movement of the mobile node between two consecutive time steps. C is the interference input matrix describing the mapping of the randomness and the velocity of the mobile node, and Δt is the sampling period. The process driving noise $\omega(k)$ due to the acceleration of the mobile node is assumed to have a zero-mean, white Gaussian with a covariance matrix $Q(k)$. Assume that there are M beacon nodes around the mobile node to detect the signal of the mobile node. We can let $D(k) = [d_1(k), d_2(k), \ldots, d_M(k)]^T$ denote the range measurements based on the TOA data between the mobile node and M beacon nodes at time step k. Then:

$$D(k) = \boldsymbol{h}(\boldsymbol{x}(k)) + v(k),\ k = 1,\ \ldots,\ K, \tag{2}$$

where $\boldsymbol{h}(\boldsymbol{x}(k)) = [h_1(\boldsymbol{x}(k)), h_2(\boldsymbol{x}(k)), \ldots, h_M(\boldsymbol{x}(k))]^T$ and the Euclidean distance between the mobile node and the mth beacon node with the position $\left(x_{bn,m}, y_{bn,m}\right)$ at time step k is:

$$h_m(\boldsymbol{x}(k)) = \sqrt{\left(x(k) - x_{bn,m}\right)^2 + \left(y(k) - y_{bn,m}\right)^2},\ m = 1,\ \ldots,\ M. \tag{3}$$

The noise vector $v(k) = [v_1(k), v_2(k), \ldots, v_M(k)]^T$ contains random variables with a variance describing Gaussian sensor noise due to the NLOS propagation. The measurement covariance matrix is:

$$R(k) = E\left\{[\boldsymbol{v}(k) - E\{\boldsymbol{v}(k)\}][\boldsymbol{v}(k) - E\{\boldsymbol{v}(k)\}]^T\right\}, \tag{4}$$

which is defined as:

$$R(k) = diag\left[\sigma_1^2, \sigma_2^2, \ldots, \sigma_M^2\right], \tag{5}$$

where the elements σ_m^2 in $R(k)$ are defined as:

$$\sigma_m^2 = \begin{cases} \sigma_L^2 & \text{if LOS condition} \\ \sigma_L^2 + \sigma_{NLOS}^2 & \text{if NLOS condition} \end{cases}. \tag{6}$$

We assume that the sensor noise variance σ_L^2, and process covariance matrix $Q(k)$ are known. σ_{NLOS}^2 is unknown.

2.2. A Brief Introduction of Existing Methods

The interacting multiple model (IMM) algorithm is used to track and locate in NLOS environments. In Vaghefi and Buehrer [8], the IMM algorithm uses two Kalman filters to smooth the TOA range measurements in both the LOS model and the NLOS model. In order to reduce the NLOS errors, the TOA range measurements smoothed by the Kalman filter is used to subtract the NLOS mean error in the NLOS model. The distance estimates between the mobile node and the beacon nodes are determined by combining the state estimate in the LOS model with the state estimate in the NLOS model, which eliminates the NLOS errors. Then, the location of the mobile node is determined using a geometric method based on the smoothed distance estimates. According to the idea of data fusion, Chen et al. [29] proposed an IMM algorithm based on TOA and RSS data fusion to find the location of a mobile node. Almost all IMM algorithms need to presuppose NLOS statistical errors to solve the NLOS interference problem, but in practice, the NLOS statistical errors are unknown. The robust IMM (RIMM) algorithm proposed in Li et al. [30] does not need prior knowledge of the NLOS error. In the

LOS model, EKF is used to estimate the location of a mobile node. In the NLOS model, REKF is used to transform the EKF equation into a linear regression problem, which is solved using robust techniques in Hammes and Zoubir [31]. Then, the state vectors of the two models are weighted using likelihood function values to determine the final state vectors of the mobile node.

EKF can achieve accurate positioning in LOS environments, but the positioning accuracy is not high in NLOS environments. Therefore, it is necessary to improve the filters of the IMM algorithm to improve the positioning accuracy. The RIMM algorithm is used to improve the filter of the NLOS model in the IMM algorithm. The MPDA algorithm in Hammes et al. [32] has a high positioning accuracy and can mitigate NLOS errors. The MPDA algorithm is a sub-optimal filtering algorithm based on the Bayesian formula, which divides the range measurements between the beacon nodes and the mobile node into different groups. Each group obtains the corresponding position estimation via a least-squares method and optimizes the position estimation by using the Gauss–Newton iteration method. The optimized location estimation is identified using NLOS detection, and the location estimation that does not fall into the validation gate is discarded. The location estimation that falls into the validation gate is weighted by the corresponding association probability to determine the location of the mobile node. Choosing an appropriate validation gate is a prerequisite for realizing the probability data association. The common validation gate forms are a rectangular gate, elliptic gate, sector gate, etc. Among them, the elliptic gate is the most widely used. The positioning accuracy of the MPDA algorithm is very high, but it is sensitive to outliers. These outliers often fail to locate, so it is necessary to abandon the outliers to ensure the positioning accuracy of the MPDA algorithm. Compared with the MPDA algorithm, the EKF algorithm has better robustness.

3. Proposed Method

3.1. General Concept

As shown in Figure 2, we assumed that the initial values of the state estimation and covariance matrix are known. The initial values of the model probability and the transition matrix are given according to empirical knowledge, and the mixed probability, mixed state estimation, and mixed covariance matrix are calculated to complete the interactive process, and then model matching is conducted. The proposed algorithm uses an extended Kalman filter in the LOS model and proposes an improved modified probabilistic data association filter in the NLOS model. The two filters work in parallel to achieve the model-matching process. In the NLOS model, the improved modified probabilistic data association filter first divides the measured values into N different groups. Each group gets a corresponding position estimation via a least-squares estimation. The N different position estimates are obtained for NLOS recognition through a validation gate, where the position estimates falling in the validation gate around the predicted position estimates are the correct position estimates. Furthermore, the position estimates that do not fall in the validation gate are discarded. The correlation probability corresponding to the correct position estimation is calculated, the correct position estimation with the corresponding correlation probability is weighted, and the updated position estimation is produced. Through the NLOS recognition, the position estimates with a large error are discarded to reduce the NLOS errors. When the NLOS interference is relatively serious, there will be no correct position estimation, which will have no significant impact on the positioning results. If there is no correct position estimation after successive attempts, a positioning failure will occur. Therefore, the NLOS tracer method is proposed to record the situation of incorrect position estimation. When no correct position estimation occurs the first time, the predicted state is regarded as the updated state. If no correct position estimation occurs twice or more times, the EKF will be used to update the position. The model probability is updated according to the result of model matching, and the updated model probability is used to weight the updated state estimation of the corresponding model to obtain the final state estimation.

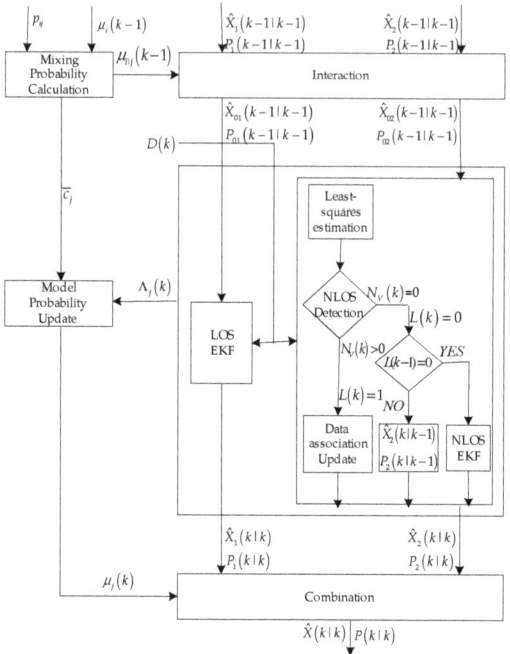

Figure 2. The flowchart for the proposed algorithm. EKF: extended Kalman filter.

3.2. Interaction

We determine the initial value of the prior probability $\mu_i(k-1)$, the Markov transition probabilities p_{ij}, the prior state $\hat{X}_i(k-1|k-1)$, and covariance matrix $P_i(k-1|k-1)$ of the proposed algorithm. However, the transition probabilities p_{ij} are not known in practice, so the initial of the transition probabilities p_{ij} are chosen based on the prior knowledge. Before the interaction, determination of the mixing probabilities $\mu_{i|j}(k-1|k-1)$ is required. The mixing probability $\mu_{i|j}(k-1|k-1)$ is given as:

$$\mu_{i|j}(k-1|k-1) = \left(1/\bar{c}_j\right)p_{ij}\mu_i(k-1),\ k = 1,\ \ldots,\ K. \tag{7}$$

The normalization factor $\bar{c}_j$ is:

$$\bar{c}_j = \sum_i p_{ij}\mu_i(k-1), \tag{8}$$

where, the mixed probabilities $\mu_{i|j}(k-1|k-1)$ obtained are applied to the interaction process. Furthermore:

$$\hat{X}_{0j}(k-1|k-1) = \sum_i \hat{X}_i(k-1|k-1)\mu_{i|j}(k-1|k-1), \tag{9}$$

$$\widetilde{X}_{ij}(k-1|k-1) = \hat{X}_i(k-1|k-1) - \hat{X}_{0j}(k-1|k-1), \tag{10}$$

$$P_{0j}(k-1|k-1) = \sum_i \mu_{i|j}(k-1|k-1)\left\{P_i(k-1|k-1) + \widetilde{X}_{ij}(k-1|k-1)\widetilde{X}_{ij}^T(k-1|k-1)\right\}, \tag{11}$$

where $\hat{X}_{0j}(k-1|k-1)$ is the mixed state estimate of the LOS/NLOS model at time step $k-1$ and $P_{0j}(k-1|k-1)$ is the mixed state covariance matrix estimate of the LOS/NLOS model at time step $k-1$. Model matching of the proposed algorithm is performed after the interaction is completed.

3.3. Model Matching

Prediction: The mixed state estimates and the mixed covariance matrix estimates obtained after the interaction are provided as initial estimates to the two model-matched filters for prediction:

$$\hat{X}_j(k|k-1) = A\hat{X}_{0j}(k-1|k-1) \tag{12}$$

$$P_j(k|k-1) = AP_{0j}(k-1|k-1)A^T + CQ(k)C^T \tag{13}$$

where $\hat{X}_j(k|k-1)$ is the state prediction of the LOS/NLOS model at time step k and $P_j(k|k-1)$ is the error covariance matrix prediction of the LOS/NLOS model at time step k.

Update: When $j = 1$, the state estimate update and the error covariance matrix estimate update are performed using the extended Kalman filter. Furthermore:

$$H_1(k) = \left.\frac{\partial h(X(k))}{\partial X(k)}\right|_{X(k)=\hat{X}_1(k|k-1)}, \tag{14}$$

$$v_1(k) = D(k) - h\left(\hat{X}_1(k|k-1)\right), \tag{15}$$

$$S_1(k) = H_1(k)P_1(k|k-1)H_1^T(k) + R_1^*(k), \tag{16}$$

where $H_1(k)$ is the Jacobian matrix at time step k. $v_1(k)$ denotes the innovation at time step $k.S_1(k|k-1)$ is the innovation covariance matrix. $R_1^*(k)$ is the covariance matrix of the measurement errors vector. Furthermore:

$$K_1(k) = P_1(k|k-1)H_1^T(k)S_1^{-1}(k|k-1), \tag{17}$$

$$\hat{X}_1(k|k) = \hat{X}_1(k|k-1) + K_1(k)v_1(k), \tag{18}$$

$$P_1(k|k) = (I_4 - K_1(k)H_1(k))P_1(k|k-1), \tag{19}$$

$$\Lambda_1(k) = N(v_1(k); 0, S_1(k)), \tag{20}$$

where $K_1(k)$ is the extended Kalman gain at time k, $\hat{X}_1(k|k)$ is the updated state estimate, $P_1(k|k)$ denotes the updated error covariance matrix, and $\Lambda_1(k)$ is the likelihood function.

When $j = 2$, the improved MPDA filter is used to update the state estimation prediction and error covariance matrix prediction in the NLOS model. The proposed method divides the range measurements into $N = \begin{pmatrix} M \\ 3 \end{pmatrix}$ different subgroups, which are used together with the positions of the corresponding beacon nodes to determine the position of the mobile node, and each subgroup uses the least squares estimation to obtain the corresponding position estimation $z_n(k)$. The position estimation prediction $\hat{z}(k|k-1)$ of the mobile node is given as:

$$\hat{z}(k|k-1) = B\hat{X}_2(k|k-1), \tag{21}$$

$$v_{2,n}(k) = z_n(k) - \hat{z}(k|k-1),\ n = 1, \ldots, N, \tag{22}$$

where $B = \begin{bmatrix} 1 & 0 & 0 & 0 \\ 0 & 1 & 0 & 0 \end{bmatrix}$ is the observation matrix, $v_{2,n}(k)$ is the innovation for the position estimate from subgroup n at time step k. Then, we perform the NLOS detection for the position estimation obtained to determine the correct position estimation.

NLOS detection: If all the innovations do not exist as NLOS errors, then there are:

$$v_{2,n}(k) \sim N(0, S_{2,n}(k)),\ n = 1, \ldots, N. \tag{23}$$

In order to validate Equation (21), we define N hypotheses and alternatives:

$$\zeta_{0,n} : v_{2,n}(k) \sim N(0, S_{2,n}(k)),\ n = 1, \ldots, N \tag{24}$$

$$\zeta_{1,n} :\ \text{not}\ \zeta_{0,n},\ n = 1, \ldots, N \tag{25}$$

If the beacon nodes from subgroup n are in the LOS condition, the hypothesis $\zeta_{0,n}$ holds true, and the position estimate $z_n(k)$ from subgroup n falls into the validation region; otherwise, the hypothesis $\zeta_{1,n}$ holds true. The test statistic $T_n(k)$ can be expressed as:

$$T_n(k) = v_{2,n}^T(k) S_{2,n}^{-1}(k) v_{2,n}(k). \tag{26}$$

$T_n(k)$ is verified with the validation gate γ to determine whether the position estimate $z_n(k)$ is within the validation region. If $T_n(k)$ is not larger than the validation gate γ, the hypothesis $\zeta_{0,n}$ holds true, and $z_n(k)$ is accurate. If $T_n(k)$ is larger than the validation gate γ, the hypothesis $\zeta_{0,n}$ is rejected, and $z_n(k)$ is discarded due to it being an NLOS error. Furthermore:

$$S_{2,n}(k) = BP_2(k|k-1)B^T + \sigma_G^2 \left(H_{2,n}^T(z_n) H_{2,n}(z_n)\right)^{-1}, \tag{27}$$

$$H_{2,n}(z_n) = \left.\frac{\partial h_{2,n}\left([x, y]^T\right)}{\partial [x, y]}\right|_{x=\hat{x}, y=\hat{y}}, \tag{28}$$

where $S_{2,n}(k)$ is the innovation covariance matrix for the position estimation from subgroup n at time step k. The number of $z_n(k)$ that falls into the validation region is calculated and is referred to as $N_V(k) (0 \le N_V(k) \le N)$. In order to judge whether $N_V(k) = 0$ is continuous or not, we propose the NLOS tracer method, which is used to record $N_V(k) = 0$. Therefore, we define:

$$L(k) = \begin{cases} 1 & \text{if } N_V(k) > 0 \\ 0 & \text{if } N_V(k) = 0 \end{cases}, \tag{29}$$

When $N_V(k)$ is larger than zero, we need to calculate the association probabilities used to weight the position estimation $z_n(k)$, which are also called the posterior probabilities. The premise of accurately calculating the association probabilities are to determine the threshold value of the validation region, and the selection of the threshold is related to the tracking threshold probability P_G. The tracking threshold probability P_G denotes the probability that the position estimation $z_n(k)$ determined from LOS beacon nodes with smallest error covariance falls into the valid region. Furthermore:

$$\int_0^{\gamma} f_{x^2(2)}(x) dx = P_G = 1 - P_{FA}, \tag{30}$$

where $f_{x^2(2)}(\cdot)$ is the chi-square probability density function with two degrees of freedom, and P_{FA} is the preset false alarm rate. Giving the probability P_G, the threshold value γ can be determined using the chi-square distribution table.

Data Association: In order to facilitate the calculation of the association probabilities, we define the following association events, following the approach taken in [33]:

$\theta_l(k)$: $\{z_l(k)$ is determined from the LOS beacon nodes with the smallest error covariance, $l = 1, \ldots, N_V(k)\}$.

$\theta_0(k)$: {none of the position estimates $z_l(k)$ at time step k stems from the LOS beacon nodes}.

The associated probabilities are:

$$\beta_l(k) = \Pr\left\{\theta_l(k) \middle| Z^k\right\} = \Pr\left\{\theta_l(k) \middle| Z(k), N_V(k), Z^{k-1}\right\},\ l = 1, \ldots, N_V(k), \tag{31}$$

where $Z(k) = \{Z_l(k)\}_{l=1}^{N_V(k)}$ and Z^k is the cumulative set of the position estimation, i.e., $Z^k = \{Z(i)\}_{i=1}^{k}$.

Since the associated probabilities are calculated based on the Bayesian formula, assuming that the innovations $v_{2,n}(k)$ are independent, Equation (31) can be rewritten as:

$$\beta_l(k) = \frac{1}{c} f\left[Z(k)\middle|\theta_l(k), N_V(k), Z^{k-1}\right] \times \Pr\left\{\theta_l(k)\middle|N_V(k), Z^{k-1}\right\},\ l = 0,\ 1,\ \ldots,\ N_V(k), \tag{32}$$

where c is the normalization factor and $f(\cdot)$ is the joint probability density function for the position estimation.

Assuming that the probability density function of the correct position estimates obeys the Gaussian distribution, and the probability density function of the inaccurate position estimate obeys the uniform distribution, the probability density function for the correct position estimates is:

$$f\left[z_l(k)\middle|\theta_l(k), N_V(k), Z^{k-1}\right] = P_G^{-1} N\left(z_l(k); \hat{z}(k|k-1), S_{2,l}(k)\right), \tag{33}$$

$$= P_G^{-1} N\left(v_{2,l}(k); 0, S_{2,l}(k)\right), \tag{34}$$

$$= P_G^{-1} \frac{\exp\left\{-\frac{1}{2} v_{2,l}^T(k) S_{2,l}^{-1}(k) v_{2,l}(k)\right\}}{2\pi\left|S_{2,l}(k)\right|^{0.5}}, \tag{35}$$

where $\left|S_{2,l}(k)\right|$ denotes the determinant of matrix $S_{2,l}(k)$. The probability density function of the inaccurate position estimate is:

$$f\left[z_l(k)\middle|\theta_l(k), N_V(k), Z^{k-1}\right] = V_l^{-1}(k), \tag{36}$$

$$V_l(k) = \gamma\pi\left|S_{2,l}(k)\right|^{0.5}, \tag{37}$$

where $V_l(k)$ is the area of the validation region [33] of the $N_V(k)$ accepted hypothesis, and the prior probabilities in Equation (32) are:

$$\Pr\left\{\theta_l(k)\middle|N_V(k), Z^{k-1}\right\} = \frac{P_d P_G}{N_V(k)},\ l = 1,\ \ldots,\ N_V(k), \tag{38}$$

$$\Pr\left\{\theta_l(k)\middle|N_V(k), Z^{k-1}\right\} = 1 - P_d P_G,\ l = 0, \tag{39}$$

where the prior probabilities of the correct position estimates are given by Equation (35), and the prior probability of the inaccurate position estimate is given by Equation (36). The detection probability P_d represents the probability that the position estimation that falls into the verification area can be correctly detected. The joint probability density function for the correct position estimates is:

$$f\left[Z(k)\middle|\theta_l(k), N_V(k), Z^{k-1}(k)\right] = \prod_{i=1\quad i\neq l}^{N_V(k)} \frac{N\left(v_{2,l}(k); 0, S_{2,l}(k)\right)}{V_i(k) P_G},\ l = 1,\ \ldots,\ N_V(k). \tag{40}$$

The joint probability density function for an inaccurate position estimation is:

$$f\left[Z(k)\middle|\theta_l(k), N_V(k), Z^{k-1}(k)\right] = \prod_{i=1}^{N_V(k)} V_i^{-1}(k),\ l = 0. \tag{41}$$

Multiplying the probability density function of the position estimates by the prior probabilities gives:

$$\beta_l'(k) = \frac{N\big(v_{2,l}(k); 0, S_{2,l}(k)\big)}{P_G}\left(\frac{P_d P_G}{N_V(k)}\right) \prod_{i=1 \quad i \neq l}^{N_V(k)} V_i^{-1}(k),\ l = 1,\ \ldots,\ N_V(k), \tag{42}$$

$$\beta_0'(k) = (1 - P_d P_G) \prod_{i=1}^{N_V(k)} V_i^{-1}(k),\ l = 0, \tag{43}$$

Thus, the association probabilities of the correct position estimate are modeled as:

$$\beta_l(k) = \frac{\beta_l'(k)}{\beta_0'(k) + \sum\limits_{l=1}^{N_V(k)} \beta_l'(k)}. \tag{44}$$

The association probability of an inaccurate position estimation is modeled as:

$$\beta_0(k) = \frac{\beta_0'(k)}{\beta_0'(k) + \sum\limits_{l=1}^{N_V(k)} \beta_l'(k)}. \tag{45}$$

When updating, we need to obtain the innovation covariance matrix $S_2(k)$ and the Kalman gain $K_2(k)$:

$$S_2(k) = BP_2(k|k-1)B^T + \sigma_G^2 I_2, \tag{46}$$

$$K_2(k) = P_2(k|k-1)B^T S_2^{-1}(k). \tag{47}$$

The updated state estimation $\hat{X}_2(k|k)$ is given as:

$$\hat{X}_2(k|k) = \hat{X}_2(k|k-1) + K_2(k) \sum_{l=1}^{N_V(k)} \beta_l(k) v_{2,l}(k). \tag{48}$$

The error covariance matrix $P_2(k|k)$ update is modeled as:

$$P_2(k|k) = \beta_0(k) P_2(k|k-1) + (1 - \beta_0(k)) P_c(k) + \widetilde{P}(k), \tag{49}$$

$$P_c(k) = (I_4 - K_2(k)B) P_2(k|k-1), \tag{50}$$

$$\widetilde{P}(k) = K_2(k) \left[\sum_{l=1}^{N_V(k)} \beta_l(k) v_{2,l}(k) v_{2,l}^T(k) - v_2(k) v_2^T(k) \right] K_2^T(k), \tag{51}$$

$$v_2(k) = \sum_{l=1}^{N_V(k)} \beta_l(k) v_{2,l}(k), \tag{52}$$

$$\Lambda_2(k) = \beta_l(k) \Lambda_{2,l}(k) \tag{53}$$

$$= \beta_l(k) N\big(v_{2,l}(k); 0, S_{2,l}(k)\big), \tag{54}$$

where $P_c(k)$ is the posterior covariance matrix of the standard Kalman filter and $\widetilde{P}(k)$ with the weighted innovation $v_2(k)$ corrects for the measurement uncertainty. $\Lambda_2(k)$ is the likelihood function when $N_V(k)$ is larger than zero at time step k.

When $N_V(k)$ is zero at time step k, the algorithm judges whether $N_V(k)$ is also zero at time step $k-1$. The proposed algorithm records the situation of $N_V(k)$ at time step $k-1$ using the NLOS tracer method. If $L(k-1)$ is 1, it indicates that $N_V(k) = 0$ appears separately, and the prediction of the state estimation and error covariance matrix are used as the updated state estimation and error covariance matrix:

$$\hat{X}_2(k|k) = \hat{X}_2(k|k-1), \tag{55}$$

$$P_2(k|k) = P_2(k|k-1), \tag{56}$$

$$S_2(k) = BP_2(k|k)B^T + 3\sigma_G^2 I_2, \tag{57}$$

$$\Lambda_2(k) = \frac{1}{2\pi|S_2(k)|^{0.5}}. \tag{58}$$

If $L(k-1)$ is zero, it indicates that $N_V(k) = 0$ appears successively. Thus, we use the extended Kalman filter to update the state estimate and the error covariance matrix estimate in the NLOS model:

$$H_2(k) = \left.\frac{\partial h(X(k))}{\partial X(k)}\right|_{X(k)=\hat{X}_2(k|k-1)}, \tag{59}$$

$$v_2(k) = D(k) - h\left(\hat{X}_2(k|k-1)\right), \tag{60}$$

$$S_2(k) = H_2(k)P_2(k|k-1)H_2(k) + R_2^*(k), \tag{61}$$

$$K_2(k) = P_2(k|k-1)H_2^T(k)S_2^{-1}(k|k-1), \tag{62}$$

$$\hat{X}_2(k|k) = \hat{X}_2(k|k-1) + K_2(k)v_2(k), \tag{63}$$

$$P_2(k|k) = (I_4 - K_2(k)H_2(k))P_2(k|k-1), \tag{64}$$

$$\Lambda_2(k) = N(v_2(k); 0, S_2(k)). \tag{65}$$

3.4. Model Probability Update

The model probabilities are updated using:

$$\mu_j(k) = (1/c)\Lambda_j(k)\bar{c}_j, \tag{66}$$

$$c = \sum_j \Lambda_j(k)\bar{c}_j, \tag{67}$$

where $\mu_j(k)$ is the updated model probability and c is the normalization factor.

3.5. Combination

The combination output estimation results are as follows:

$$\hat{X}(k|k) = \sum_j \hat{X}_j(k|k)\mu_j(k), \tag{68}$$

$$P(k|k) = \sum_j \left\{P_j(k|k) + \left[\hat{X}_j(k|k) - X(k|k)\right] \times \left[\hat{X}_j(k|k) - X(k|k)\right]^T\right\}\mu_j(k|k). \tag{69}$$

4. Experiment and Result Analysis

This part mainly discusses the simulation results of the experiment. In this paper, six beacon nodes were randomly deployed in a 100×100 area, and the mobile node ran along a fixed trajectory shown in Figure 3 with a moving step of 100. The sampling period was $\Delta t = 0.5$ s and the mobile node initial state $X_j(0) = \begin{bmatrix} 1\,\text{m} & 20\,\text{m} & 1\,\frac{\text{m}}{\text{s}} & 0.5\,\frac{\text{m}}{\text{s}} \end{bmatrix}^T$. The initial of the error covariance matrix was set to $P_j(0) = I_4$, the Markov transition matrix initial value was $p = \begin{bmatrix} 0.995 & 0.005 \\ 0.005 & 0.995 \end{bmatrix}$, and the measurement noise covariance matrix was $R_j^*(0) = \sigma_m^2 I_M$. The tracking threshold probability P_G was 0.99, and the detection probability P_d was 0.9. In order to simulate an NLOS environment, a probability value was generated randomly. Comparing this probability value with the NLOS probability threshold, if the probability that was generated randomly was less than the NLOS probability, the mobile node and the corresponding beacon nodes were considered to be in the NLOS condition. The simulation experiments were carried out under the conditions of NLOS errors obeying Gaussian, uniform, and exponential distributions. In this paper, we compared the proposed algorithm with the EKF [34], IMM-EKF [30], and MPDA [35]. The simulation results were obtained using 1000 Monte Carlo runs, and the root mean square error (RMSE) and the error cumulative distribution function (CDF) of the average positioning errors were used as the performance indicators for the evaluation algorithm:

$$RMSE = \sqrt{\frac{1}{MC}\frac{1}{K}\prod_{j=1}^{MC}\prod_{k=1}^{K}\left(\left(\hat{x}_j(k) - x_j(k)\right)^2 + \left(\hat{y}_j(k) - y_j(k)\right)^2\right)}, \tag{70}$$

$$ALE = \frac{1}{MC \cdot K}\sum_{j=1}^{MC}\sum_{k=1}^{K}\sqrt{(\hat{x}_j(k) - x_j(k))^2 + (\hat{y}_j(k) - y_j(k))^2}, \tag{71}$$

where ALE is the average localization error, which represents the average Euclidean distance between the position estimate and the true position. $K = 100$ is the number of moving steps, $MC = 1000$ is the number of time the Monte Carlo simulation ran. $\left(\hat{x}_j(k), \hat{y}_j(k)\right)$ is the position estimate of the interactive output in the *jth* Monte Carlo run and $\left(x_j(k), y_j(k)\right)$ is the true position of the mobile terminal during the *jth* Monte Carlo run.

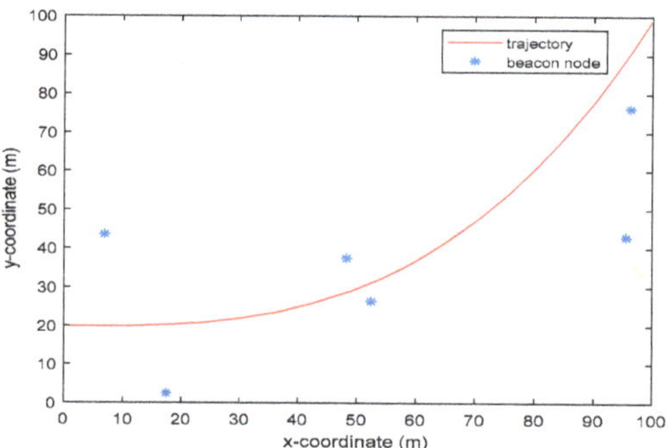

Figure 3. The trajectory of the mobile node.

4.1. Gaussian Distribution

It was assumed that the NLOS errors obeyed the Gaussian distribution $N\left(\mu_{NLOS}, \sigma_{NLOS}^2\right)$, the measurement noise obeyed the Gaussian distribution $N\left(0, \sigma_L^2\right)$, and the range measurements were irrelevant. The default parameters of the Gaussian distribution simulation experiment are shown in Table 1.

Table 1. Gaussian distribution parameter.

Parameter	Symbol	Default Values
The number of beacon nodes	M	6
NLOS error probability	P_{NLOS}	0.5
The measurements' noise	$N\left(0,\sigma_L^2\right)$	$N\left(0,1^2\right)$
The NLOS errors	$N\left(\mu_{NLOS},\sigma_{NLOS}^2\right)$	$N\left(5,6^2\right)$

As shown in Figure 4a, with the number of beacon nodes increasing from 4 to 9, the RMSEs of the four algorithms decreased. This indicates that increasing the number of beacon nodes was helpful for reducing the NLOS errors and improving the positioning accuracy of the algorithms. When the number of beacon nodes varied from 4 to 7, the RMSEs of the IMM-EKF, MPDA, and the proposed algorithm decreased rapidly. When the number of beacon nodes increased to 7, the RMSEs of the MPDA and the proposed algorithm decreased slowly, and the RMSE of the IMM-EKF hardly changed. The RMSE of the EKF decreased slowly. From Figure 4a, we can see that the positioning accuracy of the proposed algorithm was better than that of the EKF, IMM-EKF, and MPDA. When the number of beacon nodes was 9, the proposed algorithm had a higher positioning accuracy than the EKF, IMM-EKF, and MPDA at about 65.70%, 59.97%, and 25.16%, respectively.

Figure 4b shows the impact of the NLOS errors probability on RMSE. Compared with the EKF and IMM-EKF algorithms, the RMSE of the proposed algorithm increased slowly when the probability of the NLOS error was less than 0.4. However, when the probability of the NLOS error was greater than 0.4, the RMSE of the proposed algorithm increased rapidly with the increase of the NLOS errors probability. The change of root mean square error of the MPDA algorithm was similar to that of the proposed algorithm, but its growth rate was faster than that of the proposed algorithm. When compared with EKF, IMM-EKF, and MPDA, the proposed algorithm had a higher positioning accuracy.

Figure 4c shows the relationship between the RMSE and the mean value of the NLOS errors. As the mean value of the NLOS error varied from 3 to 10, the RMSE of the EKF, IMM-EKF, and MPDA increased. Although the RMSE of the proposed algorithm also increased, its RMSE was always smaller than that of the EKF, IMM-EKF, and MPDA. From Figure 5, we can see that the growth rate of the root mean square error of the proposed algorithm was obviously less than that of other three algorithms. When the mean value of NLOS error was 3, the proposed algorithm improved the positioning accuracy by about 47.22%, 32.72%, and 28.98% compared with the EKF, IMM-EKF, and MPDA, respectively. When the mean value of the NLOS error was 10, the proposed algorithm had a higher positioning accuracy than the EKF, IMM-EKF, and MPDA at about 35.76%, 36.67%, and 30.06%, respectively.

Figure 4d shows the impact of the standard deviation of NLOS errors on the RMSE. The RMSE of the EKF clearly increased with the increase of the standard deviation, which led to a rapid decline in its positioning accuracy. When the standard deviation was 10, the RMSE of the EKF, IMM-EKF, MPDA, and the proposed algorithm were 5.3948 m, 3.9240 m, 3.2254 m, and 2.0257 m, respectively. Compared with the EKF, IMM-EKF, and MPDA, the proposed algorithm improved the positioning accuracy by about 62.45%, 48.38%, and 37.20%, respectively.

Figure 4e shows the error cumulative distribution function of the localization errors, which shows that the 90th percentile of the errors of the proposed algorithm was less than 3.263 m. In contrast, the 90th percentile of errors of the EKF, IMM-EKF, and MPDA were 6.668 m, 5.494 m, and 4.191 m, respectively.

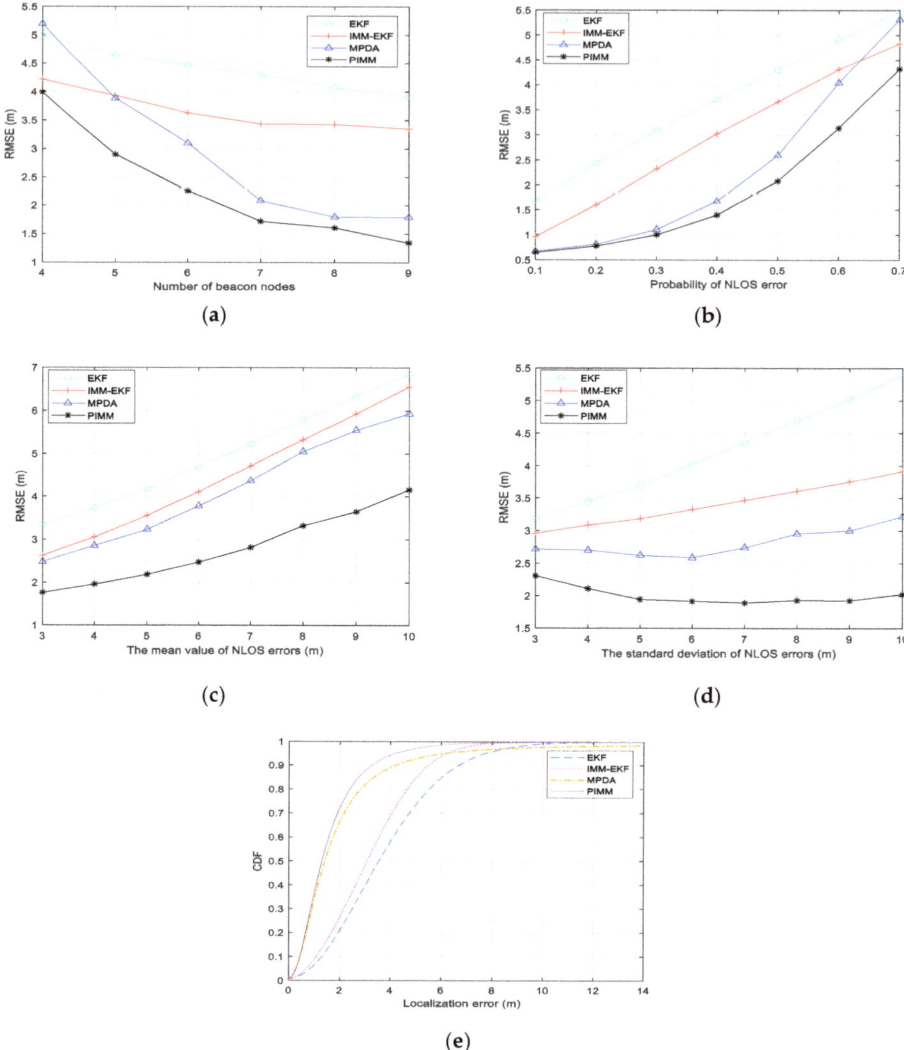

Figure 4. (**a**) Performance comparison between the EKF, interacting multiple model (IMM)-EKF, modified probabilistic data association (MPDA), and the proposed algorithm (PIMM) under a different number of beacon nodes M, where $P_{NLOS} = 0.5$, $N(0, 1^2)$, and $N(5, 6^2)$. (**b**) Performance comparison between the EKF, IMM-EKF, MPDA, and the proposed algorithm under different NLOS error probabilities P_{NLOS}, where $M = 6$, $N(0, 1^2)$, and $N(5, 6^2)$. (**c**) Performance comparison between the EKF, IMM-EKF, MPDA, and the proposed algorithm under different mean values μ_{NLOS} of the Gaussian distribution, where $M = 6$, $\sigma_{NLOS} = 6$,$P_{NLOS} = 0.5$, and $N(0, 1^2)$. (**d**) Performance comparison between the EKF, IMM-EKF, MPDA, and the proposed algorithm under different standard deviations σ_{NLOS} of the Gaussian distribution, where $M = 6$, $P_{NLOS} = 0.5$, $\mu_{NLOS} = 5$, and $N(0, 1^2)$. (**e**) The cumulative distribution function (CDF) of the localization error.

4.2. Uniform Distribution

It was assumed that the measurements' noise obeyed a Gaussian distribution $N\left(0, \sigma_L^2\right)$ and the NLOS errors obeyed a uniform distribution $U(a, b)$. The default parameters for the uniform distribution simulation experiment are shown in Table 2.

Table 2. Uniform distribution parameter.

Parameter	Symbol	Default Values
The number of beacon nodes	M	6
The NLOS error probability	P_{NLOS}	0.5
The measurements' noise	$N\left(0, \sigma_L^2\right)$	$N\left(0, 1^2\right)$
The NLOS errors	$U(a, b)$	$U(0, 14)$

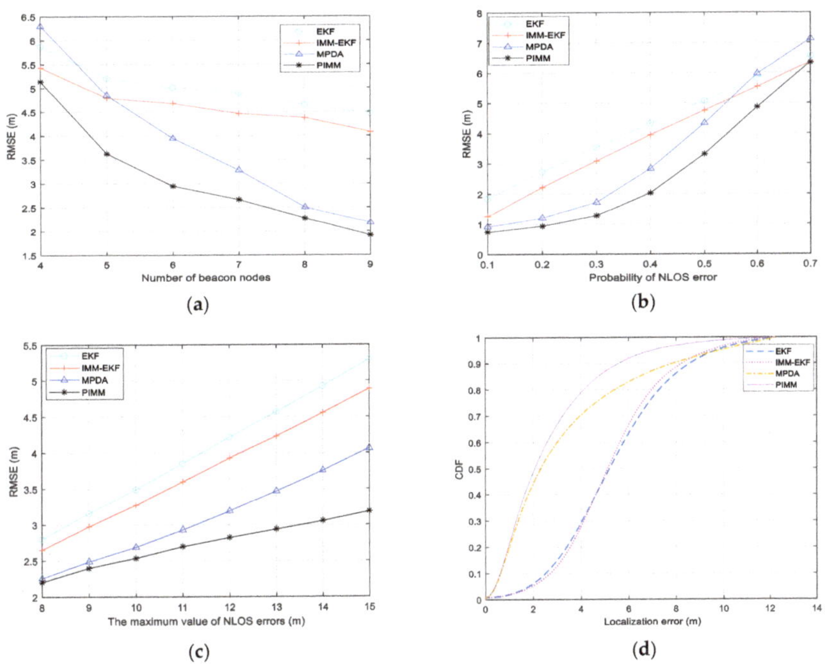

Figure 5. (**a**) Performance comparison between the EKF, IMM-EKF, MPDA, and the proposed algorithm under a different number of beacon nodes M, where $P_{NLOS} = 0.5$, $N\left(0, 1^2\right)$, and $U(0, 14)$. (**b**) Performance comparison between the EKF, IMM-EKF, MPDA, and the proposed algorithm under different NLOS error probabilities P_{NLOS}, where $M = 6$, $N\left(0, 1^2\right)$, and $U(0, 14)$. (**c**) Performance comparison between the EKF, IMM-EKF, MPDA, and the proposed algorithm under different maximum values b of uniform distribution, where $M = 6$, $P_{NLOS} = 0.5$, and $N\left(0, 1^2\right)$. (**d**) The CDF of the localization error.

Figure 5a shows the relationship between the RMSE and the number of beacon nodes when the NLOS errors took on a uniform distribution. It can be seen that the number of beacon nodes varied from 4 to 9, and the RMSEs of the EKF and IMM-EKF algorithms decreased slowly in a similar trend. The change in the number of beacon nodes had a great impact on the positioning results of the MPDA and the proposed algorithm. The proposed algorithm had a higher positioning accuracy than the EKF, IMM-EKF, and MPDA with about 39.70%, 34.81%, and 18.33%, respectively, on average.

As shown in Figure 5b, increasing the probability of the NLOS error aggravated the interference of the NLOS factors for positioning, which resulted in increasing the RMSE of the EKF, IMM-EKF, MPDA, and the proposed algorithm. We can see that the RMSE of the proposed algorithm increased slowly with the increase of the probability of the NLOS errors when the probability of the NLOS errors was not more than 0.4, but the growth rate increased gradually. On the whole, the proposed algorithm showed a significant improvement in positioning accuracy compared with the EKF and IMM-EKF algorithms when the NLOS error probability was not higher than 0.6. Compared with the MPDA, the proposed algorithm had a higher positioning accuracy.

Figure 5c shows how the RMSE varied with the maximum value of the NLOS errors. The RMSEs of the EKF, IMM-EKF, MPDA, and the proposed algorithm increased almost linearly as the maximum value of the NLOS errors gradually increased. Compared with the EKF and IMM-EKF algorithms, the proposed algorithm had a greatly improved positioning accuracy. When the maximum of the NLOS error was 15 m, the proposed algorithm had a higher positioning accuracy than the EKF, IMM-EKF, and MPDA with about 39.90%, 34.89%, and 21.67%, respectively. It can be seen from Figure 5c that the proposed algorithm significantly outperformed the other three algorithms in terms of inhibiting NLOS errors.

The error cumulative distribution function of the localization errors for the same example is depicted in Figure 5d, which shows that the 90th percentile of the proposed algorithm was less than 5.555 m. In contrast, the 90th percentile of the EKF, IMM-EKF, and MPDA were 8.566 m, 8.228 m, and 7.829 m, respectively.

4.3. Exponential Distribution

It was assumed that the measurements' noise obeyed a Gaussian distribution $N\left(0, \sigma_L^2\right)$ and the NLOS errors obeyed an exponential distribution $E(\lambda)$. The default parameters of the exponential distribution are shown in the Table 3.

Table 3. Exponential distribution parameter.

Parameter	Symbol	Default Values
The number of beacon nodes	M	6
The NLOS error probability	P_{NLOS}	0.5
The measurements noise	$N\left(0, \sigma_L^2\right)$	$N\left(0, 1^2\right)$
The NLOS errors	$E(\lambda)$	$E(8)$

We investigated how the positioning accuracy of the EKF, IMM-EKF, MPDA, and the proposed algorithm varied with the number of beacon nodes when the NLOS errors were exponential distribution (Figure 6a). As the number of beacon nodes increased, the RMSEs of the EKF, IMM-EKF, MPDA, and the proposed algorithm all decreased, but the difference was that the RMSEs of the EKF and IMM-EKF decreased slowly and the improvement of the positioning accuracy was not high, while the RMSEs of the MPDA and the proposed algorithms decreased rapidly and the positioning accuracy improved significantly. In the case of a few beacon nodes, the positioning accuracy of the proposed algorithm was higher than that of the EKF, IMM-EKF, and MPDA algorithms. In the case of more beacon nodes, the positioning accuracy of the proposed algorithm was slightly higher than that of the MPDA, and far higher than that of the EKF and IMM-EKF. Compared with the EKF, IMM-EKF, and MPDA algorithms, the proposed algorithm improved the positioning accuracy by 63.60%, 59.57%, and 20.27%, respectively, on average.

We also explored the influence of different NLOS errors probabilities on the positioning results of the EKF, IMM-EKF, MPDA, and the proposed algorithm when NLOS errors were given as an exponential distribution. From Figure 6b, we can see that the positioning accuracy of the EKF, IMM-EKF, MPDA, and the proposed algorithm declined with the increasing probability of NLOS errors. With the increase of the NLOS error probability, the RMSEs of the EKF and IMM-EKF increased almost linearly, and the RMSE growth rate of the MPDA and the proposed algorithm gradually accelerated. When the NLOS error was less than 0.3, the positioning accuracy of the proposed algorithm was slightly higher than that of the MPDA, but significantly higher than that of the EKF and IMM-EKF. When the probability of the NLOS errors was relatively large, the proposed algorithm had a better localization performance than the EKF, IMM-EKF, and MPDA. The proposed algorithm had a higher positioning accuracy than the EKF, IMM-EKF, and MPDA with about 65.83%, 53.01%, and 18.56%, respectively, on average.

In Figure 6c, the influence of changing the parameter λ of exponential distribution on the localization accuracy is shown. We can see that when the parameter λ changed from 3 to 10, the RMSEs of the EKF and IMM-EKF algorithm increased rapidly, while the RMSE of the MPDA algorithm grew slowly. The change of the parameter λ from 3 to 10 had no significant impact on the localization accuracy of the proposed algorithm. Compared with the EKF, IMM-EKF, and MPDA algorithms, the proposed algorithm improved the positioning accuracy by at least 36.16%, 33.35%, and 11.43%, respectively.

The error cumulative distribution function of the localization errors is depicted in Figure 6d. It shows that the 90th percentile of the proposed algorithm was less than 3.882 m. In contrast, the 90th percentile of the EKF, IMM-EKF, and MPDA were achieved at 12.22 m, 10.54 m, and 5.065 m, respectively.

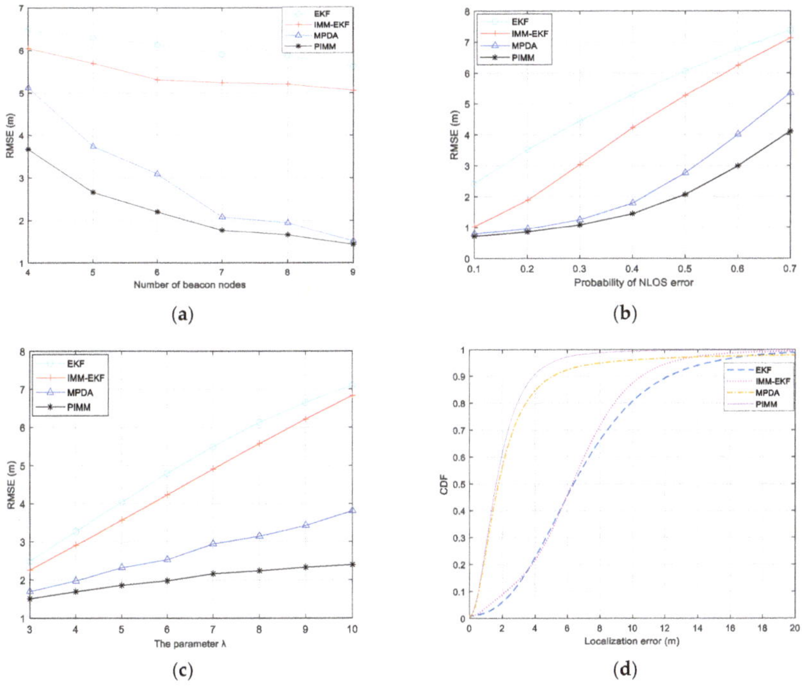

Figure 6. (**a**) Performance comparison between the EKF, IMM-EKF, MPDA, and the proposed algorithm under different numbers of beacon nodes M, where $P_{NLOS} = 0.5$, $N(0, 1^2)$, and $E(8)$. (**b**) Performance comparison between the EKF, IMM-EKF, MPDA, and the proposed algorithm under different NLOS error probabilities P_{NLOS}, where $M = 6$, $N(0, 1^2)$, and $E(8)$. (**c**) Performance comparison between the EKF, IMM-EKF, MPDA, and the proposed algorithm under different parameters (λ) the exponential distribution, where $M = 6$, $P_{NLOS} = 0.5$, and $N(0, 1^2)$. (**d**) The CDF of the localization error.

4.4. Experimental Results

In order to further verify the positioning accuracy of the proposed algorithm, we carried out experiments in a real environment. Ultra-wideband (UWB) technology was used to transmit signals between beacon nodes and a mobile node to obtain the range measurements. The time-based UWB positioning system makes use of the accurate TOA of the information exchange between devices [36] and has a high positioning accuracy. Therefore, in recent years, UWB has been widely used in indoor positioning. As shown in Figure 7a, there were six beacon nodes and a mobile node moved uniformly along the trajectory shown. In order to avoid the reflection of UWB signal from the ground, the mobile node was moved to 1.2 m above the ground. The room was 10 meters long and 7 meters wide. Because there were many obstacles in the room, the measurements were prone to be disturbed by NLOS factors, resulting in larger errors. Six beacon nodes sampled mobile nodes every 0.6 m for a total of 16 times. Each beacon node took 20 measurements at each sampling position, and the average value was taken as the final measurement.

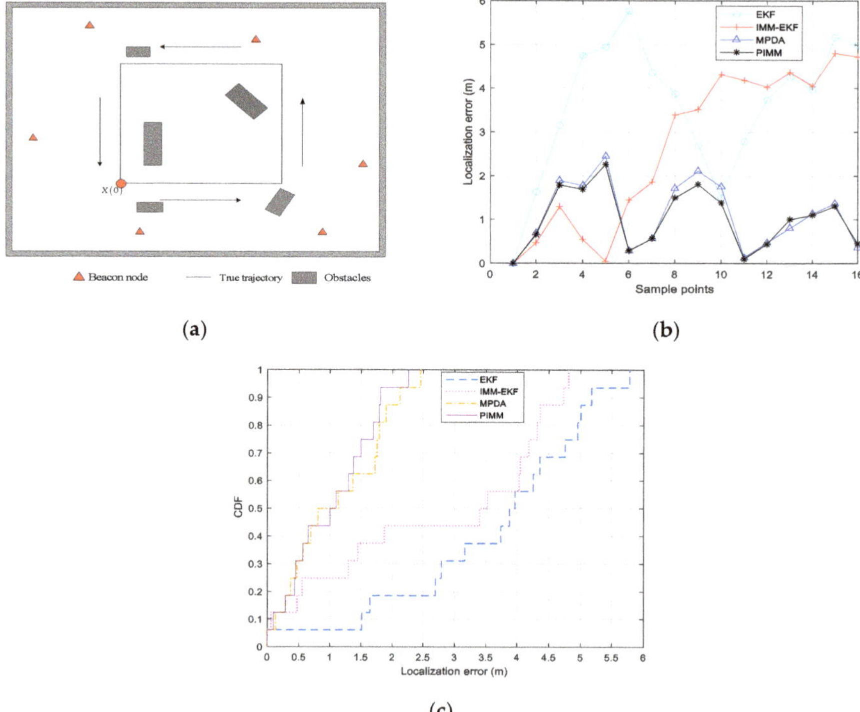

Figure 7. (**a**) The deployment of beacon nodes and the true trajectory of the mobile node. (**b**) The localization error at each sample point. (**c**) The CDF of the localization error.

The initial state of the mobile node was set $X(0) = \begin{bmatrix} 1.2\,\text{m} & 1.8\,\text{m} & 0.6\,\text{m/s} & 0\,\text{m/s} \end{bmatrix}^T$, the sampling period Δt was 1s, and other parameters were consistent with the simulation. The error distribution of each sample point is shown in Figure 7b, and the CDF of the localization error is shown in Figure 7c. Because there were some obstacles in the experimental environment, and the interference of NLOS factors was serious, the localization errors of the EKF and IMM-EKF were relatively large in most cases. Compared with the EKF and IMM-EKF, the localization errors of the MPDA and the proposed algorithm were smaller at most sample points, but the positioning accuracy of the proposed algorithm was slightly higher than that of the MPDA.

5. Conclusions

This paper proposed an NLOS error mitigation algorithm for tracking a mobile node based on the TOA measurements in a mixed LOS/NLOS environment. An improved MPDA algorithm with a stronger robustness was proposed in this paper. We applied the improved MPDA filter to the IMM-EKF algorithm framework. In this paper, an extended Kalman filter was used in the LOS model, and an improved MPDA filter was used in parallel with the NLOS model. The second filter discarded the position estimates with large errors through NLOS recognition to mitigate the NLOS errors. After updating the model probability of each model, the state estimation calculated using the two filters was based on the updated model probability weighted combination to obtain the final state estimation. The simulation and experimental results showed that the proposed algorithm outperformed the EKF, IMM-EKF, and MPDA in an NLOS environment. The proposed algorithm had a high positioning accuracy in the case of a small NLOS error probability. In the case of a high probability of NLOS error, the positioning accuracy of the proposed algorithm decreased greatly. In the future, the intended further improvement involved increasing the positioning accuracy of the proposed algorithm in the case of a high probability of NLOS error.

Author Contributions: Methodology, L.C. and M.X.; data curation, M.X. and Z.L.; data curation, M.X. and Y.W. All authors have read and agreed to the published version of the manuscript.

Funding: This work was funded by the National Natural Science Foundation of China under grant no. 61803077, Natural Science Foundation of Hebei Province under grant no. F2016501080 and no. F2015501097, and Fundamental Research Funds for the Central Universities under grant no. N172304024.

Conflicts of Interest: The authors declare no conflict of interest.

References

1. Wylie, M.P.; Holtzman, J. The non-line of sight problem in mobile location estimation. In Proceedings of the IEEE ICUPC-5th International Conference on Universal Personal Communications, Cambridge, MA, USA, 2 October 1996; pp. 827–831.
2. Su, Z.; Shao, G.; Liu, H. Semidefinite Programming for NLOS Error Mitigation in TDOA Localization. *IEEE Commun. Lett.* **2018**, *22*, 1430–1433. [CrossRef]
3. Tomic, S.; Beko, M. A bisection-based approach for exact target localization in NLOS environments. *Signal Process.* **2018**, *143*, 328–335. [CrossRef]
4. Yang, X.; Zhao, F.; Chen, T. NLOS identification for UWB localization based on import vector machine. *Int. J. Electron. Commun.* **2018**, *87*, 128–133. [CrossRef]
5. Zhang, Z.; Zhou, X.; Wang, Z.; Yan, H. Adaptive Consensus-Based Distributed Target Tracking With Dynamic Cluster in Sensor Networks. *IEEE Trans. Cybern.* **2019**, *49*, 1580–1591. [CrossRef]
6. Li, Y.Y.; Qi, G.Q.; Sheng, A.D. Performance Metric on the Best Achievable Accuracy for Hybrid TOA/AOA Target Localization. *IEEE Commun. Lett.* **2018**, *22*, 1474–1477. [CrossRef]
7. Sun, Y.M.; Zhou, Z.P.; Tang, S.L. 3D Hybrid TOA-AOA Source Localization Using an Active and a Passive Station. In Proceedings of the 2016 IEEE 13th International Conference on Signal Processing (ICSP), Chengdu, China, 6–10 November 2016; pp. 257–260.
8. Vaghefi, R.M.; Buehrer, R.M. Cooperative Localization in NLOS Environments Using Semidefinite Programming. *IEEE Commun. Lett.* **2015**, *19*, 1382–1385. [CrossRef]
9. Vaghefi, R.M.; Buehrer, R.M. Cooperative Source Node Tracking in Non-Line-of-Sight Environments. *IEEE Trans. Mob. Comput.* **2018**, *16*, 1287–1299. [CrossRef]
10. Chen, P.C. A non-line-of-sight error mitigation algorithm in location estimation. In Proceedings of the IEEE Wireless Communications Networking Conference (WCNC), New Orleans, LA, USA, 21–24 September 1999; Volume 1, pp. 316–320.
11. Park, C.H.; Chang, J.H. Robust Time-of-arrival Source Localization Employing Error Covariance of Sample Mean and Sample Median in Line-of-sight/Non-line-of-sight Mixture Environments. *EURASIP J. Adv. Signal Process.* **2016**, *89*, 1–11. [CrossRef]

12. Jiao, L.; Xing, J.; Zhang, X. LCC-RWGH: A NLOS Error Mitigation Algorithm for Localization in Wireless Sensor Network. In Proceedings of the 2007 IEEE International Conference on Control and Automation, Guangzhou, China, 30 May–1 June 2007; pp. 1354–1359.
13. Xing, J.; Zhang, J.; Jiao, L. A Robust Wireless Sensor Network Localization Algorithm in NLOS Environment. In Proceedings of the 2007 IEEE International Conference on Control and Automation, Guangzhou, China, 30 May–1 June 2007; pp. 3244–3249.
14. Wu, S.; Xu, D.; Wang, H. Adaptive NLOS Mitigation Location Algorithm in Wireless Cellular Network. *Wirel. Pers. Commun.* **2015**, *84*, 1–14. [CrossRef]
15. Li, X. An Iterative NLOS Mitigation Algorithm for Location Estimation in Sensor Networks. *Mob. Wirel. Commun.* **2006**, *10*, 1–5.
16. Horiba, M.; Okamoto, E.; Shinohara, T. An improved NLOS detection scheme for hybrid-TOA/AOA-based localization in indoor environments. In Proceedings of the 2013 IEEE International Conference on Ultra-Wideband (ICUWB), Sydney, NSW, Australia, 15–18 September 2013; Ieice Technical Report, Volume 113, pp. 37–42.
17. Horiba, M.; Okamoto, E.; Shinohara, T. An improved NLOS detection scheme using stochastic characteristics for indoor localization. In Proceedings of the 2015 International Conference on Information Networking (ICOIN), Siem Reap, Cambodia, 12–14 January 2015; pp. 478–482.
18. Zhang, S.J.; Gao, S.C.; Wang, G.; Li, Y.M. Robust NLOS Error Mitigation Method for TOA-Based Localization via Second-Order Cone Relaxation. *IEEE Commun. Lett.* **2015**, *19*, 2210–2213. [CrossRef]
19. Wang, G.; Chen, H.; Li, Y.; Ansari, N. NLOS Error Mitigation for TOA-based Localization via Convex Relaxation. *IEEE Trans. Wirel. Commun.* **2014**, *13*, 4119–4131. [CrossRef]
20. Tomic, S.; Beko, M.; Dinis, R.; Montezuma, P. A Robust Bisection-based Estimator for TOA-based Target Localization in NLOS Environments. *IEEE Commun. Lett.* **2017**, *21*, 2488–2491. [CrossRef]
21. Liu, D.; Xu, Y.; Huang, X. Identification of Location Spoofing in Wireless Sensor Networks in Non-Line-of-Sight Conditions. *IEEE Trans. Ind. Inform.* **2018**, *14*, 2375–2384. [CrossRef]
22. Han, K.; Xing, H.S.; Deng, Z.L.; Du, Y.C. A RSSI/PDR-Based Probabilistic Position Selection Algorithm with NLOS Identification for Indoor Localisation. *ISPRS Int. J. Geo-Inf.* **2018**, *7*, 232. [CrossRef]
23. Benedetto, F.; Giunta, G.; Vandendorpe, L. LOS/NLOS detection by the normalized Rayleigh-ness test. Presented at the 17th European Signal Processing Conference (EUSIPCO), Glasgow, UK, 24–28 August 2009.
24. Almazrouei, E.; Sindi, N.A.; Al-Araji, S.R. Measurement and analysis of NLOS identification metrics for WLAN systems. In Proceedings of the 2014 IEEE 25th International Symposium on Personal. Indoor and Mobile Radio Communications, Washington, DC, USA, 2–5 September 2014; pp. 280–284.
25. Xiao, Z.; Wen, H.; Markham, A. Non-Line-of-Sight Identification and Mitigation Using Received Signal Strength. *IEEE Trans. Wirel. Commun.* **2015**, *14*, 1689–1702. [CrossRef]
26. Venkatraman, S.; Caffery, J. Statistical approach to non-line-of-sight BS identification. In Proceedings of the IEEE 5th International Symposium on Wireless Personal Multimedia Communications, Honolulu, HI, USA, 27–30 October 2002; Volume 1, pp. 296–300.
27. Zhang, J.; Salmi, J.; Lohan, E.S. Analysis of Kurtosis-Based LOS/NLOS Identification Using Indoor MIMO Channel Measurement. *IEEE Trans. Veh. Technol.* **2013**, *62*, 2871–2874. [CrossRef]
28. Xiao, F.; Guo, Z.; Zhu, H. Real-time LOS/NLOS identification with WiFi. In Proceedings of the IEEE International Conference Communications (ICC), Paris, France, 21–25 May 2017; pp. 1–7.
29. Chen, B.S.; Yang, C.Y.; Liao, F.K.; Liao, J.-F. Mobile location estimator in a rough wireless environment using extended Kalman based IMM and data fusion. *IEEE Trans. Veh. Technol.* **2009**, *58*, 1157–1169. [CrossRef]
30. Li, W.L.; Jia, Y.M.; Du, J.P. TOA-based cooperative localization for mobile stations with NLOS mitigation. *J. Frankl. Inst.* **2016**, *353*, 1297–1312. [CrossRef]
31. Hammes, U.; Zoubir, A.M. Robust MT Tracking Based on M-Estimation and Interacting Multiple Model Algorithm. *IEEE Trans. Signal Process.* **2011**, *59*, 3398–3409. [CrossRef]
32. Hammes, U.; Wolsztynski, E.; Zoubir, A.M. Robust tracking and geolocation for wireless networks in NLOS environments. *IEEE J. Sel. Top. Signal Process.* **2009**, *3*, 889–901. [CrossRef]
33. Kirubarajan, T.; Bar-Shalom, Y. Probabilistic data association techniques for target tracking in clutter. *Proc. IEEE.* **2004**, *92*, 536–557. [CrossRef]
34. Cui, W.; Li, B.; Zhang, L.; Meng, W. Robust Mobile Location Estimation in NLOS Environment Using GMM, IMM, and EKF. *IEEE Syst. J.* **2019**, *13*, 3490–3500. [CrossRef]

35. Hammes, U.; Zoubir, A.M. Robust mobile terminal tracking in NLOS environments based on data association. *IEEE Trans. Signal Process.* **2010**, *51*, 5872–5882. [CrossRef]
36. Kim, D.H.; Kwon, G.R.; Pyun, J.Y. NLOS Identification in UWB channel for Indoor Positioning. Presented at the 15th IEEE Annual Consumer Communications & Networking Conference (CCNC), Las Vegas, NV, USA, 12–15 January 2018.

Article

A Wi-Fi FTM-Based Indoor Positioning Method with LOS/NLOS Identification

Minghao Si [1,2], Yunjia Wang [1,2,*], Shenglei Xu [2], Meng Sun [2] and Hongji Cao [2]

[1] Key Laboratory of Land Environment and Disaster Monitoring, MNR, China University of Mining and Technology, Xuzhou 221116, China; hmsi@cumt.edu.cn

[2] School of Environmental Science and Spatial Informatics, China University of Mining and Technology, Xuzhou 221116, China; cumtxsl@163.com (S.X.); msun@cumt.edu.cn (M.S.); hjcao@cumt.edu.cn (H.C.)

* Correspondence: wyjc411@163.com; Tel.: +86-132-252-31855

Received: 30 December 2019; Accepted: 28 January 2020; Published: 2 February 2020

Abstract: In recent years, many new technologies have been used in indoor positioning. In 2016, IEEE 802.11-2016 created a Wi-Fi fine timing measurement (FTM) protocol, making Wi-Fi ranging more robust and accurate, and providing meter-level positioning accuracy. However, the accuracy of positioning methods based on the new ranging technology is influenced by non-line-of-sight (NLOS) errors. To enhance the accuracy, a positioning method with LOS (line-of-sight)/NLOS identification is proposed in this paper. A Gaussian model has been established to identify NLOS signals. After identifying and discarding NLOS signals, the least square (LS) algorithm is used to calculate the location. The results of the numerical experiments indicate that our algorithm can identify and discard NLOS signals with a precision of 83.01% and a recall of 74.97%. Moreover, compared with the traditional algorithms, by all ranging results, the proposed method features more accurate and stable results for indoor positioning.

Keywords: indoor positioning; Wi-Fi fine timing measurement; NLOS identification; Gaussian model

1. Introduction

The importance of indoor positioning is reflected by the increasing requirements of location-based services (LBSs). Under such a background, global researchers have proposed a lot of indoor positioning methods based on different technologies, such as wireless fidelity (Wi-Fi) [1,2], Bluetooth [3], radio frequency identification (RFID) [4], computer vision [5], ultra-wideband (UWB) [6], infrared [7], and inertial navigation systems (INSs) [8], micro-electro-mechanical systems (MEMSs) [9], visible light [10,11], geomagnetic fields [12,13], and pseudolites [14,15], among others.

Among these methods, methods based on Wi-Fi have attracted much attention due to their wide distribution. They can be defined as fingerprint-based methods [16] and distance-based methods. The former can reach an accuracy of about 2 m and needs to construct a fingerprint database at the offline stage, which costs more labor and time. The latter can calculate the distance between the transceivers according to the received signal strength indicator (RSSI), then use the least square (LS), trilateral positioning, or other algorithms to derive the location. However, the RSSI, when collected at a fixed place, could be influenced by device heterogeneity and non-line-of-sight (NLOS) errors, so it is difficult to use widely.

This phenomenon did not change until 2016. The fine timing measurement (FTM) protocol has been standardized by the Institute of Electrical and Electronics Engineers (IEEE) 802.11-2016 and can provide meter-level positioning accuracy [17] through time of flight (TOF) echo technology [18]. However, just like many other ranging measurements, such as UWB [19,20] and Global Positioning System (GPS) [21,22], one of the major challenges for positioning is the mitigation of NLOS effects. If the direct path between a fixed terminal (FT) and a mobile terminal (MT) is obstructed, the time of arrival (TOA) of the signal to the FT will be delayed, which will introduce a positive bias. The use of such TOA estimates may significantly degrade accuracy during the positioning process. Hence, it is necessary to identify NLOS signals.

NLOS identification has been extensively discussed for UWB signals [23], while it has rarely been discussed for Wi-Fi signals, as only RSSIs rather than more detailed channel state information (CSI) data can be collected from Wi-Fi devices with mobile phones. Therefore, elucidating how to identify NLOS signals using RSSI has become one of the major difficulties of this work. Joan Borras [24] formulated a theoretical framework where various theoretical tests were developed for known and unknown models of NLOS errors. Ke Han [25] calculated the distance between the coordinates of the FT and that of MT, which were obtained via pedestrian dead reckoning (PDR). Then, the obtained values were used to build two Gaussian models to identify line-of-sight (LOS) signals and NLOS signals, respectively. Considering that RSSIs will attenuate as they pass through different materials, this method may perform poorly in some environments. Jo [26] determined LOS signals according to the difference of RSSI between a 2.4 GHz signal and that of a 5 GHz signal. However, the new FTM protocol that supports Wi-Fi devices can only send single-frequency signals, so this method is not suitable. Except for identification methods based on received multipath signals, some methods have been designed using the overall mobile network as their basis. Chen [27] used various combinations of ranging results to evaluate the positioning and residual of an MT. The combination with a smaller residual has a larger chance of obtaining a higher accuracy. Some other NLOS error mitigation methods using the mobile network have been reported in [28–34].

This study proposes a Wi-Fi FTM-based indoor positioning method that uses LOS/NLOS identification. A Gaussian model has been built in order to identify NLOS signals. LOS signals are used to calculate the locations of an MT with an LS algorithm. Based on the newest Wi-Fi technology, this method is suitable for future Wi-Fi devices. Moreover, the identification model built by this method is more suitable for complex indoor environments.

The contributions of this work are summarized as follows:

(1) In order to discard the NLOS error, an identification model is proposed. Differing from other Gaussian-based models, which have established two models for LOS and NLOS, the model presented here only needs to establish one model in order to adapt to the complex indoor environment.
(2) To solve the difficulty of setting parameters, we have researched and fit the relation between RSSI and ranging results, achieving adaptive settings.

The remainder of this paper is organized as follows: Section 2 introduces the theoretical framework, including the round-trip time (RTT)-based ranging model and the propagation and attenuation of Wi-Fi signals. Section 3 introduces the proposed method, including the construction of the Gaussian model for identification and the iterative least squares algorithm. Section 4 describes the experimental results by the proposed method. Section 5 presents a conclusion and points out areas for future work.

2. Theoretical Framework

2.1. RTT-Based Ranging Model

Here, TOF echo technology is used in the Wi-Fi FTM protocol and makes the RTT based ranging more convenient [18]. This new technology determines the distance according to the RTT of a Wi-Fi FTM pulse from an MT to an FT. Accordingly, it can eliminate the time synchronization errors in the time of arrival (TOA)/time difference of arrival (TDOA)-based ranging, with no need to synchronize the times of the FT and MT. The whole procedure of ranging is shown in Figure 1. It shows that a mobile phone is used as an MT to send an FTM request to FTs supporting the FTM protocol. After an FT receives the request, an Acknowledgement (ACK) signal is sent from the FT to the MT. After that, several FTM feedback signals are sent from the FT to the MT, where the mean RTT can then be calculated. The time information can be expressed as follows:

$$t_{ACK} = t_{request} + \frac{2}{c}\|p_{MT} - p_{FT}\|_2 + t_D + e_{NLOS} \tag{1}$$

where $t_{request}$ and t_{ACK} are the times for when the request was sent from the MT and when the ACK signal arrives at MT; p_{MT} is the location of the MT when the FTM request is sent; p_{FT} is the location of the FT when the MT receives the FTM request; t_D is standard time deviation between the MT and the FT; e_{NLOS} is the time delay error caused by the NLOS effect; and c is the speed of light. Hence, the distance can be calculated by Equation (2):

$$\rho = \|p_{MT} - p_{FT}\|_2 = d_{RTT} - d_D - d_{NLOS} \tag{2}$$

where d_{RTT} is the ranging result; d_D is the ranging error caused by standard time deviation, as shown in Equation (3); and d_{NLOS} is the ranging error caused by the NLOS effect. The ranging error d_D not only includes a fixed time delay error, a device error, and a start error, but also is influenced by the distance of pulse propagation and other environmental factors [35], which can be expressed as follows:

$$d_D = e_n + \varphi(s) + \varepsilon \tag{3}$$

where e_n is the constant error caused by the fixed time delay error, the device error, and the start error; $\varphi(s)$ is the error formed by different ranging results, which can be seen as a polynomial function related to the ranging results; and ε is the system noise. It can be seen that the standard time deviation error of the equipment needs to be determined and calibrated to provide more precise ranging results.

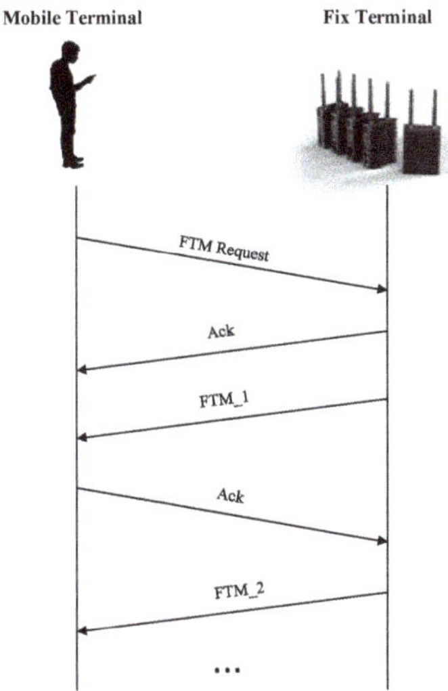

Figure 1. Procedure of Wi-Fi fine timing measurement (FTM).

2.2. Propagation of Wi-Fi Signals

With wave-particle duality, electromagnetic waves are the method of Wi-Fi propagation, which exhibit both particle properties and wave properties during propagation [36,37]. In a wide space, electromagnetic waves are only transmitted directly without obstacles. While in an indoor environment, the propagation characteristics will change due to the existence of various objects, causing reflection and diffraction, etc. In addition, electromagnetic waves have a certain penetration ability and can spread across walls. As a result, the actual propagation process of electromagnetic waves is varied and complex.

The propagation of electromagnetic waves follows the 'skin effect' [38]. That is, when there is an alternating current or alternating electromagnetic field in a conductor, the current distributes unevenly and concentrates in the 'skin' part of the conductor, that is, the current is concentrated on the thin layer of the outer surface of the conductor. As a result, the resistance of the conductor is increased, and its power loss is also increased. When an electromagnetic wave carrying a signal passes through some media, it is restricted by the skin effect, such that a part of the electromagnetic wave stays on the surface of the media and cannot completely pass through the media. The better the conductivity of the medium, the more obvious the effect of blocking in the electromagnetic wave is, and the more obvious the attenuation of the signal intensity is. Table 1 presents the signal attenuation of electromagnetic waves with a frequency of 2.4 GHz when crossing various common obstacles in indoor buildings. It can be obtained through observation that the better the conductivity of an obstacle is, the more obvious signal attenuation will be.

Table 1. Signal attenuation through different media.

Obstacle	Reinforced Concrete Wall	Plasterboard Wall	Wooden Door	Metal Door	Glass Window
Signal Attenuation Value (dBm)	15–16	3–5	3–5	10–12	6–8

Electromagnetic waves will wear out when propagating in air. The loss in free space is mainly caused by the air's own propagation and the scattering and refraction of fine dust particles. In the case of no obstacles in the air, the propagation of electromagnetic waves can be expressed as per Equation (4) [39], where d is the distance from the MT to the FT; d_0 is the reference distance; $L(d)$ is the received power of the MT when the distance from the FT is d; $L(d_0)$ is the received power when the distance from the FT to the MT is 1 m; η is the attenuation coefficient; and k is the shadowing factor, obeying a Gaussian distribution with a mean of 0. Since the model is in the case of LOS, k is 0.

$$L(d) = L(d_0) + 10 \cdot \eta \cdot lg(\frac{d}{d_0}) + k \quad (4)$$

3. The Proposed Indoor Localization Method

3.1. Overall Structure of the Proposed Method

Figure 2 shows the procedures of the proposed positioning method. The ranging results and RSSIs are collected from the FTs and then input into the LOS/NLOS identification model. Then, the model identifies NLOS signals and only outputs the LOS ranging results. If the number of LOS ranging results is greater than the minimum number required for positioning, the LOS ranging results are used as the input of the single point positioning to calculate the location of the MT. If the number of LOS ranging results is less than the minimum number, all the ranging results, including LOS and NLOS, will be used as the input of the single point positioning to derive the location.

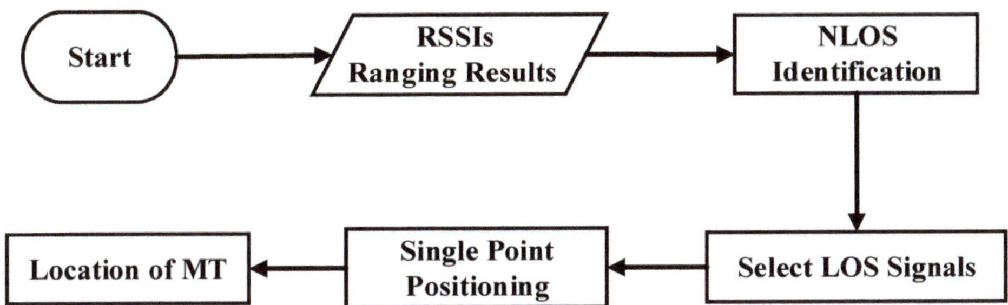

Figure 2. Flow chart of the proposed positioning method. RSSIs, received signal strength indicators; NLOS, non-line-of-sight; LOS, line-of-sight; MT, mobile terminal.

3.2. The Identification Model of LOS/NLOS

Figure 3 shows the ranging results (the red solid line and the dotted line) and RSSIs (the blue solid line and the dotted line) collected at different distances from FTs under both NLOS and LOS conditions. Here, we have placed Wi-Fi devices in an open corridor and in a closed room, then collected RSSIs and ranging results at 1 m intervals in the corridor. The sampling time was 30 s, the sampling frequency was 5 Hz, and the sampling device was a Pixel 3 mobile phone. The solid line denotes the results from a corridor and the dotted line denotes the results from a closed room. It can be seen that the LOS and NLOS ranging results are similar at any distance, but the RSSIs are quite different. This is due to the principle of electromagnetic wave propagation. RSSIs mainly reflect the amplitude of the electromagnetic wave. When electromagnetic waves travel through walls, some particles will not pass through walls, and the amplitude will significantly reduce, such that the RSSI will also be reduced [38]. The ranging results mainly reflect the time of propagation. When the electromagnetic waves pass through the wall, their speed will decrease. However, due to the small thickness of the wall and the high electromagnetic velocity, the impact on the time of propagation and ranging result is small. Under such conditions, the ranging results are not obvious when compared with those of an RSSI. The LOS/NLOS identification model was designed based on this phenomenon.

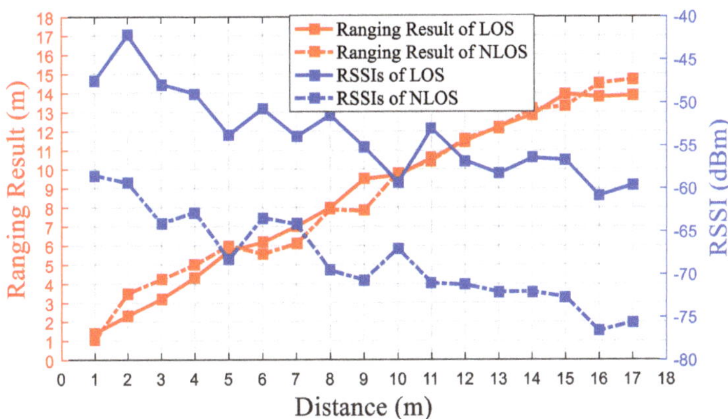

Figure 3. Ranging results and the received signal strength indicators (RSSIs) of line-of-sight (LOS/NLOS) signals.

According to Section 2.2, the attenuation of the RSSIs was greatly affected by the conductivity of obstacles. In the complex indoor environment, it is difficult to model the relationship between the distance and RSSI. Therefore, differing from the literature [24,25,40], where two models have been built under LOS and NLOS conditions, respectively, this paper's identification model was built in the line-of-sight condition, which could output probability with a large difference according to the condition of the signal.

Most existing works [41–43] use a Gaussian distribution to model the randomness of RSSI, and consequently, in this paper, a Gaussian model was established to calculate the probability of LOS by Equation (5), as shown in Figure 4. The parameters of the model were determined by ranging results, and the inputs of the model were the RSSIs of the FTs. According to the maximum likelihood method, μ is the mean of the RSSI and σ is the variance of the RSSI in Equations (6) and (7), where $RSSI_d^i$ is the RSSI collected at the distance d; N is the number of RSSI collected. The determination of μ and σ is described below in detail. After the parameters were determined, the probability could be determined by inputting the RSSI into the Gaussian model. Then, the probability was compared with the threshold θ. If the probability is greater than the threshold, the current signal is a LOS signal, and if it is less than, it is not.

$$P_{LOS}(RSSI) = \frac{A}{\sqrt{2\pi}\sigma} * exp\left(-\frac{(RSSI-\mu)^2}{2\sigma^2}\right) \tag{5}$$

$$\mu = \frac{1}{N}\sum_{i=1}^{N} RSSI_d{}^{(i)} \tag{6}$$

$$\sigma^2 = \frac{1}{N}\sum_{i=1}^{N} \left(RSSI_d{}^{(i)} - \overline{RSSI_d}\right)^2 \tag{7}$$

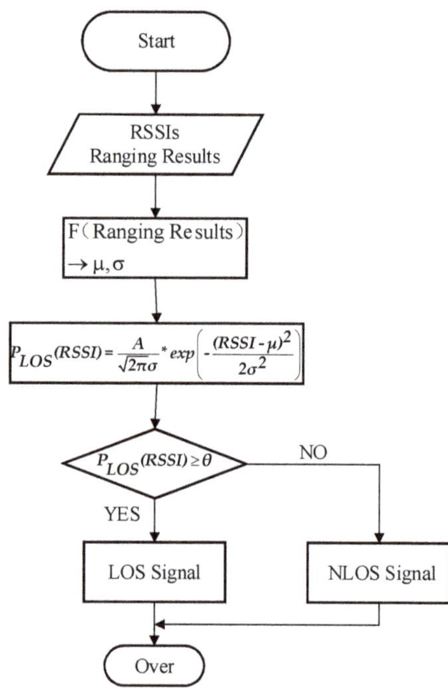

Figure 4. Flow chart of the LOS/NLOS identification process.

3.2.1. The Determination of d

In the paper, it was assumed that the ranging results could reflect the true distance to a certain extent. Therefore, we needed to evaluate the accuracy of the ranging results based on Wi-Fi RTT.

As shown in Figure 5, a tester collected the ranging results of the FTs at each grid point. Here, the grid side length was 1 m, the sampling time was 30 s, and the sampling frequency was 5 Hz. The MT and FTs were basically at the same height during sampling. The ranging accuracy is shown in Figure 6. The red and blue solid lines are the ranging errors of the LOS signals and NLOS signals, respectively. Through calculation, the average ranging error of LOS signals was 0.87 m and the variance was 0.36 m. The average ranging error of the NLOS signals was 1.16 m and the variance was 0.54m. It can be seen that RTT-based ranging results can accurately reflect the true distance between the transceivers. At the same time, the ranging accuracy and stability of the LOS signals were better than those of the NLOS signals.

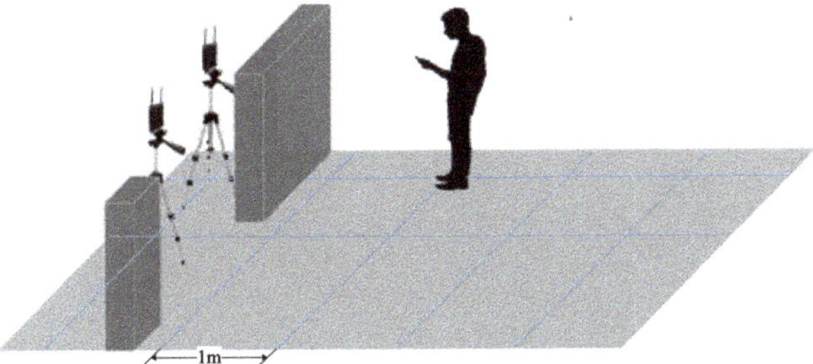

Figure 5. Experiment for estimating the accuracy of the ranging results.

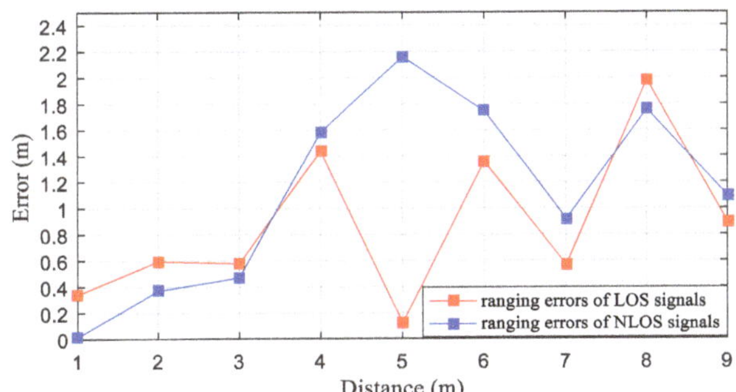

Figure 6. Ranging errors of the LOS/NLOS signals.

3.2.2. The Determination of μ

μ is the mean of RSSIs collected when the distance between an MT and FT is determined. According to the propagation model of electromagnetic waves, in Equation (5), the true distance is exponential with RSSI, so the exponential model was used to fit the ranging results and RSSIs.

A Wi-Fi device was arranged at a fixed point as the FT. The experimenter walked in a straight line with the mobile phone within 23 m from the FT. One thousand, eight hundred groups of ranging results and RSSIs were collected and used as training data. The collected ranging results and RSSIs were used as training data. The fitting of the single exponential model is shown in Figure 7a. It can be seen that fitting over 15 m performs poorly. Here, the mean square error (MSE) was used to evaluate the fitting, calculated as per Equation (8), where N is the number of RSSI used for the fitting; $RSSI^{(i)}$ is the i-th RSSI for the fitting; and $RSSI_{fit}^{(i)}$ is the i-th fitting result. By calculating the MSE of the fitting from 0 m to a different distance, as shown in Figure 7b, it was found that within 0–15 m, as the distance increases, the training data increase, and the MSE decreases. Between 15 and 20 m, the MSE becomes larger, and the fitting effect is not good. Over 20 m, the MSE tends to be stable. It can be concluded that the RSSIs collected at the distance over 15 m have a different tendency from those collected within 15 m.

Therefore, the ranging result RSSI model was changed into a double exponential model to calculate μ, as shown in Equation (9). The data in the range of 0–15 m and 15–25 m were fitted, respectively. The result of fitting using the double exponential model is shown in Figure 7c. The MSE of the fitting, using the double exponential from 0 m to different distances, is shown in Figure 7d. It can be seen that the MSE at different distances is lower than that of the single exponential, except at a distance of 4 m. The average MSE of the single exponential model was 4.85 dBm^2, and that of the improved double exponential model was 4.15 dBm^2. Therefore, the double exponential model is more in line with the ranging result–RSSI relationship.

$$MSE = \frac{1}{N}\sum_{i=1}^{N}(RSSI^{(i)} - RSSI_{fit}^{(i)}) \tag{8}$$

$$\mu = \begin{cases} a_1 + b_1 * lg(d) & d < 15m \\ a_2 + b_2 * lg(d) & d \geq 15m \end{cases} \tag{9}$$

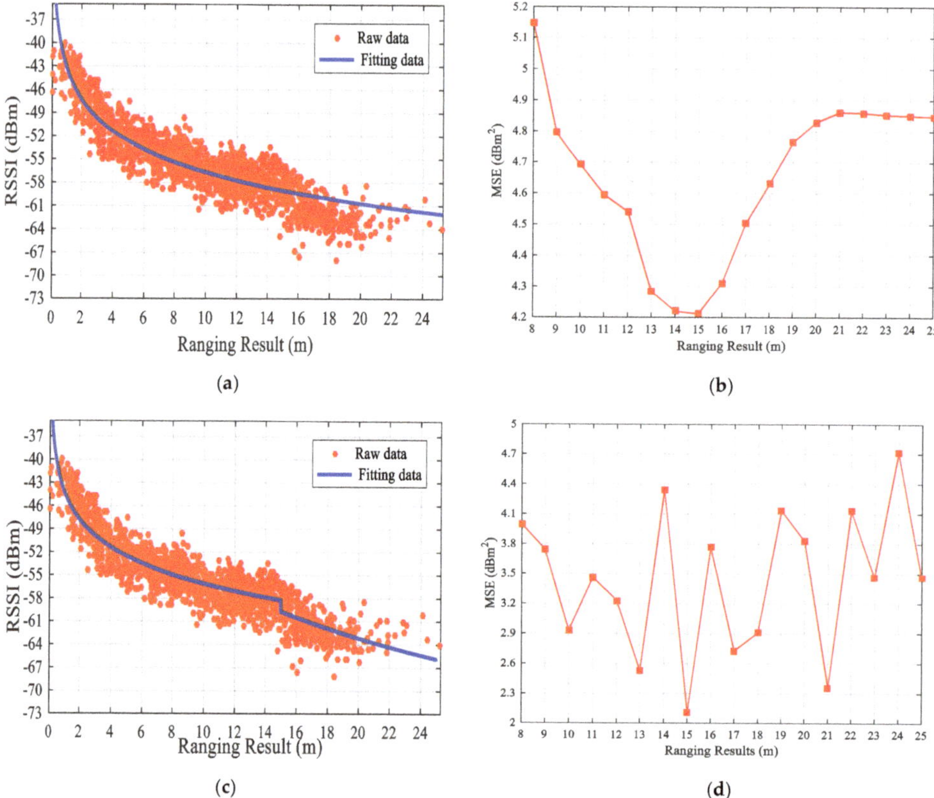

Figure 7. Experimental results of the fitting. (**a**) Comparison of the fitting data using the single exponential model and the raw RSSI data. (**b**) Mean square error (MSE) of the fitting, from 0 m to different distances, using the single exponential model. (**c**) Comparison of the fitting of data using the double exponential model and the raw RSSI data. (**d**) MSE of the fitting, from 0 m to different distances, using the single exponential model.

3.2.3. The Determination of σ

σ is the variance of RSSIs collected at a fixed point. In the literature [24,25], the variances of RSSIs collected at different distances have traditionally been set as a fixed value. According to the analysis in the second section, we assumed that the variances of RSSIs at different distances were different. For this reason, the variances were calculated, as shown in Table 2. It can be seen that the variance decreased by about 1 dBm^2 from 0 to 6 m, while it nearly did not change from 6 to 18 m. According to the trend of change, the exponential function was used to fit the variances and ranging results. To verify if the function was suitable, the sum of squares due to error (SSE) and coefficient of determination (R-square) of the fitting result was compared with that found by other functions, including the Fourier, Gaussian, linear fitting, polynomial, rational, and sum of sine functions, shown in the Table 3. SSE and R-square can be calculated by Equation (10) and Equation (11), where $\overline{RSSI}$ is the average of RSSI. The smaller the SSE or the bigger the R-square, the better the fitting results will be. It can be seen that from both the SSE and R-square that the exponential function is more suitable.

$$SSE = \sum_{i=1}^{N} (RSSI^{(i)} - RSSI_{fit}^{(i)}) \tag{10}$$

$$R - square = \frac{\sum_{i=1}^{N} (\overline{RSSI} - RSSI_{fit}^{(i)})}{\sum_{i=1}^{N} (\overline{RSSI} - RSSI^{(i)})} \tag{11}$$

Table 2. Variance of RSSIs at different distances.

Distance Range (m)	0–3	3–6	6–9	9–12	12–15	15–18
σ (dBm)	3.48	2.34	1.66	1.69	1.74	1.74

Table 3. Sum of squares due to error (SSE) and coefficient of determination (R-square) values of the different fitting functions.

Fitting Function	SSE (dBm^2)	R-Square Value
Our Model	72.77	0.48
Fourier	85.70	0.40
Linear Fitting	99.85	0.30
Polynomial	115.13	0.19
Rational	92.79	0.35
Sum of Sine	115.14	0.19

For the fitting, a large amount of RSSIs with the same ranging results should be obtained. However, due to different ranging results, a credible variance was hard to achieve. The ranging results were divided into several 0.5 m-long ranges. Next, all the training data were put into the corresponding range according to the ranging results. Then, the average of the ranging results and the variances of RSSIs in each range were calculated and used as the input of the fitting. The variances were calculated by the following two methods:

(a) The variances of RSSIs were directly calculated, according to Equation (7).

(b) All the RSSIs were counted in a certain interval and the probabilities of each RSSI were calculated. Then, the Gaussian regression was performed on the probability and the corresponding RSSI. σ of the Gaussian model was taken as the σ of the current distance.

After the σ values were determined, the ranging results and σ values were fitted to generate a ranging result-based σ model. Figure 8 shows the fitting results of methods A and B. The red dot is the variance calculated by method A, while the blue point is the variance calculated by the other method. The fitting results and raw data are shown in Figure 8a, and the residuals of the two methods are shown in Figure 8b. It can be seen that variances calculated by method B fluctuate largely, which might be caused by over-fitting. In the range of 0–4 m, the variance of method B is higher than 15. Table 1 shows that the attenuation of RSSI caused by the reinforced concrete is about 15 dBm, so that when the variance is higher than 15, the output probability of LOS and NLOS is similar, which may make the set of threshold and identification difficult. Furthermore, as shown in Figure 8b, the fitting result of method A is better than method B, as the sum of square due to error (SSE) of method A is much smaller than that of method B. Hence, method A was used to calculate the variance used for the later fitting.

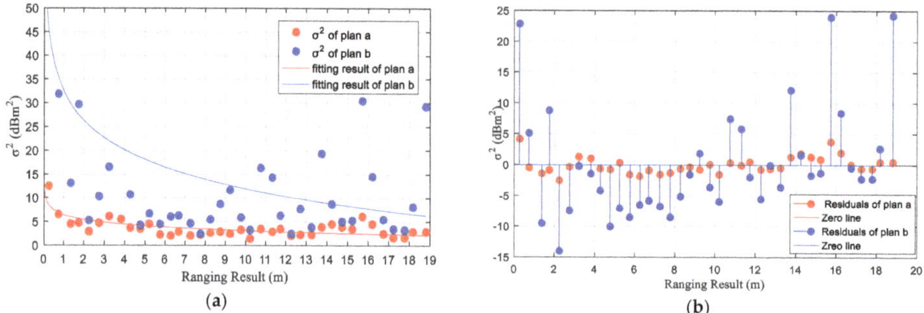

Figure 8. Fitting results of the two methods. (**a**) Fitting result and the raw data of methods A and B. (**b**) Residuals of the fitting plans of methods A and B.

3.3. Single Point Positioning

If the number of the ranging results was more than 2 after filtering, the LS algorithm was used for the single point positioning. The observation equation for the ranging results can be expressed as follows:

$$d_{RTT}^{i} = \rho^{i} + d_{D}^{i} + d_{NLOS}^{i} \tag{12}$$

In the equation, d_{RTT}^{i} is the ranging result collected from the i-th FT, ρ^{i} is the Euclidean distance between the i-th FT and the MT, d_{NLOS}^{i} is the ranging error caused by the NLOS error from the i-th FT; and d is the ranging error caused by the standard time deviation. According to Equation (12), we can derive the error observation as follows:

$$V^{i} = \rho^{i} + d_{D}^{i} + d_{NLOS}^{i} - d_{RTT}^{i} \tag{13}$$

Assuming that the approximate coordinate of MT is X = (X_0, Y_0) and that the Euclidean distance from the coordinate of the i-th FT to the approximate coordinate of MT is ρ0, the Taylor series of the above formula is:

$$V^{i} = \rho_{0}^{i} + \frac{X_0 - X_{FT}^{i}}{\rho_{0}^{i}} dX + \frac{Y_0 - Y_{FT}^{i}}{\rho_{0}^{i}} dY + d_{D}^{i} + d_{NLOS}^{i} - d_{RTT}^{i} \tag{14}$$

If we set the following:

$$L^{i} = d_{RTT}^{i} - \rho_0 - d_{D}^{i} - d_{NLOS}^{i} \tag{15}$$

Then the equation could be changed as follows:

$$V^{i} = \frac{X_0 - X_{FT}^{i}}{\rho_{0}^{i}} dX + \frac{Y_0 - Y_{FT}^{i}}{\rho_{0}^{i}} dY - L^{i} \tag{16}$$

where dX and dY are the corrections of the approximate coordinates of MT, and X^i_{FT} and Y^i_{FT} are the two coordinate values of the i-th FT. Thus, Equation (14) can be expressed in a matrix form as follows:

$$\underset{m\times 1}{V} = \underset{m\times 2}{A} \times \underset{2\times 1}{\Delta} \times \underset{m\times 1}{L} \tag{17}$$

where m is the number of ranging results after filtering; V is the vector of the correction of ranging results; L is observation error constant matrix; Δ is an unknown parameter in Equation (15); and A is the cosine coefficient matrix in Equation (19).

$$\Delta = [dX, dY] \tag{18}$$

$$A^i = [\frac{X_0 - X^i_{FT}}{\rho_0}, \frac{Y_0 - Y^i_{FT}}{\rho_0}] \tag{19}$$

The error observation equation can be established using the ranging results from different FTs, and then the parameters of the coordinate can be obtained by solving the equation with the least square method, as shown in Equation (20).

$$\Delta = (A^T A)^{-1} \times A^T \times L \tag{20}$$

The pseudocode of single point positioning is shown in Figure 9. The observation equation was constructed using the inputs to calculate Δ. Through iteration, Δ gradually approached 0, where the approximate coordinate of MT was gradually steady. When Δ was smaller than a threshold or the iteration number was more than a threshold, the iteration could meet the condition of stopping.

```
Input    :Distance matrix  D=[d^1_RTT, d^2_RTT, ..., d^m_RTT] , whose elements d^i_RTT is ranging results from MT to FTs
          Coordinates matrix  C={c^1, c^2, ..., c^m} , whose elements c^i = (X^i_FT, Y^i_FT)  is the coordinates of FTs
          Initial approximate coordinate  x=[X_0, Y_0]
Process  :
          repeat
              for all d^i_RTT ∈ D
                  construct the approximate distance matrix  R=[ρ^1_0, ρ^2_0, ..., ρ^m_0],  ρ^i_0 = ||c^i − x||_2 ;
                  construct the constant matrix L, according to the equation (4);
                  construct the cosine coefficient matrix A , according to the equation (6) ;
              end for
              calculate the unknown parameter Δ , according to the equation (7);
              update the approximate coordinare  x=x+Δ ;
          until  meet the condition of stop
Output   :x, the final approximate coordinate of MT
```

Figure 9. Single point positioning algorithm.

4. Experiment

4.1. Evaluation of the Identification Model

The performance of the identification model was evaluated. A Wi-Fi device was set in the corridor, and the experimenter walked at a constant speed in the corridor and room separately with a mobile phone to collect the RSSIs and ranging results. The data collected in the corridor were seen as LOS data, while those collected in the room were seen as NLOS data. A group of data consisted of an RSSI and a ranging result. A total of 8360 groups of data were collected, among which 1800 groups were used for training and 6560 groups were used for testing. The training data acquisition time was only one day away from that of the testing data, and the training data did not participate in the test.

The identification of NLOS/LOS was a binary problem, so precision and recall were utilized to evaluate the model. They were calculated by Equations (21) and (22), respectively, where the TP (true positive) is the number of LOS signals that were accurately identified, FP (false positive) refers to the number of NLOS signals identified as LOS signals, and FN (false negative) points to the number of LOS signals identified as NLOS signals. All the data were divided into different ranges according to the ranging results in order to evaluate the precision and recall at different distances. Here, 3280 groups of LOS data and 3280 groups of NLOS data were used to test the precision, while 3280 groups of LOS data were used to test the recall. The model output the LOS/NLOS recognition results, then TP, FP, and FN were counted to calculate the precision and recall of each interval. In this experiment, the threshold θ of the model was set as 0.7.

$$P = \frac{TP}{TP + FP} \tag{21}$$

$$R = \frac{TP}{TP + FN} \tag{22}$$

Figure 10 shows the precision at different ranges. Through calculation, the average precision was 97.08% and the MSE was 0.07. Figure 11 shows the cumulative distribution of precision. All precisions were higher than 70%, 9.09% of which were lower than 85%, and 15.15% of which were lower than 95%. It can be seen that the precision of the model is poor at some ranges, which might be caused by the poor fitting of the ranging result and RSSI.

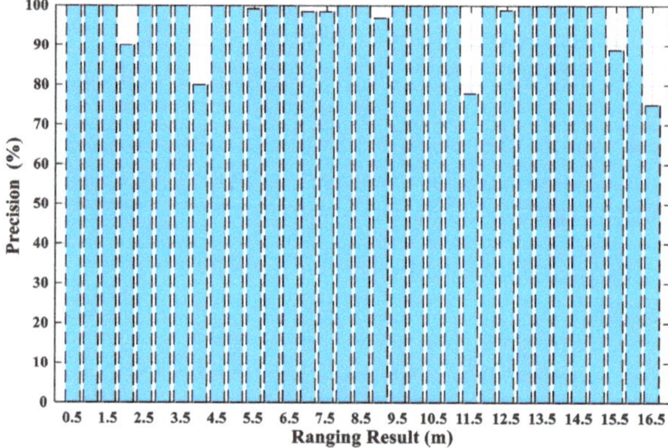

Figure 10. Precision at different ranging results.

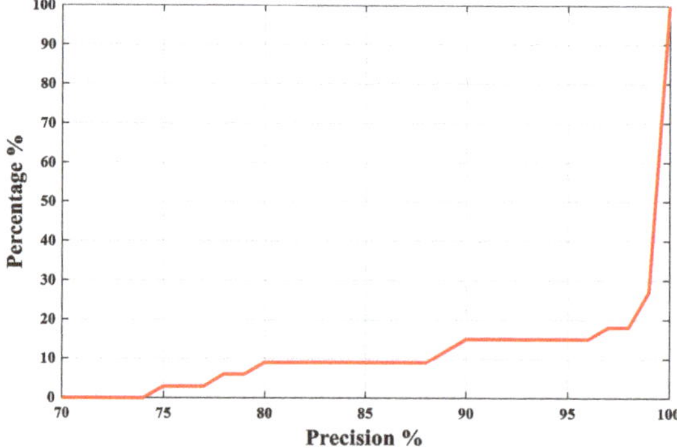

Figure 11. Cumulative distribution function of the precisions.

The recalls and mean probabilities at different ranges are shown in Figure 12, where the value of the blue column is roughly the same as that of the red line. It can be seen that the set of θ could distinguish LOS signals and NLOS signals. However, through calculation, the average recall was only 74.97% and the MSE was 0.2. Figure 13 shows the cumulative distribution of recalls. Here, 24.24% of the recalls were lower than 60%, and 51.52% of them were lower than 80%, indicating that the model often misjudges some LOS signals.

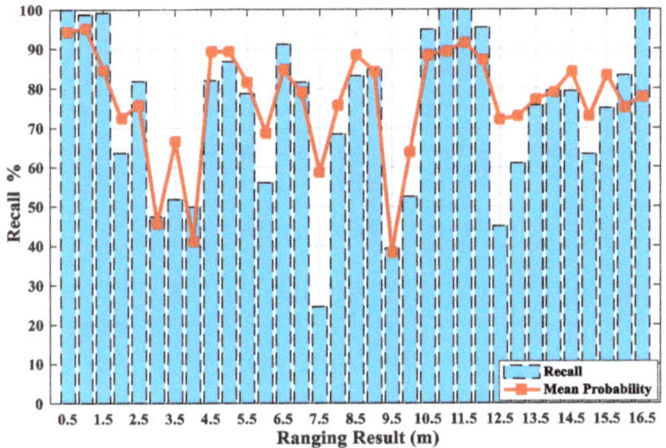

Figure 12. Recall of different ranging results.

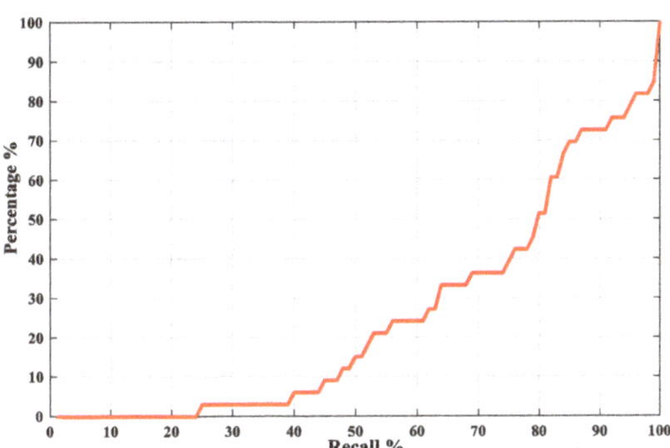

Figure 13. Cumulative distribution function of precisions.

4.2. Evaluation of Proposed Positioning Method

The experimental environment is shown in Figure 14, which includes two rooms and a corridor adjacent to each other. Both of the rooms were 19.05 m long and 5.83 m wide, while the corridor was 19.51 m long and 1.74 m wide. Totally, 10 FTs were positioned on a tripod with a height of 1.3 m. Among them, 8 were placed in two rooms on average, where 3 were positioned in the corridor, marked with a red five-pointed star. Here, 197 testing points were set, and marked with blue squares, with an interval of 1.2 m. The sampling time of each point was 30 s and the sampling frequency was 1 Hz. Two experiments were conducted to evaluate the proposed positioning method.

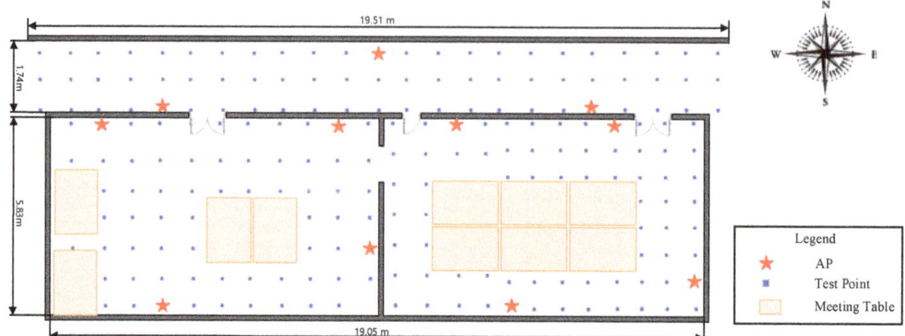

Figure 14. Experimental area.

In the first experiment, the performance of the method was evaluated when a different number of FTs were set in the NLOS condition. A different number of FTs, from 0 to 7, were set outside the eastern room in turn. The ranging results and RSSIs from all FTs were collected at 67 test points in the eastern room. Then, the data were imported into MATLAB. The location was calculated using single point positioning with identification and discarding (SPP-IAD) NLOS, and that of single point positioning (SPP).

Figure 15 shows the accuracy of the two methods. In the absence of NLOS FTs, the error of SPP-IAD is greater than that of SPP. This is because the identification model discards some ranging results from LOS FTs. However, when adding NLOS FTs, the error of SPP-IAD is smaller than SPP. As the number of NLOS FTs increases, the accuracy of SPP-IAD is basically unchanged, and the error of SPP increases. Hence, the accuracy of SPP-IAD is better than SPP when NLOS FTs are present.

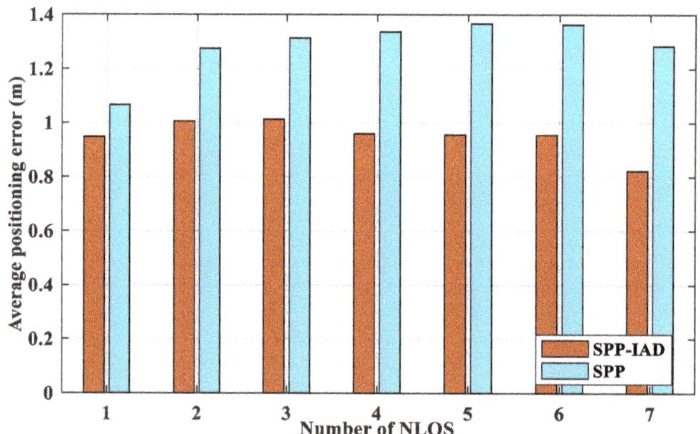

Figure 15. Average positioning error with different numbers of NLOS signals.

In the second experiment, the ranging results and RSSIs at all the test points were collected and the proposed method was used to carry out the positioning. Here, the maximum error (ME), error mean (EM), and root mean square error (RMSE) were used as indicators of the method.

Figure 16 shows the EM of the SPP-IAD and SPP at each test point. Among them, 18 EMs of SPP-IAD are larger than that of the SPP, and the maximum difference is 0.3 m.

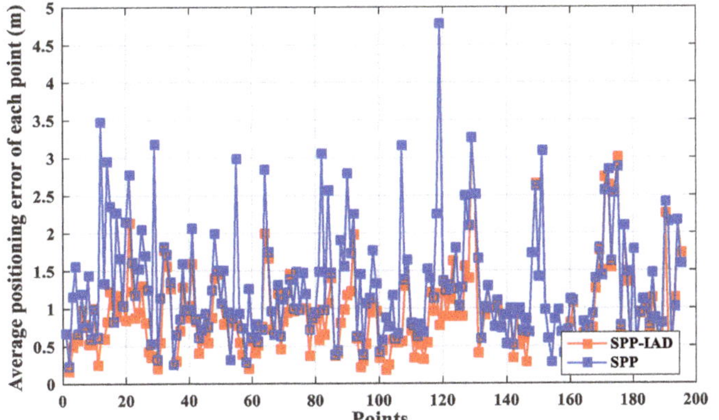

Figure 16. Average positioning error of each testing point.

As shown in Table 4, the SPP-IAD was implemented with a ME of 6.862 m, an EM of 0.932 m, and an RMSE of 0.712 m, while the other has greater error than the proposed method. Compared with SPP, it achieves a ME improvement of 1.041 m (13.17%), an EM improvement of 0.337 m (36.04%), and an RMSE improvement of 0.456 m (64.04%).

Table 4. Statistical results of positioning errors. ME, maximum error; EM, error mean; RMSE, root mean square error.

Method	ME (m)	EM (m)	RMSE (m)
SPP-IAD	6.862	0.932	0.712
SPP	7.903	1.272	1.268

The Cumulative Distribution Functions (CDFs) of the positioning errors of SPP-IAD and SPP are shown in Figure 17. Obviously, the positioning errors of SPP-IAD are smaller than those of SPP with different y-values.

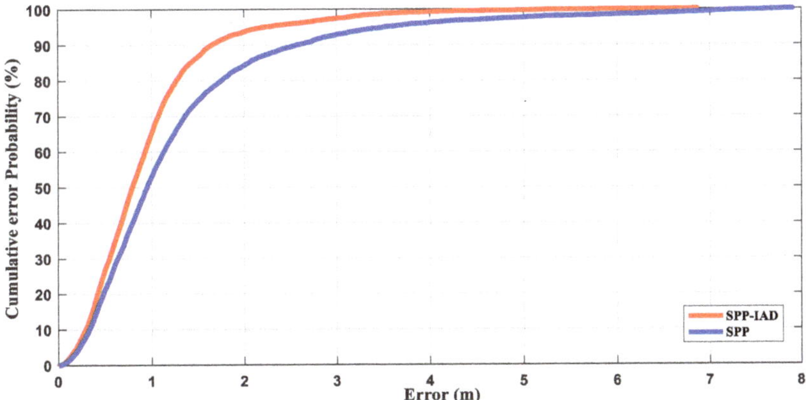

Figure 17. Cumulative probability distribution of positioning errors related to single point positioning with identification and discarding (SPP-IAD) and single point positioning (SPP).

From the two experiments, it can be concluded that the performance of SPP-IAD is better than that of SPP, both in the positioning accuracy and stability, under the condition NLOS FTs.

5. Discussion

In order to reduce the positioning errors caused by NLOS, this paper has proposed a LOS/NLOS identification model which is suitable for complex indoor environments, and the parameters of the model have been adaptively determined. After identifying and discarding NLOS ranging results, the LS-based single point positioning algorithm was used. It has been deduced that the average precision of the model was 97.08% and the average recall was 74.97%. The average error of the positioning results was 0.935 m. The accuracy found here is better than that found without identification and discarded in the case of a different number of NLOS FTs. In conclusion, the proposed method can eliminate some NLOS ranging results and effectively improve the positioning accuracy. However, the recall of the model is not high, which may cause some LOS ranging results to be eliminated during positioning and affect the positioning accuracy. In order to solve this problem, map matching [44,45] can be added to the proposed method to reduce misjudgment.

Author Contributions: Conceptualization, M.S.; Formal analysis, M.S. and S.X.; Funding acquisition, Y.W.; Methodology, M.S.; Project administration, Y.W.; Data curation, H.C. and M.S.; Writing—original draft preparation, M.S. All authors have read and agreed to the published version of the manuscript.

Funding: This research was funded by the National Key Research and Development Program of China under grant number 2016YFB0502102.

Acknowledgments: In this section you can acknowledge any support given which is not covered by the author contribution or funding sections. This may include administrative and technical support, or donations in kind (e.g., materials used for experiments).

Conflicts of Interest: The authors declare no conflict of interest.

References

1. Bi, J.; Wang, Y.; Li, X.; Qi, H.; Cao, H.; Xu, S. An Adaptive Weighted KNN Positioning Method Based on Omnidirectional Fingerprint Database and Twice Affinity Propagation Clustering. *Sensors* **2018**, *18*, 2502. [CrossRef] [PubMed]
2. Doiphode, S.R.; Bakal, J.W.; Gedam, M. A hybrid indoor positioning system based on Wi-Fi hotspot and Wi-Fi fixed nodes. In Proceedings of the 2016 IEEE International Conference on Engineering and Technology (ICETECH), Coimbatore, India, 17–18 March 2016; pp. 56–60.
3. Cao, H.; Wang, Y.; Bi, J.; Qi, H. An Adaptive Bluetooth/Wi-Fi Fingerprint Positioning Method based on Gaussian Process Regression and Relative Distance. *Sensors* **2019**, *19*, 2784. [CrossRef] [PubMed]
4. Jiang, H.; Peng, C.; Sun, J. Deep Belief Network for Fingerprinting-Based RFID Indoor Localization. In Proceedings of the ICC 2019—2019 IEEE International Conference on Communications (ICC), Shanghai, China, 20–24 May 2019; pp. 1–5.
5. Sevrin, L.; Noury, N.; Abouchi, N.; Jumel, F.; Massot, B.; Saraydaryan, J. Characterization of a multi-user indoor positioning system based on low cost depth vision (Kinect) for monitoring human activity in a smart home. In Proceedings of the 2015 37th Annual International Conference of the IEEE Engineering in Medicine and Biology Society (EMBC), Milan, Italy, 25–29 August 2015; pp. 5003–5007.
6. Tian, Q.; Wang, K.I.; Salcic, Z. A resetting approach for INS and UWB sensor fusion using Particle Filter for pedestrian tracking. *IEEE Trans. Instrum. Meas.* **2019**. [CrossRef]
7. Cahyadi, W.A.; Chung, Y.H.; Adiono, T. Infrared Indoor Positioning Using Invisible Beacon. In Proceedings of the 2019 Eleventh International Conference on Ubiquitous and Future Networks (ICUFN), Zagreb, Croatia, 2–5 July 2019; pp. 341–345.
8. Fan, Q.; Sun, B.; Sun, Y.; Zhuang, X. Performance Enhancement of MEMS-Based INS/UWB Integration for Indoor Navigation Applications. *IEEE Sensors J.* **2017**, *17*, 3116–3130. [CrossRef]
9. Smidla, J.; Simon, G. Accelerometer-based event detector for low-power applications. *Sensors* **2013**, *13*, 13978–13997. [CrossRef]
10. Zhuang, Y.; Hua, L.; Qi, L.; Yang, J.; Cao, P.; Cao, Y.; Wu, Y.; Thompson, J.; Haas, H. A Survey of Positioning Systems Using Visible LED Lights. *IEEE Commun. Surv. Tutor.* **2018**, *20*, 1963–1988. [CrossRef]
11. Simon, G.; Zachár, G.; Vakulya, G. Lookup: Robust and Accurate Indoor Localization Using Visible Light Communication. *IEEE Trans. Instrum. Meas.* **2017**, *66*, 2337–2348. [CrossRef]

12. Bhattarai, B.; Yadav, R.K.; Gang, H.; Pyun, J. Geomagnetic Field Based Indoor Landmark Classification Using Deep Learning. *IEEE Access* **2019**, *7*, 33943–33956. [CrossRef]
13. Sun, M.; Wang, Y.; Xu, S.; Cao, H.; Si, M. Indoor Positioning Integrating PDR/Geomagnetic Positioning Based on the Genetic-Particle Filter. *Appl. Sci.* **2020**, *10*, 668. [CrossRef]
14. Fujii, K.; Yonezawa, R.; Sakamoto, Y.; Schmitz, A.; Sugano, S. A combined approach of Doppler and carrier-based hyperbolic positioning with a multi-channel GPS-pseudolite for indoor localization of robots. In Proceedings of the 2016 International Conference on Indoor Positioning and Indoor Navigation (IPIN), Alcala de Henares, Spain, 4–7 October 2016; pp. 1–7.
15. Barnes, J.; Rizos, C.; Wang, J.; Small, D.; Voigt, G.; Gambale, N. Locata: A new positioning technology for high precision indoor and outdoor positioning. In Proceedings of the 2003 International Symposium on GPS\GNSS, Portland, OR, USA, 9–12 September 2003; pp. 9–18.
16. Davidson, P.; Piche, R. A Survey of Selected Indoor Positioning Methods for Smartphones. *IEEE Commun. Surv. Tutor.* **2017**, *19*, 1347–1370. [CrossRef]
17. IEEE Standard for Information technology—Telecommunications and information exchange between systems Local and metropolitan area networks—Specific requirements—Part 11: Wireless LAN Medium Access Control (MAC) and Physical Layer (PHY) Specifications. *IEEE Std 802.11-2016 (Revision of IEEE Std 802.11-2012)* **2016**, 1–3534. [CrossRef]
18. Rea, M.; Fakhreddine, A.; Giustiniano, D.; Lenders, V. Filtering Noisy 802.11 Time-of-Flight Ranging Measurements from Commoditized WiFi Radios. *IEEE/ACM Trans. Netw.* **2017**, *25*, 2514–2527. [CrossRef]
19. Dai, P.; Yang, Y.; Zhang, C.; Bao, X.; Zhang, H.; Zhang, Y. Analysis of Target Detection Based on UWB NLOS Ranging Modeling. In Proceedings of the 2018 Ubiquitous Positioning, Indoor Navigation and Location-Based Services (UPINLBS), Wuhan, China, 22–23 March 2018; pp. 1–6.
20. Gao, D.; Li, A.; Fu, J. Analysis of positioning performance of UWB system in metal NLOS environment. In Proceedings of the 2018 Chinese Automation Congress (CAC), Xi'an, China, 30 November–2 December 2018; pp. 600–604.
21. Kbayer, N.; Sahmoudi, M. Performances Analysis of GNSS NLOS Bias Correction in Urban Environment Using a Three-Dimensional City Model and GNSS Simulator. *IEEE Trans. Aerosp. Electron. Syst.* **2018**, *54*, 1799–1814. [CrossRef]
22. Egea-Roca, D.; Seco-Granados, G.; López-Salcedo, J.A. Transient change detection for LOS and NLOS discrimination at GNSS signal processing level. In Proceedings of the 2016 International Conference on Localization and GNSS (ICL-GNSS), Barcelona, Spain, 28–30 June 2016; pp. 1–6.
23. Yang, X. NLOS Mitigation for UWB Localization Based on Sparse Pseudo-Input Gaussian Process. *IEEE Sens. J.* **2018**, *18*, 4311–4316. [CrossRef]
24. Borras, J.; Hatrack, P.; Mandayam, N.B. Decision theoretic framework for NLOS identification. In Proceedings of the VTC '98. 48th IEEE Vehicular Technology Conference, Pathway to Global Wireless Revolution, Ottawa, ON, USA, 21–21 May 1998; pp. 1583–1587.
25. Han, K.; Xing, H.; Deng, Z.; Du, Y. A RSSI/PDR-Based Probabilistic Position Selection Algorithm with NLOS Identification for Indoor Localisation. *ISPRS Int. J. Geo-Inf.* **2018**, *7*, 232. [CrossRef]
26. Jo, H.; Kim, S. Indoor Smartphone Localization Based on LOS and NLOS Identification. *Sensors* **2018**, *18*, 3987. [CrossRef]
27. CHEN, P.C. A non-line-of-sight error mitigation algorithm in location estimation. In Proceedings of the IEEE Wireless Communications and Networking Conference, New Orleans, LA, USA, 21–24 September 2000; pp. 1525–3511.
28. Casas, R.; Marco, A.; Guerrero, J.J.; Falc, J. Robust estimator for non-line-of-sight error mitigation in indoor localization. *EURASIP J. Adv. Signal Process* **2006**, *2006*, 156. [CrossRef]
29. Chao-Lin, C.; Kai-Ten, F. An efficient geometry-constrained location estimation algorithm for NLOS environments. In Proceedings of the 2005 International Conference on Wireless Networks, Communications and Mobile Computing, Maui, HI, USA, 13–16 June 2005; Volume 241, pp. 244–249.
30. Li, C.; Weihua, Z. Nonline-of-sight error mitigation in mobile location. *IEEE Trans. Wirel. Commun.* **2005**, *4*, 560–573. [CrossRef]
31. Riba, J.; Urruela, A. A non-line-of-sight mitigation technique based on ML-detection. In Proceedings of the 2004 IEEE International Conference on Acoustics, Speech, and Signal Processing, Montreal, QC, Canada, 17–21 May 2004; pp. 153–156.

32. Venkatesh, S.; Buehrer, R.M. A linear programming approach to NLOS error mitigation in sensor networks. In Proceedings of the 5th International Conference on Information Processing in Sensor Networks, Nashville, TN, USA, 19–21 April 2006; pp. 301–308.
33. Xin, W.; Zongxin, W.; Dea, B.O. A TOA-based location algorithm reducing the errors due to non-line-of-sight (NLOS) propagation. *IEEE Trans. Veh. Technol.* **2003**, *52*, 112–116. [CrossRef]
34. Chan, Y.T.; Hang, H.Y.C.; Ching, P.C. Exact and approximate maximum likelihood localization algorithms. *IEEE Trans. Veh. Technol.* **2006**, *55*, 10–16. [CrossRef]
35. Zhang, R.; Hoflinger, F.; Reindl, L. Inertial Sensor Based Indoor Localization and Monitoring System for Emergency Responders. *IEEE Sens. J.* **2013**, *13*, 838–848. [CrossRef]
36. Tonomura, A.; Endo, J.; Matsuda, T.; Kawasaki, T.; Ezawa, H. Demonstration of single-electron buildup of an interference pattern. *Am. J. Phys.* **1989**, *57*, 117–120. [CrossRef]
37. Schmidt, H.T.; Fischer, D.; Berenyi, Z.; Cocke, C.L.; Gudmundsson, M.; Haag, N.; Johansson, H.A.; Källberg, A.; Levin, S.B.; Reinhed, P. Evidence of wave-particle duality for single fast hydrogen atoms. *Phys. Rev. Lett.* **2008**, *101*, 083201. [CrossRef] [PubMed]
38. Andersson, R.; Syk, M. Electromagnetic pulse forming of carbon steel sheet metal. In Proceedings of the 3rd International Conference on High Speed Forming, Dortmund, Germany, 1–6 June 2008; pp. 249–260.
39. Molisch, A.F.; Cassioli, D.; Chong, C.C.; Emami, S.; Fort, A.; Kannan, B.; Karedal, J.; Kunisch, J.; Schantz, H.G.; Siwiak, K.; et al. A Comprehensive Standardized Model for Ultrawideband Propagation Channels. *IEEE Trans. Antennas Propag.* **2006**, *54*, 3151–3166. [CrossRef]
40. Ke, W.; Wu, L. Mobile Location with NLOS Identification and Mitigation Based on Modified Kalman Filtering. *Sensors* **2011**, *11*, 1641–1656. [CrossRef]
41. Zeng, C.; Zhao, S.; Zhong, Y.; Yuan, Z.; Luo, X. An improved method for indoor positioning of WIFI based on location fingerprint. In Proceedings of the 2018 7th International Conference on Digital Home (ICDH), Guiliin, China, 30 November–1 December 2018; pp. 280–285.
42. Haeberlen, A.; Flannery, E.; Ladd, A.M.; Rudys, A.; Wallach, D.S.; Kavraki, L.E. Practical robust localization over large-scale 802.11 wireless networks. In Proceedings of the 10th Annual International Conference on Mobile Computing and Networking, Philadelphia, PA, USA, 26 September–1 October 2004; pp. 70–84.
43. Luo, J.; Zhan, X. Characterization of Smart Phone Received Signal Strength Indication for WLAN Indoor Positioning Accuracy Improvement. *J. Netw.* **2014**, *9*, 1061–1065. [CrossRef]
44. Zampella, F.; Ruiz, A.R.J.; Granja, F.S. Indoor Positioning Using Efficient Map Matching, RSS Measurements, and an Improved Motion Model. *IEEE Trans. Veh. Technol.* **2015**, *64*, 1304–1317. [CrossRef]
45. Perttula, A.; Leppäkoski, H.; Kirkko-Jaakkola, M.; Davidson, P.; Collin, J.; Takala, J. Distributed Indoor Positioning System with Inertial Measurements and Map Matching. *IEEE Trans. Instrum. Meas.* **2014**, *63*, 2682–2695. [CrossRef]

Article

Identification of NLOS and Multi-Path Conditions in UWB Localization Using Machine Learning Methods

Cung Lian Sang *, Bastian Steinhagen, Jonas Dominik Homburg, Michael Adams, Marc Hesse and Ulrich Rückert

Cognitronics and Sensor Systems Group, CITEC, Bielefeld University, 33619 Bielefeld, Germany; bsteinhagen@techfak.uni-bielefeld.de (B.S.); jhomburg@techfak.uni-bielefeld.de (J.D.H.); madams@techfak.uni-bielefeld.de (M.A.); mhesse@techfak.uni-bielefeld.de (M.H.); rueckert@techfak.uni-bielefeld.de (U.R.)

* Correspondence: csang@techfak.uni-bielefeld.de; Tel.: +49-521-106-67368

Received: 2 April 2020; Accepted: 2 June 2020; Published: 8 June 2020

Abstract: In ultra-wideband (UWB)-based wireless ranging or distance measurement, differentiation between line-of-sight (LOS), non-line-of-sight (NLOS), and multi-path (MP) conditions is important for precise indoor localization. This is because the accuracy of the reported measured distance in UWB ranging systems is directly affected by the measurement conditions (LOS, NLOS, or MP). However, the major contributions in the literature only address the binary classification between LOS and NLOS in UWB ranging systems. The MP condition is usually ignored. In fact, the MP condition also has a significant impact on the ranging errors of the UWB compared to the direct LOS measurement results. However, the magnitudes of the error contained in MP conditions are generally lower than completely blocked NLOS scenarios. This paper addresses machine learning techniques for identification of the three mentioned classes (LOS, NLOS, and MP) in the UWB indoor localization system using an experimental dataset. The dataset was collected in different conditions in different scenarios in indoor environments. Using the collected real measurement data, we compared three machine learning (ML) classifiers, i.e., support vector machine (SVM), random forest (RF) based on an ensemble learning method, and multilayer perceptron (MLP) based on a deep artificial neural network, in terms of their performance. The results showed that applying ML methods in UWB ranging systems was effective in the identification of the above-three mentioned classes. Specifically, the overall accuracy reached up to 91.9% in the best-case scenario and 72.9% in the worst-case scenario. Regarding the F1-score, it was 0.92 in the best-case and 0.69 in the worst-case scenario. For reproducible results and further exploration, we provide the publicly accessible experimental research data discussed in this paper at PUB (Publications at Bielefeld University). The evaluations of the three classifiers are conducted using the open-source Python machine learning library scikit-learn.

Keywords: UWB; NLOS identification; multi-path detection; NLOS and MP discrimination; machine learning; SVM; random forest; multilayer perceptron; LOS; DWM1000; indoor localization

1. Introduction

Indoor localization systems enable several potential applications in diverse fields. A few examples where positioning is crucial include tracking valuable assets and personal devices in IoT, ambient assisted living systems in smart homes and hospitals, logistics, autonomous driving systems, customer tracking systems in shopping and public areas, positioning systems in industrial environments, and mission-critical systems such as an application for firefighters and soldiers [1–3]. Among several technologies available for indoor localization described in the literature, ultra-wideband (UWB) technology [1,4,5] plays an increasingly important role in precise indoor localization systems due to its fine ranging resolution and obstacle-penetration capabilities [2,3,6].

In wireless ranging systems including UWB technology, the distance between the transmitter and receiver is estimated by measuring the time-of-flight (TOF) between the two transceivers and multiplying it by the speed of light [7,8]. However, the ranging algorithm assumes that the *TOF*signal is always in a direct line-of-sight (LOS) condition. Therefore, non-line-of-sight (NLOS) [9–11] and multi-path (MP) [3] conditions cause a positive bias in the estimated distances. Figure 1 expresses an abstract view of the LOS, NLOS, and MP conditions in typical wireless communications. The figure shows how a signal sent from a tag device (green pyramid shape in the middle) can be received in different scenarios at the anchor nodes (yellow pyramid shapes). In Figure 1, we define two possible MP conditions in wireless communication. The first condition is clear because the first path signal is completely blocked by the obstacle, and the only received signal in the measurement is based on the bounded signal from the transmitter. However, distance measurement in wireless communication could also be distorted by multiple reflections of signals even if there is no direct obstacle between the transceivers. For instance, wireless measurement is conducted in a narrow corridor, tunnel, etc. We confirmed the error caused by such MP conditions using UWB in our previous work [8]. The research data of the mentioned work are publicly available in [12]. Similar results based on UWB were also reported in [13]. Therefore, differentiation between the LOS, NLOS, and MP conditions in wireless ranging systems is important for precise localization systems.

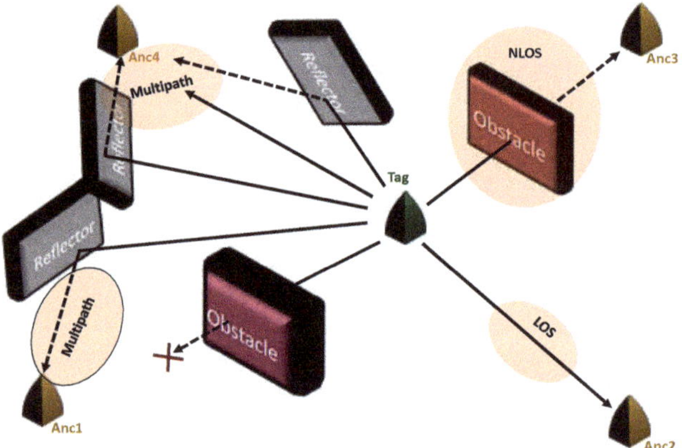

Figure 1. Illustration of LOS, NLOS and multi-path (MP) scenarios in a UWB-based ranging system.

This paper discusses the vital role of classifying the LOS, NLOS, and MP scenarios in UWB ranging system using machine learning (ML) approaches. By understanding the defined three classes, a positioning algorithm [6,14] can mitigate the biases caused by NLOS and MP conditions, i.e., by giving different weights to each class. We proposed such a mitigation technique in our previous work [2]. The common identification and mitigation techniques for the NLOS condition in UWB can be found in [9,15,16] and the references provided therein.

In fact, the multi-class identification of UWB measurement data (LOS, NLOS, and MP) in the real world is challenging in indoor environments because a variety of physical effects can distort the direct path LOS signal in different ways [3,17] (Figure 1). This includes walls, furniture, humans, the orientation of the UWB antenna, etc. Therefore, machine learning methods are attractive solutions for solving such a problem.

Identification and mitigation techniques of the NLOS condition in UWB, or wireless communications in general, using ML methods are not new. It has received significant interest recently years [9–11,17–23]. However, the major contributions in the literature address the binary classification between the LOS and NLOS in the UWB ranging system.

In contrast, this paper addresses machine learning techniques for direct identification of the three mentioned classes (LOS, NLOS, and MP) in a UWB indoor localization system using experimental data collected in seven different environments in two different test scenarios (Section 4). Using the collected real measurement data, we compare three machine learning methods, i.e., support vector machine (SVM), random forest (RF), and multilayer perceptron (MLP), in terms of their prediction accuracy, training time, and testing time. The classifiers are chosen by bearing in mind that the evaluated ML models can be used in a low cost and power efficient real-time system such as microcontroller-based platforms [24].

For the sake of reproducible results and further exploration, we provide all the experimental research data and the corresponding source codes as the Supplementary Data of this manuscript in PUB (Publications at Bielefeld University) [25], which is publicly available. The evaluation of the algorithms is conducted using the open-source Python machine learning library scikit-learn [26].

2. Problem Description

The primary goal of the identification process in wireless communications is to detect the existence of an NLOS and/or MP condition in a communication between a transmitter and a receiver. This process is crucial because the multi-path effects and the NLOS conditions strongly influence the accuracy of the measured distances in wireless communications. As an example, Figure 2d compares the error of measured distances in static scenarios in the LOS, NLOS, and MP conditions in UWB based on our experimental data. Figure 2a–c also illustrates the comparison of the conventional identification techniques for the three mentioned classes (LOS, NLOS, and MP) based on the first-path (FP) power level and the channel impulse response (CIR) (a more detailed description is given in Section 3.1).

The experimental evaluation results in Figure 2d suggest that the magnitude of the error in NLOS and MP conditions is considerably larger than the LOS condition compared to the ground truth reference. Moreover, the error introduced by the MP condition is significantly lower than the completely blocked NLOS condition, where the signal needs to penetrate the obstacle to reach the receiver. Indeed, this depends on the materials and other factors of the obstacles [3]. However, the result in Figure 2d, which is blocked by a human in this experiment, indicates that the NLOS condition introduces the highest impact on the measured distance errors. This motivated us to classify the UWB ranging systems into three classes (LOS, NLOS, MP) to improve the location accuracy in the UWB localization system. The classified ranging information is applicable in any positioning algorithm [6,14] to mitigate the biases effectively [2,18,20] caused by the NLOS and MP conditions. It should be noted that the measured distances in Figure 2d were conducted in the static scenario at approximately a 6 m distance between the anchor and tag for the three classes. The ground truth references of the distance were measured using a laser distance meter, CEM iLDM-150 (http://www.cem-instruments.in/product.php?pname=iLDM-150), which provided an accuracy of ± 1.5 mm according to the datasheet of the manufacturer. Regarding the ranging errors in UWB in a static scenario, a more rigorous evaluation was conducted in our previous work [8], where its corresponding experimental research data were given in [12]. The result of MP conditions in Figure 2d was based on a scenario when there was no obstacle between the transceivers, as illustrated in Figure 1 (Section 1). To be precise, the measurement was conducted in indoor environments where multiple reflections from walls occurred in a narrow corridor. One of the reasons the error occurred in the MP condition in this scenario was because of the preamble accumulation time delay (PATD) in the coherent receiver of the UWB chip [8]. PATD is affected by the presence of multi-path in wireless measurements. It is more notable when the arrivals of the reflected signal are within the chip period of the first path signal [8]. Moreover, it is worth mentioning that the error deviations in Figure 2d correspond to a single range (a tag to an anchor) in the measurement. In general, at least three ranges (usually more in multilateration methods) of such measures are necessary for UWB localization [6,14]. This implies that the combination of such errors in a ranging phase contributes to a significant impact on the overall system performance.

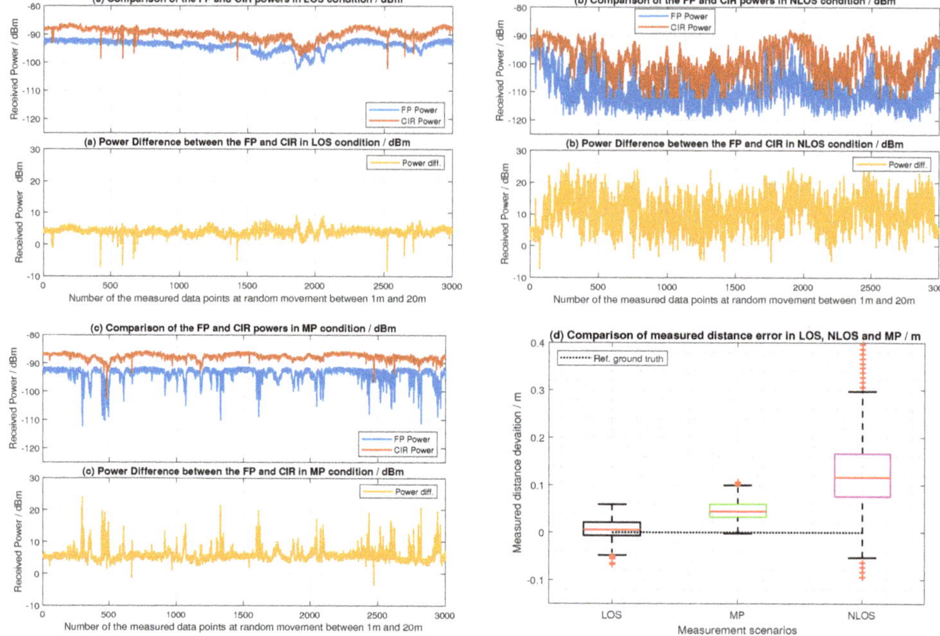

Figure 2. Illustration of the LOS, NLOS, and MP conditions in UWB ranging systems: (**a**–**c**) Comparison of the FP power, the CIR power, and the difference between the two power levels (FP and CIR power) in three scenarios: (**a**) LOS, (**b**) NLOS, and (**c**) MP conditions. The measurement was conducted for the three scenarios for random movement between 1 m and 24 m distances. (**d**) Comparison of the measured distance errors in the three mentioned conditions (LOS, NLOS, and MP) in static scenarios.

3. Related Works

In the literature, the problem solving strategy for ranging errors in UWB due to the effects of NLOS and/or MP can be coarsely classified into two steps [9,15]: (i) the NLOS identification process [16,27,28] and (ii) the NLOS mitigation process [9,16,29]. This paper is solely focused on the former case. In fact, there exists a method that bypasses the identification process and directly mitigates the ranging error using channel statistics and SVM as an ML-based classifier [18,30]. However, this method restricts the flexibility to choose different positioning algorithms in the latter case since the mitigation technique is limited to a few compatible algorithms.

The common approach is to detect the non-direct path signal (i.e., NLOS and/or MP) and use the detected information to modify the location algorithm in order to mitigate the biases caused by the NLOS and/or MP conditions [9,10,17,18,20,21,30]. In this manuscript, we divided the related works into two subsections: (i) conventional approaches without using ML techniques (Section 3.1) and (ii) ML-based approaches (Section 3.2). Our proposed technique regarding the multi-classes identification process in UWB was based on the ML-based approach.

3.1. Conventional NLOS Identification Techniques in UWB

As already mentioned in Section 1, identification of NLOS and LOS in UWB communications is not new. There have been several proposals in the literature [9,16,27,28,31–33] to identify and mitigate the NLOS conditions in UWB. However, the identification process of the MP condition is usually ignored in the literature, although the effects of the MP conditions in UWB ranging systems were acknowledged as important aspects in [3,8,13,21]. Conventionally, the NLOS detection in UWB is always regarded as a binary classification problem. The traditional NLOS detection methods can be coarsely categorized as:

- Identification of the NLOS situation based on a binary hypothesis test [27]
- NLOS detection based on the change of the signal-to-noise ratio (SNR) [28]
- NLOS identification based on channel impulse response [9,31]
- NLOS detection techniques based on the multi-path channel statistics such as the kurtosis, the mean excess delay spread, and the root mean squared delay spread [16,33]
- Detection of the NLOS condition using the received signal strength (RSS) [32,33]

In brief, the conventional NLOS identification approaches mainly rely on the statistical condition of the received signals in UWB communications. Figure 2a–c demonstrates these scenarios by comparing the first-path (FP) signal power and the channel impulse response (CIR) power for the three conditions (LOS, NLOS, and MP). Among the mentioned NLOS detection methods, the threshold approach presented by Decawave in [34] has been widely used in different UWB applications and system implementations [23,34,35]. This is accomplished by taking the difference between the estimated total received (RX) power and the first-path (FP) power using the following equations [34] (Figure 2a–c):

$$FP\ Power\ Level = 10 \cdot log_{10}(\frac{F_1^2 + F_2^2 + F_3^2}{N^2}) - A \tag{1}$$

where F_1, F_2, and F_3 are the first, second, and third harmonics of the first-path signal amplitudes for signal propagation through wireless media as in the multi-path, NLOS, and/or LOS scenarios [34,35]. N is the value of the preamble accumulation count reported in the DW1000 chip from Decawave. A is a predefined constant value, which has 133.77 for a pulse repetition frequency (PRF) of 16 MHz and 121.74 for a PRF of 64 MHz.

The estimated received power (RX) level can be defined as:

$$RX\ Power\ Level = 10 \cdot log_{10}(\frac{C \cdot 2^{17}}{N^2}) - A \tag{2}$$

where C is the value of the channel impulse response power reported in the DW1000 chip.

Therefore, the metric that specifies the conditions of LOS and NLOS in the threshold method can be achieved by computing the difference between the received and first-path power [34] as:

$$Threshold\ Power = RX\ Power\ Level - FP\ Power\ Level \tag{3}$$

In the conventional threshold approach, the measured distance is classified as a LOS when the threshold power using (3) is less than 6 dBm and defined as an NLOS when it is more than 10 dBm [34]. This is a sub-optimal acceptable solution as our particular experimental evaluation results show in Figure 2. That is, the mean value of the threshold power including its standard deviation in the LOS condition using (3) is 4.12 ± 1.13 dBm (Figure 2a) and in the the NLOS condition is 10.75 ± 5.51 dBm (Figure 2b). However, the solution is not optimal as much fluctuation can occur as described in the experimental measurement data (Figure 2a–c). The condition is harsher to solve in MP condition, where the first-path signals are hard to distinguish clearly from the received signal (Figure 2c).

Nevertheless, the classification of the three mentioned classes is not straightforward. The complexity of the classification problem increases especially in indoor environments because of several factors such as material characteristics [36], the refractive index of different materials, and so on [1,3]. Moreover, the phenomenon of the multi-path effects and NLOS depends on the properties of the medium through which the signal travels, the location (dimension of the places and rooms) where the signals are measured, the presence of other objects within the measured environment, the orientation of the UWB antenna, etc. Therefore, ML approaches have been regarded as attractive strategies for solving this complex task in recent years (Section 3.2).

3.2. Identification of the NLOS and MP Conditions in the Literature Based on Machine Learning Techniques

One of the earlier ML-based NLOS identifications in UWB was conducted in [9] using SVM as a classifier. In that paper, the identification process was considered as a binary classification problem (LOS vs. NLOS) showing that the ML approaches outperformed the traditional parametric techniques and signal processing approaches from the literature.

Consequently, several investigations of the NLOS identification process in UWB were examined in the literature using different ML techniques as a classifier such as SVM in [9,10,17,21], MLP in [23,37,38], boosted decision tree (BDT) in [38], recursive decision tree in [35], and other ML techniques such as kernel principal component analysis in [19], etc. Moreover, the unsupervised machine learning technique called "expectation maximization for Gaussian mixture models" was recently applied in [39] to classify the LOS and NLOS conditions in UWB measurement. Likewise, deep learning approaches such as the convolutional neural network (CNN) were also explored to distinguish the NLOS condition from LOS in UWB ranging [13,22]. In CNN-based deep learning approaches, the authors generally modified the existing CNN network such as GoogLeNet [22], VGG-architecture (i.e., VGG-16 or VGG-19) [13,22], AlexNet [22], etc., to be usable for the low cost UWB systems. The reported overall accuracy ranges started from 60% (using a typical ML technique such as SVM) up to 99% (using the CNN approach). In all of the above-mentioned approaches, the focus is solely on detecting the NLOS condition in UWRranging, i.e., the binary classification between LOS and NLOS.

Moreover, the performance comparison of different ML techniques for the identification of NLOS in UWB ranging was conducted in [11,17,38]. The main purpose of these analyses was to compare the impact of model selection in ML-based system applications in UWB. In [38], the performance comparison of two ML methods namely MLP and BDT was carried out for the binary classification (LOS vs. NLOS). The resultant report concluded that the BDT outperformed the MLP. Likewise, the comparison of five classifiers using MATLAB (i.e., SVM, k-nearest neighbor (KNN), binary decision tree, Gaussian process (GP), generalized linear model) was performed in [11]. The authors concluded that KNN and GP performed better than the other three models. Similarly, the authors in [17] evaluated three ML models (SVM, MLP, RF) to classify the LOS and NLOS in narrowband wireless communications (i.e., not specifically for UWB systems in this case). The authors reported that RF and MLP performed better than SVM in all of their evaluations.

In contrast to the binary classification between LOS and NLOS in UWB, the binary classification of the MP from LOS conditions was investigated in [13]. The author reported that MP effects could cause an error in UWB ranging from a few centimeters up to 60 cm. Similar deviation of the error in the MP condition can be seen in our experimental evaluation presented in Figure 2d.

Throughout the literature, the problem has been treated as a binary classification problem or hypothesis test (i.e., LOS vs. NLOS or LOS vs. MP). To the best of the authors' knowledge, only two papers addressed UWB-based ranging errors as a multi-class problem [11,21]. The first paper was based on a two-step identification process [21] using SVM as a classifier. In that paper, the LOS and NLOS were identified in the first step. Then, further classification (MP vs. NLOS) was categorized in the second step if NLOS was detected in the first step. The second paper categorized the NLOS conditions into two types (soft-NLOS vs. hard-NLOS) in addition to LOS while ignoring the MP effects [11]. The differentiation between the two NLOS types was primarily based on the material of the obstacles, which the UWB signal was passing through by penetration. The authors used two types of walls in their evaluation to classify a soft-NLOS and a hard-NLOS.

In contrast with the above-mentioned approaches, we performed a direct identification of the multi-class classification for UWB ranging systems in this paper. The classified classes were the LOS, NLOS, and MP conditions. Based on the measurement data, we performed three ML models (Section 5), namely SVM, RF, and MLP, to compare their performances (Sections 6 and 7). The experimental research data utilized in this paper [25] are provided on a public archive for reproducible results and further exploration.

4. Measurement Scenarios and Data Preparation

In this section, we describe the experimental setup of the evaluations (Section 4.1), the data collection processes including labeling and data separation (Section 4.2), and the feature extraction based on the collected data (Section 4.3) for the three evaluated ML models.

4.1. Experimental Setup

For the UWB data measurement process in the experimental evaluations, we used a DWM1000 module [34] manufactured by Decawave as the UWB hardware and the STM32 development board (NUCLEO-L476RG) [40] manufactured by STMicroelectronics as the main microcontroller (MCU). Table 1 provides the hardware configurations used in the experimental evaluations.

Table 1. Configurations of the primary hardware used in the experimental evaluation.

Types of Hardware	Properties	Values
UWB module	Module name	DWM1000
	Data rate	6.8 Mbps
	Center frequency	3993.6 MHz
	Bandwidth	499.2 MHz
	Channel	2
	Pulse-repetition frequency (PRF)	16 MHz
	Reported precision	10 cm
	manufacturer	Decawave
Microcontroller (MCU)	Module type	STM32L476RG
	Development board	NUCLEO-L476RG
	Manufacturer	STMicroelectronics

Our previous work [7,8] pointed out that the alternative double-sided two-way ranging (AltDS-TWR) method outperformed other available TWR methods in the literature in different tested scenarios. Therefore, we applied AltDS-TWR as a wireless ranging method in our evaluations. Furthermore, AltDS-TWR operated well without the need to use high precision external oscillators in the MCU [8]. Hence, the built-in high-speed internal (HSI) clock source (16 MHz) from the MCU was applied to all of the evaluation results presented in this article. According to the data sheet [41], the HSI has an accuracy of $\pm 1\%$ using the factory-trimmed RC oscillator.

During measurement, one of the transceivers among the two was connected to a computer for logging the data via a serial USART port. Both transceivers were executed with a two-way ranging software provided by Decawave for production testing of their evaluation kit (EVK1000), which is available on Decawave's website (https://www.decawave.com/software/). We modified the library of this software to extract all required features provided in Section 4.3. Then, we logged and saved the extracted features into a file for each trial in our measurement campaign. To avoid the effect of Fresnel zones in our measurement results, the antenna height was always maintained at 1.06 m in one of our UWB devices, i.e., the static one that recorded the measurement data via PC.

4.2. Data Collection Process

The required data for experimental evaluations presented in Section 7 were collected in three scenarios (two small rooms, a hall, and four corridors) at seven different places in indoor environments (Figure 3). The two rooms were the (6 m $\times$ 6 m) laboratory environment and approximately the (8 m $\times$ 6 m) communication room in which different items of furniture were placed. The collected data in narrow corridors were intended for MP conditions, where the direct LOS could not be distinguished because of multiple signal reflections from the narrow walls. Figure 2c illustrates a concrete example of this MP condition in terms of the FP and RX powers. In all cases, the data were collected for both static and dynamic conditions. In the dynamic case, the device attached to the PC stayed static, while another device was held by a human during random walks. In the static scenario, the two transceivers were at a vertical position of 90° pointing to the antenna of the DWM1000 module as an upward position without any rotation. However, the antenna of the device held by a human during the dynamic condition was randomly rotated between 0° and 180° in some cases of the data collection process. Moreover, the NLOS conditions by blocking the communication between two transceivers using a human as an obstacle were conducted in all cases depicted in Figure 3. Besides, a thick concrete wall, pieces of concrete block, and a mixture of wood and metal were also applied as parts of the obstacles for NLOS conditions in the two small rooms and their environments.

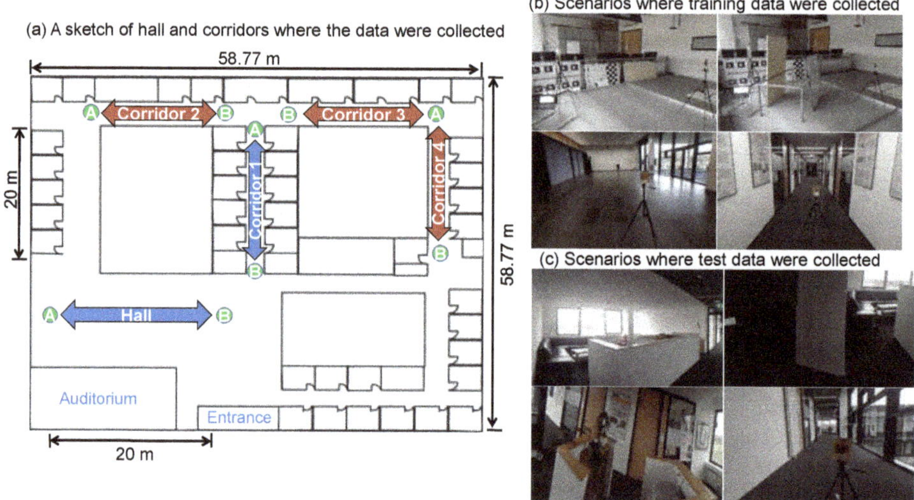

Figure 3. Illustration of the scenarios where training and test data were collected for evaluation: (**a**) A sketch of the building where the experimental data for training and test were collected. (**b**) Training data were collected for LOS, NLOS (including human blocking) and MP conditions in a laboratory, a large hall, and a corridor (blue color in (**a**)). (**c**) Similar to (**b**), test data were collected in a different room (including different types of furniture and NLOS human blocking) and a different corridor (red in (**a**)).

4.2.1. Labeling the Measured Data and Dealing with the Class Imbalance Case

The class labels (LOS, NLOS, and MP) were manually annotated in the data preprocessing phase after the measurement campaign. During the data measurement process, a block of observations for each trial regarding the three categories (LOS, NLOS, MP) was carried out and saved individually into the PC. The block referred to a collection of data belonging to the same class that was saved separately in the computer as a single file. For instance, we collected the random movement data in Corridor 2 (Figure 3) for 5 min using a data update rate of the system as 20 Hz. At the same time, we made sure that there were no obstacles between the two transceivers, and the antenna was held still at 90° during the process. Then, we annotated the block of this whole measured data as an MP class to be used in our evaluation. In this way, the whole block of that data in each evaluated scenario could easily be labeled as an LOS, NLOS, or MP.

The initial UWB dataset achieved from the measurement was imbalanced for the three demanded classes (LOS, NLOS, and MP), which is a typical phenomenon in data collection. Class imbalance refers to a scenario where the number of observations in each class is not the same in the measurement. In other words, the number of samples in one class or more classes is significantly lower or higher than those belonging to the other classes. There are several techniques to deal with the imbalanced data in classification problems including resampling techniques and algorithms [42].

We chose a random undersampling technique [42] in our evaluation to balance the three mentioned classes equally. This ensured that no artificial data points were created outside of the measured experimental data. The undersampling was performed by setting the class belonging to the smallest number of observations as a base class. Then, the classes belonging to higher samples of observation were reduced to balance with the total number of the base class by randomly selecting their elements.

4.2.2. Separation of the Training, Validation, and Test Dataset

Two independent test datasets were used in our experimental evaluations. The first test dataset was separated from the measurement environments provided in Figure 3b, i.e., the environments from which the training data came. This is the typical scenario for the presentation of classification results in most UWB-based literature [10,11,21]. In some cases in the literature, the test dataset was separately collected by intentionally switching the subject of the experiment, i.e., a person who carried the UWB device in [11]. However, the environment of the measurement stayed unchanged in the evaluation. As already mentioned in Section 2, the propagation of the measured UWB signal can be affected by several environmental factors such as the refractive index of materials, the placement of objects in the measured circumstances, etc. To examine this incident in our results, we collected a second test dataset that was independent and different from the training environments (Figure 3c). This second dataset was solely split out for the purpose of testing in our evaluation. The results using both test datasets are discussed in Section 7.

For training the evaluated ML models including the validation process, more data were gathered in the laboratory room, the hall, and the first corridor (Figure 3b). In particular, one-hundred eighty-five-thousand seven-hundred ninety observations in total were gathered after balancing the three classes in this scenario. This meant each class belonged to 61,930 data points. For testing purposes, thirty percent of the data points were left out by random shuffling in each trial conducted in Section 7. The results using this test scenario in Figure 3b are expressed as a scenario when test and training are in the same condition (Section 7).

In contrast, the measurement campaign, particularly for testing purposes, was conducted in different scenarios from the training. These measurements were carried out in a different room with various items of furniture and three different corridors (Figure 3c). The results achieved from these second test scenarios are expressed in Section 7 as a scenario where the test and training conditions were different. The total number of 36,015 data points, 12,005 for each class, was used for conducting this test scenario in our evaluation after balancing equally the three classes.

4.3. Feature Extraction

In total, twelve features were extracted from the DWM1000 UWB modules manufactured by Decawave [34] using the configuration described in Table 1. The extracted features were based on the typical parameters that are necessary in the traditional NLOS identification methods as expressed in Equations (1)–(3). This meant that no extra burden was involved by using these extracted features in our ML application. For the sake of completeness, two more features namely standard and maximum noises supported by the DW1000 module were included in the feature extraction of our evaluation. Therefore, the full features extracted and saved during the experimental evaluation were:

1. the reported measured distance
2. the compound amplitudes of multiple harmonics in the FP signal
3. the amplitude of the first harmonic in the FP signal
4. the amplitude of the second harmonic in the FP signal
5. the amplitude of the third harmonic in the FP signal

6. the amplitude of the channel impulse response (CIR)
7. the preamble accumulation count reported in the DW1000 chip module
8. the estimated FP power level using (1)
9. the estimated RX power level using (2)
10. the difference between the FP and RX power level using (3)
11. the standard noise reported in the DW1000 chip module
12. the maximum noise reported in the DW1000 chip module

Regarding the above-mentioned 12 features, we would like to mention that the feature extraction was solely based on the DW1000 chip as the UWB hardware, which was manufactured by Decawave.

5. Machine Learning Models for Identification of the LOS, NLOS, and MP Conditions

Three machine learning models (SVM, RF, and MLP) were evaluated in this paper. SVM was regarded as a baseline model in the evaluation since it was the most commonly and frequently used model for the UWB-based identification of NLOS conditions in the literature [9,10,17,21]. The configuration and setup for each classifier are discussed in the subsequent subsections.

The training and test times of each classifier reported in this section were based on a single concurrent CPU core without using any parallel computing devices such as the GPU. The evaluation was done on the same machine for all classifiers. The reported results for all classifiers in this section (SVM in Section 5.1, RF in Section 5.2, and MLP in Section 5.3) were based on 10 iterations of randomly splitting the measured training, validation, and test data. The training and test datasets used in this section were the random splitting of the data collected in Figure 3b. This explained the test datasets collected in Figure 3c, which were used only in the evaluation results presented in Section 7. The extracted features used for all classifiers in this section were based on the discussion and selection observed in Section 6. The reported training and test times per sample (mean value) for each classifier were estimated in two steps. First, we estimated the total amount of time it took for the whole dataset in the training and test phases using the corresponding training and test dataset. Then, the measured time was divided by the total number of samples to get the mean value per sample.

Generally, several parameters in the ML classifiers were tuned to achieve the optimized results. Moreover, each classifier had its specific hyper-parameters, which were not compatible with one another. Therefore, a direct comparison using exactly the same parameters for all classifiers was impossible. For the sake of simplicity and better representation of the results, the comparison is done by choosing the most important and influential parameters for each classifier in this section. This implied that the kernel type was chosen for SVM, and the number of decision trees in the forest was selected for RF. For MLP, the number of hidden layers including the total number of neurons in each layer was evaluated to choose the best option for the given problem.

For the reproducible results, the parameters for each classifier such as the activation function, optimizer, earlier stopping criteria for the training, learning rate, etc., were based on the default setting of the scikit-learn [26] library if nothing is explicitly mentioned in the following. The applied stable version of the scikit-learn library was 0.23.1 at the time of writing this paper.

5.1. Support Vector Machine Classifier for the UWB Localization System

SVM is a supervised machine learning technique suitable for solving both classification and regression problems [43,44]. It is strongly based on the framework of statistical learning theory [45]. SVM has also been recognized as one of the most frequently used classification techniques in the machine learning community in the past due to its robustness and superior performance without the need to tune several parameters compared to deep neural networks [9]. In short, SVM takes the data as an input and determines a hyper-plane that separates the data into predefined classes. The hyper-plane was established in the SVM algorithm by maximizing the margin between the separable classes as wide as possible. Table 2 presents the comparison of four kernel types in SVM using the UWB measurement data and extracted features examined in Section 4.

Table 2. Comparison of the SVM configurations based on the kernel functions.

Kernel Types	Mean Accuracy with std (%)	Mean Training Time per Sample (ms)	Mean Test Time per Sample (ms)
Radial basis function (RBF)	**82.96** ± 0.14	2.06 ± 0.18	0.99 ± 0.01
Linear function	72.59 ± 0.25	**1.92** ± 0.08	**0.53** ± 0.01
3rd order polynomial function	70.82 ± 0.19	3.05 ± 0.09	0.80 ± 0.02
Sigmoid function	50.59 ± 3.05	3.01 ± 0.27	1.59 ± 0.09

The bold text and numbers in the table refer to the chosen kernel type for the evaluation and the best performance scores for each metric respectively.

The choice of the kernel types in SVM had a strong influence on its accuracy regarding our particular measurement of UWB data. The results in Table 2 show that the radial basis function (RBF) kernel reached the highest accuracy with 82.96%, while the sigmoid function provided the poorest with 50.59%. Both linear and third-order polynomial functions had comparable results. In terms of training and test times, the linear function achieved the lowest time per sample while the sigmoid function showed the worst performance with the highest time per sample. In all circumstances, the training and test times in SVM were in the order of milliseconds. This meant that SVM had the poorest performance in terms of test time compared to RF (Section 5.2) and MLP (Section 5.3).

5.2. Random Forrest Classifier for the UWB Localization System

According to the original paper in [46], random forests (RF) are a combination of decision tree predictors in the forest such that each tree depends on the values of a random vector, which is sampled with the independent and identical distribution for all the trees. In brief, RF is built upon multiple decision trees and merges them to get a more accurate and stable prediction as its final output. Two significant advantages of RF are (i) the reduction in over-fitting by averaging several trees and (ii) the low risk of prediction error since RF typically makes the wrong prediction only when more than half of the base classifiers (decision trees) are wrong. The disadvantage, though, is that RF is typically more complex and computationally expensive than the simple decision tree algorithm. In general, the more trees in the forest, the better the prediction. However, this flexibility comes with the cost of the processing time (training and test times), as described in Table 3.

Table 3. Comparison of the RF configurations based on the numbers of decision trees in the forest.

No. of Decision Trees in the Forest	Mean Accuracy with std (%)	Mean Training Time per Sample (μs)	Mean Test Time per Sample (μs)
5 decision trees	90.91 ± 0.18	**4.84** ± 0.25	**1.26** ± 0.02
10 decision trees	91.55 ± 0.09	9.38 ± 0.27	2.33 ± 0.08
20 decision trees	91.83 ± 0.10	18.68 ± 0.52	4.66 ± 0.08
30 decision trees	91.89 ± 0.11	27.30 ± 0.41	6.82 ± 0.13
50 decision trees	91.99 ± 0.11	45.44 ± 0.62	11.30 ± 0.27
100 decision trees	92.07 ± 0.09	90.42 ± 1.39	22.43 ± 0.34
200 decision trees	92.12 ± 0.13	179.85 ± 3.10	44.85 ± 0.47
500 decision trees	**92.13** ± 0.12	460.04 ± 10.27	113.39 ± 1.44

The bold text and numbers in the table refer to the chosen decision trees in the forest for the evaluation and the best performance scores for each metric respectively.

The prediction accuracy in RF increased steadily as the number of decision trees in the forest increased (Table 3). However, the improvement slowed down when the number of trees in the forest was more than 50 in this particular UWB data. In contrast, the training and test time keep increasing linearly by the increase of the decision trees in the forest. This implied that the training and test times (the smaller the magnitude of the metric, the better the performance) were negatively affected by the growth of trees in the forest. Therefore, a trade-off between the accuracy by growing trees in the forest and the efficiency of the test time should be thoroughly made. In terms of training time, RF performed the fastest among the three classifiers compared to SVM (Section 5.1) and MLP (Section 5.3), i.e., the training time per sample in RF was in the order of microseconds.

5.3. Multi-Layer Perceptron Classifier for the UWB Localization System

MLP is a type of deep feedforward artificial neural networks, which contains at least three layers (an input layer, a hidden layer, and an output layer) in a single network [47]. Typically, the neurons in the hidden and output layers of the MLP use nonlinear activation functions such as sigmoid, ReLU, and Softmax. The term deep is usually applied when there is more than one hidden layer in the network. MLP utilizes the backpropagation algorithm [48] for training the network. In this paper, the MLP classifier is configured using the rectified linear unit (ReLU) as the activation function for the hidden layers, the Softmax function as the output layer, and the Adam (adaptive moment estimation) as an optimization algorithm. The maximum number of epochs was set to 500, allowing early stopping if the training loss had not improved for 10 consecutive epochs.

The evaluations of MLP were conducted for up to 6 fully-connected hidden layers in two conditions using 50 and 100 neurons in each hidden layer (Table 4). The results showed that there was a significant increase in the overall accuracy by adding a second and third hidden layer to the network. However, the improvement was meager when more than 3 hidden layers were used for a network. In terms of the number of neurons per layer in the network, the use of 100 neurons in each hidden layer constantly beat the use of 50 up to four consequent layers. However, the difference could not be clearly distinguished when more than 4 layers were used in the network.

In terms of the processing time (training and test times), adding more hidden layers and more neurons in the network had a negative impact on the performance (the last two columns in Table 4), i.e., the lower the processing time, the better the performance. Therefore, a trade-off between the accuracy and processing time was necessary to have an efficient performance. The results in Table 4 suggest that the use of 3 hidden layers in which each contained 100 neurons seemed a good choice for solving the evaluated UWB-based multi-class classification problem.

Table 4. Comparison of the MLP configurations based on the number of hidden layers and neurons.

No. of Neurons in each Hidden Layers	No. of Hidden Layers	Mean Accuracy with std (%)	Mean Training Time per Sample (ms)	Mean Test Time per Sample (μs)
50	1	84.93 ± 0.26	$\mathbf{0.68} \pm 0.27$	$\mathbf{1.26} \pm 0.06$
	2	88.77 ± 0.53	1.45 ± 0.07	3.23 ± 0.55
	3	90.45 ± 0.51	2.57 ± 0.25	5.46 ± 0.20
	4	90.95 ± 0.35	2.84 ± 0.63	8.79 ± 0.41
	5	$\mathbf{91.04} \pm 0.66$	3.50 ± 1.07	12.20 ± 0.26
	6	90.61 ± 1.26	3.70 ± 0.74	15.25 ± 3.28
100	1	85.88 ± 0.12	$\mathbf{1.02} \pm 0.01$	$\mathbf{2.51} \pm 0.05$
	2	89.78 ± 0.59	3.75 ± 0.19	12.20 ± 1.53
	3	91.36 ± 0.54	5.68 ± 1.23	18.12 ± 1.71
	4	$\mathbf{91.38} \pm 0.44$	5.08 ± 1.67	23.32 ± 3.75
	5	90.85 ± 0.67	7.90 ± 2.60	29.24 ± 0.94
	6	91.33 ± 0.44	9.42 ± 3.90	32.47 ± 1.81

The bold numbers in the table refer to the chosen configuration of MLP for the evaluation and the best performance scores for each metric.

5.4. Section Summary

In summary, RF was the fastest among the three methods for the given dataset in terms of the training time (Figure 4). This was because the training time per sample data in RF only took in the order of microseconds. Meanwhile, SVM and MLP were in the order of several milliseconds depending on the configuration and setup. In terms of test time, SVM performed worst among the three with a test time in the order of milliseconds, while RF and MLP are in the order of microseconds.

Taking into consideration the evaluated results presented in this section, we established the configurations of three classifiers for further processing in Section 7. The summarized overview is represented in Figure 4. The selected configurations were the radial basis function kernel approach for SVM, 50 independent decision tree estimators for RF, and three hidden fully-connected layers with 100 neurons in each layer for the MLP network.

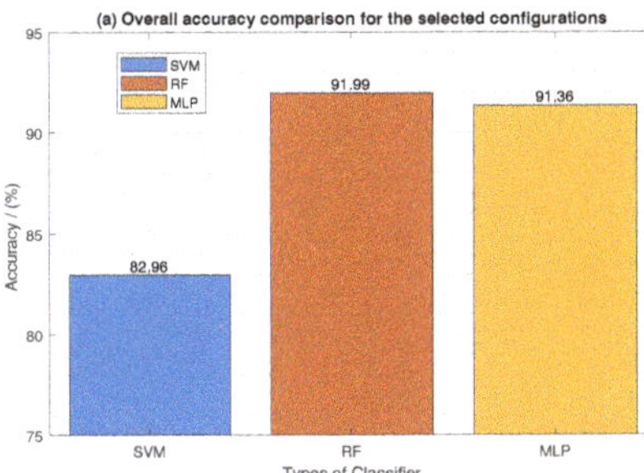

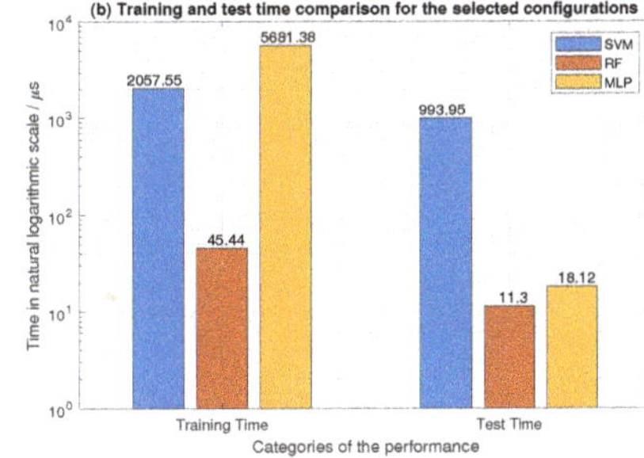

Figure 4. Summary of the chosen configurations for the three classifiers, which is the RBF kernel for SVM, 50 estimators for RF, and three hidden layers with 100 neurons in each for MLP. (**a**) Overall accuracy comparison for the selected configurations (**b**) Training and test time comparison for the selected configurations.

6. Data Preprocessing and Feature Selection

This section examines the impact of feature selections (Section 6.1) and features scaling, i.e., the standardization technique in this manuscript (Section 6.2), for the three evaluated ML models. The experimental results presented in Section 7 were performed based on the outcomes of this section.

6.1. The Impact of Feature Extraction in the Evaluated Machine Learning Models

Based on the extracted features defined in Section 4.3, the performance comparison of feature extractions for five categories is illustrated in Figure 5. The five categories were built upon: (i) 12 extracted features (i.e., the full features in our evaluation), (ii) 10 features excluding standard and maximum noises, (iii) 5 features, i.e., the reported distance, the CIR, and the first, second, and third harmonics of the FP, (iv) 3 features, i.e., reported distance, the CIR, and the first harmonics of FP, and (v) 2 features, i.e., the CIR and first harmonic of FP.

Regarding feature extractions of the UWB measurement data, Figure 5 indicates that a notable degradation in accuracy occurred for the three ML models when two features were applied in the evaluation. The rest of the categories (starting from three to 12 features) provided more or less comparable results. Moreover, we noticed during the evaluation that using the reported distance as a feature played an important role in feature extractions for UWB data. Furthermore, this was also the metric that we were most interested in estimating the position in UWB. We also observed that the contribution of the amplitudes of the three harmonics (first, second, and third) in the FP signal implied comparable impacts. This implied that picking any one of them as a feature provided an equivalent performance in the case of feature reductions. The amplitude of the CIR is undoubtedly an important feature in UWB, which represents a vital role in the identification of NLOS in the conventional technique using (2).

The empirical results in Figure 5 suggested that the most suitable choice for the evaluation in terms of minimum features and optimal performance was to use three features. This was because there were no significant gaps between the use of three to 12 features in terms of accuracy (Figure 5).

The accuracy in this context refers to the overall probability of the defined three classes, which was accurately estimated during the measurement, i.e., independent of the specific LOS, NLOS, and MP conditions. As a rule of thumb, fewer features in the model typically allow less computation and better resource efficiency, especially for MCU-based platforms. In terms of the test time, the performance of MLP degraded when less than five features were applied. However, there were no notable differences in SVM and RF when more than 2 features were used in the evaluation (Figure 5). Therefore, the experimental evaluation results presented in Section 7 were based on three features, specifically the reported distance, the amplitude of the CIR signal, and the amplitude of the first harmonics in the FP signal.

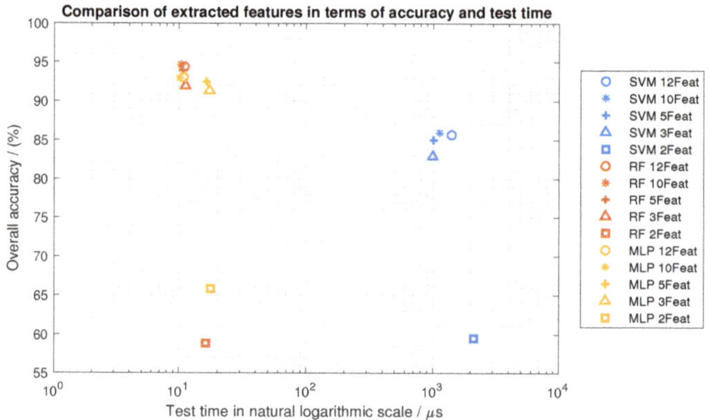

Figure 5. Performance comparison of three ML models (SVM, MLP, and RF) using different extracted features at training and test data collected for the same scenarios.

6.2. The Impact of Feature Scaling in the Evaluated Machine Learning Models

Feature scaling is a model dependent parameter in ML. It is a technique to normalize or standardize the range of independent variables or features of input measured data in a preprocessing step. This typically allows a faster training time and a better performance in many ML models. This section briefly reveals the effects of feature scaling in the three evaluated ML models. Besides, there exist ML models where their performance is not affected by the feature scaling in the preprocessing of the input data. A good representative of such a model in our evaluation was the RF classifier (Figure 6). In this paper, the feature scaling was performed using the standardization technique. This typically means rescaling the data in preprocessing to have a mean of zero and a standard deviation of one (unit variance).

Figure 6a depicts the impact of feature scaling for the three ML models in terms of the overall accuracy. Scaling the input data in the preprocessing phase had a notable impact on the SVM and MLP classifiers in terms of accuracy. In SVM, the overall accuracy was improved from 67.95% to 82.95% by scaling the features of the input data. Similarly, the accuracy of the MLP was increased from 85.70% to 91.36%. However, the RF gave equivalent outcomes in both scaled and unscaled features.

In terms of training time using three features in each model, SVM learned significantly faster when feature scaling was used (Figure 6b). To be precise, the training time of SVM reduced from $4195.14 \pm 210.38\,\mu s$ to $1977.12 \pm 185.45\,\mu s$ by scaling the features. However, RF and MLP did not show obvious improvement for training time in our small-scale three feature evaluation. Specifically, the training time of RF for the scaled and unscaled features was $44.97 \pm 2.38\,\mu s$ and $47.98 \pm 4.74\,\mu s$, respectively. Likewise, the training time of MLP for the scaled and unscaled features was $5562.66 \pm 1313.85\,\mu s$ and $5737.97 \pm 1949.05\,\mu s$, respectively.

In terms of test time, feature scaling hurt the performance of MLP (Figure 6b). Specifically, the value of the test time (the smaller, the better) in MLP degraded from 10.41 ± 0.54 μs to 17.58 ± 1.27 μs by feature scaling. On the contrary, the performance of the test time in SVM improved when feature scaling was applied, i.e., 1839.53 ± 53.70 μs for unscaled features and 956.28 ± 13.94 μs for scaled features. Again, RF did not show any significant improvements except a small variation in its standard deviation, which implied 11.48 ± 0.99 μs for unscaled features and 11.48 ± 1.42 μs for scaled features, respectively.

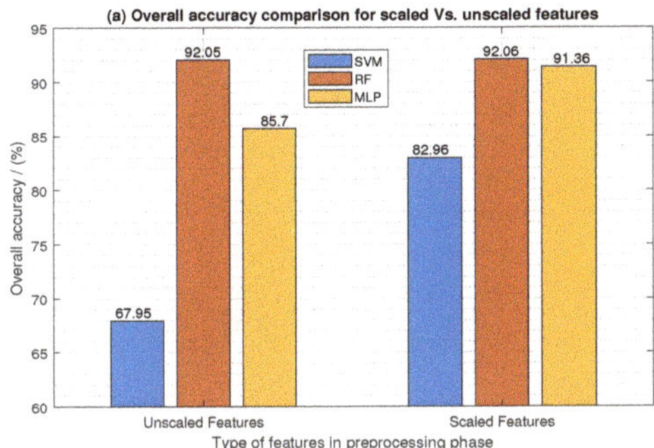

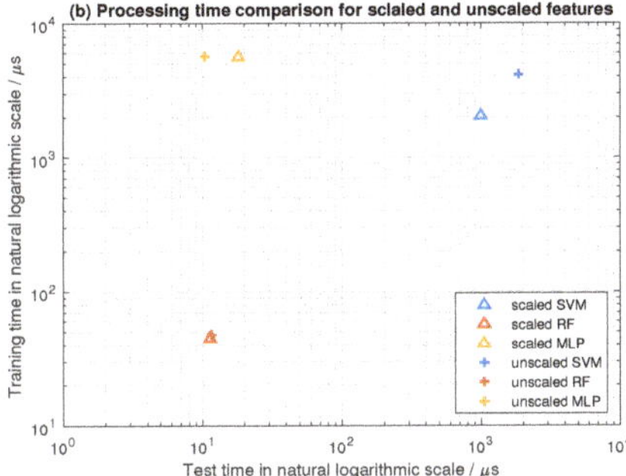

Figure 6. Comparison of the overall accuracy, training, and test times for the scaled vs. unscaled features in the preprocessing phase. (**a**) Overall accuracy comparison for scaled Vs. unscaled features (**b**) Processing time comparison for scaled and unscaled features.

7. Evaluation Results

This section examines the experimental evaluation results of three ML classifiers based on two quantitative metrics: (i) an F1-score, which was used to compare the performance of the three evaluated classifiers in this paper (Section 7.1), and (ii) a confusion matrix that gave an insightful representation of the reported results for each individual classifier (Section 7.2).

7.1. Performance Comparison of the Three Classifiers Using the Macro-Averaging F1-Score as a Metric

To give an overview of the actual state in each trial conducted 10 times, we use the macro-averaging F1-score to compare the performance of the three classifiers in this section. The F1-score, i.e., in contrast to the overall accuracy score in the confusion matrix (Section 7.2), is extensively used to quantify the classifier's performance in ML because it takes into account both the precision and recall to compute the decisive score [11,49]. It is the harmonic mean of the precision and recall, which can be expressed for a binary classification as:

$$F1 = 2 \cdot \frac{Precision \cdot Recall}{Precision + Recall} \tag{4}$$

For multi-class classification, there are two typical ways (macro-averaging and micro-averaging) to compute the overall F1-score for a classifier [49]. We applied the macro-averaging technique in our evaluation, which treated all the classes equally. Based on the mentioned macro-averaging F1-score, Figure 7 compares the experimental evaluation results of the three classifiers in two test environments (the scenario that was the same as vs. different from the training state) defined in Section 4.2.2. The solid lines denote the results of the test dataset when the data in the test state were collected in a different environment from the training state. The dotted lines show the results of the test dataset achieved from the same scenario as the training state.

In general, a significant gap between the two test scenarios was discovered in the experimental evaluation results of the three classifiers (Figure 7). The figure reveals that impressive outcomes were achieved in RF and MLP classifiers when the test and training states were conducted in the same environments. However, the performance of SVM was relatively low in this scenario compared to RF and MLP. In contrast, the performance of all classifiers was notably degraded when the test environment was different from the training state. Specifically, the resultant mean of the SVM classifier based on the macro-averaging F1-score was reduced from 0.83 (when the training and test scenarios were the same) to 0.75 (when the training and test scenarios were different). Similarly, the performance of the RF decreased from 0.92 to 0.73. The MLP classifier showed a degradation from 0.91 to 0.72. The results showed that an immediate conclusion and judgment on the choice of classifiers based on a single test scenario or environment could be misleading. The core reason was that the estimated precision and characteristics of the measured wireless signal, i.e., the UWB ranging data in our evaluation, are affected by a variety of physical impacts in indoor environments, as previously mentioned several times in this manuscript.

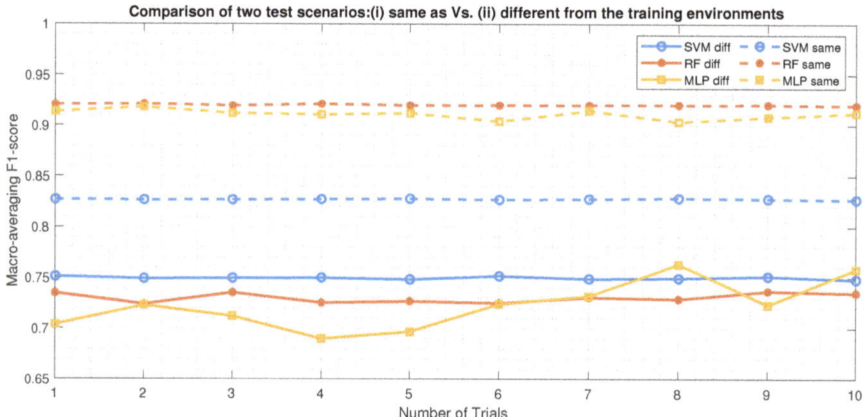

Figure 7. Performance comparison of the three evaluated classifiers based on the macro-averaging F1-score in two different scenarios: (i) dotted lines denote that the training and test data came from the same environment; (ii) solid lines denote that the training and test data came from different environments.

It is interesting to see that SVM stood out to be the best classifier in our evaluation when the test scenario was different from the training state (Figure 7). However, it had the poorest performance among the three classifiers when the same environments of training and test were applied. Moreover, the outcomes of SVM were consistent in all of the evaluated trials in both of the scenarios. Indeed, the outcomes of RF were also quite stable across the whole trials compared to the MLP classifier. However, many fluctuations were evident in the predicted outputs of the MLP, especially in the condition where the training and test environments were different.

In all experiments, the lowest F1-score was 0.69 in Trial No. 4 when MLP was used as a classifier, and the highest score reached 0.92 using RF. This outcome showed that the ML-based classifications, regardless of the type of the classifier, were more effective in the multi-class identification of UWB data than the traditional approaches described in Section 2.

7.2. Result Representation of the Three Evaluated Classifiers Using the Confusion Matrix

To examine the study more extensively, a comparative analysis of the two test scenarios for each classifier was conducted using the confusion matrix in this section. In the confusion matrix (Figures 8–10), the output class in the Y-axis refers to the prediction of the classifier, and the target class in the X-axis refers to the true reference class. The overall accuracy of the classifier is given in the bottom right corner of each confusion matrix. The last column in each category of the confusion matrix indicates the precision (positive predictive value) and its counterpart the false discovery rate (FDR) of the classifier. Likewise, the last row in each category gives the recall (sensitivity or true positive rate) and its complement the false-negative rate (FNR). The correct predictions for each category are expressed in the diagonal of the confusion matrix. The values in the off-diagonal correspond to the Type-I and Type-II errors. For a scenario where training and test datasets were collected in different environments, the confusion matrices presented in this section were based on the mode of each classifier (i.e., we chose the most frequently predicted class in each trial as our estimator output). For a scenario where the training and test environments were the same, the confusion matrix was based on Trial No. 5 out of the 10 trials described in Section 7.1. The reason was that the random splitting of the test dataset for the true class in this scenario was different for each trial. Moreover, all trials in this scenario gave comparable results as reported in Figure 7.

7.2.1. Comparative Analysis of the Two Test Scenarios for SVM Classifier

An insightful comparison of the two test scenarios based on the confusion matrix for SVM is presented in Figure 8. The result showed that the overall accuracy of the SVM significantly dropped, i.e., from 82.8% to 75.4%, when the tested dataset was different from the training state. We observed that this was the cause of a significant decrease in the identification process of the MP condition. By comparing the two test scenarios in Figure 8a,b, the predicted accuracy of the MP conditions in SVM declined from 28.6% to 19.0%. Meanwhile, there existed no sharp deviations in the predicted accuracy of the LOS and NLOS conditions in both test scenarios. This increased the misclassification rate of the LOS and NLOS conditions as an MP in the SVM classifier.

To be precise, a significant misclassification rate of the MP condition as an NLOS was detected in the evaluated results, i.e., the value rose from 0.6% (when the training and test were in the same condition) to 5.8% (when the training and test were in different conditions), as provided in Figure 8. The misclassification rate of LOS as MP was also quite high, i.e., it increased from 4.2% to 8.5%. The main reason could be the data collected for MP conditions when the two transceivers were too close to each other. In that case, it was acceptable to interpret the received signal in MP condition as an LOS.

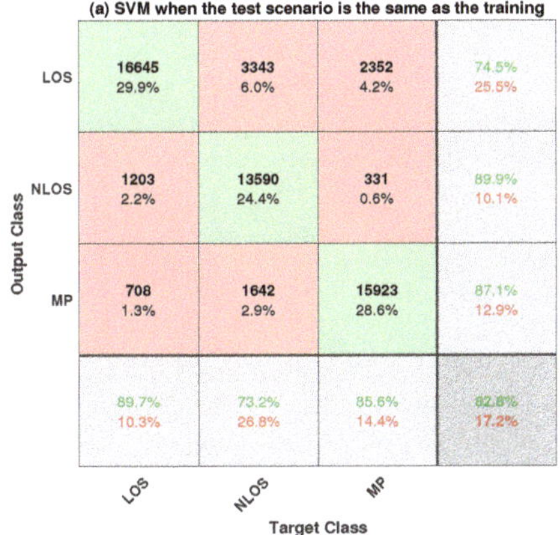

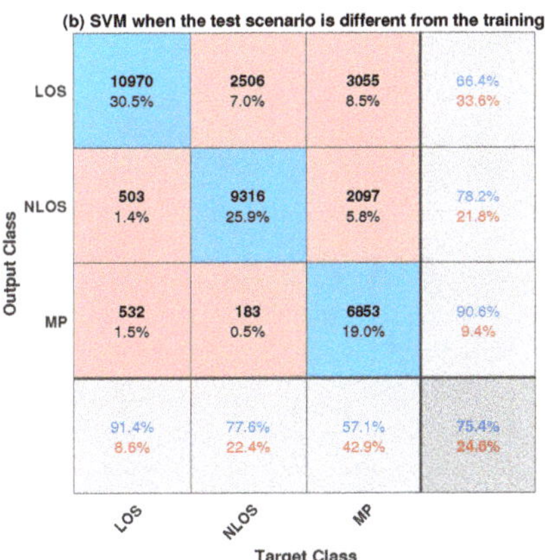

Figure 8. Comparison of the multi-label classification results for SVM using the confusion matrix in two different scenarios: (**a**) the test dataset was obtained from the same environments as in the training state; (**b**) the test dataset was collected in the environments different from the training state. In the evaluation, the radial basis function was used as the kernel for the SVM classifier.

Regarding the NLOS condition in both test scenarios, we observed quite a good outcome in the predicted accuracy, precision, and recall of the SVM classifier (Figure 8). This result was crucial because the main impact on the performance of the UWB localization algorithm was the NLOS condition (Section 2). The misclassification of the LOS as an NLOS did not produce a severe consequence on the overall performance of the UWB system. The reason was that the location algorithm assigned different weights to the classification results, and giving LOS as a smaller weight did not hurt the system performance.

7.2.2. Comparative Analysis of the Two Test Scenarios for the RF Classifier

Similar to the SVM classifier, the overall accuracy of the RF classifier degraded strikingly from 91.9 to 73.5 when the test and training environments were different (Figure 9). Again, the cause of this significant decrease in RF was evident in the predicted accuracy of the MP conditions, i.e., it reduced from 30.4% to 19.6%. Unlike the SVM classifier, the predicted accuracy of the RF classifier in both the LOS and NLOS conditions noticeably declined as well. This implied the predicted accuracy of LOS degraded from 30.9% to 26.9% while NLOS decreased from 30.6% to 27.0%. This pushed the overall accuracy of the RF classifier to be worse than the SVM in the scenario where the training and test environments were different.

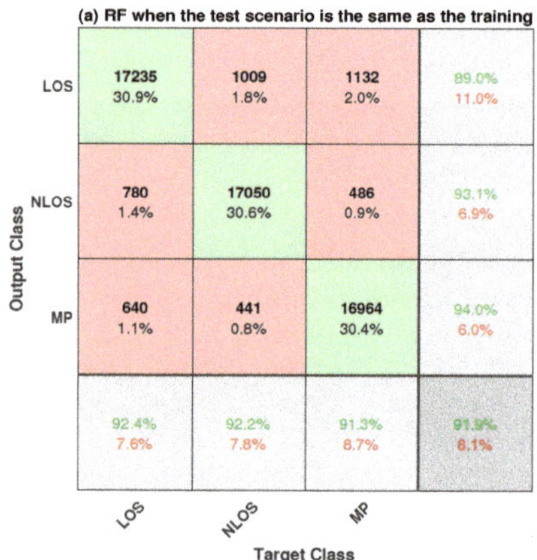

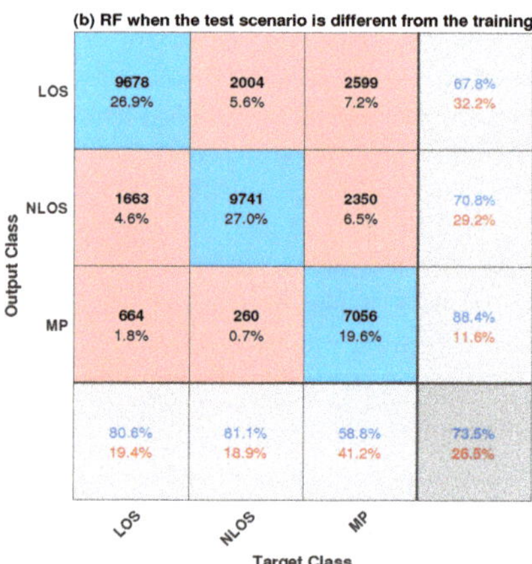

Figure 9. Comparison of the multi-class classification results for the RF classifier using the confusion matrices in two scenarios: (**a**) the test dataset obtained from the same conditions as the training data; (**b**) the test dataset collected in a different condition from the training. In the evaluation, fifty decision trees were used in the forest of the applied RF classifier.

Similar to the aforementioned reason and condition in the SVM classifier, the misclassification rate of the MP condition as either the LOS or NLOS was quite high in the RF classifier as well (Figure 9). To give the exact values, the misclassification rate of the MP condition as NLOS increased from 0.9% to 6.5%. Similarly, the misclassification of the MP as LOS grew from 2.0% to 7.2%. In the RF classifier, the misclassification rate of the NLOS as an LOS was also noticeably high, i.e., it increased from 1.8% to 5.6%. Again, the predicted accuracy of the NLOS condition in both of the two test scenarios was quite satisfying; that is, 30.6% out of 100% for the three classes in the scenario when the training and test data were in the same environments. Similarly, it was 27.0% when the data for the training and test environments were different.

7.2.3. Comparative Analysis of the Two Test Scenarios for the MLP Classifier

Figure 10 compares the multi-class classification results of two test scenarios using MLP as a classifier. Similar to the previously mentioned two classifiers, the overall accuracy of the MLP considerably declined from 91.2% to 72.9% when the test environments of the UWB were different from the training state. Repeatedly, this was caused by the false discovery rate of the MP conditions in the measured UWB data. Specifically, the predicted accuracy of MP condition in the MLP classifier decreased from 30% to 19.8% out of 100% for the three classes. This outcome showed that the MP condition was the challenging class to identify throughout our evaluation in all classifiers (Figures 8–10).

The decrease of performance in a particular class causes an increase in the false discovery rate of other classes. Specifically, for our evaluation in MLP, the misclassification rate of MP as LOS increased from 2.2% to 7.7%, while MP as NLOS increased from 1.1% to 5.9% (Figure 10). In fact, the predicted accuracy of both LOS and NLOS also declined in the MLP classifier, i.e., from 30.9% to 26.7% in the LOS condition and from 30.2% to 26.4%.

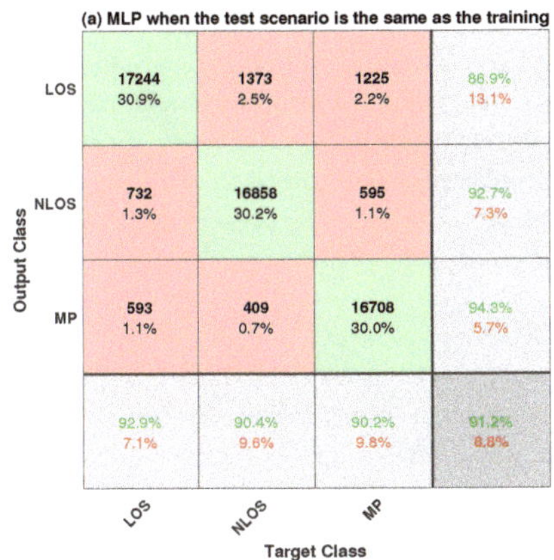

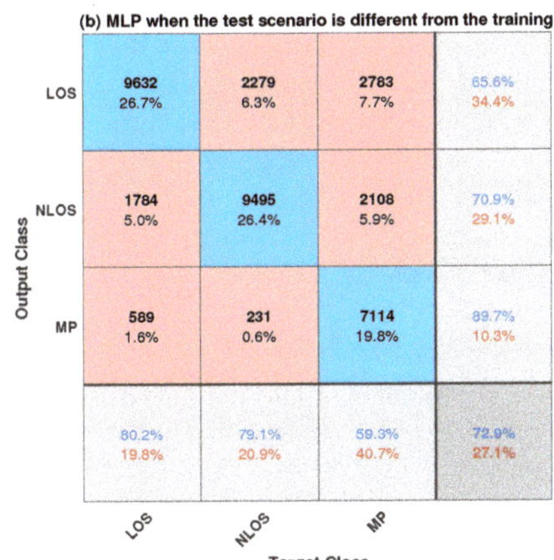

Figure 10. Comparison of the multi-class classification results for the MLP classifier using confusion matrices in two scenarios: (**a**) the test dataset obtained from the same conditions as the training phase; (**b**) the test dataset collected in a different condition from the training. In the evaluation, three fully-connected hidden layers with 100 neurons in each layer were used for the MLP classifier.

7.3. Summary of the Experimental Evaluation Results

In this paper, the multi-label classification results of the UWB data were quantified based on two metrics, i.e., the F1-score and confusion matrix. The F1-score is typically used in the literature to quantify the performance of ML-based classifiers because it provides a convenient single value score [11,42,49]. However, it can sometimes overlook the insightful information of some classes. Table 5 gives the typical summary of the two evaluated scenarios using both the F1-score and overall accuracy. The individual F1-score for each class (LOS, NLOS, and MP) is also given in the table.

Table 5. Summary of the results based on the F1-scores and overall accuracy.

Scenarios	Classifiers	Individual F1-Scores			Macro-Averaging	Overall
		LOS	NLOS	MP	F1-Scores	Accuracy (%)
Training and test environments are different	SVM	0.77	**0.78**	0.70	**0.75**	**75.35**
	RF	0.74	**0.76**	0.71	0.73	73.52
	MLP	0.72	**0.75**	0.71	0.73	72.86
Training and test environments are the same	SVM	0.81	0.81	**0.86**	0.83	82.80
	RF	0.91	**0.93**	**0.93**	**0.92**	**91.90**
	MLP	0.90	**0.92**	**0.92**	0.91	91.20

The bold numbers in the table refer to the best performance scores for each class (LOS, NLOS, and MP) and each classifier based on the metrics and scenarios.

Based on the data given in Table 5, both the macro-averaging F1-score and overall accuracy showed that the SVM classifier gave the best performance when the training and test environments were different. The consistent outcome in SVM also revealed the reason why it is one of the most frequently used classifiers in the literature [9,10,17,21]. In contrast, RF performed the best among the three evaluated classifiers regarding the same training and test environments.

However, it was hard to clarify using Table 5 that the predicted accuracy of MP condition was significantly low compared to the other two classes in all evaluated classifiers in a scenario when the training and test were in different environments. This phenomenon could be detected using the confusion matrix as previously described in Section 7.2.

8. Discussions

The identification process of the LOS, NLOS, and MP conditions in a wireless ranging system, especially in UWB, is crucial because they strongly influence the quality and accuracy of the actual measurement. For that reason, several contributions have been proposed in the literature to identify these conditions as already mentioned in Section 1. However, most of the contributions treated the issue as a binary classification problem (Section 2). To the best of our knowledge, only two papers [11,21] addressed the UWB-based classification process as a multi-class problem. In this paper, we defined three classes in UWB measurement data (LOS, NLOS, and MP conditions as presented in Section 2) and evaluated three ML classifiers (SVM, RF, and MLP as presented in Section 5) to identify these three defined classes.

Two metrics, F1-score and confusion matrix, were used for evaluating the performance of each classifier. Apart from these scores, the training and test times for each classifier were also given in our evaluation. As a matter of fact, this type of metric (training and test times) is typically ignored in the literature. However, it is undeniable that the magnitude of a test time in a certain classifier is usually vital in the overall performance of the system. Our results presented in Section 5 revealed that SVM had the poorest performance among the three classifiers in terms of the test time. In contrast, SVM gave the best performance in terms of the F1-score when different environments of training and test states were applied (Section 7). The two mentioned results showed us the interesting contradiction of two metrics in the SVM classifier, which could be an important factor in the system implementation of some ML applications.

The evaluation results based on the F1-score and overall accuracy pointed out that the measured environments in UWB had a strong effect on the performance of the three classifiers (Section 7). This referred to the striking degradation of the performance in all classifiers when different environments of training and test were applied in the evaluation. One may argue that this was caused by the overfitting of the model in the training state. However, this outcome occurred in all three evaluated classifiers regardless of the hyperparameters used in each classifier. Indeed, this was a part of the generalization problem caused by the inadequate representation of the conditions in the data. However, the data collection process in UWB, especially for the three mentioned classes, was quite time-consuming, elaborate, and costly because of the nature of the wireless signal. This meant the evaluated outcomes would be affected differently in several different ways by the conditions of the materials, types of walls, types of furniture, etc., in the measured environment. Our attempt was to provide the feasibility of the ML approach in the multi-class classification of the UWB measurement data in contrast to the conventional technique given in Section 2. The results based on the F1-score in our experiments showed 0.69 in the worst-case scenario and 0.92 in the best-case scenario. The outcomes were quite satisfying and promising.

In general, the classification results of UWB in the literature were usually reported using single value metrics, i.e., the overall accuracy score [21,23] and F1-score [11]. In some cases, the recall and true negative rate were applied in addition to the accuracy score, as in [38]. Typically, in a binary classification problem, the receiver operating characteristic (ROC) curve [50] and cumulative distribution function [9,18] are widely used. Those scores and metrics are particularly helpful for the comparative analysis of different classifiers. However, the insightful details of the actual conditions are usually overlooked or missed in some cases. Therefore, we used the complete confusion matrix in this paper to examine the individual outcomes of each class in all evaluated classifiers (Section 7.2). Using the confusion matrix, we observed that the predicted accuracy of the MP condition significantly dropped in all classifiers when different environments of training and test were used in the data. This condition cannot be clearly evident using other above-mentioned metrics. This incident also proved that the identification of the MP condition was more challenging than the other two classes (NLOS and LOS). Moreover, we observed that the predicted accuracy of the NLOS condition was generally quite good in all of the evaluations regardless of the classifiers. At the same time, the misclassification rate of NLOS as either an LOS or MP condition was also moderately low. This criterion is crucial because the highest error rate in UWB measurements is usually caused by the NLOS condition (Section 2). In contrast, the misidentification of LOS as an NLOS condition in a certain model did not hurt the overall system performance of the UWB application.

9. Conclusions

In this paper, the multi-label (LOS, NLOS, and MP) identification for UWB ranging was conducted using three machine learning techniques (SVM, RF, MLP) as a classifier. This was in contrast to the typical binary classification approaches in the literature. The experimental evaluation results based on the F1-score proved that ML-based classifiers could identify the defined three classes with a high score, i.e., 0.69 in the worst-case scenario and 0.92 in the best-case scenario. However, it was unreasonable to single out the best classifier out of the three because their performances depended extensively on the environmental changes and the metrics used to quantify them. Moreover, the insightful results based on the confusion matrix revealed that the MP condition was the most challenging to identify among the three classes. It was also evident in the confusion matrices that the predicted accuracy of the NLOS condition was quite high throughout all the evaluated experiments.

As future work, the NLOS condition could be further classified into two classes based on the study conducted in [11]. This would lead the defined classes of UWB data to become four, i.e., LOS, MP, soft-NLOS, and hard-NLOS. Indeed, this also means much more data would need to be collected to identify these four conditions in several environments using several different materials and objects. Besides, the experimental evaluation will be conducted in the actual hardware of the microcontroller-based platform instead of a PC.

Author Contributions: Conceptualization, C.L.S.; methodology, C.L.S. and J.D.H.; validation, B.S., J.D.H., M.A., M.H., and U.R.; software, C.L.S., B.S., and M.A.; writing, original draft preparation, C.L.S.; writing, review and editing, B.S., J.D.H., M.A., M.H., and U.R.; visualization, C.L.S., J.D.H., M.A., and M.H.; project administration, M.H. and U.R.; supervision, U.R.; funding acquisition, U.R. All authors read and agreed to the published version of the manuscript.

Funding: We acknowledge support for the article processing charge by the Deutsche Forschungsgemeinschaft and the Open Access Publication Fund of Bielefeld University.

Acknowledgments: This work was supported by the Cluster of Excellence Cognitive Interaction Technology "CITEC" (EXC 277) at Bielefeld University, funded by the German Research Foundation (DFG). Author Cung Lian Sang was partly supported in this work by the German Academic Exchange Service (DAAD). The authors are responsible for the contents of this publication. We also sincerely thank the anonymous reviewers for their valuable suggestions and feedback.

Conflicts of Interest: The authors declare no conflict of interest. The funders had no role in the design of the study; in the collection, analyses, or interpretation of the data; in the writing of the manuscript; nor in the decision to publish the results.

Abbreviations

The following abbreviations are used in this manuscript:

AltDS-TWR	Alternative double-sided two-way ranging
BDT	Boosted decision tree
CIR	Channel impulse response
CNN	Convolutional neural network
FP	First-path
GP	Gaussian process
HSI	High speed internal (clock)
IoT	Internet of Things
KNN	K-nearest neighbor
LOS	Line-of-sight
MCU	Microcontroller unit
ML	Machine learning
MLP	Multi-layer perceptron
MP	Multi-path
NLOS	Non-line-of-sight
PATD	Preamble accumulation time delay
PUB	Publications at Bielefeld University
RBF	Radial basis function
RF	Random forest
RSS	Received signal strength
RX	Received or receiver
SNR	Signal-to-noise ratio
SVM	Support vector machine
TOF	Time-of-flight
UWB	Ultra-wideband

References

1. Win, M.Z.; Dardari, D.; Molisch, A.F.; Wiesbeck, W.; Zhang, J. History and Applications of UWB [Scanning the Issue]. *Proc. IEEE* **2009**, *97*, 198–204. [CrossRef]
2. Sang, C.L.; Adams, M.; Korthals, T.; Hörmann, T.; Hesse, M.; Rückert, U. A Bidirectional Object Tracking and Navigation System using a True-Range Multilateration Method. In Proceedings of the 2019 International Conference on Indoor Positioning and Indoor Navigation (IPIN), Pisa, Italy, 30 September–3 October 2019; pp. 1–8. [CrossRef]
3. Dardari, D.; Conti, A.; Ferner, U.; Giorgetti, A.; Win, M.Z. Ranging With Ultrawide Bandwidth Signals in Multipath Environments. *Proc. IEEE* **2009**, *97*, 404–426. [CrossRef]
4. Win, M.Z.; Scholtz, R.A. Impulse radio: How it works. *IEEE Commun. Lett.* **1998**, *2*, 36–38. [CrossRef]
5. Win, M.Z.; Scholtz, R.A. Ultra-wide bandwidth time-hopping spread-spectrum impulse radio for wireless multiple-access communications. *IEEE Trans. Commun.* **2000**, *48*, 679–689. [CrossRef]
6. Sang, C.L.; Adams, M.; Hesse, M.; Hörmann, T.; Korthals, T.; Rückert, U. A Comparative Study of UWB-based True-Range Positioning Algorithms using Experimental Data. In Proceedings of the 2019 16th Workshop on Positioning, Navigation and Communications (WPNC), Bremen, Germany, 23–24 October 2019; pp. 1–6. [CrossRef]
7. Sang, C.L.; Adams, M.; Hörmann, T.; Hesse, M.; Porrmann, M.; Rückert, U. An Analytical Study of Time of Flight Error Estimation in Two-Way Ranging Methods. In Proceedings of the 2018 International Conference on Indoor Positioning and Indoor Navigation (IPIN), Nantes, France, 24–27 September 2018; pp. 1–8. [CrossRef]
8. Lian Sang, C.; Adams, M.; Hörmann, T.; Hesse, M.; Porrmann, M.; Rückert, U. Numerical and Experimental Evaluation of Error Estimation for Two-Way Ranging Methods. *Sensors* **2019**, *19*, 616. [CrossRef]
9. Maranò, S.; Gifford, W.M.; Wymeersch, H.; Win, M.Z. NLOS identification and mitigation for localization based on UWB experimental data. *IEEE J. Sel. Areas Commun.* **2010**, *28*, 1026–1035. [CrossRef]

10. Li, W.; Zhang, T.; Zhang, Q. Experimental researches on an UWB NLOS identification method based on machine learning. In Proceedings of the 2013 15th IEEE International Conference on Communication Technology, Guilin, China, 17–19 November 2013; pp. 473–477. [CrossRef]
11. Barral, V.; Escudero, C.J.; García-Naya, J.A.; Maneiro-Catoira, R. NLOS Identification and Mitigation Using Low-Cost UWB Devices. *Sensors* **2019**, *19*, 3464. [CrossRef]
12. Lian Sang, C.; Adams, M.; Hörmann, T.; Hesse, M.; Porrmann, M.; Rückert, U. Supplementary Experimental Data for the Paper entitled Numerical and Experimental Evaluation of Error Estimation for Two-Way Ranging Methods. *Sensors* **2019**, *19*, 616. [CrossRef]
13. Schmid, L.; Salido-Monzú, D.; Wieser, A. Accuracy Assessment and Learned Error Mitigation of UWB ToF Ranging. In Proceedings of the 2019 International Conference on Indoor Positioning and Indoor Navigation (IPIN), Pisa, Italy, 30 September–3 October 2019; pp. 1–8. [CrossRef]
14. Shen, G.; Zetik, R.; Thoma, R.S. Performance comparison of TOA and TDOA based location estimation algorithms in LOS environment. In Proceedings of the 2008 5th Workshop on Positioning, Navigation and Communication, Hannover, Germany, 27 March 2008; pp. 71–78. [CrossRef]
15. Khodjaev, J.; Park, Y.; Saeed Malik, A. Survey of NLOS identification and error mitigation problems in UWB-based positioning algorithms for dense environments. *Ann. Telecommun. Ann. Des Télécommun.* **2010**, *65*, 301–311. [CrossRef]
16. Guvenc, I.; Chong, C.; Watanabe, F. NLOS Identification and Mitigation for UWB Localization Systems. In Proceedings of the 2007 IEEE Wireless Communications and Networking Conference, Kowloon, China, 11–15 March 2007; pp. 1571–1576. [CrossRef]
17. Huang, C.; Molisch, A.F.; He, R.; Wang, R.; Tang, P.; Ai, B.; Zhong, Z. Machine Learning-Enabled LOS/NLOS Identification for MIMO System in Dynamic Environment. *IEEE Trans. Wirel. Commun.* **2020**, 1. [CrossRef]
18. Wymeersch, H.; Marano, S.; Gifford, W.M.; Win, M.Z. A Machine Learning Approach to Ranging Error Mitigation for UWB Localization. *IEEE Trans. Commun.* **2012**, *60*, 1719–1728. [CrossRef]
19. Savic, V.; Larsson, E.G.; Ferrer-Coll, J.; Stenumgaard, P. Kernel Methods for Accurate UWB-Based Ranging With Reduced Complexity. *IEEE Trans. Wirel. Commun.* **2016**, *15*, 1783–1793. [CrossRef]
20. Zandian, R.; Witkowski, U. NLOS Detection and Mitigation in Differential Localization Topologies Based on UWB Devices. In Proceedings of the 2018 International Conference on Indoor Positioning and Indoor Navigation (IPIN), Nantes, France, 24–27 September 2018; pp. 1–8. [CrossRef]
21. Kolakowski, M.; Modelski, J. Detection of direct path component absence in NLOS UWB channel. In Proceedings of the 2018 22nd International Microwave and Radar Conference (MIKON), Poznan, Poland, 14–17 May 2018; pp. 247–250. [CrossRef]
22. Niitsoo, A.; Edelhäusser, T.; Mutschler, C. Convolutional Neural Networks for Position Estimation in TDoA-Based Locating Systems. In Proceedings of the 2018 International Conference on Indoor Positioning and Indoor Navigation (IPIN), Nantes, France, 24–27 September 2018; pp. 1–8. [CrossRef]
23. Cwalina, K.K.; Rajchowski, P.; Blaszkiewicz, O.; Olejniczak, A.; Sadowski, J. Deep Learning-Based LOS and NLOS Identification in Wireless Body Area Networks. *Sensors* **2019**, *19*, 4229. [CrossRef] [PubMed]
24. Leech, C.; Raykov, Y.P.; Ozer, E.; Merrett, G.V. Real-time room occupancy estimation with Bayesian machine learning using a single PIR sensor and microcontroller. In Proceedings of the 2017 IEEE Sensors Applications Symposium (SAS), Glassboro, NJ, USA, 13–15 March 2017; pp. 1–6. [CrossRef]
25. Lian Sang, C.; Steinhagen, B.; Homburg, J.D.; Adams, M.; Hesse, M.; Rückert, U. Supplementary Research Data for the Paper entitled Identification of NLOS and Multi-path Conditions in UWB Localization using Machine Learning Methods. *Appl. Sci.* **2020**. [CrossRef]
26. Pedregosa, F.; Varoquaux, G.; Gramfort, A.; Michel, V.; Thirion, B.; Grisel, O.; Blondel, M.; Prettenhofer, P.; Weiss, R.; Dubourg, V.; et al. Scikit-learn: Machine Learning in Python. *J. Mach. Learn. Res.* **2011**, *12*, 2825–2830.
27. Borras, J.; Hatrack, P.; Mandayam, N.B. Decision theoretic framework for NLOS identification. In Proceedings of the VTC '98. 48th IEEE Vehicular Technology Conference. Pathway to Global Wireless Revolution (Cat. No.98CH36151), Ottawa, ON, Canada, 21 May 1998, Volume 2; pp. 1583–1587, doi:10.1109/VETEC.1998.686556. [CrossRef]
28. Schroeder, J.; Galler, S.; Kyamakya, K.; Jobmann, K. NLOS detection algorithms for Ultra-Wideband localization. In Proceedings of the 2007 4th Workshop on Positioning, Navigation and Communication, Hannover, Germany, 22 March 2007; pp. 159–166. [CrossRef]

29. Denis, B.; Daniele, N. NLOS ranging error mitigation in a distributed positioning algorithm for indoor UWB ad-hoc networks. In Proceedings of the International Workshop on Wireless Ad-Hoc Networks, Oulu, Finland, 31 May–3 June 2004; pp. 356–360. [CrossRef]
30. Wu, S.; Zhang, S.; Xu, K.; Huang, D. Neural Network Localization With TOA Measurements Based on Error Learning and Matching. *IEEE Access* **2019**, *7*, 19089–19099. [CrossRef]
31. Maali, A.; Mimoun, H.; Baudoin, G.; Ouldali, A. A new low complexity NLOS identification approach based on UWB energy detection. In Proceedings of the 2009 IEEE Radio and Wireless Symposium, San Diego, CA, USA, 18–22 January 2009; pp. 675–678. [CrossRef]
32. Xiao, Z.; Wen, H.; Markham, A.; Trigoni, N.; Blunsom, P.; Frolik, J. Non-Line-of-Sight Identification and Mitigation Using Received Signal Strength. *IEEE Trans. Wirel. Commun.* **2015**, *14*, 1689–1702. [CrossRef]
33. Gururaj, K.; Rajendra, A.K.; Song, Y.; Law, C.L.; Cai, G. Real-time identification of NLOS range measurements for enhanced UWB localization. In Proceedings of the 2017 International Conference on Indoor Positioning and Indoor Navigation (IPIN), Sapporo, Japan, 18–21 September 2017; pp. 1–7. [CrossRef]
34. Decawave. DW1000 User Manual: How to Use, Configure and Program the DW1000 UWB Transceiver. 2017. Available online: https://www.decawave.com/sites/default/files/resources/dw1000_user_manual_2.11.pdf (accessed on 5 June 2020).
35. Musa, A.; Nugraha, G.D.; Han, H.; Choi, D.; Seo, S.; Kim, J. A decision tree-based NLOS detection method for the UWB indoor location tracking accuracy improvement. *Int. J. Commun. Syst.* **2019**, *32*, e3997. [CrossRef]
36. Smith, D.R.; Pendry, J.B.; Wiltshire, M.C.K. Metamaterials and Negative Refractive Index. *Science* **2004**, *305*, 788–792. [CrossRef]
37. Bregar, K.; Hrovat, A.; Mohorcic, M. NLOS Channel Detection with Multilayer Perceptron in Low-Rate Personal Area Networks for Indoor Localization Accuracy Improvement. In Proceedings of the 8th Jožef Stefan International Postgraduate School Students' Conference, Ljubljana, Slovenia, 31 May–1 June 2016; Volume 31.
38. Krishnan, S.; Xenia Mendoza Santos, R.; Ranier Yap, E.; Thu Zin, M. Improving UWB Based Indoor Positioning in Industrial Environments Through Machine Learning. In Proceedings of the 2018 15th International Conference on Control, Automation, Robotics and Vision (ICARCV), Singapore, 18–21 November 2018; pp. 1484–1488. [CrossRef]
39. Fan, J.; Awan, A.S. Non-Line-of-Sight Identification Based on Unsupervised Machine Learning in Ultra Wideband Systems. *IEEE Access* **2019**, *7*, 32464–32471. [CrossRef]
40. STMicroelectronics. UM1724 User Manual: STM32 Nucleo-64 Boards (MB1136). 2019. Available online: https://www.st.com/resource/en/user_manual/dm00105823-stm32-nucleo-64-boards-mb1136-stmicroelectronics.pdf (accessed on 5 June 2020).
41. STMicroelectronics. STM32L476xx Datasheet—Ultra-low-power Arm Cortex-M4 32-bit MCU+FPU, 100DMIPS, up to 1MB Flash, 128 KB SRAM, USB OTG FS, LCD, ext. SMPS. 2019. Available online: https://www.st.com/resource/en/datasheet/stm32l476je.pdf (accessed on 5 June 2020).
42. Charte, F.; Rivera, A.J.; del Jesus, M.J.; Herrera, F. Addressing imbalance in multilabel classification: Measures and random resampling algorithms. *Neurocomputing* **2015**, *163*, 3–16. doi:10.1016/j.neucom.2014.08.091. [CrossRef]
43. Cortes, C.; Vapnik, V. Support-vector networks. *Mach. Learn.* **1995**, *20*, 273–297. [CrossRef]
44. Suykens, J.; Vandewalle, J. Least Squares Support Vector Machine Classifiers. *Neural Process. Lett.* **1999**, *9*, 293–300.:1018628609742. [CrossRef]
45. Smola, A.J.; Schölkopf, B. A tutorial on support vector regression. *Stat. Comput.* **2004**, *14*, 199–222.:STCO.0000035301.49549.88. [CrossRef]
46. Breiman, L. Random Forests. *Mach. Learn.* **2001**, *45*, 5–32.:1010933404324. [CrossRef]
47. Hastie, T.; Tibshirani, R.; Friedman, J. *The Elements of Statistical Learning*; Springer Series in Statistics; Springer New York Inc.: New York, NY, USA, 2001.
48. Rumelhart, D.E.; Hinton, G.E.; Williams, R.J. Learning representations by back-propagating errors. *Nature* **1986**, *323*, 533–536. [CrossRef]
49. Sokolova, M.; Lapalme, G. A systematic analysis of performance measures for classification tasks. *Inf. Process. Manag.* **2009**, *45*, 427–437. [CrossRef]

50. Choi, J.; Lee, W.; Lee, J.; Lee, J.; Kim, S. Deep Learning Based NLOS Identification With Commodity WLAN Devices. *IEEE Trans. Veh. Technol.* **2018**, *67*, 3295–3303. [CrossRef]

Sample Availability: The experimental research data and the corresponding source code used in this paper are publicly available in PUB—Publication at Bielefeld University [25].

Article

Bearing Estimation for Indoor Localization Systems Using Planar Circular Photodiode Arrays

Gergely Zachár [1], Gergely Vakulya [1] and Gyula Simon [2,*]

1 ElectrIT, 8200 Veszprém, Hungary; zachar@electrit.hu (G.Z.); vg@electrit.hu (G.V.)
2 Alba Regia Technical Faculty, Óbuda University, 8000 Székesfehérvár, Hungary
* Correspondence: simon.gyula@amk.uni-obuda.hu

Received: 30 April 2020; Accepted: 21 May 2020; Published: 26 May 2020

Abstract: An inexpensive bearing estimation sensor and a corresponding method has been recently proposed for indoor localization applications. The system utilizes modulated infrared LED sources as light beacons, and a planar circular photodiode array (PCPA) as sensor. The PCPA measures the light intensity of the beacons in multiple channels, from which the bearings of the beacons can be estimated using least squares (LS) method, with an accuracy in the range of 1 degree. In this paper a novel estimation method is proposed, which provides fast bearing estimates from the PCPA measurements using the frequency domain. The computational complexity of the novel method is orders of magnitude less than that of the LS solution, at the price of a slight decrease in accuracy. The performance of the PCPA is analyzed in the presence of reflections and the tilting of the sensor. The results demonstrate that the effect of reflections can be significant, while the tilting of the sensor has only a minor effect on the bearing estimation. The applicability of the measurement device in indoor localization applications is illustrated by real localization examples.

Keywords: bearing estimation; azimuth estimation; signal processing; position estimation; photodiode array

1. Introduction

Most indoor localization systems utilize some beaconing infrastructure (e.g., radio, light, sound sources) along with corresponding sensors to estimate various properties of the beacons (e.g., distance between the sensor and the beacon, angle of arrival and direction of the beacon, view angles between beacons, etc.). From the measured properties various location estimation methods can be used to produce the location estimates for the tracked object, which can be either a sensor or a beacon. In the first case, usually, multiple beacons are deployed at known positions and the sensor's position is unknown, while in the latter case, multiple sensors are utilized with known positions to measure the unknown position of a beacon.

In this paper we focus on systems using angle of arrival (AoA) or angle difference of arrival (ADoA) measurements [1]. Some of these localization systems use acoustic sources [2,3], an interesting example being weapon localization where the beacon is the muzzle blast of the weapon [2], but the majority of AoA/ADoA systems utilize optical sources, e.g., blinking LEDs [4–13]. In such systems bearings are measured between the sources and the sensors, and based upon the angle measurements the unknown location is estimated using various means, e.g., least squares estimates [4], consensus functions [5], or exhaustive search [4].

The direction of a beacon can be measured in three dimensions (3D), when two angles, the azimuth and the elevation, are determined, or in two dimensions (2D) only, when one angle, the bearing (i.e., azimuth), is measured. The 3D approach allows localization in 3D with arbitrary sensor orientation, while 2D measurements can be utilized when certain constraints are fulfilled, namely in this case the sensor orientation must be either measured or known (typical use case is when the sensor is heading upwards) [4,5]. The main advantage of the 2D approach is that the computational complexity can be decreased significantly [6]. The measurement method proposed in this paper provides 2D bearing measurements.

For the measurement of bearings of light sources image sensors (e.g., cameras) can be utilized. The tilting of the sensor can be either measured by auxiliary sensors [4,5] or can be estimated from the image itself [7,8].

Light sensors (e.g., photodiodes) offer a more cost-effective solution; such devices utilize the directional sensitivity of the light sensors. The quadrant photodiode angular diversity aperture (QADA), which utilizes four specially arranged photodiodes on a plane, with an auxiliary aperture was proposed [9]. Traditional photodiodes (PDs), arranged in various geometrical formations were also proposed to measure bearings: in [10] and [11] the PDs were attached on the edges of a cube, while the planar circular photodiode array (PCPA) utilizes PDs placed on the surface of a cylinder [12,13].

In this paper the performance properties of the PCPA are analyzed. The PCPA was proposed with a corresponding least squares (LS) bearing estimation in [12]. The hardware architecture was improved in [13]. In this paper a novel computationally efficient bearing estimation, using frequency domain, is proposed. The performance of the measurement methods in the presence of various physical error sources is analyzed. The results demonstrate that the effect of reflections can be significant, while the tilting of the sensor has only minor effect on the bearing estimation. The applicability of the measurement device in indoor localization applications is illustrated by real localization examples.

2. The Planar Circular Photodiode Array (PCPA)

2.1. Background of Operation

The operation of the proposed method is based on the directional sensitivity of the photodiodes. Figure 1 shows the measured normalized angular response of a photodiode, type SFH205F [14], as a function of angle φ of the incoming light. The left-hand side is shown in a polar system, while the right-hand side in a Cartesian coordinate system. Notice that the maximum sensitivity value was set to unity, hence the name normalized. Apart from the direction, the detected light intensity (i.e., the current flowing through the photodiode) also depends on the emitted light intensity and the distance between the source and the sensor.

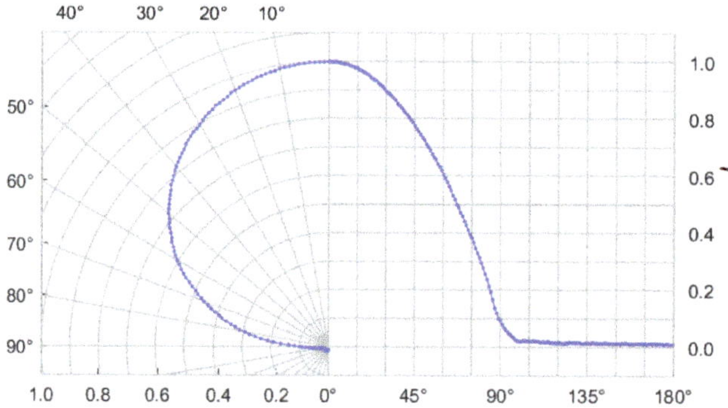

Figure 1. The normalized radial sensitivity of a photodiode.

Using multiple photodiodes, the dependence of the distance and the source light intensity can be cancelled, thus angle φ can be measured. Our goal was to create a sensor and a corresponding measurement method which can determine the angle of the incoming light in 2 dimensions. Thus it is supposed that the light source and the sensor are placed (approximately) on the same plane. The proposed measurement setup will be introduced in the next Section.

2.2. Measurement Method

The measurement setup is shown in Figure 2. The system contains beacons $B^{(1)}, B^{(2)}, \ldots, B^{(K)}$, which are infrared transmitters, each of which having a unique blinking frequency, which allows the separation of the beacon signals in the measurements. The PCPA consists of N PDs arranged evenly along a cylinder, facing outwards, thus the angle between neighboring channels is $\Delta\varphi = 360/N$. The beacons and the PCPA are assumed to be approximately on the same plane.

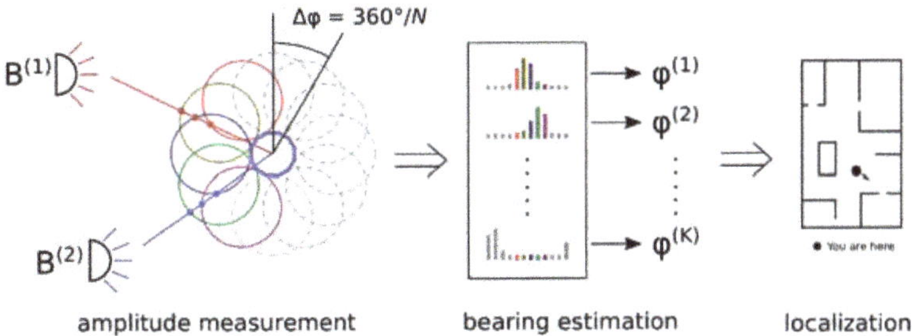

Figure 2. The bearing measurement process using planar circular photodiode array (PCPA). Different colors represent different channels; left: sensitivity curves, middle: measured amplitudes.

Notice that each PCPA channel measures the additive incoming light intensity, coming from all beacons. The beacons' signals are separated in the frequency domain by the digital signal processing unit [12] and the detected amplitude of each beacon is measured by each channel. Using the measured intensities the bearing $\varphi^{(k)}$, $k = 1, 2, \ldots, K$ of each beacon $B^{(k)}$ is determined. A least squares (LS) and a fast frequency domain (FD) method will be introduced in Sections 2.3 and 2.4, respectively.

2.3. Least-Squares Estimatior

The least-squares angle estimate is reviewed here based on the results of [12] and [13]. The incoming light intensities from a particular light source are measured and the N-element measurement vector denoted with Ψ, as follows:

$$\Psi = [\Psi_0, \Psi_1, \ldots \Psi_{N-1}]^T, \tag{1}$$

where Ψ_i is the measured light intensity in channel i and $(.)^T$ is the transpose operator.

A calibration matrix $R_{N \times M}$ is created where N is the number of channels in the PCPA and M is the number of calibration measurements. Row n of R corresponds to a particular PCPA channel n, and the values in this row contain the measured channel sensitivity values, sampled in M equidistant angles φ_m, $0 \leq m < M - 1$, where

$$\varphi_m = m \cdot 360/M. \tag{2}$$

Notice that Figure 1 contains such a normalized measurement for one channel. See Section 4.1 for further multichannel measurement examples.

On the other hand, columns of R correspond to directions φ_m, as follows:

$$R = \begin{bmatrix} R_0 & R_1 & \ldots & R_{M-1} \end{bmatrix}, \tag{3}$$

where column R_m contains intensity values $r_{n,m}$ for each channel n, measured using a light source with bearing φ_m :

$$R_m = \begin{bmatrix} r_{0,m} & r_{1,m} & \ldots & r_{N-1,m} \end{bmatrix}^T \tag{4}$$

The least squares bearing estimate will be the angle φ_m for which the shape of R_m is closest to the shape of Ψ, in the least squares sense. Recall that the measured light intensity depends not only on the bearing but also on the distance between the light source and the sensor, and also on the power of the light source. Thus we use a scaling in order to define the distance between R_m and Ψ, as follows:

$$E_m(\lambda) = \| R_m - \lambda\Psi \|_2 . \tag{5}$$

In (5) E_m is the squared error between the calibration vector R_m and the measured vector Ψ, and λ is the unknown scaling factor. The least squares cost function for angle φ_m is defined as the smallest possible value of $E_m(\lambda)$, as follows:

$$e_m = \min_{\lambda} E_m(\lambda) \tag{6}$$

As a first step, we determine the optimal value for λ, given the value of m. From (2) and (3) it directly follows that

$$E_m(\lambda) = \sum_{n=0}^{N-1} (r_{n,m} - \lambda\Psi_n)^2 = \sum_{n=0}^{N-1} \left(r_{n,m}^2 + (\lambda\Psi_n)^2 - 2\lambda r_{n,m}\Psi_n \right) \tag{7}$$

At the minimum of E_m the derivative $\frac{dE_m}{d\lambda}$ is zero:

$$\frac{dE_m}{d\lambda} = \frac{d}{d\lambda}\left(\sum_{n=0}^{N-1} r_{n,m}^2 + \lambda^2 \sum_{n=0}^{N-1} (\Psi_n)^2 - 2\lambda \sum_{n=0}^{N-1} r_{n,m}\Psi_n \right) = 2\lambda \sum_{n=0}^{N-1} (\Psi_n)^2 - 2\sum_{n=0}^{N-1} r_{n,m}\Psi_n = 0 \tag{8}$$

From (8), the optimum value of λ_m is the following:

$$\lambda_m = \frac{\sum_{n=0}^{N-1} r_{n,m}\Psi_n}{\sum_{n=0}^{N-1} (\Psi_n)^2} = \frac{(R_m)^T\Psi}{\Psi^T\Psi}, \tag{9}$$

Using the optimal value of λ_m, the cost function (6) now can be calculated as follows:

$$e_m = \min_{\lambda} E_m(\lambda) = E_m(\lambda_m) = \| R_m - \lambda_m\Psi \|_2 . \tag{10}$$

Notice that each index m corresponds to bearing φ_m. Thus index m_{opt}, for which the cost function (9) is minimal, defines the best bearing estimate. The optimal bearing index m_{opt} is the following:

$$m_{opt} = \operatorname*{argmin}_{m} e_m, \tag{11}$$

and, using (2), is the bearing estimate is

$$\hat{\varphi} = \varphi_{m_0} = m_0 \cdot \frac{360}{M}. \tag{12}$$

Notices:

(1) The resolution of the bearing estimate (11) is $d\hat{\varphi} = 360/M$, i.e., the higher M is, the better the resolution. Observing Figure 1, it is apparent that the sensitivity curves are smooth, which enables interpolation between measured values. The sensitivity value $r_{n,m'}$ for an arbitrary angle φ', for which $\varphi_m < \varphi' < \varphi_{m+1}$, can be calculated as follows:

$$r_{n,m'} = r_{n,m} + (r_{n,m+1} - r_{n,m}) \frac{\varphi' - \varphi_m}{\varphi_{m+1} - \varphi_m}. \tag{13}$$

Using (13), the size of the stored calibration matrix can be kept low, while the resolution can arbitrarily be increased.

(2) The computational complexity of the least squares estimate as follows: the calculation of λ_m requires approx. $8N$ operations (multiplications, divisions, additions, and subtractions). The brute force minimization requires the calculation of $8NM$ calculations, while a gradient search requires $8NL$ operations, where L is the number of gradient search steps. In practice L may be smaller than M, but the number of calculations is still at least an order of magnitude higher than $8N$. In the current implementation the calculation of a bearing estimate requires 150 ms in a PC-class computer. Thus the implementation of the LS estimator is troublesome on low-scale microcontrollers.

2.4. Fast Angle Estimation in the Frequency Domain

To overcome the high computational complexity of the LS estimator, described in Section 2.3, a fast computation method will be proposed, where the angle estimation is made in the frequency domain.

The following assumptions are made:

- The relative sensitivity characteristics of the individual PCPA channels are almost identical. The universal normalized sensitivity characteristic is $S_{rel}(\varphi)$, where S_{rel} is periodic with 2π;
- The sensitivity characteristics are even symmetrical.

The justification of the assumptions can be seen in the measurements shown in Figure 3. Figure 3a shows the normalized sensitivity characteristics of 16 PCPA channels (these are in fact the radial sensitivities of the applied individual photodiodes), while Figure 3b shows their maximum difference between the measured normalized sensitivities. As the figure shows, the difference is indeed small; this is expected since the same type of photodiode was used in each channel. Figure 3a also shows that the sensitivity curve is indeed even symmetrical.

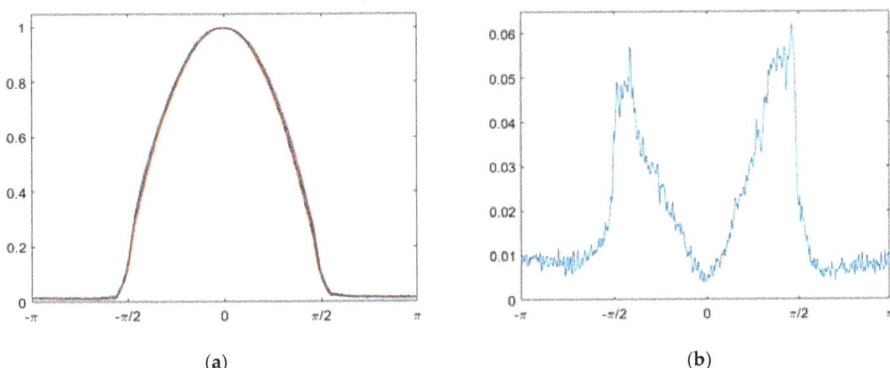

Figure 3. Sensitivity of 16 PCPA channels as a function of the direction of incoming light: (**a**) normalized sensitivity of all PCPA channels (each color shows a different channel), (**b**) maximum difference between the normalized channel sensitivities.

Since the channels are rotated by $\Delta\varphi = 360/N$ (see Figure 2), the incoming light angle, coming from the same source, changes by $\Delta\varphi$ at each channel. Thus the Ψ_i $(i = 0, 1, \ldots, N-1)$ values, provided by a measurement, can be considered as equidistant samples of the $S_{rel}(\varphi)$ sensitivity curve, as follows:

$$\Psi_i = S_{rel}(\varphi + i\Delta\varphi). \tag{14}$$

The measurement, modeled as a sampling process, is illustrated in Figure 4, for incoming angle values of $\varphi = 0$ and $\varphi = -120$, for the photodiode shown in Figure 1. The N–element sample vector, as a function of φ, is denoted by $\Psi(\varphi)$. The angle estimation should determine the value of φ, from the measured vector $\Psi(\varphi)$.

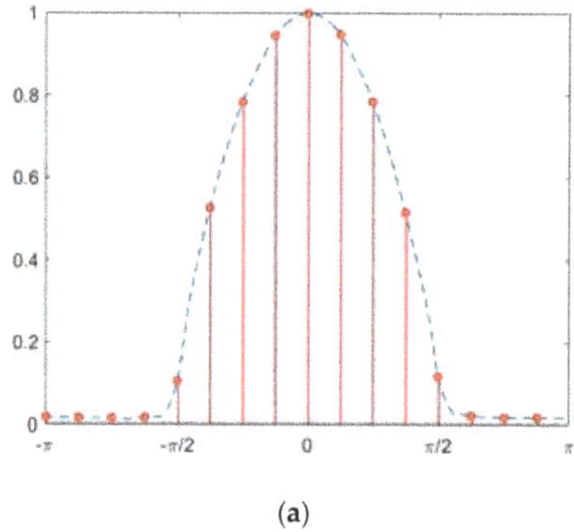

(**a**)

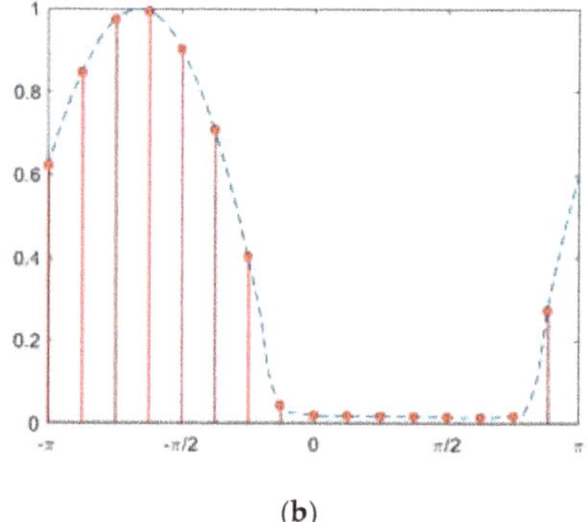

(**b**)

Figure 4. Measurement values, as samples of the sensitivity curve. (**a**) $\varphi = 0$, (**b**) $\varphi = -120$. Blue line: continuous sensitivity curve, red dots: samples.

The case when the light source is in direction zero, i.e., $\varphi = 0$, as in Figure 4a is examined. $S(0)$ is the DFT of the sampled sensitivity curve $\Psi(0)$, as follows:

$$S(0) = DFT(\Psi(0)). \tag{15}$$

Since $\Psi(0)$ is even symmetric, $S(0)$ is real, i.e., the angle of all harmonic components in $S(0)$, including the fundamental harmonic $S_1(0)$, is zero. In considering a case of an incoming light with arbitrary angle of φ, where the DFT of $\Psi(\varphi)$ is $S(\varphi)$:

$$S(\varphi) = DFT(\Psi(\varphi)) \tag{16}$$

Using the shift property of the DFT, the value of the fundamental harmonic $S_1(\varphi)$ of $S(\varphi)$ can be expressed as follows:

$$S_1(\varphi) = e^{-j\varphi} S_1(0). \tag{17}$$

Since $S_1(0)$ is real, the angle of $S_1(\varphi)$ is

$$\angle S_1(\varphi) = -\varphi. \tag{18}$$

Thus, the frequency domain (FD) angle estimation can be carried out in the following way: First calculate the first harmonic component of the DFT of the measured vector Ψ, as follows:

$$S_1 = \sum_{k=0}^{N-1} \Psi_k w^k. \tag{19}$$

where

$$w^k = e^{-j\frac{2\pi}{N}k} \tag{20}$$

Then the angle estimate $\hat{\varphi}$ is the following:

$$\hat{\varphi} = -\angle S_1. \tag{21}$$

The computational demand of the fast method is only approx. $2N$ (real) operations, which is significantly smaller, than that of the least squares estimate.

3. Measurement System

The architecture of the measurement system is shown in Figure 5. The beacons generate infrared light, blinking with unique frequencies. The beacons contain a PIC12F1572 microcontroller, a driver stage and a high power infrared LED. The illustration of Figure 5 contains $K = 4$ beacons.

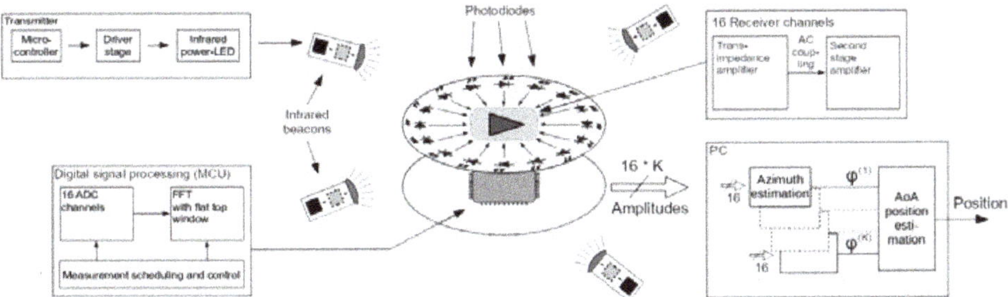

Figure 5. The measurement system containing the beacons, the 16-channel PCPA with the digital signal processing unit, the bearing estimation and the final position estimation units.

The current implementation of the PCPA consists of $N = 16$ photodiodes. These SFH205F types have a cylindrical front side [14]. Each of them is followed by a two-stage transimpedance amplifier, providing fourth order lowpass filtering and 130 dBOhm gain. After multiplexing, the amplified and filtered signals are sampled and converted by the internal analog-digital converter of the PIC32MZ2048EFM064 microcontroller with 40 kHz/channel. The spectra of the 16 signals are calculated by the microcontroller with fast Fourier transform using a flat top window and a transform size of 1024. For each beacon, the amplitude, corresponding to the beacon frequency, is measured on each channel, thus the PCPA provides $16K$ amplitude values with an update frequency of 39 Hz.

The second processing step is currently performed on a PC, where the bearing estimates are computed using either the LS or the FD estimate, as described in Sections 2.3 and 2.4, respectively. The current LS estimator can provide bearing values with 6 Hz on a PC, while the FD estimator's update frequency is higher than 1000 Hz. The execution time of the FD estimator, implemented and optimized on the microcontroller, is approx.. 100 µs.

The bearing measurements are then used to estimate the position of the sensor. Using the known beacon locations and the measured bearings the position estimation can be computed by any AoA/ADoA method, e.g., exhaustive search [4], consensus functions [5], geometric methods [6], or LS methods [4] may be utilized The location estimates on the PC can be calculated to provide an update for each bearing measurement (i.e., with 39 Hz).

The photo of the PCPA is shown in Figure 6. The PCPA sensor (photodiodes) is a cylinder with diameter of 38 mm and height of 20 mm. In the current implementation the analog and digital electronic circuits are located on two circular PCBs with diameter of 100 mm (no attempt to optimize the size was made). The photo also shows the turntable, used in the measurements.

Both the beacons and the sensor are of low cost: the bill of material (BOM) cost is USD 14 per beacon (including main power supply), and USD 30 per receiver.

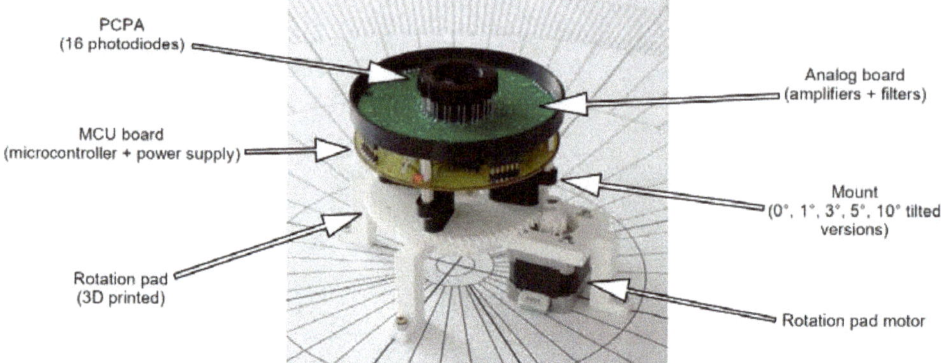

Figure 6. Photo of the PCPA deployed on the turntable, used in the automatic measurements.

4. Measurements and Evaluation

In this section measurements are presented to analyze the sensitivity of the PCPA to two practically important error sources. The first error source is reflections: the beacon signals may be reflected from various surfaces near the PCPA, thus not only the line of sight signals, but other disturbing components will be measured, causing error. The second possible error source is the tilting of the sensor: in practical scenarios the sensor's alignment may not be perfect, thus the not homogenous vertical sensitivity of the applied PDs may cause error.

In the subsequent sections, first the reference measurement is introduced. The impact of the error sources will be analyzed on

- raw measurements, i.e., the measured light intensities (Section 4.2),
- bearing estimations (Section 4.3),
- and finally, localization results in some illustrative scenarios (Section 4.4).

The generic test setup can be seen on Figure 7. The center piece is the PCPA itself, which was placed on a turntable. With this turntable the measurement device could be accurately rotated around the vertical axis. Ideally the PCPA's normal vector is vertical, but to investigate the effect of tilting, the PCPA may be mounted on the turntable with various tilting angles, denoted by α in Figure 7. The turntable allowed the measurement of the radiant intensity characteristics automatically with a resolution of 1°.

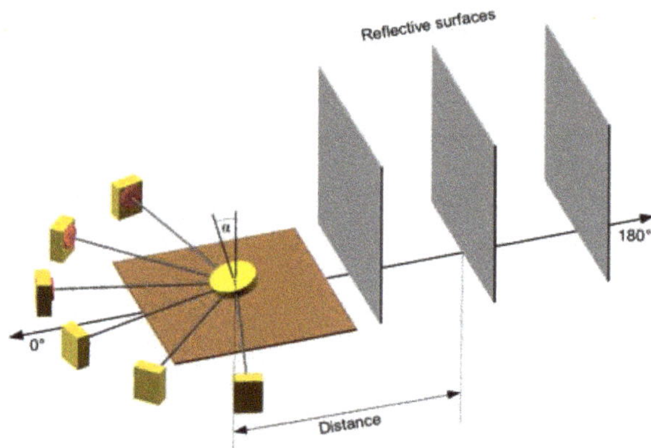

Figure 7. The measurement setup with the light sources, PCPA, and reflective surfaces.

The beacons were deployed on one side of the PCPA, around direction zero, as shown in Figure 7. On the other side of the PCPA (in direction 180°) various reflective surfaces were placed at different distances from the PCPA. The reflective materials, including white paper, black paper, aluminum foil, and mirror, were placed on 80 × 80 cm wooden surfaces.

It is important to note, that the measurements were carried out in a 5 m × 4 m office, which was furnished and had several large white areas on the walls. Thus some of the measurements may include the influence of the reflections of the environment.

4.1. Reference Measurement

The LS estimator requires the reference matrix R, which contains the reference intensity characteristics of each channel. Recall that each row of R corresponds to a channel, while each column corresponds to a bearing. The values of R can be produced by a measurement, carried out as follows. A beacon was placed in front of the PCPA, at direction of 0°. To minimalize the effects of reflections a black foam was placed on the opposite side of the PCPA (between directions of (−90°)–90°, and the beacon was placed in a tube pointing to the measurement device.

During the measurement, the PCPA was rotated with the turntable with a resolution of 1°, while the light intensity was averaged from 20 measurements at each angular position. For better visibility, the data corresponding to each channel were normalized. The results of the measurements can be seen for all the 16 channels, and the three selected channels in Figure 8.

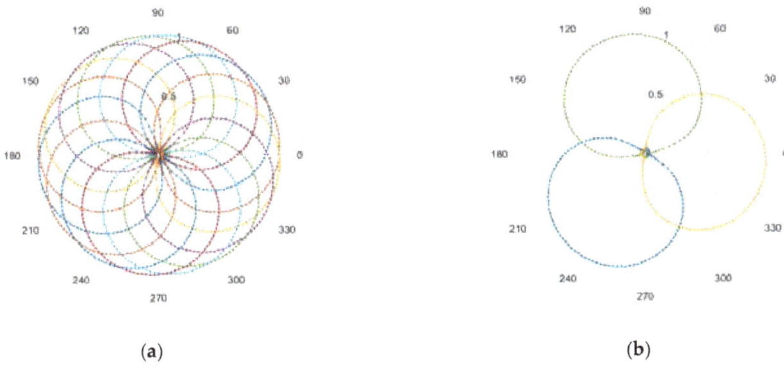

Figure 8. Radiant sensitivity of the measurement channels. The plot shows the normalized reference matrix of a 16-channel PCPA. Different colors represent different rows of the matrix, corresponding to different channels; the mth value in a row is plotted at angle φ_m. (**a**) All 16 channels of the PCPA. (**b**) For better, visibility 3 selected channels.

Note, that the resolution of the reference matrix R can be increased with interpolation, as described in Section 2.3. Additionally, note that the resulting characteristics are relatively "smooth" and almost identical (see also Figure 3, showing the same measurements).

4.2. Light Intensity

In the case of the reference measurement we strove to eliminate environmental factors, mainly reflections. However, it is interesting and also crucial to know what kinds of deviations are produced in the measurements in non-ideal conditions. In this chapter the effect of reflective materials and the tilt of the PCPA are be analyzed on the recorded light intensities.

4.2.1. Effects of Reflective Surfaces

For examining the effect of reflecting surfaces, a series of measurements were conducted utilizing the test setup depicted in Figure 7. In the measurements four beacons were placed at ±22.5° and ±67.5°, and an aluminum foil was used as a reflective material. The reflective surface was placed at 0.5, 1.0, and 1.5 m from the center of the PCPA. During the measurements, the PCPA made a full rotation with a resolution of 2°, and the light intensity values were recorded at each angular position.

In Figure 9 the results of the experiments can be seen, where the blue lines represent the normalized light intensity values of the measurement channel facing 0°. The angular placements of the beacons are represented with the red dots while the reflective surface is always at 180°, denoted by a black arc; the length of the arc representing the viewing angle of the surface (the higher the distance between the surface and the PCPA the smaller the viewing angle). The first row of the figure represents the reference measurement where no artificial reflective material was present (but some unwanted reflection from the walls still may be present here).

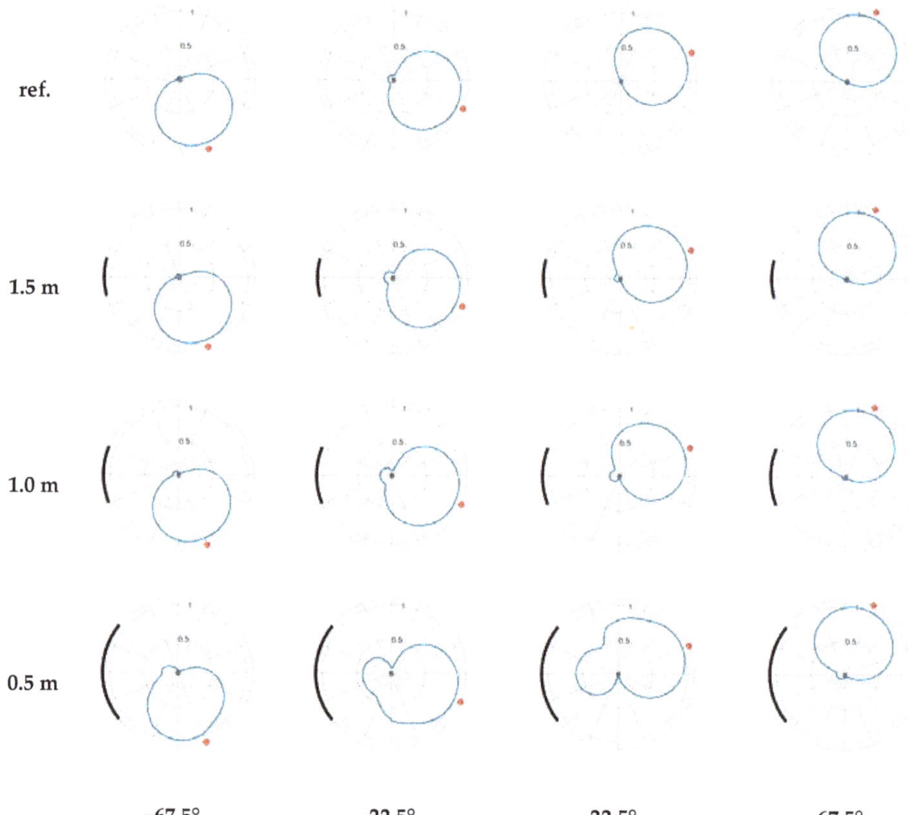

Figure 9. Comparison of light intensity measurements with 4 beacons. From top to bottom: no reflective surface, aluminum foil at 1.5, 1.0, and 0.5 m. The red dot represents the angular placement of the beacon, black arcs denote the reflective surfaces.

In Figure 9 several trends can be clearly seen: with decreasing distance between the PCPA and the reflective surface the deviations from the ideal (reference) characteristic are increasing. The most prominent abnormality is the bulb-like shape, created by the reflection of the beacon. In these angular positions the measured values should be zero because the photodiode at these positions is facing backwards, thus no direct line-of-sight is present.

It is important to note that the measurement shown in Figure 9 is an extreme and conspicuous case of reflectivity, where a large piece (80 cm) of aluminum foil was used. For comparison, Figure 10 shows measurements of different materials with the same setup at a distance of 0.5 m. From the measurements it shows that the aluminum foil may have greater impact than the mirror (at least in the setup shown in Figure 10), probably due to the combined effect of reflection and scattering. Black paper has a minor, but still visible, effect, while white paper also shows significant impact.

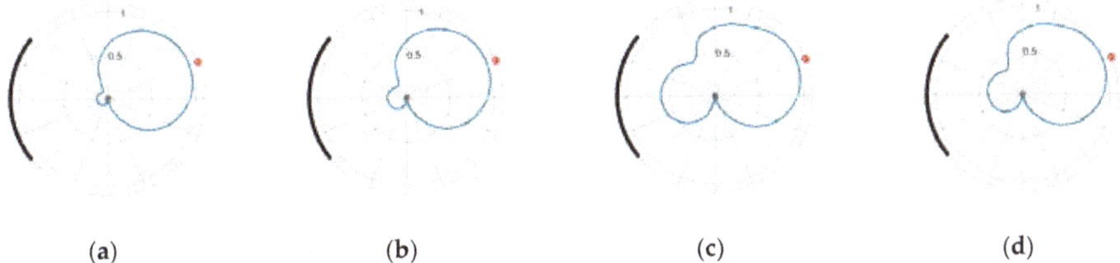

(a) (b) (c) (d)

Figure 10. Comparison of light intensity measurements with different reflective materials at 0.5 m: (**a**) black paper, (**b**) white paper, (**c**) aluminum foil, (**d**) mirror. Blue line: measured intensity, red dot: angular placement of the beacon.

4.2.2. Effects of PCPA Tilt

In principle, the PCPA was designed with the assumption that both the beacons and the utilized photodiodes are aligned on a plane, i.e., the normal vector of the PCPA is vertical. However, in real world applications (e.g., localization of a mobile robot platform) the PCPA may be tilted. Thus it is important to analyze the deviations caused by a small amount of tilt.

In our experiments we utilized the setup and rotation pad depicted on Figure 6. The PCPA was assembled with multiple mounts, which provided a tilt of 0°, 1°, 3°, 5°, and 10°.

The results of the measurements can be seen on Figure 11, where the maximum absolute amplitude errors are plotted relative to the 0° tilt measurement values. The exact utilized error function was derived as follows. The measured not-normalized sensitivity characteristics of channel m, measured at bearing φ, using tilt angle of α, are denoted by $S_m(\varphi, \alpha)$. The reference measurement is $S_m(\varphi, 0)$, which is shown in Figure 8. The measurement error $\Delta S_{tilt}(\varphi, \alpha)$ at bearing angle φ, caused by the tilting of the sensor, is defined as follows:

$$\Delta S_{tilt}(\varphi, \alpha) = \max_m \left(abs \left(\frac{S_m(\varphi, 0)}{\max(S_m(\varphi, 0))} - \frac{S_m(\varphi, \alpha)}{\max(S_m(\varphi, 0))} \right) \right) \tag{22}$$

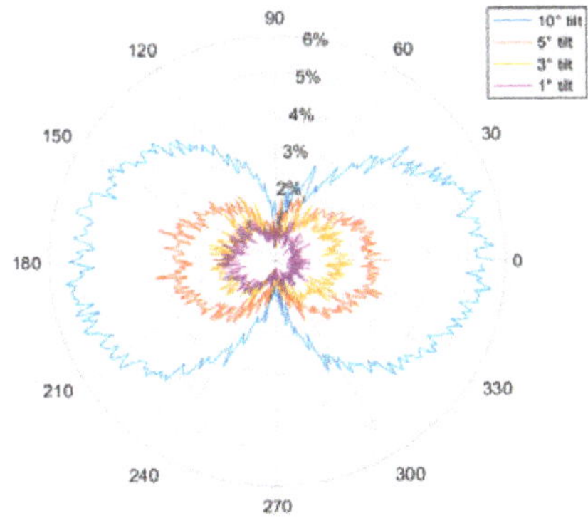

Figure 11. Maximum absolute error of the relative amplitude measurements with different amount of sensor tilt.

Notice that in (22) the first term is the normalized reference sensitivity characteristics of channel m, while the second term is measured with the tilted sensor using the same (reference) normalization factor. Thus (22) provides a metric to characterize the measurement error, due to tilting. Figure 11 shows the measured ΔS_{tilt} values for $\alpha = 1, 3, 5, 10$.

As Figure 11 shows, the largest error is expected in directions toward the tilt, i.e., in our case towards directions 0° and 180°. The error is small in directions perpendicular to the tilting direction, in this case 90° and 270°. It should be noted that the error never reaches 0 and a small asymmetry is also present, due to the measurement and processing noise and a minimal reflection from the environment. Additionally, it is notable that the $\alpha = 10$ tilt for the PCPA is an exaggerated case: most of the intended applications for the device provide less deviation (e.g., in case of industrial mobile robots the tilt is limited by the maximum inclination of ramps).

Based on the results shown in Figure 11, it is evident that the effect of the sensor tilt is negligible to the measured amplitude values (mostly below 5%) compared to the deviations caused by reflective surfaces (e.g., Figures 9 and 10, where the deviation can be as much as 10%–660%).

It is important to note, that the relatively low amplitude variations caused by the tilt of the PCPA is due to the shape of the photodiodes. The selected type has a cylindrical front side, in contrast to the most common half sphere shape, providing smaller vertical sensitivity.

4.3. Bearing Estimation

The next stage of the processing chain is the bearing estimation phase. In this section, the effect of the reflections and the tilt of the PCPA will be examined to the angle estimation. In the evaluation, both the LS and the FD estimation methods will be utilized and the results will be compared.

4.3.1. Effects of Reflective Surfaces

Measurements with reflective surfaces were conducted utilizing a measurement setup similar to the one shown in Figure 7, but in this case six beacons were used at the angular positions of ±15°, ±45°, and ±75°. During the measurements full rotations were made with resolution of 2°. As a reflective surface, white paper and aluminum foil were assembled on an 80 × 80 cm board. The distances of the reflective materials in the measurements were set to 0.5, 1.0, and 1.5 m.

The measured angle estimation errors are shown in Figure 12, while the numerical values can be seen in Table 1.

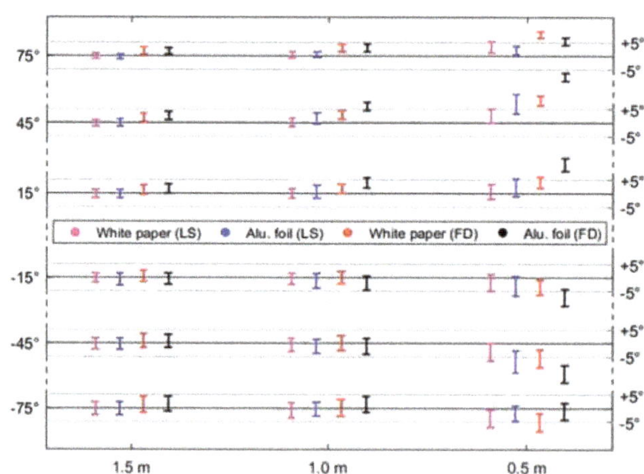

Figure 12. Bearing estimation errors with different reflective materials, different distances, and both with the least squares (LS) and frequency domain (FD) angle estimation method. The whiskers correspond to $\pm 3\sigma$.

Table 1. The mean and standard deviation of the bearing estimation errors.

Reflective Material/Method	1.5 m	1.0 m	0.5 m
White paper (LS)	0.8°/0.6°	1.0°/0.6°	3.0°/0.9°
Alu. Foil (LS)	0.9°/0.6°	1.5°/0.7°	4.3°/1.1°
White paper (FD)	1.7°/0.7°	2.0°/0.7°	6.2°/0.8°
Alu. Foil (FD)	2.0°/0.7°	3.5°/0.7°	9.3°/0.8°

As the results show, by decreasing the distance between the PCPA and the reflective material the error of the estimated angle increases. The proximity of the reflective surface increases the bias of the calculated angle, the direction of the error also depending on the angle between the beacon and the reflective material. The FD estimator clearly shows higher bias compared to the LS estimator, while the standard deviation of the measurements are similar. In worst case (aluminum foil at 0.5 m), the mean error was as much as 4.3° for the LS, and 9.3° for the FD estimator. For realistic cases (distance 1.5 m) the error is around 1° and 2° for the LS and FD estimators, respectively.

4.3.2. Effects of PCPA Tilt

To analyze the angle estimation error caused by the tilt of the PCPA we conducted multiple measurements. In our experiments we utilized the setup and rotation pad depicted on Figures 6 and 7. The PCPA was assembled with multiple mounts, which provided tilt values of 0°, 1°, 3°, 5°, and 10°.

The bearing estimation errors can be seen in Figure 13, where the errors are plotted for both the LS and FD estimator. The errors were defined as the differences between the bearing estimates using the 1°, 3°, 5°, and 10° tilted measurements, and the bearing estimates using the 0° tilt reference measurement.

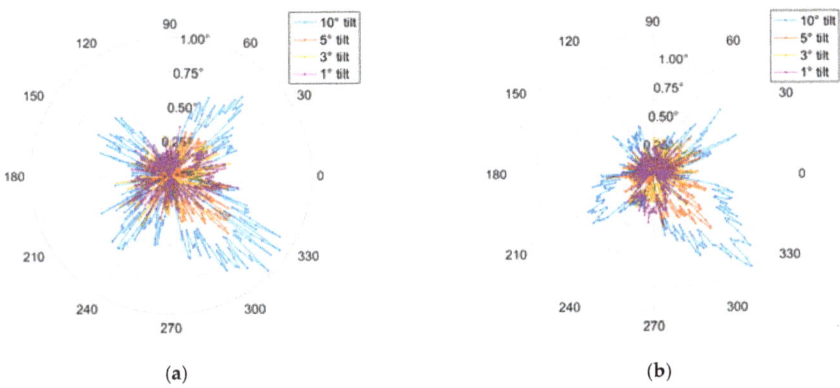

Figure 13. Measured relative angle estimation error with different amount of tilt. (**a**) LS estimator, (**b**) FD estimator.

As Figure 13 shows, the angle errors due to the tilt of the PCPA are negligible, usually below 0.25° (except the 10° case), which is the same magnitude as the measurement error of the current system. Note, that reflective surfaces cause several times higher angle estimation inaccuracies.

4.4. Localization Measurements

To illustrate the potential application of the proposed bearing measurement method an indoor measurement setup was built. In this section, the performance of the system is analyzed through the localization accuracies with various amounts of reflections and tilt.

The system was installed in a 5 m by 4 m furnished office room, with several white areas on the walls. The measurements utilized five or six beacons as anchor points. The nine reference positions were on the grid shown in the center of Figure 14. At each reference point the sensor was rotated, taking 72 independent measurements. From the measured bearing values, the sensor position was estimated using a least-squares ADoA-based position estimator.

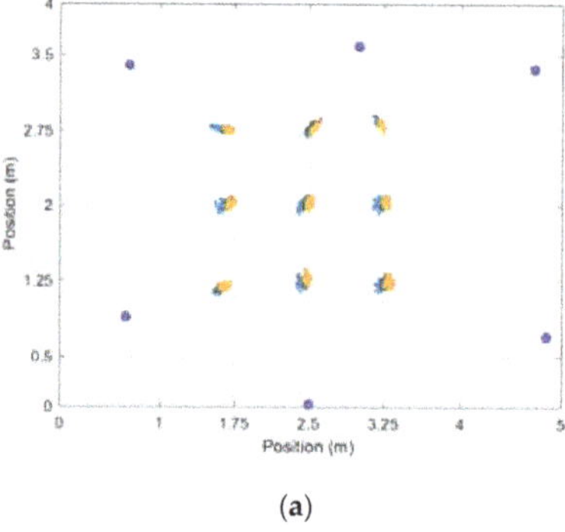

(**a**)

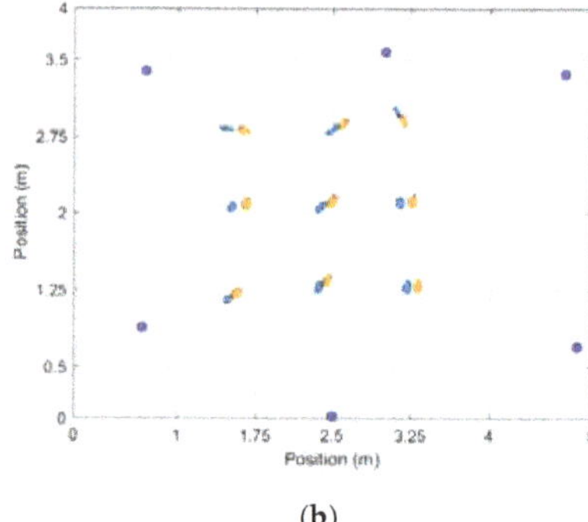

(**b**)

Figure 14. Localization with various amount of reflections, utilizing 6 beacons: (**a**) LS estimator, (**b**) FD estimator. Dark big blue dots: beacon positions, light blue dots: localization with reflective surface next to the PCPA, yellow dots: localization without reflective surface.

4.4.1. Localization with Various Amounts of Reflections

In our first experiment, six beacons were utilized. At each reference point, two independent measurement runs were made: a "normal" measurement with no artificial reflections, and a measurement with white paper placed next to the PCPA (50 cm to the right) as a reflective surface. The placement of the beacons and the measurement results can be seen in Figure 14, utilizing the LS and the FD methods for bearing estimation.

The statistical results of the experiments are presented in Table 2, where the localization accuracies are also calculated for a case where only four beacons are present. In these scenarios the measurements of the top and bottom beacons are excluded from the localization.

Table 2. Mean and standard deviation of the localization error (units in cm).

Setup	4 Beacon (LS)	4 Beacons (FD)	6 Beacon (LS)	6 Beacon (FD)
normal	8.37/5.64	16.37/9.80	6.63/3.47	14.36/3.92
with reflections	15.02/8.02	28.78/11.34	9.93/5.52	18.39/8.56

As can be seen in Figure 14 and Table 2, the localization accuracy based on the LS estimator outperforms the FD estimator in similar cases. This is probably due to the bias caused by the reflections of the environment, to which the FD estimator is more sensitive than the LS estimator (see Figure 12).

It is noteworthy that the white paper placed right to the reference positions are clearly biased the position estimates to the left. Without artificial reflections, which can be considered "normal" operation condition, the accuracy of the systems is around 7–8 cm for the LS, and 14–16 cm for the FD estimator. The effect of artificial reflections is clearly higher when the number of beacons is only four (since the redundancy is small), resulting in 15–28 cm of mean error; while for the six-beacon case the effect is smaller (due to higher redundancy), increasing the error to 10–18 cm.

4.4.2. Localization Width Tilted PCPA

Another localization experiment was conducted utilizing five beacons, with a similar beacon arrangement as can be seen in Figure 14. The purpose of the measurements was to quantify the localization error caused by the tilt of the PCPA. Table 3 summarizes the statistical results of the experiments.

Table 3. Mean and standard deviation of the localization error with 5 beacons (units in cm).

Setup	LS Estimator	FD Estimator
0° tilt	5.27/3.05	10.19/3.77
10° tilt	5.36/3.18	9.93/3.76

As the data in Table 3 indicates, the tilted and non-tilted measurements are almost identical. This indicates that the utilized hardware, more precisely the cylindrical shape of the photodiode is highly tolerant to the tilt of the PCPA.

5. Conclusions

Planar circular photodiode arrays provide a cost-efficient method to measure bearings for indoor localization. In the measurements presented here, modulated infrared LED sources are used for the beacons, and the sensor is a set of photodiodes (channels), equally arranged on the surface of a cylinder. The measurement process provides amplitude estimates for each beacon at each channel, from which the bearing estimation process provides the azimuth estimates.

A novel fast frequency domain bearing estimation method was proposed. The advantage of the proposed solution is that it has much lower computational complexity than the least squares method. The disadvantage of the fast solution is that the accuracy is somewhat decreased: according to the measurements, the fast method is more sensitive to reflections.

The application of the proposed system requires that the transmitters and the PCPA be at the same plane. Tilting of the sensor, or uncontrolled elevation of the transmitters may cause additional error. According to measurements, the effect of this error source is negligible if the tilting of the sensor or, equivalently, the elevation of the beacon, remains below 10°.

The effect of reflections was also studied. The results show that the method is sensitive to reflections, reflective surfaces in the vicinity of 1 m can cause severe degradation of the estimates. Real measurement results indicate that in realistic circumstances the accuracy of the bearing estimation is approximately 1° for the least squares and 2° for the fast method.

The applicability of the measurement method for indoor localization was also demonstrated. In a 5 m by 4 m office the mean localization error was 8 cm and 16 cm for the least squares and the fast methods, respectively, using four beacons.

Author Contributions: Conceptualization, G.Z. and G.S.; methodology, G.S., G.Z., and G.V.; software, G.Z. and G.V.; validation, G.Z., G.V. and G.S.; writing—original draft preparation, G.Z., G.V., G.S.; writing—review and editing, G.S. All authors have read and agreed to the published version of the manuscript.

Funding: This research received no external funding.

Conflicts of Interest: The authors declare no conflict of interest.

References

1. Bergen, M.H.; Schaal, F.S.; Klukas, R.; Cheng, J.; Holzman, J.F. Toward the implementation of a universal angle-based optical indoor positioning system. *Front. Optoelectron.* **2018**, *11*, 116–127. [CrossRef]
2. Serrenho, F.G.; Apolinário, J.A., Jr.; Ramos, A.L.L.; Fernandes, R.P. Gunshot airborne surveillance with rotary wing UAV-embedded microphone array. *Sensors* **2019**, *19*, 4271. [CrossRef] [PubMed]
3. Cui, X.; Yu, K.; Zhang, S.; Wang, H. Azimuth-only estimation for TDOA-based direction finding with three-dimensional acoustic array. *IEEE Trans. Instrum. Meas.* **2020**, *69*, 985–994. [CrossRef]
4. Huynh, P.; Yoo, M. VLC-based positioning system for an indoor environment using an image sensor and an accelerometer sensor. *Sensors* **2016**, *16*, 783. [CrossRef] [PubMed]
5. Simon, G.; Zachár, G.; Vakulya, G. Lookup: Robust and accurate indoor localization using visible light communication. *IEEE Trans. Instrum. Meas.* **2017**, *66*, 2337–2348. [CrossRef]

6. Rátosi, M.; Simon, G. Real-time localization and tracking using visible light communication. In Proceedings of the 2018 International Conference on Indoor Positioning and Indoor Navigation (IPIN), Nantes, France, 24–27 September 2018; pp. 1–8.
7. Zhu, B.; Cheng, J.; Wang, Y.; Yan, J.; Wang, J. Three-dimensional vlc positioning based on angle difference of arrival with arbitrary tilting angle of receiver. *IEEE J. Sel. Areas Commun.* **2018**, *36*, 8–22. [CrossRef]
8. Li, Y.; Ghassemlooy, Z.; Tang, X.; Lin, B.; Zhang, Y. A VLC smartphone camera based indoor positioning system. *IEEE Photonics Technol. Lett.* **2018**, *30*, 1171–1174. [CrossRef]
9. Cincotta, S.; Neild, A.; He, C.; Armstrong, J. Visible light positioning using an aperture and a quadrant photodiode. In Proceedings of the 2017 IEEE Globecom Workshops (GC Wkshps), Singapore, 4–8 December 2017; pp. 1–6.
10. Arafa, A.; Jin, X.; Klukas, R. Wireless indoor optical positioning with a differential photosensor. *IEEE Photonics Technol. Lett.* **2012**, *24*, 1027–1029. [CrossRef]
11. Arafa, A.; Dalmiya, S.; Klukas, R.; Holzman, J. Angle-of-arrival reception for optical wireless location technology. *Opt. Express* **2015**, *23*, 7755–7766. [CrossRef] [PubMed]
12. Zachár, G.; Vakulya, G.; Simon, G. Azimuth estimation for indoor localization using redundant planar circular photodiode array. In Proceedings of the 10th International Conference on Indoor Positioning and Indoor Navigation (IPIN 2019), Pisa, Italy, 30 September–3 October 2019; pp. 1–8. Available online: http://ceur-ws.org/Vol-2498/short20.pdf (accessed on 14 April 2020).
13. Zachár, G.; Vakulya, G.; Simon, G. Bearing estimation using planar circular photodiode arrays. In Proceedings of the 2020 IEEE International Instrumentation and Measurement Technology Conference, Torino, Italy, 22–25 May 2020.
14. Osram. SFH 205 F Silicon PIN Photodiode with Daylight Blocking Filter Datasheet. Available online: https://dammedia.osram.info/media/resource/hires/osram-dam-5488340/SFH%20205%20F_EN.pdf (accessed on 14 April 2020).

Article

Single-Camera Trilateration

Yu Zhou [1,*], Wenfei Liu [2], Xionghui Lu [2] and Xu Zhong [2]

1 Department of Engineering, College of Engineering, State University of New York Polytechnic Institute, Utica, NY 13502, USA

2 Department of Mechanical Engineering, College of Engineering and Applied Sciences, Stony Brook University, Stony Brook, NY 11794, USA; wenfliu@ic.sunysb.edu (W.L.); luxionghui@gmail.com (X.L.); zhongxu2@gmail.com (X.Z.)

* Correspondence: zhouy2@sunypoly.edu

Received: 2 October 2019; Accepted: 4 December 2019; Published: 9 December 2019

Abstract: This paper presents a single-camera trilateration scheme which estimates the instantaneous 3D pose of a regular forward-looking camera from a single image of landmarks at known positions. Derived on the basis of the classical pinhole camera model and principles of perspective geometry, the proposed algorithm estimates the camera position and orientation successively. It provides a convenient self-localization tool for mobile robots and vehicles equipped with onboard cameras. Performance analysis has been conducted through extensive simulations with representative examples, which provides an insight into how the input errors and the geometric arrangement of the camera and landmarks affect the performance of the proposed algorithm. The effectiveness of the proposed algorithm has been further verified through an experiment.

Keywords: localization; trilateration; mobile robots

---

1. Introduction

This paper introduces a single-camera trilateration scheme that determines the instantaneous 3D pose (position and orientation) of a regular forward-looking camera, which can be moving, based on a single image of landmarks taken by the camera. It provides a convenient self-localization tool for mobile robots and other vehicles with onboard cameras.

Self-localization is a fundamental problem in mobile robotics. This refers to how a mobile robot localizes itself in its environment. Tremendous effort has been made to study this topic. The approach of this paper is inspired by the principle of trilateration which, given the correspondence of external references, is the most adopted external reference-based localization technique for mobile robots.

Trilateration refers to positioning an object based on the measured distances between the object and multiple references at known positions [1,2], i.e., locating an object by solving a system of equations in the form of:

$$(\mathbf{p}_i - \mathbf{p}_0)^T(\mathbf{p}_i - \mathbf{p}_0) = r_i^2 \tag{1}$$

where $\mathbf{p}_0$ denotes the unknown position of the object, $\mathbf{p}_i$ the known position of the ith reference point, and r_i the measured distance between $\mathbf{p}_0$ and $\mathbf{p}_i$. (Even when more than three references are involved in positioning, we still call it "trilateration" instead of "multilateration", because "multilateration" has been used to name the positioning process based on the difference in the distances from three or more references [3].)

Equation (1) represents the classical ranging-based trilateration problem which depends on measuring the distances between the target object and references. Existing algorithms include: closed-form solutions which have low time complexity, such as the algebraic algorithm proposed by Fang [4] and Ziegert and Mize [5] which refers to the base plane, that by Manolakis [6] which adapts to an arbitrary frame of reference (a few typos in [6] were fixed by Rao [7]), that by Coope [8] which finds the intersection points of n spheres in P^n based on Gaussian elimination, and the line intersection algorithm proposed by Pradhan et al. [9], as well as the geometric solution developed by Thomas and Ros using Cayley–Menger determinants [2], etc.; and numerical methods which provide optimal estimation by minimizing the residue of (1) in some form, such as least-squares estimation [8,10,11], Taylor-series [12], the extended Kalman filtering (EKF) [13–15], linear algebraic techniques [8,16], and particle filtering [17,18], etc. Our work in [19] proposed a closed-form algorithm for the non-linear least-squares trilateration formulation that is dominantly addressed with numerical methods, which provides near optimal estimation with low time complexity. Correspondingly, range sensors, which provide direct distance measurements based on the time of flight of the signals, have been dominantly used in trilateration systems, such as those based on ultrasonic [20–34], radio frequency (RF) [4,14,15,17,18,35–45], laser [5], and infrared [46] signals, just to name a few. Existing ranging-based trilateration systems mostly adopt a GPS (Global Positioning System)-like configuration with either "fixed transmitters + onboard receiver" or "onboard transmitter + fixed receivers". They require installing and maintaining relevant signaling devices in the target environment. The accuracy of the time-of-flight measurement depends on the accuracy of synchronization among the transmitters and receivers. Trilateration for moving objects needs to accommodate the time difference among sequentially received ranging signals and the effect of movement. The signal delay caused by obstacles, which is not easy to detect and filter, is a common source of localization inaccuracy. An onboard ranging-based self-trilateration system, which transmits ranging signals and positions itself based on reflected signals, is highly challenging to realize, due to the difficulty in identifying references from only the reflected signals.

Meanwhile, cameras have become major onboard sensors for autonomous mobile robots in various navigation tasks, due to the comprehensive amount of information contained in images. Correspondingly, vision-based trilateration is considered a natural extension of the classical ranging-based trilateration. Stereovision-based trilateration systems have been proposed [47–49]. Compared with ranging-based trilateration which depends on signaling between the separate transmitter and receiver, vision-based trilateration determines the ranges to landmarks from onboard-taken camera images, and thus achieves self-trilateration given the correspondence of landmarks. Vision-based trilateration provides a complementary scheme which naturally avoids some aforementioned common issues with ranging-based trilateration, such as synchronization among signaling devices and obstacle-caused signal delay. Because a camera can capture multiple landmarks at the same time, a camera-based trilateration system can thus obtain simultaneous distance measurements from multiple landmarks. One intrinsic challenge for stereovision is to match the landmark projections in the two simultaneous images, in particular when the numbers of observed landmarks are different in the two images, the two cameras view different subsets of landmarks (which are only partly overlapped), or landmarks are projected close to each other on the same image. Mismatching in landmark projections will result in invalid landmark reconstruction, and hence may cause serious errors in localization. This problem can be avoided if trilateration can be achieved using one camera.

Thus, by comparing it with the ranging-based and stereovision-based trilateration schemes, we feel that single-camera trilateration has certain advantages in the category of trilateration techniques for mobile robots and vehicles. However, there has been a lack of systematic discussion on single-camera trilateration so far, with very limited work using single-camera ranging to assist localization, e.g., as a constraint to correct the odometry for 2D robot self-localization [50]. With the intention to fully explore the capability of single-camera trilateration as an independent 3D self-localization scheme, in this paper we propose a general algorithm for single-camera trilateration that can determine the instantaneous 3D pose of a moving camera from a single image of three or more landmarks at known positions, and give a systematic discussion on the formulation and performance analysis of single-camera trilateration. Minimizing the requirements on the hardware system, the proposed algorithm is based on the classical pinhole model of the regular forward-looking camera. When the camera is fixed on a mobile robot, the self-localization of the camera is equivalent to the self-localization of the robot, where the pose of the camera and that of the robot are different by a fixed camera-robot transformation which can be obtained by the robot hand-eye calibration [51,52]; when the camera is movable on the robot, e.g., pan and tilt, the robot pose can be calculated from the camera pose by incorporating the camera-robot transformation. The proposed scheme extends the classical ranging-based trilateration formulation to encompass the geometry of single-camera imaging and defines an integrated trilateration procedure that fits with the nature of single-camera imaging. It also makes the estimation of the camera orientation a natural part of the complete trilateration process, by taking advantage of the geometric information contained in the image.

As the ranging-based trilateration requires the input of the global positions of reference objects, the camera-based trilateration requires the correspondence between observed landmarks and a map of known landmarks. Focusing on discussing the proposed single-camera trilateration algorithm itself in this paper, we refer to the approaches in the literature [53–57] for the solutions of this correspondence problem. Minimally speaking, the proposed single-camera trilateration algorithm is highly applicable to indoor environments with identity-encoded artificial landmarks, such as shape patterns and color patterns, and outdoor environments with salient natural landmarks, such as uniquely-shaped mountain peaks and buildings. It can also be incorporated into a simultaneous localization and mapping (SLAM) process which generates a map of those distinct landmarks while locating the mobile robot with respect to the map.

We initially came up with the concept of single-camera trilateration and carried out some preliminary work based on simulation in [58]. This current paper will present a fully developed single-camera trilateration algorithm with comprehensive performance analysis. Compared with our preliminary work, the work of this current paper made highly significant and extensive further development. The specific contributions on top of our preliminary work include:

(1) The proposed single-camera trilateration algorithm is fully developed with more systematic derivation and well-adopted notations.

(2) The algorithm is developed with high level of completion. In particular, in the preliminary work, the camera orientation estimation algorithm only estimated the directional vector of the camera optical axis; the current work largely enhances the camera orientation estimation by providing a complete estimation of the 3D rotation matrix of the camera frame relative to the environment frame.

(3) A comprehensive performance analysis of the proposed single-camera trilateration algorithm is carried out. In the preliminary work, the performance analysis of the algorithm considered mainly the effect of the landmark position errors in the environment and image; in the current work, a much more extensive performance analysis is carried out with the consideration of not only the effect of the landmark errors but also that of the errors in camera parameters. Moreover, along with the new camera orientation estimation algorithm, the corresponding performance indices are re-defined, and an enhanced performance analysis is carried out. In addition, the performance analysis in the current work is carried out with a much larger number of simulation data.

(4) An experiment is carefully set up and carried out to further verify the effectiveness of the proposed single-camera trilateration algorithm, and check the estimation error under the combined effect of various input errors. Our preliminary work was simulation-based. The experimental work tests the proposed algorithm in the physical environment, which is a significant addition to the simulation-based performance analysis.

The rest of this paper is laid out as follows. Section 2 will present the derivation of the proposed algorithm. Section 3 will present the error analysis of the proposed algorithm. Section 4 will present the experimental verification. Section 5 will summarize this work.

2. Single-Camera Trilateration Algorithm

2.1. Notation and Assumptions

The proposed single-camera trilateration algorithm is derived on the basis of the classical pinhole camera model (Figure 1). Here, $X_gY_gZ_g$ denotes the global frame of reference attached to the environment in which the camera is moving, and a vector defined in $X_gY_gZ_g$ is labeled with a superscript "g", e.g., ${}^g\mathbf{v}$; $X_cY_cZ_c$ denotes the camera frame with the origin at the optical center $\mathbf{p}_0$ and Z_c axis pointing along the optical axis from $\mathbf{p}_0$ towards the image center $\mathbf{q}_0$, and a vector defined in $X_cY_cZ_c$ is labeled with a superscript "c", e.g., ${}^c\mathbf{v}$; UV denotes the image frame with the origin at the upper left corner of the image, and a vector defined in UV is labeled with a superscript "m", e.g., ${}^m\mathbf{v}$; UVW is the corresponding 3D extension of UV. The vectors in $X_gY_gZ_g$, $X_cY_cZ_c$ and UVW are 3-dimensional, while those in UV are 2-dimensional. A position defined in $X_gY_gZ_g$ or $X_cY_cZ_c$ is in the unit of millimeters, while that defined in UV or UVW is in the unit of pixels.

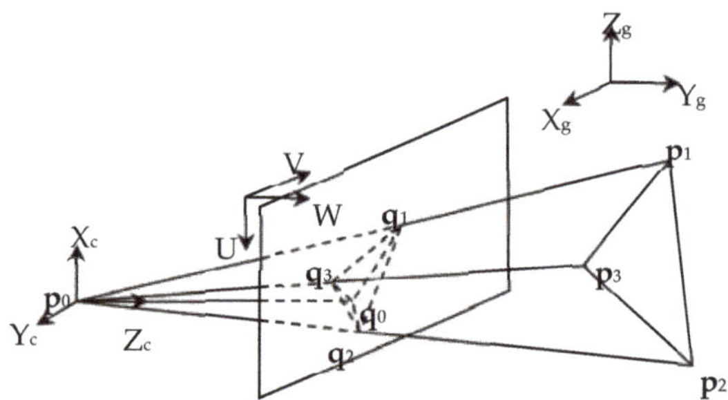

Figure 1. Pinhole camera model. This figure is adapted from Figure 1 in [58]. (© (2007) IEEE. Reprinted, with permission, from [Liu W.; Zhou Y. Recovering the position and orientation of a mobile robot from a single image of identified landmarks. In Proceedings of 2007 IEEE/RSJ International Conference on Intelligent Robots and Systems, 2007; pp. 1065–1070.]).

In particular, $\mathbf{p}_0$ denotes the optical center of the camera, and its positions defined in $X_gY_gZ_g$ and $X_cY_cZ_c$ are denoted by ${}^g\mathbf{p}_0$ and ${}^c\mathbf{p}_0$, respectively. $\mathbf{q}_0$ denotes the image center (also known as the principal point)—The intersection between the optical axis and image plane (which are perpendicular to each other), and its position defined in UV is denoted by ${}^m\mathbf{q}_0$. The distance between $\mathbf{p}_0$ and $\mathbf{q}_0$ is known as the focal length f. Assuming that there are N landmarks captured by the camera (where $N' \geq 3$ for 3D trilateration according to the principle of trilateration), we denote the reference point of the ith landmark as $\mathbf{p}_i$, where $i \in \{1,2, \ldots ,N\}$, and its positions defined in $X_gY_gZ_g$ and $X_cY_cZ_c$ are denoted by ${}^g\mathbf{p}_i$ and ${}^c\mathbf{p}_i$, respectively. Moreover, $\mathbf{q}_i$ denotes the projection of $\mathbf{p}_i$ on the image plane, and its positions defined in $X_cY_cZ_c$ and UV are denoted by ${}^c\mathbf{q}_i$ and ${}^m\mathbf{q}_i$, respectively.

As a necessary input to the proposed single-camera trilateration algorithm, the set of intrinsic parameters of the involved camera can be retrieved from an off-line camera calibration process, a classic topic which has been addressed extensively in computer vision literature such as [59–61]. A well-adopted camera calibration toolbox can also be found at [62]. Using the calibration toolbox, we can determine the set of intrinsic camera parameters, including the focal length ${}^m\mathbf{f} = [f_u,f_v]^T$, image center ${}^m\mathbf{q}_0 = [u_0,v_0]^T$, and distortion coefficients, as well as their estimation uncertainty. In particular, $f_u = fD_us_u, f_v = fD_v$, where f is the physical focal length (mm), D_u the pixel size in the U direction (pixel/mm), D_v the pixel size in the V direction, and s_u the scale factor. It is known that f, D_u, D_v and s_u are in general impossible to calibrate separately [60–62]. Therefore, instead of using f, D_u, D_v and s_u separately, we use f_u and f_v in our algorithm.

The pinhole model is established on the basis of the calibrated camera parameters and corrected image. In the following derivation, we assume that the image distortions have been corrected according to the calibrated distortion coefficients. We also assume that at least three landmarks have been captured, segmented, identified and located at ${}^{m}\mathbf{q}_i$ in UV. Relevant image processing techniques can be found in existing literature [59,63,64]. Moreover, we assume that the positions of the involved landmarks in the global frame $X_gY_gZ_g$, ${}^{g}\mathbf{p}_i$, have been retrieved from a map of known landmarks using existing methods [53–57].

The camera position is defined by the position of its optical center $\mathbf{p}_0$ in $X_gY_gZ_g$, ${}^{g}\mathbf{p}_0$, and the camera orientation is defined by the rotation matrix of $X_cY_cZ_c$ with respect to $X_gY_gZ_g$, ${}^{g}\mathbf{R}_c$. The proposed algorithm will estimate ${}^{g}\mathbf{p}_0$ and ${}^{g}\mathbf{R}_c$, which will be discussed in the following two subsections respectively. The following relationships obtained from the pinhole model will be used in the derivation:

$$ {}^{g}\mathbf{p}_i = {}^{g}\mathbf{R}_c{}^{c}\mathbf{p}_i + {}^{g}\mathbf{p}_0, \tag{2} $$

$$ {}^{c}\mathbf{p}_i = {}^{c}\mathbf{q}_i\frac{{}^{c}z_i}{f}, \tag{3} $$

$$ {}^{c}\mathbf{q}_i = \begin{bmatrix} \frac{u_i-u_0}{D_u s_u} & \frac{v_i-v_0}{D_v} & f \end{bmatrix}^T, \tag{4} $$

where u_i and v_i denote the U and V components of ${}^{m}\mathbf{q}_i$ respectively.

2.2. Camera Position Estimation

Assuming that $N \geq 3$ landmarks are chosen for localizing the camera, we have a system of N independent trilateration equations in $X_gY_gZ_g$ in the form of:

$$ ({}^{g}\mathbf{p}_i - {}^{g}\mathbf{p}_0)^T({}^{g}\mathbf{p}_i - {}^{g}\mathbf{p}_0) = r_i^2, \forall i \in \{1,\ldots,N\}, \tag{5} $$

which is in fact Equation (1) defined in $X_gY_gZ_g$. In principle, given ${}^{g}\mathbf{p}_i$ and r_i, one can estimate ${}^{g}\mathbf{p}_0$ by solving Equation (5). Representing the standard ranging-based trilateration problem, Equation (5) can be solved using one of the existing closed-form or numerical algorithms in the literature [2,4–16,19]. Here, ${}^{g}\mathbf{p}_i$ are known as the pre-mapped landmark positions, but r_i are yet to be determined.

In order to determine r_i, we can write down a system of totally N $(N-1)/2$ equations as:

$$ r_i^2 + r_j^2 - 2r_ir_j\cos\theta_{ij} = r_{ij}^2, \forall i,j \in \{1,\ldots,N\}, i \neq j, \tag{6} $$

each of which is defined in a triangle $\Delta\mathbf{p}_0\mathbf{p}_i\mathbf{p}_j$ in $X_gY_gZ_g$ according to the law of cosine. Here, θ_{ij} denotes the vertex angle associated with $\mathbf{p}_0$ in $\Delta\mathbf{p}_0\mathbf{p}_i\mathbf{p}_j$ (which is also known as the visual angle that $\mathbf{p}_i\mathbf{p}_j$ subtends at $\mathbf{p}_0$), and r_{ij} the distance between $\mathbf{p}_i$ and $\mathbf{p}_j$. In principle, given r_{ij} and $\cos\theta_{ij}$, one can estimate r_i by solving Equation (6). At least two general schemes can be adopted to obtain r_i:

(1) Solving N equations for N variables: first, choose a subset of N independent equations from Equation (6) which contain N variables $\{r_i|i = 1,\ldots,N\}$; next, by combining the N equations and eliminating $N-1$ variables $\{r_i|i = 2,\ldots,N\}$, obtain a 2^{N-1}-degree polynomial equation of ${r_1}^2$, i.e.,

$$ \sum_{k=0}^{2^{N-1}} c_k(r_1^2)^k = 0; \tag{7} $$

then, solve Equation (7) for r_1, and substitute r_1 into the chosen N equations to solve for $\{r_i|i = 2,\ldots,N\}$ iteratively. When $N = 3$, (7) is a 4-degree polynomial equation which can be solved in the closed form. When $N > 3$, (7) is a polynomial equation of degree 8 or greater. Although it has been proven (by Niels Henrik Abel in 1824) that there cannot be any closed-form formula for such a polynomial equation, a number of numerical algorithms are available, such as Laguerre's method [65], the Jenkins–Traub method [66], the Durand–Kerner method [67], Aberth method [68], splitting circle method [69] and the Dandelin–Gräffe method [70]. The roots of a polynomial can also be found as the eigenvalues of its companion matrix [71–73].

(2) Solving an associated optimization problem: A nonlinear least-squares problem can be formulated to estimate the optimal r_i that minimizes the residue of Equation (6), i.e.,

$$\mathbf{r} = \underset{\mathbf{r}}{\operatorname{argmin}} \sum_{\substack{i,j=1 \\ i \neq j}}^{N} \left(r_i^2 + r_j^2 - 2r_i r_j \cos\theta_{ij} - r_{ij}^2\right)^2, \tag{8}$$

where $\mathbf{r}$ denotes the vector of $\{r_i|i = 1, \ldots, N\}$. Search-based optimization algorithms are commonly used to solve this category of problems, including both local optimization algorithms, e.g., the steepest descent method and the Newton–Raphson method (also known as Newton's method), and global optimization algorithms, e.g., the simulated annealing and the genetic algorithm [74–78].

To solve Equations (6)–(8), we need to know r_{ij} and $\cos\theta_{ij}$. Here, $r_{ij} = \sqrt{({}^g\mathbf{p}_i - {}^g\mathbf{p}_j)^T({}^g\mathbf{p}_i - {}^g\mathbf{p}_j)}$, but $\cos\theta_{ij}$ are yet to be determined.

In order to determine $\cos\theta_{ij}$, we have the following equation in the triangle $\Delta\mathbf{p}_0\mathbf{q}_i\mathbf{q}_j$ in $X_cY_cZ_c$ according to the law of cosine:

$$\cos\theta_{ij} = \frac{\rho_i^2 + \rho_j^2 - \rho_{ij}^2}{2\rho_i\rho_j}, \tag{9}$$

where $\rho_i = \sqrt{{}^c\mathbf{q}_i^T\,{}^c\mathbf{q}_i}$ denotes the distance between $\mathbf{p}_0$ and $\mathbf{q}_i$, and $\rho_{ij} = \sqrt{({}^c\mathbf{q}_i - {}^c\mathbf{q}_j)^T({}^c\mathbf{q}_i - {}^c\mathbf{q}_j)}$ denotes the distance between $\mathbf{q}_i$ and $\mathbf{q}_j$. Substituting Equation (4) into Equation (9), we obtain:

$$\cos\theta_{ij} = \frac{a_{ij}}{b_{ij}}, \tag{10}$$

where

$$a_{ij} = [(\frac{u_i - u_0}{f_u})^2 + (\frac{v_i - v_0}{f_v})^2 + 1] + [(\frac{u_j - u_0}{f_u})^2 + (\frac{v_j - v_0}{f_v})^2 + 1] - [(\frac{u_i - u_j}{f_u})^2 + (\frac{v_i - v_j}{f_v})^2],$$

$$b_{ij} = 2\sqrt{(\frac{u_i - u_0}{f_u})^2 + (\frac{v_i - v_0}{f_v})^2 + 1}\sqrt{(\frac{u_j - u_0}{f_u})^2 + (\frac{v_j - v_0}{f_v})^2 + 1}.$$

Since ${}^m\mathbf{q}_0 = [u_0, v_0]^T$ and ${}^m\mathbf{f} = [f_u, f_v]^T$ are determined through camera calibration and ${}^m\mathbf{q}_i$ are obtained by segmenting the visual landmarks from the image, $\cos\theta_{ij}$ can be calculated from Equation (10) directly.

Based on the above derivation, it is clear that the camera can be positioned by reversing the above steps. That is, first calculate $\cos\theta_{ij}$ from Equation (10), next estimate r_i from Equation (6), and then estimate ${}^g\mathbf{p}_0$ from Equation (5).

2.3. Camera Orientation Estimation

Knowing ${}^g\mathbf{p}_0$ and the set of ${}^g\mathbf{p}_i$, we obtain from Equation (2):

$${}^g\Delta\mathbf{P} = {}^g\mathbf{R}_c\,{}^c\mathbf{P} \tag{11}$$

where ${}^{g}\Delta\mathbf{P}$ is a 3 × N matrix with the ith column as ${}^{g}\Delta\mathbf{p}_i = {}^{g}\mathbf{p}_i - {}^{g}\mathbf{p}_0$, and ${}^{c}\mathbf{P}$ is a 3 × N matrix with the ith column as ${}^{c}\mathbf{p}_i$. Here, ${}^{c}\mathbf{p}_i$ can be determined by substituting Equation (4) into Equation (3):

$$ {}^{c}\mathbf{p}_i = {}^{c}z_i \begin{bmatrix} \frac{u_i-u_0}{f_u} & \frac{v_i-v_0}{f_v} & 1 \end{bmatrix}^T, \tag{12} $$

where ${}^{c}z_i$ can be calculated as

$$ {}^{c}z_i = \frac{fr_i}{\rho_i} = \frac{r_i}{\sqrt{\left(\frac{u_i-u_0}{f_u}\right)^2 + \left(\frac{v_i-v_0}{f_v}\right)^2 + 1}}. \tag{13} $$

Given ${}^{g}\Delta\mathbf{P}$ and ${}^{c}\mathbf{P}$, ${}^{g}\mathbf{R}_c$ can in principle be calculated from Equation (11) as

$$ {}^{g}\mathbf{R}_c = {}^{g}\Delta\mathbf{P}{}^{g}\Delta\mathbf{P}^T \left({}^{c}\mathbf{P}{}^{g}\mathbf{P}^T\right)^{-1}. \tag{14} $$

However, due to input errors, such as those in reference positioning, camera calibration and image segmentation, the resulting ${}^{g}\mathbf{R}_c$ may not strictly be a rotation matrix. In particular, the orthonormality of the matrix may not be guaranteed. Thus, instead of using Equation (14), we go through the following process to obtain a valid rotation matrix.

We define an optimal approximation of ${}^{g}\mathbf{R}_c$ as the one that minimizes the residue of Equation (11), i.e.,

$$ \begin{aligned} {}^{g}\mathbf{R}_{c,opt} &= \underset{{}^{g}\mathbf{R}_c}{\operatorname{argmin}} \sum_{i=1}^{N} \left({}^{g}\mathbf{R}_c{}^{c}\mathbf{p}_i - {}^{g}\Delta\mathbf{p}_i\right)^T \left({}^{g}\mathbf{R}_c{}^{c}\mathbf{p}_i - {}^{g}\Delta\mathbf{p}_i\right) \\ &= \underset{{}^{g}\mathbf{R}_c}{\operatorname{argmin}} Tr\left[\left({}^{g}\mathbf{R}_c{}^{c}\mathbf{P} - {}^{g}\Delta\mathbf{P}\right)^T \left({}^{g}\mathbf{R}_c{}^{c}\mathbf{P} - {}^{g}\Delta\mathbf{P}\right)\right] \\ &= \underset{{}^{g}\mathbf{R}_c}{\operatorname{argmin}} Tr\left({}^{c}\mathbf{P}^{T c}\mathbf{P} + {}^{g}\Delta\mathbf{P}^{T g}\Delta\mathbf{P} - 2{}^{g}\Delta\mathbf{P}^{T g}\mathbf{R}_c{}^{c}\mathbf{P}\right) \\ &= \underset{{}^{g}\mathbf{R}_c}{\operatorname{argmax}} Tr\left({}^{g}\Delta\mathbf{P}^{T g}\mathbf{R}_c{}^{c}\mathbf{P}\right) \end{aligned} . \tag{15} $$

where Tr ($\mathbf{M}$) denotes the trace of the matrix $\mathbf{M}$. Since $Tr\,(\mathbf{M}_1\mathbf{M}_2) = Tr\,(\mathbf{M}_2\mathbf{M}_1)$ for two matrices $\mathbf{M}_1$ and $\mathbf{M}_2$, we have:

$$ \begin{aligned} {}^{g}\mathbf{R}_{c,opt} &= \underset{{}^{g}\mathbf{R}_c}{\operatorname{argmax}} Tr\left({}^{c}\mathbf{P}{}^{g}\Delta\mathbf{P}^{T g}\mathbf{R}_c\right) \\ &= \underset{{}^{g}\mathbf{R}_c}{\operatorname{argmax}} Tr\left(\mathbf{ADB}^{T g}\mathbf{R}_c\right) \\ &= \underset{{}^{g}\mathbf{R}_c}{\operatorname{argmax}} Tr\left(\mathbf{DB}^{T g}\mathbf{R}_c\mathbf{A}\right) \end{aligned} , \tag{16} $$

where $\mathbf{ADB}^T$ is the singular value decomposition of the 3 × 3 matrix ${}^{c}\mathbf{P}{}^{g}\Delta\mathbf{P}^T$ in which $\mathbf{A}$ and $\mathbf{B}$ are orthogonal matrices and $\mathbf{D}$ is an diagonal matrix with non-negative real numbers on the diagonal. Since $\mathbf{B}^{T g}\mathbf{R}_c\mathbf{A}$ is an orthogonal matrix and $\mathbf{D}$ is a non-negative diagonal matrix, we must have:

$$ \max Tr\left(\mathbf{DB}^{T g}\mathbf{R}_c\mathbf{A}\right) = Tr(\mathbf{D}), \tag{17} $$

and correspondingly

$$ \mathbf{B}^{T g}\mathbf{R}_c\mathbf{A} = \mathbf{I}. \tag{18} $$

Then, we obtain from Equation (18)

$$ {}^{g}\mathbf{R}_c = \mathbf{BA}^T, \tag{19} $$

which is guaranteed to be an orthogonal matrix.

Based on the above derivation, it is clear that the camera orientation can be determined by first calculating ${}^{c}\mathbf{p}_i$ from Equation (12), next conducting the singular value decomposition of the matrix ${}^{c}\mathbf{P}{}^{g}\Delta\mathbf{P}^T$, and then getting ${}^{g}\mathbf{R}_c$ from Equation (19).

2.4. Algorithm Summary

Following the discussions in the above two subsections, we summarize the proposed single-camera trilateration algorithm as Algorithm 1. It estimates the instantaneous 3D pose of a camera, which can be moving, from a single image of landmarks at known positions. Strictly speaking, Algorithm 1 provides a general framework. In particular, Equation (6) in Step (2) and Equation (5) in Step (3) can be solved by a variety of algorithms specific for those steps, as we pointed out.

Algorithm 1: *Single-Camera Trilateration in P^3*

Input: The camera parameters ${}^{m}\mathbf{q}_0$ (image center) and ${}^{m}\mathbf{f}$ (focal length), the positions of a set of $N \geq 3$ landmarks in $X_gY_gZ_g$, $\{{}^{g}\mathbf{p}_i | i \in Z^+, 1 \leq i \leq N\}$, and their corresponding projections in UV, $\{{}^{m}\mathbf{q}_i | i \in Z^+, 1 \leq i \leq N\}$.

Output: The position ${}^{g}\mathbf{p}_0$ and orientation ${}^{g}\mathbf{R}_c$ of the camera in $X_gY_gZ_g$.

(1) Calculate $\cos\theta_{ij}$ from Equation (10);
(2) Solve Equation (6) for $\{r_i | i \in Z^+, 1 \leq i \leq N\}$;
(3) Solve Equation (5) for ${}^{g}\mathbf{p}_0$;
(4) Calculate $\{{}^{c}\mathbf{p}_i | i \in Z^+, 1 \leq i \leq N\}$ from Equation (12);
(5) Conduct the singular value decomposition of the matrix ${}^{c}\mathbf{P}^{g}\Delta\mathbf{P}^T$;
(6) Calculate ${}^{g}\mathbf{R}_c$ from Equation (19);
(7) Return ${}^{g}\mathbf{p}_0$ and ${}^{g}\mathbf{R}_c$.

---

3. Performance Analysis

The input to the proposed single-camera trilateration algorithm consists of the global positions of the involved landmarks ${}^{g}\mathbf{p}_i$, their positions in the image ${}^{m}\mathbf{q}_i$, the camera focal length ${}^{m}\mathbf{f}$ and image center ${}^{m}\mathbf{q}_0$. In practice, the uncertainty in camera calibration and inaccuracy of image segmentation cause errors in ${}^{m}\mathbf{f}$, ${}^{m}\mathbf{q}_0$ and ${}^{m}\mathbf{q}_i$, and imperfect mapping brings errors to ${}^{g}\mathbf{p}_i$. In this section, we analyze the effect of these input errors on the accuracy of the estimation output of the camera position ${}^{g}\mathbf{p}_0$ and orientation ${}^{g}\mathbf{R}_c$.

3.1. Performance Indices

We define an input vector $\mathbf{x}$ which contains all the above input parameters, i.e., $\mathbf{x} = [{}^{g}\mathbf{p}_i{}^{T}\,{}^{m}\mathbf{q}_i{}^{T}\,{}^{m}\mathbf{f}^{T}\,{}^{m}\mathbf{q}_0{}^{T}]^T$. Then ${}^{g}\mathbf{p}_0$ is a function of $\mathbf{x}$, i.e., ${}^{g}\mathbf{p}_0 = {}^{g}\mathbf{p}_0(\mathbf{x})$. Denoting the actual value and random error of $\mathbf{x}$ as $\overline{\mathbf{x}}$ and $\delta\mathbf{x}$ respectively, we have $\mathbf{x} = \overline{\mathbf{x}} + \delta\mathbf{x}$. Correspondingly, the actual value and output estimate of ${}^{g}\mathbf{p}_0$ can be written as ${}^{g}\mathbf{p}_0(\overline{\mathbf{x}})$ and ${}^{g}\mathbf{p}_0(\mathbf{x}) = {}^{g}\mathbf{p}_0(\overline{\mathbf{x}} + \delta\mathbf{x})$, respectively, and the estimation error of ${}^{g}\mathbf{p}_0$ is $\delta{}^{g}\mathbf{p}_0(\mathbf{x}) = {}^{g}\mathbf{p}_0(\overline{\mathbf{x}} + \delta\mathbf{x}) - {}^{g}\mathbf{p}_0(\overline{\mathbf{x}})$.

We denote the errors of input quantities ${}^{g}\mathbf{p}_i$, ${}^{m}\mathbf{q}_i$, ${}^{m}\mathbf{f}$ and ${}^{m}\mathbf{q}_0$ as $\delta{}^{g}\mathbf{p}_i$, $\delta{}^{m}\mathbf{q}_i$, $\delta{}^{m}\mathbf{f}$ and $\delta{}^{m}\mathbf{q}_0$ respectively. Adopting the classical scheme of error analysis used in [2,6], we evaluate the effects of $\delta{}^{g}\mathbf{p}_i$, $\delta{}^{m}\mathbf{q}_i$, $\delta{}^{m}\mathbf{f}$ and $\delta{}^{m}\mathbf{q}_0$ on $\delta{}^{g}\mathbf{p}_0$, respectively. For any of these input error vectors (generally denoted as $\delta\mathbf{v}$), we assume that the vector components are zero-mean random variables and uncorrelated with one another with a common standard deviation $\delta\mathbf{v}$. We adopt two well-accepted performance indices [2,6], the normalized total bias B_p which represents the sensitivity of the systematic position estimation error to the input error, and the normalized total standard deviation error S_p which represents the sensitivity of the position estimation uncertainty to the input error:

$$B_p(\delta\mathbf{v}) = \frac{\left|E(\delta{}^{g}\mathbf{p}_0)\right|}{\sigma_v^2}, \tag{20}$$

$$S_p(\delta\mathbf{v}) = \frac{\sqrt{Tr[\mathrm{var}(\delta{}^{g}\mathbf{p}_0)]}}{\sigma_v}, \tag{21}$$

where |**a**|denotes the norm of a vector **a**, *Tr* (**M**) denotes the trace of a matrix **M**, E(.) denotes the mean of a random variable, and var(.) denotes the variance of a random variable.

Similar to ${}^{g}\mathbf{p}_0$, ${}^{g}\mathbf{R}_c$ is also a function of **x**, i.e., ${}^{g}\mathbf{R}_c = {}^{g}\mathbf{R}_c(\mathbf{x})$. To make the presentation consistent, we discuss the orientation estimation error in the vector form, and correspondingly define an error metric of ${}^{g}\mathbf{R}_c$ as:

$$\delta^{g}\boldsymbol{\theta}_c = [\delta\theta_1\ \delta\theta_2\ \delta\theta_3]^T, \tag{22}$$

where

$$\delta\theta_i(\mathbf{x}) = \arccos[\mathbf{n}_i(\overline{\mathbf{x}})^T\mathbf{n}_i(\overline{\mathbf{x}}+\delta\mathbf{x})], \tag{23}$$

in the unit of degrees. Here $\mathbf{n}_i$ denotes the *i*th column of ${}^{g}\mathbf{R}_c$ which corresponds to the x, y or z directional vector of the camera frame with respect to the global frame. As a result, $\delta^{g}\boldsymbol{\theta}_c$ shows the angular difference between the actual and erroneous camera orientations. Similar to Equations (20) and (21), we define the corresponding performance indices—the normalized total bias B_θ and the normalized total standard deviation error S_θ as:

$$\mathrm{B}_\theta(\delta\mathbf{v}) = \frac{|E(\delta^{g}\boldsymbol{\theta}_c)|}{\sigma_v^2}, \tag{24}$$

$$\mathrm{S}_\theta(\delta\mathbf{v}) = \frac{\sqrt{Tr[\mathrm{var}(\delta^{g}\boldsymbol{\theta}_c)]}}{\sigma_v}, \tag{25}$$

where B_θ represents the sensitivity of the systematic orientation estimation error to the input error, and S_θ represents the sensitivity of the orientation estimation uncertainty to the input error.

In the following performance analysis, we will report the variation of B_p, S_p, B_θ and S_θ across the simulated spaces of ${}^{g}\mathbf{p}_0$ under the representative σ_v values of $\delta^{g}\mathbf{p}_i$, $\delta^{m}\mathbf{q}_i$, $\delta^{m}\mathbf{f}$ and $\delta^{m}\mathbf{q}_0$.

3.2. Simulation Settings

A performance analysis has been conducted on the proposed single-camera trilateration algorithm with representative examples in which a moving camera is localized in $\mathbf{P}^3$ based on three landmarks, simulated in Matlab.

In the simulated global frame of reference, the landmarks (reference points) are placed at ${}^{g}\mathbf{p}_1 = [\ 500\quad 0\quad 0\ ]^T$, ${}^{g}\mathbf{p}_2 = [\ -250\quad 250\sqrt{3}\quad 0\ ]^T$ and ${}^{g}\mathbf{p}_3 = [\ -250\quad -250\sqrt{3}\quad 0\ ]^T$. They form an equilateral triangle on the XY plane, inscribed in a circle centered at the origin of the global frame with a radius of r = 500 length units (Figure 2).

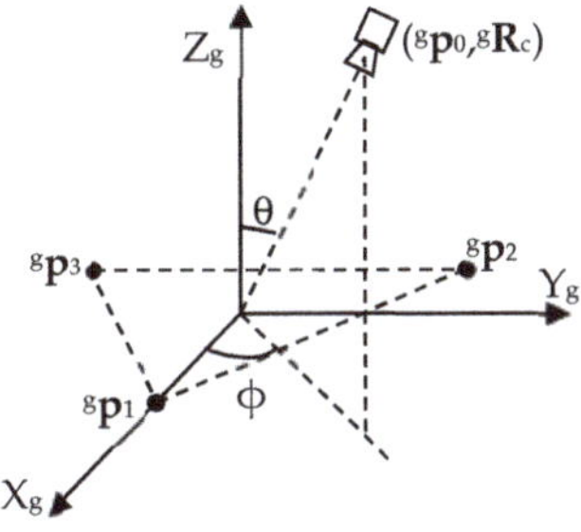

Figure 2. Simulation settings. This figure is adapted from Figure 3 in [58]. (© (2007) IEEE. Reprinted, with permission, from [Liu W.; Zhou Y. Recovering the position and orientation of a mobile robot from a single image of identified landmarks. In Proceedings of the 2007 IEEE/RSJ International Conference on Intelligent Robots and Systems, 2007; pp. 1065–1070.]).

To study the effect of the geometric arrangement of the camera relative to the landmarks on the performance indices, we let the camera move on a data acquisition sphere S centered at the origin of the global frame with a radius of R, and keep the optical axis of the camera pointing towards the origin. We define a tilt angle θ as the angle between Z_g and $-\mathbf{n}_3$, and a pan angle ϕ as the angle between X_g and the orthogonal projection of $-\mathbf{n}_3$ on the XY plane (Figure 2). To study the effect of the observation distance on the performance indices, we test the algorithm with different values of R. To keep the inscribing circle of the landmarks in the camera field of view all the time, we use $R \geq 3000$. It means that the required maximal angle of view is about 19° (corresponding to R = 3000 and $\theta = 0°$), which is a mild requirement met by a large number of lenses. For a specific R, by changing the tilt angle θ and pan angle ϕ, we move the camera on the sphere, and check the variation of the performance indices on the sphere.

The intrinsic camera parameters used in the simulation are obtained by calibrating a real camera (a MatrixVision BlueFox−120 640 × 480 CCD (Charged Coupled Device) grayscale camera with a Kowa 12 mm lens) using the camera calibration toolbox in [62] and a printed black/white checkerboard pattern. In particular, ${}^{m}\mathbf{f} = [1627.5609, 1629.9348]^T$ and ${}^{m}\mathbf{q}_0 = [333.9088, 246.3799]^T$.

The standard deviations of input errors are chosen mainly according to the calibration results of the above camera and lens. Specifically, using the camera calibration toolbox in [62] to calibrate the camera and lens, we obtained the estimation error for ${}^{m}\mathbf{f}$ as [1.1486, 1.1069]. Since the estimation errors are approximately three times the standard deviation of $\delta^{m}\mathbf{f}$ [62], a representative standard deviation of the $\delta^{m}\mathbf{f}$ components can be taken as $\sigma_f = 0.4$. Meanwhile, the estimation error for ${}^{m}\mathbf{q}_0$ was obtained as [1.9374, 1.5782], which means that a representative standard deviation of the $\delta^{m}\mathbf{q}_0$ components can be taken as $\sigma_c = 0.6$. In addition, we take $\sigma_p = 25$ as a representative standard deviation of the $\delta^{g}\mathbf{p}_i$ components, which corresponds to 5% of the base radius r. Besides, $\sigma_q = 0.5$ is taken as a representative standard deviation of the $\delta^{m}\mathbf{q}_i$ components based on a conservative estimation of the average image segmentation operation.

For performance analysis, we check the effect of $\delta^{g}\mathbf{p}_i$, $\delta^{m}\mathbf{q}_i$, $\delta^{m}\mathbf{f}$ and $\delta^{m}\mathbf{q}_0$ separately on $\delta^{g}\mathbf{p}_0$ and $\delta^{g}\boldsymbol{\theta}_c$. For each of these input error vectors, we assume that the vector components are zero-mean random variables following the corresponding Gaussian distribution and uncorrelated with one another with the same standard deviation; we discretize the data acquisition sphere S with a set of equally gapped θ and ϕ; for each point (θ, ϕ) on S, 10,000 samples of the erroneous input are generated according to the random distribution; the corresponding erroneous output estimates are obtained using the proposed single-camera trilateration algorithm; by comparing the erroneous estimates with the correct camera pose corresponding to (θ, ϕ) on S, the statistics of the output error are generated, and then the performance indices are calculated.

In order to optimally estimate ${}^{g}\mathbf{p}_0$ and r_i, we carry out the steps (2) and (3) of Algorithm 1 by solving the corresponding least-squares problems of Equations (5) and (6) respectively. The Newton–Raphson method [74] is used to search for the solutions, which has a sufficient rate of convergence. To avoid being trapped in local minima, for continuous camera position estimation, the previous camera position is used as the initial guess to estimate the current camera position, which guarantees the convergence of the solution to the actual camera position.

3.3. Trends of Performance Indices under Representative Input Error Standard Deviations

To evaluate the impact of $\delta^{g}\mathbf{p}_i$, $\delta^{m}\mathbf{q}_i$, $\delta^{m}\mathbf{f}$ and $\delta^{m}\mathbf{q}_0$ on $\delta^{g}\mathbf{p}_0$ and $\delta^{g}\boldsymbol{\theta}_c$, B_p, S_p, B_θ and S_θ are at first estimated with representative input error standard deviations across the data acquisition sphere S with R = 3000. Specifically, $B_p(\delta^{g}\mathbf{p}_i)$, $S_p(\delta^{g}\mathbf{p}_i)$, $B_\theta(\delta^{g}\mathbf{p}_i)$ and $S_\theta(\delta^{g}\mathbf{p}_i)$ are estimated with $\sigma_p = 25$ (Figure 3); $B_p(\delta^{m}\mathbf{f})$, $S_p(\delta^{m}\mathbf{f})$, $B_\theta(\delta^{m}\mathbf{f})$ and $S_\theta(\delta^{m}\mathbf{f})$ are estimated with $\sigma_f = 0.4$ (Figure 4); $B_p(\delta^{m}\mathbf{q}_0)$, $S_p(\delta^{m}\mathbf{q}_0)$, $B_\theta(\delta^{m}\mathbf{q}_0)$ and $S_\theta(\delta^{m}\mathbf{q}_0)$ are estimated with $\sigma_c = 0.6$ (Figure 5); and $B_p(\delta^{m}\mathbf{q}_i)$, $S_p(\delta^{m}\mathbf{q}_i)$, $B_\theta(\delta^{m}\mathbf{q}_i)$ and $S_\theta(\delta^{m}\mathbf{q}_i)$ are estimated with $\sigma_q = 0.5$ (Figure 6). Since S is vertically symmetrical to the base plane (XY plane) and horizontally symmetrical to the vertical planes containing the medians of the equilateral base triangle, we only need to display the variation of B_p, S_p, B_θ and S_θ across a patch constrained by

$\theta \in [0°, 90°]$ and $\phi \in [0°, 60°]$, while the values of these performance indices at any point on S outside the patch can be obtained according to the symmetry. Each sub-figure in Figures 3–6 presents the variation of a performance index, under a specific input error standard deviation, with respect to the observation direction (defined by the tilt angle θ and pan angle ϕ) at a specific distance. Each sub-figure includes one surface plot and one contour plot, both generated using Matlab. The surface plot presents a 3D surface which visualizes the values of the corresponding performance index as heights and colors above a horizontal grid defined by θ and ϕ. The general trend of the performance index's variation is reflected by the variation in the height and color of the surface. Meanwhile, the contour plot presents a set of isolines (contour lines) at different value levels of the corresponding performance index which varies across θ and ϕ. The contour lines are labeled with the associated values of the performance index in order to provide more numerical details of the variation. To match with the aforementioned data acquisition patch on S, the $\theta - \phi$ grid in each plot is presented in the form of a 2D sector, where the tilt angle of the observation direction θ varies in the radial direction of the sector from 0° to 90° and the pan angle ϕ varies along the arc from 0° to 60°.

For the effect of the landmark position error $\delta^g \mathbf{p}_i$:

(1) As shown in Figure 3a, $B_p(\delta^g \mathbf{p}_i)$ in general decreases as θ increases from 0° towards 90°, although its maximum may not take place exactly at $\theta = 0°$. However, it forms a ridge when θ approaches 90°, which is lower than the crest near $\theta = 0°$. When $\sigma_p = 25$, the ratio between the two peaks is about 6.73.

(2) As shown in Figure 3b–d respectively, $S_p(\delta^g \mathbf{p}_i)$, $B_\theta(\delta^g \mathbf{p}_i)$ and $S_\theta(\delta^g \mathbf{p}_i)$ in general decrease as θ increases from 0° to 90°, although their maxima may not be exactly at $\theta = 0°$.

For the effect of the imaging-related errors $\delta^m \mathbf{q}_i$, $\delta^m \mathbf{f}$ and $\delta^m \mathbf{q}_0$:

(1) As shown in Figures 4a and 5a, as θ increases from 0°, $B_p(\delta^m \mathbf{f})$ and $B_p(\delta^m \mathbf{q}_0)$ increase at first to form a high peak, and then decrease generally as θ increases further towards 90°. However, they reach a second, local peak when θ approaches 90°, which is much lower than the first peak. Compared with the first peak, the second peak is far less significant: when $\sigma_f = 0.4$, the ratio between the peaks of B_p $(\delta^m \mathbf{f})$ is about 1095.84; when $\sigma_c = 0.6$, the ratio between the peaks of B_p $(\delta^m \mathbf{q}_0)$ is about 1525.50.

(2) As shown in Figure 4b–d, Figures 5b–d and 6, as θ increases from 0°, $B_p(\delta^m \mathbf{q}_0)$, $S_p(\delta^m \mathbf{f})$, $S_p(\delta^m \mathbf{q}_0)$, $S_p(\delta^m \mathbf{q}_i)$, $B_\theta(\delta^m \mathbf{f})$, $B_\theta(\delta^m \mathbf{q}_0)$, $B_\theta(\delta^m \mathbf{q}_i)$, $S_\theta(\delta^m \mathbf{f})$, $S_\theta(\delta^m \mathbf{q}_0)$ and $S_\theta(\delta^m \mathbf{q}_i)$ all increase at first to form a high peak, and then decreases as θ increases further towards 90°.

The general decreasing trend of these performance indices towards $\theta = 90°$ implies that the camera pose estimation is less sensitive to the input errors as the observations of landmarks are made more parallel and closer to the base plane. Moreover, the non-circular patterns from the contour plots in Figures 3–6 reflect the influence of the geometric arrangement of the landmarks.

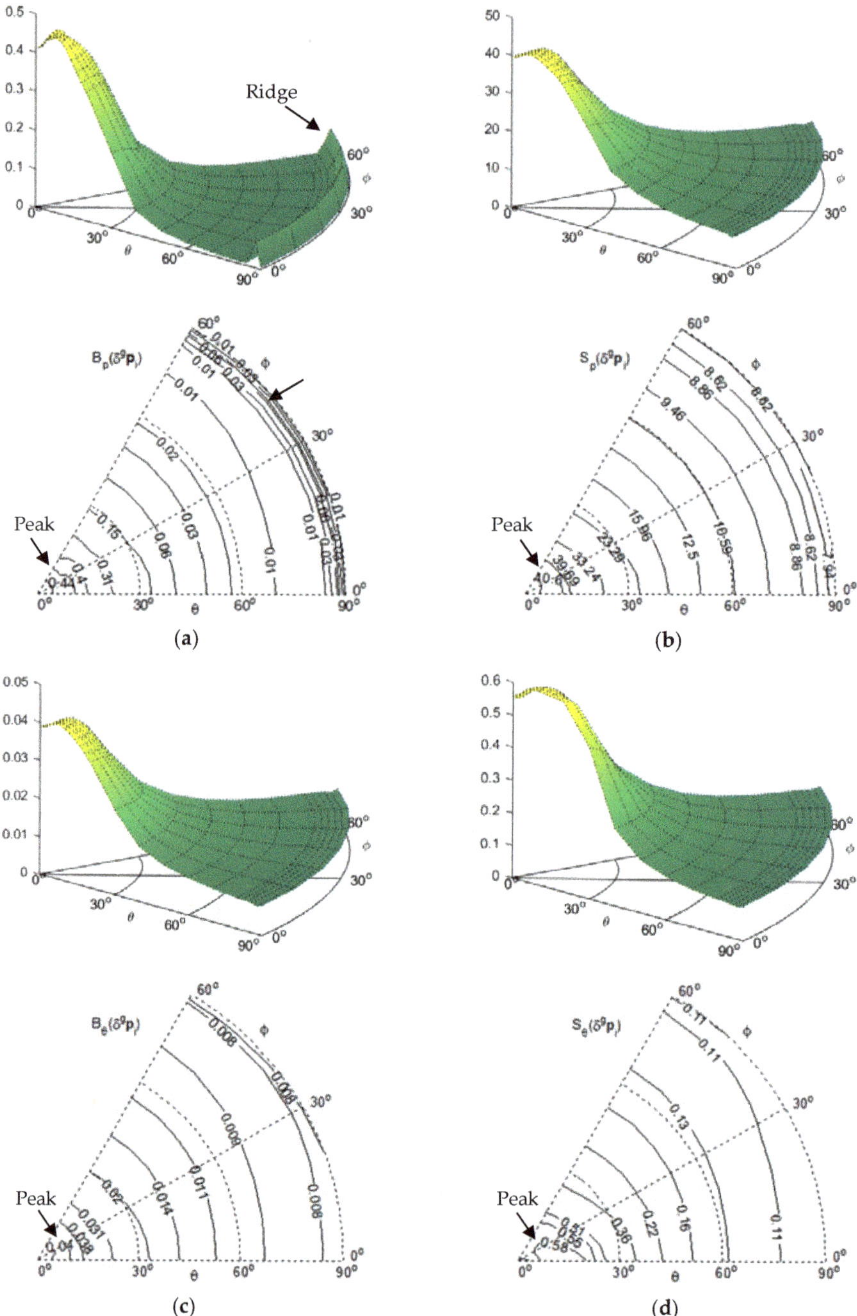

Figure 3. $B_p(\delta^g \mathbf{p}_i)$, $S_p(\delta^g \mathbf{p}_i)$, $B_\theta(\delta^g \mathbf{p}_i)$ and $S_\theta(\delta^g \mathbf{p}_i)$ with $\sigma_p = 25$ and $R = 3000$. (**a**) Surface plot and contour plot of $B_p(\delta^g \mathbf{p}_i)$; (**b**) Surface plot and contour plot of $S_p(\delta^g \mathbf{p}_i)$; (**c**) Surface plot and contour plot of $B_\theta(\delta^g \mathbf{p}_i)$; (**d**) Surface plot and contour plot of $S_\theta(\delta^g \mathbf{p}_i)$.

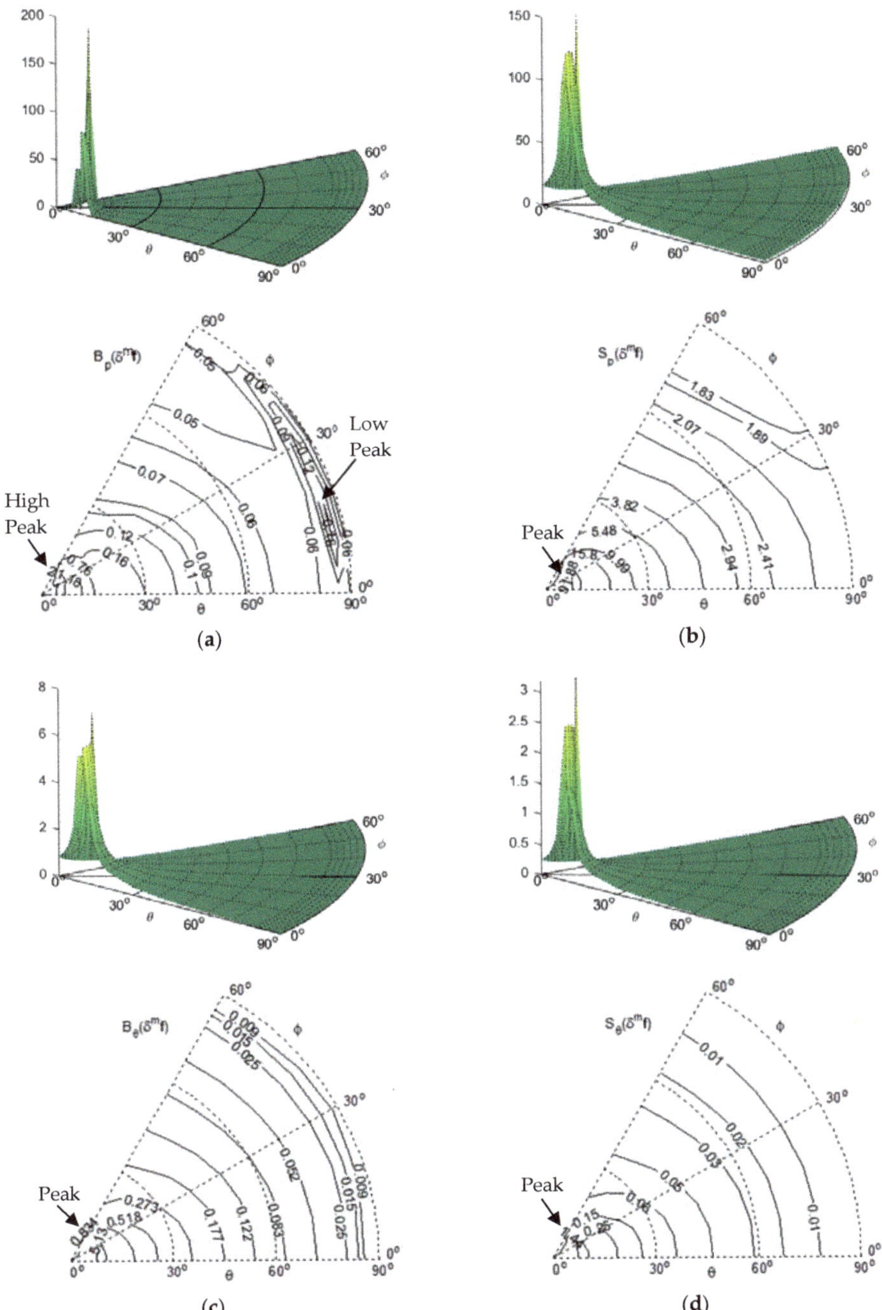

Figure 4. $B_p(\delta^m\mathbf{f})$, $S_p(\delta^m\mathbf{f})$, $B_\theta(\delta^m\mathbf{f})$ and $S_\theta(\delta^m\mathbf{f})$ with $\sigma_f = 0.4$ and R = 3000. (**a**) Surface plot and contour plot of $B_p(\delta^m\mathbf{f})$; (**b**) Surface plot and contour plot of $S_p(\delta^m\mathbf{f})$; (**c**) Surface plot and contour plot of $B_\theta(\delta^m\mathbf{f})$; (**d**) Surface plot and contour plot of $S_\theta(\delta^m\mathbf{f})$.

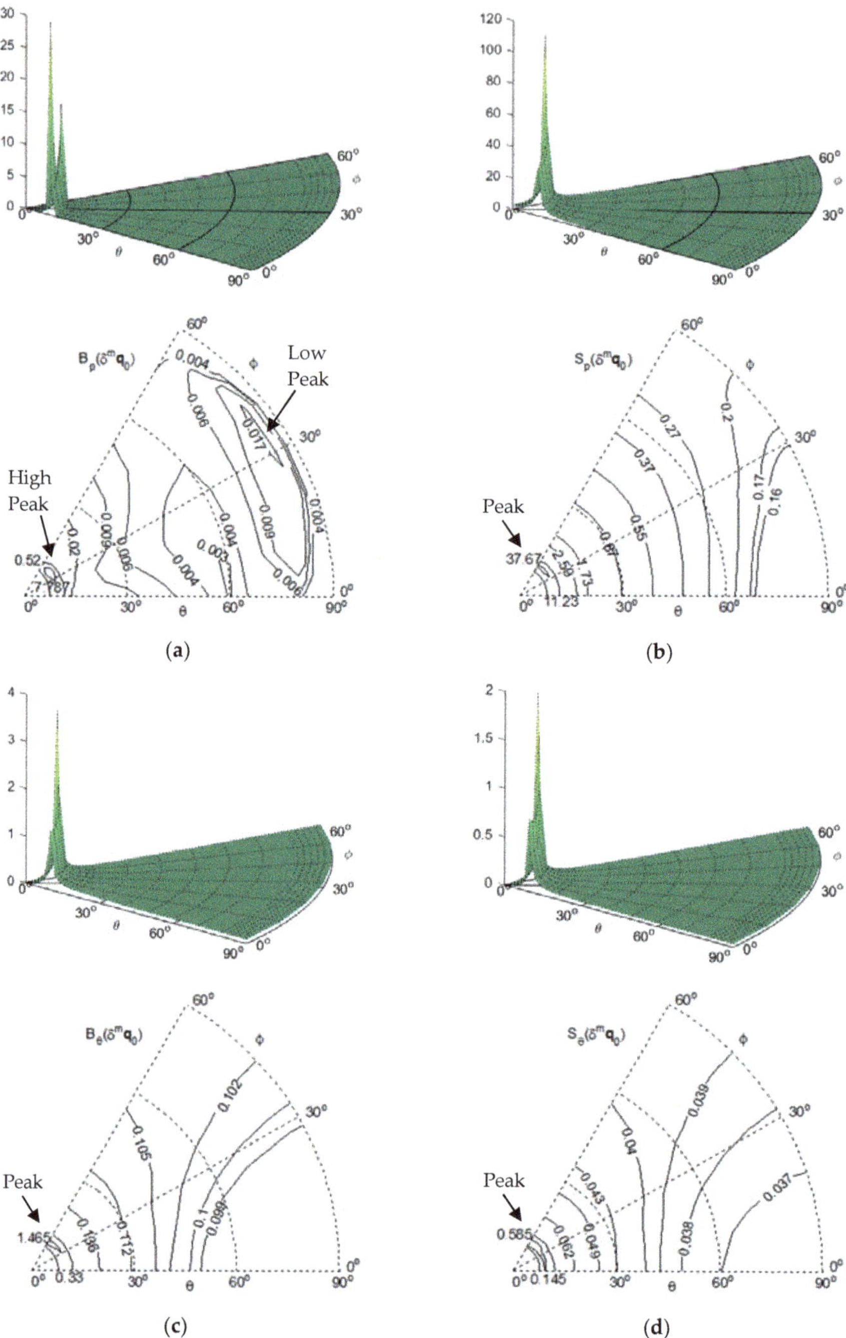

Figure 5. $B_p(\delta^m \mathbf{q}_0)$, $S_p(\delta^m \mathbf{q}_0)$, $B_\theta(\delta^m \mathbf{q}_0)$ and $S_\theta(\delta^m \mathbf{q}_0)$ with $\sigma_c = 0.6$ and R = 3000. (**a**) Surface plot and contour plot of $B_p(\delta^m \mathbf{q}_0)$; (**b**) Surface plot and contour plot of $S_p(\delta^m \mathbf{q}_0)$; (**c**) Surface plot and contour plot of $B_\theta(\delta^m \mathbf{q}_0)$; (**d**) Surface plot and contour plot of $S_\theta(\delta^m \mathbf{q}_0)$.

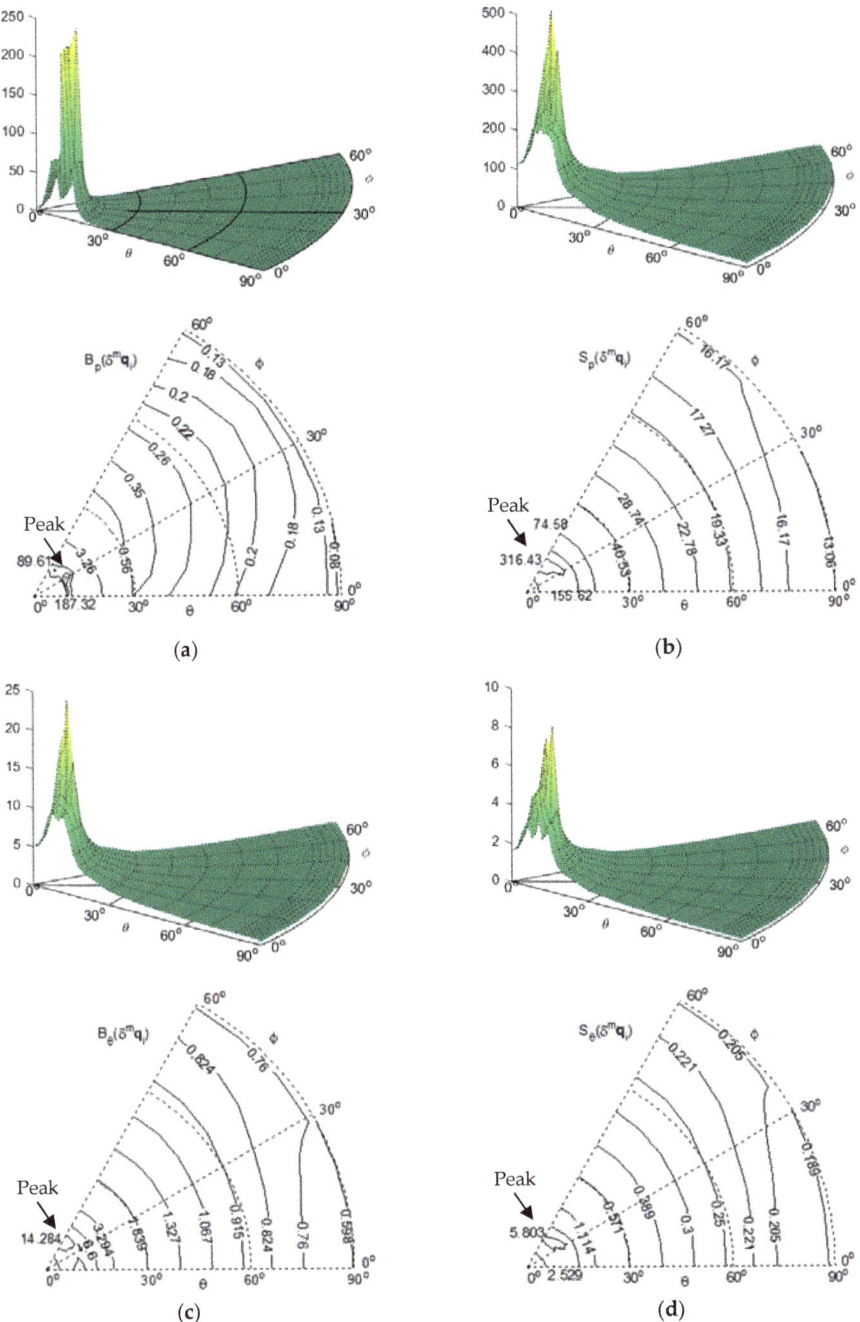

Figure 6. $B_p(\delta^m \mathbf{q}_i)$, $S_p(\delta^m \mathbf{q}_i)$, $B_\theta(\delta^m \mathbf{q}_i)$ and $S_\theta(\delta^m \mathbf{q}_i)$ with $\sigma_c = 0.5$ and R = 3000. (**a**) Surface plot and contour plot of $B_p(\delta^m \mathbf{q}_i)$; (**b**) Surface plot and contour plot of $S_p(\delta^m \mathbf{q}_i)$; (**c**) Surface plot and contour plot of $B_\theta(\delta^m \mathbf{q}_i)$; (**d**) Surface plot and contour plot of $S_\theta(\delta^m \mathbf{q}_i)$.

3.4. Variation of Performance Indices under Variation of Input Error Standard Deviations

To examine the variation of B_p, S_p, B_θ and S_θ under the variation of the input error standard deviations, we compare the values of these performance indices obtained with a set of different values of the input error standard deviations on the data acquisition sphere S with R = 3000. Specifically, $B_p(\delta^g \mathbf{p}_i)$, $S_p(\delta^g \mathbf{p}_i)$, $B_\theta(\delta^g \mathbf{p}_i)$ and $S_\theta(\delta^g \mathbf{p}_i)$ are estimated with $\sigma_p = \{25,50,75\}$ (Figure 7); $B_p(\delta^m \mathbf{f})$, $S_p(\delta^m \mathbf{f})$, $B_\theta(\delta^m \mathbf{f})$ and $S_\theta(\delta^m \mathbf{f})$ are estimated with $\sigma_f = \{0.4,0.6,0.8\}$ (Figure 8); $B_p(\delta^m \mathbf{q}_0)$, $S_p(\delta^m \mathbf{q}_0)$, $B_\theta(\delta^m \mathbf{q}_0)$ and $S_\theta(\delta^m \mathbf{q}_0)$ are estimated with $\sigma_c = \{0.6,0.8,1.0\}$ (Figure 9); and $B_p(\delta^m \mathbf{q}_i)$, $S_p(\delta^m \mathbf{q}_i)$, $B_\theta(\delta^m \mathbf{q}_i)$ and $S_\theta(\delta^m \mathbf{q}_i)$ are estimated with $\sigma_q = \{0.5,0.7,0.9\}$ (Figure 10). For the convenience of visualization, we compute the mean of each performance index across ϕ at each θ, plot the curve of variation for this mean value with respect to θ, and include the standard deviation across ϕ as the vertical error bar at each θ. Each vertical error bar has a length equal to the standard deviation both above and below the specific mean value at a specific θ. Moreover, to accommodate the large span between the maxima and minima in these performance indices, we plot the data as logarithmic scale for the vertical axis.

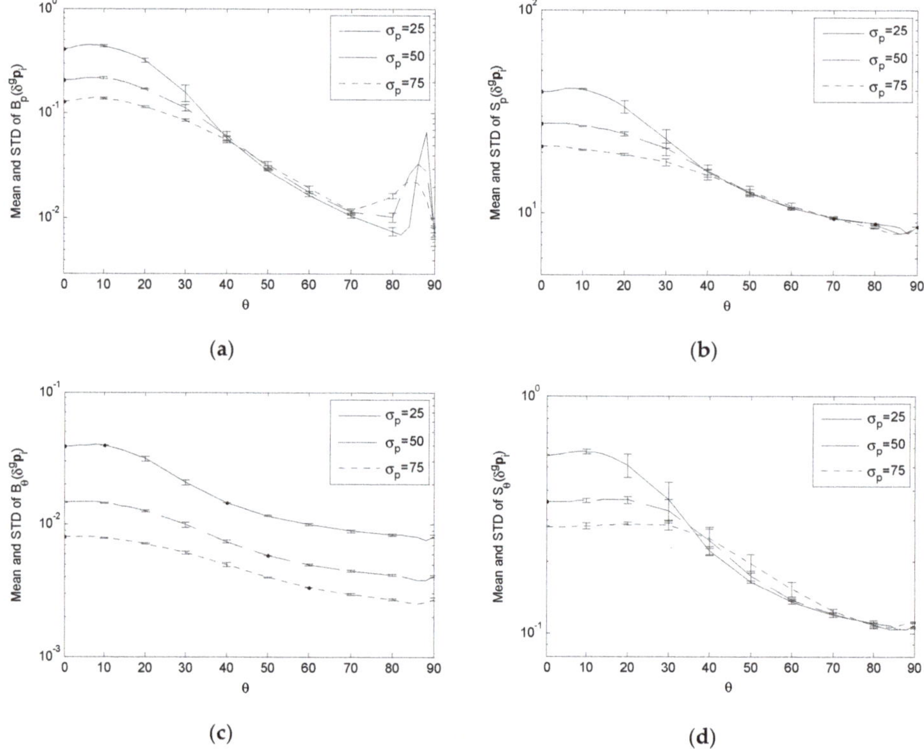

Figure 7. Comparing $B_p(\delta^g \mathbf{p}_i)$, $S_p(\delta^g \mathbf{p}_i)$, $B_\theta(\delta^g \mathbf{p}_i)$ and $S_\theta(\delta^g \mathbf{p}_i)$ with σ_p = {25, 50, 75} and R = 3000. (**a**) Variation of $B_p(\delta^g \mathbf{p}_i)$; (**b**) Variation of $S_p(\delta^g \mathbf{p}_i)$; (**c**) Variation of $B_\theta(\delta^g \mathbf{p}_i)$; (**d**) Variation of $S_\theta(\delta^g \mathbf{p}_i)$.

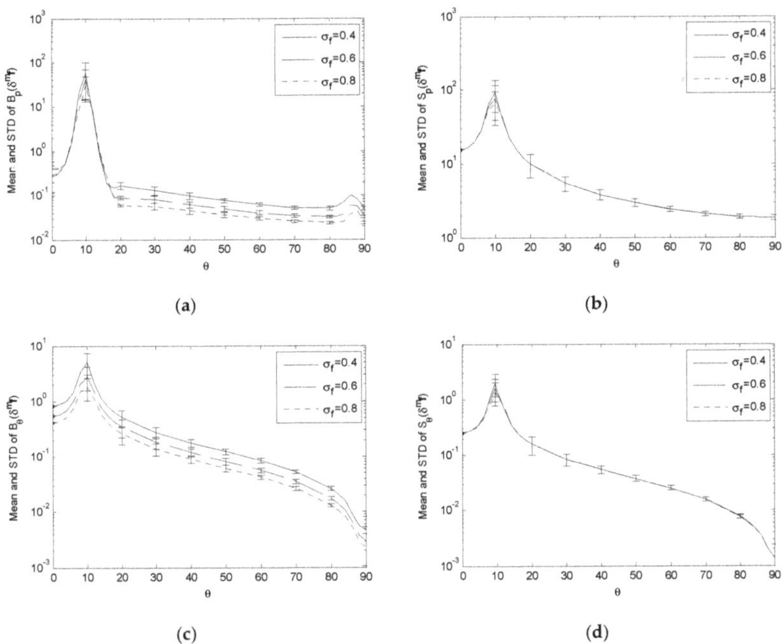

Figure 8. Comparing $B_p(\delta^m\mathbf{f})$, $S_p(\delta^m\mathbf{f})$, $B_\theta(\delta^m\mathbf{f})$ and $S_\theta(\delta^m\mathbf{f})$ with σ_f = {0.4,0.6,0.8} and R = 3000. (**a**) Variation of $B_p(\delta^m\mathbf{f})$; (**b**) Variation of $S_p(\delta^m\mathbf{f})$; (**c**) Variation of $B_\theta(\delta^m\mathbf{f})$; (**d**) Variation of $S_\theta(\delta^m\mathbf{f})$.

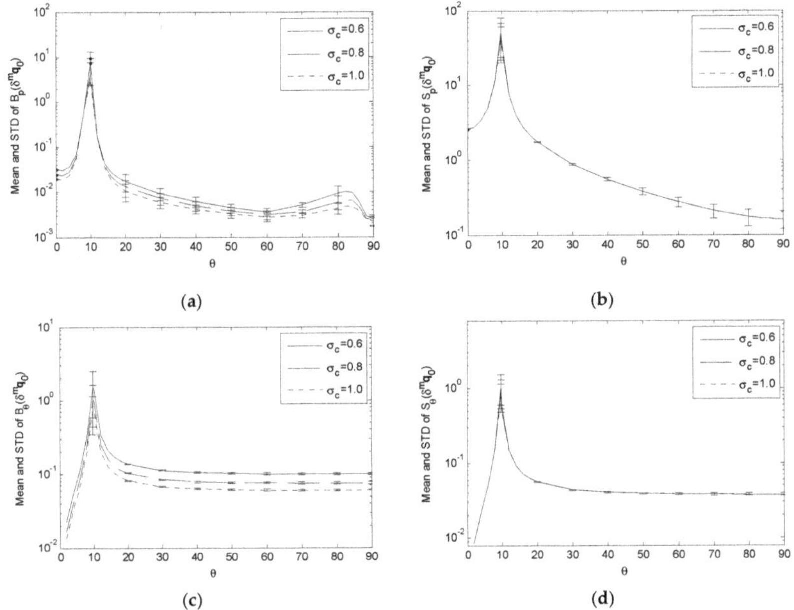

Figure 9. Comparing $B_p(\delta^m\mathbf{q}_0)$, $S_p(\delta^m\mathbf{q}_0)$, $B_\theta(\delta^m\mathbf{q}_0)$ and $S_\theta(\delta^m\mathbf{q}_0)$ with σ_c = {0.6,0.8,1.0} and R = 3000. (**a**) Variation of $B_p(\delta^m\mathbf{q}_0)$; (**b**) Variation of $S_p(\delta^m\mathbf{q}_0)$; (**c**) Variation of $B_\theta(\delta^m\mathbf{q}_0)$; (**d**) Variation of $S_\theta(\delta^m\mathbf{q}_0)$.

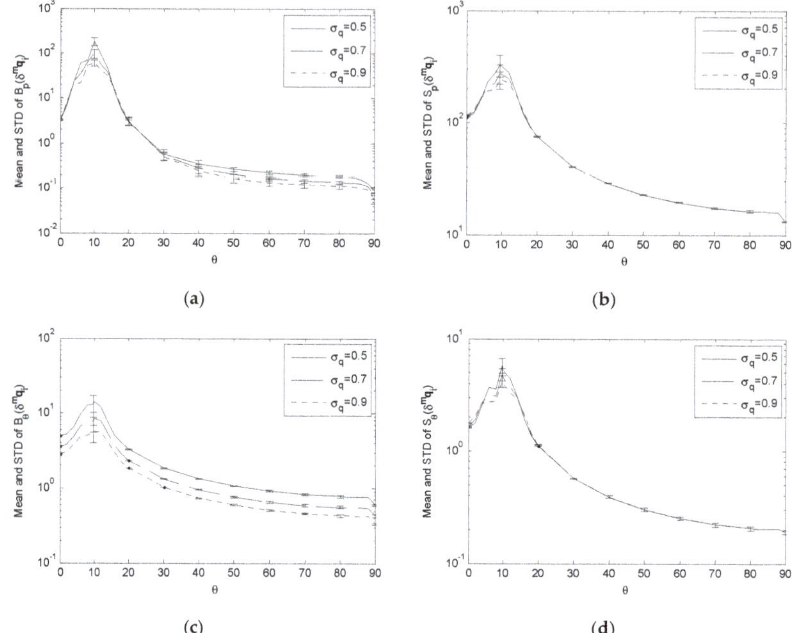

Figure 10. Comparing $B_p(\delta^m\mathbf{q}_i)$, $S_p(\delta^m\mathbf{q}_i)$, $B_\theta(\delta^m\mathbf{q}_i)$ and $S_\theta(\delta^m\mathbf{q}_i)$ with $\sigma_q = \{0.5,0.7,0.9\}$ and R = 3000. (**a**) Variation of $B_p(\delta^m\mathbf{q}_i)$; (**b**) Variation of $S_p(\delta^m\mathbf{q}_i)$; (**c**) Variation of $B_\theta(\delta^m\mathbf{q}_i)$; (**d**) Variation of $S_\theta(\delta^m\mathbf{q}_i)$.

For the effect of $\delta^g\mathbf{p}_i$, the simulation results show that, as σ_p increases,

(1) $B_p(\delta^g\mathbf{p}_i)$, $S_p(\delta^g\mathbf{p}_i)$ and $S_\theta(\delta^g\mathbf{p}_i)$ have lower peaks, resulting in flatter variation across S (Figure 7a,b,d);

(2) $B_\theta(\delta^g\mathbf{p}_i)$ decreases across S (Figure 7c).

For the effect of $\delta^m\mathbf{q}_i$, $\delta^m\mathbf{f}$ and $\delta^m\mathbf{q}_0$, the simulation results show that, as σ_q, σ_f and σ_c increase,

(1) $B_p(\delta^m\mathbf{f})$, $B_p(\delta^m\mathbf{q}_0)$, $B_p(\delta^m\mathbf{q}_i)$, $B_\theta(\delta^m\mathbf{f})$, $B_\theta(\delta^m\mathbf{q}_0)$ and $B_\theta(\delta^m\mathbf{q}_i)$ tend to have lower crests and decrease across the rest of S (Figures 8, 9 and 10a,c);

(2) $S_p(\delta^m\mathbf{f})$, $S_p(\delta^m\mathbf{q}_0)$, $S_p(\delta^m\mathbf{q}_i)$, $S_\theta(\delta^m\mathbf{f})$, $S_\theta(\delta^m\mathbf{q}_0)$ and $S_\theta(\delta^m\mathbf{q}_i)$ tend to have lower crests and are essentially independent of the variation of σ_f, σ_c and σ_q for the rest of S (Figures 8, 9 and 10b,d).

The decrease in these performance indices across the whole or portion of S, corresponding to the increase in the input error standard deviations, means that the proposed single-camera trilateration algorithm has a reduced sensitivity to higher input uncertainty under the corresponding geometric arrangements of the camera and landmarks.

3.5. Variation of Performance Indices under Variation of Observation Distance

To examine the variation of B_p, S_p, B_θ and S_θ under the variation of R, we compare the values of these performance indices obtained with representative input error standard deviations on different data acquisition spheres with different radii R = {3000, 4000, 5000} (Figures 11–14). For the convenience of comparison, we adopt the same scheme of data plotting as Figures 7–10.

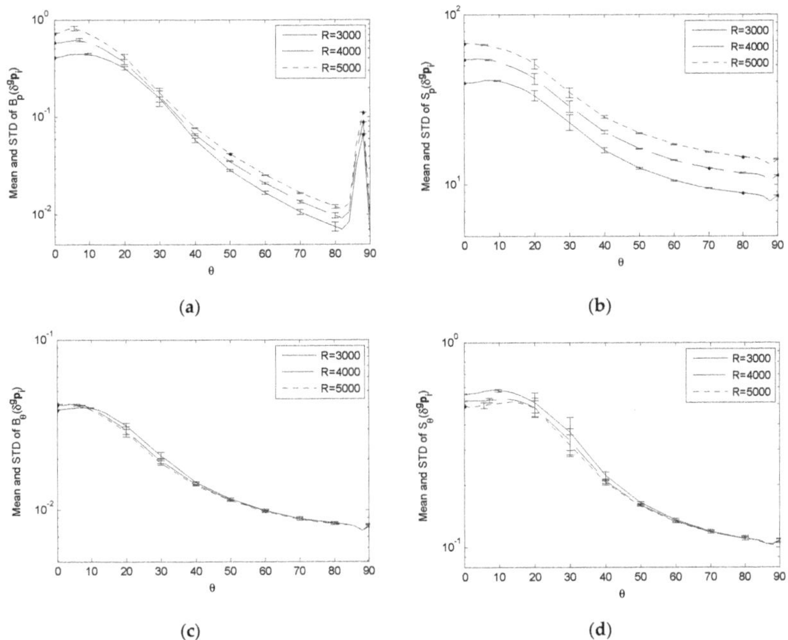

Figure 11. Comparing $B_p(\delta^g \mathbf{p}_i)$, $S_p(\delta^g \mathbf{p}_i)$, $B_\theta(\delta^g \mathbf{p}_i)$ and $S_\theta(\delta^g \mathbf{p}_i)$ with $\sigma_p = 25$ and R = {3000, 4000, 5000}. (**a**) Variation of $B_p(\delta^g \mathbf{p}_i)$; (**b**) Variation of $S_p(\delta^g \mathbf{p}_i)$; (**c**) Variation of $B_\theta(\delta^g \mathbf{p}_i)$; (**d**) Variation of $S_\theta(\delta^g \mathbf{p}_i)$.

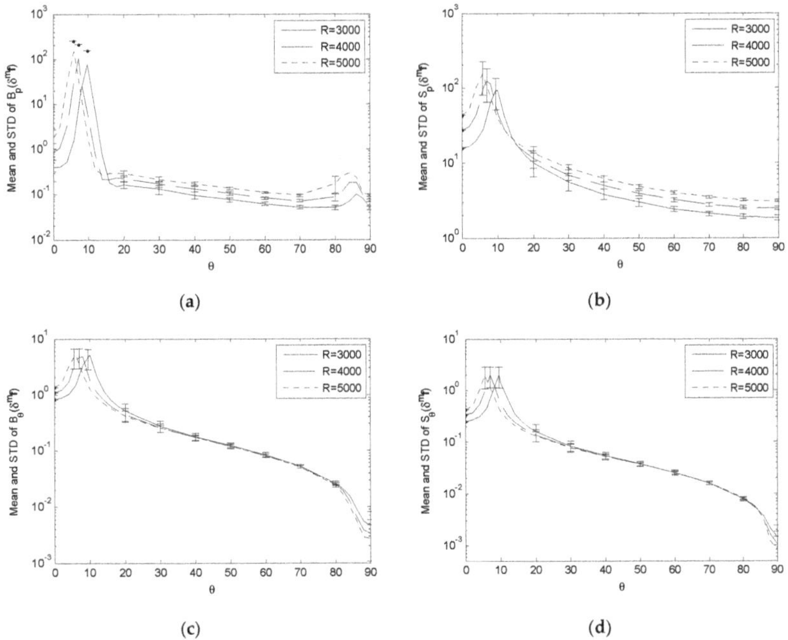

Figure 12. Comparing $B_p(\delta^m \mathbf{f})$, $S_p(\delta^m \mathbf{f})$, $B_\theta(\delta^m \mathbf{f})$ and $S_\theta(\delta^m \mathbf{f})$ with $\sigma_f = 0.4$ and R = {3000, 4000, 5000}. (**a**) Variation of $B_p(\delta^m \mathbf{f})$; (**b**) Variation of $S_p(\delta^m \mathbf{f})$; (**c**) Variation of $B_\theta(\delta^m \mathbf{f})$; (**d**) Variation of $S_\theta(\delta^m \mathbf{f})$.

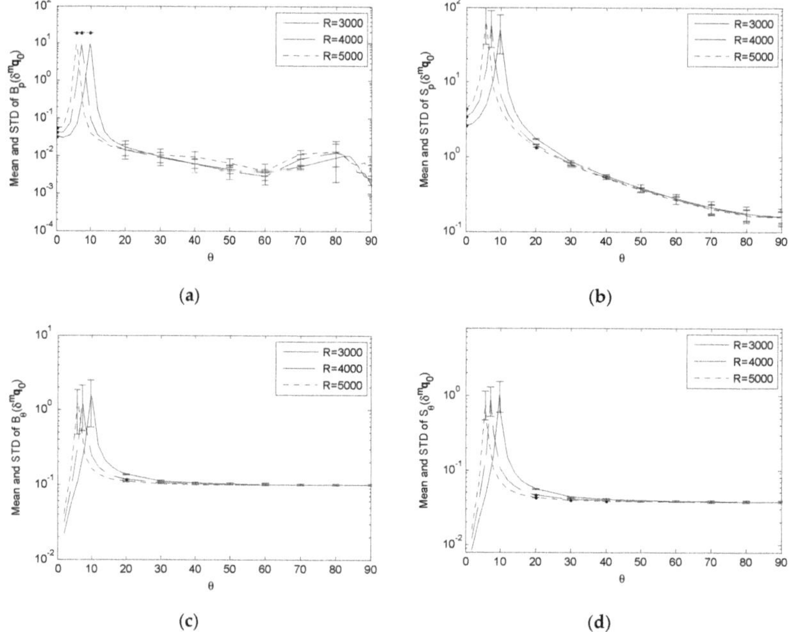

Figure 13. Comparing $B_p(\delta^m \mathbf{q}_0)$, $S_p(\delta^m \mathbf{q}_0)$, $B_\theta(\delta^m \mathbf{q}_0)$ and $S_\theta(\delta^m \mathbf{q}_0)$ with $\sigma_c = 0.6$ and R = {3000, 4000, 5000}. (**a**) Variation of $B_p(\delta^m \mathbf{q}_0)$; (**b**) Variation of $S_p(\delta^m \mathbf{q}_0)$; (**c**) Variation of $B_\theta(\delta^m \mathbf{q}_0)$; (**d**) Variation of $S_\theta(\delta^m \mathbf{q}_0)$.

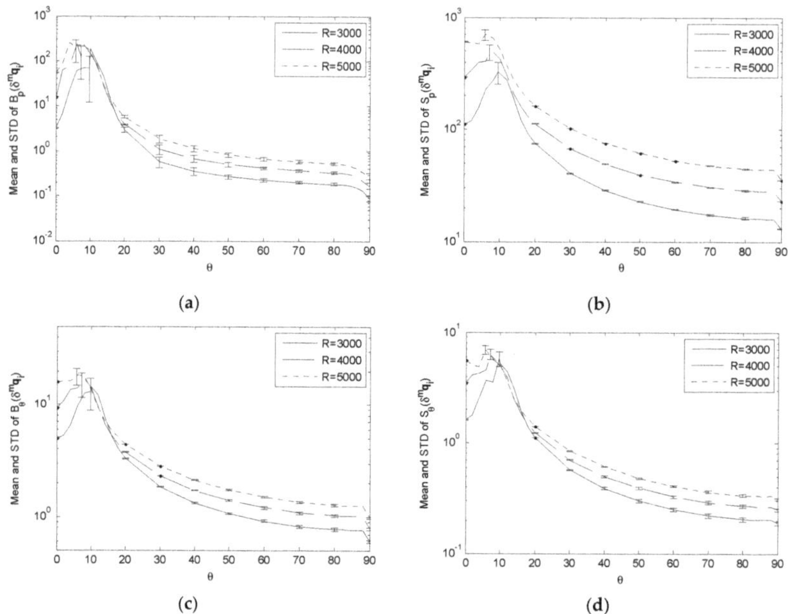

Figure 14. Comparing $B_p(\delta^m \mathbf{q}_i)$, $S_p(\delta^m \mathbf{q}_i)$, $B_\theta(\delta^m \mathbf{q}_i)$ and $S_\theta(\delta^m \mathbf{q}_i)$ with $\sigma_q = 0.5$ and R = {3000, 4000, 5000}. (**a**) Variation of $B_p(\delta^m \mathbf{q}_i)$; (**b**) Variation of $S_p(\delta^m \mathbf{q}_i)$; (**c**) Variation of $B_\theta(\delta^m \mathbf{q}_i)$; (**d**) Variation of $S_\theta(\delta^m \mathbf{q}_i)$.

For the effect of $\delta^g \mathbf{p}_i$, the simulation results show that, as R increases,

(1) $B_p(\delta^g \mathbf{p}_i)$ and $S_p(\delta^g \mathbf{p}_i)$ increases across S (Figure 11a,b);
(2) $B_\theta(\delta^g \mathbf{p}_i)$ and $S_\theta(\delta^g \mathbf{p}_i)$ have very small changes and can be considered being essentially independent of the variation of R (Figure 11c,d).

It means that a longer observation distance tends to increase the sensitivity of the positioning bias and uncertainty to the landmark position error, while it has little impact on the corresponding estimation bias and uncertainty of the camera orientation.

For the effect of $\delta^m \mathbf{q}_i$, $\delta^m \mathbf{f}$ and $\delta^m \mathbf{q}_0$, the simulation results show that, as R increases,

(1) $B_p(\delta^m \mathbf{f})$, $S_p(\delta^m \mathbf{f})$, $B_p(\delta^m \mathbf{q}_i)$, $S_p(\delta^m \mathbf{q}_i)$, $B_\theta(\delta^m \mathbf{q}_i)$ and $S_\theta(\delta^m \mathbf{q}_i)$ in general have an "increasing and shifting" trend, i.e., their variation can be considered as increasing across θ with the peaks shifting towards $\theta = 0°$ (Figure 12a,b and Figure 14a–d);
(2) $B_\theta(\delta^m \mathbf{f})$, $S_\theta(\delta^m \mathbf{f})$, $B_p(\delta^m \mathbf{q}_0)$, $S_p(\delta^m \mathbf{q}_0)$, $B_\theta(\delta^m \mathbf{q}_0)$ and $S_\theta(\delta^m \mathbf{q}_0)$ essentially have their peaks shift towards $\theta = 0°$ (Figure 12c,d and Figure 13a–d).

This means that a longer observation distance tends to increase the sensitivity of the positioning bias and uncertainty to the focal length error, and the sensitivity of the positioning and orienting bias and uncertainty to the landmark segmentation error while, except for causing the peaks to shift, it has little impact on the orienting bias and uncertainty due to the focal length error, and the positioning and orienting bias and uncertainty due to the image center error.

3.6. Error-Prone Regions in Performance Indices

Figure 4, Figure 5, Figure 6, Figure 8, Figure 9, Figure 10, Figures 12–14 indicate that a sharp high peak consistently appears in B_p, S_p, B_θ and S_θ of $\delta^m \mathbf{f}$, $\delta^m \mathbf{q}_i$ and $\delta^m \mathbf{q}_0$. Each performance index increases to form a high peak as θ increases from 0°, and in general decreases as θ further increases towards 90°. (Though $B_p(\delta^m \mathbf{f})$ and $B_p(\delta^m \mathbf{q}_0)$ have a second peak later on, it is not so significant as the first peak). The high peak stands out sharply from its neighborhood. On the other side, the crests and transitions in Figures 3, 7 and 11, which show the effect of landmark position errors on the performance indices, are relatively mild.

This high peak is of particular interest. It reflects:

(1) The camera localization algorithm is more sensitive to the errors from the calibration of the camera parameters and segmentation of the landmarks from the image than the errors in positioning the landmarks in the environment.
(2) The neighborhood of a high peak represents a range of camera orientations (relative to the base plane) from which the camera localization is highly sensitive to the errors in camera parameters and image segmentation, resulting in high estimation bias and uncertainty.

Thus, improving the accuracy of camera calibration and image segmentation will help to improve the accuracy of camera localization.

Moreover, knowing the values of θ corresponding to the peaks will help to avoid those error-prone regions in practice. The simulation results reveal that:

(1) Given a specific geometric arrangement of the landmarks and a specific observation distance, the peaks of all these performance indices appear at the same θ value, as shown in Figures 8–10;
(2) The θ value at which the peaks appear shifts as the observation distance changes, as shown in Figures 12–14.

This means that the θ value corresponding to the peaks in the above performance indices is mostly determined by the geometric arrangement of the landmarks and camera.

Although a closed-form estimation of the peak θ value is not available yet, a close approximation can be found through simulations. Specifically in our example where the three reference points define an equilateral triangle, the simulation results show that the peaks occur when the camera is located above the inscribing circle of the base triangle. That is, the θ value corresponding to the peaks is well approximated as:

$$\theta = \arcsin(\frac{r}{R}), \tag{26}$$

where r denotes the base radius. This relationship has been verified by comparing the numerically estimated peak locations in these performance indices with θ calculated from Equation (26) (Table 1). Since the θ value at which the peaks appear is consistent across B_p, S_p, B_θ and S_θ of $\delta^m\mathbf{f}$, $\delta^m\mathbf{q}_i$ and $\delta^m\mathbf{q}_0$, only $B_p(\delta^m\mathbf{f})$ is reported here as a representative. Table 1 shows a high consistency between the numerically estimated θ and that calculated from Equation (26). Since the high peaks in these performance indices mean high bias and uncertainty in camera pose estimation, tracking θ at which the observation is made can alert the localization process the error-prone region in practice.

The above discussion shows that the singular cases represented by the sharp high peaks existing in the plots of performance analysis are related to both the spatial relationship between the camera and landmarks and the usage of camera imaging as the input. Equation (26) provides an effective tool for predicting/estimating potentially singular observation positions. Ultimately, a closed-form representation of the performance indices will lead to analytical analysis of the singular cases. This will be explored in our future work.

Table 1. θ Corresponding to the high peak in B_p ($\delta^m\mathbf{f}$) [1].

R	θ from Simulation	$\theta = \arcsin(\frac{r}{R})$
2000	14.47	14.48
3000	9.50	9.59
4000	7.21	7.18
5000	5.64	5.74
6000	4.92	4.78
7000	3.96	4.10
8000	3.64	3.58

[1] In this table, $r = 500$, $\sigma_f = 0.4$, and the unit of θ is degree.

4. Experimental Test

An experiment was also carried out to further verify the effectiveness of the proposed single-camera trilateration algorithm, and check the estimation error under the combined effect of various input errors.

A landmark pattern was designed as shown in Figure 15a. It consists of a 679 mm × 679 mm square-shaped checker board pattern and a circle co-centered with the checker board and with a diameter of 679 mm. There are 12 landmark dots equally spaced on the circle, as shown in Figure 15b. The pattern was hung under the ceiling of an indoor experimental space with the base plane parallel to the flat floor. We used the base plane as the XY plane of the global frame of reference, with the Z axis pointing downwards and the center of the pattern as the origin of the frame.

(a)

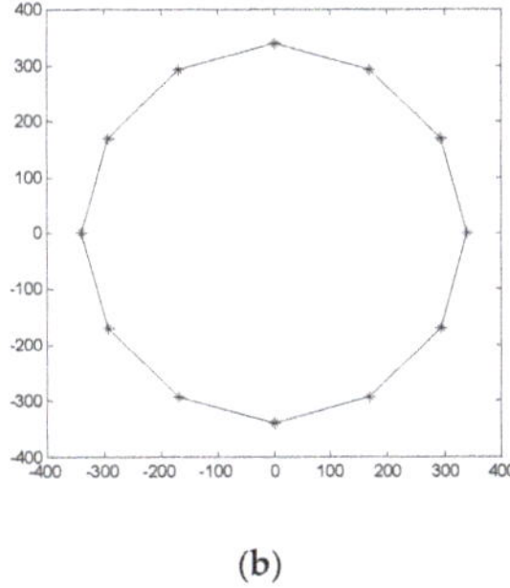

(b)

Figure 15. Landmark pattern used in the experiment: (**a**) landmark pattern; (**b**) landmark dot arrangement.

The exact camera, which was calibrated to provide the realistic values of camera parameters for the above simulations, was installed on a Directed Perception PTU-D46-17 controllable pan-tile unit on the top of a Pioneer 3-DX mobile robot (Figure 16) such that the camera was kept on a plane parallel to the base plane at a distance of 1600 mm.

Figure 16. Onboard experimental system.

We moved the camera (by moving the mobile robot and adjusting the pan-tilt unit) to a number of locations at different observation distances R (which were defined in Section 3.2), one location at one distance. At each location (each R), we adjusted the tilt angle θ and pan angle ϕ (which were defined in Section 3.2) of the camera using the pan-tile unit such that the center of the landmark pattern fell on the image center of the camera, and thus kept the camera oriented towards the landmark pattern. At each location, an image of the landmark pattern was taken by the camera, and stored in the onboard laptop computer.

The collected landmark images were processed after the data acquisition process. Focusing on evaluating the proposed single-camera trilateration algorithm itself, we segmented the landmarks from each image manually after correcting the image distortion. Because the camera optical center is not a physical point on the camera, it is highly difficult, if not impossible, to directly measure the position of the camera optical center and the orientation of the camera optical axis. Thus it is highly difficult to evaluate the localization accuracy of the proposed algorithm according to the golden truth of the camera poses in the global frame. Instead, we chose to compare the camera localization results from the proposed single-camera trilateration algorithm with those obtained from camera calibration (using the camera calibration toolbox in [62]). The camera calibration process calibrated both the intrinsic parameters of the camera, which were input to the proposed algorithm for camera pose estimation, and the extrinsic parameters associated with each calibration image of the checker board pattern, which were used as the reference camera pose estimate to reflect the localization accuracy of the proposed algorithm.

To evaluate the accuracy of the proposed single-camera localization algorithm at different observation distances, we would like to obtain a collection of camera localization results corresponding to different pan angles ϕ at each observation distance R so that we can derive the statistics of localization accuracy at each R, where, as we adjusted the pan/tile unit to align the camera optical axis with the center of the landmark pattern at each R, the tilt angle θ is fixed at each R. Due to the size limitation on the experimental space, it was difficult for us to pan the camera relative to the Z axis of the global frame all around as we kept increasing R. Instead, we took advantage of the design of our landmark pattern. With 12 landmark dots equally spaced on the circle as shown in Figure 15b, we have 12 equally spaced landmark sets labeled as (1, 5, 9), (2, 6, 10), (3, 7, 11), (4, 8, 12), (5, 9, 1), (6, 10, 2), (7, 11, 3), (8, 12, 4), (9, 1, 5), (10, 2, 6), (11, 3, 7) and (12, 4, 8). As mentioned above, at each R (and θ), we took one camera image of the landmark pattern. By using different 3-landmark sets from the same image, we in fact localize the camera corresponding to different ϕ. For example, using landmark set (2, 6, 10) is equivalent to panning the camera from landmark set (1, 5, 9) about the global Z direction by $\Delta\phi = 30°$. In this way, instead of physically panning the camera, by using different landmark sets to localize the camera, we equivalently obtained the camera localization results corresponding to different ϕ. On the other side, at each R (and θ), the reference camera pose estimate corresponding to each landmark set was obtained by rotating the calibrated camera pose, which was calculated based on the image of the checker board pattern at R (and θ) using the camera calibration toolbox and corresponds to the landmark set (1,5,9), about the global Z axis with a proper $\Delta\phi$.

The localization results from the proposed single-camera localization algorithm are compared with those from the camera calibration process in Table 2. The calibrated camera pose is used as the reference estimate, and the corresponding calibration error reported by the camera calibration toolbox is presented as the standard deviation of the camera pose calibration. The difference between the camera pose estimated using the proposed algorithm and the calibrated camera pose is recorded as $\Delta\mathbf{p}_0$ (position estimation difference) and $\Delta\boldsymbol{\theta}_c$ (orientation estimation difference), and the mean and standard deviation of these differences are reported. The results in Table 2 show that on average the proposed algorithm attains a localization estimation very close to that attained with camera calibration. The average difference in position estimation is only about 6 mm at an observation distance of R = 4.3 m (corresponding to R_h = 4 m in Table 2), and is smaller with shorter observation distances; the average difference in orientation estimation remains lower than 0.3 degree at an observation distance of $R \leq 4.3$ m. While camera calibration cannot be used for real-time localization of a moving camera, the proposed algorithm, which provides a comparable estimation of the camera pose on the go, presents an effective online localization approach for many mobile robot applications. We also notice that the difference in camera position estimation between the two compared methods and the associated standard deviation increase as the observation distance increases. This is mainly because the effective size of the pixels increases as the distance increases. Since the proposed algorithm depends on the segmented landmarks from the same image, the camera positioning error clearly increases as the observation distance and thus the effective pixel size increase; while the camera position estimation error with camera calibration increases much slower, because the calibration process takes advantage of all the images taken at different distances and all the grid corners of the checker board pattern to average out the effect of the increasing effective pixel size. On the other side, the difference in camera orientation estimation between the two compared methods and the associated standard deviation do not appear to be affected as the observation distance increases, due to the cancellation between the effect of the observation distance and that of the effective pixel size. Other factors contributing to the localization error of both methods include the calibration error of the camera intrinsic parameters, and the positioning errors of the landmarks and grid corners due to the printing inaccuracy of the landmark pattern and manual measurement inaccuracy.

Table 2. Camera pose estimation error—the proposed algorithm versus camera calibration [1].

	Errors of Camera Pose Calibration		Errors of Single-Camera Trilateration			
R_h (mm)	$\lvert STD(\delta^g p_0)\rvert$ (mm)	$\lvert STD(\delta^g \theta_c)\rvert$ (deg)	$\lvert E(\Delta^g p_0)\rvert$ (mm)	$\lvert STD(\Delta^g p_0)\rvert$ (mm)	$\lvert E(\Delta^g \theta_c)\rvert$ (deg)	$\lvert STD(\Delta^g \theta_c)\rvert$ (deg)
2000	4.5627	0.1806	1.3539	10.9633	0.0838	0.3564
2500	5.5080	0.1448	1.4588	12.4060	0.2302	0.2878
3000	6.3335	0.1348	5.0328	17.5257	0.2874	0.3189
3500	7.2251	0.1273	4.4541	29.3335	0.1970	0.3084
4000	8.2313	0.1228	6.1605	31.9332	0.2548	0.4927

[1] In this table, R_h denotes the horizontal distance between the camera and the Z axis of the global frame which passes through the center of the base triangle, E(.) denotes mean, STD(.) denotes standard deviation, and |.| denotes the norm of a vector.

In addition, to test the speed of the proposed single-camera trilateration algorithm, we implemented a compiled standalone executable (exe file) of the algorithm on a laptop PC with a 1.66 GHz Intel Core 2 CPU, when processing the experimental data. The average estimation speed was about 0.06762 s. In our implementation, the main time-consuming steps were to solve Equations (5) and (6) by using the Newton–Raphson method to search for the solutions of the associated least-squares optimization problems in order to obtain an optimal camera pose estimation.

5. Conclusions and Future Work

This paper has introduced an effective single-camera trilateration scheme for regular forward-looking cameras. Based on the classical pin-hole model and principles of perspective geometry, the position and orientation of such a camera can be calculated successively from a single image of a few landmarks at known positions. The proposed scheme is an advantageous addition to the category of trilateration techniques:

(1) The camera position is trilaterated from multiple simultaneously captured landmarks, which naturally resolves the common concern of the synchronization among multiple distance measurements in ranging-based trilateration systems.
(2) The estimation of the camera orientation becomes an integrated part of the proposed trilateration scheme, taking advantage of the geometrical information contained in a single image of multiple landmarks.
(3) The instantaneous camera pose is estimated from only one single image, which naturally resolves the issue of mismatching among a pair of images in the stereovision-based trilateration.

The proposed single-camera trilateration scheme provides a convenient tool for the self-localization of mobile robots and other vehicles which are equipped with cameras. Depending on the identification of the landmarks involved, the proposed algorithm targets environments with identifiable landmarks, such as indoor and outdoor environments with artificial landmarks, metropolitan environments, and natural environments with distinct landmarks.

The performance of the proposed algorithm has been analyzed in this work based on extensive simulations with a representative set of geometric arrangements among the landmarks and camera and a representative set of input errors. In practice, simulations with the settings of a targeted environment and the input errors of a targeted camera will provide an effective way to predict the algorithm performance in the specific environment and system. The simulation results will provide valuable guidance for the implementation of the proposed algorithm, and facilitate mobile robots to achieve accurate self-localization by avoiding sensitive regions of the performance indices.

Nevertheless, a closed-form solution will further reduce the time complexity of the proposed algorithm, and a closed-form estimation of the performance indices will enable more efficient performance analysis and prediction of the proposed algorithm. As pointed out, existing closed-form solutions can be adopted into the steps (2) and (3) of the proposed framework (Algorithm 1) to solve the 3-landmark case. However, it is challenging to find a closed-form solution which is globally optimal and applies to the general case of 3D trilateration based on more than 3 landmarks. This will be considered in our future work.

Moreover, we target the development of an automatic localization system for mobile robot self-localization based on the proposed single-camera trilateration scheme. The experimental system shown in Figure 16 has provided the necessary onboard hardware for the proposed scheme. Our future work will focus on the development of the software system and supporting vision algorithms.

In addition, over the years of research and development, numerous localization algorithms/approaches in different categories have been proposed in the literature, from deterministic to probabilistic, from ranging-based to image-based. The work of this paper is carried out within the scope of trilateration, with the intention to provide a convenient addition to this category of localization approaches. Meanwhile, the development in other categories of localization approaches are highly significant, e.g., the approaches in the big category of computer vision-based localization [79–81]. Our future work on localization may expand into those categories, and comparison study among related approaches is expected.

Author Contributions: Conceptualization of the presented approach, funding acquisition, research supervision, derivation and finalization of the complete algorithm of position and orientation estimation, performance analysis, paper writing, Y.Z.; derivation of the position estimation algorithm of the presented approach, W.L.; experimental study, X.L. and X.Z.

Funding: This research was partly supported by National Science Foundation, grant number1360873.

Conflicts of Interest: The authors declare no conflict of interest.

References

1. Borenstein, J.; Everett, H.R.; Feng, L.; Wehe, D. Mobile robot positioning: Sensors and techniques. *J. Robot. Syst.* **1997**, *14*, 231–249. [CrossRef]
2. Thomas, F.; Ros, L. Revisiting trilateration for robot localization. *IEEE Trans. Robot.* **2005**, *21*, 93–101. [CrossRef]
3. Multilateration. Available online: http://en.wikipedia.org/wiki/Multilateration (accessed on 5 December 2019).
4. Fang, B. Trilateration and extension to global positioning system navigation. *J. Guid.* **1986**, *9*, 715–717. [CrossRef]
5. Ziegert, J.; Mize, C.D. The laser ball bar: A new instrument for machine tool metrology. *Precis. Eng.* **1994**, *16*, 259–267. [CrossRef]
6. Manolakis, D.E. Efficient solution and performance analysis of 3-D position estimation by trilateration. *IEEE Trans. Aerosp. Electron. Syst.* **1996**, *32*, 1239–1248. [CrossRef]
7. Rao, S.K. Comments on efficient solution and performance analysis of 3-D position estimation by trilateration. *IEEE Trans. Aerosp. Electron. Syst.* **1998**, *34*, 681.
8. Coope, I.D. Reliable computation of the points of intersection of n spheres in Rn. *Aust. N. Z. Ind. Appl. Math. J.* **2000**, *42*, C461–C477. [CrossRef]
9. Pradhan, S.; Hwang, S.S.; Cha, H.R.; Bae, Y.C. Line intersection algorithm for the enhanced TOA trilateration technique. *Int. J. Hum. Robot.* **2014**, *11*, 144203. [CrossRef]
10. Navidi, W.; Murphy, W.S., Jr.; Hereman, W. Statistical methods in surveying by trilateration. *Comput. Stat. Data Anal.* **1998**, *27*, 209–217. [CrossRef]
11. Hu, W.C.; Tang, W.H. Automated least-squares adjustment of triangulation-trilateration figures. *J. Surv. Eng.* **2001**, *127*, 133–142. [CrossRef]
12. Foy, W.H. Position-location solutions by Taylor-series estimation. *IEEE Trans. Aerosp. Electron. Syst.* **1976**, *AES-12*, 187–194. [CrossRef]
13. Pent, M.; Spirito, M.A.; Turco, E. Method for positioning GSM mobile stations using absolute time delay measurements. *Electron. Lett.* **1997**, *33*, 2019–2020. [CrossRef]

14. Cho, H.; Kim, S.W. Mobile robot localization using biased chirp-spread-spectrum ranging. *IEEE Trans. Ind. Electron.* **2010**, *57*, 2826–2835.
15. Fujii, K.; Arie, H.; Wang, W.; Kaneko, Y.; Sakamoto, Y.; Schmitz, A.; Sugano, S. Improving IMES localization accuracy by integrating dead reckoning information. *Sensors* **2016**, *16*, 163. [CrossRef] [PubMed]
16. Zhou, Y. An efficient least-squares trilateration algorithm for mobile robot localization. In Proceedings of the 2009 IEEE/RSJ International Conference on Intelligent Robots and Systems, St. Louis, MO, USA, 10–15 October 2009; pp. 3474–3479.
17. Hsu, C.C.; Yeh, S.S.; Hsu, P.L. Particle filter design for mobile robot localization based on received signal strength indicator. *Trans. Inst. Meas. Control* **2016**, *38*, 1311–1319. [CrossRef]
18. Ullah, Z.; Xu, Z.; Lei, Z.; Zhang, L. A robust localization, slip estimation, and compensation system for WMR in the indoor environments. *Symmetry* **2018**, *10*, 149. [CrossRef]
19. Zhou, Y. A closed-form algorithm for the least-squares trilateration problem. *Robotica* **2011**, *29*, 375–389. [CrossRef]
20. Cheng, X.; Shu, H.; Liang, Q.; Du, D.H. Silent positioning in underwater acoustic sensor networks. *IEEE Trans. Veh. Technol.* **2008**, *57*, 1756–1766. [CrossRef]
21. Ward, A.; Jones, A.; Hopper, A. A new location technique for the active office. *IEEE Pers. Commun.* **1997**, *4*, 42–47. [CrossRef]
22. Priyantha, N.B.; Chakraborty, A.; Balakrishnan, H. The Cricket location-support system. In Proceedings of the 6th ACM/IEEE International Conference on Mobile Computing and Networking, Boston, MA, USA, 6–10 August 2000; pp. 32–43.
23. Harter, A.; Hopper, A.; Steggles, P.; Ward, A.; Webster, P. The anatomy of a context-aware application. *Wirel. Netw.* **2002**, *8*, 187–197. [CrossRef]
24. Urena, J.; Hernandez, A.; Jimenez, A.; Villadangos, J.M.; Mazo, M.; Garcia, J.C.; Garcia, J.J.; Alvarez, F.J.; De Marziani, C.; Perez, M.C.; et al. Advanced sensorial system for an acoustic LPS. *Microprocess. Microsyst.* **2007**, *31*, 393–401. [CrossRef]
25. Ningrum, E.S.; Hakkun, R.Y.; Alasiry, A.H. Tracking and formation control of leader-follower cooperative mobile robots based on trilateration data. *EMITTER Int. J. Eng. Technol.* **2015**, *3*, 88–98. [CrossRef]
26. Petković, M.; Sibinović, V.; Popović, D.; Mitić, V.; Todorović, D.; Đorđević, G.S. Analysis of two low-cost and robust methods for indoor localisation of mobile robots. *Facta Univ. Ser. Electron. Energetics* **2017**, *30*, 403–416. [CrossRef]
27. Dell'Erba, R. Determination of spatial configuration of an underwater swarm with minimum data. *Int. J. Adv. Robot. Syst.* **2015**, *12*, 1–18. [CrossRef]
28. Ureña, J.; Hernández, A.; García, J.; Villadangos, J.M.; Pérez, M.C.; Gualda, D.; Álvarez, F.J.; Aguilera, T. Acoustic local positioning with encoded emission beacons. *Proc. IEEE* **2018**, *106*, 1042–1062. [CrossRef]
29. Park, J.; Lee, J. Beacon selection and calibration for the efficient localization of a mobile robot. *Robotica* **2014**, *32*, 115–131. [CrossRef]
30. Park, J.; Choi, M.; Zu, Y.; Lee, J. Indoor localization system in a multi-block workspace. *Robotica* **2010**, *28*, 397–403. [CrossRef]
31. Kim, S.J.; Kim, B.K. Accurate hybrid global self-localization algorithm for indoor mobile robots with two-dimensional isotropic ultrasonic receivers. *IEEE Trans. Instrum. Meas.* **2011**, *60*, 3391–3404. [CrossRef]
32. Martín, A.J.; Alonso, A.H.; Ruíz, D.; Gude, I.; De Marziani, C.; Pérez, M.C.; Álvarez, F.J.; Gutiérrez, C.; Ureña, J. EMFi-based ultrasonic sensory array for 3D localization of reflectors using positioning algorithms. *IEEE Sens. J.* **2015**, *15*, 2951–2962. [CrossRef]
33. Sanchez, A.; de Castroa, A.; Elvira, S.; Glez-de-Rivera, G.; Garrido, J. Autonomous indoor ultrasonic positioning system based on a low-cost conditioning circuit. *Measurement* **2012**, *45*, 276–283. [CrossRef]
34. Eom, W.; Park, J.; Lee, J. Hazardous area navigation with temporary beacons. *Int. J. Control Autom. Syst.* **2010**, *8*, 1082–1090. [CrossRef]
35. El-Rabbany, A. *Introduction to GPS: The Global Positioning System*; Artech House: Norwood, MA, USA, 2002.
36. Ashkenazi, V.; Park, D.; Dumville, M. Robot positioning and the global navigation satellite system. *Ind. Robot* **2000**, *27*, 419–426. [CrossRef]
37. Folster, F.; Rohling, H. Data association and tracking for automotive radar networks. *IEEE Trans. Intell. Transp. Syst.* **2005**, *6*, 370–377. [CrossRef]

38. Bahl, P.; Padmanabhan, V.N. User location and tracking in an in-building radio network. In *Microsoft Research Technical Report MSR-TR-99-12*; Microsoft Corporation: Redmond, WA, USA, 1999.
39. Hightower, J.; Borriello, G.; Want, R. SpotON: An indoor 3D location sensing technology based on RF signal strength. In *UW CSE Technical Report 2000-02-02*; University of Washingtion: Seattle, WA, USA, 2000.
40. Ahmad, F.; Amin, M.G. Noncoherent approach to through-the-wall radar localization. *IEEE Trans. Aerosp. Electron. Syst.* **2006**, *42*, 1405–1419. [CrossRef]
41. González, J.; Blanco, J.L.; Galindo, C.; Ortiz-de-Galisteo, A.; Fernandez-Madrigal, J.A.; Moreno, F.A.; Martínez, J.L. Mobile robot localization based on Ultra-Wide-Band ranging: A particle filter approach. *Robot. Auton. Syst.* **2009**, *57*, 496–507. [CrossRef]
42. Vougioukas, S.G.; He, L.; Arikapudi, R. Orchard worker localisation relative to a vehicle using radio ranging and trilateration. *Biosyst. Eng.* **2016**, *147*, 1–16. [CrossRef]
43. Booma, G.; Umamakeswari, A. An analytical approach for effective location marking of landmine using robot. *Res. J. Pharm. Biol. Chem. Sci.* **2015**, *6*, 521–529.
44. Cotera, P.; Velazquez, M.; Cruz, D.; Medina, L.; MBandala, M. Indoor robot positioning using an enhanced trilateration algorithm. *Int. J. Adv. Robot. Syst.* **2016**, *13*, 1–8. [CrossRef]
45. Reiser, D.; Paraforos, D.S.; Khan, M.T.; Griepentrog, H.W.; Vazquez-Arellano, M. Autonomous field navigation, data acquisition and node location in wireless sensor networks. *Precis. Agric.* **2017**, *18*, 279–292. [CrossRef]
46. Gorostiza, E.M.; Galilea, J.L.L.; Meca, F.J.M.; Monzú, D.S.; Zapata, F.E.; Puerto, L.P. Infrared sensor system for mobile-robot positioning in intelligent spaces. *Sensors* **2011**, *11*, 5416–5438. [CrossRef]
47. Bais, A.; Sablatnig, R. Landmark based global self-localization of mobile soccer robots. *Lect. Notes Comput. Sci.* **2006**, *3852*, 842–851.
48. Zhou, Y.; Liu, W.; Huang, P. Laser-activated RFID-based indoor localization system for mobile robots. In Proceedings of the 2007 IEEE International Conference on Robotics and Automation, Roma, Italy, 10–14 April 2007; pp. 4600–4605.
49. Spacek, L.; Burbridge, C. Instantaneous robot self-localization and motion estimation with omnidirectional vision. *Robot. Auton. Syst.* **2007**, *55*, 667–674. [CrossRef]
50. Stroupe, A.W.; Sikorski, K.; Balch, T. Constraint-based landmark localization. *Lect. Notes Comput. Sci.* **2003**, *2752*, 8–24.
51. Mao, J.; Huang, X.; Jiang, L. A flexible solution to AX=XB for robot hand-eye calibration. In Proceedings of the 10th WSEAS International Conference on Robotics, Control and Manufacturing Technology, Hangzhou, China, 11–13 April 2010; pp. 118–122.
52. Shah, M.; Eastman, R.D.; Tsai, H. An overview of robot-sensor calibration methods for evaluation of perception systems. In Proceedings of the Workshop on Performance Metrics for Intelligent Systems, College Park, MD, USA, 20–22 March 2012; pp. 15–20.
53. Zitova, B.; Flusser, J. Image registration methods: A survey. *Image Vis. Comput.* **2003**, *21*, 977–1000. [CrossRef]
54. De Jong, F.; Caarls, J.; Bartelds, R.; Jonker, P. A two-tiered approach to self-localization. *Lect. Notes Comput. Sci.* **2002**, *2377*, 405–410.
55. Bandlow, T.; Klupsch, M.; Hanek, R.; Schmitt, T. Fast image segmentation, object recognition and localization in a robocup scenario. *Lect. Notes Comput. Sci.* **2000**, *1856*, 111–128.
56. Choi, W.; Ryu, C.; Kim, H. Navigation of a mobile robot using mono-vision and mono-audition. In Proceedings of the IEEE International Conference on Systems, Man, and Cybernetics, Tokyo, Japan, 12–15 October 1999; pp. 686–691.
57. Atiya, S.; Hager, G.D. Real-time vision-based robot localization. *IEEE Trans. Robot. Autom.* **1993**, *9*, 785–800. [CrossRef]
58. Liu, W.; Zhou, Y. Recovering the position and orientation of a mobile robot from a single image of identified landmarks. In Proceedings of the 2007 IEEE/RSJ International Conference on Intelligent Robots and Systems, San Diego, CA, USA, 29 October–2 November 2007; pp. 1065–1070.
59. Forsyth, D.A.; Ponce, J. *Computer Vision: A Modern Approach*; Prentice Hall: Upper Saddle River, NJ, USA, 2002.
60. Tsai, R.Y. A versatile camera calibration technique for high-accuracy 3D machine vision metrology using off-the-shelf TV cameras and lenses. *IEEE J. Robot. Autom.* **1987**, *3*, 323–344. [CrossRef]

61. Heikkila, J.; Silven, O. A four-step camera calibration procedure with implicit image correction. In Proceedings of the 1997 IEEE Computer Society Conference on Computer Vision and Pattern Recognition, San Juan, PR, USA, 17–19 June 1997; pp. 1106–1112.
62. Camera Calibration Toolbox for Matlab. Available online: http://www.vision.caltech.edu/bouguetj/calib_doc/ (accessed on 5 December 2019).
63. Shapiro, L.G.; Stockman, G.C. *Computer Vision*; Prentice Hall: Upper Saddle River, NJ, USA, 2001.
64. Gonzalez, R.C.; Woods, R.E. *Digital Image Processing*, 3rd ed.; Prentice Hall: Upper Saddle River, NJ, USA, 2007.
65. Acton, F.S. *Numerical Methods that Work*; Harper & Row: Manhattan, NY, USA, 1970.
66. Press, W.H.; Teukolsky, S.A.; Vetterling, W.T.; Flannery, B.P. *Numerical Recipes in C++: The Art of Scientific Computing*, 2nd ed.; Cambridge University Press: New York, NY, USA, 2002.
67. Petkovic, M.S.; Carstensen, C.; Trajkovic, M. Weierstrass formula and zero-finding methods. *Numer. Math.* **1995**, *69*, 353–372. [CrossRef]
68. Bini, D.A. Numerical computation of polynomial zeros by means of Aberth's method. *Numer. Algorithms* **1996**, *13*, 179–200. [CrossRef]
69. Pan, V.Y. Univariate polynomials: Nearly optimal algorithms for numerical factorization and root-finding. *J. Symb. Comput.* **2002**, *33*, 701–733. [CrossRef]
70. Malajovich, G.; Zubelli, J.P. Tangent Graeffe Iteration. *Numer. Math.* **2001**, *89*, 749–782. [CrossRef]
71. Toh, K.; Trefethen, L.N. Pseudozeros of polynomials and pseudospectra of companion matrices. *Numer. Math.* **1994**, *68*, 403–425. [CrossRef]
72. Edelman, A.; Murakami, H. Polynomial roots from companion matrix eigenvalues. *Math. Comput.* **1995**, *64*, 763–776. [CrossRef]
73. Jonsson, G.F.; SVavasis, S. Solving polynomials with small leading coefficients. *SIAM J. Matrix Anal. Appl.* **2005**, *26*, 400–414. [CrossRef]
74. Spall, J.C. *Introduction to Stochastic Search and Optimization: Estimation, Simulation, and Control*; John Wiley & Sons: Hoboken, NJ, USA, 2005.
75. Kirkpatrick, S.; Gelatt, C.D.; Vecchi, M.P. Optimization by simulated annealing. *Science* **1983**, *220*, 671–680. [CrossRef] [PubMed]
76. Coley, D.A. *An Introduction to Genetic Algorithms for Scientists and Engineers*; World Scientific: Hackensack, NJ, USA, 1999.
77. Back, T.; Fogel, D.B.; Michalewicz, Z. *Evolutionary Computation 1: Basic Algorithms and Operators*; Institute of Physics: London, UK, 2000.
78. Back, T.; Fogel, D.B.; Michalewicz, Z. *Evolutionary Computation 2: Advanced Algorithms and Operators*; Institute of Physics: London, UK, 2000.
79. Eggert, D.W.; Lorusso, A.; Fisher, R.B. Estimating 3-D rigid body transformations: A comparison of four major algorithms. *Mach. Vis. Appl.* **1997**, *9*, 272–290. [CrossRef]
80. Guan, J.; Deboeverie, F.; Slembrouck, M.; van Haerenborgh, D.; van Cauwelaert, D.; Veelaert, P.; Philips, W. Extrinsic calibration of camera networks using a sphere. *Sensors* **2015**, *15*, 18985–19005. [CrossRef]
81. Penne, R.; Ribbens, B.; Roios, P. An exact robust method to localize a known sphere by means of one image. *Int. J. Comput. Vis.* **2019**, *127*, 1012–1024. [CrossRef]

Article

Identification of Markers in Challenging Conditions for People with Visual Impairment Using Convolutional Neural Network

Mostafa Elgendy [1,2,*], Tibor Guzsvinecz [1] and Cecilia Sik-Lanyi [1]

1 Department of Electrical Engineering and Information Systems, University of Pannonia, 8200 Veszprém, Hungary; guzsvinecz@virt.uni-pannon.hu (T.G.); lanyi@almos.uni-pannon.hu (C.S.-L.)
2 Department of Computer Science, Faculty of Computers and Informatics, Benha University, Benha 13511, Egypt
* Correspondence: mostafa.elgendy@virt.uni-pannon.hu; Tel.: +36-88-624-000/6188

Received: 17 October 2019; Accepted: 23 November 2019; Published: 26 November 2019

Abstract: People with visual impairment face a lot of difficulties in their daily activities. Several researches have been conducted to find smart solutions using mobile devices to help people with visual impairment perform tasks. This paper focuses on using assistive technology to help people with visual impairment in indoor navigation using markers. The essential steps of a typical navigation system are identifying the current location, finding the shortest path to the destination, and navigating safely to the destination using navigation feedback. In this research, the authors proposed a system to help people with visual impairment in indoor navigation using markers. In this system, the authors have re-defined the identification step to a classification problem and used convolutional neural networks to identify markers. The main contributions of this paper are: (1) A system to help people with visual impairment in indoor navigation using markers. (2) Comparing QR codes with Aruco markers to prove that Aruco markers work better. (3) Convolutional neural network has been implemented and simplified to detect the candidate markers in challenging conditions and improve response time. (4) Comparing the proposed model with another model to prove that it gives better accuracy for training and testing.

Keywords: smartphone; assistive technology; visually impaired; navigational system; indoor navigation; computer vision; markers

---

1. Introduction

People with visual impairment (PVI) have limitations in the function of their visual system. These limitations prevent them from seeing and performing daily activities such as navigation or shopping [1–5]. Therefore, it is important to develop solutions to help PVI improve their mobility, protect them from injury, and encourage them to interact socially [6]. PVI also face challenges when navigating in shops, and to access and identify products [7]. While outdoor navigation is already solved by the use of a global positioning system, indoor navigation is still problematic [8–11].

As a result, multiple researches have been carried out to help PVI in indoor navigation and identifying products in shops or other objects using Wi-Fi, Bluetooth, RFID, NFC, and tags. These solutions differ in the types of the signals used, the positioning method, and the accuracy [12–19]. These solutions fall into three categories: Tag Based Systems: Such as radio-frequency identification (RFID) and near field communication (NFC), which use wireless components to transfer data from a tag attached to an object for the purpose of automatic identification and tracking [20,21]. Computer Vision Based Systems: These systems install or attach unique visual tags such as quick response (QR) codes or augmented reality (AR) markers to help in the navigation and recognition process [22–25]. Other systems do not require tags to be installed as they use computer vision (CV) techniques to analyze images and extract some features from them which help in navigating and identifying objects [26–33]. Hybrid Systems: These systems take the strengths of two or more systems and combine them into a new system to deliver better accuracy and performance [34–36]. In this research, the authors have focused on CV techniques.

CV non-tag-based systems are cost-effective, as they need little or no infrastructure and most of them can be easily installed on smartphones. However, they have several limitations. Their performance may be unpredictable in real-world environments, as image resolution may be affected by some factors including motion blur and changes in certain conditions, such as illumination, orientation, and scale. These systems also use extensive computational power which is not suitable for real-time usage. It is also difficult for PVI to capture photos in good quality [26,37]. CV tag-based systems rely on detectable visual landmarks such as QR codes, barcodes, and AR markers, which improve precision, robustness, and speed [5,37]. However, CV tag-based techniques require a prior installation of tags in the correct place and building a map to store these locations. Moreover, if many tagged items are installed in a small area, PVI may be confused by receiving information about all of them at the same time. Furthermore, tags not related to the system may confuse PVI. These tags are susceptible to damage caused by movement through the supply chain or by the weather. Moreover, it is difficult for a smartphone camera to detect CV tags if the PVI are moving fast, and the recognition rate decreases as the distance between the reader and the tags increases [38]. Based on our evaluation of the available technologies, the authors have concentrated on CV tag-based techniques.

Our goal is to build an indoor navigation system for PVI using CV tag-based techniques. A wide range of tags may be used, but QR codes and square markers are the most popular ones as they provide four correspondence points which are enough to perform camera position estimation. However, captured images in many real-life applications may have some problems such as blur due to motion, distortion of marker, and marker occlusion which makes identification difficult. The authors proposed a navigation system to improve marker identification in challenging conditions using convolutional neural network (CNN). By using CNN, the identification problem is converted to a typical classification problem.

This manuscript presents a system to help PVI navigate using markers. In this system, markers have been printed on papers and installed in indoor locations. When any marker is detected during navigation using the smartphone camera, the system uses this marker as a starting position. Then, it searches for the shortest path from that node to the destination node. During navigation, when these markers are detected, navigation commands are sent as feedback to help PVI. The main contributions of this article are the following:

- Building a system to help PVI in indoor navigation using markers.
- Making a comparison between multiple CV tags to select which is the best. The results proved that Aruco markers are the most fit for purpose of all markers.
- Formulating the marker identification problem of the CV tag-based system as a classification problem and solving it using the convolutional neural network (CNN). CNN has been implemented to detect candidate markers in challenging conditions. The authors have tested this system on real examples and achieved a significantly high accuracy in marker identification.
- Simplifying the CNN model to improve response time and making it suitable for real-time usage.
- Comparing the model of the authors with another model presented in [39] to evaluate the accuracy of training and testing.

The next part of this article is structured in the following way. Section 2 reviews the most relevant related works. Section 3 discusses the design of the proposed system. Section 4 presents the experimentation carried out, while section five draws some conclusions and future works.

2. Related Work

The main task of indoor navigation is to find the location of PVI and allow them to navigate safely in indoor environments such as public buildings and shopping malls. In recent years, smartphones have become useful because of their integrated cameras and various sensors. These technologies allow developers to build many applications to help PVI to navigate indoors and avoid obstacles [40]. Additionally, multiple researchers used the field of CV to improve the quality of these applications by improving accuracy and allowing them to be suitable for real-time use. In the marker-based method, the system consists of a mobile device with a camera, markers, and a server. The camera is used to scan markers, while the server is used for storing information such as the map [14]. In this section, the authors have focused on the available CV solutions based on QR codes and markers.

2.1. QR Codes

Ebsar is an Arabic system which provides indoor navigation for PVI by preparing the building and then guiding them using navigation commands [41]. At first, this system constructed a graph where each node represents a place or a checkpoint and a QR code is generated for each. Each edge is labelled with the number of steps and the direction between nodes connected to it. To start navigating, it seeks the nearest node to the PVI's location. After finding this node, it searches for the shortest path from that node to the destination node. During navigation, it provides Arabic voice feedback to the PVI using Google Glass. This system used Google glasses to facilitate detecting QR codes and communicating with the user, which makes it inexpensive and allows PVI to navigate without having to use their hands. However, this system requires an internet connection to download the building graphs from the server. Moreover, it is better to use markers which can be detected from a long distance rather than a QR code. Adding haptic feedback enables operation in a noisy environment.

AssisT-In used QR codes to help cognitively impaired people navigate inside new and complex environments [42]. QR codes are installed inside the building. Then, PVI use one of them as a starting point by scanning it. After determining the desired destination, the system calculates the shortest path to reach it. The system starts navigation by guiding PVI from the start node to subsequent nodes until reaching the destination node. When PVI scan a QR code during navigation, the system provides a feedback to them as a text message in the voice of a virtual pet such as a cartoon dog. If PVI scan a QR code not belonging to AssisT-In, it notifies them to keep searching for the correct QR code. However, it is difficult for PVI to capture good quality photos with their smartphone camera as most of the photos may be blurry. If more than one QR code is detected at the same time, it is better to select the one to use based on the distance and not select it randomly.

Zhang et al. proposed a navigation approach using a mobile robot in an indoor environment based on QR codes as landmarks [38]. These QR codes are placed in a distribution such as a grid pattern at the ceiling and the system constructs a map for them. Furthermore, an industrial camera is added to the robot to rapidly identify these QR codes. With this configuration, the camera can detect at least one QR code in its field of view and can estimate the position of the robot. The proposed recognition algorithm can localize the robot accurately and is suitable for real-time tasks. However, the robot failed to recognize QR codes in a completely dark environment. It is also hard to identify QR codes in adverse conditions such as a blurring effect due to motion and occlusion.

A smartphone system was developed to help PVI navigate in unknown indoor places using QR codes [43]. It starts by determining the type of current position, then fetching the environmental information from colour QR codes using a simple CV algorithm based on their colour and edges. When in motion, the change in location is computed continuously using two inertial sensors and PVI routes are recorded to guide them during the return route. During navigation, the system provides PVI with feedbacks using beeping or Text-to-Speech (TTS). The proposed method combined spatial language and virtual sounds to provide productive feedback which leads to better performance and reduces navigation errors. This system only requires minor modifications to the environment such as installing QR codes. Coloured QR codes facilitate separating and identifying them from the background.

However, objects only within 2.5 m were detected, which needed improvement. In adverse conditions, such as the blurring effect of motion, the system has difficulties identifying QR codes.

An android navigation application was introduced for PVI using QR codes that utilizes the smartphone's camera [44]. QR codes intended to be used by PVI are installed on the floor. Initially, the current location is defined by scanning one of the existing QR codes. Then, it finds the shortest and most optimal path to the PVI's destination. During navigation, any deviation from the predefined path is detected and corrected. All the instructions are given in an audio form to the PVI. This application provides automatic navigation on pre-defined paths for PVI and does not require any additional hardware. This application is capable of scanning QR codes of different sizes and in different challenging environments. However, instructions in audio and haptic form should be added to increase performance and reduce navigation errors. Moreover, it is better to use markers which can be detected from a long distance rather than a QR code.

Blind shopping is a solution which offers a better shopping experience for PVI with features including product search and navigation inside the store using voice messages [45]. The system combined an RFID reader on the tip of a white cane with mobile technology to identify RFID tags and navigate inside the shop. The system provided a web-based management part for configuration, generating QR codes for product shelves and RFID tag markers attached to the supermarket floor. It also, gives navigation feedback to PVI using voice commands via their smartphone. However, a Wi-Fi connection is required to retrieve the data from an online database. RFID tags and QR codes cannot be detected from a long distance.

2.2. Markers

Square markers are square shaped tags, as shown in Figure 1. They have a thick black border and the inner region contains images or binary codes represented in the form of grids of black and white regions. The reason for using thick black border is to ensure quick detection on any surface. For example, when considering Aruco markers [25], which have been used in our system, it is a square marker where the detection process is performed in two main steps: Detection of the marker candidates and codification analysis.

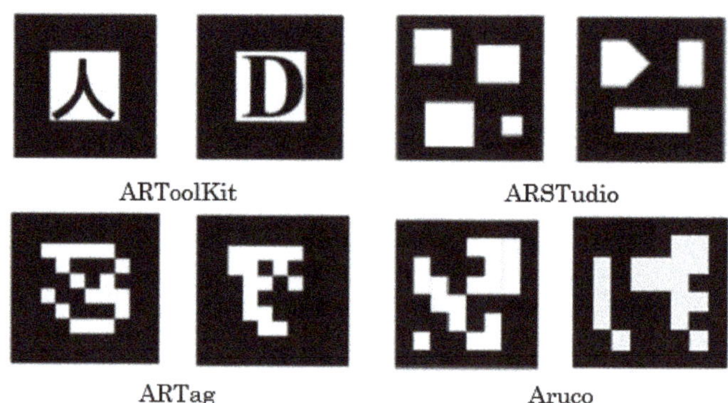

Figure 1. Examples of square markers.

Dash et al. proposed an AR system to be used in kindergarten to learn the alphabet by detecting the markers that may be present in a scene using CNN [35]. Markers have been printed on papers within a rectangular box. Then, children can show them in front of the camera attached to automatically render the virtual object over the marker with appropriate position and orientation. This system achieved high accuracy in marker identification and augmentation of the virtual objects making the system resistant to environmental noises and position variation. However, it may fail to detect markers from a long distance.

Delfa et al. proposed an approach for indoor localization and navigation using Bluetooth and an embedded camera on the smartphone [46,47]. It operates in two modes: Background and foreground mode. The background mode gives a low-level accuracy position estimation using Bluetooth. The foreground mode provides high accuracy by using the smartphone's camera to detect visual tags deployed onto the floor at known points. The system can detect tags in real-time to estimate the PVI position with a high level of accuracy and navigate towards the target. The marker's colours are chosen to be different from the colour of the floor to enhance the speed and efficiency. These colours guarantee high contrast between the floor's colour and the tag's border. However, the system failed to detect markers from a long distance and cannot detect more than one tag at the same time. Moreover, adding haptic feedback enables it to work in noisy places.

Bacik et al. presented an autonomous flying quadrocopter using a single onboard camera and augmented reality markers for localization and mapping [48]. The system is capable of estimating the position of the quadrocopter using a coordinate system defined by the first detected marker. To improve the robustness of marker-based navigation, it used a fuzzy control to achieve a fully autonomous flight. However, the precision of the mapping approach and the response time requires improvement. The system also fails to detect markers from a long distance.

3. Materials and Methods

Navigating and tracking PVI and objects inside buildings have many obstacles. While Global Positioning System (GPS) is used for outdoor navigation with good results, it is not suitable for indoor use, as it may give inaccurate results. For example, it is impossible to use a GPS to automatically determine which floor the user is currently on in a tall building. Using the mobile device's accelerometer and compass to track users showed better results and proved to be more available than GPS tracking but, requires a wireless connection or General Packet Radio Service (GPRS) service to function. Tags can be used to ensure that even when the PVI deviate from the correct path, they will immediately allow the system to adjust the exact location and continue navigating. This manuscript proposed a system to help PVI navigate indoors using markers. At first, tags are printed on multiple pieces of paper and placed in the specified location of interest. Then, a graph is created where nodes represent the markers' position. Edges are labelled with the number of steps and the direction between nodes are connected to it. The system starts by requesting the PVI to select their starting point based on the surrounding tags. When a marker is detected by the smartphone camera, the system will use this marker as a starting position.

To start navigation, the system asks the PVI to choose their destination using voice commands. Then, it searches in the database for the shortest path from this point to the destination. This path is a list of checkpoints that the PVI should pass to arrive at their destination. During navigation, continuous guidance is given to them when moving from one point to the next. Figure 2 shows a diagram to illustrate these steps.

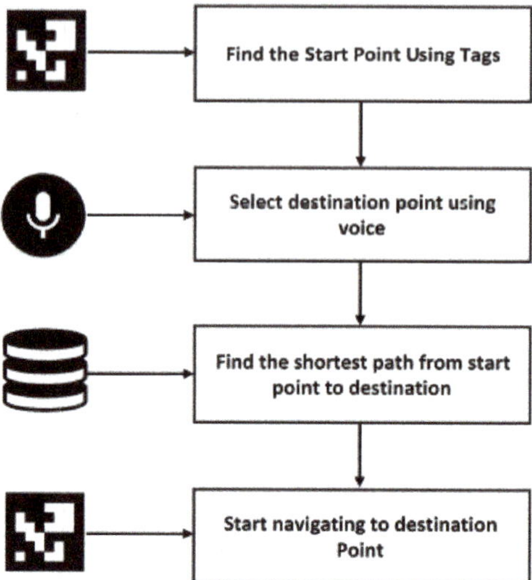

Figure 2. Components of the proposed system.

The authors found that the essential components of this prototype are: 1. How to accurately find PVI location at any time based on the installed tags so, they can continue navigating safely. 2. How to navigate from the starting point to the destination based on these tags. Therefore, the authors divided the proposed system into two parts: Identifying PVI location and navigating to a destination.

3.1. *Navigating to a Destination*

The proposed system functions through two major phases: (a) Preparing the building for PVI by installing markers at the specified locations and building a map for them. (b) Guiding PVI during navigation using TTS. The following sections describe these phases in detail.

3.1.1. Building a Map

Before PVI navigate inside the building, a sighted person should construct a floor plan. This person moves around the building to explore the available paths of the interest points such as labs and classrooms. After that, markers are generated and attached to the walls at those points to accurately guide PVI. Figure 3 shows a floor plan of the fourth floor of the Faculty of Information Technology at the University of Pannonia in Veszprem, Hungary. As shown in the Figure 3, the interest points are marked with red circles. Then, the authors built an internal map using a graph to store the relations between these points. Our prototype used the number of steps between markers during map construction. To do this, the authors tracked a sighted person's movements between markers to use them on the graph edges. Figure 4 shows a constructed graph representation for the fourth-floor plan. Finally, nodes and edges are stored in a database to be used by PVI during navigation.

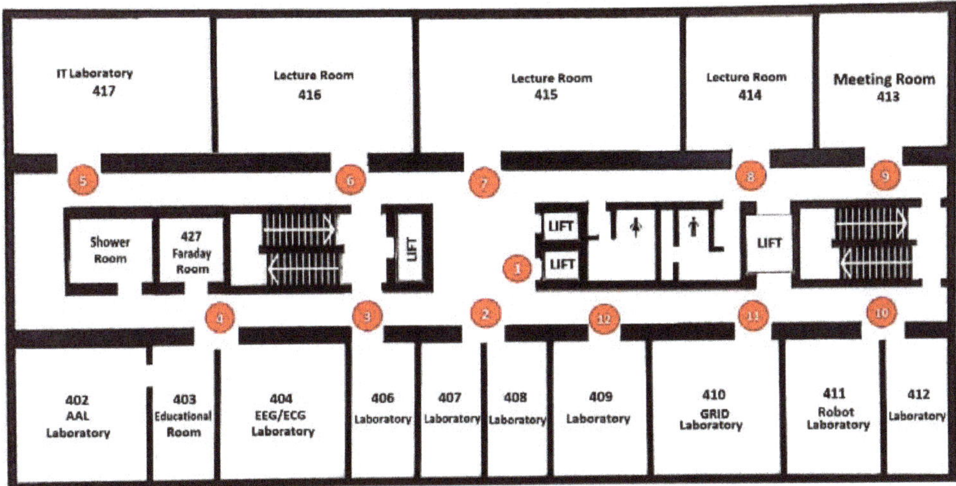

Figure 3. The plan of the fourth floor's corridor of the Faculty of Information Technology at the University of Pannonia.

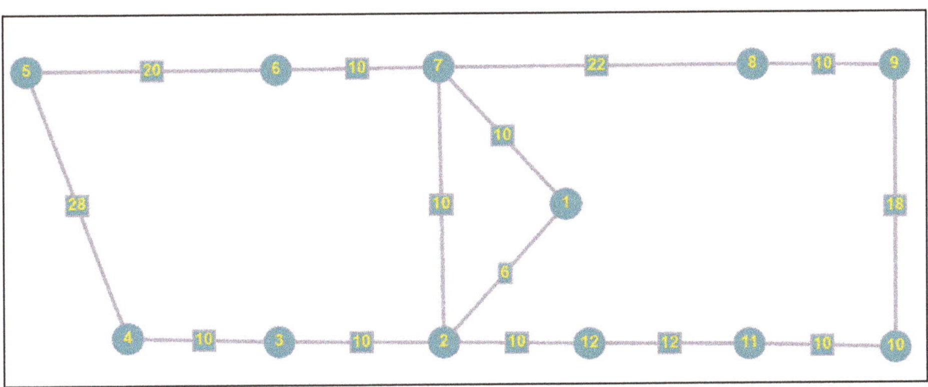

Figure 4. A graph of the fourth floor's corridor at the same faculty.

3.1.2. Navigation

Tracking the exact location of the PVI indoors is impeded by many obstacles. While a GPS is used widely in outdoor tracking, it is inaccurate when used for tracking objects indoors. Using the accelerometer and the compass (or magnetometer) to track users proved to work better than GPS tracking, which requires a wireless connection to function. At first, the PVI search for a marker around them to use it as a starting point. Once a marker is detected, it tells them that the initial location is identified. After that, it asks PVI to specify the desired target location. Then, it estimates the shortest path from the start point to the target location and instructs the PVI to start walking in the appropriate direction. Several techniques may be implemented, but the authors have chosen the Dijkstra algorithm. When walking, it guides them to the destination using navigation commands. When the PVI reach the first location of the path and eventually scan the marker placed on the wall, our system guides them to the next maker on the graph. This process is repeated until the PVI arrive at their destination. The main feature of the navigation module is to give continuous navigation commands to reach the destination.

3.2. Identifying PVI Location

A typical CV tag-based system to identify PVI location consists of tags, a database, a camera, a processing unit, and the use of audio and haptic feedback [41]. Identifying markers is the most crucial part of our system. They help find the destination successfully, as long as they are identified accurately. So, the authors put together the following three research questions.

RQ1: Which tag gives better accuracy based on the chosen criteria?
RQ2: Are there any difficulties or problems with identifying tags?
RQ3: Is this solution suitable for real-time usage?

To answer these questions, the authors have divided the implementation of this system into three steps:

1. Comparing QR code with Aruco markers to select which tag is the best.
2. Solving the problem of detecting markers in challenging conditions using the convolutional neural network (CNN) model.
3. Simplifying the CNN model to minimize the response time.

3.2.1. Comparing QR code with Aruco markers

To select which tag is the best, the authors have compared them based on certain criteria [5]. The first criterion is the cost of applying the technology to any solution. It is shown that CV tag techniques can be used without any cost except for printing the QR codes or AR markers and putting them in the correct place. When using a barcode, there is no need to print them, as they are already placed on each product. Tag-based techniques can be used at a low cost, as shops only need the RFID or NFC tags to be installed in the correct places. If CV techniques are used, high-quality equipment, such as cameras, are required for satisfactory results. The second criterion is the equipment needed to detect and identify products or places. For CV tag-based solutions, PVI only need their smartphone cameras to detect and identify items. For non-CV tag-based techniques, some solutions only need a smartphone's camera, while others need high-quality cameras to capture high resolution images and machines with powerful processors for computation. In tag-based techniques, PVI need RFID readers or smartphones supporting NFC technology. The third criterion is the number of items to be scanned simultaneously. Only RFID readers, AR markers, and CV techniques can scan multiple items at the same time, which is useful in some situations, such as if PVI want to identify and count the items in their shopping cart. The fourth criterion is whether the PVI must be in the line of sight with the identified products or not. For tag-based solutions, PVI do not have to be in the line of sight of items, and the PVI can identify them in any direction. In tag-based solutions, there is no need for RFID tags and NFC tags to be in the line of sight, while CV solutions depend on some other parameters, such as the tag size in QR codes or barcodes, and the marker or camera parameters for CV techniques. The fifth criterion is the storage capacity of each solution. Only some tags, such as RFID, NFC, and QR codes, have a storage capacity, while others, such as AR markers and barcodes, do not have any storage capacity. The last criterion is the detection range, in tag-based solutions, tags must be within 3 m for RFID tags and within 10 cm for NFC tags. In CV tag-based solutions, tags can be detected from a distance which depends on tag size and image resolution. Based on this evaluation, the authors found that QR codes and markers were the most suitable markers to select.

So, the authors have compared QR codes with markers to select which give more accuracy and put together the following hypotheses:

H1: *Aruco markers are better than QR codes.*

H2: *Aruco markers can be detected from long distances.*

The authors have developed two applications to identify the location of PVI with the architecture shown in Figure 5.

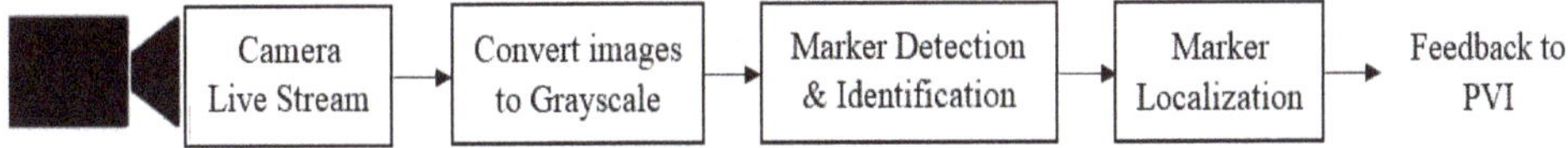

Figure 5. Main components of the comparison application.

An important task after the detection and identification of markers is to obtain the camera's position from them. For the position estimation, the authors need to know the parameters of the camera which came from the calibration. These parameters are the camera matrix and the distortion coefficients. The camera has some parameters such as extrinsic, intrinsic, and distortion coefficients. To estimate these parameters, it is important to have 3D world points of a real scene and the corresponding 2D image points which are called camera calibration. Camera calibration estimates the parameters of the lens and sensor of an image which makes it important for AR applications. This can be done using multiple images of a pattern such as a checkerboard, square grid, circle hexagonal grid or circle regular grid. So, before using the two applications, the authors made camera calibration to obtain the camera position from markers. These applications work the following way: At first, the application opens the camera to get a live stream of images. Then, it converts the image to grayscale and sends it to the desired library to detect and identify the marker. After that, it calculates the distance to the marker and gives feedback to the PVI using voice commands. QR codes were used as markers in the first application, while Aruco markers were used instead of QR codes in the second one.

3.2.2. Detecting Markers in Challenging Conditions

After comparison, the authors found that Aruco markers are better and give more accuracy than QR codes. So, the authors decided to use Aruco markers in the proposed system. The authors also found that both cannot be detected in challenging conditions such as long distances, blurring effects, distortion of marker, and occlusion. To solve this problem, the authors converted the identification part of the problem to a classification one and used CNN to identify the markers as shown in Figure 6. The authors formed the following hypothesis:

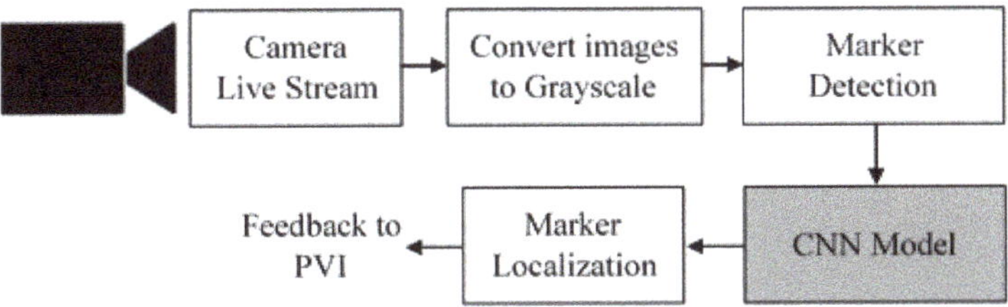

Figure 6. Main components of the application using CNN.

H3: *CNN models can be used to detect Aruco markers in challenging conditions.*

To detect markers in the mentioned challenging conditions, the authors converted the identification component in the architecture shown in Figure 5 to a classification one and CNN is used to identify the markers. The architecture consists of two main units, the first unit marked in white is used to extract markers from captured frames, while the second one is marked in grey and used for classification. Following the image conversion to greyscale and detecting markers, a CNN model is used to identify them. This model returns the correct ID of detected markers or if it fails to do so, it determines that no marker is available. In a typical marker-based application, image frames are affected by various noisy conditions such as blurring and distortion of markers, etc. Such noises affect the accuracy of marker identification. The authors have created a dataset that covers most of the challenges above. A dataset is created corresponding to each class of challenging conditions and different orientations. This dataset contains 40 classes of 10 markers and one to indicate no markers [37]. As shown in Figure 7, the authors have created different effects by randomly applying various transformations to the original images. For each class, the authors generated 500 samples. So, a total of 14,000 images were created for the 28 classes. To identify that no markers were visible in the image captured by the camera, the authors assigned the null class with images taken from a dataset [35]. 75% of the input samples from each class were used for training and rest for validation.

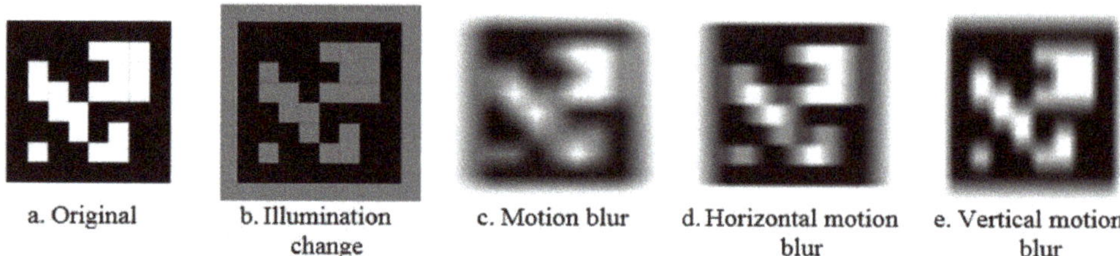

Figure 7. A few samples of makers with illumination change and motion blur: (**a**) Original; (**b**) Illumination change; (**c**) Motion blur; (**d**) Horizontal motion blur; (**e**) Vertical motion blur.

CNN automatically learns the most efficiently from raw data instead of using traditional machine learning techniques. A typical CNN has a convolutional layer, which is the basis. It also contains an activation function for transforming the summed weighted input from the node into the activation of the node or output for that input. Another so called pooling layer is used, which performs a downsampling operation along the spatial dimension. These layers together are used for effective feature extraction and are called feature extraction unit (FEU). For classification, CNN also contains the FC layer. As shown in Figure 8, the authors used three FEUs in the proposed model. In the first FEU, the authors used a convolutional layer with a dimension of $20 \times 5 \times 5$ and a sigmoid activation function. Then, max-pooling with a pooling filter size of 2 and stride of 2 is used to down sample the outputs by a scale factor of 2. The second and the third FEU are of the same structure as the first one with a slight variation. The authors used a convolutional layer with a dimension of $50 \times 5 \times 5$ in the second FEU, while a convolutional layer with a dimension of $200 \times 5 \times 5$ was used in the third one.

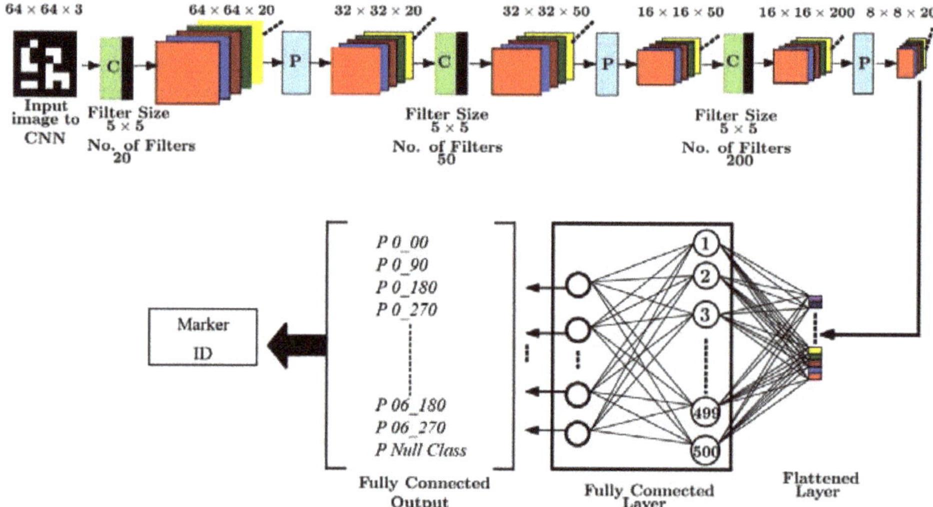

Figure 8. The proposed convolutional neural network (CNN) architecture used in training of markers.

For classification, the output of the last feature extraction unit is converted into a vector through flattening and is provided as an input to the fully connected layer. The output of the fully connected layer is converted into probabilities corresponding to 29 classes of the classifier. The authors used a fully connected layer after the flatten operation contained 500 components which were mapped to 29 number of classes. During the training stage, the Adam optimizer with a learning rate of has been used over the cost function to determine the optimum value of the weight parameters. The results showed that Aruco markers can be detected in challenging conditions using our CNN model. However, this model still failed to detect markers under occlusive conditions. Moreover, the time required for this model to detect markers should be minimized.

3.2.3. Simplified CNN Model to Minimize the Response Time

The results showed that Aruco markers can be detected in adverse conditions, albeit the execution time is not suitable for real-time usage. To minimize the time required to detect markers, the authors formulated the following hypothesis:

H4: *CNN models can be simplified by removing some convolutional layers to improve the response time.*

To minimize the response time, the authors have simplified the convolutions layers of the CNN model and used the same parameters for training and validation. As shown in Figure 9, the authors used two FEUs in the simplified model. In the first FEU, the authors used a convolutional layer with a dimension of 20 × 5 × 5 and a sigmoid activation function; then Max pooling was used with a pooling filter size of 2 and stride of 2 to downsample the layer's outputs by a scale factor of 2. In the second FEU, a convolutional layer with a dimension of 200 × 5 × 5 was used. Dropout layers are placed to prevent an overfitting problem. The simplified model can detect Aruco markers very well in adverse conditions and the execution time is minimized to be suitable for real-time identification of markers. However, this model still failed to detect markers under occlusive conditions.

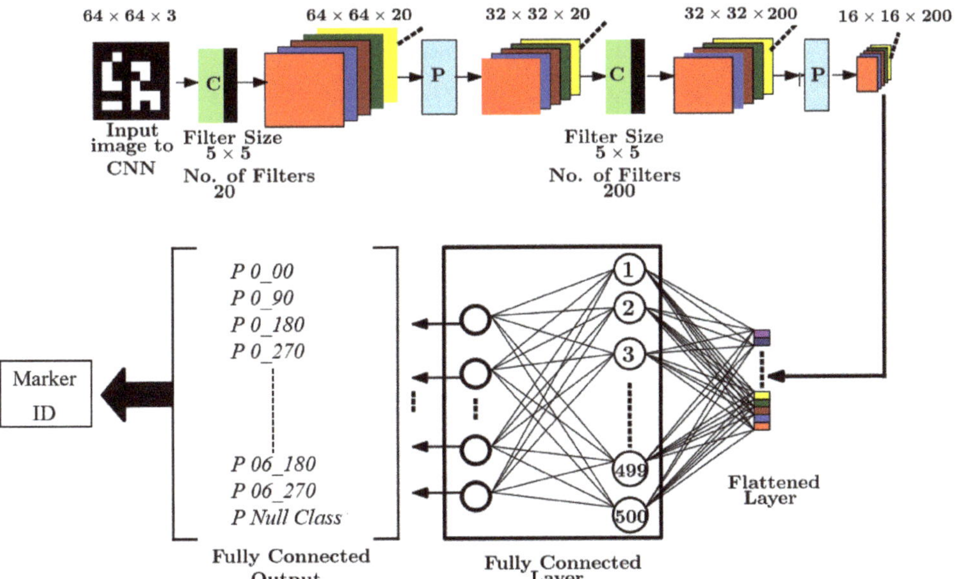

Figure 9. The proposed simplified CNN architecture.

4. Results and Discussion

The authors evaluated their system in two steps: Compared QR code with Aruco markers then, evaluated the performance of CNN in challenging conditions. The authors used HTC Desire 826 (HTC, Taiwan) with 2 GB RAM, octa-core (4 x 1.7 GHz Cortex-A53 and 4 x 1.0 GHz Cortex-A53) CPU and Adreno 405 GPU for the first task. For the second one, a DELL INSPIRON N5110 computer (DELL, U.S.) with Intel Core i7-2630 QM 2.00 GHz CPU, 6 MB cache, quad-core, 8 GB RAM, and a 1 GB NVIDIA GeForce GT 525 M graphics card was used.

4.1. Comparing QR Code with Aruco Markers

In the first application, QR codes with dimension of 10 × 10 cm are printed and installed in the environment at regular intervals. Each QR code stores information about the current location. For pre-processing, the authors used an open-source CV library called OpenCV that implement many algorithms for image processing and CV. The authors also used an open-source library called Zxing for detecting and identifying QR codes. When PVI use this application, it activates the camera to capture photos until a QR code is detected. The position details stored in the detected QR Code and the distance are given to the PVI as voice commands.

The second application is the same as the first one, but it uses Aruco markers with the same dimension instead of QR codes. The authors also used an open-source library called Aruco library for detecting and identifying markers. After testing the two applications in different situations, the authors made a comparison to select which one is the best, as shown in Figure 10.

Aruco Markers	1 Meter	2 Meters	3 Meters	4 Meters
# of scanned items	Multiple	Multiple	Multiple	Multiple
30 degree	√	√	√	√
60 degree	√	√	√	√
80 degree	√	√	√	√
Moving fast	√	x	x	x
Line of sight	x	x	x	x
Blur	x	x	x	x
Camera out of focus	x	x	x	x
Occlusion	x	x	x	x

(a)

QR Codes	1 Meter	2 Meters	3 Meters	4 Meters
# of scanned items	Multiple	Multiple	x	x
30 degree	√	√	x	x
60 degree	√	√	x	x
80 degree	x	x	x	x
Moving fast	x	x	x	x
Line of sight	x	x	x	x
Blur	x	x	x	x
Camera out of focus	x	x	x	x
Occlusion	x	x	x	x

(b)

Figure 10. Evolution of quick response (QR)codes and Aruco markers: (**a**) Results of evaluating Aruco markers in different situations from various distances (1, 2, 3, and 4 m); (**b**) results of evaluating QR codes in different situations from various distances (1, 2, 3, and 4 m).

After testing these two applications in different situations, the authors found that Aruco markers can be detected from distances up to 4 m while QR codes were limited to 2 m only. To detect Aruco markers from long and short distances, there is no need for the camera to be in their line of sight. For QR codes, they cannot be detected from farther than 2 m and whether the camera was in their line of sight was irrelevant. From these results, the authors have deduced that Aruco markers are better than QR codes, which lead to the verification H1. So, the authors formulate Thesis T1: Aruco markers work better than QR codes. As a result, the authors selected Aruco markers as tags for our prototype.

4.2. Navigation Using Aruco Markers

After selecting Aruco markers for our system, the authors built maps for the first and the fourth floor. Depending on which floor the PVI wish to navigate, the right map is loaded from the database once the application has been launched. At that point, the prototype is controlled by the user via voice commands. Figure 11 showed screenshots of the system. As shown in part (a) of the Figure 11, it asks the PVI to select the starting point by pressing on the screen. As shown in part (b), it launches the mobile camera and guides them to search for any markers to be used as a starting point. As shown in part (c), the third step is to select the destination using voice commands. For example, PVI say "lab 417" to be their destination. The system calculates the shortest path from the starting point to the destination using the Dijkstra algorithm. Then, it launches the smartphone camera and starts guiding the PVI to the next point using voice commands as shown in part (d). The authors used the commands shown in Table 1.

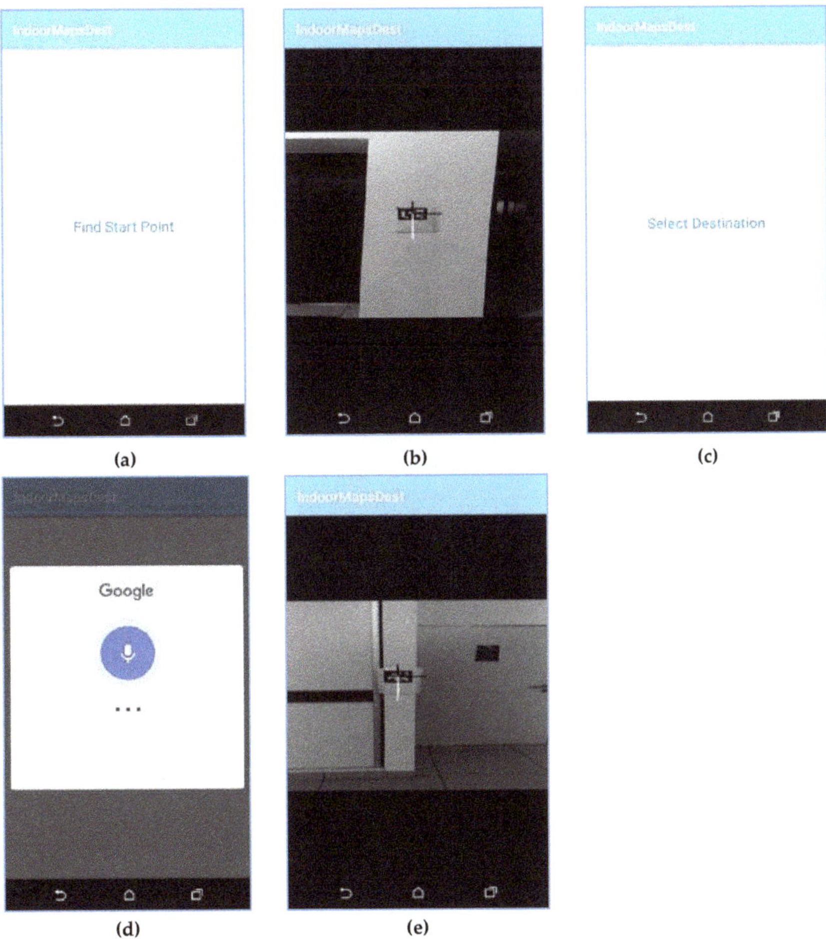

Figure 11. Screenshots of our prototype: (**a**) User presses this button to select the starting point; (**b**) User moves the smartphone's camera left and right to detect the starting point; (**c**) User presses this button to select the destination point; (**d**) User selects the destination using voice commands to start navigation; (**e**) User moves the smartphone camera left and right to detect the next point until reaching destination.

Table 1. Commands given by the PVI and navigation commands by the prototype.

Command Type	Name	Description
PVI's voice commands	"Go to" + destination	The PVI order the prototype to lead them toward the predefined destinations.
	"Start"	The PVI order the prototype to go to the start activity to select the start point.
	"Exit"	The PVI order the prototype to exit.
Navigation instructions	"Incorrect destination, you should press on the screen and select it again"	The prototype informs the user that they should provide another destination.
	"Go straight" + number of steps	The prototype directs the user to go straight for number of steps.
	"Turn left", "Turn right"	The prototype directs the user to turn left or right.
	"You have detected your next point, so, you should go straight to reach it"	The prototype informs the user that the next point is detected, and the user should approach it.
	"You have passed this point successfully"	The prototype informs the user that they passed this point successfully and have started navigating to the next point.
	"You have reached your destination so, go straight to it"	Once the user reaches the desired destination, the prototype informs him.

For example, PVI select the elevator as a starting point which is stored in the database as a node 1. They also choose their destination as "lab 417" which is stored as node 5. Using our system, the shortest path between node 1 and node 5 is calculated as shown in Figure 12. To reach laboratory number 417 (node 5), PVI start walking from the elevator (node 1). They go from node 1 to the next node, which is node 7. At node 7, navigation commands guide them to turn left and move straight for ten steps to reach node 6. Finally, PVI walk for twenty steps to reach node 5, which is the destination.

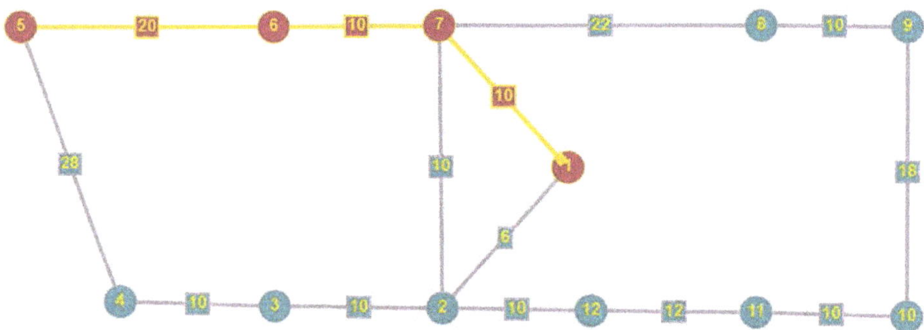

Figure 12. Path to destination example.

The authors divided testing of the prototype into two test cases. The first case was to test it with blindfolded people or PVI, collection of the feedback and updating it. The second case is to evaluate the prototype again after applying modifications using the PVI's comments. The prototype was tested on the corridor of the first and the fourth floor (Video S1, Video S2). In the beginning, a short introduction to the case study was provided to the participants. The authors trained the users for 30 min to know how to use the prototype for navigating from one place to another. The goal was to test whether the prototype was easy to use or not. It also tested if the users could effectively interpret the feedbacks or not. The authors made sure that there were no obstacles on the way to the destination. During navigation, the user held the smartphone in his hands roughly at chest level with the screen facing towards him. The smartphone was held in portrait orientation while slightly tilted at an angle nearly perpendicular to the horizontal plane. As shown in Figure 13, using this angle is enough for covering the walking area in front of the PVI and identifying markers. For a hands-free option, the smartphone may also be mounted on the user's chest. Audio feedback is provided to the user via headphones connected to the smartphone or by a smartphone's speaker.

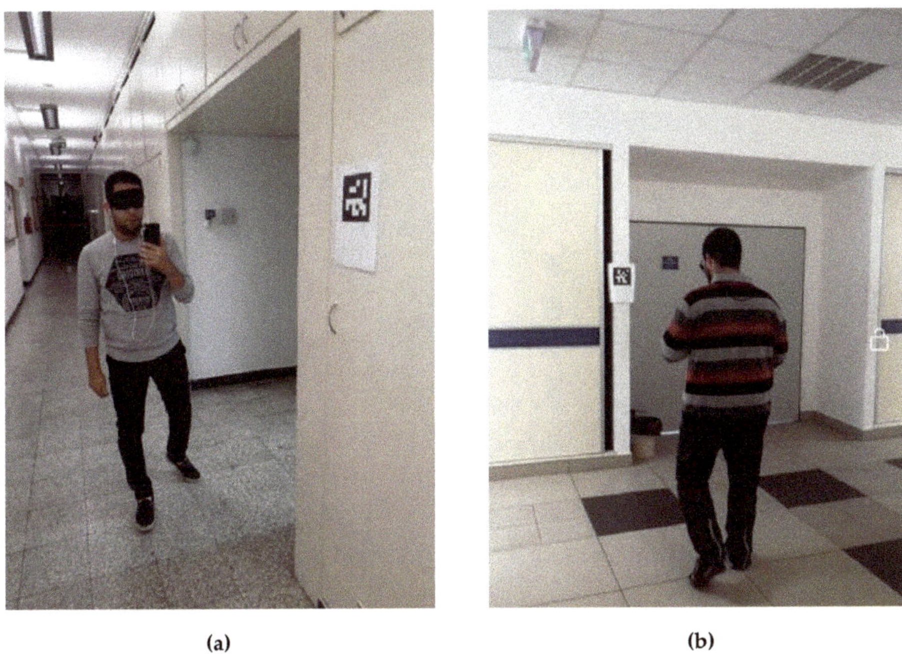

Figure 13. Screenshots of our prototype: (**a**) A screenshot of a blindfolded person while testing our prototype on the first floor; (**b**) a screenshot of a PVI while testing our prototype on the fourth floor.

4.2.1. First Test Case

After learning how to use the prototype, they have tested it several times by selecting a start point and destination. The prototype assists them in moving from the starting point to the destination using navigation feedback. During the process, the authors have discovered some problems: 1. Sometimes PVI failed to understand the feedbacks and as a result, the feedback needed to be improved. 2. PVI can hardly detect markers because they were placed higher than the view of the camera. So, installing markers at a lower position is necessary. 3. PVI move their hands rapidly during navigation which causes images to be captured with occlusion. 4. PVI cannot detect markers because they are moving their hands a lot and tags move out of the smartphone's camera view. 5. PVI take shorter steps compared to blindfolded participants, so the authors should calculate the number of steps based on PVI rather than the blindfolded individuals. 6. PVI occasionally create situations that cannot be managed by our prototype. For example, if their next point is node 7 and they go in the wrong direction leading to another point. The prototype should check whether this point is node 6 or not. If it is node 6, the prototype should continue navigating, because node 6 in the graph is the next point to destination after the node. However, if it was another point, the prototype should ask the PVI to go back and search for node 7 again.

4.2.2. Second Test Case

The authors tried to solve the problem occurring during the first test case. For the first problem, they improved the feedback based on the comments of the PVI. As shown in Figure 14, markers should be installed in a different style to solve the second problem. Instead of adding one marker at each interest point, eight markers are installed with the same ID. This implementation makes detection easier and solved the third and fourth problems. It also helps PVI of different heights to detect markers easily. The authors count the steps in the same way as the PVI walk to solve the fifth problem. The authors tried to manage all situations and conditions raised during the testing phase of the prototype. Users have tested the prototype several times by selecting a starting point and destination. PVI found it easier to detect markers faster than before. Using this arrangement of markers can be detected easily while moving their hands rapidly. Additionally, they found the audio feedback to be satisfactory.

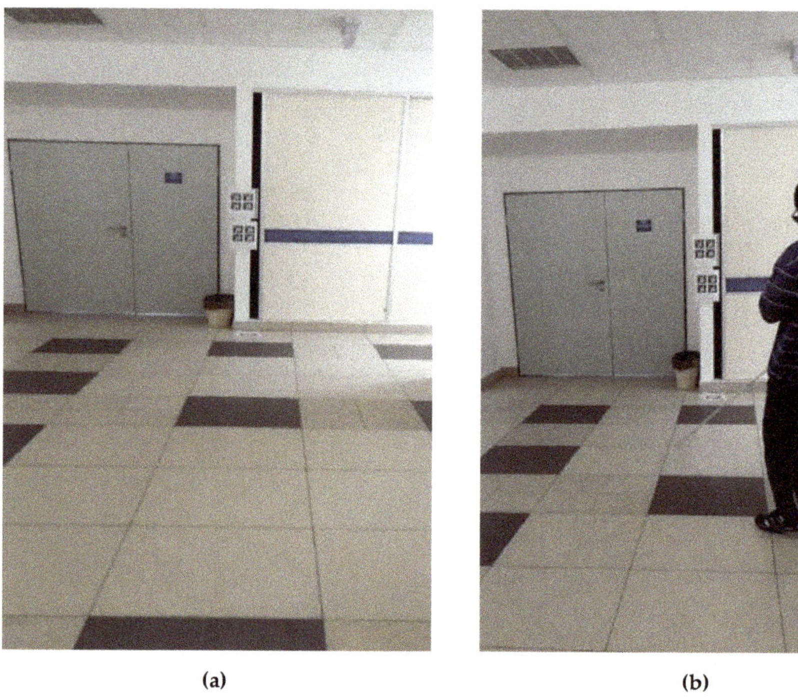

Figure 14. Screenshots of our prototype: (**a**) A screenshot of markers after updates; (**b**) a screenshot of PVI while testing our prototype on the fourth floor.

4.3. *CNN Performance Evaluation in Challenging Conditions*

As discussed, the authors trained the CNN model using the created dataset. The training was conducted in batches with size of 32. The Keras framework for deep learning was used for implementing the CNN model. The original size of the markers in the synthetic data set were 512 × 512. For training the CNN model, the original images were resized to 64 × 64 and were given as an input to the first convolution layer. The entire dataset was divided into two parts: 75% for training and 25% for validation. Training and validation datasets were used to determine the layer parameters during the training phase of the model. Additionally, the authors tested the model with new data that had not been encountered before. The test performance of the model was carried out on the trained model. The output was produced with 29 class values. For all the experimental results, the training phase of the CNN model was carried out for 100 epochs. Experimental results were obtained for all classes. The authors also trained the model with the dataset proposed to find which marker produces better results. The training and validation plots for accuracy and loss are shown in Figures 15 and 16.

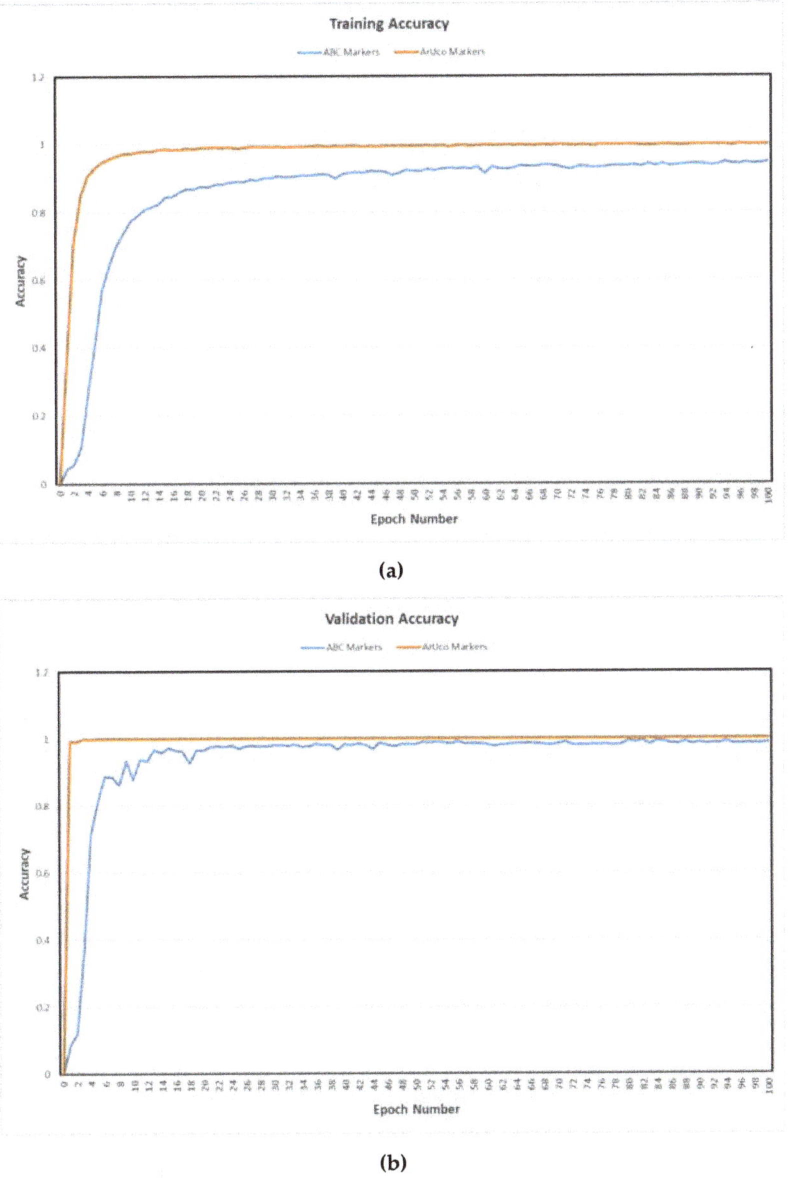

Figure 15. Comparative accuracy graphs after applying our model on two datasets where our dataset is drawn in red and the other dataset is drawn in blue: (**a**) Results of the training accuracy of the two datasets; (**b**) results of the validation accuracy of the two datasets.

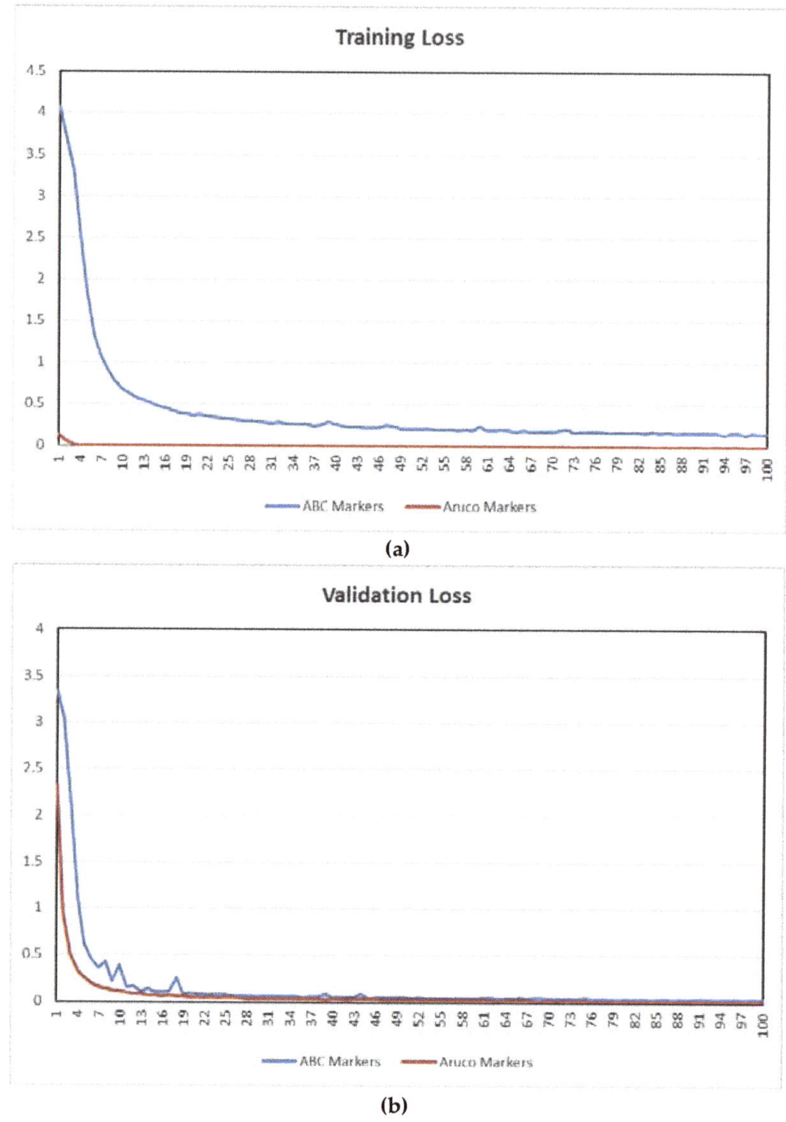

Figure 16. Comparative loss graphs after applying our model on two datasets where our dataset is drawn in red and the other dataset is drawn in blue: (**a**) Results of the training loss of the two datasets; (**b**) results of the validation loss of the two datasets.

The results show that the proposed model can detect Aruco markers very well in challenging conditions as it provides approximately 97% accuracy for training and 99.97% accuracy for testing. When applying it to the other dataset, it gave 86% for training accuracy and 94.74% for testing accuracy. From these results, the authors improved the detection in challenging conditions and verify H3. The authors also improved the detection of markers from a long distance and verify H2. So, the authors formulate Thesis T2 and Thesis T3. T2: Aruco markers can be detected from long distances. T3: CNN models can be used to detect Aruco markers in challenging conditions such as blurring effect due to motion. However, this model still failed to detect markers under occlusive conditions. Moreover, the time required for this model to detect markers should be minimized.

4.4. Simplified CNN Model Suitable for Real-Time

The authors simplified the CNN model by applying only two sequences of FEU instead of three. The authors used the same parameters and data of the CNN model for training, validating, and testing. Figures 17 and 18 show the training and validation accuracy graphs of the proposed model. The training and validation accuracy of the simplified model is better than the accuracy of the model proposed in [39]. It also shows that the training and validation curves of our two models are close to each other.

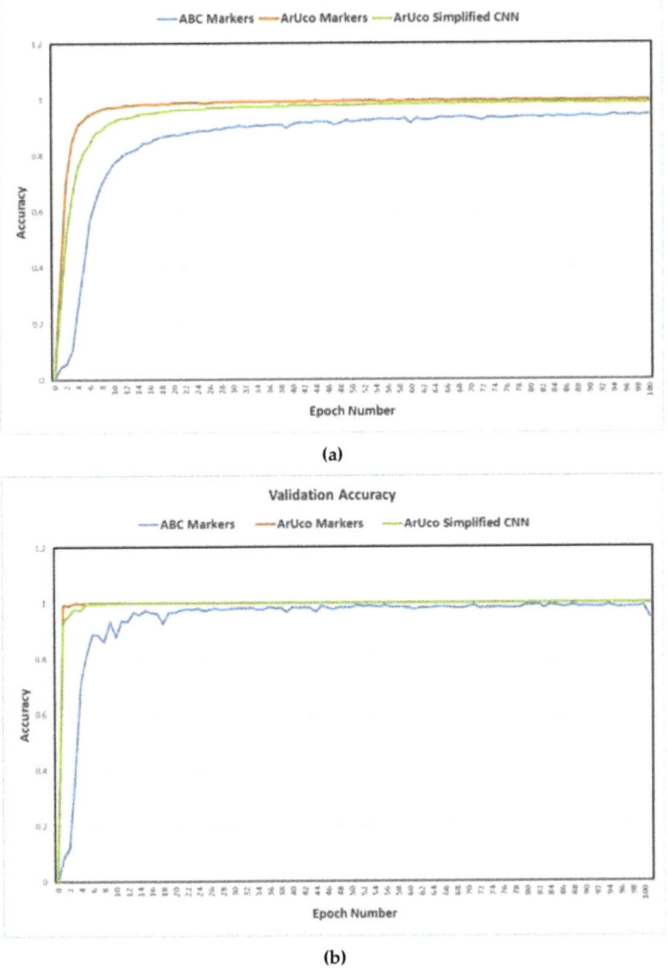

Figure 17. Comparative accuracy graphs after applying our first model to our dataset which is drawn in red and applying the simplified models to the two datasets where our dataset is drawn in green and the other dataset is drawn in blue: (**a**) Results of the training accuracy of the two datasets; (**b**) results of the validation accuracy of the two datasets.

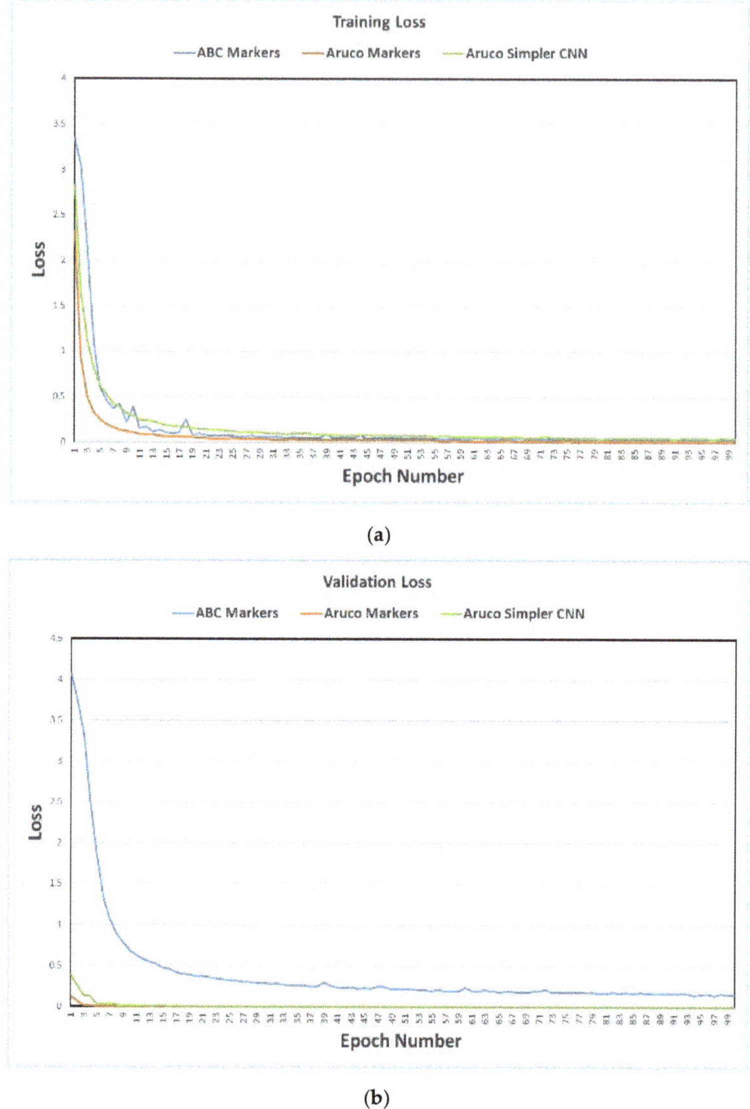

Figure 18. Comparative loss graphs after applying our first model to our dataset which is drawn in red and applying the simplified models to the two datasets where our dataset is drawn in green and the other dataset is drawn in blue: (**a**) Results of the training loss of the two datasets; (**b**) results of the validation loss of the two datasets.

The execution time for detecting markers in the simplified model is better than the complex one. Thus, this model is suitable for real-time identification of markers. The authors have simplified the convolutional layers of the CNN model and used the same parameters for training and validation. The simplified model can detect Aruco markers very well in challenging conditions as it gives approximately 95.5% accuracy for training and 99.82% accuracy for testing. Furthermore, the training and testing accuracy of the simplified model is better than the accuracy of the other dataset. The training and testing curves of our models are also close to each other. The execution time for detecting markers in the simplified model is better than the complex one. So, this simplified model minimized the response time and led to the verification H4. The authors formulate T4. T4: CNN models can be simplified by removing some convolutional layers to improve the response time. However, this model still failed to detect markers under occlusive conditions.

As a summary, the authors found the following results based on the above experiments. Aruco markers are selected to be used based on a comparison with QR codes in different situations. A navigation system is proposed to help PVI in navigation indoors using Aruco markers. After testing the prototype by PVI and blindfolded individuals, feedback and comments were discussed. Based on the discussion, the prototype was updated and tested again by PVI. The authors solved the problem of detecting markers in challenging conditions by using CNN and compared the results to another model. The results showed that our CNN gives better accuracy. The CNN model was simplified to be suitable for real-time usage.

5. Conclusions

The goal was to design a navigation system for PVI using makers. First, a sighted user is required to walk through the building to indicate different points of interest. Then, markers are printed and placed on walls at the specified locations. A map is built with a graph to be used for finding the shortest path to the destination. PVI use a smartphone to find the current location based on markers around them. Finally, they are guided to the destination using voice feedback. Our evaluation showed that Aruco markers are better for localization than QR codes as they can be detected from distances twice as long (compared to QR codes). Moreover, they can be detected in adverse conditions using CNN and the time for identification can be minimized by simplifying the CNN.

For future work, the prototype should be improved to add automatic map construction. The authors identified markers in challenging conditions using CNN architecture. This work has only dealt with the identification steps, while detection has been done using a method based on image thresholding and rectangle extraction. The authors can use CNN's models such as you only look once (YOLO) or single shot detector (SSD) to fully perform the detection and identification steps as they can be trained to automatically to it. The authors did not analyze the behavior of their proposal against occlusions, which are indeed a common problem when markers are used. Consequently, a set of experiments to evaluate our proposal against the occlusion problem and its solutions, in case of bad performance, could be of great interest.

Supplementary Materials: The following are available online at http://www.mdpi.com/2076-3417/9/23/5110/s1, Video S1: Blindfolded_Test.mp4. Video S2: PVI_Test.mp4.

Author Contributions: The work presented in this paper corresponds to a collaborative development by all authors. C.S.-L. supervised the overall project. M.E. participated in all design tasks of the system. M.E. and T.G. were involved in the development of the proposed system. C.S.-L. and T.G. were involved in the analysis of the different components of the system. All authors have written and revised the manuscript.

Funding: This research received no external funding.

Acknowledgments: The authors would like to kindly thank Kristof Bakonyvari, the participant of the PVI community and his assisted guides, Tibor Tuza and Efraim Hege, students of the University of Pannonia who participated in the testing phase of the project. Moreover, the authors thank the help of Mohamed Issa, as a blindfolded tester.

Conflicts of Interest: The authors declare no conflict of interest.

References

1. Giudice, N.A. Navigating without vision: Principles of blind spatial cognition. In *Handbook of Behavioral and Cognitive Geography*; Edward Elgar Publishing: Cheltenham, UK, 2018; pp. 260–288.
2. Legge, G.E.; Granquist, C.; Baek, Y.; Gage, R. Indoor Spatial Updating with Impaired Vision. *Investig. Opthalmol. Vis. Sci.* **2016**, *57*, 6757. [CrossRef]

3. Schinazi, V.R.; Thrash, T.; Chebat, D. Spatial navigation by congenitally blind individuals. *Wiley Interdiscip. Rev. Cogn. Sci.* **2016**, *7*, 37–58. [CrossRef]
4. Yuan, C.W.; Hanrahan, B.V.; Lee, S.; Rosson, M.B.; Carroll, J.M. Constructing a holistic view of shopping with people with visual impairment: A participatory design approach. *Univers. Access Inf. Soc.* **2019**, *18*, 127–140. [CrossRef]
5. Elgendy, M.; Sik-Lanyi, C.; Kelemen, A. Making Shopping Easy for People with Visual Impairment Using Mobile Assistive Technologies. *Appl. Sci.* **2019**, *9*, 1061. [CrossRef]
6. Bhowmick, A.; Hazarika, S.M. An insight into assistive technology for the visually impaired and blind people: State-of-the-art and future trends. *J. Multimodal User Interfaces* **2017**, *11*, 149–172. [CrossRef]
7. Kostyra, E.; Żakowska-Biemans, S.; Śniegocka, K.; Piotrowska, A. Food shopping, sensory determinants of food choice and meal preparation by visually impaired people. Obstacles and expectations in daily food experiences. *Appetite* **2017**, *113*, 14–22. [CrossRef] [PubMed]
8. Tapu, R.; Mocanu, B.; Zaharia, T. DEEP-SEE: Joint Object Detection, Tracking and Recognition with Application to Visually Impaired Navigational Assistance. *Sensors* **2017**, *17*, 2473. [CrossRef] [PubMed]
9. Velázquez, R.; Pissaloux, E.; Rodrigo, P.; Carrasco, M.; Giannoccaro, N.; Lay-Ekuakille, A. An Outdoor Navigation System for Blind Pedestrians Using GPS and Tactile-Foot Feedback. *Appl. Sci.* **2018**, *8*, 578. [CrossRef]
10. Domingo, M.C. An overview of the Internet of Things for people with disabilities. *J. Netw. Comput. Appl.* **2012**, *35*, 584–596. [CrossRef]
11. Dias, M.B.; Steinfeld, A.; Dias, M.B. Indoor Wayfinding and Navigation. Karimi, H.A., Ed.; CRC Press: Boca Raton, FL, USA, 2015.
12. Tapu, R.; Mocanu, B.; Zaharia, T. Wearable assistive devices for visually impaired: A state of the art survey. *Pattern Recognit. Lett.* **2018**, in press. [CrossRef]
13. Petsiuk, A.L.; Pearce, J.M. Low-Cost Open Source Ultrasound-Sensing Based Navigational Support for the Visually Impaired. *Sensors* **2019**, *19*, 3783. [CrossRef] [PubMed]
14. Sakpere, W.; Adeyeye Oshin, M.; Mlitwa, N.B. A State-of-the-Art Survey of Indoor Positioning and Navigation Systems and Technologies. *S. Afr. Comput. J.* **2017**, *29*, 145–197. [CrossRef]
15. Jafri, R.; Ali, S.A.; Arabnia, H.R.; Fatima, S. Computer vision-based object recognition for the visually impaired in an indoors environment: A survey. *Vis. Comput.* **2014**, *30*, 1197–1222. [CrossRef]
16. Elmannai, W.; Elleithy, K. Sensor-Based Assistive Devices for Visually-Impaired People: Current Status, Challenges, and Future Directions. *Sensors* **2017**, *17*, 565. [CrossRef] [PubMed]
17. Real, S.; Araujo, A. Navigation Systems for the Blind and Visually Impaired: Past Work, Challenges, and Open Problems. *Sensors* **2019**, *19*, 3404. [CrossRef]
18. Chen, S.; Yao, D.; Cao, H.; Shen, C. A Novel Approach to Wearable Image Recognition Systems to Aid Visually Impaired People. *Appl. Sci.* **2019**, *9*, 3350. [CrossRef]
19. Elgendy, M.; Sik Lanyi, C. Review on Smart Solutions for People with Visual Impairment. In *Lecture Notes in Computer Science (Including Subseries Lecture Notes in Artificial Intelligence and Lecture Notes in Bioinformatics)*; Springer Science+Business Media: Berlin, Germany, 2018; Volume 10896, pp. 81–84.
20. Valero, E.; Adán, A.; Cerrada, C. Evolution of RFID Applications in Construction: A Literature Review. *Sensors* **2015**, *15*, 15988–16008. [CrossRef]
21. Ozdenizci, B.; Coskun, V.; Ok, K. NFC Internal: An Indoor Navigation System. *Sensors* **2015**, *15*, 7571–7595. [CrossRef]
22. Garrido-Jurado, S.; Muñoz-Salinas, R.; Madrid-Cuevas, F.J.; Marín-Jiménez, M.J. Automatic generation and detection of highly reliable fiducial markers under occlusion. *Pattern Recognit.* **2014**, *47*, 2280–2292. [CrossRef]
23. Marchand, E.; Uchiyama, H.; Spindler, F. Pose Estimation for Augmented Reality: A Hands-On Survey. *IEEE Trans. Vis. Comput. Graph.* **2016**, *22*, 2633–2651. [CrossRef]
24. Olson, E. AprilTag: A robust and flexible visual fiducial system. In Proceedings of the 2011 IEEE International Conference on Robotics and Automation, Shanghai, China, 9–13 May 2011; pp. 3400–3407.
25. Garrido-Jurado, S.; Muñoz-Salinas, R.; Madrid-Cuevas, F.J.; Medina-Carnicer, R. Generation of fiducial marker dictionaries using Mixed Integer Linear Programming. *Pattern Recognit.* **2016**, *51*, 481–491. [CrossRef]

26. Acuna, R.; Li, Z.; Willert, V. MOMA: Visual Mobile Marker Odometry. In Proceedings of the IPIN 2018—9th International Conference on Indoor Positioning and Indoor Navigation, Nantes, France, 24–27 September 2018; pp. 206–212.
27. Cortés, X.; Serratosa, F. Cooperative pose estimation of a fleet of robots based on interactive points alignment. *Expert Syst. Appl.* **2016**, *45*, 150–160. [CrossRef]
28. Engel, J.; Schöps, T.; Cremers, D. LSD-SLAM: Large-Scale Direct monocular SLAM. In *Lecture Notes in Computer Science (Including Subseries Lecture Notes in Artificial Intelligence and Lecture Notes in Bioinformatics)*; Springer: Cham, Switzerland, 2014; Volume 8690, pp. 834–849.
29. Mur-Artal, R.; Montiel, J.M.M.; Tardos, J.D. ORB-SLAM: A Versatile and Accurate Monocular SLAM System. *IEEE Trans. Robot.* **2015**, *31*, 1147–1163. [CrossRef]
30. Zhong, S.; Liu, Y.; Chen, Q. Visual orientation inhomogeneity based scale-invariant feature transform. *Expert Syst. Appl.* **2015**, *42*, 5658–5667. [CrossRef]
31. Yi, C.; Tian, Y.; Arditi, A. Portable Camera-Based Assistive Text and Product Label Reading from Hand-Held Objects for Blind Persons. *IEEE/ASME Trans. Mechatron.* **2014**, *19*, 808–817. [CrossRef]
32. Mekhalfi, M.L.; Melgani, F.; Zeggada, A.; De Natale, F.G.B.; Salem, M.A.-M.; Khamis, A. Recovering the sight to blind people in indoor environments with smart technologies. *Expert Syst. Appl.* **2016**, *46*, 129–138. [CrossRef]
33. Takizawa, H.; Yamaguchi, S.; Aoyagi, M.; Ezaki, N.; Mizuno, S. Kinect cane: An assistive system for the visually impaired based on the concept of object recognition aid. *Pers. Ubiquitous Comput.* **2015**, *19*, 955–965. [CrossRef]
34. Wong, E.J.; Yap, K.M.; Alexander, J.; Karnik, A. HABOS: Towards a platform of haptic-audio based online shopping for the visually impaired. In Proceedings of the 2015 IEEE Conference on Open Systems (ICOS), Bandar Melaka, Malaysia, 24–26 August 2015; pp. 62–67.
35. Zientara, P.A.; Lee, S.; Smith, G.H.; Brenner, R.; Itti, L.; Rosson, M.B.; Carroll, J.M.; Irick, K.M.; Narayanan, V. Third Eye: A Shopping Assistant for the Visually Impaired. *Computer* **2017**, *50*, 16–24. [CrossRef]
36. Jafri, R.; Ali, S.A. A Multimodal Tablet–Based Application for the Visually Impaired for Detecting and Recognizing Objects in a Home Environment. In *Lecture Notes in Computer Science (Including Subseries Lecture Notes in Artificial Intelligence and Lecture Notes in Bioinformatics)*; Springer: Cham, Switzerland, 2014; Volume 8547, pp. 356–359.
37. Mondéjar-Guerra, V.; Garrido-Jurado, S.; Muñoz-Salinas, R.; Marín-Jiménez, M.J.; Medina-Carnicer, R. Robust identification of fiducial markers in challenging conditions. *Expert Syst. Appl.* **2018**, *93*, 336–345. [CrossRef]
38. Zhang, H.; Zhang, C.; Yang, W.; Chen, C.-Y. Localization and navigation using QR code for mobile robot in indoor environment. In Proceedings of the 2015 IEEE International Conference on Robotics and Biomimetics (ROBIO), Zhuhai, China, 6–9 December 2015; pp. 2501–2506.
39. Dash, A.K.; Behera, S.K.; Dogra, D.P.; Roy, P.P. Designing of marker-based augmented reality learning environment for kids using convolutional neural network architecture. *Displays* **2018**, *55*, 46–54. [CrossRef]
40. AL-Madani, B.; Orujov, F.; Maskeliūnas, R.; Damaševičius, R.; Venčkauskas, A. Fuzzy Logic Type-2 Based Wireless Indoor Localization System for Navigation of Visually Impaired People in Buildings. *Sensors* **2019**, *19*, 2114. [CrossRef] [PubMed]
41. Al-Khalifa, S.; Al-Razgan, M. Ebsar: Indoor guidance for the visually impaired. *Comput. Electr. Eng.* **2016**, *54*, 26–39. [CrossRef]
42. Torrado, J.C.; Montoro, G.; Gomez, J. Easing the integration: A feasible indoor wayfinding system for cognitive impaired people. *Pervasive Mob. Comput.* **2016**, *31*, 137–146. [CrossRef]
43. Ko, E.; Kim, E.Y. A Vision-Based Wayfinding System for Visually Impaired People Using Situation Awareness and Activity-Based Instructions. *Sensors* **2017**, *17*, 1882.
44. Idrees, A.; Iqbal, Z.; Ishfaq, M. An efficient indoor navigation technique to find optimal route for blinds using QR codes. In Proceedings of the 2015 IEEE 10th Conference on Industrial Electronics and Applications (ICIEA), Auckland, New Zealand, 15–17 June 2015; pp. 690–695.
45. López-de-Ipiña, D.; Lorido, T.; López, U. Indoor Navigation and Product Recognition for Blind People Assisted Shopping. In *Lecture Notes in Computer Science (including subseries Lecture Notes in Artificial Intelligence and Lecture Notes in Bioinformatics)*; Springer: Cham, Switzerland, 2011; Volume 6693, pp. 33–40.

46. La Delfa, G.C.; Catania, V.; Monteleone, S.; De Paz, J.F.; Bajo, J. Computer Vision Based Indoor Navigation: A Visual Markers Evaluation. In *Advances in Intelligent Systems and Computing*; Springer: Berlin, Germany, 2015; Volume 376, pp. 165–173.
47. Delfa, G.L.; Catania, V. Accurate indoor navigation using smartphone, bluetooth low energy and visual tags. In Proceedings of the 2nd Conference on Mobile and Information Technologies in Medicine, Prague, Czech Republic, 21 October–21 November 2014; pp. 5–8.
48. Bacik, J.; Durovsky, F.; Fedor, P.; Perdukova, D. Autonomous flying with quadrocopter using fuzzy control and ArUco markers. *Intell. Serv. Robot.* **2017**, *10*, 185–194. [CrossRef]

Article

Scene Description for Visually Impaired People with Multi-Label Convolutional SVM Networks

Yakoub Bazi [1,*], Haikel Alhichri [1], Naif Alajlan [1] and Farid Melgani [2]

1 Department of Computer Engineering, College of Computer and Information Sciences, King Saud University, Riyadh 11543, Saudi Arabia; hhichri@ksu.edu.sa (H.A.); najlan@ksu.edu.sa (N.A.)
2 Department of Information Engineering and Computer Science, University of Trento, via Sommarive 9, 38123 Trento, Italy; melgani@disi.unitn.it
* Correspondence: ybazi@ksu.ed.sa; Tel.: +9665-0646-9808

Received: 21 October 2019; Accepted: 21 November 2019; Published: 23 November 2019

Abstract: In this paper, we present a portable camera-based method for helping visually impaired (VI) people to recognize multiple objects in images. This method relies on a novel multi-label convolutional support vector machine (CSVM) network for coarse description of images. The core idea of CSVM is to use a set of linear SVMs as filter banks for feature map generation. During the training phase, the weights of the SVM filters are obtained using a forward-supervised learning strategy unlike the backpropagation algorithm used in standard convolutional neural networks (CNNs). To handle multi-label detection, we introduce a multi-branch CSVM architecture, where each branch will be used for detecting one object in the image. This architecture exploits the correlation between the objects present in the image by means of an opportune fusion mechanism of the intermediate outputs provided by the convolution layers of each branch. The high-level reasoning of the network is done through binary classification SVMs for predicting the presence/absence of objects in the image. The experiments obtained on two indoor datasets and one outdoor dataset acquired from a portable camera mounted on a lightweight shield worn by the user, and connected via a USB wire to a laptop processing unit are reported and discussed.

Keywords: visually impaired (VI); computer vision; deep learning; multi-label convolutional support vector machine (M-CSVM)

---

1. Introduction

Chronic blindness may occur as an eventual result of various causes, such as cataract, glaucoma, age-related macular degeneration, corneal opacities, diabetic retinopathy, trachoma, and eye conditions in children (e.g., caused by vitamin A deficiency) [1]. Recent factsheets from the World Health Organization, as per October 2018 [1], indicate that 1.3 billion people suffer from some form of vision impairment, including 36 million people who are considered blind. These facts highlight an urgent need to improve the quality of life for people with vision disability, or at least to lessen its consequences.

Towards achieving the earlier endeavor, assistive technology ought to exert an essential role. On this point, the latest advances gave rise to several designs and prototypes, which can be regarded from two distinct but complementary perspectives, namely (1) assistive mobility and obstacle avoidance, and (2) object perception and recognition. The first perspective enables the visually impaired (VI) persons to navigate more independently, while the latter emphasizes consolidating their comprehension of the nature of the nearby objects, if any.

Navigation-focused technology constitutes the bulk of the literature. Many works have been carried out making use of ultrasonic sensors to probe the existence of nearby obstacles via transmitting and subsequently receiving ultrasonic waves. The time consumed during this process is commonly termed time of flight (TOF). In [2] for instance, a prototype that consists of a guide cane, around a housing, a wheelbase, and a handle is presented. The housing is surrounded by ten ultrasonic sensors, eight of which are spaced by 15° and set up on the frontal part, and the other two sensors are placed on the edge to sense lateral objects. The user can use a mini joystick to control the preferred direction and maneuver the cane in order to inspect the area. If an obstacle is sensed, an obstacle avoidance algorithm (installed on a computer) estimates an alternative obstacle-free path and steers the cane through by applying a gentle force felt by the user on the handle. A similar concept was suggested in [3]. Another prototype, which consists of smart clothing equipped with a microcontroller, ultrasonic sensors, and alerting vibrators was proposed in [4]. The sensors are adopted to detect potential obstacles, whilst a neuro-fuzzy-based controller estimates the obstacle's position (left, right, and front) and indicates navigation tips such as 'turn left' or 'turn right'. A similar concept was presented in [5]. Another ultrasonic-based model was designed in [6].

The common downsides of the prototypes discussed earlier are their size and power consumption render them impractical for daily use by a VI individual. Alternative navigation models suggest the use of the Global Positioning System (GPS) to determine the location of the VI users to instruct them towards a predefined destination [7]. These latter models, however, can be accurate to determine the location of the user but are unable to tackle the issue of obstacle avoidance.

Regarding the recognition aspect, the literature reports some contributions, which are mostly based on computer vision. In [8], a banknote recognition system for the VI, which makes use of speeded-up robust features (SURF), was presented. Diego et al. [9] proposed a supported supermarket shopping design, which considers a camera-based product recognition by means of QR codes that are placed on the shelves. Another product barcode detection, as well as reading, was developed in [10]. In [11] another travel assistant was proposed. It considers the text zones displayed on public transportation buses (at bus stops) to extract information related to the bus line number. The proposed prototype elaborates a given image (acquired by a portable camera) and subsequently notifies by voice the outcome to the user. In another work [12], assisted indoor staircases detection (within 1–5 m ahead) was suggested. In [13] an algorithm meant to aid VI people was proposed to read texts encountered in natural scenes. Door detection in unfamiliar environments was also considered in [14]. Assisted indoor scene understanding for single object detection and recognition was also suggested in [15] by means of the scale invariant feature transform. In [16], the authors proposed a system called Blavigator to assist VI, which was composed of an interface, Geographic Information System (GIS), navigation, object recognition and a central decision module. In a known location, this system uses object recognition algorithms to provide contextual feedback to the user in order to validate the positioning module and the navigation system for the VI. In another work, [17] the authors proposed a vision-based wayfinding aid for blind people to access unfamiliar indoor environments to find different rooms such as office, and laboratory and building amenities such as exits and elevators. This system incorporates object detection with text recognition.

The recognition part of VI assistance technology is mostly approached from a single object perspective (i.e., single-label classification where the attention is focused on single objects). This approach, despite its usefulness in carrying out information to the VI person, remains rather limited since multiple objects may be encountered in daily life. Thereupon, broadening the emphasis to multiple objects by solving the problem from a multi-label classification perspective seems to offer a more significant alternative in terms of usefulness to the VI user. The concept of multi-label classification was adopted recently in some works based on handcrafted features, and was commonly termed 'coarse' scene multi-labeling/description [18,19]. It has demonstrated its effectiveness in detecting the presence of multiple objects of interest across an indoor environment in a very low processing time. Similar designs were envisioned in other areas such as remote sensing [20] to address the issue of object detection in high-resolution images.

Recently deep convolutional neural networks (CNNs) have achieved impressive results in applications such as image classification [21–25], object detection [26–29], and image segmentation [30,31]. These networks have the ability to learn richer representations in a hierarchical way compared to handcrafted-based methods. Modern CNNs are made up of several alternating convolution blocks with repeated structures. The whole architecture is trained end-to-end using the backpropagation algorithm [32].

Usually, CNNs perform well for datasets with large labeled data. However, they are prone to overfitting when dealing with datasets with very limited labeled data as in the context of our work. In this case, it is has been shown in many studies that it is more appealing to transfer knowledge from CNNs (such as AlexNet [25], VGG-VD [33], GoogLeNet [24], and ResNet [23]) pre-trained on an auxiliary recognition task with very large labeled data instead of training from scratch [34–37]. The possible knowledge transfer solutions include fine-tuning of the labeled data of the target dataset or to exploit the network feature representations with an external classifier. We refer the readers to [35] where the authors presents several factors affecting the transferability of these representations.

In this paper, we propose an alternative solution suitable for datasets with limited training samples mainly based on convolutional SVM networks (CSVMs). Actually, SVMs are among the most popular supervised classifiers available in the literature. They rely on the margin maximization principle, which makes them less sensitive to overfitting problems. They have been widely used for solving various recognition problems. Additionally, they are also commonly placed on the top of a CNN feature extractor for carrying out the classification task [35]. In a recent development, these classifiers have been extended to act as convolutional filters for the supervised generation of features maps for single object detection in remote sensing imagery [38]. Compared to standard CNNs, CSVM introduces a new convolution trick based on SVMs and does not rely on the backpropagation algorithm for training. Basically, this network is based on several alternating convolution and reduction layers followed by a classification layer. Each convolution layer uses a set of linear SVMs as filter banks, which are convolved with the feature maps produced by the precedent layer to generate a new set of feature maps. For the first convolution layer, the SVM filters are convolved with the original input images. Then the SVM weights of each convolution layer are computed in a supervised way by training on patches extracted from the previous layer. The feature maps produced by the convolution layers are then fed to a pooling layer. Finally, the high-level representations obtained by the network are fed again to a linear SVM classifier for carrying out the classification task. In this work, we extend them to the case of multi-label classification. In particular, we introduce a novel multi-branch CSVM architecture, where each branch will be used to detect one object in the image. We exploit the correlation between the objects present in the image by fusing the intermediate outputs provided by the convolution layers of each branch by means of an opportune fusion mechanism. In the experiments, we validate the method on images obtained from different indoor and outdoor spaces.

The rest of this paper is organized as follows. In Section 2, we provide a description of the proposed multi-label CSVM (M-CSVM) architecture. The experimental results and discussions are presented in Sections 3 and 4, respectively. Finally, conclusions and future developments are reported in Section 5.

2. Proposed Methodology

Let us consider a set of M training RGB images $\{X_i, y_i\}_{i=1}^{M}$ of size $r \times c$ acquired by a portable digital camera mounted on a lightweight shield worn by the user, and connected via a USB wire to a laptop processing unit, where $X_i \in \mathcal{R}^{r \times c \times 3}$, and (r, c) refer the number of rows and columns of the images. Let us assume also $y_i = [y_1, y_2, \ldots, y_K]^T$ is its corresponding label vector, where K represents the total number of targeted classes. In a multi-label setting, the label $y_i = 1$ is set to 1 if the corresponding object is present; otherwise, it is set to 0. Figure 1 shows a general view of the proposed M-CSVM classification system, which is composed from K branches. In next sub-sections, we detail the convolution and fusion layers, which are the main ingredient of the proposed method.

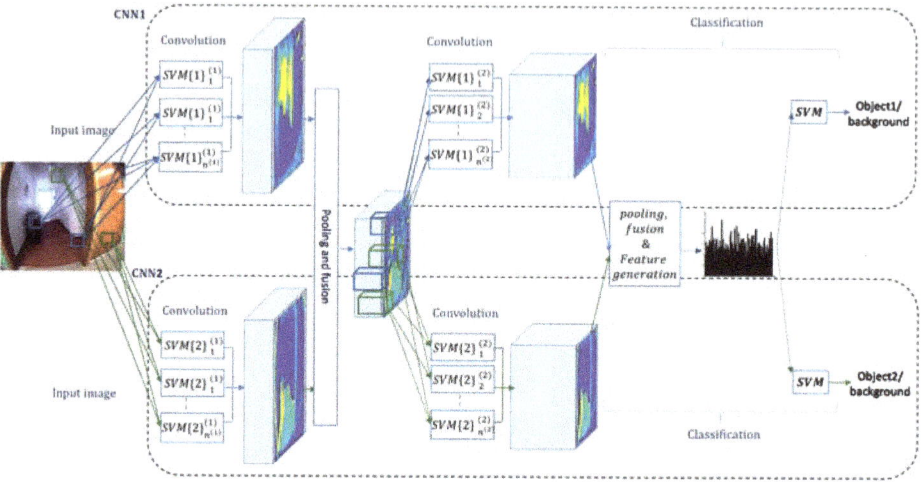

Figure 1. Proposed multi-label convolutional support vector machine (M-CSVM) architecture for detecting the presence of multiple objects in the image.

2.1. A. Convolution Layer

In this section, we present the SVM convolution technique for the first convolution layer. The generalization to subsequent layers is straightforward. In a binary classification setting, the training set $\{X_i, y_i\}_{i=1}^{M}$ is supposed to be composed of M positive and negative RGB images and the corresponding class labels are set to $y_i \in \{+1, -1\}$. The positive images contain the object of interest, whereas the negatives ones represent background. From each image X_i, we extract a set of patches of size $h \times h \times 3$ and represent them as feature vectors x_i of dimension d, with $d = h \times h \times 3$. After processing the M training images, we obtain a large training set $Tr^{(1)} = \{x_i, y_i\}_{i=1}^{m^{(1)}}$ of size $m^{(1)}$ as shown in Figure 2.

Next, we learn a set of SVM filters on different sub-training sets $Tr_{sub}^{(1)} = \{x_i, y_i\}_{i=1}^{l}$ of size l randomly sampled from the training set $Tr^{(1)}$. The weight vector $w \in \mathcal{R}^d$ and bias $b \in \mathcal{R}$ of each SVM filter are determined by optimizing the following problem [39,40]:

$$\min_{w,b} w^T w + C \sum_{i=1}^{l} \xi(w, b; x_i, y_i) \tag{1}$$

where C is a penalty parameter, which can be estimated through cross-validation. As loss function, we use $\xi(w, b; x_i, y_i) = \max\left(1 - y_i\left(w^T x_i + b\right), 0\right)$ referred as the hinge loss. After training, we represent the weights of the SVM filters as $\left\{w_k^{(1)}\right\}_{k=1}^{n^{(1)}}$, where $w_k^{(1)} \in \mathcal{R}^{h \times h \times 3}$ refers to kth-SVM filter weight matrix, while $n^{(1)}$ is the number of filters. Then, the complete weights of the first convolution layer are grouped into a filter bank of four dimensions $W^{(1)} \in \mathcal{R}^{h \times h \times 3 \times n^{(1)}}$.

In order to generate the feature maps, we simply convolve each training image $\{X_i\}_{i=1}^{M}$, with the obtained filters as is usually done in standard CNN to generate a set of 3D hyper-feature maps $\left\{H_i^{(1)}\right\}_{i=1}^{M}$. Here $H_i^{(1)} \in \mathcal{R}^{r^{(1)} \times c^{(1)} \times n^{(1)}}$ is the new feature representation of image X_i composed of $n^{(1)}$ feature maps (Figure 3). To obtain the kth feature map $h_{ki}^{(1)}$, we convolve the kth filter with a set of sliding windows of size $h \times h \times 3$ (with a predefined stride) over the training image X_i as shown in Figure 4:

$$h_{ki}^{(1)} = f\left(X_i * w_k^{(1)}\right), \; k = 1, \ldots, n^{(1)} \tag{2}$$

where $*$ is the convolution operator and f is the activation function.

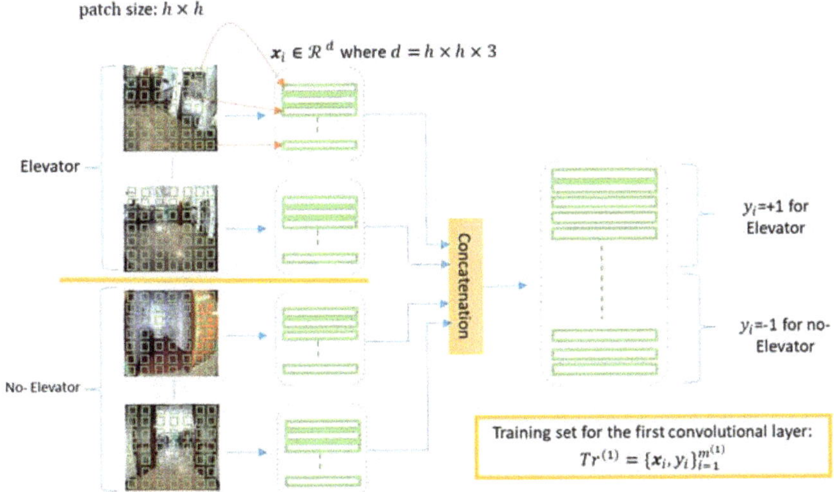

Figure 2. Training set generation for the first convolution layer.

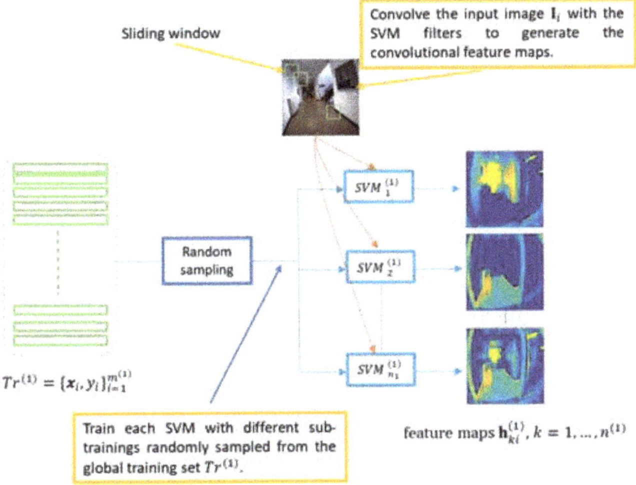

Figure 3. Supervised feature map generation.

In the following algorithm, we present the implementation of this convolutional layer. The generalization to subsequent convolution layer is simply made by considering the obtained feature maps as new input to the next convolution layer.

Algorithm 1 Convolution layer

Input: Training images $\{X_i, y_i\}_{i=1}^{M}$; SVM filters: $n^{(1)}$; filter parameters (width h, and stride);
l : size of sampled training set for generating a single feature map.
Output: Feature maps: $H^{(1)} \in \mathcal{R}^{r^{(1)} \times c^{(1)} \times n^{(1)} \times M}$
1: $Tr^{(1)} = \text{global\_training}\left(\{I_i, y_i\}_{i=1}^{M}\right)$;
2: $W^{(1)} = \text{learn\_SVM\_filters}\left(Tr^{(1)}, n^{(1)}, l\right)$
3: $H^{(1)} = \text{convolution}\left(\{X_i\}_{i=1}^{M}, W^{(1)}\right)$;
4: $H^{(1)} = ReLU\left(H^{(1)}\right)$

2.2. B. Fusion Layer

In a multi-label setting, we run multiple CSVMs depending on the number of objects. Each CSVM will apply a set of convolutions on the image under analysis as shown in Figure 1. Then each convolution layer is followed by a spatial reduction layer. This reduction layer is similar to the spatial pooling layer in standard CNNs. It is commonly used to reduce the spatial size of the feature maps by selecting the most useful features for the next layers. It takes small blocks from the resulting features maps and sub-samples them to produce a single output from each block. Here, we use the average pooling operator for carrying out reduction. Then, to exploit the correlation between objects present in the image, we propose to fuse the feature maps provided by each branch. In particular, we opt for the max-pooling strategy in order to highlight the different detected objects by each branch. Figure 4 shows an example of fusion process for two branches. The input image contains two objects, Laboratories (object1) and Bins (object2). The first CSVM tries to highlight the first object, while the second one is devoted for the second object. The output maps provided by the pooling operation are fused using the max-rule in order to get a new feature-map where the two concerned objects are highlighted as can be seen in Figure 4. We recall that the feature maps obtained by this operation will be used as input to the next convolution layer for each branch.

Algorithm 2 Pooling and fusion

Input: Feature maps: $H^{(i)}$ $i = 1, \ldots, K$ produced by the ith CSVM branch
Output: Fusion result: $H^{(f1)}$
1: Apply an average pooling to each $H^{(i)}$ to generate a set of activation maps of reduced spatial size.
2: Fuse the resulting activation using max rule to generate the feature map $H^{(f1)}$. These maps will be used as a common input to the next convolution layers in each CSVM branch.

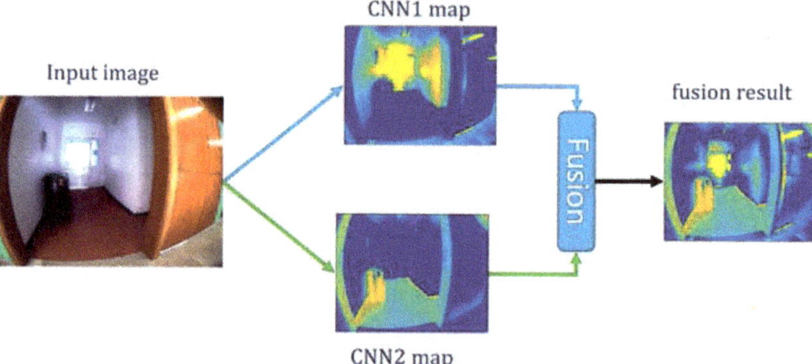

Figure 4. Example of fusion of output maps of two convolutional neural networks (CNNs).

2.3. C. Feature Generation and Classification

After applying several convolutions, reduction, and fusion layers, the high-level reasoning of the network is done by training K binary SVM classifiers to detect the presence/absence of the objects in the image. If we let $\left\{H_i^{(L)}, y_i\right\}_{i=1}^{M}$ be the hyper-feature maps obtained by the last computing layer (convolution or reduction depending on the network architecture) and, if we suppose also that each hyper-feature map $H_i^{(L)}$ is composed of $n^{(L)}$ feature maps, then a possible solution for extracting the high-level feature vector $z_i \in \mathcal{R}^{n^{(L)}}$ of dimension for the training image $\mathbf{X}_i$ could be simply done by computing the mean or max value for each feature map.

3. Experimental Results

3.1. Dataset Description

In the experiments, we evaluate the proposed method on three datasets taken by a portable camera mounted on a lightweight shield worn by the user, and connected via a USB wire to a laptop processing unit. This system incorporates navigation and recognition modules. In a first step, the user runs the application to load the offline-stored information related to recognition and navigation. Then he can control this system using verbal commands as shown in Figure 5a. For the sake of clarity, we provide also a general view of the application, where the user asks to go to the 'elevator'. Upon the arrival to the desired destination using a path planning module, the prototype notifies the user that the destination is reached. Figure 5b shows the current view of the camera, where the destination (elevators) is displayed. The system also features a virtual environment emulating the real movement of the user within the indoor space. As can be seen, the user is symbolized by the black top-silhouette, emitting two lines, the blue line refers to the user's current frontal view; the green point refers to the destination estimated by the path planning module; while the red dot highlights the final destination. The interface displays also markers displayed as thick lines laying on the walls for helping in the localization. In our work, we have used this system to acquire different images used for developing the recognition module based on M-CSVM.

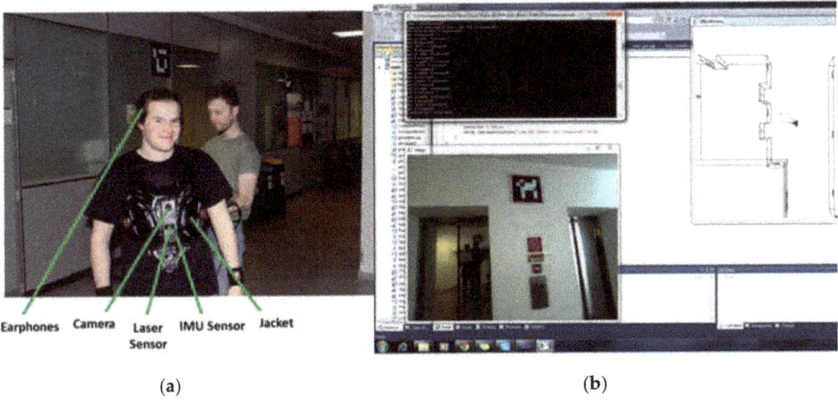

(a) (b)

Figure 5. Overview of the system developed for visually impaired (VI) people. (**a**) Camera mounted on the chest, tablet, and headphones for voice communication; and (**b**) graphical interface.

The first and second datasets acquired by this system are composed of images of size 320 × 240 pixels. Both datasets have been taken at two different indoor spaces of the faculty of science of University of Trento (Italy). The first dataset contains 58 training and 72 testing images, whereas the second dataset contains 61 training images and 70 testing images. On the other the side, the third dataset is related to outdoor environment and was acquired over different locations across the city of Trento. The locations were selected based on their importance as well as the density of people frequenting them. The dataset comprises two hundred (200) images of size 275 × 175 pixels, which were equally divided into 100 training and testing images, respectively. It is noteworthy that the training images for all datasets were selected in such a way as to cover all the predefined objects in the considered indoor and outdoor environments. To this end, we have selected the objects deemed to be the most important ones in the considered spaces. Regarding the first dataset, 15 objects were considered as follows: 'External Window', 'Board', 'Table', 'External Door', 'Stair Door', 'Access Control Reader', 'Office', 'Pillar', 'Display Screen', 'People', 'ATM', 'Chairs', 'Bins', 'Internal Door', and 'Elevator'. Whereas, for the second set, the list was the following: 'Stairs', 'Heater', 'Corridor', 'Board', 'Laboratories', 'Bins', 'Office', 'People', 'Pillar', 'Elevator', 'Reception', 'Chairs', 'Self Service', 'External Door', and 'Display Screen'. Finally, for the last dataset, a total of 26 objects were defined as follows: 'People', 'Building', 'Bar(s)', 'Monument(s)', 'Chairs/Benches', 'Green ground', 'Vehicle(s)', 'Stairs', 'Walk path/Sidewalk', 'Fence/Wall', 'Tree(s)/Plant(s)', 'Garbage can(s)', 'Bus stop', 'Crosswalk', 'River', 'Roundabout', 'Pole(s)/Pillar(s)', 'Shop(s)', 'Supermarket(s)', 'Pound/Birds', 'Underpass', 'Bridge', 'Railroad', 'Admiration building', 'Church', and 'Traffic signs'. Figure 6 shows sample images from each dataset.

Figure 6. Example of images taken by a portable camera: (**a**) dataset 1, (**b**) dataset 2, and (**c**) dataset 3.

In the experiments, we assessed the performances of the method in terms of sensitivity (SEN) and specificity (SPE) measures:

$$\text{SEN} = \frac{\text{True Positive}}{\text{True Positive} + \text{False Negative}} \quad (3)$$

$$\text{SPE} = \frac{\text{True Negative}}{\text{True Negative} + \text{False Positive}} \quad (4)$$

The sensitivity expresses the classification rate of real positive cases, i.e., the efficiency of the algorithm towards detecting existing objects. The specificity, on the other hand, underlines the tendency of the algorithm to detect the true negatives, i.e., the non-existing objects. We also propose to compute the average of the two earlier measures as follows:

$$\text{AVG} = \frac{\text{SEN} + \text{SPE}}{2} \tag{5}$$

3.2. Results

The architecture of the proposed CSVM involves many parameters. To identify a suitable architecture, we propose to investigate three main parameters, which are: the number of layers of the network, number of feature maps generated by each convolution layer and the spatial sizes of the kernels. To compute the best parameter values, we use a cross-validation technique with a number of folds equal to 3. Due to the limited number of training samples and the large number of possible combinations, we set the maximum number of possible layers to 3, the maximum number of SVMs in each layer to 512 with step of 2^i ($i = 0, \ldots, 9$) and maximum kernel size for each layer is fixed to 10% of the size of the current map (we consider the minimum between the height and the width as the considered size) with step of 2. The obtained best values of the parameters by cross-validation are listed in Table 1. This table indicates that only one layer is enough for the first dataset, whereas the two other datasets require two layers to get the best performances. Concerning the number of SVMs and the related spatial size, dataset 3 presents the simplest architecture with just one SVM at the first layer and two SVMs at the second one with spatial size of 3. It is worth recalling that in the experiments.

Table 1. Best parameter values of the convolutional support vector machine (CSVM) for each dataset.

Dataset	Layer 1		Layer 2	
	Number of SVMs	**Spatial Size**	**Number of SVMs**	**Spatial Size**
Dataset 1	512	7	/	/
Dataset 2	32	7	256	3
Dataset 3	1	3	2	3

In order to evaluate our method, we compare it with results obtained using three different pre-trained CNNs, which are ResNet [23], GoogLeNet [24] and VGG16 [33]. All the results in terms of accuracies are reported in Table 2.

Table 2. Classification obtained by M-CSVM compared to CNNs for: (**a**) dataset 1, (**b**) dataset 2, and (**c**) dataset 3.

(a)

Method	SEN (%)	SPE (%)	AVG (%)
ResNet	71.16	93.84	82.50
GoogLeNet	78.65	94.34	86.49
VGG16	74.53	94.58	84.55
Ours	89.14	84.26	86.70

(b)

Methods	SEN (%)	SPE (%)	AVG (%)
ResNet	89.54	96.38	92.96
GoogLeNet	83.63	96.86	90.25
VGG16	81.81	96.14	88.98
Ours	93.64	92.17	92.90

(c)

Methods	SEN (%)	SPE (%)	AVG (%)
ResNet	64.17	92.40	78.29
GoogLeNet	62.50	93.27	77.88
VGG16	64.32	93.98	79.15
Ours	80.79	82.27	81.53

From these tables, it can be seen that in eight cases out of nine, our proposed method by far outperforms the different pre-trained CNNs. In seven cases the improvement is clearly important (more than 2%). Only in one case (ResNet, dataset 2) does a CNN method give a slightly better result than our method (92.96% compared to 92.90%).

4. Discussion

Besides the classification accuracies, another important performance parameter is the runtime. Table 3 shows the training time of the proposed method for the three datasets. It can be seen clearly that the training of M-CSVM is fast and needs just few seconds to few minutes in the worse case. In details, dataset 1 presents the highest runtime (76 s) which is due to the high number of filters used for this dataset (512), while training of dataset 3 is much faster (just 8 s) due to the simplicity of the related network (see Table 1).

Table 3. Training time of the proposed M-CSVM.

Dataset	Runtime (s)
Dataset 1	76
Dataset 2	42
Dataset 3	8

Regarding runtime at the prediction phase, which includes the feature extraction and classification, the M-CSVM method presents different runtime for the three datasets depending on the complexity of the adopted architecture. For instance, as we can see in Table 4, the highest runtime is with the first dataset 1 with 0.200 s per image, which is due to the high number of filters adopted for it (512). In contrast, the third dataset requires only 0.002 s to extract features and estimate the classes for each image. This short time is due to the small number of SVMs used in this network for this dataset. It is also important to mention that the runtime provided by our method outperforms the three pre-trained CNNs on two datasets (datasets 2 and 3), especially for the dataset 3 where the difference is significant. Except for dataset 1, where GoogLeNet is slightly faster due to complexity of the related network.

Table 4. Comparison of average runtime per image for: (**a**) dataset 1; (**b**) dataset 2, and (**c**) dataset 3.

(**a**)

Method	Runtime (s)
ResNet	0.207
GoogLeNet	0.141
VGG16	0.291
Ours	0.206

(**b**)

Method	Runtime (s)
ResNet	0.208
GoogLeNet	0.144
VGG16	0.291
Ours	0.115

(**c**)

Method	Runtime (s)
ResNet	0.207
GoogLeNet	0.145
VGG16	0.300
Ours	0.002

5. Conclusions

In this paper, we have presented a novel M-CSVM method for describing the image content for VI people. This method has the following important proprieties: (1) it allows SVMs to act as convolutional filters; (2) it uses a forward supervised learning strategy for computing the weights of the filters; and (3) it estimates each layer locally, which reduces the complexity of the network. The experimental results obtained on the three datasets with limited training samples confirm the promising capability of the proposed method with respect to state-of-the-art methods based on pre-trained CNNs. For future developments, we plan to investigate architectures based on residual connections such as in modern networks, and to explore uses of advanced strategies based on reinforcement learning for finding an optimized M-CVSM architecture. Additionally, we plan to extend this method to act as a detector by localizing the detected object in the image.

Author Contributions: Y.B. implemented the method and wrote the paper. H.A., N.A., and F.M. contributed to the analysis of the experimental results and paper writing.

Funding: This work was supported by NSTIP Strategic Technologies Programs, number 13-MED-1343-02 in the Kingdom of Saudi Arabia.

Acknowledgments: This work was supported by NSTIP Strategic Technologies Programs number: 13-MED-1343-02 in the Kingdom of Saudi Arabia.

Conflicts of Interest: The authors declare no conflict of interest.

References

1. Blindness and Vision Impairment. Available online: https://www.who.int/news-room/fact-sheets/detail/blindness-and-visual-impairment (accessed on 12 September 2019).
2. Ulrich, I.; Borenstein, J. The GuideCane-applying mobile robot technologies to assist the visually impaired. *IEEE Trans. Syst. Man Cybern. Part A Syst. Hum.* **2001**, *31*, 131–136. [CrossRef]
3. Shoval, S.; Borenstein, J.; Koren, Y. The NavBelt-a computerized travel aid for the blind based on mobile robotics technology. *IEEE Trans. Biomed. Eng.* **1998**, *45*, 1376–1386. [CrossRef]
4. Bahadir, S.K.; Koncar, V.; Kalaoglu, F. Wearable obstacle detection system fully integrated to textile structures for visually impaired people. *Sens. Actuators A Phys.* **2012**, *179*, 297–311. [CrossRef]
5. Shin, B.-S.; Lim, C.-S. Obstacle Detection and Avoidance System for Visually Impaired People. In Proceedings of the Haptic and Audio Interaction Design, Seoul, Korea, 29–30 November 2007; Oakley, I., Brewster, S., Eds.; Springer: Berlin/Heidelberg, Germany, 2007; pp. 78–85.
6. Bousbia-Salah, M.; Bettayeb, M.; Larbi, A. A Navigation Aid for Blind People. *J. Intell. Robot. Syst.* **2011**, *64*, 387–400. [CrossRef]
7. Brilhault, A.; Kammoun, S.; Gutierrez, O.; Truillet, P.; Jouffrais, C. Fusion of Artificial Vision and GPS to Improve Blind Pedestrian Positioning. In Proceedings of the 2011 4th IFIP International Conference on New Technologies, Mobility and Security, Paris, France, 7–10 February 2011; pp. 1–5.
8. Hasanuzzaman, F.M.; Yang, X.; Tian, Y. Robust and Effective Component-Based Banknote Recognition for the Blind. *IEEE Trans. Syst. Man Cybern. Part C (Appl. Rev.)* **2012**, *42*, 1021–1030. [CrossRef]
9. López-de-Ipiña, D.; Lorido, T.; López, U. BlindShopping: Enabling Accessible Shopping for Visually Impaired People through Mobile Technologies. In Proceedings of the Toward Useful Services for Elderly and People with Disabilities, Montreal, QC, Canada, 20–22 June 2011; Abdulrazak, B., Giroux, S., Bouchard, B., Pigot, H., Mokhtari, M., Eds.; Springer: Berlin/Heidelberg, Germany, 2011; pp. 266–270.
10. Tekin, E.; Coughlan, J.M. An Algorithm Enabling Blind Users to Find and Read Barcodes. *Proc. IEEE Workshop Appl. Comput. Vis.* **2009**, *2009*, 1–8.
11. Pan, H.; Yi, C.; Tian, Y. A primary travelling assistant system of bus detection and recognition for visually impaired people. In Proceedings of the 2013 IEEE International Conference on Multimedia and Expo Workshops (ICMEW), San Jose, CA, USA, 15–19 July 2013; pp. 1–6.
12. Jia, T.; Tang, J.; Lik, W.; Lui, D.; Li, W.H. Plane-based detection of staircases using inverse depth. In Proceedings of the Australasian Conference on Robotics and Automation (ACRA), Wellington, New Zealand, 3–5 December 2012.

13. Chen, X.R.; Yuille, A.L. Detecting and reading text in natural scenes. In Proceedings of the 2004 IEEE Computer Society Conference on Computer Vision and Pattern Recognition (CVPR), Washington, DC, USA, 27 June–2 July 2004; Volume 2, p. II.
14. Yang, X.; Tian, Y. Robust door detection in unfamiliar environments by combining edge and corner features. In Proceedings of the 2010 IEEE Computer Society Conference on Computer Vision and Pattern Recognition-Workshops, San Francisco, CA, USA, 13–18 June 2010; pp. 57–64.
15. Wang, S.; Tian, Y. Camera-Based Signage Detection and Recognition for Blind Persons. In Proceedings of the Computers Helping People with Special Needs, Linz, Austria, 11–13 July 2012; Miesenberger, K., Karshmer, A., Penaz, P., Zagler, W., Eds.; Springer: Berlin/Heidelberg, Germany, 2012; pp. 17–24.
16. Fernandes, H.; Costa, P.; Paredes, H.; Filipe, V.; Barroso, J. Integrating Computer Vision Object Recognition with Location Based Services for the Blind. In Proceedings of the Universal Access in Human-Computer Interaction. Aging and Assistive Environments, Heraklion, Greece, 22–27 June 2014; Stephanidis, C., Antona, M., Eds.; Springer International Publishing: Cham, Switzerland, 2014; pp. 493–500.
17. Tian, Y.; Yang, X.; Yi, C.; Arditi, A. Toward a computer vision-based wayfinding aid for blind persons to access unfamiliar indoor environments. *Mach. Vis. Appl.* **2013**, *24*, 521–535. [CrossRef]
18. Malek, S.; Melgani, F.; Mekhalfi, M.L.; Bazi, Y. Real-Time Indoor Scene Description for the Visually Impaired Using Autoencoder Fusion Strategies with Visible Cameras. *Sensors* **2017**, *17*, 2641. [CrossRef]
19. Mekhalfi, M.L.; Melgani, F.; Bazi, Y.; Alajlan, N. Fast indoor scene description for blind people with multiresolution random projections. *J. Vis. Commun. Image Represent.* **2017**, *44*, 95–105. [CrossRef]
20. Moranduzzo, T.; Mekhalfi, M.L.; Melgani, F. LBP-based multiclass classification method for UAV imagery. In Proceedings of the 2015 IEEE International Geoscience and Remote Sensing Symposium (IGARSS), Milan, Italy, 26–31 July 2015; pp. 2362–2365.
21. Mariolis, I.; Peleka, G.; Kargakos, A.; Malassiotis, S. Pose and category recognition of highly deformable objects using deep learning. In Proceedings of the 2015 International Conference on Advanced Robotics (ICAR), Istanbul, Turkey, 27–31 July 2015; pp. 655–662.
22. He, K.; Zhang, X.; Ren, S.; Sun, J. Spatial Pyramid Pooling in Deep Convolutional Networks for Visual Recognition. In Proceedings of the Computer Vision–ECCV 2014, Zurich, Switzerland, 6–12 September 2014; Springer: Cham, Switzerland, 2014; pp. 346–361.
23. He, K.; Zhang, X.; Ren, S.; Sun, J. Deep Residual Learning for Image Recognition. In Proceedings of the 2016 IEEE Conference on Computer Vision and Pattern Recognition (CVPR), Las Vegas, NV, USA, 27–30 June 2016; pp. 770–778.
24. Szegedy, C.; Liu, W.; Jia, Y.; Sermanet, P.; Reed, S.; Anguelov, D.; Erhan, D.; Vanhoucke, V.; Rabinovich, A. Going deeper with convolutions. In Proceedings of the 2015 IEEE Conference on Computer Vision and Pattern Recognition (CVPR), Boston, MA, USA, 7–12 June 2015; pp. 1–9.
25. Krizhevsky, A.; Sutskever, I.; Hinton, G.E. ImageNet Classification with Deep Convolutional Neural Networks. In Proceedings of the 25th International Conference on Neural Information Processing Systems-Volume 1, Lake Tahoe, Nevada, 3–6 December 2012; Curran Associates Inc.: Brooklyn, NY, USA, 2012; pp. 1097–1105.
26. Ren, S.; He, K.; Girshick, R.; Zhang, X.; Sun, J. Object Detection Networks on Convolutional Feature Maps. *IEEE Trans. Pattern Anal. Mach. Intell.* **2017**, *39*, 1476–1481. [CrossRef] [PubMed]
27. Girshick, R.; Donahue, J.; Darrell, T.; Malik, J. Region-Based Convolutional Networks for Accurate Object Detection and Segmentation. *IEEE Trans. Pattern Anal. Mach. Intell.* **2016**, *38*, 142–158. [CrossRef] [PubMed]
28. Ren, S.; He, K.; Girshick, R.; Sun, J. Faster R-CNN: Towards Real-Time Object Detection with Region Proposal Networks. *IEEE Trans. Pattern Anal. Mach. Intell.* **2017**, *39*, 1137–1149. [CrossRef] [PubMed]
29. Ouyang, W.; Zeng, X.; Wang, X.; Qiu, S.; Luo, P.; Tian, Y.; Li, H.; Yang, S.; Wang, Z.; Li, H.; et al. DeepID-Net: Object Detection with Deformable Part Based Convolutional Neural Networks. *IEEE Trans. Pattern Anal. Mach. Intell.* **2017**, *39*, 1320–1334. [CrossRef] [PubMed]
30. Farabet, C.; Couprie, C.; Najman, L.; LeCun, Y. Learning Hierarchical Features for Scene Labeling. *IEEE Trans. Pattern Anal. Mach. Intell.* **2013**, *35*, 1915–1929. [CrossRef] [PubMed]
31. Sun, W.; Wang, R. Fully Convolutional Networks for Semantic Segmentation of Very High Resolution Remotely Sensed Images Combined With DSM. *IEEE Geosci. Remote Sens. Lett.* **2018**, *15*, 474–478. [CrossRef]
32. Srivastava, N.; Hinton, G.; Krizhevsky, A.; Sutskever, I.; Salakhutdinov, R. Dropout: A Simple Way to Prevent Neural Networks from Overfitting. *J. Mach. Learn. Res.* **2014**, *15*, 1929–1958.

33. Simonyan, K.; Zisserman, A. Very Deep Convolutional Networks for Large-Scale Image Recognition. *arXiv* **2014**, arXiv:1409.1556.
34. Chatfield, K.; Simonyan, K.; Vedaldi, A.; Zisserman, A. Return of the Devil in the Details: Delving Deep into Convolutional Nets. *arXiv* **2014**, arXiv:1405.3531.
35. Azizpour, H.; Razavian, A.S.; Sullivan, J.; Maki, A.; Carlsson, S. Factors of Transferability for a Generic ConvNet Representation. *IEEE Trans. Pattern Anal. Mach. Intell.* **2016**, *38*, 1790–1802. [CrossRef]
36. Nogueira, R.F.; de Alencar Lotufo, R.; Machado, R.C. Fingerprint Liveness Detection Using Convolutional Neural Networks. *IEEE Trans. Inf. Forensics Secur.* **2016**, *11*, 1206–1213. [CrossRef]
37. Gao, C.; Li, P.; Zhang, Y.; Liu, J.; Wang, L. People counting based on head detection combining Adaboost and CNN in crowded surveillance environment. *Neurocomputing* **2016**, *208*, 108–116. [CrossRef]
38. Bazi, Y.; Melgani, F. Convolutional SVM Networks for Object Detection in UAV Imagery. *IEEE Trans. Geosci. Remote Sens.* **2018**, *56*, 3107–3118. [CrossRef]
39. Fan, R.-E.; Chang, K.-W.; Hsieh, C.-J.; Wang, X.-R.; Lin, C.-J. Liblinear: A Library for Large Linear Classification. *J. Mach. Learn. Res.* **2008**, *9*, 1871–1874.
40. Chang, K.; Lin, C. Coordinate Descent Method for Large-scale L2-loss Linear Support Vector Machines. *J. Mach. Learn. Res.* **2008**, *9*, 1369–1398.

MDPI
St. Alban-Anlage 66
4052 Basel
Switzerland
Tel. +41 61 683 77 34
Fax +41 61 302 89 18
www.mdpi.com

Applied Sciences Editorial Office
E-mail: applsci@mdpi.com
www.mdpi.com/journal/applsci

www.ingramcontent.com/pod-product-compliance
Lightning Source LLC
LaVergne TN
LVHW071626170726
843515LV00009B/2489

* 9 7 8 3 0 3 6 5 1 4 8 3 3 *